Progress in Mathematical Physics
Volume 25

Editors-in-Chief
Anne Boutet de Monvel, *Université Paris VII Denis Diderot*
Gerald Kaiser, *The Virginia Center for Signals and Waves*

Editorial Board
D. Bao, *University of Houston*
C. Berenstein, *University of Maryland, College Park*
P. Blanchard, *Universität Bielefeld*
A.S. Fokas, *Imperial College of Science, Technology and Medicine*
C. Tracy, *University of California, Davis*
H. van den Berg, *Wageningen University*

Sergiu Klainerman
Francesco Nicolò

The Evolution Problem in General Relativity

Birkhäuser
Boston • Basel • Berlin

Sergiu Klainerman
Princeton University
Department of Mathematics
Princeton, NJ 08544
U.S.A.

Francesco Nicolò
Università degli studi di Roma "Tor Vergata"
Dipartimento di Matematica
Facoltà di Scienze, M.F.N.
Roma, 00100
Italy

Library of Congress Cataloging-in-Publication Data

Klainerman, Sergiu, 1950
The evolution problem in general relativity / Sergiu Klainerman and Francesco Nicolò.
p. cm.– (Progress in mathematical physics ; v. 25)
Includes bibliographical references and index.
ISBN 0-8176-4254-4 (alk. paper) — ISBN 3-7643-4254-4 (alk. paper)
1. General relativity (Physics) 2. Evolution equations. 3. Mathematical physics. I. Nicolò, Francesco, 1943- II. Title. III. Series.

QC173.6 .K57 2002
530.11–dc21 2002074351
CIP

AMS Subject Classifications: Primary: 83C05; Secondary: 83C20, 35L70, 35L15

Printed on acid-free paper.

Birkhäuser

ISBN 0-8176-4254-4 SPIN 10842860
ISBN 3-7643-4254-4

Reformatted from authors' files by TEXniques, Cambridge, MA.
Printed in the United States of America.

9 8 7 6 5 4 3 2 1

Birkhäuser Boston • Basel • Berlin
A member of BertelsmannSpringer Science+Business Media GmbH

Contents

Preface

The main goal of this work is to revisit the proof of the global stability of Minkowski space by D. Christodoulou and S. Klainerman, [Ch-Kl]. We provide a new self-contained proof of the main part of that result, which concerns the full solution of the radiation problem in vacuum, for arbitrary asymptotically flat initial data sets. This can also be interpreted as a proof of the global stability of the external region of Schwarzschild spacetime.

The proof, which is a significant modification of the arguments in [Ch-Kl], is based on a *double null foliation* of spacetime instead of the mixed *null-maximal foliation* used in [Ch-Kl]. This approach is more naturally adapted to the radiation features of the Einstein equations and leads to important technical simplifications.

In the first chapter we review some basic notions of differential geometry that are systematically used in all the remaining chapters. We then introduce the Einstein equations and the initial data sets and discuss some of the basic features of the initial value problem in general relativity. We shall review, without proofs, well-established results concerning local and global existence and uniqueness and formulate our main result.

The second chapter provides the technical motivation for the proof of our main theorem. We start by reviewing the standard proof of local existence and uniqueness for systems of nonlinear wave equations. We then discuss methods for proving global existence results, by stressing the importance of symmetries. We also emphasize the importance of a structural condition, called the null condition in establishing global results in $3+1$ dimensions. The cancellation that results when this formal condition is adopted illustrates the advantage of working with null frames. An essential result is the derivation of uniform decay estimates for linearized equations using only energy inequalities and the symmetries of Minkowski spacetime.

We proceed to show how the same method can be used to derive full decay estimates for the Weyl fields which satisfy the linear Bianchi equations in flat spacetime. The latter provides a crucial stepping stone to the Einstein equations. Finally we provide the reader with a detailed discussion of the basic ideas in the proof of the main theorem. All the remaining chapters, with the exception of the last, are dedicated to the proof of our main theorem. The proof is essentially self-contained, except for several topics which are treated in [Ch-Kl] and to which we provide ample reference.

In the last chapter we derive the most important consequences of our main theorem. In particular we give a rigorous derivation of the Bondi mass law and discuss the asymptotic properties of our spacetime. Due to our approach, based on the double null foliation, we are able to provide a straightforward definition of the outgoing null infinity. This makes the derivation of our asymptotic results simpler and more intuitive than the corresponding ones in the last chapter of [Ch-Kl]. In particular we are able to give a simple derivation of the connection between the Bondi mass and the ADM mass.

S. Klainerman
F. Nicolò

Acknowledgments: We want to thank D. Christodoulou for discussing with us many important ideas concerning our work. While we regret his decision to discontinue the original collaboration we would like to acknowledge his essential role in the original setup of our proof of the main theorem. This is particularly true in connection with Chapter 7 in which we follow his suggestions concerning the formulation of the last slice problem and the proof of its local existence. We are also happy to acknowledge a set of personal notes regarding the setup of the double null foliation. Their content is reflected in Section 3.1 of our book.

The Evolution Problem in General Relativity

1

Introduction

In the first part of this chapter we briefly present the main notions of differential geometry that we are going to use systematically throughout the book. We also describe, in some detail, results connected to the symmetry properties of Einstein spacetimes. The second part of the chapter is devoted to the introduction of the *initial value problem* in general relativity and to give a short survey of the main results on this subject. Finally we give a first description, in broad terms, of the central results of this book, and we compare them with the result of D. Christodoulou and S. Klainerman in [Ch-Kl].

1.1 Generalities about Lorentz manifolds

1.1.1 Lorentz metric, vector and tensor fields, covariant derivative, Lie derivative

A Lorentz manifold, or simply a spacetime, consists of a pair $(\mathcal{M}, \mathbf{g})$ where $\mathcal{M}$ is an orientable $(n+1)$-dimensional manifold whose points correspond to physical events and $\mathbf{g}$ is a Lorentz metric defined on it, that is, a smooth, nondegenerate 2-covariant symmetric tensor field of signature $(n, 1)$.[1] This means that at each point $p \in \mathcal{M}$ one can choose a basis of $n+1$ vectors, $\{e_{(\alpha)}\}$, belonging to the tangent space $T\mathcal{M}_p$, such that

$$\mathbf{g}(e_{(\alpha)}, e_{(\beta)}) = \eta_{\alpha\beta} \tag{1.1.1}$$

for all $\alpha, \beta = 0, 1, \ldots, n$, where η is the diagonal matrix with entries $-1, 1, \ldots, 1$. If X is an arbitrary vector at p expressed in terms of the basis $\{e_{(\alpha)}\}$ as $X = X^\alpha e_{(\alpha)}$, we have

$$\mathbf{g}(X, X) = \eta_{\alpha\beta} X^\alpha X^\beta = -(X^0)^2 + (X^1)^2 + (X^2)^2 + \cdots + (X^n)^2. \tag{1.1.2}$$

[1] We assume that our reader is already familiar with the basic concepts of differential geometry such as manifolds, tensor fields, covariant, Lie and exterior differentiation. For a short introduction to these concepts, see Chapter 1 of [Haw-El].

The primary example of a spacetime is Minkowski spacetime, the spacetime of special relativity. Minkowski spacetime plays the same role in Lorentzian geometry as Euclidean space does in Riemannian geometry. In this case the manifold $\mathcal{M}$ is diffeomorphic to R^{n+1} and there exist globally defined systems of coordinates, x^α, relative to which the metric takes the diagonal form $-1, 1, \dots, 1$. All such systems are related through Lorentz transformations and are called inertial. We shall denote a Minkowski spacetime of dimension $n+1$ by (M^{n+1}, η).

In view of (1.1.2) the Lorentz metric divides the vectors in the tangent space $T\mathcal{M}_p$ at each p into timelike, null or spacelike according to whether the quadratic form

$$(X, X) = g_{\mu\nu} X^\mu X^\nu \tag{1.1.3}$$

is, respectively, negative, zero or positive. The set of null vectors N_p forms a double cone called the null cone of the corresponding point p. The set of timelike vectors I_p forms the interior of this cone. The vectors in the union of I_p and N_p are called causal. The set S_p of spacelike vectors is the complement of $I_p \cup N_p$.

Together with the orthonormal frames we will use in the following the null frames,[2] $\{e_3, e_4, e_a\}$, satisfying

$$\begin{aligned} \mathbf{g}(e_3, e_3) = \mathbf{g}(e_4, e_4) = 0\ ,\ \ \mathbf{g}(e_3, e_4) = -2 \\ \mathbf{g}(e_3, e_a) = \mathbf{g}(e_4, e_a) = 0\ ,\ \ \mathbf{g}(e_a, e_b) = \delta_{ab}, \end{aligned}$$

where the e_a are orthonormal spacelike vectors with $a = 1, \dots, n-2$.

We conclude this introductory section by stating, without proof, a proposition that shows already at this level the fundamental role played by the null cones in Lorentzian geometry. For its proof, see Chapter 3 of [Haw–El].

Proposition 1.1.1 *The specification of the null cones N_p uniquely determines the metric up to a factor of proportionality. In other words any two Lorentz metrics on $\mathcal{M}$ which have the same null structure are conformally equivalent.*

Notation: Throughout the book we use the following notational conventions:

1. We use boldface characters to denote the spacetime metric $\mathbf{g}$, the Riemann curvature tensor $\mathbf{R}$, its conformal part $\mathbf{C}$, as well as the connection $\mathbf{D}$.

2. Their components relative to arbitrary frames are also denoted by boldface characters. Thus given a frame $\{e_\alpha\}$ we write $\mathbf{g}_{\alpha\beta} = \mathbf{g}(e_\alpha, e_\beta)$ $\mathbf{R}_{\alpha\beta\gamma\delta} = \mathbf{R}(e_\alpha, e_\beta, e_\gamma, e_\delta)$ and, for an arbitrary tensor T,

$$\begin{aligned} &T_{\alpha\beta\gamma\delta\dots} \equiv T(e_\alpha, e_\beta, e_\gamma, e_\delta, \dots) \\ &\mathbf{D}_\alpha \mathbf{D}_\beta \dots \mathbf{D}_\delta T_{\epsilon\dots\lambda} \equiv (\mathbf{D}.\mathbf{D}.\ \dots \mathbf{D}.T)(e_\alpha, e_\beta, \dots, e_\delta, e_\epsilon, \dots, e_\lambda)\ . \end{aligned}$$

3. We do not use boldface characters for the components of tensors relative to an arbitrary system of coordinates. Thus, for instance, in (1.1.3) $g_{\mu\nu} = \mathbf{g}(\frac{\partial}{\partial x^\mu}, \frac{\partial}{\partial x^\nu})$.

---

[2]We often write e_α instead of $e_{(\alpha)}$ to simplify notation.

4. To denote the indices associated with noncoordinate frames, we use the first Greek letters $\alpha, \beta, \gamma, \delta, \ldots$. The Greek letters $\mu, \nu, \rho, \sigma, \ldots$ refer to spacetime coordinate frames.[3]

5. When we refer to tensor quantities defined on a spacelike three-dimensional hypersurface, Σ, we use the Latin letters $i, j, l, k, \ldots$. In this case it will be clear from the text which kind of components we are using.

6. When we consider tensors restricted to two-dimensional surfaces, S, diffeomorphic to S^2, we use the Latin letters $a, b, c, d, \ldots$ only to indicate their components with respect to an adapted orthonormal frame $\{e_a\}$. We point out explicitly the cases in which the components are written with respect to an arbitrary frame or a set of coordinates of S. We will, however, in the sequel, restrict ourselves mainly to orthonormal or null frames and, of course, to dimension $n = 3$.

Now we present the properties of the three fundamental operators of differential geometry on a Riemann or Lorentz manifold: the exterior derivative, the Lie derivative and its connection with its associated covariant derivative.

The exterior derivative

Given a scalar function f its differential df is the 1-form defined by

$$df(X) = X(f),$$

for any vector field X. This definition can be extended to all differential forms on $\mathcal{M}$ in the following way.

Definition 1.1.1

(i) The exterior derivative d is a linear operator defined from the space of all k-forms to that of $(k+1)$-forms on $\mathcal{M}$. Thus for all k-forms A, B and real numbers λ, μ

$$d(\lambda A + \mu B) = \lambda dA + \mu dB.$$

(ii) For any k-form A and arbitrary form B

$$d(A \wedge B) = dA \wedge B + (-1)^k A \wedge dB.$$

(iii) For any form A

$$d^2 A = 0\,.$$

We recall that, if Φ is a smooth map defined from $\mathcal{M}$ to another manifold $\mathcal{M}'$, then

$$d(\Phi^* A) = \Phi^*(dA)\,.$$

Finally if A is a 1-form and X, Y arbitrary vector fields, we have the equation

$$dA(X,Y) = \Big(X(A(Y)) - Y(A(X)) - A([X,Y]) \Big),$$

which can be easily generalized to arbitrary k-forms; see [Sp], Vol. I, Chapter 7, Theorem 13.

---

[3]When no confusion arises, we sometimes also use the first Greek letters for the coordinate components.

The Lie derivative

Consider an arbitrary vector field X. In the local coordinates x^μ, the flow of X is given by the system of differential equations

$$\frac{dx^\mu}{dt} = X^\mu(x^0(t), \ldots, x^n(t)) .$$

The corresponding curves, $x^\mu(t)$, are the integral curves of X. For each point $p \in \mathcal{M}$ there exists an open neighborhood $\mathcal{U}$, a small $\epsilon > 0$ and a family of diffeomorphisms $\Phi_t : \mathcal{U} \to \mathcal{M},\ |t| \le \epsilon$, obtained by taking each point in $\mathcal{U}$ to a parameter distance t along the integral curves of X. We use these diffeomorphisms to construct, for any given tensor T at p, the family of tensors $(\Phi_t)_* T$ at$\Phi_t(p)$.[4]

Definition 1.1.2 *The Lie derivative $\mathcal{L}_X T$ of a tensor field T with respect to X is*

$$\mathcal{L}_X T|_p \equiv \lim_{t\to 0} \frac{1}{t}\left(T|_p - (\Phi_t)_* T|_p\right) .$$

It has the following properties:

(i) $\mathcal{L}_X$ maps linearly (p, q)-tensor fields into tensor fields of the same type.

(ii) $\mathcal{L}_X$ commutes with contractions.

(iii) For any tensor fields S, T,

$$\mathcal{L}_X(S \otimes T) = \mathcal{L}_X S \otimes T + S \otimes \mathcal{L}_X T .$$

If X is a vector field we easily see that

$$\mathcal{L}_X Y = [X, Y] .$$

If A is a k-form we have, as a consequence of the commutation formula of the exterior derivative with the pullback Φ^*,

$$d(\mathcal{L}_X A) = \mathcal{L}_X(dA).$$

We remark that the Lie bracket of two coordinate vector fields vanishes, $[\frac{\partial}{\partial x^\mu}, \frac{\partial}{\partial x^\nu}] = 0$. The converse is also true; (see [Sp], Vol. I, Chapter 5).

Proposition 1.1.2 *If $X_{(0)}, \ldots, X_{(k)}$ are linearly independent vector fields in a neighborhood of a point p and the Lie bracket of any two of them is zero, then there exists a coordinate system x^μ around p such that $X_{(\rho)} = \frac{\partial}{\partial x^\rho}$ for each $\rho = 0, \ldots, k$.*

The above proposition is the main step in the proof of the Frobenius Theorem. To state the theorem we recall the definition of a k-distribution in $\mathcal{M}$. This is an arbitrary smooth assignment of a k-dimensional plane π_p at every point in a domain $\mathcal{U}$ of $\mathcal{M}$. The distribution is said to be involute if, for any vector fields X, Y on $\mathcal{U}$ with $X|_p, Y|_p \in \pi_p$, for

---

[4]Given a diffeomorphism Φ, Φ_* is the linear map $T\mathcal{M}_p \to T\mathcal{M}_{\Phi(p)}$ defined on vector fields as $(\Phi_* X)(f)|_p \equiv X(f(\Phi(p)))$ and on 1-forms as the inverse of the pullback associated with Φ.

any $p \in \mathcal{U}$, we have $[X, Y]|_p \in \pi_p$. This is clearly the case for integrable distributions.[5] Indeed if $X|_p, Y|_p \in T\mathcal{N}_p$ for all $p \in \mathcal{N}$, then X, Y are tangent to $\mathcal{N}$ and so is their commutator $[X, Y]$. The Frobenius Theorem establishes that the converse is also true, that is, being in involution is also a sufficient condition for the distribution to be integrable.[6]

Theorem 1.1.1 **(Frobenius Theorem)** *A necessary and sufficient condition for a distribution $(\pi_p)_{p\in\mathcal{U}}$ to be integrable is that it be involute.*

The connection and the covariant derivative

Definition 1.1.3 *A connection* $\mathbf{D}$ *at a point $p \in \mathcal{M}$ is a rule that assigns to each $X \in T\mathcal{M}_p$ a differential operator $\mathbf{D}_X$.*[7] *This operator maps vector fields Y into vector fields $\mathbf{D}_X Y$ in such a way that, with $\alpha, \beta \in C$ and f, g scalar functions on $\mathcal{M}$,*

(i) $\mathbf{D}_{fX+gY} Z = f\mathbf{D}_X Z + g\mathbf{D}_Y Z;$
(ii) $\mathbf{D}_X(\alpha Y + \beta Z) = \alpha\mathbf{D}_X Y + \beta\mathbf{D}_X Z;$
(iii) $\mathbf{D}_X fY = X(f)Y + f\mathbf{D}_X Y.$

Therefore, at a given point p,

$$\mathbf{D}Y \equiv Y^{\alpha}_{;\beta}\theta^{(\beta)} \otimes e_{(\alpha)}, \tag{1.1.4}$$

where the $\theta^{(\beta)}$ are the 1-forms of the basis dual with respect to the orthonormal frame $e_{(\beta)}$.[8] On the other side, from (iii),

$$\mathbf{D}fY = df \otimes Y + f\mathbf{D}Y,$$

so that

$$\mathbf{D}Y = \mathbf{D}(Y^{\alpha}e_{(\alpha)}) = dY^{\alpha} \otimes e_{(\alpha)} + Y^{\alpha}\mathbf{D}e_{(\alpha)},$$

and finally[9]

$$\mathbf{D}Y = \left(e_{(\beta)}(Y^{\alpha}) + Y^{\gamma}\theta^{(\alpha)}(\mathbf{D}_{e_{(\beta)}}e_{(\gamma)})\right)\theta^{(\beta)} \otimes e_{(\alpha)}. \tag{1.1.5}$$

Therefore

$$Y^{\alpha}_{;\beta} = \left(e_{(\beta)}(Y^{\alpha}) + \Gamma^{\alpha}_{\beta\gamma}Y^{\gamma}\right),$$

and the connection is determined by its connection coefficients,

$$\Gamma^{\alpha}_{\beta\gamma} = \theta^{(\alpha)}(\mathbf{D}_{e_{(\beta)}}e_{(\gamma)}), \tag{1.1.6}$$

which, in a coordinate basis, are the usual Christoffel symbols

$$\Gamma^{\mu}_{\nu\rho} = dx^{\mu}(\mathbf{D}_{\frac{\partial}{\partial x^{\nu}}}\frac{\partial}{\partial x^{\rho}}) \, .$$

---

[5] Recall that a distribution π on $\mathcal{U}$ is said to be integrable if through every point $p \in \mathcal{U}$ there passes a unique submanifold $\mathcal{N}$ of dimension k such that $\pi_p = T\mathcal{N}_p$.

[6] For a proof see [Sp], Vol. I, Chapter 6.

[7] Recall that the notion of affine connection does not depend on the metric of $\mathcal{M}$.

[8] Note that $Y^{\alpha}_{;\beta} = \theta^{(\alpha)}(\mathbf{D}_{e_{(\beta)}}Y)$.

[9] We use, from previous definitions, $df(\cdot) = e_{(\alpha)}(f)\theta^{(\alpha)}(\cdot)$.

Finally

$$\mathbf{D}_X Y = \big(X(Y^\alpha) + \Gamma^\alpha_{\beta\gamma} X^\beta Y^\gamma\big) e_{(\alpha)}, \tag{1.1.7}$$

and, in the particular case of a coordinate frame,

$$\mathbf{D}_X Y = \left(X^\mu \frac{\partial Y^\nu}{\partial x^\mu} + \Gamma^\nu_{\rho\sigma} X^\rho Y^\sigma\right) \frac{\partial}{\partial x^\nu}. \tag{1.1.8}$$

Definition 1.1.4 *The Levi-Civita connection on $\mathcal{M}$ is the unique connection on $\mathcal{M}$ that satisfies* $\mathbf{Dg} = 0$.

Thus for any three vector fields X, Y, Z

$$Z(\mathbf{g}(X, Y)) = \mathbf{g}(\mathbf{D}_Z X, Y) + \mathbf{g}(X, \mathbf{D}_Z Y),$$

and relative to a system of coordinates, x^μ, the Christoffel symbols of the connection are given by the standard formula

$$\Gamma^\mu_{\rho\nu} = \frac{1}{2} g^{\mu\tau} \left(\partial_\rho g_{\nu\tau} + \partial_\nu g_{\tau\rho} - \partial_\tau g_{\nu\rho}\right) .$$

The Levi-Civita connection is torsion free namely

$$\mathbf{D}_X Y - \mathbf{D}_Y X = [X, Y] .$$

This allows one to connect it to the Lie derivative. Thus if T is a k-covariant tensor we have, in a coordinate basis,

$$(\mathcal{L}_X T)_{\sigma_1 \ldots \sigma_k} = X^\mu T_{\sigma_1 \ldots \sigma_k ; \mu} + X^\mu_{\ ;\sigma_1} T_{\mu \sigma_2 \ldots \sigma_k} + \ldots . + X^\mu_{\ ;\sigma_k} T_{\sigma_1 \ldots \sigma_{k-1} \mu} .$$

The covariant derivative is also connected to the exterior derivative according to the following simple formula. If A is a k-form, we have[10] $A_{[\sigma_1 \ldots \sigma_k ; \mu]} = A_{[\sigma_1 \ldots \sigma_k , \mu]}$ and

$$dA = \sum A_{\sigma_1 \ldots \sigma_k ; \mu} dx^\mu \wedge dx^{\sigma_1} \wedge dx^{\sigma_2} \wedge \cdots \wedge dx^{\sigma_k} .$$

Definition 1.1.5 *Given a smooth curve $\mathbf{x} : [0, 1] \to \mathcal{M}$, parametrized by t, let $T = \left(\frac{\partial}{\partial t}\right)_{\mathbf{x}}$ be the corresponding tangent vector field. A vector field X, defined on the curve, is said to be parallel transported along it if $\mathbf{D}_T X = 0$. Let the curve have the parametric equations $x^\nu = x^\nu(t)$. Then $T^\mu = \frac{dx^\mu}{dt}$ and the components $X^\mu = X^\mu(\mathbf{x}(t))$ satisfy the system of ordinary differential equation*

$$\frac{\mathbf{D}}{dt} X^\mu \equiv \frac{dX^\mu}{dt} + \Gamma^\mu_{\rho\sigma}(\mathbf{x}(t)) \frac{dx^\rho}{dt} X^\sigma = 0 .$$

---

[10] $A_{\sigma_1 \ldots \sigma_k , \mu}$ denotes $\frac{\partial}{\partial x^\mu} A_{\sigma_1 \ldots \sigma_k}$. $[\sigma_1 \ldots \sigma_k ; \mu]$ or $[\sigma_1 \ldots \sigma_k , \mu]$ refer to the usual antisymmetrization with respect to all indices.

The curve is said to be geodesic if, at every point of the curve, $\mathbf{D}_T T$ is tangent to the curve, $\mathbf{D}_T T = \lambda T$. In this case one can reparametrize the curve such that, relative to the new parameter s, the tangent vector $S = \left(\frac{\partial}{\partial s}\right)_{\mathbf{x}}$ satisfies $\mathbf{D}_S S = 0$. Such a parameter is called an *affine parameter*. The affine parameter is defined up to a transformation $s = as' + b$ for a, b constants. Relative to an affine parameter s and arbitrary coordinates x^μ the geodesic curves satisfy the equations

$$\frac{d^2 x^\mu}{ds^2} + \Gamma^\mu_{\rho\sigma} \frac{dx^\rho}{ds} \frac{dx^\sigma}{ds} = 0.$$

A geodesic curve parametrized by an affine parameter is simply called a geodesic. Timelike geodesics correspond to histories of particles freely falling in the gravitational field represented by the connection coefficients. In this case the affine parameter s is called the proper time of the particle.

Given a point $p \in \mathcal{M}$ and a vector X in the tangent space $T_p\mathcal{M}$, let $\mathbf{x}(t)$ be the unique geodesic starting at p with "velocity" X. We define the exponential map

$$\exp_p : T_p\mathcal{M} \to \mathcal{M} ,$$

where, for each $X \in T_p\mathcal{M}$, $\exp_p(X)$ is the point in $\mathcal{M}$ a unit parameter distance along the geodesic $\mathbf{x}(t)$ from p.

This map may not be defined for all $X \in T_p\mathcal{M}$. The theorem of existence for systems of ordinary differential equations implies that the exponential map is defined in a neighbourhood of the origin in $T_p\mathcal{M}$. If the exponential map is defined for all $T_p\mathcal{M}$ for every point p, the manifold $\mathcal{M}$ is said to be geodesically complete. In general if the connection is a C^r connection[11] there exists an open neighborhood $\mathcal{U}_0$ of the origin in $T_p\mathcal{M}$ and an open neighborhood $\mathcal{V}_p$ of the point p in $\mathcal{M}$ such that the map $\exp_p$ is a C^r diffeomorphism of $\mathcal{U}_0$ onto $\mathcal{V}_p$. The neighborhood $\mathcal{V}_p$ is called a normal neighborhood of p.[12]

1.1.2 Riemann curvature tensor, Ricci tensor, Bianchi identities

In flat spacetime if we parallel transport a vector along any closed curve we obtain the vector we started with. This fails in general because the second covariant derivatives of a vector field do not commute. This lack of commutation is measured by the Riemann curvature tensor,

$$\mathbf{R}(X,Y)Z = \mathbf{D}_X(\mathbf{D}_Y Z) - \mathbf{D}_Y(\mathbf{D}_X Z) - \mathbf{D}_{[X,Y]}Z, \tag{1.1.9}$$

or, written in components relative to an arbitrary frame,

$$\mathbf{R}^\alpha{}_{\beta\gamma\delta} = \theta^{(\alpha)}\big((\mathbf{D}_\gamma \mathbf{D}_\delta - \mathbf{D}_\delta \mathbf{D}_\gamma)e_{(\beta)}\big) . \tag{1.1.10}$$

Relative to a coordinate system x^μ and written in terms of the $g_{\mu\nu}$ components, the Riemann components have the expression

$$R^\mu{}_{\nu\rho\sigma} = \frac{\partial \Gamma^\mu_{\sigma\nu}}{\partial x^\rho} - \frac{\partial \Gamma^\mu_{\rho\nu}}{\partial x^\sigma} + \Gamma^\mu_{\rho\tau}\Gamma^\tau_{\sigma\nu} - \Gamma^\mu_{\sigma\tau}\Gamma^\tau_{\rho\nu}. \tag{1.1.11}$$

---

[11] A C^r connection is such that if Y is a C^{r+1} vector field, then $\mathbf{D}Y$ is a C^r vector field.

[12] For a more general discussion of the exponential map see [Sp], Vol. I and [Haw-El].

The fundamental property of the curvature tensor, first proved by Riemann, states that if $\mathbf{R}$ vanishes identically in a neighborhood of a point p one can find families of local coordinates such that, in a neighborhood of p, $g_{\mu\nu} = \eta_{\mu\nu}$.[13]

The trace of the curvature tensor relative to the metric $\mathbf{g}$ is a symmetric tensor called the Ricci tensor,[14]

$$\mathbf{R}_{\alpha\beta} = \mathbf{g}^{\gamma\delta}\mathbf{R}_{\alpha\gamma\beta\delta}.$$

The scalar curvature is the trace of the Ricci tensor

$$\mathbf{R} = \mathbf{g}^{\alpha\beta}\mathbf{R}_{\alpha\beta}.$$

The Riemann curvature tensor of an arbitrary spacetime $(\mathcal{M}, \mathbf{g})$ has the following symmetry properties: [15]

$$\begin{aligned} &\mathbf{R}_{\alpha\beta\gamma\delta} = -\mathbf{R}_{\beta\alpha\gamma\delta} = -\mathbf{R}_{\alpha\beta\delta\gamma} = \mathbf{R}_{\gamma\delta\alpha\beta} \\ &\mathbf{R}_{\alpha\beta\gamma\delta} + \mathbf{R}_{\alpha\gamma\delta\beta} + \mathbf{R}_{\alpha\delta\beta\gamma} = 0. \end{aligned} \tag{1.1.12}$$

It also satisfies the second Bianchi identities, which we refer to here as *the Bianchi equations,* and in a given frame has the form

$$\mathbf{D}_{[\epsilon}\mathbf{R}_{\gamma\delta]\alpha\beta} = 0. \tag{1.1.13}$$

The traceless part of the curvature tensor, $\mathbf{C}$, has the following expression

$$\begin{aligned} \mathbf{C}_{\alpha\beta\gamma\delta} \quad = \quad & \mathbf{R}_{\alpha\beta\gamma\delta} - \frac{1}{n-1}\left(\mathbf{g}_{\alpha\gamma}\mathbf{R}_{\beta\delta} + \mathbf{g}_{\beta\delta}\mathbf{R}_{\alpha\gamma} - \mathbf{g}_{\beta\gamma}\mathbf{R}_{\alpha\delta} - \mathbf{g}_{\alpha\delta}\mathbf{R}_{\beta\gamma}\right) \\ + \quad & \frac{1}{n(n-1)}(\mathbf{g}_{\alpha\gamma}\mathbf{g}_{\beta\delta} - \mathbf{g}_{\alpha\delta}\mathbf{g}_{\beta\gamma})\mathbf{R}. \end{aligned} \tag{1.1.14}$$

Observe that $\mathbf{C}$ verifies all the symmetry properties of the Riemann tensor:

$$\begin{aligned} &\mathbf{C}_{\alpha\beta\gamma\delta} = -\mathbf{C}_{\beta\alpha\gamma\delta} = -\mathbf{C}_{\alpha\beta\delta\gamma} = \mathbf{C}_{\gamma\delta\alpha\beta} \\ &\mathbf{C}_{\alpha\beta\gamma\delta} + \mathbf{C}_{\alpha\gamma\delta\beta} + \mathbf{C}_{\alpha\delta\beta\gamma} = 0 \end{aligned} \tag{1.1.15}$$

and, in addition, $\mathbf{g}^{\alpha\gamma}\mathbf{C}_{\alpha\beta\gamma\delta} = 0$.

We say that two metrics $\mathbf{g}$ and $\hat{\mathbf{g}}$ are conformal if $\hat{\mathbf{g}} = \lambda^2\mathbf{g}$ for some nonzero differentiable function λ. Then the following theorem holds; see [Haw-El], Chapter 2.

Theorem 1.1.2 *Let* $\hat{\mathbf{g}} = \lambda^2\mathbf{g}$, $\hat{\mathbf{C}}$ *the Weyl tensor relative to* $\hat{\mathbf{g}}$ *and* $\mathbf{C}$ *the Weyl tensor relative to* $\mathbf{g}$. *Then*

$$\hat{\mathbf{C}}^{\alpha}_{\ \beta\gamma\delta} = \mathbf{C}^{\alpha}_{\ \beta\gamma\delta}\ ,$$

which shows that $\mathbf{C}$ *is conformally invariant.*

---

[13] For a thorough discussion and proof of this fact we refer to [Sp], Vol. II.

[14] In an arbitrary frame $\mathbf{R}_{\alpha\beta} \equiv \mathrm{Ricci}(e_{(\alpha)}, e_{(\beta)})$, $\mathbf{g}^{\alpha\beta} \equiv \mathbf{g}^{-1}_{\alpha\beta}$.

[15] The second ones are called the first Bianchi identities.

1.1.3 Isometries and conformal isometries, Killing and conformal Killing vector fields

Definition 1.1.6 *A diffeomorphism $\Phi : \mathcal{U} \subset \mathcal{M} \to \mathcal{M}$ is said to be a conformal isometry if, at every point p, $\Phi^*\mathbf{g} = \Lambda^2\mathbf{g}$, that is,*

$$(\Phi^*\mathbf{g})(X, Y)|_p = \mathbf{g}(\Phi_* X, \Phi_* Y)|_{\Phi(p)} = \Lambda^2\mathbf{g}(X, Y)|_p$$

with $\Lambda \neq 0$. If $\Lambda = 1$, Φ is called an isometry of $\mathcal{M}$.

Definition 1.1.7 *A vector field K which generates a one-parameter group of isometries, respectively, conformal isometries, is called a Killing, (resp. conformal Killing) vector field.*

Let K be such a vector field and Φ_t the corresponding one-parameter group. Since the $(\Phi_t)_*$ are conformal isometries, we infer that $\mathcal{L}_K\mathbf{g}$ must be proportional to the metric $\mathbf{g}$. Moreover, $\mathcal{L}_K\mathbf{g} = 0$ if K is a Killing vector field.

Definition 1.1.8 *Given an arbitrary vector field X we denote ${}^{(X)}\pi$ the deformation tensor of X defined by the formula*

$${}^{(X)}\pi_{\alpha\beta} = (\mathcal{L}_X g)_{\alpha\beta} = \mathbf{D}_\alpha X_\beta + \mathbf{D}_\beta X_\alpha \ .$$

${}^{(X)}\pi$ measures, in a precise sense, how much the diffeomorphism generated by X differs from an isometry or a conformal isometry. The following proposition holds; see [Haw-El], Chapter 2, p. 43.

Proposition 1.1.3 *The vector field X is Killing if and only if ${}^{(X)}\pi = 0$. It is conformal Killing if and only if ${}^{(X)}\pi$ is proportional to* $\mathbf{g}$.

Remark: One can choose local coordinates $x^0, x^1, \ldots, x^n$ such that $X = \frac{\partial}{\partial x^0}$. It then immediately follows that, relative to these coordinates, the metric $\mathbf{g}$ is independent of x^0.

Proposition 1.1.4 *On any spacetime $\mathcal{M}$ of dimension $n+1$, there can be no more than $\frac{1}{2}(n+1)(n+2)$ linearly independent Killing vector fields.*

Proof. Proposition 1.1.4 is an easy consequence of the following relation, valid for an arbitrary vector field X, obtained by a straightforward computation and the use of the Bianchi identities,

$$\mathbf{D}_\beta\mathbf{D}_\alpha X_\lambda = \mathbf{R}_{\lambda\alpha\beta\delta}X^\delta + {}^{(X)}\Gamma_{\alpha\beta\lambda}, \tag{1.1.16}$$

where

$${}^{(X)}\Gamma_{\alpha\beta\lambda} = \frac{1}{2}\left(\mathbf{D}_\beta\pi_{\alpha\lambda} + \mathbf{D}_\alpha\pi_{\beta\lambda} - \mathbf{D}_\lambda\pi_{\alpha\beta}\right) \tag{1.1.17}$$

and $\pi \equiv {}^{(X)}\pi$ is the X deformation tensor. In fact, if X is a Killing vector field equation (1.1.16) becomes

$$\mathbf{D}_\beta(\mathbf{D}_\alpha X_\lambda) = \mathbf{R}_{\lambda\alpha\beta\delta}X^\delta, \tag{1.1.18}$$

which implies that any Killing vector field is completely determined by the $\frac{1}{2}(n+1)(n+2)$ values of X and $\mathbf{D}X$ at a given point. Then the argument goes in this way: let p, q be two points connected by a curve $x(t)$ with tangent vector T. Let $L_{\alpha\beta} \equiv \mathbf{D}_\alpha X_\beta$. Along $x(t)$, X, L satisfy the system of differential equations

$$\frac{\mathbf{D}}{dt}X = T \cdot L \ , \ \frac{\mathbf{D}}{dt}L = \mathbf{R}(\cdot, \cdot, X, T).$$

Therefore the values of X, L along the curve are uniquely determined by their values at p.

The spacetime that possesses the maximum number of Killing and conformal Killing vector fields is the Minkowski spacetime $\mathbf{M}^{n+1}$. Let us review its associated isometries and conformal isometries.

Let x^μ be an inertial, positively oriented coordinate system. We have the following:

1. Translations: for any given vector $a = (a^0, a^1, \ldots, a^n) \in \mathbf{M}^{n+1}$,

$$x^\mu \to x^\mu + a^\mu;$$

2. Lorentz rotations: given any $\Lambda = \Lambda^\rho_\sigma \in \mathbf{O}(1, n)$,

$$x^\mu \to \Lambda^\mu_\nu x^\nu;$$

3. Scalings: given any real number $\lambda \neq 0$,

$$x^\mu \to \lambda x^\mu;$$

4. Inversions: consider the transformation $x^\mu \to I(x^\mu)$, where

$$I(x^\mu) = \frac{x^\mu}{(x, x)},$$

is defined for all points $x \in \mathbf{M}^{n+1}$ such that $(x, x) \neq 0$.

The first two sets of transformations are isometries of $\mathbf{M}^{n+1}$, and the group generated by them is called the Poincarè group. The last two types of transformations are conformal isometries. The group generated by all the above transformations is called the conformal group. In fact the Liouville theorem, whose infinitesimal version will be proved below, states that the conformal group is the group of all the conformal isometries of $\mathbf{M}^{n+1}$.

Let us list the Killing and conformal Killing vector fields that generate the above transformations.

1. The generators of translations in the x^μ directions, $\mu = 0, 1, \ldots, n$:

$$T_\mu = \frac{\partial}{\partial x^\mu}.$$

2. The generators of the Lorentz rotations in the (μ, ν) plane:

$$L_{\mu\nu} = x_\mu \partial_\nu - x_\nu \partial_\mu.$$

3. The generator of the scaling transformations:

$$S = x^\mu \partial_\mu .$$

4. The generators of the inverted translations:[16]

$$K_\mu = 2x_\mu x^\rho \frac{\partial}{\partial x^\rho} - x^\rho x_\rho \frac{\partial}{\partial x^\mu} .$$

We also list below the commutator relations between these vector fields,

$$\begin{aligned}
&[L_{\alpha\beta}, L_{\gamma\delta}] = \eta_{\alpha\gamma} L_{\beta\delta} - \eta_{\beta\gamma} L_{\alpha\delta} + \eta_{\beta\delta} L_{\alpha\gamma} - \eta_{\alpha\delta} L_{\beta\gamma} \\
&[L_{\alpha\beta}, T_\gamma] = \eta_{\alpha\gamma} T_\beta - \eta_{\beta\gamma} T_\alpha \\
&[T_\alpha, T_\beta] = 0 \\
&[T_\alpha, S] = T_\alpha \\
&[T_\alpha, K_\beta] = 2(\eta_{\alpha\beta} S + L_{\alpha\beta}) \\
&[L_{\alpha\beta}, S] = [K_\alpha, K_\beta] = 0 \\
&[L_{\alpha\beta}, K_\gamma] = \eta_{\alpha\gamma} K_\beta - \eta_{\beta\gamma} K_\alpha .
\end{aligned} \tag{1.1.19}$$

Denoting by $\mathcal{P}(1, n)$ the Lie algebra generated by the vector fields T_α, $L_{\beta\gamma}$ and by $\underline{\mathcal{K}}(1, n)$ the Lie algebra generated by all the vector fields T_α, $L_{\beta\gamma}$, S, K_δ, we state the following version of the Liouville theorem.

Theorem 1.1.3

(i) $\mathcal{P}(1, n)$ is the Lie algebra of all Killing vector fields in $\mathbf{M}^{n+1}$.
(ii) If $n > 1$, $\underline{\mathcal{K}}(1, n)$ is the Lie algebra of all Killing and conformal Killing vector fields in $\mathbf{M}^{n+1}$.
(iii) If $n = 1$, all conformal Killing vector fields in $\mathbf{M}^{1+1}$ have the expression

$$f(x^0 + x^1)(\partial_0 + \partial_1) + g(x^0 - x^1)(\partial_0 - \partial_1),$$

where f, g are arbitrary smooth functions of one variable.

Proof. The proof of (i) follows immediately, as a particular case, from Proposition 1.1.4. From (1.1.16), with $\mathbf{R} = 0$, and X Killing, we have

$$D_\mu D_\nu X_\lambda = 0 .$$

Therefore, there exist constants $a_{\mu\nu}$, b_μ such that $X^\mu = a_{\mu\nu} x^\nu + b_\mu$. Since X is Killing, $D_\mu X_\nu = -D_\nu X_\mu$ which implies $a_{\mu\nu} = -a_{\nu\mu}$. Consequently X can be written as a linear combination, with real coefficients, of the vector fields T_α, $L_{\beta\gamma}$.

Now let X be a conformal Killing vector field. There exists a function Ω such that

$$^{(X)}\pi_{\mu\nu} = \Omega \eta_{\mu\nu} . \tag{1.1.20}$$

[16] Observe that the vector fields K_μ can be obtained by applying I_* to the vector fields T_μ.

From (1.1.16) and (1.1.17) it follows that

$$D_\mu D_\nu X_\lambda = \frac{1}{2}\left(\eta_{\nu\lambda}\partial_\mu\Omega + \eta_{\mu\lambda}\partial_\nu\Omega - \eta_{\nu\mu}\partial_\lambda\Omega\right). \tag{1.1.21}$$

Taking the trace with respect to μ, ν on both sides of (1.1.21) and (1.1.20) we infer that

$$\begin{aligned} \Box X_\lambda &= -\frac{n-1}{2}\Omega_{,\lambda} \\ D^\mu X_\mu &= \frac{n+1}{2}\Omega. \end{aligned} \tag{1.1.22}$$

and applying D^λ to the first equation and $\Box$ to the second and subtracting, we obtain

$$\Box\Omega = 0. \tag{1.1.23}$$

Applying D_μ to the first equation of (1.1.22) and using (1.1.23) we obtain

$$\begin{aligned} (n-1)D_\mu D_\lambda\Omega &= \frac{n-1}{2}(D_\mu D_\lambda\Omega + D_\lambda D_\mu\Omega) = -\Box(D_\mu X_\lambda + D_\lambda X_\mu) \\ &= -(\Box\Omega)\eta_{\mu\lambda} = 0. \end{aligned} \tag{1.1.24}$$

Hence for $n \neq 1$, $D_\mu D_\lambda\Omega = 0$. This implies that Ω must be a linear function of x^μ. We can therefore find a linear combination, with constant coefficients, $cS + d^\alpha K_\alpha$ such that the deformation tensor of $X - (cS + d^\alpha K_\alpha)$ must be zero. This is the case because ${}^{(S)}\pi = 2\eta$ and ${}^{(K_\mu)}\pi = 4x_\mu\eta$. Therefore $X - (cS + d^\alpha K_\alpha)$ is Killing which, in view of (i), proves part (ii) of the result.

Part (iii) can be easily derived by solving (1.1.20). Indeed posing $X = a\partial_0 + b\partial_1$, we obtain $2D_0X_0 = -\Omega$, $2D_1X_1 = \Omega$ and $D_0X_1 + D_1X_0 = 0$. Hence a, b satisfy the system

$$\frac{\partial a}{\partial x^0} = \frac{\partial b}{\partial x^1}, \quad \frac{\partial b}{\partial x^0} = \frac{\partial a}{\partial x^1}.$$

Hence the 1-form $adx^0 + bdx^1$ is exact, $adx^0 + bdx^1 = d\phi$, and $\frac{\partial^2\phi}{\partial x^{0^2}} = \frac{\partial^2\phi}{\partial x^{1^2}}$; that is, $\Box\,\phi = 0$. In conclusion

$$X = \frac{1}{2}\left(\frac{\partial\phi}{\partial x^0} + \frac{\partial\phi}{\partial x^1}\right)(\partial_0 + \partial_1) + \frac{1}{2}\left(\frac{\partial\phi}{\partial x^0} - \frac{\partial\phi}{\partial x^1}\right)(\partial_0 - \partial_1),$$

which proves the result.

1.2 The Einstein equations

The Einstein equations link the metric $g_{\mu\nu}$ to the matter fields ψ, with energy-momentum tensor $T(\psi)_{\mu\nu}$, by

$$\begin{aligned} G_{\mu\nu} &= 8\pi T_{\mu\nu} \\ F(\psi) &= 0, \end{aligned} \tag{1.2.1}$$

where $G_{\mu\nu} \equiv R_{\mu\nu} - \frac{1}{2} g_{\mu\nu} R$ is the Einstein tensor and R, the scalar curvature, is the trace of the Ricci tensor, $R = g^{\mu\nu} R_{\mu\nu}$. The second line of (1.2.1) summarizes the dynamical equations of the matter fields. As a consequence of the twice contracted Bianchi identities the energy momentum tensor $T_{\mu\nu}$ satisfies the local conservation laws

$$D^\mu T_{\mu\nu} = 0\,.$$

It is also important to emphasize that a solution of the coupled Einstein field matter equations is, in fact, a class of equivalence of solutions. More precisely, if Φ is a diffeomorphism of $\mathcal{M}$, then $\{\mathcal{M}, \mathbf{g}, \psi\}$ and $\{\mathcal{M}, \Phi^*\mathbf{g}, \Phi^*\psi\}$ describe the "same" solution of the Einstein equations.

1.2.1 The initial value problem, initial data sets and constraint equations

The general formulation of the initial value problem is given in the definitions below.

Definition 1.2.1 *An initial data set is given by a set $\{\Sigma, \overline{g}, \overline{k}, \overline{\psi}\}$ where Σ is a three-dimensional manifold, $\overline{\psi}$ is the prescribed matter fields on Σ, $\overline{g}$ is a Riemannian metric,*[17] *and $\overline{k}$ is a covariant symmetric tensor field satisfying the constraint equations:*[18]

$$\begin{aligned} \nabla^j \overline{k}_{ij} - \nabla_i \mathrm{tr}\, \overline{k} &= 8\pi j_i \\ \overline{R} - |\overline{k}|^2 + (\mathrm{tr}\, \overline{k})^2 &= 16\pi\rho. \end{aligned} \tag{1.2.2}$$

Two initial data sets $\{\Sigma, \overline{g}_1, \overline{k}_1, \overline{\psi}_1\}$ and $\{\Sigma, \overline{g}_2, \overline{k}_2, \overline{\psi}_2\}$ are said to be equivalent if there exists a diffeomorphism χ of Σ such that $\bar{g}_1 = \chi^* \bar{g}_2$, $\bar{k}_1 = \chi^* \bar{k}_2$, $\bar{\psi}_1 = \chi^* \bar{\psi}_2$.

Definition 1.2.2 (The Cauchy problem) *To solve the Einstein matter field equations with a given initial data set means finding a four-dimensional manifold $\mathcal{M}$, a Lorentz metric g and fields ψ satisfying the coupled Einstein matter equations as well as an imbedding*

$$i : \Sigma \to \mathcal{M},$$

such that $i^*(g) = \overline{g}$, $i^*(k) = \overline{k}$, $i^*(\psi) = \overline{\psi}$ where g is the induced metric and k is the second fundamental form (the extrinsic curvature) of the submanifold $i(\Sigma) \subset \mathcal{M}$.[19] *The constraint equations for $\overline{g}$ and $\overline{k}$ are thus the pullback of the Codazzi and Gauss equations induced on $i(\Sigma)$. Two equivalent initial data sets are supposed to lead to equivalent solutions.*

Definition 1.2.3 *The spacetime manifold $\mathcal{M}$ defined above is called a development of the initial data set $\{\Sigma, \overline{g}, \overline{k}, \overline{\psi}\}$.*

Definition 1.2.4 *If $i(\Sigma)$ is a Cauchy surface,*[20] *the spacetime $\mathcal{M}$ is called a Cauchy development of the initial data set $\{\Sigma, \overline{g}, \overline{k}, \overline{\psi}\}$.*

---

[17] The differentiability class of g and k will be discussed later on.

[18] Here $j_i = T_{0i}(\overline{\psi})$, $\rho = T_{00}(\overline{\psi})$. $\overline{R}_{ij}$ denotes the Ricci curvature of Σ with the metric $\overline{g}$, $\overline{R}$ is the scalar curvature.

[19] The second fundamental form of the hypersurface $i(\Sigma)$, immersed in a manifold $(\mathcal{M}, \mathbf{g})$ is given by $k = -\frac{1}{2}\mathcal{L}_N \mathbf{g}$, where N is the unit normal vector field of $i(\Sigma)$. In $\mathcal{M}$ arbitrary coordinates, $k_{\mu\nu} = -\Pi_\mu^\rho \Pi_\nu^\sigma D_\rho n_\sigma$, where n is the covariant unit normal of $i(\Sigma)$, $n_\mu = g_{\mu\nu} N^\nu$, Π_μ^ρ is the projection operator on $T(i(\Sigma))$ and D is the covariant derivative.

[20] $i(\Sigma)$ is a Cauchy surface if every non-spacelike inextendible curve in $\mathcal{M}$ intersects it exactly once. In this case the spacetime $\mathcal{M}$ is globally hyperbolic; see [Haw-El], Chapter 6.

From now on we will restrict ourselves to the Einstein equations in the vacuum case. In other words we assume everywhere that $T(\psi) = 0$. Therefore the Einstein equations take the form

$$R_{\mu\nu} = 0\,.$$

1.3 Local existence for Einstein's vacuum equations

1.3.1 Reduction to the nonlinear wave equations

As discussed above a solution of the Einstein equations consists of an equivalence class of spacetimes $(\mathcal{M}, \mathbf{g})$ relative to diffeomorphisms of $\mathcal{M}$. Thus to construct a solution of the Einstein equations we have the freedom to choose an appropriate member of that class. Typically this means choosing a special coordinate system such as the wavelike coordinate system. Expressed relative to wavelike coordinates, the Einstein vacuum equations take the form of a system of second-order quasilinear wave equations, hyperbolic in the sense of Leray, [Le], to which one can apply well-established analytic methods; see Chapter 2. Using this formalism, Y. Choquet-Bruhat [Br1], [Br2] was the first to prove a comprehensive local existence and uniqueness result for the initial value problem in general relativity. Later the result was revisited and improved by many authors; see in particular the result of T. Hughes, T. Kato and J. Marsden, [Hu-Ka-Ms] and of A. Fisher and J. Mardsen [F-Ms1], [F-Ms2], who expressed the reduced equations, (1.3.1), in the form of a symmetric hyperbolic system to which they could apply the general theory developed by T. Kato; see also [Ch-Mu]. Recently, these classical results were improved even further in [Kl-Rodn1]–[Kl-Rodn3], see also [Kl-Rodn] and a short dicussion of this at the end of Section 2.1.1.

In what follows we shall give a short review of the wavelike coordinates and the derivation of the reduced Einstein equations. Let $\hat{g}$ be a given Lorentz metric on $\mathcal{M}$. For a metric g on $\mathcal{M}$, we introduce[21]

$$V^\alpha = g^{\mu\nu}(\Gamma^\alpha_{\mu\nu} - \hat{\Gamma}^\alpha_{\mu\nu}),$$

and define, with $\hat{D}$ the covariant derivative associated to the metric $\hat{g}$,

$$R^{(h)}_{\alpha\beta} = R_{\alpha\beta} - \frac{1}{2}(\hat{D}_\alpha V_\beta + \hat{D}_\beta V_\alpha)\,.$$

A simple calculation shows that

$$R^{(h)}_{\alpha\beta} = -\frac{1}{2}\,\Box_g g_{\alpha\beta} + H_{\alpha\beta}(g, \partial g),$$

with H a quadratic expression depending on g and its first derivatives[22] and

$$\Box_g g_{\alpha\beta} = g^{\mu\nu}\partial_\mu\partial_\nu g_{\alpha\beta}\,.$$

---

[21] Although $\Gamma^\alpha_{\mu\nu}$ and $\hat{\Gamma}^\alpha_{\mu\nu}$ are not tensors, their difference is.

[22] Of course, H also depends on $\hat{g}$, $\partial\hat{g}$ and $\partial^2\hat{g}$.

The condition $V^\alpha = 0$ is satisfied if and only if the identity map

$$\mathrm{Id} : (\mathcal{M}, g) \to (\mathcal{M}, \hat{g})$$

is a wave map. This means that the coordinates x^α, $\{x^0 = t, x^1, x^2, x^3\}$, have to satisfy the equation

$$\Box_g x^\alpha + \hat{\Gamma}^\alpha_{\beta\gamma} \partial_\lambda x^\beta \partial^\lambda x^\gamma = -(\hat{\Gamma}^\alpha - \Gamma^\alpha) = 0.$$

In that case, the vacuum equations are *reduced* to $R^{(h)}_{\alpha\beta} = 0$, that is, to

$$\Box_g g_{\alpha\beta} = 2H_{\alpha\beta}(g, \partial g), \tag{1.3.1}$$

which is a *weakly coupled* system[23] of nonlinear wave equations. The general case described above was used in the work of H. Friedrich; see [Fr3] and also [Fr4].

If the background metric is the Minkowski metric, $\hat{g}_{\alpha\beta} = \eta_{\alpha\beta}$, we have $\hat{\Gamma}^\alpha_{\mu\nu} = 0$ and $V^\alpha = g^{\mu\nu}\Gamma^\alpha_{\mu\nu} = \Gamma^\alpha$. Condition (1.3.1) reduces to the more familiar wavelike coordinate condition

$$\Box_g x^\alpha = 0, \qquad \alpha = 0, 1, 2, 3. \tag{1.3.2}$$

In what follows we restrict ourselves to the choice $\hat{g}_{\alpha\beta} = \eta_{\alpha\beta}$. To construct solutions of the Einstein equations one solves the reduced equations (1.3.1), subject to initial conditions satisfying the constraint equations $n^\nu G_{\nu\mu} = 0$, where n is the unit normal to Σ and $G_{\mu\nu}$ is the Einstein tensor. Observe that in the constraint equations the second derivatives with respect to t of the metric are absent. Moreover, if we choose Σ as the hyperplane $t = 0$ with $g_{\mu\nu}$ a Gaussian metric on it, $g_{00}(0, x) = -1$, $g_{0i}(0, x) = 0$, and we define $k_{ij} = -\frac{1}{2}\frac{\partial g_{ij}}{\partial t}$, the constraint equations coincide with equations (1.2.2) with $\rho = j_i = 0$.

The main goal of the approach described above is, therefore, to reduce the general system of Einstein equations $R(g)_{\mu\nu} = 0$ to the hyperbolic system

$$R^{(h)}(g)_{\mu\nu} = 0. \tag{1.3.3}$$

This has to be connected with the initial value formulation. In this respect the crucial observation is that if the constraints and the condition $\Gamma^\alpha = 0$ are satisfied by the initial data, then they are automatically propagated by the solutions of the reduced equations (1.3.1). The precise statement is given by the following proposition whose proof is in [Ch-Mu] and [F-Ms1]; see also [Br1].

Proposition 1.3.1 *Let $g_{\mu\nu}$ be the components of a metric tensor g written in a specific set of coordinates, x^μ, with $x^0 = t$, $x = (x^1, x^2, x^3)$, such that:*

(i) On $\mathcal{M}$ it satisfies the reduced Einstein equations $R^{(h)}_{\mu\nu}(g) = 0$. [24]

(ii) On $\Sigma \equiv \Sigma_{t=0}$ it satisfies the initial conditions

$$\left(g_{\mu\nu}(0, x), \frac{\partial g_{\mu\nu}}{\partial t}(0, x)\right) = (\phi_{\mu\nu}(x), \psi_{\mu\nu}(x)),$$

[23] The system has a diagonal structure with respect to the highest order terms.

[24] Here on $\mathcal{M}$ we indicate the region of R^{3+1} where the reduced equations are satisfied.

where $\{\phi_{\mu\nu}(x), \psi_{\mu\nu}(x)\}$ *satisfy the conditions* $\overline{\Gamma}^{\alpha}(0,x) = 0$ *and the constraints* $\bar{g}^{0\nu}\overline{G}_{\nu\mu}(x) = 0$.[25] *Then* $\Gamma^\alpha(x^\lambda) = 0$ *on all of* $\mathcal{M}$ *and, therefore,* $g_{\mu\nu}$ *is also a solution of the Einstein equations* $R_{\mu\nu}(g) = 0$.

The proof of the proposition is achieved in two steps which we sketch below; see also [Hu-Ka-Ms], [Wa2],

Step 1: Let $\{\Sigma, \bar{g}, \bar{k}\}$ be an initial data set. Let us require that the coordinate system be also Gaussian on Σ, adding therefore to $\bar{g}_{ij} \equiv g_{ij}(0,x)$, $g_{00}(0,x) = -1,\ g_{0i}(0,x) = 0$. There exists a coordinate transformation ξ, $x^\lambda \to x'^\lambda = \xi^\lambda(x^\alpha)$, such that on Σ

$$
\begin{aligned}
&(\xi^0(0,x), \xi^j(0,x)) = (0, x^j)\\
&\left(\frac{\partial \xi^0}{\partial t}(0,x), \frac{\partial \xi^j}{\partial t}(0,x)\right) = (1,0)\\
&\frac{\partial^2 \xi^\lambda}{\partial t^2}(0,x) = -\Gamma^\lambda(0,x).
\end{aligned}
\tag{1.3.4}
$$

In the new coordinates x' we have

$$
\begin{aligned}
&g'_{\mu\nu}(0,x) = g_{\mu\nu}(0,x)\ ,\ k'_{ij}(0,x) = k_{ij}(0,x)\\
&k'_{0i}(0,x) = -\frac{1}{2}\frac{\partial g'_{0i}}{\partial t'}(0,x) = k_{0i}(0,x) - \frac{1}{2}g_{ij}(0,x)\Gamma^j(0,x)\\
&k'_{00}(0,x) = -\frac{1}{2}\frac{\partial g'_{00}}{\partial t'}(0,x) = k_{00}(0,x) + \Gamma^0(0,x),
\end{aligned}
\tag{1.3.5}
$$

and, from the transformation rule of Γ^α,

$$
\Gamma'^\alpha = \Gamma^\rho \frac{\partial \xi^\alpha}{\partial x^\rho} - g^{\rho\sigma}\frac{\partial^2 \xi^\alpha}{\partial x^\rho \partial x^\sigma} = -\Box_g \xi^\alpha,
\tag{1.3.6}
$$

it follows that $\Gamma'^\alpha(0,x) = 0$. Moreover the conditions $\frac{\partial \Gamma'^\mu}{\partial t} = 0$ on Σ are automatically fulfilled from the constraint equations $G^0_\mu = 0$ when g satisfies the reduced equations.

Step 2: $\Gamma^\alpha(0,x) = 0$ and $\frac{\partial \Gamma^\mu}{\partial t}(0,x) = 0$ imply $\Gamma^\alpha = 0$ on all of $\mathcal{M}$. This is achieved by observing that the twice contracted Bianchi equations lead to the following system of linear equations satisfied by the Γ^α

$$
\frac{1}{2}g^{\beta\gamma}\frac{\partial^2 \Gamma^\mu}{\partial x^\beta \partial x^\gamma} + A^{\beta\mu}_\nu(g, \partial g)\frac{\partial \Gamma^\nu}{\partial x^\beta} = 0,
$$

where $g_{\rho\sigma}$ is a solution of the reduced equations (1.3.3). The uniqueness properties of the initial value problem for such systems proves the result.

Therefore, given Proposition 1.3.1, to reduce the solution of the Einstein vacuum equations to the solution of the reduced system (1.3.1), we need initial data that satisfy the constraint equations and also the conditions $\Gamma^\alpha = 0$.

[25] Consistently with the initial data $\bar{g}, \bar{k}$ on Σ, $\overline{\Gamma}^\alpha = \bar{g}^{\mu\nu}\Gamma^\alpha_{\mu\nu}(\bar{g}, \bar{k})$ and $\overline{G}_{\nu\mu}$ is the expression of the Einstein tensor $G_{\nu\mu}$ on Σ relative to $\bar{g}, \bar{k}$.

Let $\{\Sigma, \bar{g}, \bar{k}\}$ be an initial data set and let the initial conditions for the reduced system (1.3.1) given by

$$g_{\mu\nu}(0,x) = \phi_{\mu\nu}\ , \quad \frac{\partial g_{\mu\nu}}{\partial t}(0,x) = \psi_{\mu\nu}. \tag{1.3.7}$$

We have to connect the latter, $\{\phi_{\mu\nu}, \psi_{\mu\nu}\}$, to the former, $\{\bar{g}_{ij}, \bar{k}_{ij}\}$. To achieve that we restrict to a Gaussian coordinate system requiring that $\phi_{00} = -1$, $\phi_{0i} = 0$. Then from the first line of (1.3.5) we obtain immediately that

$$\phi_{ij} = \bar{g}_{ij},\ \ \psi_{ij} = -2\bar{k}_{ij}\ .$$

The remaining data ψ_{00}, ψ_{0i} are determined from (1.3.4) and the next two lines of (1.3.5). The result is

$$\psi_{00} = -4tr_{\bar{g}}\bar{k}_{ij}\ ,\ \ \psi_{0i} = \Delta_{\bar{g}}x_i. \tag{1.3.8}$$

Proof. The third line of (1.3.5) can be rewritten as

$$\psi_{00} = \frac{\partial g_{00}}{\partial t}(0,x) - 2\Gamma^0(0,x) = -4\mathrm{tr}_{\bar{g}}\bar{k}, \tag{1.3.9}$$

where the last equality in (1.3.9) comes from the explicit expression of $\Gamma^0(0,x)$, namely, $\Gamma^0(0,x) = \frac{1}{2}\frac{\partial g_{00}}{\partial t}(0,x) + 2\mathrm{tr}_{\bar{g}}\bar{k}$. The second relation follows from the explicit computation of the second line of (1.3.5) which gives

$$\psi_{0i} = g_{ij}(0,x)^{(3)}\Gamma^j(0,x) = -\Delta_{\bar{g}}\xi^i,$$

where ${}^{(3)}\Gamma^j$ is the contracted Christoffel symbol relative to the Riemannian metric of Σ, $\bar{g}_{ij}$. The last equality arises from equation (1.3.4) and the definition $\Delta_{\bar{g}}\xi^i \equiv \bar{g}^{ls}\frac{\partial^2}{\partial x^l \partial x^s}\xi^i - {}^{(3)}\Gamma^j \frac{\partial}{\partial x^l}\xi^i$.

In view of the fact that $G^0{}_\mu$ does not depend on $\partial_t g_{0\mu}$ it follows immediately that the constraint equations are also satisfied for this choice of initial data.

1.3.2 Local existence for the Einstein vacuum equations using wave coordinates

Before stating the local existence and uniqueness theorem for the Einstein equations,[26] we recall the definition of local Sobolev spaces H^s_{loc}.

Definition 1.3.1 *Given a three-dimensional Riemannian manifold Σ and an integer $s \geq 0$,[27] we say that $f \in H^s_{loc}$ if, for any compact subset $K \subset \Sigma$, we have $\int_K |D^s f|^2 < \infty$.*

Theorem 1.3.1 (Local existence) *Let $\{\Sigma, \bar{g}, \bar{k}\}$ be an initial data set and assume that Σ admits a locally finite C^1 covering by open coordinate charts $\{U_\alpha\}$[28] such that $(\bar{g}, \bar{k}) \in$*

[26] This version is due to Hughes, Kato and Mardsen, see [Hu-Ka-Ms].

[27] The definition can be extended to the noninteger s with the help of the Fourier transform.

[28] This means that any point in Σ has a neighborhood which intersects only a finite number of the open sets U_α. The sets U_α are related by C^1 coordinate transformations.

$H^s_{loc}(U_\alpha) \times H^{s-1}_{loc}(U_\alpha)$ with some $s > \frac{5}{2}$. There exists a globally hyperbolic development $(\mathcal{M}, \mathbf{g})$ of $\{\Sigma, \bar{g}, \bar{k}\}$ for which Σ is a Cauchy hypersurface. Moreover $\mathbf{g}$ is uniquely determined, up to an H^{s+1} coordinate transformation, by $\{\bar{g}, \bar{k}\}$.

Remark: The result $s > \frac{5}{2}$ has been recently significantly improved to $s > 2$; see [Kl-Rodn]. The proof, however, is based on microlocal techniques which are beyond the scope of this book. We have already indicated how the existence part of Theorem 1.3.1 can be reduced to the study of the initial value problem for (1.3.1). In Section 2.1.1. we shall analyze in detail the main analytic ideas for constructing solutions to hyperbolic systems such as (1.3.1).

Sketch of the proof. According to the discussion in the previous section, it suffices to prove the existence of solutions to the reduced Einstein equations (1.3.1). By a simple domain of dependence argument it suffices to consider that the initial data are supported in a fixed coordinate patch. Thus, the problem is reduced to the initial value problem for systems of nonlinear wave equations. We shall discuss this issue in more detail in Section 1 of the next chapter. For a detailed account of the proof of Theorem 1.3.1 we refer the reader to [Hu-Ka-Ms]; previous proofs of the local existence theorem are in [Br1] and [F-Ms1]; for a survey see [Fr-Re]. The result applies, in particular, to asymptotically flat initial data sets.[29] More precise information concerning the behavior at spacelike infinity for asymptotically flat initial data sets can be derived by using weighted Sobolev spaces; see [Ch-Mu].

Concerning uniqueness we observe that the Cauchy development $(\mathcal{M}, \mathbf{g})$ described above is not unique. In fact a coordinate transformation $z = \sigma(x)$, which on Σ takes the form

$$\sigma^\mu(x) = x^\mu \,, \quad \frac{\partial \sigma^\mu}{\partial x^0}(x) = \delta^\mu_0 \,, \quad \frac{\partial^2 \sigma^\mu}{\partial x^{02}}(x) = 0 \tag{1.3.10}$$

changes neither Γ^α on Σ_0 nor the other initial conditions. Therefore, if ξ is a transformation to wavelike coordinates that on Σ connects $\{\Sigma, \bar{g}, \bar{k}\}$ with the initial conditions $(g_{\mu\nu}(0,x) = \phi_{\mu\nu}, \frac{\partial g_{\mu\nu}}{\partial t}(0,x) = \psi_{\mu\nu})$ (see (1.3.7)), then $\{\sigma \circ \xi\}$ is too.

Let g_{re} be the solution of the reduced equations (1.3.3) with initial data $(\phi_{\mu\nu}, \psi_{\mu\nu})$, then $g(x) = \xi^* g_{re}(\xi(x))$ and $\tilde{g}(x) = (\sigma \circ \xi)^* g_{re}(\sigma \circ \xi(x))$ are different solutions of the vacuum Einstein equations, $R_{\mu\nu} = 0$, corresponding to the same initial data set.

To prove the uniqueness in the sense stated in the theorem, we also have to show that any two developments of the Einstein equations corresponding to the same initial conditions [30] are connected by a coordinate transformation, a diffeomorphism. The idea of the proof is very simple: If $\tilde{g}$ and g are two Einstein metrics corresponding to the same initial data set, then on Σ they share the same Γ^α. We then define, according to (1.3.7), the coordinate transformations $\tilde{\xi}$ and ξ, respectively, and check that the two sets of initial conditions for the reduced equations (1.3.1) coincide. Therefore the spacetime metrics $g, \tilde{g}$ produce two solutions, $\xi^{*-1} g$, $\tilde{\xi}^{*-1}\tilde{g}$, of the reduced Einstein equations with the same

[29] For these initial data sets one can derive a uniform existence time. The uniformity of time can be made precise by using the geodesic distance function from a point of the spacetime to the initial hypersurface.

[30] The same applies to equivalent initial data sets.

initial conditions and satisfying $\Gamma^\alpha = 0$. In view of the uniqueness results for hyperbolic systems, the two solutions coincide. Then the composition of the transformations $\tilde{\xi}^{-1} \circ \xi$ gives the diffeomorphism we are looking for.

1.3.3 General foliations of the Einstein spacetime

We recall the following result due to R. Geroch, [Ge], see also [Haw-El] Chapter 6.

Theorem 1.3.2 *Assume that the spacetime $(\mathcal{M}, \mathbf{g})$ is globally hyperbolic,*[31] *then $(\mathcal{M}, \mathbf{g})$ can be foliated along a timelike direction and is diffeomorphic to $R \times \mathcal{S}$ where $\mathcal{S}$ is a three-dimensional Riemannian manifold.*

Sketch of the proof. To construct the diffeomorphism $\mathcal{T} : R \times \mathcal{S} \to \mathcal{M}$ one proceeds in the following steps.

1. One shows first that there exists a continuous function $t(\cdot)$ on $\mathcal{M}$ such that

$$\Sigma_a = \{p \in \mathcal{M} | t(p) = a\}$$

is a Cauchy hypersurface. We identify $\mathcal{S}$ with Σ_0. Then one proves, using a smoothing procedure in the definition of the time function, see [Se], that there exists a global C^2 time function t whose level sets are Cauchy hypersurfaces.

2. One defines on $\mathcal{M}$ a timelike vector field V such that, in view of the properties of the Cauchy hypersurfaces, its integral curves $\Psi(s; p)$ solution of

$$\frac{d\Psi^\mu}{ds} = V^\mu(\Psi(s))$$

define a map $\beta : q \in \mathcal{M} \to \beta(q) \in \mathcal{S}$ through the relation

$$q = \Psi(\bar{s}(q); \beta(q)) .$$

3. One defines the diffeomorphism $\mathcal{T}$ through the relation

$$\mathcal{T}^{-1}(q) = (t(q), \beta(q)) .$$

Therefore the diffeomorphism $\mathcal{T}$ is specified once the function $t(q)$ and the vector field

$$V|_q = \left.\frac{\partial \Psi^\mu}{\partial s}\right|_q \frac{\partial}{\partial x^\mu}$$

are defined on on $\mathcal{M}$. In general, the vector field V is not orthogonal to the hypersurface Σ_t. To decompose it into its orthogonal and tangent components, let us introduce the timelike vector field orthogonal to Σ_t,

$$F = \left(g^{\mu\nu} \frac{\partial t}{\partial x^\nu}\right) \frac{\partial}{\partial x^\mu}, \tag{1.3.11}$$

[31] A set $\mathcal{N}$ is *globally hyperbolic* if the strong causality assumption holds on $\mathcal{N}$ and for any two points $p, q \in \mathcal{N}$, $J^+(p) \cap J^-(q)$ is compact and contained in $\mathcal{N}$. For a thorough discussion of the properties of globally hyperbolic spacetimes see [Haw-El], Chapter 6.

and define the future directed unit normal vector field to Σ_t,

$$N = \frac{1}{(-\mathbf{g}(F, F))^{\frac{1}{2}}} F. \tag{1.3.12}$$

Then one decomposes V into components parallel and orthogonal to Σ_t, $V = V_{\|} + V_{\perp}$ where

$$\begin{aligned} V_{\perp}{}^{\mu} &= -(V^{\nu}N_{\nu})N^{\mu} = \frac{\mathbf{g}(V, F)}{\mathbf{g}(F, F)} F^{\mu} \\ V_{\|}{}^{\mu} &= V^{\mu} - \frac{\mathbf{g}(V, F)}{\mathbf{g}(F, F)} F^{\mu} = g^{\mu\nu}(g_{\nu\rho} + N_{\nu}N_{\rho})V^{\rho} = h^{\mu}_{\rho} V^{\rho}, \end{aligned} \tag{1.3.13}$$

and $h^{\mu}_{\nu} \equiv (g^{\mu}_{\nu} + N^{\mu}N_{\nu})$ is the projection tensor on Σ_t. The function

$$\Phi(q) \equiv (-\mathbf{g}(q)(F, F))^{-\frac{1}{2}} \tag{1.3.14}$$

is called the *lapse function* and the vector field tangent to Σ_t

$$X^{\nu}(q) \equiv \frac{V_{\|}{}^{\nu}(q)}{\mathbf{g}(q)(V, F)}, \tag{1.3.15}$$

is called the *shift vector.*

Lemma 1.3.1 *The following relation holds*

$$\frac{\partial}{\partial t} = \Phi N + X .$$

Proof. Observe that from

$$t = t(q) = t(x^{\mu}(q)) = t(\Psi^{\mu}(s(t, p); p),$$

we have

$$1 = \frac{\partial t}{\partial t} = \frac{\partial t}{\partial x^{\mu}} \frac{\partial \psi^{\mu}}{\partial s} \frac{\partial s}{\partial t} = \mathbf{g}(F, V) \frac{\partial s}{\partial t}. \tag{1.3.16}$$

Therefore

$$\begin{aligned} \frac{\partial}{\partial t} &= \frac{\partial x^{\mu}}{\partial t} \frac{\partial}{\partial x^{\mu}} = \frac{\partial s}{\partial t} \frac{\partial \Psi^{\mu}}{\partial s} \frac{\partial}{\partial x^{\mu}} = \frac{1}{\mathbf{g}(F, V)} V_{\perp} + \frac{1}{\mathbf{g}(F, V)} V_{\|} \\ &= \frac{F}{\mathbf{g}(F, F)} + X = \Phi N + X . \end{aligned}$$

The next lemma, whose proof is in the appendix to this chapter, gives the explicit form of the metric components in the coordinates $(t(q), \beta(q))$.

Lemma 1.3.2 *Choosing as coordinates of the generic point $q \in \mathcal{M}$*

$$x^0(q) = t(q) \ \text{ and } \ \tilde{x}^i(q) = \beta^i(q) ,$$

we may express the metric tensor $\mathbf{g}$ *as*

$$\mathbf{g}(q)(\cdot, \cdot) = -\Phi^2(q)dt^2 + g_{ij}(q)(d\tilde{x}^i + X^i dt)(d\tilde{x}^j + X^j dt),$$

where

$$-\Phi^2(q) = \mathbf{g}(q)(F, F)^{-1} \ , \ \ g_{0i}(q) = X_i \ .$$

In what follows we assume our spacelike foliation to be given by the level hypersurfaces of the time function t. Let (g_{ij}, k_{ij}) be the induced metric and the second fundamental form on Σ_t, with k given by $k_{ij} = -\frac{1}{2}(\mathcal{L}_N g)_{ij}$. Consider a frame $\{e_0 = N, e_1, e_2, e_3\}$ satisfying $[\partial_t, e_i] = 0$. We obtain the following evolution and constraint equations

Evolution equations:

$$\begin{aligned}\partial_t g_{ij} &= -2\Phi k_{ij} + \mathcal{L}_X g_{ij} \\ \partial_t k_{ij} &= -\nabla_i \nabla_j \Phi + \Phi(R_{ij} + \mathrm{tr}k k_{ij} - 2k_{im}k^m_{\ j}) + \mathcal{L}_X k_{ij}\end{aligned} \tag{1.3.17}$$

Constraint equations:

$$\begin{aligned}&R - |k|^2 + (\mathrm{tr}k)^2 = 0 \\ &\nabla_i \mathrm{tr}k - \nabla^j k_{ij} = 0\end{aligned} \tag{1.3.18}$$

with R_{ij} denoting the Ricci curvature of the induced Σ_t metric.[32][33]

1.3.4 *Maximal foliations of Einstein spacetime*

Let us recall that in a Lorentz manifold a maximal hypersurface is one that is spacelike and maximizes the volume among all possible compact perturbations of it. It satisfies the equation $\mathrm{tr}\, k = 0$. The constraint equations for the level hypersurfaces of a maximal foliation take, in this case, the form

$$\begin{aligned}&R - |k|^2 = 0 \\ &\nabla^j k_{ij} = 0 \ , \quad \mathrm{tr}k = 0.\end{aligned} \tag{1.3.19}$$

1.3.5 *A proof of local existence using the maximal foliation*

We review the proof of local existence and uniqueness for the Einstein vacuum equations in the maximal foliation. The specific gauge conditions are $X = 0$, $\mathrm{tr}k = 0$. Thus the equations, (1.3.17), (1.3.18) take the form

Evolution equations:

$$\begin{aligned}\frac{\partial g_{ij}}{\partial t} &= -2\Phi k_{ij} \\ \frac{\partial k_{ij}}{\partial t} &= -\nabla_j \nabla_i \Phi + \Phi(R_{ij} - 2k_{il}k^l_{\ j}),\end{aligned} \tag{1.3.20}$$

---

[32] See [An-Mon].

[33] If the Einstein spacetime is not a vacuum one the previous equations take the form

$$\begin{aligned}&\partial_t g_{ij} = -2\Phi k_{ij} + \mathcal{L}_X g_{ij} \\ &\partial_t k_{ij} = -\nabla_i \nabla_j \Phi + \Phi(-\mathbf{R}_{ij} + R_{ij} + \mathrm{tr}k k_{ij} - 2k_{im}k^m_{\ j}) + \mathcal{L}_X k_{ij} \\ &R - |k|^2 + (\mathrm{tr}k)^2 = 2\mathbf{R}_{TT} + \mathbf{R} \ , \quad \nabla_i \mathrm{tr}k - \nabla^j k_{ij} = \mathbf{R}_{Ti}.\end{aligned}$$

Constraint equations:

$$
\begin{aligned}
&R - |k|^2 = 0 \\
&\mathrm{div} k = 0\,, \quad \mathrm{tr}\, k = 0,
\end{aligned} \tag{1.3.21}
$$

Lapse equation:[34]

$$
\triangle \Phi = |k|^2 \Phi. \tag{1.3.22}
$$

It is easy to check that the evolution–lapse equations preserve the constraint equations (see Proposition 1.3.1). In other words it suffices to assume that the constraints are satisfied by the initial data set $\{\Sigma, g_0, k_0\}$.[35] We can then try to solve the evolution equations for g, k coupled with the elliptic equation satisfied by Φ. This system is however not in standard hyperbolic form. This is due not only to the fact that the lapse equation is elliptic but also, ignoring Φ, to the fact that the evolution system for g, k is not hyperbolic. Indeed, the principal part of the Ricci tensor R_{ij}, expressed relative to the metric g_{ij}, is not elliptic. This problem can be overcome by differentiating the evolution equation for k_{ij} with respect to t.

The detailed proof is given in [Ch-Kl], Chapter 10. The final result is as follows.

Theorem 1.3.3 *Let $\{\Sigma, g_0, k_0\}$ be an initial data set satisfying the following conditions:*

(i) $\{\Sigma, g_0\}$ is a complete Riemannian manifold diffeomorphic to R^3.

(ii) The isoperimetric constant $\mathcal{I}(\Sigma, g_0)$ is finite, where $\mathcal{I}$ is defined to be

$$
\sup_S \frac{V(S)}{A(S)^{3/2}}.
$$

with S an arbitrary surface in Σ, $A(S)$ its area and $V(S)$ the enclosed volume.

(iii) The Ricci curvature $Ric(g_0)$, relative to the distance function d_0 from a given point O, satisfies

$$
Ric(g_0) \in H_{2,1}(\Sigma, g_0).
$$

(iv) k is a 2-covariant symmetric trace-free tensorfield on Σ satisfying

$$
k \in H_{3,1}(\Sigma, g_0),
$$

where for a given tensor field h, $\|h\|_{H_{s,\tau}(\Sigma,g_0)}$ denotes the norm

$$
\|h\|_{H_{s,\tau}(\Sigma,g_0)} = \left(\sum_{i=0}^{s} \int \sigma_0^{2\tau+2i} |\nabla_0^i h|^2 d\mu_{g_0} \right)^{1/2}
$$

and $\sigma_0 = \sqrt{1 + d_0^2}$.[36]

(v) (g_0, k_0) satisfy the constraint equations on Σ.

---

[34] In the asymptotically flat case one has to normalize Φ by the condition $\Phi \to 1$ at spacelike infinity.

[35] $\Sigma \equiv i(\Sigma)\,, i^*(g_0) = \bar{g}\,,\ i^*(k_0) = \bar{k}$.

[36] These weighted Sobolev norms, see [Ch-Mu] and [Br-Ch2], give more control on the behavior of solutions at spacelike infinity. In particular they prove an H^s version of the propagation of the asymptotic flatness condition.

Then there exists a unique, local-in-time smooth development, foliated by a normal, maximal time foliation t with a range in some interval $[0, t_*]$ and with $t = 0$ corresponding to the initial slice Σ. Moreover

$$\begin{aligned} g(t) - g_0 &\in C^1([0, t_*];\ H_{3,1}(\Sigma, g_0)) \\ k(t) &\in C^0([0, t_*];\ H_{3,1}(\Sigma, g_0)). \end{aligned}$$

1.3.6 Maximal Cauchy developments

We recall the general result of Y. Choquet-Bruhat and R. Geroch, [Br-Ge] concerning the existence and uniqueness of a maximal Cauchy development of an initial data set. Without going into details it is intuitively clear what it means for one Cauchy development to be an extension of another.[37]

An extension is called proper if it is strictly larger than the development it extends. A Cauchy development that has no proper extensions is called maximal. The following theorem states precisely the result of Y. Choquet-Bruhat and R. Geroch, but with better differentiability conditions due to S.W. Hawking and G.F.R. Ellis, see [Haw-El], Chapter 7.

Theorem 1.3.4 *Let $\{\Sigma, \bar{g}, \bar{k}\}$ be an arbitrary $\mathcal{H}^s$ initial data set with $s \geq 4$. There exists a unique, future, maximal globally hyperbolic vacuum extension (MGHVE) $(\mathcal{M}^*, g^*)$. Moreover, the development can be represented by $\mathcal{M}^* = [0, 1) \times \Sigma$ and $g^*(t, \cdot) \in C^0\Big([0, 1); \mathcal{H}^s(\Sigma)\Big) \cap C^1\Big([0, 1); \mathcal{H}^{s-1}(\Sigma)\Big)$.*

Although the result of Bruhat and Geroch can be deduced quite easily from Theorem 1.3.1, it is conceptually very important because it allows us to associate, *to any initial data set*, a unique maximal globally hyperbolic spacetime. Thus any construction, obtained by an evolutionary approach from initial data, must necessarily be included in the corresponding MGHVE spacetime which should, therefore, be viewed as our main object of study.

1.3.7 Hawking–Penrose singularities, the cosmic censorship

Soon after the formulation of the theory of general relativity it was realized that the Einstein equations could lead to the formation of singularities.[38] A standard example is given (see [Haw-El], Chapter 5), by the Friedman or Robertson–Walker spacetime with positive curvature, which evolves from the "big bang" singularity to the "big crunch" singularity. Therefore the question of whether singularities generally occurr in vacuum Einstein spacetimes has been an important and open question for years. This problem is considered, basically, satisfactorily settled by Hawking and Penrose in their famous singularity theorems; see [Haw-El].

[37] A development $(\mathcal{M}', \mathbf{g}')$ of $\{\Sigma, \bar{g}, \bar{k}\}$ is an extension of the development $(\mathcal{M}, \mathbf{g})$ of the same initial data set if there exists a diffeomorphism ψ from $\mathcal{M}$ to an open subset $\mathcal{U} \subset \mathcal{M}'$ which maps the metric $\mathbf{g}$ to the restriction of the metric $\mathbf{g}'$ on $\mathcal{U}$, $\psi(\mathcal{M}) = \mathcal{U}$, $\psi^*(\mathbf{g}'|_{\mathcal{U}}) = \mathbf{g}$. For a precise statement of the Bruhat–Geroch result see also [Br-Y].

[38] See [Chr] for a review of the problem.

Rephrased in the language of the initial value problem, the question is that of the time-like and null geodesic completeness of the maximal future Cauchy vacuum development. The singularity theorems answer this question in the negative. In particular, we recall the Penrose Theorem [Pe3] which, in the vacuum Einstein case, can be stated in the following way.

Theorem: *If in the initial data set $\{\Sigma, \overline{g}, \overline{k}\}$ Σ is noncompact and, moreover, contains a closed trapped surface $\mathcal{S}$,*[39] *then the corresponding maximal future development is incomplete.*

The singularity theorems motivated some efforts to try to formulate precise statements about the predictive power of the Einstein equations and the nature of the singularities. Along this line of thought Penrose proposed two "cosmic censorship conjectures".[40]

Penrose's first conjecture, called the "weak cosmic censorship", can be formulated in many ways. The version we state here[41] makes use of the following result, a direct corollary of the result proved in this book.[42]

Corollary 1.3.5 *For any asymptotically flat initial data set $\{\Sigma, \overline{g}, \overline{k}\}$ with maximal future development $(\mathcal{M}, \mathbf{g})$, one can find a suitable domain Ω_0 with compact closure in Σ such that the boundary of its domain of influence $I^+(\Omega_0)$ in $\mathcal{M}$ has complete null generating geodesics.*

The above corollary can be used to introduce the concept of complete future null infinity.[43]

Definition 1.3.2 *The maximal future Cauchy development $(\mathcal{M}, \mathbf{g})$ of an asymptotically flat initial data set possesses a complete future null infinity if, for any positive real number A, we can find a domain Ω containing the set Ω_0 of the previous corollary, such that the boundary $\mathcal{D}^-$ of the domain of dependance of Ω in $\mathcal{M}$ has the property that each of its null generating geodesics has a total affine length, in $\mathcal{D}^- \cap I^+(\Omega_0)$, greater than or equal to A.*

Weak cosmic censorship (WCC): *Generic asymptotically flat initial data have maximal future developments possessing a complete future null infinity.*

Remarks: This conjecture asserts that, for all but possibly an exceptional set of initial conditions, no singularities may be observed from infinity. In other words the singularities in general relativity must be hidden, generically, by regions of spacetime called black holes, in which all future causal geodesics remain necessarily trapped. So far the only

---

[39] A closed trapped surface is a C^2 compact, unbounded, spacelike two-dimensional surface such that a displacement of $\mathcal{S}$ in $\mathcal{M}$ along the congruence of the future null-outgoing directions decreases, pointwise, the area element.

[40] There are many references on this subject; see for instance [An-Mon] and the detailed discussion in [Chr].

[41] Due to D. Christodoulou [Ch6].

[42] A proof of Corollary 1.3.5 can be also derived indirectly from [Ch-Kl]. The result proved in this book avoids however a great deal of work.

[43] This concept is usually defined in the general relativity literature through the concept of a regular conformal compactification of a spacetime, by attaching a boundary at infinity. (The notion of conformal compactification, due to Penrose, is discussed in [Haw-El] and [Wa2].) The definition given here, due to [Ch6], avoids the technical issue of the specific degree of smoothness of the compactification.

satisfactory rigorous proof of the conjecture, due to Christodoulou, was obtained for the special case of spherically symmetric solutions of the Einstein equations coupled with a scalar field; see [Ch5]. Christodoulou had previously proved the existence of naked singularities for his model [Ch4] and thus had to show that the WCC conjecture holds true only in a generic sense.[44]

The weak cosmic censorship conjecture does not preclude, however, the possibility that singularities may be visible by local observers. This could lead to the paradoxical situation in which predicted outcomes of observations made by such observers are not unique. Since predictability is a fundamental requirement of classical physics it seems reasonable to want it to be valid throughout spacetime. Predictability is known to fail, however, within the black hole of a Kerr solution. In that case the maximal development of any complete spacelike hypersurface has a future boundary called a Cauchy horizon upon which the Kerr solution is perfectly smooth and yet beyond which there are many possible smooth extensions. This failure of predictability is due to a global pathology of the geometry of characteristics and not to a loss of local regularity. To avoid this pathology and ensure uniqueness we want the maximal development of generic initial data to be inextendible. Motivated by these considerations Penrose introduced strong cosmic censorship, which forbids such undesirable features of singularities.

Strong cosmic censorship (SCC): *Generic initial data sets have maximal Cauchy future developments that are locally inextendible, in a continuous manner, as Lorentz manifolds. In other words every maximal Hausdorff development of a generic initial data set, compact or asymptotically flat, is a Cauchy development.*

Remarks: The formulation above leaves open the sense in which the maximal future developments are inextendible. One possibility could be that, disregarding some possible exceptional initial conditions, the maximal future development of an initial data set is such that, along any future, inextendible, timelike geodesics of finite length,[45] the spacetime curvature components expressed relative to a parallel-transported orthonormal frame along the geodesic must become infinite as the value of the arc length tends to its limiting value. The precise definition of extendibility, however, is a subtle issue that will probably only be settled together with a complete solution of the conjecture.

Finally, note that invalidity of SCC implies the existence of Cauchy horizons, which suggests that the uniqueness of $(\mathcal{M}, \mathbf{g})$ is lost beyond them.

1.3.8 The C-K Theorem and the Main Theorem

We conclude this first chapter by stating the two theorems we intend to discuss. The rest of the book is devoted to the proof of the second theorem, but at the end of Chapter 3, a proof of the first theorem is also given.

We start by giving preliminary definitions of "asymptotically flat initial data sets" and of "strong asymptotically flat initial data sets," which enter into the statements of both theorems. [46]

---

[44] See also [Chr], Section 1.4 for a review of "nongeneric" examples of naked singularity formation.

[45] That is, geodesics of bounded proper time.

[46] More precise definitions are given in Chapter 3, where the Main Theorem is stated in detail.

Definition 1.3.3 *We say that a data set* $\{\Sigma_0, g, k\}$ *is asymptotically flat if there exists a coordinate system* (x^1, x^2, x^3) *defined outside a sufficiently large compact set such that, relative to this coordinate system,*

$$g_{ij} = (1 + \frac{2M}{r})\delta_{ij} + o(r^{-1})$$
$$k_{ij} = o(r^{-2}). \tag{1.3.23}$$

Definition 1.3.4 *An initial data set* $\{\Sigma_0, g, k\}$ *is strongly asymptotically flat if there exists a coordinate system* (x^1, x^2, x^3) *defined outside a sufficiently large compact set such that, relative to this coordinate system,*

$$g_{ij} = (1 + 2M/r)\delta_{ij} + o_4(r^{-\frac{3}{2}})$$
$$k_{ij} = o_3(r^{-\frac{5}{2}}). \tag{1.3.24}$$

We also introduce the following functional associated to any asymptotically flat initial data set,

$$J_0(\Sigma_0, g, k) = \sup_{\Sigma_0}\left((d_0^2+1)^3|\mathrm{Ric}|^2\right) + \int_{\Sigma_0}\sum_{l=0}^{3}(d_0^2+1)^{l+1}|\nabla^l k|^2 + \int_{\Sigma_0}\sum_{l=0}^{1}(d_0^2+1)^{l+3}|\nabla^l B|^2, \tag{1.3.25}$$

where d_0 is the geodesic distance from a fixed point O on Σ_0 and B is the Bach tensor.[47]

C–K Theorem (Global stability of Minkowski space using a maximal foliation) *There exists an* ϵ *sufficiently small such that if* $J_0(\Sigma_0, g, k) \leq \epsilon$, *then the initial data set* $\{\Sigma_0, g, k\}$, *strongly asymptotically flat and maximal, has a unique, globally hyperbolic, smooth, geodesically complete solution.*[48] *This development is globally asymptotically flat which means that the Riemann curvature tensor tends to zero along any causal or space-like geodesic. Moreover, there exists a global maximal time function* t *and an optical function* u *defined everywhere outside an "internal region."*[49] *The null-outgoing foliation defined by* u *corresponds to the propagation properties of the spacetime.*

Main Theorem (Global stability using a double null foliation) *Consider an initial data set* $\{\Sigma_0, g, k\}$, *strongly asymptotically flat and maximal, and assume* $J_0(\Sigma_0, g, k)$ *is bounded. Then, given a sufficiently large compact set* $K \subset \Sigma_0$ *such that* $\Sigma_0 \setminus K$ *is diffeomorphic to* R^3/B_1 *and under additional smallness assumptions that are made precise in Section 3.7, there exists a unique development* $(\mathcal{M}, \mathbf{g})$ *with the following properties:*[50]

---

[47] See Chapter 3, Section 3.6 and also [Ch-Kl] for the definition of B and discussions about the quantity J_0.

[48] This coincides with the maximally hyperbolic development of Choquet-Bruhat and Geroch, [Br-Ge].

[49] See details in [Ch-Kl].

[50] This development coincides, roughly speaking, with the complement of the domain of influence of the compact set K. This means, in particular, that for any point $p \in (\mathcal{M}, g)$ any causal curve passing through it intersects $\Sigma_0 \setminus K$ once and only once.

(i) $(\mathcal{M}, \mathbf{g})$ *can be foliated by a double null foliation* $\{C(\lambda)\}$ *and* $\{\underline{C}(\nu)\}$ *whose outgoing leaves* $C(\lambda)$ *are complete.* [51]

(ii) We have detailed control of all the quantities associated with the double null foliations of the spacetime. We also have detailed control of the asymptotic behavior of the Riemann curvature tensor along the null-outgoing and spacelike geodesics.

(iii)If $J(\Sigma_0, g, k)$ *is small, we can extend* $(\mathcal{M}, \mathbf{g})$ *to a smooth, complete solution compatible with the global stability of Minkowski space.*

In this work we only provide complete proofs for (i) and (ii); see Section 3.7 for a complete discussion of our results.

1.4 Appendix

Proof of Lemma 1.3.2

The proof goes along the following three steps.

Step 1: Using $(t, \tilde{x}^i)$, the adapted coordinates of $R \times \mathcal{S}$, we compute $\tilde{\mathbf{g}}(t, p) \equiv \mathcal{T}^*\mathbf{g}(q)$, the metric induced on $R \times \mathcal{S}$. The metric is defined through $\mathcal{T}$ in the following way,

$$\tilde{\mathbf{g}}(t, p)((a, \tilde{Y}), (b, \tilde{Y})) = \mathbf{g}(q)(\mathcal{T}_*(a, \tilde{Y}), \mathcal{T}_*(b, \tilde{Y})).$$

Step 2: Let us consider the tangent space of $R \times \Sigma_0$:

$$T(R \times \Sigma_0) = TR \times T\Sigma_0$$
$$T(R \times \Sigma_0)_{(t,p)} = R_t \times (T\Sigma_0)_p \quad , \ p \in \Sigma_0.$$

The generic vector of $T(R \times \Sigma_0)$ is

$$(a, \tilde{X})_{(t,p)} \equiv \left(a \frac{\partial}{\partial t}\bigg|_t , \tilde{X}^i_p \frac{\partial}{\partial \tilde{x}^i}\bigg|_p \right)$$

and[52]

$$\mathcal{T}_*(a, \tilde{X})_{(t,p)} \in (T\mathcal{M})_q \ , \ \mathcal{T}_*(0, \tilde{X})_{(t,p)} \in (T\Sigma_t)_q,$$

where

$$t(q) = t \ , \ \Psi(\bar{s}, p) = q \ .$$

We derive

$$\mathcal{T}_*(1, 0)_{(t,p)} = \frac{V^\mu_q}{\mathbf{g}(F, V)} \frac{\partial}{\partial x^\mu}\bigg|_q \ , \ \mathcal{T}_*(0, \tilde{Y})_{(t,p)} = \tilde{Y}^i_p \frac{\partial \Psi^\mu}{\partial \tilde{x}^i}\bigg|_{(\bar{s},p)} . \tag{1.4.1}$$

[51] This definiton means that the null geodesics generating $C(\lambda)$ can be indefinitely extended toward the future.

[52] Observe that $\mathcal{T}_*(a, \tilde{0})_{(t,p)}$ does not belong to $(T\Sigma_t)^{\perp}_q$.

Step 3: Combining Step 1 and Step 2 we obtain

$$
\begin{aligned}
\tilde{g}_{00}(t,p) &= -(-\mathbf{g}(q)(F,F))^{-1} + \mathbf{g}(q)(X,X) \\
\tilde{g}_{0i}(t,p) &= \mathbf{g}(q)(X, \frac{\partial \Psi}{\partial \tilde{x}^i}) = X_i \\
\tilde{g}_{ij}(t,p) &= \mathbf{g}(q)\left(\frac{\partial \Psi}{\partial \tilde{x}^i}\frac{\partial \Psi}{\partial \tilde{x}^j}\right),
\end{aligned} \tag{1.4.2}
$$

which completes the proof of the lemma.

The more delicate part to prove is equation (1.4.1) of Step 2. Let f be a function on $\mathcal{M}$. Then the vector $Y_q \equiv \mathcal{T}_*(a, \tilde{Y})_{(t,p)}$ applied to f gives

$$
\begin{aligned}
Y_q(f) &= Y^\mu_q \frac{\partial f}{\partial x^\mu}\bigg|_q = (a, \tilde{Y})_{(t,p)}(f \circ \mathcal{T}) \\
&= a\frac{\partial}{\partial t}(f \circ \mathcal{T})\big|_{(t,p)} + \tilde{Y}^i_p \frac{\partial}{\partial \tilde{x}^i}(f \circ \mathcal{T})\big|_{(t,p)},
\end{aligned} \tag{1.4.3}
$$

where

$$
a\frac{\partial}{\partial t}(f \circ \mathcal{T})|_{(t,p)} = a\frac{\partial f}{\partial x^\mu}\bigg|_q \frac{\partial \mathcal{T}^\mu}{\partial t}\bigg|_{(t,p)}
$$

and, using (1.3.16),

$$
\frac{\partial \mathcal{T}^\mu}{\partial t}\bigg|_{(t,p)} = \frac{\partial \Psi^\mu}{\partial s}\bigg|_{\bar{s}} \frac{\partial s}{\partial t}\bigg|_{(t,p)} = \frac{1}{\mathbf{g}(F,V)} V^\mu_q .
$$

Moreover,

$$
\frac{\partial}{\partial \tilde{x}^i}(f \circ \mathcal{T})\bigg|_{(t,p)} = \frac{\partial f}{\partial x^\mu}\bigg|_q \frac{\partial \Psi^\mu}{\partial \tilde{x}^i}\bigg|_{(\bar{s},p)},
$$

so that finally

$$
\begin{aligned}
\mathcal{T}_*(1, \tilde{0})_{(t,p)} &= \frac{V_q}{\mathbf{g}(F,V)} = \frac{V^\mu_q}{\mathbf{g}(F,V)}\frac{\partial}{\partial x^\mu}\bigg|_q = \frac{\partial \Psi^\mu}{\partial s}\bigg|_{(\bar{s},p)} \frac{\partial}{\partial x^\mu}\bigg|_q \\
\mathcal{T}_*(0, \frac{\partial}{\partial \tilde{x}^i})_{(t,p)} &= \frac{\partial \Psi}{\partial \tilde{x}^i}\bigg|_q = \frac{\partial \Psi^\mu}{\partial \tilde{x}^i}\bigg|_{(\bar{s},p)} \frac{\partial}{\partial x^\mu}\bigg|_q .
\end{aligned}
$$

Therefore,

$$
Y = \left(a\frac{V^\mu_q}{\mathbf{g}(F,V)} + \tilde{Y}^i_p \frac{\partial \Psi^\mu}{\partial \tilde{x}^i}\bigg|_{(\bar{s},p)}\right) \frac{\partial}{\partial x^\mu}\bigg|_q .
$$

Step 3 is then simply achieved with the following substitutions:

$$
\begin{aligned}
\tilde{g}_{00}(t,p) &= \tilde{\mathbf{g}}(t,p)\left((1,\tilde{0}),(1,\tilde{0})\right) = \frac{\mathbf{g}(q)(V,V)}{\mathbf{g}(q)(V,F)^2} \\
&= \frac{1}{\mathbf{g}(q)(V,F)^2}\left(\mathbf{g}(q)(V_\perp, V_\perp) + \mathbf{g}(q)(V_{||}, V_{||})\right) \\
&= -(-\mathbf{g}(q)(F,F))^{-1} + \mathbf{g}(q)(X,X)
\end{aligned}
$$

$$\tilde{g}_{ij}(t,p) = \tilde{\mathbf{g}}(t,p)\left((0,\frac{\partial}{\partial \tilde{x}^i}),(0,\frac{\partial}{\partial \tilde{x}^j})\right) = \mathbf{g}(q)\left(\frac{\partial \Psi}{\partial \tilde{x}^i}\frac{\partial \Psi}{\partial \tilde{x}^j}\right).$$

Finally[53]

$$\begin{aligned}\tilde{g}_{0i}(t,p) &= \tilde{\mathbf{g}}(t,p)((1,\tilde{0}),(0,\frac{\partial}{\partial \tilde{x}^j})) = \mathbf{g}(q)\left(\frac{V}{\mathbf{g}(V,F)},\frac{\partial \Psi}{\partial \tilde{x}^i}\right)\\ &= \frac{1}{\mathbf{g}(V,F)}\mathbf{g}(q)(V_{||},\frac{\partial \Psi}{\partial \tilde{x}^i}) = \mathbf{g}(q)(X,\frac{\partial \Psi}{\partial \tilde{x}^i}) = X_i \ .\end{aligned}$$

[53]The vector components X_i are relative to the $(t,\tilde{x}^i)$ coordinates.

2

Analytic Methods in the Study of the Initial Value Problem

The goal of this chapter is to introduce the reader to the global analytic methods that play a fundamental role in the remaining chapters of the book. We start with a discussion of local and global existence results for systems of nonlinear wave equations. As we have pointed out in the previous sections, the Einstein vacuum equations can be reduced to such systems of partial differential equations with the help of wavelike coordinates. Thus the general framework of systems of nonlinear wave equations provides a very convenient first introduction to some of the basic analytic tools in the study of the evolution problem in general relativity.

2.1 Local and global existence for systems of nonlinear wave equations

2.1.1 Local existence for nonlinear wave equations

Recall that, written relative to a system of wavelike coordinates, the Einstein equations take on the reduced form

$$\frac{1}{2} g^{\alpha\beta} \frac{\partial^2 g_{\mu\nu}}{\partial x^\alpha \partial x^\beta} = H_{\mu\nu}(g, \partial g),$$

where H is a quadratic expression relative to the first derivatives of g. Writing $g_{\alpha\beta} = \eta_{\alpha\beta} + u_{\alpha\beta}$, with η the Minkowski metric and u a small perturbation, we derive a system of equations of the form

$$\Box u = N(u, \partial u, \partial^2 u) \tag{2.1.1}$$

where $u = (u^{(1)}, \ldots, u^{(k)})$ is a vector in R^k. We shall denote by ∂ the spacetime gradient $\partial = (\partial_0, \partial_1, \ldots, \partial_n)$, by D the space gradient $D = (\partial_1, \ldots, \partial_n)$ and by $\Box = \Box_\eta = \eta^{\alpha\beta}\partial_\alpha\partial_b$, the D'Alembertian with respect to the Minkowski metric of R^{n+1}. The nonlinear part N of the Einstein equations consists of a large number of terms that can be organized into two categories:

1. Terms that can be written as a product of a real analytic function of u, a component of u and a second partial derivative of a component of u. Schematically,
$$F(u) \cdot u \cdot \partial^2 u;$$
2. Terms that can be written as a product of a real analytic function of u and a product of first derivatives of two components of u. Schematically,
$$F(u) \cdot \partial u \cdot \partial u.$$

From the point of view of proving local and global existence results terms of the first type are considerably more difficult to treat. It makes sense, therefore, to start with a treatment of equations that contain only terms of the second type. In doing this we shall make, for the sake of clarity, two more simplifications.

We will assume that the nonlinearity is quadratic in the first derivatives of u, that is, $F(u)$ constant and u a scalar function. Both simplifications are irrelevant insofar as the main ideas of the proof are concerned. Indeed it will be clear from our discussion how to extend the proof to the general case. In fact we shall see that an appropriate modification of the argument presented below will also be used in the global theory. We therefore consider an equation of the form

$$\Box u = N = \partial u \cdot \partial u. \tag{2.1.2}$$

We solve (2.1.2) subject to the initial conditions at $t = 0$,

$$u(0, x) = f(x), \quad \partial_t u(0, x) = g(x). \tag{2.1.3}$$

The solution of equations (2.1.2), (2.1.3) can be expressed in the form

$$u = u^0 + \Box^{-1} N. \tag{2.1.4}$$

Here u^0 is a solution of the homogeneous equation

$$\Box u^0 = 0, \tag{2.1.5}$$

subject to the initial conditions (2.1.3) and, for an arbitrary spacetime function F, $\Box^{-1}F$ is defined as the unique solution v of

$$\Box v = F, \tag{2.1.6}$$

subject to zero initial data, that is, $v = \partial_t v = 0$ at $t = 0$.

In view of the classical contraction argument finding a "local-in-time" solution u of (2.1.2) amounts to finding a $T > 0$ and a space of functions $X = X(T)$, defined on the time slab $[0, T] \times R^n$, in which we can apply a contraction mapping. In other words one has to find a space X that satisfies the following properties:

1. The homogeneous solution u^0 belongs to X.
2. If $u \in X$, then $\mathcal{R}[u] = \square^{-1}N(u) \in X$.
3. The mapping $\mathcal{R}: u \longrightarrow \square^{-1}N(u)$ is a contraction.

To achieve (2) and (3) we need "good estimates" for the inhomogeneous equation (2.1.6). More precisely, since the nonlinear term N depends on the derivatives of u, we need estimates that gain a derivative. This means that we need estimates for the first derivatives of u, in an appropriate norm, in terms of estimates for N itself. The energy estimate is precisely such an estimate.[1]

Lemma 2.1.1 *Let u be a general solution of the inhomogeneous equation $\square u = F$. Define*

$$Q[u](t) = \left(\frac{1}{2}\int_{\Sigma_t} |\partial u|^2 dx\right)^{\frac{1}{2}}, \tag{2.1.7}$$

where $|\partial u|^2 = |\partial_0 u|^2 + |\partial_1 u|^2 + \cdots + |\partial_n u|^2$. Then

$$Q[u](t) \leq Q[u](0) + \int_0^t \|F(s)\|_{L^2} ds. \tag{2.1.8}$$

In the case of the homogeneous wave equation $\square u = 0$ we have the "energy identity"

$$Q[u](t) = Q[u](0). \tag{2.1.9}$$

Both inequalities follow easily by multiplying the wave equation by $\partial_t u$ and then integrating on the spacetime slab $[0, T] \times R^n$ where we perform a simple integration by parts argument.

It thus makes sense to ask whether the space of functions u endowed with the norm $\sup_{t\in[0,T]} Q[u](t)$ satisfies the above properties. The answer is clearly negative; property 2 fails due to the lack of sufficient differentiability. The problem is that we cannot bound $\|(\partial u)^2\|_{L^2}$ in terms of the energy norm $\|\partial u\|^2_{L^2}$. However the following modification works: consider the operators $D^I = \partial_1^{i_1} \cdots \partial_n^{i_n}$ with $I = (i_1, \ldots, i_n)$ and $|I| = i_1 + \cdots + i_n$. Let us define, for $i \geq 0$

$$Q_i[u](t) = \left(\sum_{|I|\leq i} Q^2[D^I u](t)\right)^{\frac{1}{2}}. \tag{2.1.10}$$

In view of the fact that D^I commutes with $\square$ we have for the solutions of $\square u = F$

$$Q_i[u](t) \leq Q_i[u](0) + \int_0^t \|F(s)\|_{H^i}, \tag{2.1.11}$$

[1] It is very important to remark that, for dimension n greater than or equal to 2, the energy estimate is, in fact, the only L^p-type estimate with this property. This is easily seen in the case of norms that are L^p in space and uniform in time. The case of general local spacetime L^p norms is harder (see T. Wolff [W].)

where $H^i = H^i(R^n)$ denotes the Sobolev space[2] of functions f in R^n endowed with the norm

$$\|f\|_{H^i} = \Big(\sum_{|I| \le i} \int |D^I f(x)|^2 dx \Big)^{\frac{1}{2}}. \tag{2.1.12}$$

Moreover, for the solutions of the homogeneous problem (2.1.5),

$$Q_i[u^0](t) = Q_i[u^0](0). \tag{2.1.13}$$

Motivated by this we define on the slab $[0, T] \times R^n$ the function space $X = X(T; s)$ of functions $u \in C^1\Big([0, T]; H^{s-1}(R^n)\Big) \cap C^0\Big([0, T]; H^s(R^n)\Big)$ endowed with the norm

$$\|u\|_X = \sup_{[0,T]} Q_{s-1}[u]. \tag{2.1.14}$$

We claim that for $s > \frac{n}{2} + 1$, the space X satisfies both properties 2 and 3. The first property is obviously true. The second follows from Proposition 2.1.1.

Proposition 2.1.1 *For $s > \frac{n}{2}$ the Sobolev space $H^s = H^s(R^n)$ forms an algebra, that is,*

$$\|f \cdot g\|_{H^s} \le c\|f\|_{H^s} \cdot \|g\|_{H^s}.$$

In fact let $v = \Box^{-1} N(u)$. In view of (2.1.11) and (2.1.13) and using Proposition 2.1.1 we derive, for $s \ge \frac{n}{2} + 1$,

$$\|v\|_X \le Q_{s-1}[u^0] + cT\|u\|_X^2. \tag{2.1.15}$$

To prove the contraction property (3) we restrict ourselves to the ball

$$\|u\|_X \le \Delta,$$

with Δ sufficiently large so that $Q_{s-1}[u^0] \le \frac{1}{2}\Delta$. Then we choose T sufficiently small, proportional to Δ^{-1}, such that $\Delta \le Q_{s-1}[u^0] + cT\Delta^2$. With this choice of T and Δ the operator $\mathcal{R}$ maps the ball $\|u\|_X \le \Delta$ into itself. Finally, using the same argument as in the derivation of (2.1.15) we show that

$$\|\mathcal{R}[u_1] - \mathcal{R}[u_2]\|_X \le cT\Delta\|u_1 - u_2\|_X. \tag{2.1.16}$$

Therefore, for a small $T > 0$, we infer that the map $\mathcal{R}$ is a contraction, which proves, therefore, the following theorem:

---

[2]Sobolev-type spaces play an important role in the subject because of the energy-type inequalities of Lemma 3.1. For a useful monograph on the subject see [Ad].

Theorem 2.1.1 *Assume that $f \in H^s(R^n)$, $g \in H^{s-1}(R^n)$, with (f, g) the initial data (2.1.3). Then, if $s \geq s_0$ for a fixed $s_0 > \frac{n}{2}+1$, there exists a time $T > 0$, depending only on the size of $\|f\|_{H^{s_0}(R^n)} + \|g\|_{H^{s_0-1}(R^n)}$, and a unique solution $u \in C^1\Big([0, T]; H^{s-1}(R^n)\Big) \cap C^0\Big([0, T]; H^s(R^n)\Big)$ that satisfies (2.1.2) and initial conditions (2.1.3).*

Proof: The proof of Proposition (2.1.1) is standard. It can, for example, be easily derived by Fourier transform methods; it can also be derived from the following more general Moser-type estimates; see [Ho].

Proposition 2.1.2 *For every $s \geq 0$ the space $H^s(R^n) \cap L^\infty(R^n)$ forms an algebra. Moreover we have the estimate*

$$||f \cdot g||_{H^s} \leq c\,(||f||_{L^\infty}||g||_{H^s} + ||g||_{L^\infty}||f||_{H^s})\,. \tag{2.1.17}$$

We also have the following classical version of the Sobolev inequality,

Proposition 2.1.3 *The Sobolev space $H^s(R^n)$, for $s > \frac{n}{2}$, is contained in the space of bounded continuous functions in R^n and we have the estimate*

$$||f||_{L^\infty} \leq c||f||_{H^s}. \tag{2.1.18}$$

Using Proposition 2.1.2 to estimate the term $\|\partial u \cdot \partial u\|_{H^i}$ on the right-hand side of the inequality (2.1.11) and applying the standard Gronwall inequality,[3] we derive the following a priori estimate for solutions u of (2.1.2)

$$Q_s[u](t) \leq Q_s[u](0) \exp\left(c_s \int_0^t \|\partial u(t')\|_{L^\infty} dt'\right). \tag{2.1.19}$$

This estimate can be used to prove the following characterization of the maximal time of existence in Theorem 2.1.1.

Theorem 2.1.2 *Under the same assumptions as in the previous theorem, the unique solution u can be extended to any slab $[0, T] \times R^n$ as long as*

$$\int_0^T ||\partial u(s)||_{L^\infty} ds < \infty\,.$$

Both Theorems 2.1.1 and 2.1.2 are valid for more general equations. In fact the argument presented above extends easily over equations of the type $\Box u = F(u)\partial u \cdot \partial u$. To treat the general quasilinear case, the previous approach has to be somewhat modified. The idea is to appropriately modify the energy norm (2.1.7) so that we can still rely on

[3]The Gronwall inequality is a basic estimate used to study the dependence on initial conditions that will be used repeatedly in this book; see for instance [Ho], Chapter 6.

energy-type estimates as we have done before. Consider, for example, scalar equations of the form

$$A^{\alpha\beta}(u)\partial_\alpha\partial_\beta u = N(u, \partial u), \tag{2.1.20}$$

with $A^{\alpha\beta}$ a Lorentz metric depending on u. We define the mapping

$$u \to \mathcal{R}(u) = v, \tag{2.1.21}$$

where v is the unique solution of the linear wave equation

$$A^{\alpha\beta}(u)\partial_i\partial_j v = N(u, \partial u), \tag{2.1.22}$$

subject to the given initial conditions. We only need to prove then that the mapping $\mathcal{R}$ is a contraction. This can be done by following precisely the same steps as before. The only modification is in the definition of the energy integral norm, which is now defined with the help of the energy-momentum tensor $T_{\alpha\beta} = \partial_\alpha v \partial_\beta v - A_{\alpha\beta}(u)A^{\gamma\delta}(u)\partial_\gamma v \partial_\delta v$ associated with solutions v of the linear equation (2.1.22). The integral norm $Q[v](t)$ is defined on spacelike hypersurfaces Σ_t by integrating the energy density $T(n, n)$, where n is the future unit normal to Σ_t. One then proceeds precisely as in the case of the simple model equation (2.1.2) described above and shows that the results of both Theorems 2.1.1 and 2.1.2 hold true for general equations of the type (2.1.1).

Remark 2.1.1 The basic building blocks in the proof of the local existence Theorem 2.1.1 are:

1. basic energy estimate; see (2.1.8)
2. higher energy estimates; see (2.1.11)
3. Sobolev inequality; see (2.1.18)
4. bootstrap estimate; see (2.1.15)
5. contraction estimate; see (2.1.16).

These elements are typical to all classical local existence results,[4] and we shall also encounter them, in a modified form, in the global theory.

[4]We call classical results those based on energy estimates and Sobolev inequalities. These types of results require $s_0 > \frac{n}{2} + 1$ as in Theorem 2.1.1. To do better one needs to replace the Sobolev inequalities by the far more refined Strichartz and bilinear estimates; see the review article [Kl-Se] for a discussion of such results for semilinear equations of type (2.1.2). The case of quasilinear equations, such as the reduced Einstein equations, is far more difficult and has only recently been addressed; see [Ba-Ch1], [Ta], [Kl5], [Kl-Rodn] and [Kl-Rodn1]-[Kl-Rodn3]. The low regularity result in [Kl-Rodn] mentioned above requires Strichartz-type inequalities for linear wave operators with rough coefficients.

2.1.2 Global existence for nonlinear wave equations

In trying to prove a global result for the Einstein equations it makes sense to start with wavelike coordinates and to study the question of existence of global smooth solutions for the reduced system (1.3.1). In the spirit of our discussion in Subsection 2.1.1 we first look at the scalar model equation

$$\Box u = \partial u \cdot \partial u, \tag{2.1.23}$$

subject to the initial conditions $u = f$, $\partial_t u = g$ at $t = 0$ and ask whether the local solutions can be continued for infinite time. In Theorem 2.1.2 we have shown that the solution given by the local existence theorem can be extended in any interval of time $[0, T]$ for which

$$\int_0^T ||\partial u(s)||_{L^\infty} ds < \infty.$$

In fact in the case of equations of the type (2.1.23) we had the precise estimate (2.1.19),

$$Q_s[u](T) \leq Q_s[u](0) \exp\left(c_s \int_0^T ||\partial u(t)||_{L^\infty} dt\right),$$

with Q_s the energy-type norms introduced in (2.1.10). This suggests that to obtain a global solution we have to control the asymptotic behavior of the L^∞ norm of ∂u. If u is a solution of the linear wave equation

$$\begin{aligned} &\Box u = 0 \\ &u(0) = f, \ \partial_t u(0) = g, \end{aligned} \tag{2.1.24}$$

it is possible to show from the explicit form of the fundamental solution that, as $|t|$ goes to infinity,

$$||u(t)||_{L^\infty} \leq C|t|^{-\frac{n-1}{2}}, \tag{2.1.25}$$

where C depends in a specific way on the data f and g. This method of deriving the asymptotic behavior of u, based on the explicit form of the fundamental solution, is very cumbersome in applications to nonlinear problems. It would be particularly difficult to implement for quasilinear wave equations such as (1.3.1).

Another method for deriving the asymptotic behavior of solutions to (2.1.24) is the *conformal method* introduced by Penrose [Pe2], to obtain the asymptotic behavior of linear massless field equations. This technique was later developed by Christodoulou [Ch1], [Ch2] and Friedrich [Fr1], [Fr2], [Fr3]. The problem with the conformal method is that it requires a lot of decay of the initial data f, g at spacelike infinity, which is incompatible, in the case of the Einstein vacuum equations, with the long range properties of asymptotically flat initial data sets.

In what follows we give a short outline of a different method (see [Kl4], [Kl3] and [Ho])[5] of deriving not only the uniform asymptotic behavior but also the propagation

[5] See also [Kl5] for new applications of this technique to Strichartz-type inequalities and improved regularity results for quasilinear wave equations.

properties of solutions to the linear wave equation based on the conformal symmetries of the Minkowski spacetime.[6] This method can be easily generalized to nonlinear situations and its main ideas will turn out to be central in our discussion of the Einstein vacuum equations.

Minkowski spacetime is equipped with a family of Killing and conformal Killing vector fields

$$\begin{aligned} T_\mu &= \partial_\mu \\ O_{\mu\nu} &= x_\mu \partial_\nu - x_\nu \partial_\mu \\ S &= t\partial_t + x^i \partial_i \\ K_0 &= -(t^2 + r^2)\partial_t - 2tx^i\partial_i \\ K_i &= 2x_i S - < x, x > \partial_i . \end{aligned} \tag{2.1.26}$$

The Killing vector fields T_μ and $O_{\mu\nu}$ commute with $\Box$ while S preserves the space of solutions since $\Box u = 0$ implies $\Box \mathcal{L}_S u = 0$ as $[\Box\,, S] = 2\Box$.[7] Based on this observation we define the following *generalized Sobolev norms*:

$$\begin{aligned} E_0[u](t) &= \|u(t,\cdot)\|_{L^2(R^3)} \\ E_k[u](t) &= \sum_{X_{i_1},\ldots,X_{i_j}} E_0[\mathcal{L}_{X_{i_1}} \mathcal{L}_{X_{i_2}} \cdots \mathcal{L}_{X_{i_j}} u](t), \end{aligned} \tag{2.1.27}$$

with the sum taken over $0 \leq j \leq k$ and over all Killing vector fields T, O as well as the scaling vector field S.

The crucial point of this method is that the generalized energy-type norms[8]

$$Q_k[u](t) \equiv E_k[Du](t) \tag{2.1.28}$$

are conserved by solutions to equation (2.1.24). The desired decay estimates of solutions to (2.1.24) can now be derived from the following global version of the Sobolev inequalities.[9] (Compare this with Proposition (2.1.3).)

Proposition 2.1.4 *Let u be an arbitrary function in R^{n+1} such that $E_s[u]$ is finite for some $s > \frac{n}{2}$. Then for $t > 0$*

$$|u(t,x)| \leq c \frac{1}{(1 + t + |x|)^{\frac{n-1}{2}} (1 + |t - |x||)^{\frac{1}{2}}} E_s[u]. \tag{2.1.29}$$

Therefore if the data f, g in (2.1.24) are such that the quantity $Q_s[u] < \infty$, it follows that for $t > 0$,

$$|Du(t,x)| \leq c \frac{1}{(1 + t + |x|)^{\frac{n-1}{2}} (1 + |t - |x||)^{\frac{1}{2}}}, \tag{2.1.30}$$

---

[6] As in Penrose's method the conformal structure is essential. However, one has the flexibilty to use it in a way that is best adapted to the problem at hand.

[7] Since u is a scalar function the Lie derivative $\mathcal{L}_S u$ is equal to Su.

[8] Hereafter $Du = (\partial_0 u, \ldots, \partial_n u)$.

[9] For details see [Kl3] and [Ho], Chapter 6.

an estimate that fits the expected propagation properties of the linear equation $\square u = 0$.

As in the derivation of the estimate (2.1.19) we can now prove an estimate of the same type expressed in terms of the new generalized energy norms defined by (2.1.28). Combining that estimate with the global Sobolev inequality (2.1.30) one derives

$$\mathcal{Q}(T) \le \mathcal{Q}(0) \exp c \left(\int_0^T (1+t)^{-\frac{n-1}{2}} \right) \mathcal{Q}(T), \tag{2.1.31}$$

where

$$\mathcal{Q}(T) = \sup_{[0,T]} Q_s[u](t), \tag{2.1.32}$$

for some $s > \frac{n}{2}$. This leads to a global bound for $\mathcal{Q}$ provided that $\mathcal{Q}(0)$ is small and $n > 3$. Therefore, for $n > 3$ and sufficiently small data, the local solution provided by Theorem 2.1.1 can be extended for all time; see [Kl4], [Ho].

For $n = 3$, the case of interest for general relativity, the estimate in (2.1.31) leads to a logarithmic divergence.[10] Nevertheless there are still interesting situations, in space dimension $n = 3$, wherein one can prove the existence of small global solutions. One favorable situation is, for instance, the case where the nonlinear part consists only of terms of order higher than quadratic, such as $Du \cdot Du \cdot Du$. A much more interesting situation, which turns out to be of great relevance in our discussion below, arises when we allow quadratic terms but require that they satisfy the *null condition*; see [Kl1], [Kl2], [Ch2] and also [Ho]. Roughly speaking, this means that the quadratic terms of the equation appear only through the intermediary of the *null quadratic forms*

$$\begin{aligned} Q_0(u,v) &= \eta^{\alpha\beta}\partial_\alpha u \partial_\beta v \\ Q_{\alpha\beta}(u,v) &= \partial_\alpha u \partial_\beta v - \partial_\beta u \partial_\alpha v \end{aligned} \tag{2.1.33}$$

with η the Minkowski metric; see [Kl2] and [Ho] for details. If the null condition is satisfied one can prove a small data global existence result even for $n = 3$.[11] The basic observation at the origin of this result has to do with the propagation properties of waves expressed relative to null frames.

Consider again the linear equation (2.1.24) and the estimate (2.1.30). The derivatives of Du, expressed relative to the standard Cartesian frame, do not behave any better along the null directions than $|t|^{-\frac{n-1}{2}}$. We get, however, a more detailed picture of the behavior of the derivatives of u by considering a null frame $\{e_3, e_4, e_a\}$[12] with null vectors $e_3 =$

---

[10]This logarithmic divergence is not an artifact of the proof. There are examples (see [John1], [John2]) of nonlinear wave equations in $n = 3$ for which all perturbations of the trivial solution form singularities in finite time. Moreover this situation is generic.

[11]The result is proved for a general class of quasilinear systems of wave equations in [Kl2] (see also [Ho]), based on the ideas sketched here, and in [Ch2] with the help of the conformal method. As we remarked above the conformal method requires more regularity for the data at spacelike infinity.

[12]An explicit expression of a null frame in the Minkowski spacetime R^{3+1} is given by e_3, e_4 as well as

$$e_\theta = \frac{1}{r}\frac{\partial}{\partial\theta} \quad e_\phi = \frac{1}{r sin\theta}\frac{\partial}{\partial\phi}$$

$\frac{\partial}{\partial t} - \frac{\partial}{\partial r}$, $e_4 = \frac{\partial}{\partial t} + \frac{\partial}{\partial r}$, and where $\{e_a\}$ is an orthonormal frame spanning the orthogonal complement of $\{e_3, e_4\}$. It is easy to prove, from (2.1.29), the following estimates for $t > 0$ and $s > \frac{n}{2} + 1$:

$$
\begin{aligned}
|D_{e_4} u(t,x)| &\leq c \frac{1}{(1+t+|x|)^{(\frac{n-1}{2}+1)}(1+|t-|x||)^{\frac{1}{2}}} E_s[u](t) \\
|D_{e_a} u(t,x)| &\leq c \frac{1}{(1+t+|x|)^{(\frac{n-1}{2}+1)}(1+|t-|x||)^{\frac{1}{2}}} E_s[u](t) \\
|D_{e_3} u(t,x)| &\leq c \frac{1}{(1+t+|x|)^{\frac{n-1}{2}}(1+|t-|x||)^{\frac{3}{2}}} E_s[u](t).
\end{aligned}
\tag{2.1.34}
$$

Thus, for $t > 0$, only D_{e_3} fails to improve. By symmetry D_{e_4} fails to improve for $t < 0$.

The null condition for systems of wave equations of type (2.1.23) simply prevents the presence of terms such as $(D_{e_3} u)^2$ and $(D_{e_4} u)^2$. This allows us to overcome the logarithmic divergence in (2.1.31) and thus prove a small data global existence result.

Remark 2.1.2 The main ingredients in the proof of global existence discussed above are:

1. generalized energy-type norms;
2. Killing and conformal Killing vector fields;
3. null frames;
4. some appropriate version of the null condition.

The last point (4) is crucial in $3+1$ dimensions because without it there is no global existence. It turns out, however, that the reduced Einstein equations (1.3.1) do not satisfy such a condition. This was first pointed out by Y. Choquet-Bruhat, [Br3] and later substantiated by L. Blanchet and T. Damour [Bl-D]. The problem is connected with the wavelike gauge itself which behaves badly on large scales. We have thus to abandon wavelike coordinates altogether.

Returning to the Einstein equations we realize that the main difficulties we face are:

1. the problem of coordinates;
2. the strong nonlinear features of the Einstein equations;
3. the long-range terms in the initial data;
4. the nontrivial propagation properties of the expected solutions.

The first two problems have already been discussed. The strongly nonlinear character of the equations requires one to rely on a quite rigid analytic approach based on energy estimates, background symmetries and some subtle cancellation properties manifest in the nonlinear structure of the equations. This last point, in particular, calls for an invariant approach. But this is not all. We can certainly not expect that the spacetime we plan to construct admits any Killing or conformal Killing vector fields. The best we can hope for is that it admits some approximate ones, namely vector fields whose deformation tensors are small in an appropriate way.[13] In trying to do this we encounter the difficulties (3) and (4). The $1/r$ decay of the metric, due to the presence of the *mass term*, has the long range effect of changing the asymptotic behavior of the null geodesics. Thus the causal structure of the spacetime we construct is not asymptotic to that of the Minkowski spacetime.

---

[13] More precisely, these vector fields must have the property that the traceless parts of their deformation tensors, as defined in (1.1.8), are asymptotically small.

To deal with these problems one has to devise a strategy that is independent, as much as possible, of a specific choice of coordinates. From this point of view it makes sense to try to derive the main propagation properties of our solutions in terms of the Riemann curvature tensor. As it will turn out the propagation properties of the Riemann curvature tensor are least sensitive to problems (3) and (4) and best suited, as a starting point, to exhibit the *null structure* properties of the Einstein equations. The key to doing so is the Bianchi equations. In the next section we shall analyze the main properties of that system of equations in Minkowski spacetime.

2.2 Weyl fields and Bianchi equations in Minkowski spacetime

We start by defining the Weyl tensor field as a tensor field with all the algebraic properties of the conformal part of the Riemann tensor field; see (1.1.14) and (1.1.15). In this section we restrict ourselves to $3+1$ dimensions. A more detailed discussion of Weyl tensor fields appears in Chapter 3, Section 3.2. We intend this section to be an introduction for it.

Definition 2.2.1 *Given a spacetime $(\mathcal{M},\mathbf{g})$, we call a Weyl field a tensor field W that satisfies the properties*

$$\begin{aligned}&W_{\alpha\beta\gamma\delta} = -W_{\beta\alpha\gamma\delta} = -W_{\alpha\beta\delta\gamma} = W_{\gamma\delta\alpha\beta}\\&W_{\alpha\beta\gamma\delta} + W_{\alpha\gamma\delta\beta} + W_{\alpha\delta\beta\gamma} = 0\\&g^{\alpha\gamma}W_{\alpha\beta\gamma\delta} = 0.\end{aligned} \tag{2.2.1}$$

We say that a Weyl tensor field is a solution of the Bianchi equations in $(\mathcal{M},\mathbf{g})$ if, relative to the Levi-Civita connection of $\mathbf{g}$, it satisfies the equation

$$\mathbf{D}_{[\lambda}W_{\gamma\delta]\alpha\beta} = 0. \tag{2.2.2}$$

When the spacetime $(\mathcal{M},\mathbf{g})$ is a solution of the Einstein vacuum equations $\mathbf{R}_{\alpha\beta} = 0$, the curvature tensor coincides with its conformal part $\mathbf{C}$ and is, therefore, a Weyl tensor field that satisfies the Bianchi equations (2.2.2).

In this section we review the main properties of Weyl tensor fields and of the Bianchi equations (2.2.2) in a fixed background space $(\mathcal{M},\mathbf{g})$; see [Ch-Kl1]. We start by recalling the definitions and properties of the Hodge duals of a given Weyl field,[14]

$${}^\star W_{\alpha\beta\gamma\delta} \equiv \frac{1}{2}\in_{\alpha\beta\mu\nu}W^{\mu\nu}{}_{\gamma\delta}\,,\quad W^\star_{\alpha\beta\gamma\delta} = W_{\alpha\beta}{}^{\mu\nu}\frac{1}{2}\in_{\mu\nu\gamma\delta}.$$

Proposition 2.2.1

i) If W is a Weyl field, then $W^\star = {}^\star W$ and ${}^\star({}^\star W) = -W$.

ii) The following four sets of equations are equivalent

$$\begin{aligned}&D_{[\sigma}W_{\gamma\delta]\alpha\beta} = 0\,,\quad D^\lambda W_{\lambda\gamma\alpha\beta} = 0\\&D_{[\sigma}{}^\star W_{\gamma\delta]\alpha\beta} = 0\,,\quad D^{\lambda\star}W_{\lambda\gamma\alpha\beta} = 0.\end{aligned}$$

[14] $\in_{\mu\nu\rho\sigma}$ are the components of the volume form of $(\mathcal{M},\mathbf{g})$ in arbitrary coordinates.

iii) The Bianchi equations (3.2.1) are conformally invariant; [15] *see [Pe1], [Pe2] and also [Ch-Kl1], [Ch-Kl].*

iv) If W is a Weyl field the modified Lie derivative[16]

$$\hat{\mathcal{L}}_X W = \mathcal{L}_X W - \frac{1}{2}{}^{(X)}[W] + \frac{3}{8}\mathrm{tr}^{(X)}\pi W \tag{2.2.3}$$

is a Weyl field and $\hat{\mathcal{L}}_X {}^\star W = {}^\star\hat{\mathcal{L}}_X W$.

v) If W satisfies the Bianchi equations and X is a conformal Killing vector field, then $\hat{\mathcal{L}}_X W$ *is also a solution of the Bianchi equations.*

These equations look complicated. Nevertheless they are quite similar to the more familiar Maxwell equations. This becomes apparent if we decompose W into its "electric" and "magnetic" parts. Given vector fields X, Y we introduce $i_{(X,Y)}$ through the relation $(i_{(X,Y)}W)_{\mu\nu} = W_{\mu\rho\nu\sigma}X^\rho Y^\sigma$, then, with $X = Y = T_0$, we define

$$E = i_{(T_0,T_0)}W\ ,\ \ H = i_{(T_0,T_0)}{}^\star W. \tag{2.2.4}$$

These two covariant, symmetric and traceless tensor fields E and H, tangent to the hyperplanes $\Sigma_t \equiv \{p \in \mathcal{M} | t(p) = t\}$, determine completely the Weyl tensor field. It is easy to write the Bianchi equations for this decomposition and obtain the following Maxwell-type equations:

$$\begin{aligned}
&\Phi^{-1}\partial_t E + \mathrm{curl} H = \rho(E, H)\\
&\Phi^{-1}\partial_t H - \mathrm{curl} E = \sigma(E, H)\\
&\mathrm{div} E = k \wedge H\\
&\mathrm{div} H = -k \wedge E,
\end{aligned}$$

where ∇ is the covariant derivative with respect to Σ_t, $(\mathrm{div}E)_i \equiv \nabla^j E_{ij}$, $(\mathrm{curl}E)_{ij} \equiv \epsilon_i^{lk}\nabla_l E_{kj}$ and the analogous expressions hold for H.[17] Moreover $(k \wedge E)_i \equiv \epsilon_i{}^{mn}k_m{}^l E_{ln}$ and the analogous expression holds for H.

This strong formal analogy with the Maxwell equations goes even further. In fact, the Bianchi equations possess an analogue of the electromagnetic tensor called the Bel–Robinson tensor (see [Bel]) which allows one to derive, in the case of Minkowski spacetime, conserved quantities.

Definition 2.2.2 *The Bel–Robinson tensor of the Weyl field W is the four-covariant tensor field*

$$Q_{\alpha\beta\gamma\delta} = W_{\alpha\rho\gamma\sigma}W_\beta{}^\rho{}_\delta{}^\sigma + {}^\star W_{\alpha\rho\gamma\sigma}{}^\star W_\beta{}^\rho{}_\delta{}^\sigma . \tag{2.2.5}$$

---

[15] This means that whenever we perform a conformal transformation Φ of the spacetime $(\mathcal{M}, \mathbf{g})$ with $\tilde{\mathbf{g}} = \Phi_*\mathbf{g} = \Lambda^2\mathbf{g}$, then $\tilde{W} = \Lambda^{-1}\Phi_* W$ is a solution of the Bianchi equations for the spacetime $(\mathcal{M}, \tilde{\mathbf{g}})$.

[16] ${}^{(X)}[W]_{\alpha\beta\gamma\delta} = {}^{(X)}\pi_\alpha^\lambda W_{\lambda\beta\gamma\delta} + {}^{(X)}\pi_\beta^\lambda W_{\alpha\lambda\gamma\delta} + {}^{(X)}\pi_\gamma^\lambda W_{\alpha\beta\lambda\delta} + {}^{(X)}\pi_\delta^\lambda W_{\alpha\beta\gamma\lambda}$, where ${}^{(X)}\pi$ is the deformation tensor relative to the vector field X.

[17] The coordinate system is chosen such that the background metric has the form $\mathbf{g}(\cdot,\cdot) = -\Phi^2 dt^2 + g_{ij}dx^i dx^j$.

The Bel–Robinson tensor has the following important properties, which recall those of the energy-momentum tensor of the Maxwell equations; see [Ch-Kl], [Ch-Kl1]:

Proposition 2.2.2
i) Q is symmetric and traceless relative to all pairs of indices.
ii) Q satisfies the following positivity condition: $Q(X_1, X_2, X_3, X_4)$ is nonnegative for any non-spacelike future directed vector fields X_1, X_2, X_3, X_4.[18]
iii) If W is a solution of the Bianchi equations, then

$$\text{Div}\, Q_{\beta\gamma\delta} = D^\alpha Q_{\alpha\beta\gamma\delta} = 0. \tag{2.2.6}$$

Proposition 2.2.3 *Let $Q(W)$ be the Bel–Robinson tensor of a Weyl field W and X, Y, Z a triplet of vector fields. We define the covariant vector field P associated to the triplet as*

$$P_\alpha = Q_{\alpha\beta\gamma\delta} X^\beta Y^\gamma Z^\delta. \tag{2.2.7}$$

Using all the symmetry properties of Q we have

$$\begin{aligned} \text{Div}\, P &= \text{Div}\, Q_{\beta\gamma\delta} X^\beta Y^\gamma Z^\delta \\ &+ \frac{1}{2} Q_{\alpha\beta\gamma\delta}\left({}^{(X)}\pi^{\alpha\beta} Y^\gamma Z^\delta + {}^{(Y)}\pi^{\alpha\gamma} X^\beta Z^\delta + {}^{(Z)}\pi^{\alpha\delta} X^\beta Y^\gamma\right). \end{aligned} \tag{2.2.8}$$

Thus, to any X, Y, Z Killing or conformal Killing vector fields we can associate a conserved quantity.

Theorem 2.2.1 *Let W be a solution of the Bianchi equations and let $X, Y, Z, V_1, \ldots, V_k$ be Killing or conformal Killing vector fields. Then*

i) $\text{Div} P = 0$ *where P is defined by (2.2.7);*

ii) The integral $\int_{\Sigma_t} Q[W](X, Y, Z, T_0) d^3x$ is finite and constant for all t provided that it is finite at $t = 0$.

iii) The integrals

$$\int_{\Sigma_t} Q[\hat{\mathcal{L}}_{V_1} \hat{\mathcal{L}}_{V_2} \ldots \hat{\mathcal{L}}_{V_k} W](X, Y, Z, T_0) d^3x$$

are finite and constant for all t provided that they are finite at $t = 0$.

2.2.1 Asymptotic behavior of the Weyl fields in Minkowski spacetime

Minkowski spacetime is equipped with the following geometric structures.

1. *Hyperplanes:* consist of the level hypersurfaces of the time function t, $\Sigma_t \equiv \{p \in \mathcal{M} | t(p) = t\}$.

2. *Canonical Null Foliation:* consists of the double family of null cones $\{C(u), \underline{C}(\underline{u})\}$ defined as the level hypersurfaces of the functions $u = t - r$ and $t + r$.

[18] If we restrict ourselves to timelike vector fields $Q(X_1, X_2, X_3, X_4)$ is positive.

$$\begin{aligned} C(u) &\equiv \{p \in \mathcal{M} | u(p) = u = t - r\}\,, \text{ (outgoing)}, \\ \underline{C}(\underline{u}) &\equiv \{p \in \mathcal{M} | \underline{u}(p) = \underline{u} = t + r\}\,, \text{ (incoming)}. \end{aligned} \tag{2.2.9}$$

3. *Canonical Sphere Foliation:* consists of the family of 2-spheres $S(t, u) = \Sigma_t \cap C(u)$, or $S(u, \underline{u}) = C(u) \cap C(\underline{u})$. For each fixed t the family $\{S(t, u)\}$ produces an S^2-foliation of the hyperplane Σ_t. This coincides, of course, with the standard foliation by the surfaces $S_{t,r} = \{(t, x) \in \Sigma_t | |x| = r\}$.

4. *Canonical Null Pair:* given by the vector fields

$$e_3 = \partial_t - \partial_r, \quad e_4 = \partial_t + \partial_r.$$

We can complete the pair e_3, e_4 to form the null frame $\{e_1, e_2, e_3, e_4\}$ at a generic point p by choosing an orthonormal frame $\{e_a\}$, $a \in \{1, 2\}$ on the tangent space to the sphere $S(t, u)$ passing through p.

5) *Conformal Structure:* Minkowski spacetime has a family of Killing and conformal Killing vector fields (see subsection 2.1.2) among which we note

$$T_0 = \frac{1}{2}(e_3 + e_4)\,, \quad S = \frac{1}{2}(ue_3 + \underline{u}e_4)\,, \quad K_0 = \frac{1}{2}(u^2 e_3 + \underline{u}^2 e_4). \tag{2.2.10}$$

T_0 corresponds to time translations, S to scaling transformations and K_0 to inverted time translations. In addition to these we shall also make use of the *rotation vector fields*:

$$^{(i)}O = \epsilon_{ijk}(x_j \partial_k - x_k \partial_j). \tag{2.2.11}$$

We next define the null components of the Weyl tensor.[19]

Definition 2.2.3 *Let e_3, e_4 be a null pair and W a Weyl field. At a given point p we introduce the following tensors defined on the tangent space to the sphere $S(t, u)$ passing through the point p,*

$$\begin{aligned} &\alpha(W)(X, Y) = W(X, e_4, Y, e_4)\,, \ \underline{\alpha}(W)(X, Y) = W(X, e_3, Y, e_3) \\ &\beta(W)(X) = \frac{1}{2} W(X, e_4, e_3, e_4)\,, \ \underline{\beta}(W)(X) = \frac{1}{2} W(X, e_3, e_3, e_4) \\ &\rho(W) = \frac{1}{4} W(e_3, e_4, e_3, e_4)\,, \ \sigma(W) = \frac{1}{4}\rho({}^\star W) = \frac{1}{4} {}^\star W(e_3, e_4, e_3, e_4) \end{aligned} \tag{2.2.12}$$

where X and Y are arbitrary vector fields tangent to $S(t, u)$.

It is easy to verify that α and $\underline{\alpha}$ are symmetric traceless tensors, β and $\underline{\beta}$ are vector fields and ρ, σ are scalar fields. The total number of independent components is, as expected, ten and they completely describe the Weyl tensor field. The Bianchi equations satisfied by W (see for instance [Ch-Kl1]) expressed in terms of these components are as follows:

[19] This null decomposition of W originates in the work of E.T. Newman, R. Penrose, [Ne-Pe2].

Bianchi equations

$$
\begin{aligned}
&\not{\mathbf{D}}_4\underline{\alpha} + \tfrac{1}{2}\mathrm{tr}\chi\,\underline{\alpha} = -\not\nabla\widehat{\otimes}\underline{\beta} &&, \ \not{\mathbf{D}}_3\underline{\beta} + 2\mathrm{tr}\underline{\chi}\ \underline{\beta} = -\not{\mathrm{d}}\mathrm{iv}\,\underline{\alpha}\\
&\not{\mathbf{D}}_4\underline{\beta} + \mathrm{tr}\chi\,\underline{\beta} = -\not\nabla\rho + {}^{\star}\not\nabla\sigma &&, \ \mathbf{D}_3\rho + \tfrac{3}{2}\mathrm{tr}\underline{\chi}\,\rho = -\not{\mathrm{d}}\mathrm{iv}\,\underline{\beta}\\
&\mathbf{D}_4\rho + \tfrac{3}{2}\mathrm{tr}\chi\rho = \not{\mathrm{d}}\mathrm{iv}\,\beta &&, \ \mathbf{D}_3\sigma + \tfrac{3}{2}\mathrm{tr}\underline{\chi}\sigma = -\not{\mathrm{d}}\mathrm{iv}\,{}^{\star}\underline{\beta}\\
&\mathbf{D}_4\sigma + \tfrac{3}{2}\mathrm{tr}\chi\sigma = -\not{\mathrm{d}}\mathrm{iv}\,{}^{\star}\beta &&, \ \not{\mathbf{D}}_3\beta + \mathrm{tr}\underline{\chi}\beta = \not\nabla\rho + {}^{\star}\not\nabla\sigma\\
&\not{\mathbf{D}}_4\beta + 2\mathrm{tr}\chi\beta = \not{\mathrm{d}}\mathrm{iv}\,\alpha &&, \ \not{\mathbf{D}}_3\alpha + \tfrac{1}{2}\mathrm{tr}\underline{\chi}\alpha = \not\nabla\widehat{\otimes}\beta
\end{aligned}
\tag{2.2.13}
$$

where, here, $\mathrm{tr}\,\chi = -\mathrm{tr}\underline{\chi} = \frac{2}{r}$, $\not{\mathbf{D}}_4$ and $\not{\mathbf{D}}_3$ are the projections on the tangent space to $S(t,u)$ of the covariant derivatives along the null directions, $\not{\mathrm{d}}\mathrm{iv}$ and $\not\nabla$ are the projections on the tangent space to $S(t,u)$, of the divergence and the covariant derivative relative to Σ_t, and $\widehat{\otimes}$ denotes twice the traceless part of the symmetric tensor product. The Hodge operator * indicates the dual of the tensor fields relative to the tangent space of $S(t,u)$.[20]

Our first goal is to show how to derive the asymptotic properties of a solution to the Bianchi equations in Minkowski spacetime for initial data at $t = 0$, compatible with the assumptions we will use later on to study the Einstein equations. From this perspective we expect that, for a given spacetime that satisfies the Einstein vacuum equations, the curvature tensor $\mathbf{R}$ behaves, on the initial hypersurface, like r^{-3} as $r \to \infty$. This is due to the presence of an ADM mass term different from zero (see [Ar-De-Mi]) in the definition of the asymptotic flatness for the initial data of the Einstein vacuum equations. Moreover, since the ADM mass is a time-independent constant, we expect that the time and angular derivatives of $\mathbf{R}$ behave better.

With this in mind it makes perfect sense to assume initial data for our Bianchi equations in Minkowski spacetime such that all the terms in the following sum are bounded at $t = 0$:

$$
\begin{aligned}
\mathcal{Q}(t) \quad = \quad & \int_{\Sigma_t} Q(\hat{\mathcal{L}}_O W)(K_0, K_0, T_0, T_0)\\
+ & \int_{\Sigma_t} Q(\hat{\mathcal{L}}_O^2 W)(K_0, K_0, T_0, T_0)\\
+ & \int_{\Sigma_t} Q(\hat{\mathcal{L}}_S \hat{\mathcal{L}}_O W)(K_0, K_0, T_0, T_0)\\
+ & \int_{\Sigma_t} Q(\hat{\mathcal{L}}_{T_0} W)(K_0, K_0, K_0, T_0)\\
+ & \int_{\Sigma_t} Q(\hat{\mathcal{L}}_O \hat{\mathcal{L}}_{T_0} W)(K_0, K_0, K_0, T_0)
\end{aligned}
\tag{2.2.14}
$$

with $\hat{\mathcal{L}}_X$ defined by (2.2.3). Note also that $|\hat{\mathcal{L}}_O f|^2 \equiv \sum_{i=1,2,3} |\hat{\mathcal{L}}_{{}^{(i)}O} f|^2$.

To understand the meaning of this quantity one should observe that $K_0 = \frac{1}{2}(u^2 e_3 + \underline{u}^2 e_4)$ and $T_0 = \frac{1}{2}(e_3 + e_4)$ are the only future-directed, causal conformal Killing vector fields

[20] Its exact definition is in Chapter 3, Section 3.1.

in Minkowski spacetime.[21] Thus the only choices for the vector fields X, Y, Z in (3.2.7), such that $Q(X, Y, Z, T_0)$ is a positive quantity consistent with our above discussion are between T_0 and K_0. In view of Theorem 2.2.1, we can conclude that $\mathcal{Q}(t)$ is conserved, and therefore bounded for all time. Combining these conservation laws with the global Sobolev inequalities (see Proposition 2.2.4 below) and taking advantage of the Bianchi equations (2.2.13), we prove the following theorem.

Theorem 2.2.2 *Assume $\mathcal{Q}$ is finite at $t = 0$. Then,*

i) In the exterior region[22] we have the following bounds for the various null components of the Weyl tensor

$$\begin{aligned}
&\sup_{Ext}|r^{\frac{7}{2}}\alpha| \leq C_0\,,\ \sup_{Ext}|r\tau_-^{\frac{5}{2}}\underline{\alpha}| \leq C_0\\
&\sup_{Ext}|r^{\frac{7}{2}}\beta| \leq C_0\,,\ \sup_{Ext}|r^2\tau_-^{\frac{3}{2}}\underline{\beta}| \leq C_0\\
&\sup_{Ext}|r^3\tau_-^{\frac{1}{2}}\sigma| \leq C_0\,,\ \sup_{Ext}|r^3\tau_-^{\frac{1}{2}}(\rho-\overline{\rho})| \leq C_0
\end{aligned} \tag{2.2.15}$$

where $\overline{\rho}$ is the average of ρ on the spheres $S(t, u)$, $\tau_-^2 = 1 + u^2$ and C_0 is a constant that depends on $\mathcal{Q}(t = 0)$;

ii) In the interior region

$$|W(t, x)| \leq c(1 + t)^{-\frac{7}{2}};$$

iii) The mass term $\overline{\rho}$ is in fact zero.[23]

Since a result analogous to this for the full Einstein equations is at the heart of the proofs of the *C–K Theorem* and of the present *Main Theorem*, we give here the main ideas of the proof of Theorem 2.2.2. From the identities

$$\begin{aligned}
Q(W)(e_3, e_3, e_3, e_3) &= 2|\underline{\alpha}|^2\\
Q(W)(e_4, e_4, e_4, e_4) &= 2|\alpha|^2\\
Q(W)(e_3, e_3, e_3, e_4) &= 4|\underline{\beta}|^2\\
Q(W)(e_3, e_4, e_4, e_4) &= 4|\beta|^2\\
Q(W)(e_3, e_3, e_4, e_4) &= 4(\rho^2 + \sigma^2),
\end{aligned} \tag{2.2.16}$$

we obtain, by a straightforward calculation,

$$\begin{aligned}
Q(W)(K_0, K_0, T_0, T_0) &= \frac{1}{8}\underline{u}^4|\alpha|^2 + \frac{1}{8}u^4|\underline{\alpha}|^2 + \frac{1}{2}(\underline{u}^4 + \frac{1}{2}\underline{u}^2u^2)|\beta|^2\\
&+ \frac{1}{2}(\underline{u}^4 + u^4 + \underline{u}^2u^2)(\rho^2 + \sigma^2) + \frac{1}{2}(u^4 + \frac{1}{2}\underline{u}^2u^2)|\underline{\beta}|^2
\end{aligned}$$

---

[21] See the Liouville theorem, Theorem (1.1.3).

[22] The exterior region refers to the set of points of $\mathcal{M}$ such that $r \geq t$. Its complement will be called the internal region or the interior.

[23] This is due to the fact that, relative to the "electromagnetic" decomposition, $\mathrm{div}\, E = 0$ and $\rho = E_{NN}$, with $N = \frac{x_i}{|x|}\partial_i$; see [Ch-Kl1]. In a general background spacetime, $\mathrm{div}\, E$ has nontrivial source terms and consequently $\overline{\rho}$ fails to be zero. The asymptotic behavior of $\bar{\rho}$ is, in fact, tied to the nontriviality of the ADM mass.

$$\begin{aligned} Q(W)(K_0, K_0, K_0, T_0) &= \frac{1}{8}\underline{u}^6|\alpha|^2 + \frac{1}{8}u^6|\underline{\alpha}|^2 + \frac{1}{4}(\underline{u}^6 + 3\underline{u}^4u^2)|\beta|^2 \\ &+ \frac{3}{4}(\underline{u}^2 + u^2)\underline{u}^2u^2(\rho^2 + \sigma^2) + \frac{1}{4}(u^6 + 3\underline{u}^2u^4)|\underline{\beta}|^2. \end{aligned} \tag{2.2.17}$$

We sketch two different methods for proving the estimates (2.2.15). The first based, on the maximal spacelike hypersurfaces $t = const$, corresponds to the method used in the proof of the *C–K Theorem* while the second, based on the null hypersurfaces $t - r = u$, $t + r = \underline{u}$, corresponds to the double null foliation approach of the *Main Theorem* .

In the first approach, based on the maximal foliation, the proof uses the conservation of the quantity $\mathcal{Q}(t)$, the null Bianchi equations (2.2.13) as well as the following form of the global Sobolev inequalities.[24]

Proposition 2.2.4 *Let F be a smooth tensor field, tangent at each point to the corresponding $S = S(t, r)$. Denote by N the exterior unit normal to S, $\nabla\!\!\!/$ the induced covariant derivative and $\nabla\!\!\!/_N F$ the projection to S of the normal derivative $\nabla_N F$. We have*

Nondegenerate version:

$$\begin{aligned} \sup_{S(t,r)}(r^{\frac{3}{2}}|F|) &\leq c\left(\int_{\Sigma_t} |F|^2 + r^2|\nabla\!\!\!/ F|^2 + r^2|\nabla\!\!\!/_N F|^2 \right. \\ &\left. + r^4|\nabla\!\!\!/^2 F|^2 + r^4|\nabla\!\!\!/\nabla\!\!\!/_N F|^2\right)^{\frac{1}{2}}, \end{aligned} \tag{2.2.18}$$

Degenerate version:

$$\begin{aligned} \sup_{S(t,r)}(r\tau_-^{\frac{1}{2}}|F|) &\leq c\left(\int_{\Sigma_t} |F|^2 + r^2|\nabla\!\!\!/ F|^2 + \tau_-^2|\nabla\!\!\!/_N F|^2 \right. \\ &\left. + r^4|\nabla\!\!\!/^2 F|^2 + r^2\tau_-^2|\nabla\!\!\!/\nabla\!\!\!/_N F|^2\right)^{\frac{1}{2}}. \end{aligned} \tag{2.2.19}$$

In what follows we use all the results mentioned above in order to derive the asymptotic properties of α. We start by applying Proposition 2.2.4 to $F = r^2\alpha$ and derive

$$\begin{aligned} \sup_{\Sigma_t}(r^{\frac{7}{2}}|\alpha|) &\leq c\left(\int_{\Sigma_t} r^4|\alpha|^2 + r^4|r\nabla\!\!\!/\alpha|^2 + r^6|\nabla\!\!\!/_N\alpha|^2 \right. \\ &\left. + r^4|r^2\nabla\!\!\!/^2\alpha|^2 + r^6|r\nabla\!\!\!/\nabla\!\!\!/_N\alpha|^2\right)^{\frac{1}{2}}. \end{aligned}$$

The integrals on the right-hand side are controlled in terms of the quantity $\mathcal{Q}$ as follows.

1. The integrals $\int_{\Sigma_t} r^4|\alpha|^2$ and $\int_{\Sigma_t} r^4|r\nabla\!\!\!/\alpha|^2$ are both bounded by $\int_{\Sigma_t} Q(\hat{\mathcal{L}}_O W)(K_0, K_0, T_0, T_0)$. This follows from eq. (2.2.17) and the following simple identity (see [Ch-Kl1]),

$$|\hat{\mathcal{L}}_O\alpha|^2 \equiv \sum_i |\hat{\mathcal{L}}_{{}^{(i)}O}\alpha|^2 = r^2|\nabla\!\!\!/\alpha|^2 + 4|\alpha|^2. \tag{2.2.20}$$

[24] See [Ch-Kl], Proposition 3.2.3.

2. The integral $\int_{\Sigma_t} r^4 |r^2 \nabla\!\!\!/^2 \alpha|^2$ is bounded by $\int_{\Sigma_t} Q(\hat{\mathcal{L}}_O^2 W)(K_0, K_0, T_0, T_0)$.

We are left with the integrals $\int_{\Sigma_t} r^4 |r \nabla\!\!\!/_N \alpha|^2$ and $\int_{\Sigma_t} r^6 |r \nabla\!\!\!/ \nabla\!\!\!/_N \alpha|^2$. We indicate how to estimate the first. The second can then be dealt with in the same way. Observe that

$$\nabla_N \alpha = D_{T_0}\alpha - D_{e_3}\alpha \ .$$

In view of this it suffices to estimate

$$\int_{\Sigma_t} r^6 |\mathbf{D}\!\!\!/_{T_0}\alpha|^2 \text{ and } \int_{\Sigma_t} r^6 |\mathbf{D}\!\!\!/_{e_3}\alpha|^2.$$

3. The integral $\int_{\Sigma_t} r^6 |\mathbf{D}\!\!\!/_{T_0}\alpha|^2$ is bounded by $\int_{\Sigma_t} Q(\hat{\mathcal{L}}_{T_0} W)(K_0, K_0, K_0, T_0)$. This can be checked again with the help of (2.2.17).

4. To bound the last integral, $\int_{\Sigma_t} r^6 |\mathbf{D}\!\!\!/_{e_3}\alpha|^2$, we have to use the null Bianchi equations, (2.2.13) to express $\mathbf{D}\!\!\!/_{e_3}\alpha$ in terms of $\frac{1}{r}\alpha$ and $\nabla\!\!\!/\beta$. It follows immediately that these integrals are bounded by $\int_{\Sigma_t} Q(\hat{\mathcal{L}}_O W)(K_0, K_0, T_0, T_0)$.

The other components of the Weyl tensor can be treated in the same manner. The results in the interior region are much easier to derive; see [Ch-Kl1].

Remark 2.2.1 The proof of the asymptotic estimates of Theorem 2.2.2 described above is based on energy-type estimates on the maximal spacelike hypersurfaces $t = \text{const}$. This is the main reason that a maximal spacelike foliation was used in [Ch-Kl].

In what follows we sketch a different approach to derive Theorem 2.2.2 using instead the double null foliation $t - r = u$, $t + r = \underline{u}$; see 2.2.9. The main idea of the new approach is to introduce some new quantities analogous to $\mathcal{Q}(t)$ (see 2.2.15) associated with both families of null hypersurfaces. We call these quantities *flux quantities* and we establish their boundedness in terms of the initial data.

Denoting by λ, ν the values taken by the functions $u(p)$, $\underline{u}(p)$ respectively, we define $V(\lambda, \nu)$ as the causal past of $S(\lambda, \nu) \equiv C(\lambda) \cap \underline{C}(\nu)$,

$$V(\lambda, \nu) = J^-(S(\lambda, \nu)).$$

We call $\mathcal{K}$ the region of the Minkowski spacetime, $V(\lambda_0, \nu_*)$, for a fixed couple (λ_0, ν_*). $\mathcal{K}$ lies in the future of the initial hypersurface $\Sigma_{t=0}$ and is foliated by the two families of null hypersurfaces $\{C(\lambda)$, $\{\underline{C}(\nu)\}$ with λ, ν varying in the finite intervals $[\lambda_1, \lambda_0]$ and $[\nu_0, \nu_*]$, respectively, where $\nu_0 = -\lambda_0$ and $\nu_* = -\lambda_1$. For simplicity we may assume $\lambda_0 = -\nu_0 = 0$. We shall also call the null hypersurface $\underline{C}(\nu_*)$ *the last slice* of the spacetime region $\mathcal{K}$ under consideration and denote it by $\underline{C}_*$.[25] $\mathcal{K}$ lies outside $I^+(0)$, the chronological future of the origin, and in the causal past of the null hypersurface $\underline{C}_*$.

Remark: In the proof of the Main Theorem, ν_* is assumed finite and the central part of the proof consists in showing that, in fact, $\nu_* = \infty$. Here we may assume directly $\nu_* = \infty$. In this case $\mathcal{K}$ is the whole complement of $I^+(0)$.

---

[25] For reasons that become clear in the next chapter.

To define the quantity analogous to the conserved quantity $\mathcal{Q}(t)$ used in the previous proof, of Theorem 2.2.2, we go back to equation (2.2.8) (see Proposition 2.2.3) which we integrate on $V(\lambda, \nu)$. If X, Y, Z are conformal Killing vector fields, we derive the identity

$$\int_{\underline{C}(\nu)\cap V(\lambda,\nu)} Q(W)(X, Y, Z, e_3) + \int_{C(\lambda)\cap V(\lambda,\nu)} Q(W)(X, Y, Z, e_4)$$
$$= \int_{\Sigma_0\cap V(\lambda,\nu)} Q(W)(X, Y, Z, T_0). \tag{2.2.21}$$

Applying this identity to $\hat{\mathcal{L}}_{T_0} W,\ \hat{\mathcal{L}}_O W,\ \hat{\mathcal{L}}_O \hat{\mathcal{L}}_{T_0} W,\ \hat{\mathcal{L}}_O^2 W$, with X, Y, Z one of the conformal timelike vector fields T_0, K_0, we are led to consider the following quantities:

$$\begin{aligned}\mathcal{Q}(\lambda, \nu) &= \mathcal{Q}_1(\lambda, \nu) + \mathcal{Q}_2(\lambda, \nu)\\ \underline{\mathcal{Q}}(\lambda, \nu) &= \underline{\mathcal{Q}}_1(\lambda, \nu) + \underline{\mathcal{Q}}_2(\lambda, \nu)\end{aligned} \tag{2.2.22}$$

where

$$\begin{aligned}\mathcal{Q}_1(\lambda, \nu) &= \int_{C(\lambda)\cap V(\lambda,\nu)} Q(\hat{\mathcal{L}}_{T_0} W)(K_0, K_0, K_0, e_4)\\ &\quad + \int_{C(\lambda)\cap V(\lambda,\nu)} Q(\hat{\mathcal{L}}_O W)(K_0, K_0, T_0, e_4)\\ \mathcal{Q}_2(\lambda, \nu) &= \int_{C(\lambda)\cap V(\lambda,\nu)} Q(\hat{\mathcal{L}}_O \hat{\mathcal{L}}_{T_0} W)(K_0, K_0, K_0, e_4)\\ &\quad + \int_{C(\lambda)\cap V(\lambda,\nu)} Q(\hat{\mathcal{L}}_O^2 W)(K_0, K_0, T_0, e_4)\end{aligned} \tag{2.2.23}$$

and

$$\begin{aligned}\underline{\mathcal{Q}}_1(\lambda, \nu) &= \int_{\underline{C}(\nu)\cap V(\lambda,\nu)} Q(\hat{\mathcal{L}}_{T_0} W)(K_0, K_0, K_0, e_3)\\ &\quad + \int_{\underline{C}(\nu)\cap V(\lambda,\nu)} Q(\hat{\mathcal{L}}_O W)(K_0, K_0, T_0, e_3)\\ \underline{\mathcal{Q}}_2(\lambda, \nu) &= \int_{\underline{C}(\nu)\cap V(\lambda,\nu)} Q(\hat{\mathcal{L}}_O \hat{\mathcal{L}}_{T_0} W)(K_0, K_0, K_0, e_3)\\ &\quad + \int_{\underline{C}(\nu)\cap V(\lambda,\nu)} Q(\hat{\mathcal{L}}_O^2 W)(K_0, K_0, T_0, e_3).\end{aligned} \tag{2.2.24}$$

In view of the identity (2.2.21) we infer that both flux quantities $\mathcal{Q}(\lambda, \nu)$ and $\underline{\mathcal{Q}}(\lambda, \nu)$ are bounded by $\mathcal{Q}(t = 0)$. Assuming that the initial data are such that $\mathcal{Q}(t = 0)$ is finite it follows that both quantities $\mathcal{Q}(\lambda, \nu)$ and $\underline{\mathcal{Q}}(\lambda, \nu)$ are finite and independent of the values of λ, ν. We have thus derived the following.

Proposition 2.2.5 *Consider the spacetime region $\mathcal{K}$, as defined in Remark 2.2.1, and assume that the data satisfy the condition $\mathcal{Q}_0 \equiv \mathcal{Q}(t = 0) < \infty$. Then the following* *quantities are uniformly bounded for all $\lambda \leq 0$ and $\nu \geq 0$:*

$$
\begin{aligned}
&\int_{C(\lambda)\cap V(\lambda,\nu)} Q(\hat{\mathcal{L}}_O W)(K_0, K_0, T, e_4)\ ,\ \int_{\underline{C}(\nu\cap V(\lambda,\nu)} Q(\hat{\mathcal{L}}_O W)(K_0, K_0, T, e_3)\\
&\int_{C(\lambda)\cap V(\lambda,\nu)} Q(\hat{\mathcal{L}}_T W)(K_0, K_0, K_0, e_4)\ ,\ \int_{\underline{C}(\nu)\cap V(\lambda,\nu)} Q(\hat{\mathcal{L}}_T W)(K_0, K_0, K_0, e_3)\\
&\int_{C(\lambda)\cap V(\lambda,\nu)} Q(\hat{\mathcal{L}}_O^2 W)(K_0, K_0, T, e_4)\ ,\ \int_{\underline{C}(\nu)\cap V(\lambda,\nu)} Q(\hat{\mathcal{L}}_O^2 W)(K_0, K_0, T, e_3) \qquad (2.2.25)\\
&\int_{C(\lambda)\cap V(\lambda,\nu)} Q(\hat{\mathcal{L}}_O \hat{\mathcal{L}}_T W)(K_0, K_0, K_0, e_4)\ ,\ \int_{\underline{C}(\nu)\cap V(\lambda,\nu)} Q(\hat{\mathcal{L}}_O \hat{\mathcal{L}}_T W)(K_0, K_0, K_0, e_3).
\end{aligned}
$$

In order to prove Theorem 2.2.2 we need, in addition to the above proposition, the following analogue of Proposition 2.2.4, whose proof in the more general situation is given in Chapter 4, Proposition 4.1.4.

Proposition 2.2.6 *Let F be a smooth tensor field, tangent at each point to the corresponding $S(\lambda, \nu)$ passing through that point. The following estimates hold uniformly with regard to $\lambda \leq 0$, $\nu \geq 0$, where $\nu_0 = \underline{u}|_{C(\lambda)\cap\Sigma_0}$, $\lambda_1 = u|_{\underline{C}(\nu)\cap\Sigma_0}$,*

$$
\begin{aligned}
\sup_{S(\lambda,\nu)}(r^{\frac{3}{2}}|F|) \leq\ & c\Bigg[\left(\int_{S(\lambda,\nu_0)} r^4|F|^4\right)^{\frac{1}{4}} + \left(\int_{S(\lambda,\nu_0)} r^4|\nabla\!\!\!\!/ F|^4\right)^{\frac{1}{4}}\\
&+\Bigg(\int_{C(\lambda)\cap V(\lambda,\nu)} |F|^2 + r^2|\nabla\!\!\!\!/ F|^2 + r^2|\mathbf{D}\!\!\!\!/_4 F|^2\\
&+r^4|\nabla\!\!\!\!/^2 F|^2 + r^4|\nabla\!\!\!\!/ \mathbf{D}\!\!\!\!/_4 F|^2\Bigg)^{\frac{1}{2}}\Bigg] \qquad (2.2.26)
\end{aligned}
$$

and[26]

$$
\begin{aligned}
\sup_{S(\lambda,\nu)}(r\tau_-^{\frac{1}{2}}|F|) \leq\ & c\Bigg[\left(\int_{S(\lambda,\nu_0)} r^2\tau_-^2|F|^4\right)^{\frac{1}{4}} + \left(\int_{S(\lambda,\nu_0)} r^2\tau_-^2|r\nabla\!\!\!\!/ F|^4\right)^{\frac{1}{4}}\\
&+\Bigg(\int_{C(\lambda)\cap V(\lambda,\nu)} (|F|^2 + r^2|\nabla\!\!\!\!/ F|^2 + \tau_-^2|\mathbf{D}\!\!\!\!/_4 F|^2\\
&+r^4|\nabla\!\!\!\!/^2 F|^2 + r^2\tau_-^2|\nabla\!\!\!\!/ \mathbf{D}\!\!\!\!/_4 F|^2\Bigg)^{\frac{1}{2}}\Bigg]. \qquad (2.2.27)
\end{aligned}
$$

The previous computation can also be done to express the sup *norms in terms of integrals along the null incoming hypersurfaces $\underline{C}(\nu)$. The results are*

$$
\begin{aligned}
\sup_{S(\lambda,\nu)}(r^{\frac{3}{2}}|F|) \leq\ & c\Bigg[\left(\int_{S(\lambda_1,\nu)} r^4|F|^4\right)^{\frac{1}{4}} + \left(\int_{S(\lambda_1,\nu)} r^4|\nabla\!\!\!\!/ F|^4\right)^{\frac{1}{4}}\\
&+\Bigg(\int_{\underline{C}(\nu)\cap V(\lambda,\nu)} |F|^2 + r^2|\nabla\!\!\!\!/ F|^2 + r^2|\mathbf{D}\!\!\!\!/_3 F|^2\\
&+r^4|\nabla\!\!\!\!/^2 F|^2 + r^4|\nabla\!\!\!\!/ \mathbf{D}\!\!\!\!/_3 F|^2\Bigg)^{\frac{1}{2}}\Bigg] \qquad (2.2.28)
\end{aligned}
$$

---

[26] We write (2.2.27) only for symmetry reasons, but we will never use it.

and

$$\begin{aligned}\sup_{S(\lambda,\nu)}(r\tau_-^{\frac{1}{2}}|F|) \;\leq\; & c\Bigg[\left(\int_{S(\lambda_1,\nu)} r^2\tau_-^2|F|^4\right)^{\frac14}+\left(\int_{S(\lambda_1,\nu)} r^2\tau_-^2|r\nabla\!\!\!/ F|^4\right)^{\frac14}\\ &+\Bigg(\int_{\underline{C}(\nu)\cap V(\lambda,\nu)} |F|^2+r^2|\nabla\!\!\!/ F|^2+\tau_-^2|\mathbf{D}_3F|^2\\ &+r^4|\nabla\!\!\!/^2F|^2+r^2\tau_-^2|\nabla\!\!\!/\mathbf{D}_3F|^2\Bigg)^{\frac12}\Bigg].\end{aligned}\tag{2.2.29}$$

We show how to use these new $\mathcal{Q}$ quantities, introduced above, and Proposition 2.2.6 to derive the asymptotic properties of $\underline{\alpha}$ in the flat case.

Asymptotic behavior of $\underline{\alpha}$:

We observe that the quantities $Q(W)(K,K,T,e_4)$, $Q(W)(K,K,K,e_4)$, for an arbitrary Weyl field W, do not involve the null component $\underline{\alpha}$ of W. This follows easily from the expressions $T_0=\frac12(e_3+e_4)$, $K_0=\frac12(u^2e_3+\underline{u}^2e_4)$ as well as equations (2.2.16). We are therefore obliged to look at the integrals along $\underline{C}(\nu)$ among those of (2.2.25). On the other hand, according to estimate (2.2.29) of Proposition 2.2.6, applied to $\tau_-^2\underline{\alpha}$, we have

$$\begin{aligned}\sup_{S(\lambda,\nu)}(r\tau_-^{\frac52}|\underline{\alpha}|) \;\leq\; & c\Bigg[\left(\int_{S(\lambda,\nu_0)} r^2\tau_-^{10}|\underline{\alpha}|^4\right)^{\frac14}+\left(\int_{S(\lambda,\nu_0)} r^2\tau_-^{10}|r\nabla\!\!\!/\underline{\alpha}|^4\right)^{\frac14}\\ &+\Bigg(\int_{\underline{C}(\nu)\cap V(\lambda,\nu)} \tau_-^4|\underline{\alpha}|^2+\tau_-^4|r\nabla\!\!\!/\underline{\alpha}|^2+\tau_-^6|\mathbf{D}_3\underline{\alpha}|^2\\ &+\tau_-^4|r^2\nabla\!\!\!/^2\underline{\alpha}|^2+\tau_-^6|r\nabla\!\!\!/\mathbf{D}_3\underline{\alpha}|^2\Bigg)^{\frac12}\Bigg].\end{aligned}\tag{2.2.30}$$

1. In view of (2.2.20) the integrals $\int_{\underline{C}(\nu)\cap V(\lambda,\nu)}\tau_-^4|\underline{\alpha}|^2$ and $\int_{\underline{C}(\nu)\cap V(\lambda,\nu)}\tau_-^4|r\nabla\!\!\!/\alpha|^2$ can be estimated by the bounded integral

$$\int_{\underline{C}(\nu)\cap V(\lambda,\nu)} Q(\hat{\mathcal{L}}_OW)(K_0,K_0,T_0,e_3).$$

2. Similarly, the integral $\int_{\underline{C}(\nu)\cap V(\lambda,\nu)}\tau_-^4|r^2\nabla\!\!\!/^2\alpha|^2$ can be estimated by the bounded integral $\int_{\underline{C}(\nu)\cap V(\lambda,\nu))} Q(\hat{\mathcal{L}}_O^2W)(K_0,K_0,T_0,e_3)$.

3. We are left with the integrals $\int_{\underline{C}(\nu)\cap V(\lambda,\nu)}\tau_-^6|\mathbf{D}_3\underline{\alpha}|^2$ and $\int_{\underline{C}(\nu)\cap V(\lambda,\nu)}\tau_-^6|r\nabla\!\!\!/\mathbf{D}_3\underline{\alpha}|^2$. Observe that $\mathbf{D}_3\underline{\alpha}=\mathbf{D}_{T_0}\underline{\alpha}-\mathbf{D}_4\underline{\alpha}$. In view of this it suffices to estimate $\int_{\underline{C}(\nu)\cap V(\lambda,\nu)}\tau_-^6|\mathbf{D}_{T_0}\underline{\alpha}|^2$ and $\int_{\underline{C}(\nu)\cap V(\lambda,\nu)}\tau_-^6|\mathbf{D}_4\underline{\alpha}|^2$.

The first integral is bounded by $\int_{\underline{C}(\nu)\cap V(\lambda,\nu)} Q(\hat{\mathcal{L}}_{T_0}W)(K_0,K_0,K_0,e_3)$. To bound the second integral, $\int_{\underline{C}(\nu)\cap V(\lambda,\nu)}\tau_-^6|\mathbf{D}_4\underline{\alpha}|^2$, we use the null Bianchi equations (2.2.13) to express $\mathbf{D}_4\underline{\alpha}$ in terms of $r^{-1}\underline{\alpha}$ and $\nabla\!\!\!/\underline{\beta}$ which are bounded by

$$\int_{\underline{C}(\nu)\cap V(\lambda,\nu} Q(\hat{\mathcal{L}}_OW)(K_0,K_0,T_0,e_3).$$

Finally, in the same way, one sees that $\int_{\underline{C}(\nu)\cap V(\lambda,\nu)} \tau_-^6 |r \nabla\!\!\!/ \mathbf{D}_3 \underline{\alpha}|^2$ is bounded by the two integrals $\int_{\underline{C}(\nu)\cap V(\lambda,\nu)} Q(\hat{\mathcal{L}}_O^2 W)(K_0, K_0, T_0, e_3)$ and $\int_{\underline{C}(\nu)\cap V(\lambda,\nu)} Q(\hat{\mathcal{L}}_O \hat{\mathcal{L}}_T W)(K_0, K_0, K_0, e_3)$.

Therefore we have obtained the asymptotic result for $\underline{\alpha}$ stated in Theorem 2.2.2. The other components of the Weyl tensor can be treated in the same manner.[27]

2.3 Global nonlinear stability of Minkowski spacetime

As we mentioned earlier, the Bianchi equations provide the keystone in the overall strategy of the proofs of both the *C–K Theorem* and the present *Main Theorem*. They allow us to introduce the crucial energy-type quantities similar to the $E_s[Du](t)$ energy-type norms (see 2.1.27) introduced in the earlier discussion concerning global solutions of nonlinear wave equations.

More importantly they allow us to make an essential conceptual linearization of the Einstein equations. This consists in the following bootstrap scheme.

1. One can first assume given the spacetime with its well-defined causal structure and study the Bianchi equations as a linear system on the given background spacetime. Unlike the case of Minkowski spacetime we do not have any symmetry at our disposal and, therefore, no conserved quantities Q. We can assume, however, and this will have to be justified as part of our overall bootstrap argument, that our background spacetime comes equipped with *approximate Killing and conformal Killing vector fields*. By this we mean vector fields X whose traceless parts of their deformation tensors are small in an appropriate way. Using this we can construct quantities analogous to the Q's introduced in Minkowski spacetime and discussed in the previous section. Instead of being conserved we need to prove that they remain bounded by an universal constant times their value on the initial hypersurface. This leads, just as in the flat case, to precise asymptotic estimates for the various components of the Riemann tensor.

2. To close the bootstrap we then proceed in the opposite way. We assume given a spacetime whose curvature tensor satisfies the asymptotic properties obtained in step (1) and deduce from them the assumptions concerning the causal structure made there.

The properties of the causal structure we construct have to be expressed relative to a foliation induced by two functions. In the *C–K Theorem*, for example, one had to rely on a time function $t(p)$ whose levels are maximal spacelike hypersurfaces and an optical function $u(p)$, whose levels are the null outgoing hypersurfaces which play the role of the null outgoing cones in Minkowski spacetime.

The optical function u is by far the more important one since all the radiation features of the Einstein equations depend heavily on it. The precise definition of u as a solution of the eikonal equation,

$$g^{\mu\nu}\partial_\mu u \partial_\nu u = 0 ,$$

allows us to treat the nontrivial asymptotic properties of the causal structure of a spacetime with nonvanishing ADM mass.

---

[27] All these estimates are discussed in full generality in Chapter 5.

The maximal foliation also seemed to be important because of the traditional role played by the time t in deriving energy estimates as, for instance, in the case of the wave equations discussed in Section 2.1. The nonlocal features of the maximal foliation lead, however, to enormous technical complications that are not intrinsic to the real problem of evolution. In the *Main Theorem* we rely instead on a double null foliation where u is defined as before and the second function $\underline{u}$ is defined symmetrically as an incoming solution of the eikonal equation whose levels are null incoming hypersurfaces. This procedure is naturally adapted to the local hyperbolic features of the Einstein equations.

2.4 Structure of the work

In view of the previous discussion the plan of the remaining six chapters is as follows:

- Chapter 3 contains all the main geometric constructions, the definitions of the main quantities $\mathcal{Q}$, a precise formulation of the *Main Theorem* and a detailed description of the strategy of its proof.

  We start with a discussion of the double null foliation, the canonical null pairs and null frames and the associated Ricci coefficients. We then present the null decomposition of the Weyl tensor followed by the structure equations and the Bianchi equations expressed relative to our null frames. The structure equations relative to a double null foliation have been previously derived by other authors; see for example [Br-D-Is-M] and the references therein.[28]

  We introduce then the important notion of the canonical double null foliation defined in terms of initial data solutions of the *last slice* and of the *initial hypersurface* problems. We also review the main properties of the Bel–Robinson tensor. This, together with the definition of the vector fields T, S, K_0, ${}^{(i)}O$, related to the analogous vector fields introduced in Minkowski space (see Subsection 2.2.1) allows us to define our main quantity $\mathcal{Q}$.

  In addition to the $\mathcal{Q}$ norm we introduce the other two fundamental family of norms, the $\mathcal{R}$ norms, which describe regularity and asymptotic properties of the null components of the Riemann tensor, and the $\mathcal{O}$ norms, which contain detailed regularity and asymptotic information for the connection coefficients.

  We introduce also a large family of norms describing the regularity and the asymptotic properties of the null components of the deformation tensors of T, S, K_0, ${}^{(i)}O$. These norms are controlled in terms of the $\mathcal{O}$ norms.

  We state precise results concerning the relationship between the $\mathcal{R}$, the $\mathcal{O}$ and the $\mathcal{Q}$ norms which form the heart of the proof of our *Main theorem.*

  These above-mentioned results require precise assumptions on the initial data. We describe the sense in which these data have to be small.

---

[28] The first systematic use of null tetrads, not necessarily tied to foliations, goes back to E.T. Newman and R. Penrose, [Ne-Pe1].

Finally we give the precise statement of the *Main Theorem* and give a detailed account of all the steps of the proof.

In the end of Chapter 3 we give, for comparison, a short review of the proof of *C–K Theorem.*

- Chapter 4 contains all the results concerning the $\mathcal{O}$ norms. They are obtained by assuming that we control the $\mathcal{R}$ norms, (*bootstrap assumption*), and are expressed in terms of the initial conditions on Σ_0 and on the last slice. The crucial and delicate issue here is to control the regularity and asymptotic behavior of the null structure coefficients with respect to that of the null components of the curvature tensor. These require subtle estimates depending heavily on the geometric properties of the null structure equations introduced in Chapter 3. Although some of the main ideas we rely on are similar to those in [Ch-Kl] we encounter many additional difficulties since we have to estimate not only the null connection coefficients associated with the null hypersurfaces $C(u)$ but also those associated with the null-incoming hypersurfaces $\underline{C}(\underline{u})$;[29] in fact, these are heavily coupled in the null structure equations.

  In the last section of this chapter we obtain the estimates of the rotation deformation tensors, based on the results of the previous sections.

- Chapter 5 is devoted to the control of the curvature tensor. Making appropriate smallness assumptions for the $\mathcal{O}$ norms of the connection coefficients, we show how to control the $\mathcal{R}$ norms in terms of the $\mathcal{Q}$ norms.

- In Chapter 6, which is central to the whole book, we establish the boundedness of the $\mathcal{Q}$ norms. This requires detailed analysis of the large number of error terms generated because of the nontriviality of the deformation tensors of the vector fields involved in the definition of the $\mathcal{Q}$ norms[30] introduced in Chapter 3, Section 3.5.1.

- In Chapter 7 we discuss the solution of the so-called *initial slice* and *final slice* problems. These are needed to define the *canonical double null foliation* of the spacetime region we construct. As in [Ch-Kl] the canonical null foliation plays a fundamental role in our approach; we explain this in more detail in Chapter 3. The solution of the initial slice problem is a simplified version of the analogous result proved in [Ch-Kl]; the final slice problem is however significantly different from the last slice problem discussed in [Ch-Kl] and we discuss it in detail.

- Chapter 8 is devoted to collecting some conclusions on the asymptotic properties of the global solutions not discussed in the Main Theorem. They do not differ significantly from those discussed in the last chapter of [Ch-Kl] except in that, due to the kind of foliations we have used here, they are obtained in a much simpler way.[31]

[29] In [Ch-Kl] these were replaced by the elliptic estimates of the geometric quantities associated to the maximal time foliation. The additional difficulties of treating the null structure equations are more than compensated by the avoidance of the very technical elliptic estimates and the gain in symmetry.

[30] Analogous to the definition 2.2.23 and 2.2.24 in the flat case.

[31] Nevertheless here we can state the connection between the Bondi and the ADM mass.

3

Definitions and Results

3.1 Connection coefficients

3.1.1 *Null second fundamental forms and torsion of a spacelike 2-surface*

Let S be a closed 2-dimensional surface embedded in a 3+1-dimensional spacetime $(\mathcal{M}, \mathbf{g})$. We assume that S has a compact filling by which we mean that there exists a Cauchy hypersurface Σ containing S such that S is the boundary of a compact region of Σ.

Let γ be the induced metric on S,

$$\gamma(X, Y) = \mathbf{g}(X, Y) \tag{3.1.1}$$

for all $X, Y \in TS$, the tangent space to S. We denote by $d\mu_\gamma$ the area element and by ϵ_{ab} its components relative to an orthonormal frame $(e_a)_{a=1,2}$. We denote by $|S|$ the area and by $r(S)$ the radius of S,

$$r(S) = \sqrt{\frac{1}{4\pi}|S|}. \tag{3.1.2}$$

Let $\nabla\!\!\!/$ be the induced connection on S and $\mathbf{K}$ its Gauss curvature. We recall that if $\mathbf{R}\!\!\!/$ is the intrinsic Riemann curvature tensor and X, Y, Z are three arbitrary vector fields tangent to S,[1]

$$\mathbf{R}\!\!\!/(X, Y)Z = (\gamma(Y, Z)X - \gamma(X, Z)Y)\,\mathbf{K}\,. \tag{3.1.3}$$

---

[1] Relative to an arbitrary orthonormal frame $(e_a)_{a=1,2}$ of S

$$\mathbf{R}\!\!\!/_{abcd} = (\delta_{ac}\delta_{bd} - \delta_{ad}\delta_{bc})\mathbf{K}.$$

At every point p in S we consider the orthogonal complement $T_pS^\perp$ relative to $T_p\mathcal{M}$. This intersects the null cone through p along two null directions. Consider the future-oriented half lines corresponding to these directions and their projections to the tangent space of Σ at p. The half line whose projection points toward the unbounded component of Σ is called future outgoing, the other one future incoming, at p. Similarly, we define the past-incoming and past-outgoing directions. The past-incoming direction at p is complementary to the future-outgoing while the past-outgoing is complementary to the future-incoming. Note also that these definitions do not depend on the particular "fillings" of S. In other words, they do not depend on the choice of the hypersurface Σ passing through S.

At any point $p \in S$ we choose e_4, e_3 to be two future-directed null vectors corresponding to the outgoing and incoming directions and subject to the normalization condition

$$g(e_4, e_3) = -2. \tag{3.1.4}$$

Here e_4 corresponds to the future-outgoing direction and e_3 to the future-incoming one.

Definition 3.1.1 *A smooth choice of such vectors is called a null pair of S.*

According to our definition a null pair is uniquely defined up to a scaling transformation

$$e_4{}' = ae_4, \quad e_3{}' = a^{-1}e_3, \tag{3.1.5}$$

for some smooth positive function a.

Definition 3.1.2 *Given a tensor U defined on $\mathcal{M}$ and tangent to S, we define $\not{\mathbf{D}}_4U$ and $\not{\mathbf{D}}_3U$ to be the projections to TS of $\mathbf{D}_4U$ and $\mathbf{D}_3U$.*

Definition 3.1.3 *Corresponding to any normalized null pair e_4, e_3 we define the null second fundamental forms of S to be the 2-covariant tensors on S*

$$\chi(X, Y) = g(\mathbf{D}_X e_4, Y), \quad \underline{\chi}(X, Y) = g(\mathbf{D}_X e_3, Y), \tag{3.1.6}$$

where X, Y are vector fields tangent to S and $\mathbf{D}$ denotes the connection on $(\mathcal{M}, \mathbf{g})$. Moreover we define the torsion of S to be the 1-form

$$\zeta(X) = \frac{1}{2} g(\mathbf{D}_X e_4, e_3). \tag{3.1.7}$$

Clearly, as $[X, Y] \in TS$, $\chi, \underline{\chi}$ are 2-covariant symmetric tensors on S.[2] Performing a scaling transformation of the form (3.1.5) we have

$$\chi' = a\chi \,, \quad \underline{\chi}' = a^{-1}\underline{\chi}. \tag{3.1.8}$$

[2]If S is the standard sphere, $(x^1)^2 + (x^2)^2 + (x^3)^2 = r^2$, on the spacelike hypersurface $x^0 = const$ in Minkowski spacetime, the standard choice of the null pair is $e_4 = \partial_t + \partial_r$, $e_3 = \partial_t - \partial_r$. In this case $\mathrm{tr}\chi = -\mathrm{tr}\underline{\chi} = \frac{2}{r}$, $\hat{\chi} = \hat{\underline{\chi}} = 0$, $\zeta = 0$ and $K = \frac{1}{r^2}$. In the Schwarzschild spacetime, where S is an orbit of the rotation group, a natural choice of a null pair is $e_4 = \Phi^{-1}(\frac{\partial}{\partial t} + \frac{\partial}{\partial r_*})$, $e_3 = \Phi^{-1}(\frac{\partial}{\partial t} - \frac{\partial}{\partial r_*})$ with $r_* \equiv r + 2m\log(\frac{r}{2m} - 1)$, $\Phi^2 = \left(1 - \frac{2m}{r}\right)$. In this case, $\chi_{ab} = \delta_{ab}\frac{\Phi}{r}$, $\underline{\chi}_{ab} = -\delta_{ab}\frac{\Phi}{r}$.

Hence $\chi, \underline{\chi}$ are uniquely defined up to a transformation of the form (3.1.8). Under the same scaling transformation the torsion ζ transforms according to the formula

$$\zeta'(X) = \zeta(X) - a^{-1}X(a). \tag{3.1.9}$$

Remark: The covariant derivatives, intrinsic to S, of χ and $\underline{\chi}$ are not invariant under the scaling transformation (3.1.5). Nevertheless the following tensors transform nicely under it:

$$\nabla\!\!\!/_X \chi + \zeta(X)\chi \ , \ \ \nabla\!\!\!/_X \underline{\chi} - \zeta(X)\underline{\chi},$$

where X is an arbitrary vector field on S. Indeed

$$\nabla\!\!\!/_X \chi' + \zeta'(X)\chi' = a\left(\nabla\!\!\!/_X \chi + \zeta(X)\chi\right)$$
$$\nabla\!\!\!/_X \underline{\chi}' - \zeta'(X)\underline{\chi}' = a^{-1}\left(\nabla\!\!\!/_X \chi - \zeta(X)\chi\right).$$

We shall call the above quantities the conformal derivatives of $\chi, \underline{\chi}$.

We denote by $\mathrm{tr}\chi$, $\mathrm{tr}\underline{\chi}$ the traces with respect to γ of $\chi, \underline{\chi}$ and by $\hat{\chi}, \hat{\underline{\chi}}$ their traceless parts

$$\begin{aligned} \hat{\chi}(X,Y) &= \chi(X,Y) - \frac{1}{2}\mathrm{tr}\chi\gamma(X,Y) \\ \hat{\underline{\chi}}(X,Y) &= \underline{\chi}(X,Y) - \frac{1}{2}\mathrm{tr}\underline{\chi}\gamma(X,Y). \end{aligned} \tag{3.1.10}$$

Observe that the product $\mathrm{tr}\chi\,\mathrm{tr}\underline{\chi}$ is independent of the choice of the null pair.

Definition 3.1.4 *Given the 1-form ξ on S we define its Hodge dual*[3]

$${}^\star\xi_a = \in_{ab} \xi^b.$$

If ξ is a symmetric traceless 2-tensor we define the following left, ${}^\star\xi$, and right, $\xi^\star$, Hodge duals

$${}^\star\xi_{ab} = \in_{ac} \xi^c{}_b \ , \ \xi^\star_{ab} = \xi_a{}^c \in_{cb} \ .$$

Remark: If ξ is a 1-form, ${}^\star({}^\star\xi) = -\xi$. If ξ is a symmetric traceless 2-tensor, the tensors ${}^\star\xi$, $\xi^\star$ are also symmetric, traceless and satisfy

$${}^\star\xi = -\xi^\star \ , \ {}^\star({}^\star\xi) = -\xi.$$

Remark: Another simple but important property is the following:[4] Let ξ, η be 2-covariant symmetric traceless tensors. Then

$$\xi_{ac}\eta_{cb} + \xi_{bc}\eta_{ca} = (\xi \cdot \eta)\delta_{ab}. \tag{3.1.11}$$

---

[3] Here a, b are just coordinate indices.

[4] It will be used in Chapter 6.

We always decompose a symmetric 2-tensor ξ into its trace, $\mathrm{tr}\xi = \delta^{ab}\xi_{ab}$, and its traceless part. If ξ_{ab} is such a tensor we write its traceless part

$$\hat{\xi}_{ab} = \xi_{ab} - \frac{1}{2}\mathrm{tr}\xi\,\delta_{ab}.$$

Given $S \subset \mathcal{M}$ and the fixed null pair $\{e_4, e_3\}$ we can associate to S two triplets $\{\mathcal{N}, L, \phi_s\}$, $\{\underline{\mathcal{N}}, \underline{L}, \underline{\phi}_s\}$ as follows: starting with the vector field e_4 given on S, we introduce the one parameter flow $\phi_s(p) = l(s; p)$ where $l(s; p)$ denotes the null geodesic parametrized by the affine parameter s[5] with initial conditions

$$l(0) = p\ ,\ (\frac{d}{ds}l)(0) = e_4|_p.$$

We define L by

$$\frac{d}{ds}l(s; p) = L(s; p)\ .$$

Clearly L satisfies

$$g(L, L) = 0, \quad \mathbf{D}_L L = 0\ .$$

The flow $\{\phi_s\}$ generates, starting from S, a family of two-dimensional surfaces $\{S_s\}$. The union of all future-outgoing null geodesics initiating at points in S forms a three-dimensional null hypersurface which we denote by $\mathcal{N}$. The diffeomorphism ϕ_t can be extended from points on S to any point q on $\mathcal{N}$: given $q = l(s, p) \in S_s$, ϕ_t moves q along $l(s, p)$, as follows:

$$\phi_t :\ \mathcal{N} \in q \longrightarrow \phi_t(q) = l(s + t; p) \in \mathcal{N}\ .$$

By replacing e_4 with e_3 we can repeat the same procedure and define the triplet $\{\underline{\mathcal{N}}, \underline{L}, \underline{\phi}_s\}$. Observe that the hypersurfaces $\mathcal{N}$ and $\underline{\mathcal{N}}$ are independent on the particular choice of the null pair.

Definition 3.1.5 *Given S, we call $\mathcal{N}$ and $\underline{\mathcal{N}}$ the null-outgoing and null-incoming hypersurfaces generated by S.*

Let X be a vector field defined on S and tangent to it, $X \in TS$. We extend X to $\mathcal{N}$ (denoting it again by X) as follows:

$$X_q \equiv d\phi_s \cdot X_p,$$

where $p \in S$, $q = \phi_s(p) \in S_s$ and $d\phi_s$ is the differential of ϕ_s. The extension is such that $\phi_{s*}X = X$ holds for any s, where ϕ_{s*} is the standard pushforward. According to the definiton of the Lie derivative, we have that on $\mathcal{N}$

$$[L, X]_q = (\mathcal{L}_L X)_q = \lim_{h\to 0}\frac{1}{h}[X_q - (\phi_{h*}X)_q] = 0\ .$$

[5]This means that the vector field L satisfies $Ls = 1$.

This implies that if we denote by ψ_t the flow generated by the extended X, the flows ψ_t and ϕ_s commute, $\psi_t \circ \phi_s = \phi_s \circ \psi_t$.

Let $\mathcal{N}$ be the corresponding null-outgoing hypersurface generated by S and let U be a tensor of type $\binom{0}{r}$ defined on $\mathcal{N}$ and tangential to each S_s.[6] We introduce the operation $\mathcal{D}U$ in the following way.

Definition 3.1.6 *At each point of S*

$$\mathcal{D}U = \frac{d}{ds}\phi_s^* U\Big|_{s=0} \tag{3.1.12}$$

where $\phi_s^* U$ is defined as usual as

$$\phi_s^* U(X_p, \ldots, Z_p) = U(d\phi_s \cdot X_p, \ldots, d\phi_s \cdot Z_p)$$

for $X_p, \ldots, Z_p$ in the tangent space to S.

The same definition can be given for $\underline{\mathcal{D}}$ by substituting L with $\underline{L}$ and $\mathcal{N}$ with $\underline{\mathcal{N}}$

$$\underline{\mathcal{D}}U = \frac{d}{ds}\underline{\phi}^*_s U\Big|_{s=0} . \tag{3.1.13}$$

The operation $\mathcal{D}$ can be trivially extended to the whole of $\mathcal{N}$ and is intrinsic to $\mathcal{N}$. Observe also that $\mathcal{D}$ is essentially the Lie derivative $\mathcal{L}_L$. In fact, if U is the restriction to $\mathcal{N}$ of a spacetime tensor field, then $\mathcal{D}U = \mathcal{L}_L U$. For example, denoting by γ the restriction of the spacetime metric $\mathbf{g}$ to the surfaces $S_s \subset \mathcal{N}$, we have

$$\begin{aligned}\mathcal{D}\gamma(X, Y) &= (\mathcal{L}_L\mathbf{g})(X, Y) = L(\mathbf{g}(X, Y)) - \mathbf{g}(\mathcal{L}_L X, Y) - \mathbf{g}(X, \mathcal{L}_L Y)\\ &= \mathbf{g}(\mathbf{D}_X L, Y) + \mathbf{g}(X, \mathbf{D}_Y L) = 2\chi(X, Y)\end{aligned}$$

and similarly for $\underline{\mathcal{D}}\gamma$. Therefore,

$$\mathcal{D}\gamma = 2\chi \ , \ \underline{\mathcal{D}}\gamma = 2\underline{\chi}. \tag{3.1.14}$$

To stress the geometric and physical importance of χ and $\underline{\chi}$, it is appropriate to recall the following properties. Let $|S|(s) = \int_S d\mu_{\gamma_s}$ be the area of S_s with γ_s the metric on S equal to the pullback by $(\phi_s)^*$ of g restricted to S_s. Then

$$\frac{d}{ds}|S|_{s=0} = \int_S \mathrm{tr}\chi \ , \ \frac{d}{d\underline{s}}|S|_{\underline{s}=0} = \int_S \mathrm{tr}\underline{\chi}. \tag{3.1.15}$$

In other words $\mathrm{tr}\chi$, $\mathrm{tr}\underline{\chi}$ measure the change of area of S in the direction of e_4, and e_3 respectively. The null second fundamental forms χ and $\underline{\chi}$ measure also the change of the length of a curve Γ on S when mapped by ϕ_s on the surface S_s. In fact let $\Gamma : t \to \Gamma(t) \in S$ and let $\Gamma_s \equiv \phi_s(\Gamma)$. The length $|\Gamma|_s$ of Γ_s satisfies the following equations, where $V = \frac{d\Gamma}{dt}$,

$$\frac{d}{ds}|\Gamma|_{s=0} = \int \frac{\chi(V, V)}{|V|^2}dt \ , \ \frac{d}{d\underline{s}}|\Gamma|_{\underline{s}=0} = \int \frac{\underline{\chi}(V, V)}{|V|^2}dt. \tag{3.1.16}$$

[6] It is enough that U be defined in a neighborhood of S.

3.1.2 Null decomposition of the curvature tensor

Consider a surface S and a fixed null pair $\{e_4, e_3\}$. Associated with this is the null frame $\{e_4, e_3, e_1, e_2\}$, where $\{e_1, e_2\}$ is an arbitrary orthonormal frame for TS. Note that the quantities we define below depend only on the choice of the null pair. We express, at each point of S, the various components of the Riemann curvature tensor of $(\mathcal{M}, \mathbf{g})$ with respect to it.

We recall that the curvature tensor has the following symmetry properties:

$$\mathbf{R}_{\alpha\beta\gamma\delta} = -\mathbf{R}_{\beta\alpha\gamma\delta} = -\mathbf{R}_{\alpha\beta\delta\gamma} = \mathbf{R}_{\gamma\delta\alpha\beta}$$
$$\mathbf{R}_{\alpha\beta\gamma\delta} + \mathbf{R}_{\alpha\gamma\delta\beta} + \mathbf{R}_{\alpha\delta\beta\gamma} = 0.$$

The curvature tensor has 20 independent components. Half of these components are taken into account by the Ricci curvature. The remaining 10 components correspond to the conformal curvature tensor $\mathbf{C}$; see (1.1.14),

$$\begin{aligned}\mathbf{C}_{\alpha\beta\gamma\delta} &= \mathbf{R}_{\alpha\beta\gamma\delta} - \frac{1}{2}\big(\mathbf{g}_{\alpha\gamma}\mathbf{R}_{\beta\delta} + \mathbf{g}_{\beta\delta}\mathbf{R}_{\alpha\gamma} - \mathbf{g}_{\beta\gamma}\mathbf{R}_{\alpha\delta} - \mathbf{g}_{\alpha\delta}\mathbf{R}_{\beta\gamma}\big)\\ &\quad + \frac{1}{6}(\mathbf{g}_{\alpha\gamma}\mathbf{g}_{\beta\delta} - \mathbf{g}_{\alpha\delta}\mathbf{g}_{\beta\gamma})\mathbf{R}.\end{aligned}$$

The conformal curvature tensor $\mathbf{C}$ is the primary example of what we call a Weyl field namely a $\binom{0}{4}$ tensor field W that satisfies all the symmetry properties of the Riemann curvature tensor

$$\begin{aligned}&W_{\alpha\beta\gamma\delta} = -W_{\beta\alpha\gamma\delta} = -W_{\alpha\beta\delta\gamma} = W_{\gamma\delta\alpha\beta}\\ &W_{\alpha\beta\gamma\delta} + W_{\alpha\gamma\delta\beta} + W_{\alpha\delta\beta\gamma} = 0\end{aligned} \tag{3.1.17}$$

and in addition,

$$g^{\alpha\gamma} W_{\alpha\beta\gamma\delta} = 0. \tag{3.1.18}$$

For a Weyl tensor field W the following definitions of left and right Hodge duals are equivalent:

$$^{\star}W_{\alpha\beta\gamma\delta} = \frac{1}{2}\epsilon_{\alpha\beta\lambda\zeta}W^{\lambda\zeta}{}_{\gamma\delta} \quad , \quad W^{\star}_{\alpha\beta\gamma\delta} = W_{\alpha\beta}{}^{\lambda\zeta}\frac{1}{2}\epsilon_{\lambda\zeta\gamma\delta}, \tag{3.1.19}$$

where $\in^{\alpha\beta\gamma\delta}$ are the components of the volume element in $\mathcal{M}$. One can easily show that $^{\star}W = W^{\star}$ is also a Weyl tensor field and that $^{\star}(^{\star}W) = -W$.

We define in the following the null components of the Weyl field relative to the null frame.

Definition 3.1.7 *Let e_4, e_3 be the null pair of the null frame. Let W be a Weyl field and introduce the following tensor fields operating, at each $p \in S$, on the subspace TS_p of the tangent space $T\mathcal{M}_p$:*

$$\alpha(W)(X, Y) = W(X, e_4, Y, e_4)$$

$$\begin{aligned}
\beta(W)(X) &= \frac{1}{2}W(X, e_4, e_3, e_4)\\
\rho(W) &= \frac{1}{4}W(e_3, e_4, e_3, e_4)\\
\sigma(W) &= \frac{1}{4}\rho({}^\star W) = \frac{1}{4}{}^\star W(e_3, e_4, e_3, e_4)\\
\underline{\beta}(W)(X) &= \frac{1}{2}W(X, e_3, e_3, e_4)\\
\underline{\alpha}(W)(X, Y) &= W(X, e_3, Y, e_3)
\end{aligned} \tag{3.1.20}$$

where X,Y are arbitrary vectors tangent to S at p. We call the set

$$\left\{\alpha(W),\ \underline{\alpha}(W),\ \beta(W),\ \underline{\beta}(W),\ \rho(W),\ \sigma(W)\right\}, \tag{3.1.21}$$

the null decomposition of W relative to e_4, e_3.

We can easily check that, in view of (3.1.18), $\alpha(W), \underline{\alpha}(W)$ are symmetric traceless tensors. Thus they have two independent components each. Together the total number of independent components of the set (3.1.21) accounts for all ten degrees of freedom of the Weyl tensor field W.

The null components of W can be expressed in terms of the null decomposition (recall that $W_{\alpha\beta\gamma\delta} \equiv W(e_\alpha, e_\beta, e_\gamma, e_\delta)$) in the following way:

$$\begin{aligned}
&W_{a33b} = -\underline{\alpha}_{ab}\ ,\ W_{a334} = 2\underline{\beta}_a\\
&W_{a44b} = -\alpha_{ab}\ ,\ W_{a443} = -2\beta_a\\
&W_{a3b4} = -\rho\delta_{ab} + \sigma\epsilon_{ab}\\
&W_{a3bc} = -{}^\star({}^\star W)_{a3bc} = \epsilon_{bc}{}^*\underline{\beta}_a\\
&W_{a4bc} = -{}^\star({}^\star W)_{a4bc} = -\epsilon_{bc}{}^*\beta_a\\
&\delta_{ab}W_{a3bc} = \underline{\beta}_c\ ,\ \delta_{ab}W_{a4bc} = -\beta_c\\
&W_{dcab}\delta_{da}\delta_{cb} = -2\rho\ ,\ W_{3434} = -4\rho\\
&W_{ab34} = 2\epsilon_{ab}\sigma
\end{aligned} \tag{3.1.22}$$

where ${}^*\alpha$, ${}^*\underline{\alpha}$, ${}^*\beta$, ${}^*\underline{\beta}$ are the Hodge duals of α, $\underline{\alpha}$, β, $\underline{\beta}$ relative to TS_p. Thus, according to Definition 3.1.4,

$$\begin{aligned}
&\alpha({}^*W) = -{}^*\alpha(W)\ ,\ \beta({}^*W) = -{}^*\beta(W)\ ,\ \underline{\alpha}({}^*W) = {}^*\underline{\alpha}(W)\\
&\underline{\beta}({}^*W) = -{}^*\underline{\beta}(W)\ ,\ \rho({}^*W) = \sigma(W)\ ,\ \sigma({}^*W) = -\rho(W)\ .
\end{aligned}$$

If we rescale the null pair,

$$e_4 \to e_4' = ae_4\ ,\ e_3 \to e_3' = a^{-1}e_3\ ,$$

the null components of W change according to

$$\begin{aligned}
&\alpha' = a^2\alpha, \quad \underline{\alpha}' = a^{-2}\underline{\alpha},\\
&\beta' = a\beta, \quad \underline{\beta}' = a^{-1}\underline{\beta},\\
&\rho' = \rho, \quad \sigma' = \sigma.
\end{aligned} \tag{3.1.23}$$

In view of this we associate to the null components of W the following weights which we refer to as the signature of the corresponding component:

$$\begin{aligned} \mathrm{sign}(\underline{\alpha}) &= -2, & \mathrm{sign}(\alpha) &= 2, \\ \mathrm{sign}(\underline{\beta}) &= -1, & \mathrm{sign}(\beta) &= 1, \\ \mathrm{sign}(\rho) &= 0, & \mathrm{sign}(\sigma) &= 0. \end{aligned} \tag{3.1.24}$$

We also remark that, under the interchange of components e_3 and e_4 in the null decomposition of W, we have

$$\alpha \to \underline{\alpha}\,,\ \beta \to -\underline{\beta}\,,\ \rho \to \rho\,,\ \sigma \to -\sigma. \tag{3.1.25}$$

Remark: Throughout the remainder of this chapter $(\mathcal{M}, \mathbf{g})$ refers to an Einstein vacuum spacetime.

3.1.3 Null structure equations of a 2-surface S

The following equations associated to a fixed 2-surface S are a subset of the set of null structure equations relative to the spacetime $(\mathcal{M}, \mathbf{g})$.

Proposition 3.1.1 *The Gauss curvature* $\mathbf{K}$ *of S, the null second fundamental forms* $\chi, \underline{\chi}$ *and the torsion* ζ *corresponding to a null pair* e_4, e_3 *satisfy the following null structure equations on S:*

i) Gauss equation:

$$\mathbf{K} = -\frac{1}{4}\mathrm{tr}\chi\,\mathrm{tr}\underline{\chi} + \frac{1}{2}\hat{\chi}\cdot\hat{\underline{\chi}} - \rho$$

ii) Null Codazzi equations:

$$\begin{aligned} \not{\mathrm{d}}\mathrm{iv}\,\hat{\chi} + \hat{\chi}\cdot\zeta &= \frac{1}{2}(\not{\nabla}\mathrm{tr}\chi + \zeta\,\mathrm{tr}\chi) - \beta \\ \not{\mathrm{d}}\mathrm{iv}\,\hat{\underline{\chi}} - \hat{\underline{\chi}}\cdot\zeta &= \frac{1}{2}(\not{\nabla}\mathrm{tr}\underline{\chi} - \zeta\,\mathrm{tr}\underline{\chi}) + \underline{\beta} \end{aligned}$$

iii) Torsion equation:

$$\mathrm{c}\not{\mathrm{u}}\mathrm{rl}\,\zeta + \frac{1}{2}\hat{\chi}\wedge\hat{\underline{\chi}} = \sigma.$$

The proof of this proposition is in the appendix to this chapter.

3.1.4 Integrable S-foliations of the spacetime

Assume that the spacetime $(\mathcal{M}, \mathbf{g})$ is foliated by a smooth codimension-two foliation whose leaves are compact spacelike 2-surfaces diffeomorphic to S^2. We shall refer to it as an S-foliation of the spacetime.

Definition 3.1.8 *A tensor field on $(\mathcal{M}, \mathbf{g})$, that is tangent at each point to the leaf of the foliation passing through that point is called S-tangent.*

At every point $p \in \mathcal{M}$ we consider the future-incoming and future-outgoing null directions normal to the leaves of the foliation[7] and choose, correspondingly, a null pair e_4, e_3. We introduce the following definitions.

Definition 3.1.9 *An "adapted" null frame consists, in addition to the null pair e_3, e_4, of an orthonormal frame $\{e_a\}_{a=1,2}$ tangent to the two-dimensional S-surfaces.*

Definition 3.1.10 *The S-foliation is said to be null-outgoing (resp. null-incoming) integrable if the distribution formed by the tangent spaces of S together with the null-outgoing (resp. null-incoming) direction is integrable. An S-foliation that is both null-outgoing and null-incoming integrable is called double null integrable.*

Proposition 3.1.2 *A null-outgoing (null-incoming) integrable foliation is locally given by the level hypersurfaces of a function $u(p)$ $(\underline{u}(p))$ that satisfies the eikonal equation*

$$g^{\mu\nu}\partial_\mu w \partial_\nu w = 0. \tag{3.1.26}$$

We shall refer to u and $\underline{u}$ as the outgoing and incoming optical functions, and we denote their level surfaces by

$$\begin{aligned} C(\lambda) &= \{p \in \mathcal{M} | u(p) = \lambda\} \\ \underline{C}(\nu) &= \{p \in \mathcal{M} | \underline{u}(p) = \nu\}. \end{aligned} \tag{3.1.27}$$

Proof: If the S-foliation is null-outgoing integrable then the distribution made by the linear span formed by TS and e_4

$$p \in \mathcal{M} \longrightarrow \Delta_p \equiv \{TS \oplus e_4\}_p$$

is integrable, which means that at each p there is a submanifold $\mathcal{N} \subset \mathcal{M}$ such that

$$T\mathcal{N}_p = \Delta_p \,.$$

The null hypersurface $\mathcal{N}$ can be expressed locally as the level hypersurface of a function u. It follows that the covariant vector n defined by $n_\mu = \partial_\mu u$ satisfies $n(e_a) = 0,\ n(e_4) = 0,\ a \in \{1, 2\}$. Therefore $g^{\mu\nu} n_\nu = g^{\mu\nu}\partial_\nu u$ is a null vector field proportional to e_4, which implies

$$g^{\mu\nu}\partial_\mu u \partial_\nu u = 0 \,.$$

Everything goes in the same way for the null-incoming integrable foliation, with the obvious substitutions, $p \in \mathcal{M} \longrightarrow \underline{\Delta}_p \equiv \{TS \oplus e_3\}_p$ and $\underline{\mathcal{N}}$ instead of $\mathcal{N}$.

The following corollary is an immediate consequence of Proposition 3.1.2.

---

[7] We will simply refer to them as the null directions of the foliation.

Corollary 3.1.1 *A double null integrable S-foliation can be locally described by the level hypersurfaces $C(\lambda)$, $\underline{C}(\nu)$ associated to an outgoing optical function u and an incoming optical function $\underline{u}$. The leaves of the foliation take the form*

$$S(\lambda,\nu)=C(\lambda)\cap\underline{C}(\nu). \tag{3.1.28}$$

Definition 3.1.11 *The pair of foliations of the spacetime $(\mathcal{M},\mathbf{g})$ defined by the null hypersurfaces $C(\lambda)$ and $\underline{C}(\nu)$ is called the "double null foliation" associated to u and $\underline{u}$.*

We associate to u, $\underline{u}$ the null geodesic vector fields

$$L^\rho\equiv-g^{\rho\mu}\partial_\mu u \ \text{ and } \ \underline{L}^\rho\equiv-g^{\rho\mu}\partial_\mu\underline{u}, \tag{3.1.29}$$

which satisfy

$$\mathbf{D}_L L=0\ ,\ \mathbf{D}_{\underline{L}}\underline{L}=0\ .$$

We refer to L, $\underline{L}$ as a null geodesic pair of the double null foliation. At each point L and $\underline{L}$ are proportional to e_4 and e_3, respectively.

Definition 3.1.12 *Given a double null foliation with associated null geodesic vector fields L, $\underline{L}$ we define its "spacetime lapse function" Ω by*

$$2\Omega^2=-\mathbf{g}(L,\underline{L})^{-1}=-(g^{\rho\sigma}\partial_\rho u\partial_\sigma\underline{u})^{-1}. \tag{3.1.30}$$

So far we have identified the null second fundamental forms and the torsion as natural geometric objects corresponding to a given two-dimensional surface S embedded in spacetime. We now look at the remaining connection coefficients associated to an arbitrary S-foliation with a fixed null pair

$$\begin{aligned}
\xi_a&=\frac{1}{2}\mathbf{g}(\mathbf{D}_{e_4}e_4,e_a)\ \ ,\ \ \underline{\xi}_a=\frac{1}{2}\mathbf{g}(\mathbf{D}_{e_3}e_3,e_a)\\
\eta_a&=-\frac{1}{2}\mathbf{g}(\mathbf{D}_{e_3}e_a,e_4)\ ,\ \underline{\eta}_a=-\frac{1}{2}\mathbf{g}(\mathbf{D}_{e_4}e_a,e_3)\\
\omega&=-\frac{1}{4}\mathbf{g}(\mathbf{D}_{e_4}e_3,e_4)\ \ ,\ \ \underline{\omega}=-\frac{1}{4}\mathbf{g}(\mathbf{D}_{e_3}e_4,e_3).
\end{aligned} \tag{3.1.31}$$

It is straightforward to check that, in the case of a double null integrable foliation, $\xi=\underline{\xi}=0$.

The fact that an S-foliation is double null integrable does not depend on the choice of the null pair $\{e_4,e_3\}$. In fact $\{e_4,e_3\}$ can be subjected to a scaling transformation

$${e_4}'=ae_4,\ \ {e_3}'=a^{-1}e_3\ .$$

In the following we make a specific choice of a null pair.

Definition 3.1.13 *Given a double null foliation with its geodesic null pair $\{L,\underline{L}\}$ and spacetime lapse function Ω we introduce*

$$e_4=\hat{N}=2\Omega L\ ,\ e_3=\underline{\hat{N}}=2\Omega\underline{L}, \tag{3.1.32}$$

which we call the "normalized null pair of the foliation" since

$$\mathbf{g}(\hat{N},\underline{\hat{N}})=4\Omega^2\mathbf{g}(L,\underline{L})=-2\ .$$

Remark: Given a double null foliation, the scalar function Ω^2 and the normalized null pair $\{\hat{N}, \underline{\hat{N}}\}$ are uniquely determined. This definition implies that

$$\mathbf{D}_{\hat{N}}\hat{N} = (\mathbf{D}_{\hat{N}} \log \Omega)\hat{N} \ , \ \mathbf{D}_{\underline{\hat{N}}}\underline{\hat{N}} = (\mathbf{D}_{\underline{\hat{N}}} \log \Omega)\underline{\hat{N}} \ .$$

Recalling the definitions of the connection coefficients given in (3.1.31) we find

$$\begin{aligned}
\underline{\xi}_a &= \frac{1}{2}\mathbf{g}(\mathbf{D}_{\underline{\hat{N}}}\underline{\hat{N}}, e_a) = \frac{1}{2}\underline{\hat{N}}(\log\Omega)\mathbf{g}(\underline{\hat{N}}, e_a) = 0\\
\xi_a &= \frac{1}{2}\mathbf{g}(\mathbf{D}_{\hat{N}}\hat{N}, e_a) = \frac{1}{2}\hat{N}(\log\Omega)\mathbf{g}(\hat{N}, e_a) = 0\\
\underline{\omega} &= \frac{1}{4}\mathbf{g}(\mathbf{D}_{\underline{\hat{N}}}\underline{\hat{N}}, \hat{N}) = -\frac{1}{2}\mathbf{D}_3(\log\Omega)\\
\omega &= \frac{1}{4}\mathbf{g}(\mathbf{D}_{\hat{N}}\hat{N}, \underline{\hat{N}}) = -\frac{1}{2}\mathbf{D}_4(\log\Omega)\\
\underline{\eta}_a &= -\zeta_a + \nabla\!\!\!\!/_a \log\Omega \ , \ \eta_a = \zeta_a + \nabla\!\!\!\!/_a \log\Omega\\
\zeta_a &= \frac{1}{2}\mathbf{g}(\mathbf{D}_{e_a}\hat{N}, \underline{\hat{N}}).
\end{aligned} \tag{3.1.33}$$

Thus all the connection coefficients of a double null integrable foliation can be expressed in terms of $\chi, \underline{\chi}, \zeta, \Omega$. We also remark that, under the interchange of components e_3 and e_4 in the connection coefficients, we have

$$(\chi, \underline{\chi}) \rightarrow (\underline{\chi}, \chi) \ , \ (\eta, \underline{\eta}) \rightarrow (\underline{\eta}, \eta) \ , \ (\omega, \underline{\omega}) \rightarrow (\underline{\omega}, \omega) \ , \ \zeta \rightarrow -\zeta \ . \tag{3.1.34}$$

The next definition introduces another important property of null-outgoing and null-incoming integrable foliations.

Definition 3.1.14 *Consider an arbitrary S-foliation and a null-outgoing vector field N normal to each S. N is said to be equivariant relative to the foliation if the leaves of the foliation are Lie transported by N. The same definition applies to a null-incoming vector field $\underline{N}$.*

In other words, if N ($\underline{N}$) is equivariant, then the 1-parameter family of diffeomorphisms $\{\phi_t\}$ ($\{\underline{\phi}_t\}$) generated by N ($\underline{N}$) maps the leaves of the foliation into themselves.

Lemma 3.1.1 *Let ϕ_t be the 1-parameter family of diffeomorphisms generated by the equivariant vector field N mapping a given 2-surface S of the foliation onto another leaf S'. Let X be a S-tangent vector field defined on $\mathcal{M}$. Then $\phi_{t*}X$ is also S-tangent, at each point, to $S' = \phi_t(S)$ and so is $\mathcal{L}_N X = [N, X]$.*

Proof: Let $p \in S$ and $q = \phi_t(p)$. Then

$$\begin{aligned}
(\phi_{t*}X)_q(f) &= d\phi_t \cdot X_p(f) = X_p(f \circ \phi_t)\\
&= \lim_{h\rightarrow 0}\frac{1}{h}[(f \circ \phi_t \circ \psi_h \circ \phi_t^{-1})(q) - f(q)],
\end{aligned} \tag{3.1.35}$$

where ψ_h is the one-parameter diffeomorphism generated by the vector field X and $(\phi_t \circ \psi_h \circ \phi_t^{-1})$ is a curve on S' whose tangent vector at q is $(\phi_{t*}X)_q$. Moreover, because

$$(\mathcal{L}_N X)_q = \lim_{h\rightarrow 0}\frac{1}{h}[X_q - (\phi_{h*}X)_q] \ ,$$

it follows that $\mathcal{L}_N X = [N, X]$ is S-tangent.

Lemma 3.1.2 *For a double null integral S-foliation the null-outgoing vector field $N = 2\Omega^2 L$ and the null-incoming vector field $\underline{N} = 2\Omega^2 \underline{L}$ are equivariant relative to it.*

Proof: Let $\{\phi_t\}$ be the one-parameter family of diffeomorphisms generated by a null-outgoing vector field N_0 such that

$$\begin{aligned} \underline{u}(\phi_t(p)) &= \underline{u}(p) + t \\ u(\phi_t(p)) &= u(p). \end{aligned}$$

Hence

$$\begin{aligned} \frac{\partial}{\partial t}\underline{u}(\phi_t(p)) = 1 &= \frac{\partial {\phi_t}^\mu}{\partial t}\frac{\partial}{\partial x^\mu}\underline{u}(p) = N_0(\underline{u}) \\ \frac{\partial}{\partial t}u(\phi_t(p)) = 0 &= \frac{\partial {\phi_t}^\mu}{\partial t}\frac{\partial}{\partial x^\mu}u(p) = N_0(u). \end{aligned} \tag{3.1.36}$$

Therefore there must exist a scalar function a such that $N_0 = aL$. Defining in the same way $\underline{\phi}_t$ as the diffeomorphism generated by the null-incoming vector field $\underline{N}_0$, it follows also that $\underline{N}_0(u) = 1,\ \underline{N}_0(\underline{u}) = 0$ and, as before, this implies the existence of a scalar function $\underline{a}$ such that $\underline{N}_0 = \underline{a}\,\underline{L}$. From (3.1.36) we have

$$1 = N_0(\underline{u}) = N_0^\mu \partial_\mu \underline{u} = -g_{\mu\nu}N_0^\mu(-g^{\nu\sigma}\partial_\sigma \underline{u}) = -g_{\mu\nu}N_0^\mu \underline{L}^\nu = -\underline{a}^{-1}\mathbf{g}(N_0, \underline{N}_0),$$

and from it,

$$\mathbf{g}(N_0, \underline{N}_0) = -\underline{a}\ .$$

Repeating the same calculation, but interchanging N_0 and $\underline{N}_0$, gives

$$-a = \mathbf{g}(\underline{N}_0, N_0) = -\underline{a}\ .$$

From $N_0 = aL,\ \underline{N}_0 = a\underline{L}$ and (3.1.30) we obtain

$$a = 2\Omega^2. \tag{3.1.37}$$

Finally, recalling equation (3.1.32), we have

$$N_0 = \Omega\hat{N} = 2\Omega^2 L = N\ ,\ \ \underline{N}_0 = \Omega\underline{\hat{N}} = 2\Omega^2\underline{L} = \underline{N}. \tag{3.1.38}$$

The next lemma is a simple generalization of equation (3.1.15). It will be systematically used in the next chapters.

Lemma 3.1.3 *Let S be a two-dimensional surface diffeomorphic to S^2. For any scalar function f, the following equations hold:*

$$\begin{aligned} \frac{d}{d\underline{u}}\int_{S(u,\underline{u})} f d\mu_\gamma &= \int_{S(u,\underline{u})}\left(\frac{df}{d\underline{u}} + \Omega \mathrm{tr}\chi f\right) d\mu_\gamma \\ \frac{d}{du}\int_{S(u,\underline{u})} f d\mu_\gamma &= \int_{S(u,\underline{u})}\left(\frac{df}{du} + \Omega \mathrm{tr}\underline{\chi} f\right) d\mu_\gamma\ . \end{aligned} \tag{3.1.39}$$

Proof: Explicit computation gives

$$\frac{d}{d\underline{u}}\int_{S(u,\underline{u})} f d\mu_\gamma = \int_{S(u,\underline{u})} \frac{df}{d\underline{u}} + \lim_{\delta\to 0}\frac{1}{\delta}\left(\int_{S(u,\underline{u}+\delta)} f d\mu_\gamma - \int_{S(u,\underline{u})} f d\mu_\gamma\right),$$

where $d\mu_\gamma$, the area form of $S(u,\underline{u})$, is the 2-form $d\mu_\gamma = \sqrt{g}dx^1 \wedge dx^2$. The null vector field $N = \Omega\hat{N}$ generates the diffeomorphism ϕ_δ sending $S(u,\underline{u})$ onto $S(u,\underline{u}+\delta)$. Let $q = \phi_\delta(p) \in S(u,\underline{u}+\delta)$ with $p \in S(u,\underline{u})$. Then the following relation holds:

$$\int_{S(u,\underline{u}+\delta)} f d\mu_\gamma = \int_{S(u,\underline{u})} f d\mu_{\Phi^*_\delta\gamma}.$$

It is easy to prove that

$$d\mu_{\Phi^*_\delta\gamma} = d\mu_\gamma + \delta\mathrm{tr}\left(\frac{1}{2}\mathcal{L}_N\gamma\right)d\mu_\gamma + O(\delta^2) = (1+\delta\Omega\mathrm{tr}\chi)d\mu_\gamma + O(\delta^2),$$

and from this relation, the first line of (3.1.39) follows. The second equation is proved in exactly the same way.

We recall the notion of *Fermi transported null frame.*

Definition 3.1.15 *Given a null pair e_3, e_4, we say that a null frame $\{e_4, e_3, e_a\}$ is Fermi transported along $C(\lambda)$ if $\mathbf{D}\!\!\!/_{e_4}e_a = 0$; we say that it is Fermi transported along $\underline{C}(\nu)$ if $\mathbf{D}\!\!\!/_{e_3}e_a = 0$.*

Remark: Since $N, \underline{N}$ do not commute we cannot simultaneously have a null frame Fermi transported along both $C(\lambda)$ and $\underline{C}(\nu)$.

3.1.5 Null structure equations of a double null foliation

We assume that the spacetime $(\mathcal{M}, \mathbf{g})$ is foliated by a smooth S-foliation whose leaves are compact spacelike 2-surfaces diffeomorphic to S^2. We consider a null frame adapted to the S-foliation, and write the null structure equations satisfied by the connection coefficients with respect to it. Denoting the null frame and its dual basis

$$\{e_\alpha\} = \{e_1, e_2, e_3, e_4\}\ ,\ \{\theta^\alpha\} = \{\theta^1, \theta^2, \theta^3, \theta^4\},$$

we define

$$\mathbf{D}_{e_\alpha}e_\beta \equiv \Gamma^\gamma_{\alpha\beta}e_\gamma\ ,\ \mathbf{R}(e_\alpha, e_\beta)e_\gamma \equiv \mathbf{R}^\delta{}_{\gamma\alpha\beta}\, e_\delta. \tag{3.1.40}$$

The connection coefficients introduced above are the nonvanishing components of $\Gamma^\gamma_{\alpha\beta}$. The connection 1-form and curvature 2-form are

$$\begin{aligned}\omega^\alpha_\beta &\equiv \Gamma^\alpha_{\gamma\beta}\theta^\gamma\\ \Omega^\alpha_\beta &\equiv \frac{1}{2}\mathbf{R}^\alpha{}_{\beta\gamma\delta}\theta^\gamma\wedge\theta^\delta.\end{aligned} \tag{3.1.41}$$

They satisfy the first and the second structure equations,[8] see [Sp],[9]

$$
\begin{aligned}
d\theta^\alpha &= -\omega^\alpha_\gamma \wedge \theta^\gamma \\
d\omega^\delta_\gamma &= -\omega^\delta_\sigma \wedge \omega^\sigma_\gamma + \Omega^\delta_\gamma \, .
\end{aligned}
\tag{3.1.42}
$$

We shall now consider the case of a double null foliation with the normalized null pair $\{\hat{N}, \underline{\hat{N}}\}$; see Definition 3.1.13. The first structure equations, written explicitly in terms of the connection coefficients, take the form (see (3.1.31)),[10]

$$
\begin{aligned}
\mathbf{D}_a e_b &= \nabla\!\!\!\!/_a e_b + \frac{1}{2}\chi_{ab} e_3 + \frac{1}{2}\underline{\chi}_{ab} e_4 \\
\mathbf{D}_a e_3 &= \underline{\chi}_{ab} e_b + \zeta_a e_3 \ , \quad \mathbf{D}_a e_4 = \chi_{ab} e_b - \zeta_a e_4 \\
\mathbf{D}_3 e_a &= \mathbf{D}\!\!\!\!/_3 e_a + \eta_a e_3 \ , \quad \mathbf{D}_4 e_a = \mathbf{D}\!\!\!\!/_4 e_a + \underline{\eta}_a e_4 \\
\mathbf{D}_3 e_3 &= (\mathbf{D}_3 \log \Omega) e_3 \ , \quad \mathbf{D}_3 e_4 = -(\mathbf{D}_3 \log \Omega) e_4 + 2\eta_b e_b \\
\mathbf{D}_4 e_4 &= (\mathbf{D}_4 \log \Omega) e_4 \ , \quad \mathbf{D}_4 e_3 = -(\mathbf{D}_4 \log \Omega) e_3 + 2\underline{\eta}_b e_b .
\end{aligned}
\tag{3.1.43}
$$

We also state the following commutation relations, which are often used in the sequel

$$
\begin{aligned}
&[\hat{N}, e_a] = \mathbf{D}\!\!\!\!/_4 e_a - \chi_{ab} e_b + (\nabla\!\!\!\!/_a \log \Omega)\hat{N} \\
&[\underline{\hat{N}}, e_a] = \mathbf{D}\!\!\!\!/_3 e_a - \underline{\chi}_{ab} e_b + (\nabla\!\!\!\!/_a \log \Omega)\underline{\hat{N}} \\
&[\hat{N}, \underline{\hat{N}}] = -(\mathbf{D}_4 \log \Omega)\underline{\hat{N}} + (\mathbf{D}_3 \log \Omega)\hat{N} - 4\zeta_b e_b ,
\end{aligned}
\tag{3.1.44}
$$

and from them

$$
\begin{aligned}
&[N, e_a] = \Omega(\mathbf{D}\!\!\!\!/_4 e_a - \chi_{ab} e_b) \\
&[\underline{N}, e_a] = \Omega(\mathbf{D}\!\!\!\!/_3 e_a - \underline{\chi}_{ab} e_b) \\
&[N, \underline{N}] = -4\Omega^2 \zeta_b e_b .
\end{aligned}
\tag{3.1.45}
$$

Recalling that all the connection coefficients of a double null foliation can be expressed relative to $\chi, \underline{\chi}, \zeta, \Omega$, the second null structure equations take the following form.

Proposition 3.1.3 (Null structure equations) *The coefficients $\chi, \underline{\chi}, \zeta, \Omega$ associated to a double null foliation and relative to the normalized null pair $\{e_4 = \hat{N}, e_3 = \underline{\hat{N}}\}$ satisfy the following equations:*

$$
\begin{aligned}
&\mathbf{D}\!\!\!\!/_3 \zeta + 2\underline{\chi} \cdot \zeta - \mathbf{D}\!\!\!\!/_3 \nabla\!\!\!\!/ \log \Omega = -\underline{\beta} \\
&\mathbf{D}\!\!\!\!/_4 \zeta + 2\chi \cdot \zeta + \mathbf{D}\!\!\!\!/_4 \nabla\!\!\!\!/ \log \Omega = -\beta \\
&\mathbf{D}\!\!\!\!/_4 \hat{\chi} + \mathrm{tr}\chi\, \hat{\chi} - (\mathbf{D}_4 \log \Omega)\hat{\chi} = -\alpha \\
&\mathbf{D}_4 \mathrm{tr}\chi + \frac{1}{2}(\mathrm{tr}\chi)^2 - (\mathbf{D}_4 \log \Omega)\mathrm{tr}\chi + |\hat{\chi}|^2 = 0
\end{aligned}
$$

---

[8] The structure equations can be stated in a more general framework for a general S-foliation, as discussed in the appendix to this chapter and even in the absence of a foliation; see the general Newman–Penrose formalism, [Ne-Pe2].

[9] With the obvious modifications due to the Lorentzian metric.

[10] Hereafter $\mathbf{D}_{e_a} = \mathbf{D}_a$ and $\mathbf{D}_{e_{(3,4)}} = \mathbf{D}_{(3,4)}$.

$$
\begin{aligned}
&\not{\mathbf{D}}_3\underline{\hat{\chi}} + \mathrm{tr}\underline{\chi}\,\underline{\hat{\chi}} - (\mathbf{D}_3\log\Omega)\underline{\hat{\chi}} = -\underline{\alpha}\\
&\mathbf{D}_3\mathrm{tr}\underline{\chi} + \frac{1}{2}(\mathrm{tr}\underline{\chi})^2 - (\mathbf{D}_3\log\Omega)\mathrm{tr}\underline{\chi} + |\underline{\hat{\chi}}|^2 = 0 \qquad (3.1.46)\\
&\not{\mathbf{D}}_4\underline{\hat{\chi}} + \frac{1}{2}\mathrm{tr}\chi\,\underline{\hat{\chi}} + \frac{1}{2}\mathrm{tr}\underline{\chi}\,\hat{\chi} + (\mathbf{D}_4\log\Omega)\underline{\hat{\chi}} + \not{\nabla}\widehat{\otimes}\zeta - \zeta\widehat{\otimes}\zeta\\
&\quad +2\zeta\widehat{\otimes}\not{\nabla}\log\Omega - (\not{\nabla}\widehat{\otimes}\not{\nabla})\log\Omega - \not{\nabla}\log\Omega\widehat{\otimes}\not{\nabla}\log\Omega = 0\\
&\mathbf{D}_4\mathrm{tr}\underline{\chi} + \frac{1}{2}\mathrm{tr}\chi\mathrm{tr}\underline{\chi} + (\mathbf{D}_4\log\Omega)\mathrm{tr}\underline{\chi} + \hat{\chi}\cdot\underline{\hat{\chi}} + 2\not{\mathrm{d}}\mathrm{iv}\,\zeta - 2\not{\Delta}\log\Omega\\
&\quad -2|\zeta|^2 - 4\zeta\cdot\not{\nabla}\log\Omega - 2|\not{\nabla}\log\Omega|^2 = 2\rho\\
&\not{\mathbf{D}}_3\hat{\chi} + \frac{1}{2}\mathrm{tr}\underline{\chi}\,\hat{\chi} + \frac{1}{2}\mathrm{tr}\chi\,\underline{\hat{\chi}} + (\mathbf{D}_3\log\Omega)\hat{\chi} - \not{\nabla}\widehat{\otimes}\zeta - \zeta\widehat{\otimes}\zeta\\
&\quad -2\zeta\widehat{\otimes}\not{\nabla}\log\Omega - (\not{\nabla}\widehat{\otimes}\not{\nabla})\log\Omega - \not{\nabla}\log\Omega\widehat{\otimes}\not{\nabla}\log\Omega = 0\\
&\mathbf{D}_3\mathrm{tr}\chi + \frac{1}{2}\mathrm{tr}\underline{\chi}\mathrm{tr}\chi + (\mathbf{D}_3\log\Omega)\mathrm{tr}\chi + \underline{\hat{\chi}}\cdot\hat{\chi} - 2\not{\mathrm{d}}\mathrm{iv}\,\zeta - 2|\zeta|^2\\
&\quad -2\not{\Delta}\log\Omega - 4\zeta\cdot\not{\nabla}\log\Omega - 2|\not{\nabla}\log\Omega|^2 = 2\rho
\end{aligned}
$$

$$
\begin{aligned}
&\not{\nabla}\mathrm{tr}\underline{\chi} - \not{\mathrm{d}}\mathrm{iv}\,\underline{\chi} + \zeta\cdot\underline{\chi} - \zeta\mathrm{tr}\underline{\chi} = -\underline{\beta}\\
&\not{\nabla}\mathrm{tr}\chi - \not{\mathrm{d}}\mathrm{iv}\,\chi - \zeta\cdot\chi + \zeta\mathrm{tr}\chi = \beta\\
&\not{\mathrm{c}}\mathrm{url}\,\zeta - \frac{1}{2}\underline{\hat{\chi}}\wedge\hat{\chi} = \sigma \qquad (3.1.47)\\
&\mathbf{K} + \frac{1}{4}\mathrm{tr}\chi\mathrm{tr}\underline{\chi} - \frac{1}{2}\underline{\hat{\chi}}\cdot\hat{\chi} = -\rho,
\end{aligned}
$$

and

$$
\frac{1}{2}(\mathbf{D}_4\mathbf{D}_3\log\Omega + \mathbf{D}_3\mathbf{D}_4\log\Omega) + (\mathbf{D}_3\log\Omega)(\mathbf{D}_4\log\Omega) + 3|\zeta|^2 - |\not{\nabla}\log\Omega|^2 = -\rho. \qquad (3.1.48)
$$

Remark: Recall that $\not{\mathbf{D}}_4, \not{\mathbf{D}}_3$ are the projections of $\mathbf{D}_4, \mathbf{D}_3$ to TS. Moreover, given U, V two covariant S-tangent vector fields, $U\widehat{\otimes}V$ is defined as twice the traceless part of their symmetric tensor product $U \otimes V$,

$$
(U\widehat{\otimes}V)_{ab} \equiv U_aV_b + U_bV_a - \delta_{ab}U\cdot V. \qquad (3.1.49)
$$

Proof: See the appendix to this chapter, Subsection 3.8.2.

3.1.6 The Einstein equations relative to a double null foliation

Among the complete set of structure equations, (3.1.46), (3.1.47), (3.1.48), we identify those that do not depend on the null components of the curvature tensor. They are the equations that correspond to $\mathbf{R}(e_\alpha, e_\beta) = 0$. In other words they can be interpreted as the

“Einstein vacuum equations”, expressed relative to the double null foliation:

$$
\left\{
\begin{aligned}
&\mathbf{D}_4\mathrm{tr}\chi + \tfrac{1}{2}(\mathrm{tr}\chi)^2 + 2\omega\mathrm{tr}\chi + |\hat{\chi}|^2 = 0\\
&\mathbf{D}_4\mathrm{tr}\underline{\chi} + \mathrm{tr}\chi\,\mathrm{tr}\underline{\chi} - 2\omega\mathrm{tr}\underline{\chi} = -2\mathbf{K} + 2\not{\mathrm{div}}\,(-\zeta + \not\nabla\log\Omega) + 2|-\zeta + \not\nabla\log\Omega|^2\\
&\not{\mathbf{D}}_4\hat{\underline{\chi}} - 2\omega\hat{\underline{\chi}} = \not\nabla\widehat{\otimes}\underline{\eta} + \underline{\eta}\widehat{\otimes}\underline{\eta} - \tfrac{1}{2}(\mathrm{tr}\underline{\chi}\,\hat{\chi} + \mathrm{tr}\chi\,\hat{\underline{\chi}})\\
&\not{\mathbf{D}}_4\zeta + \zeta\chi + \mathrm{tr}\chi\,\zeta = \not{\mathrm{div}}\,\chi - \not\nabla\mathrm{tr}\chi - \not{\mathbf{D}}_4\not\nabla\log\Omega
\end{aligned}
\right.
\tag{3.1.50}
$$

$$
\left\{
\begin{aligned}
&\mathbf{D}_3\mathrm{tr}\underline{\chi} + \tfrac{1}{2}\mathrm{tr}\underline{\chi}^2 + 2\underline{\omega}\mathrm{tr}\underline{\chi} + |\hat{\underline{\chi}}|^2 = 0\\
&\mathbf{D}_3\mathrm{tr}\chi + \mathrm{tr}\underline{\chi}\,\mathrm{tr}\chi - 2\underline{\omega}\mathrm{tr}\chi = -2\mathbf{K} + 2\not{\mathrm{div}}\,(\zeta + \not\nabla\log\Omega) + 2|\zeta + \not\nabla\log\Omega|^2\\
&\not{\mathbf{D}}_3\hat{\chi} - 2\underline{\omega}\hat{\chi} = \not\nabla\widehat{\otimes}\eta + \eta\widehat{\otimes}\eta - \tfrac{1}{2}(\mathrm{tr}\chi\,\hat{\underline{\chi}} + \mathrm{tr}\underline{\chi}\,\hat{\chi})\\
&\not{\mathbf{D}}_3\zeta + \zeta\underline{\chi} + \mathrm{tr}\underline{\chi}\,\zeta = -\not{\mathrm{div}}\,\underline{\chi} + \not\nabla\mathrm{tr}\underline{\chi} + \not{\mathbf{D}}_3\not\nabla\log\Omega,
\end{aligned}
\right.
\tag{3.1.51}
$$

and

$$
\begin{aligned}
\frac{1}{2}(\mathbf{D}_4\mathbf{D}_3\log\Omega + \mathbf{D}_3\mathbf{D}_4\log\Omega) + (\mathbf{D}_3\log\Omega)(\mathbf{D}_4\log\Omega) + 3|\zeta|^2 - |\not\nabla\log\Omega|^2&\\
= \mathbf{K} + \frac{1}{4}\mathrm{tr}\chi\,\mathrm{tr}\underline{\chi} - \frac{1}{2}\hat{\underline{\chi}}\cdot\hat{\chi}\ .&
\end{aligned}
\tag{3.1.52}
$$

Remark: The number of equations written in (3.1.50), (3.1.51), (3.1.52) is 13 instead of 10, as the independent components of the Ricci tensor. Therefore three of them are not independent. A careful look shows that the three equations $\mathbf{Ricci}(e_a, e_b) = 0$ can be written either as

$$
\begin{aligned}
&\not{\mathbf{D}}_4\hat{\underline{\chi}} - 2\omega\hat{\underline{\chi}} = \not\nabla\widehat{\otimes}\underline{\eta} + \underline{\eta}\widehat{\otimes}\underline{\eta} - \frac{1}{2}(\mathrm{tr}\underline{\chi}\,\hat{\chi} + \mathrm{tr}\chi\,\hat{\underline{\chi}})\\
&\mathbf{D}_4\mathrm{tr}\underline{\chi} + \mathrm{tr}\chi\,\mathrm{tr}\underline{\chi} - 2\omega\mathrm{tr}\underline{\chi} = -2\mathbf{K} + 2\not{\mathrm{div}}\,(-\zeta + \not\nabla\log\Omega) + 2|-\zeta + \not\nabla\log\Omega|^2,
\end{aligned}
$$

or as

$$
\begin{aligned}
&\not{\mathbf{D}}_3\hat{\chi} - 2\underline{\omega}\hat{\chi} = \not\nabla\widehat{\otimes}\eta + \eta\widehat{\otimes}\eta - \frac{1}{2}(\mathrm{tr}\chi\,\hat{\underline{\chi}} + \mathrm{tr}\underline{\chi}\,\hat{\chi})\\
&\mathbf{D}_3\mathrm{tr}\chi + \mathrm{tr}\underline{\chi}\,\mathrm{tr}\chi - 2\underline{\omega}\mathrm{tr}\chi = -2\mathbf{K} + 2\not{\mathrm{div}}\,(\zeta + \not\nabla\log\Omega) + 2|\zeta + \not\nabla\log\Omega|^2,
\end{aligned}
$$

which restores to 10 the total number of the Einstein equations.

To look at these equations as partial differential equations it is appropriate to rewrite them in terms of the $\mathcal{D}$, $\underline{\mathcal{D}}$ derivatives defined in Subsection 3.1.1. Recalling Definitions 3.1.12, 3.1.13, it follows immediately that

$$
\begin{aligned}
\mathcal{D}\mathrm{tr}\chi &= \Omega\mathbf{D}_4\mathrm{tr}\chi\ ,\ \mathcal{D}\mathrm{tr}\underline{\chi} = \Omega\mathbf{D}_4\mathrm{tr}\underline{\chi}\\
\underline{\mathcal{D}}\mathrm{tr}\chi &= \Omega\mathbf{D}_3\mathrm{tr}\chi\ ,\ \underline{\mathcal{D}}\mathrm{tr}\underline{\chi} = \Omega\mathbf{D}_3\mathrm{tr}\underline{\chi}
\end{aligned}
$$

$$\mathcal{D}\zeta = \Omega\left(\mathbf{D}\!\!\!/_4\zeta + \zeta\cdot\chi\right),\ \underline{\mathcal{D}}\zeta = \Omega\left(\mathbf{D}\!\!\!/_3\zeta + \zeta\cdot\underline{\chi}\right)$$
$$\mathcal{D}\hat{\chi} = \Omega\left(\mathbf{D}\!\!\!/_4\hat{\chi} + 2\hat{\chi}\cdot\hat{\chi}\right),\ \mathcal{D}\underline{\hat{\chi}} = \Omega\left(\mathbf{D}\!\!\!/_4\underline{\hat{\chi}} + 2\underline{\hat{\chi}}\cdot\hat{\chi}\right)$$
$$\underline{\mathcal{D}}\hat{\chi} = \Omega\left(\mathbf{D}\!\!\!/_3\hat{\chi} + 2\hat{\chi}\cdot\underline{\hat{\chi}}\right),\ \underline{\mathcal{D}}\underline{\hat{\chi}} = \Omega\left(\mathbf{D}\!\!\!/_3\underline{\hat{\chi}} + 2\underline{\hat{\chi}}\cdot\underline{\hat{\chi}}\right). \tag{3.1.53}$$

Observe also that the equations for ζ along the $C(\lambda)$ and $\underline{C}(\nu)$ null hypersurfaces can be replaced by similar equations relative to η and $\underline{\eta}$ in view of the relations, see (3.1.33),

$$\eta = \zeta + \nabla\!\!\!\!/\,\log\Omega\ ,\ \underline{\eta} = -\zeta + \nabla\!\!\!\!/\,\log\Omega. \tag{3.1.54}$$

Thus the previous equations (3.1.50) and (3.1.51) take the following form:

$$\begin{aligned}
&\mathcal{D}\mathrm{tr}\chi - (\mathcal{D}\log\Omega)\mathrm{tr}\chi + \frac{1}{2}\Omega\mathrm{tr}\chi^2 + \Omega|\hat{\chi}|^2 = 0\\
&\mathcal{D}\mathrm{tr}\underline{\chi} + (\mathcal{D}\log\Omega)\mathrm{tr}\underline{\chi} + \Omega\mathrm{tr}\chi\mathrm{tr}\underline{\chi} = 2\Omega\left(-\mathbf{K} + \mathrm{d}\!\!/\mathrm{iv}\,\underline{\eta} + |\underline{\eta}|^2\right)\\
&\mathcal{D}\underline{\hat{\chi}} + (\mathcal{D}\log\Omega)\underline{\hat{\chi}} - 2\Omega\underline{\hat{\chi}}\cdot\hat{\chi} = \Omega\left(\nabla\!\!\!\!/\,\widehat{\otimes}\underline{\eta} + \underline{\eta}\widehat{\otimes}\underline{\eta} - \frac{1}{2}(\mathrm{tr}\underline{\chi}\hat{\chi} + \mathrm{tr}\chi\underline{\hat{\chi}})\right)\\
&\mathcal{D}\eta + \Omega\mathrm{tr}\chi\eta = \Omega(\mathrm{d}\!\!/\mathrm{iv}\,\chi - \nabla\!\!\!\!/\,\mathrm{tr}\chi) + \Omega\left(\hat{\chi}\cdot\nabla\!\!\!\!/\,\log\Omega + \frac{3}{2}\mathrm{tr}\chi\nabla\!\!\!\!/\,\log\Omega\right),
\end{aligned} \tag{3.1.55}$$

and

$$\begin{aligned}
&\underline{\mathcal{D}}\mathrm{tr}\underline{\chi} - (\underline{\mathcal{D}}\log\Omega)\mathrm{tr}\underline{\chi} + \frac{1}{2}\Omega\mathrm{tr}\underline{\chi}^2 + \Omega|\underline{\hat{\chi}}|^2 = 0\\
&\underline{\mathcal{D}}\mathrm{tr}\chi + (\underline{\mathcal{D}}\log\Omega)tr\chi + \Omega\mathrm{tr}\underline{\chi}\mathrm{tr}\chi = 2\Omega\left(-\mathbf{K} + \mathrm{d}\!\!/\mathrm{iv}\,\eta + |\eta|^2\right)\\
&\underline{\mathcal{D}}\hat{\chi} + (\underline{\mathcal{D}}\log\Omega)\hat{\chi} - 2\Omega\hat{\chi}\cdot\underline{\hat{\chi}} = \Omega\left(\nabla\!\!\!\!/\,\widehat{\otimes}\eta + \eta\widehat{\otimes}\eta - \frac{1}{2}(\mathrm{tr}\chi\underline{\hat{\chi}} + \mathrm{tr}\underline{\chi}\hat{\chi})\right)\\
&\underline{\mathcal{D}}\underline{\eta} + \Omega\mathrm{tr}\underline{\chi}\underline{\eta} = \Omega(\mathrm{d}\!\!/\mathrm{iv}\,\underline{\chi} - \nabla\!\!\!\!/\,\mathrm{tr}\underline{\chi}) + \Omega\left(\underline{\hat{\chi}}\cdot\nabla\!\!\!\!/\,\log\Omega + \frac{3}{2}\mathrm{tr}\underline{\chi}\nabla\!\!\!\!/\,\log\Omega\right),
\end{aligned} \tag{3.1.56}$$

and

$$\mathcal{D}\underline{\mathcal{D}}\log\Omega + \underline{\mathcal{D}}\mathcal{D}\log\Omega = 2\Omega^2\left(\eta\cdot\underline{\eta} - 2|\eta - \nabla\!\!\!\!/\,\log\Omega|^2 + \mathbf{K} + \frac{1}{4}\mathrm{tr}\chi\mathrm{tr}\underline{\chi} - \frac{1}{2}\underline{\hat{\chi}}\cdot\hat{\chi}\right). \tag{3.1.57}$$

These equations form a closed system when supplemented by equations (3.1.14)

$$\mathcal{D}\gamma = 2\chi\ ,\ \underline{\mathcal{D}}\gamma = 2\underline{\chi}.$$

Equations (3.1.55), (3.1.56), (3.1.57) and (3.1.14) can be expressed in terms of null coordinates by supplementing u and $\underline{u}$ with angular coordinates θ, ϕ.

We start with a fixed system of coordinates θ_0, ϕ_0 defined on $S(\lambda_0, \nu_*) = \underline{C}_* \cap C_0$, where $\underline{C}_* = \underline{C}(\underline{u} = \nu_*)$, $C_0 = C(u = \lambda_0)$ and define $\mathcal{M}$ as the causal past of $S(\lambda_0, \nu_*)$. Consider any other surface $S(\lambda, \nu) \subset \mathcal{M}$. We can transport the coordinates from $S(\lambda_0, \nu_*)$ to $S(\lambda, \nu)$ in two different ways, using the flows ϕ_t and $\underline{\phi}_s$ associated to the equivariant null pair N and $\underline{N}$,

$$\begin{aligned}
\theta|_{S(\lambda,\nu)} &= \theta_0(\underline{\phi}_s \circ \phi_t)\\
\phi|_{S(\lambda,\nu)} &= \phi_0(\underline{\phi}_s \circ \phi_t)
\end{aligned} \tag{3.1.58}$$

or

$$\theta|_{S(\lambda,\nu)} = \theta_0(\phi_t \circ \underline{\phi}_s)$$
$$\phi|_{S(\lambda,\nu)} = \phi_0(\phi_t \circ \underline{\phi}_s), \tag{3.1.59}$$

where $t = \nu_* - \nu$, $s = \lambda_0 - \lambda$. Recall that $\phi_{\nu_*-\nu}$ is the diffeomorphism from $S(\lambda, \nu)$ to $S(\lambda, \nu_*)$ and $\underline{\phi}_{\lambda_0-\lambda}$ is the diffeomorphism from $S(\lambda, \nu)$ to $S(\lambda_0, \nu)$. Observe that the two definitions differ; indeed, since N and $\underline{N}$ do not commute,

$$\phi_t \circ \underline{\phi}_s \neq \underline{\phi}_s \circ \phi_t \,.$$

Since N and $\underline{N}$ are equivariant one of the two choices corresponds to $N = \frac{\partial}{\partial \underline{u}}$ and the other to $\underline{N} = \frac{\partial}{\partial u}$.[11] Choosing $N = \frac{\partial}{\partial \underline{u}}$, we infer that $\underline{N}$ must have the form $\underline{N} = \frac{\partial}{\partial u} + X$. To determine the vector field X observe that (see (3.1.45))[12]

$$[N, \underline{N}] = Z \equiv -4\Omega^2 \zeta(e_b) e_b. \tag{3.1.60}$$

On the other hand, from the previous expressions for N and $\underline{N}$,

$$[N, \underline{N}] = \left(\frac{\partial}{\partial \underline{u}} X^a\right) \frac{\partial}{\partial \omega^a}, \tag{3.1.61}$$

where $\omega^1 = \theta, \omega^2 = \phi$. Therefore in view of equation (3.1.60) we can uniquely define the vector field X by[13]

$$\frac{\partial}{\partial \underline{u}} X^a = Z^a = -4\Omega^2 \gamma^{ab} \zeta_b \;,\; X^a|_{\underline{C}_*} = 0. \tag{3.1.62}$$

With this choice of coordinates the metric $\mathbf{g}$ has the following expression

$$\mathbf{g}(\cdot\,,\cdot) = -2\Omega^2 \left(du d\underline{u} + d\underline{u} du\right) + \gamma_{ab}(d\omega^a - X^a du)(d\omega^b - X^b du). \tag{3.1.63}$$

3.1.7 *The characteristic initial value problem for the Einstein equations*

In terms of the previous choices of coordinates we can now interpret equations (3.1.55) and (3.1.56) as equations for the six unknowns γ_{ab}, X_a and Ω of the spacetime metric.

Consider two null hypersurfaces C_*, $\underline{C}_*$, both originating from a common two-dimensional surface S; see Definition 3.1.5. On $C_* \cup \underline{C}_*$ we assume prescribed γ_{ab}, X_a and Ω. How much freedom we have in assigning these six quantities will be discussed later on.

On C_*, then $\mathrm{tr}\chi$, $\hat{\chi}$ and ζ are automatically determined by (see (3.1.14) and (3.1.62)) $2\chi = \mathcal{D}\gamma$ and $Z = \mathcal{L}_N X$. Of course, in view of relations (3.1.54) and (3.1.33), η, $\underline{\eta}$ and ω are also determined on C_*.

[11] Recall that $N(\underline{u}) = 1$, $N(u) = 0$, $\underline{N}(u) = 1$, $\underline{N}(\underline{u}) = 0$.

[12] We can also write $[\mathcal{D}, \underline{\mathcal{D}}] = \mathcal{L}_{[N,\underline{N}]} = \mathcal{L}_Z$.

[13] Here a, b denote coordinates on S.

On $\underline{C}_*$, $\mathrm{tr}\underline{\chi}$, $\hat{\underline{\chi}}$, $\underline{\omega}$, η and $\underline{\eta}$ are again automatically determined from (see (3.1.14)) $2\underline{\chi} = \underline{\mathcal{D}}\gamma$ and using definitions (3.1.33) for $\underline{\omega}$, η, and $\underline{\eta}$. X is chosen to be zero, as allowed by our choice of coordinates; see (3.1.62), (3.1.63).

Moreover on $C_* \cup \underline{C}_*$ we can determine the remaining connection coefficients, $\mathrm{tr}\chi$, $\hat{\chi}$ on $\underline{C}_*$ and $\mathrm{tr}\underline{\chi}$, $\hat{\underline{\chi}}$ on C_*. In fact, using equations (3.1.55) and (3.1.56),

$$\begin{aligned}
&\mathcal{D}\mathrm{tr}\underline{\chi} + (\mathcal{D}\log\Omega)tr\underline{\chi} + \Omega\mathrm{tr}\chi\mathrm{tr}\underline{\chi} = 2\Omega\left(-\mathbf{K} + \mathrm{d}\!\!/\mathrm{iv}\,\underline{\eta} + |\underline{\eta}|^2\right)\\
&\mathcal{D}\hat{\underline{\chi}} + (\mathcal{D}\log\Omega)\hat{\underline{\chi}} - 2\Omega\hat{\underline{\chi}}\cdot\hat{\chi} = \Omega\left(\nabla\!\!\!/\widehat{\otimes}\underline{\eta} + \underline{\eta}\widehat{\otimes}\underline{\eta} - \frac{1}{2}(tr\underline{\chi}\hat{\chi} + tr\chi\hat{\underline{\chi}})\right)\\
&\underline{\mathcal{D}}\mathrm{tr}\chi + (\underline{\mathcal{D}}\log\Omega)tr\chi + \Omega\mathrm{tr}\underline{\chi}\mathrm{tr}\chi = 2\Omega\left(-\mathbf{K} + \mathrm{d}\!\!/\mathrm{iv}\,\eta + |\eta|^2\right)\\
&\underline{\mathcal{D}}\hat{\chi} + (\underline{\mathcal{D}}\log\Omega)\hat{\chi} - 2\Omega\hat{\chi}\cdot\hat{\underline{\chi}} = \Omega\left(\nabla\!\!\!/\widehat{\otimes}\eta + \eta\widehat{\otimes}\eta - \frac{1}{2}(tr\chi\hat{\underline{\chi}} + tr\underline{\chi}\hat{\chi})\right),
\end{aligned}$$

these quantities are uniquely determined on C_* and $\underline{C}_*$, respectively, in terms of their values on S. Therefore, once the γ_{ab}, X_a and Ω metric components are assigned, all the null connection coefficients are specified on $C_* \cup \underline{C}_*$.

To determine the spacetime $\mathcal{K}$, given the initial data on $C_* \cup \underline{C}_*$, we can proceed formally in the following way.[14] The incoming evolution equations (3.1.56) for χ, $\mathrm{tr}\underline{\chi}$, $\underline{\eta}$

$$\begin{aligned}
&\underline{\mathcal{D}}\mathrm{tr}\underline{\chi} - (\underline{\mathcal{D}}\log\Omega)tr\underline{\chi} + \frac{1}{2}\Omega\mathrm{tr}\underline{\chi}^2 + \Omega|\hat{\underline{\chi}}|^2 = 0\\
&\underline{\mathcal{D}}\mathrm{tr}\chi + (\underline{\mathcal{D}}\log\Omega)tr\chi + \Omega\mathrm{tr}\underline{\chi}\mathrm{tr}\chi = 2\Omega\left(-\mathbf{K} + \mathrm{d}\!\!/\mathrm{iv}\,\eta + |\eta|^2\right)\\
&\underline{\mathcal{D}}\hat{\chi} + (\underline{\mathcal{D}}\log\Omega)\hat{\chi} - 2\Omega\hat{\chi}\cdot\hat{\underline{\chi}} = \Omega\left(\nabla\!\!\!/\widehat{\otimes}\eta + \eta\widehat{\otimes}\eta - \frac{1}{2}(tr\chi\hat{\underline{\chi}} + tr\underline{\chi}\hat{\chi})\right)\\
&\underline{\mathcal{D}}\underline{\eta} + \Omega\mathrm{tr}\underline{\chi}\underline{\eta} = \Omega(\mathrm{d}\!\!/\mathrm{iv}\,\underline{\chi} - \nabla\!\!\!/\mathrm{tr}\underline{\chi}) + \Omega\left(\hat{\underline{\chi}}\cdot\nabla\!\!\!/\log\Omega + \frac{3}{2}\mathrm{tr}\underline{\chi}\nabla\!\!\!/\log\Omega\right)
\end{aligned}$$

and the equation (3.1.57), written as an evolution equation along e_3 for ω,

$$\underline{\mathcal{D}}\omega - 2\Omega\underline{\omega}\omega = \frac{3}{2}\Omega|\zeta|^2 + \Omega\zeta\cdot\nabla\!\!\!/\log\Omega - \frac{1}{2}\Omega|\nabla\!\!\!/\log\Omega|^2 - \frac{1}{2}\Omega\left(\mathbf{K} + \frac{1}{4}\mathrm{tr}\chi\mathrm{tr}\underline{\chi} - \frac{1}{2}\hat{\underline{\chi}}\cdot\hat{\chi}\right),$$

allows us to determine, from the initial conditions on C_*, χ, $\mathrm{tr}\underline{\chi}$, $\underline{\eta}$ and ω.

Therefore it remains to determine $\hat{\underline{\chi}}$, $\underline{\omega}$ and η. Then knowing η and $\underline{\eta}$, $\nabla\!\!\!/\log\Omega$ is also determined. These can be obtained from the outgoing evolution equations in (3.1.55) for $\hat{\underline{\chi}}$ and η,

$$\begin{aligned}
&\mathcal{D}\hat{\underline{\chi}} + (\mathcal{D}\log\Omega)\hat{\underline{\chi}} - 2\Omega\hat{\underline{\chi}}\cdot\hat{\chi} = \Omega\left(\nabla\!\!\!/\widehat{\otimes}\underline{\eta} + \underline{\eta}\widehat{\otimes}\underline{\eta} - \frac{1}{2}(tr\underline{\chi}\hat{\chi} + tr\chi\hat{\underline{\chi}})\right)\\
&\mathcal{D}\eta + \Omega\mathrm{tr}\chi\eta = \Omega(\mathrm{d}\!\!/\mathrm{iv}\,\chi - \nabla\!\!\!/\mathrm{tr}\chi) + \Omega\left(\hat{\chi}\cdot\nabla\!\!\!/\log\Omega + \frac{3}{2}\mathrm{tr}\chi\nabla\!\!\!/\log\Omega\right)
\end{aligned}$$

[14] See the remark at the end of this subsection.

and from equation (3.1.57), written as an evolution equation along e_4 for $\underline{\omega}$,

$$\mathcal{D}\underline{\omega}-2\Omega\omega\underline{\omega}=\frac{3}{2}\Omega|\zeta|^2-\Omega\zeta\cdot\nabla\!\!\!/\log\Omega-\frac{1}{2}\Omega|\nabla\!\!\!/\log\Omega|^2-\frac{1}{2}\Omega\left(\mathbf{K}+\frac{1}{4}\mathrm{tr}\underline{\chi}\mathrm{tr}\chi-\frac{1}{2}\hat{\chi}\cdot\hat{\underline{\chi}}\right),$$

once their initial values are given on $\underline{C}_*$.

In fact all these equations have to be solved simultaneously since they are heavily coupled.[15]

The initial data constraints

Observe that the initial metric components γ, X, Ω on $C_*\cup\underline{C}_*$ cannot be freely assigned. In fact let us recall the relations (see (3.1.14), (3.1.33) and (3.1.62),

$$\frac{\partial}{\partial\underline{u}}X^a=-4\Omega^2\zeta^a\,,\ \frac{\partial}{\partial\underline{u}}\gamma_{ab}=2\chi_{ab}\,,\ \frac{\partial}{\partial\underline{u}}\Omega=-2\Omega^2\omega, \tag{3.1.64}$$

and observe that, on C_*, $\mathrm{tr}\chi$ and ζ have to satisfy the equations[16]

$$\mathcal{D}\mathrm{tr}\chi-(\mathcal{D}\log\Omega)tr\chi+\frac{1}{2}\Omega\mathrm{tr}\chi^2+\Omega|\hat{\chi}|^2=0 \tag{3.1.65}$$
$$\mathcal{D}\zeta+\Omega\mathrm{tr}\chi\zeta=\Omega(\mathrm{d\!\!/iv}\,\chi-\nabla\!\!\!/\mathrm{tr}\chi+\chi\cdot\nabla\!\!\!/\log\Omega)+(\nabla\!\!\!/\mathcal{D}\log\Omega+2(\mathcal{D}\log\Omega)\nabla\!\!\!/\log\Omega).$$

Therefore if we assign Ω and the traceless part of $\gamma(\cdot,\cdot)$, $\hat{\gamma}_{ab}$ freely on C_*, then we can easily derive ω and $\hat{\chi}_{ab}$ on C_* as well as , using (3.1.65), $\mathrm{tr}\chi$ and ζ once they are given on S. Then, by simple integration we also know the trace of γ_{ab} and X_a on C_*, provided they are specified on S.

The situation is similar on $\underline{C}_*$. The equations analogous to (3.1.65) are

$$\underline{\mathcal{D}}\mathrm{tr}\underline{\chi}-(\underline{\mathcal{D}}\log\Omega)tr\underline{\chi}+\frac{1}{2}\Omega\mathrm{tr}\underline{\chi}^2+\Omega|\hat{\underline{\chi}}|^2=0 \tag{3.1.66}$$
$$\underline{\mathcal{D}}\zeta+\Omega\mathrm{tr}\underline{\chi}\zeta=\Omega(-\mathrm{d\!\!/iv}\,\underline{\chi}+\nabla\!\!\!/\mathrm{tr}\underline{\chi}-\underline{\chi}\cdot\nabla\!\!\!/\log\Omega).-(\nabla\!\!\!/\underline{\mathcal{D}}\log\Omega+2(\underline{\mathcal{D}}\log\Omega)\nabla\!\!\!/\log\Omega)$$

Again we can assign Ω and $\hat{\gamma}_{ab}$ freely on $\underline{C}_*$ and, knowing γ_{ab} and ζ_a on S, determine all the initial data on $\underline{C}_*$. The only relevant difference is that on $\underline{C}_*$ we can also impose $X_a=0$.

Remark: The Einstein equations written in the form (3.1.55), (3.1.56) are highly nonlinear and manifestly nonhyperbolic. The procedure we outlined above to solve them is very formal. It can only be implemented locally, using a variant of the Cauchy–Kowaleski method, in the class of analytic spacetimes. This procedure is, however, completely inadequate for studying global solutions. In fact in the proof of the Main Theorem we circumvent them completely by relying instead on the full set of structure equations (3.1.46), (3.1.47),(3.1.48), where the null components α, β, ρ, σ, $\underline{\beta}$, $\underline{\alpha}$ of the curvature tensor are treated as external sources. Indeed we show that these curvature components can be estimated, separately, from the Bianchi equations, as discussed in the remaining part of this chapter, see also the preliminary discussion in Chapter 2.

---

[15] See also A.D. Rendall, [Ren].

[16] The outgoing equation for ζ can be derived immediately from the one for η.

3.2 Bianchi equations in an Einstein vacuum spacetime

Definition 3.2.1 *A Weyl tensor field is a solution of the Bianchi equations in $(\mathcal{M}, \mathbf{g})$ if, relative to the Levi-Civita connection $\mathbf{D}$ of $\mathbf{g}$, it satisfies*

$$D_{[\sigma} W_{\gamma\delta]\alpha\beta} = 0. \tag{3.2.1}$$

Remarks:

1. If W verifies the equations (3.2.1) on a background spacetime $(\mathcal{M}, \mathbf{g})$, then it must also satisfy the compatibility condition (see [Ch8])

$$\mathbf{R}_{\mu}^{\ \alpha\beta\gamma\star} W_{\nu\alpha\beta\gamma} - \mathbf{R}_{\nu}^{\ \alpha\beta\gamma\star} W_{\mu\alpha\beta\gamma} \ .$$

2. The primary example of a solution of (3.2.1) is the Riemann curvature tensor of an Einstein vacuum spacetime $(\mathcal{M}, \mathbf{g})$.

We review the main properties of Weyl tensor fields and of the Bianchi equations (3.2.1). They are a generalization of those discussed in Chapter 2 in the case of Minkowski spacetime; see also the extended discussion in [Ch-Kl].

Definition 3.2.2 *Given a Weyl field W and X, an arbitrary vector field, we define the modified Lie derivative relative to X by*

$$\hat{\mathcal{L}}_X W = \mathcal{L}_X W - \frac{1}{2}{}^{(X)}[W] + \frac{3}{8}\mathrm{tr}\,{}^{(X)}\pi\, W, \tag{3.2.2}$$

where

$${}^{(X)}[W]_{\alpha\beta\gamma\delta} = {}^{(X)}\pi_{\alpha}^{\ \mu} W_{\mu\beta\gamma\delta} + {}^{(X)}\pi_{\beta}^{\ \mu} W_{\alpha\mu\gamma\delta} + {}^{(X)}\pi_{\gamma}^{\ \mu} W_{\alpha\beta\mu\delta} + {}^{(X)}\pi_{\delta}^{\ \mu} W_{\alpha\beta\gamma\mu} \tag{3.2.3}$$

and ${}^{(X)}\pi$ is the deformation tensor of X.

Proposition 3.2.1

i) The following four sets of equations are equivalent

$$D_{[\sigma} W_{\gamma\delta]\alpha\beta} = 0\ ,\ \ D^{\mu} W_{\mu\nu\alpha\beta} = 0$$
$$D^{\mu\star} W_{\mu\nu\alpha\beta} = 0\ ,\ \ D_{[\sigma}{}^{\star} W_{\gamma\delta]\alpha\beta} = 0\ .$$

ii) The Bianchi equations (3.2.1) are conformally covariant (see [Pe1], [Pe2] and also [Ch-Kl1], [Ch-Kl]).[17]

iii) $\hat{\mathcal{L}}_X W$ is also a Weyl field and satisfies $\hat{\mathcal{L}}_X{}^* W = {}^*\hat{\mathcal{L}}_X W$.

Proof: See [Ch-Kl], Chapter 7.

The Bianchi equations (3.2.1) look complicated. Nevertheless they are quite similar to the more familiar Maxwell equations. This is already obvious formally, but it becomes even

---

[17] This means that whenever we perform a conformal transformation ϕ of the spacetime $(\mathcal{M}, g)$ with $\tilde{g} = \phi_* g = \Lambda^2 g$, then $\tilde{W} = \Lambda^{-1}\phi_* W$ is a solution of the Bianchi equations for the spacetime $(\mathcal{M}, \tilde{g})$.

more apparent if we decompose W into its "electric" and "magnetic" parts. Given vector fields X, Y we define $i_{(X,Y)}$ through the relation $(i_{(X,Y)}W)_{\mu\nu} = W_{\mu\zeta\nu\sigma}X^\zeta Y^\sigma$. Assume the Einstein spacetime $(\mathcal{M}, \mathbf{g})$ is foliated by the spacelike hypersurfaces Σ_t and denote by T the normal vector field to Σ_t; then, with $X = Y = T$, define

$$E = i_{(T,T)}W \ , \ \ H = i_{(T,T)}{}^\star W. \tag{3.2.4}$$

These two covariant symmetric traceless tensor fields E and H, tangent to the hypersurfaces Σ_t, determine completely the Weyl tensor field. It is easy to write the Bianchi equations for this decomposition and obtain the following Maxwell-type equations:

$$\begin{aligned}
&\Phi^{-1}\partial_t E + \mathrm{curl} H = \rho(E, H)\\
&\Phi^{-1}\partial_t H - \mathrm{curl} E = \sigma(E, H)\\
&\mathrm{div} E = k \wedge H\\
&\mathrm{div} H = -k \wedge E \ ,
\end{aligned}$$

where ∇ is the covariant derivative with respect to Σ_t, k is the second fundamental form of Σ_t,

$$(\mathrm{div} E)_i = \nabla^j E_{ij} \ , \ \ (\mathrm{curl}\ E)_{ij} = \epsilon_i^{\ lk}\nabla_l E_{kj},$$

and the corresponding expressions hold for H. The explicit expressions of $\rho(E, H)$ and $\sigma(E, H)$ can be found in [Ch-Kl], p. 146.

The strong formal analogy with the Maxwell equations goes even further. In fact, the Bianchi equations possess a tensor analogous to the energy-momentum tensor, the Bel–Robinson tensor; see [Bel].

Definition 3.2.3 *The Bel–Robinson tensor of the Weyl field W is the 4-covariant tensor field*

$$Q_{\alpha\beta\gamma\delta} = W_{\alpha\rho\gamma\sigma} W_{\beta\ \delta}^{\ \rho\ \sigma} + {}^\star W_{\alpha\rho\gamma\sigma}{}^\star W_{\beta\ \delta}^{\ \rho\ \sigma}. \tag{3.2.5}$$

The Bel–Robinson tensor has the following important properties, which resemble those of the energy-momentum tensor of the Maxwell equations; see [Ch-Kl] Chapter 7.[18]

Proposition 3.2.2

i) Q is symmetric and traceless relative to all pairs of indices.

ii) Q satisfies the following positivity condition: $Q(X_1, X_2, X_3, X_4)$ is positive, unless $W = 0$, for any timelike vector fields.[19]

iii) If W is a solution of the Bianchi equations, then

$$D^\alpha Q_{\alpha\beta\gamma\delta} = 0. \tag{3.2.6}$$

Definition 3.2.4 *Given a vector field X we denote by ${}^{(X)}\pi \equiv \mathcal{L}_X g$ its deformation tensor. We denote by ${}^{(X)}\hat{\pi}_{\mu\nu} = {}^{(X)}\pi_{\mu\nu} - \frac{1}{4}g_{\mu\nu}\mathrm{tr}{}^{(X)}\pi$ its traceless part. It measures in a precise sense how much the diffeomorphism generated by X differs from an isometry or a conformal isometry.*

---

[18] Here $\alpha, \beta, \gamma, \delta$ are coordinate indices.

[19] We need this property when at most two of these vector fields are different, in which case the proof is straightforward; see [Ch-Kl].

Proposition 3.2.3 *Let $Q(W)$ be the Bel–Robinson tensor of a Weyl field W and X, Y, Z a triplet of vector fields. We define the covariant vector field P associated to the triplet as*

$$P_\alpha = Q_{\alpha\beta\gamma\delta}X^\beta Y^\gamma Z^\delta. \tag{3.2.7}$$

Using all the symmetry properties of Q we have

$$\begin{aligned}\mathbf{Div} P &= (\mathbf{Div} Q)_{\beta\gamma\delta}X^\beta Y^\gamma Z^\delta \\ &+\frac{1}{2}Q_{\alpha\beta\gamma\delta}\left({}^{(X)}\pi^{\alpha\beta}Y^\gamma Z^\delta + {}^{(Y)}\pi^{\alpha\gamma}X^\beta Z^\delta + {}^{(Z)}\pi^{\alpha\delta}X^\beta Y^\gamma\right).\end{aligned}$$

Recall (see Chapter 2, Section 2.2) that this expression can be used, if X, Y, Z are Killing or conformal Killing vector fields, to construct conserved quantities.

We assume that the Einstein vacuum spacetime, or more precisely a region of it, is foliated by a smooth double null foliation. The Bianchi equations can be expressed in terms of the null components $\alpha, \beta, \rho, \sigma, \underline{\beta}, \underline{\alpha}$ of the curvature tensor, according to the following proposition.

Proposition 3.2.4 *Expressed relatively to an adapted null frame, the Bianchi equations take the form*

$$\begin{aligned}
\underline{\alpha}_4 &\equiv \mathbf{D}\!\!\!/_4\underline{\alpha} + \frac{1}{2}\mathrm{tr}\chi\underline{\alpha} = -\nabla\!\!\!\!/\widehat{\otimes}\underline{\beta} + \left[4\omega\underline{\alpha} - 3(\hat{\underline{\chi}}\rho - {}^\star\hat{\underline{\chi}}\sigma) + (\zeta - 4\underline{\eta})\widehat{\otimes}\underline{\beta}\right]\\
\underline{\beta}_3 &\equiv \mathbf{D}\!\!\!/_3\underline{\beta} + 2\mathrm{tr}\underline{\chi}\ \underline{\beta} = -\mathrm{d}\!\!\!/\mathrm{iv}\underline{\alpha} - \left[2\underline{\omega}\underline{\beta} + (-2\zeta + \eta)\cdot\underline{\alpha}\right]\\
\underline{\beta}_4 &\equiv \mathbf{D}\!\!\!/_4\underline{\beta} + \mathrm{tr}\chi\underline{\beta} = -\nabla\!\!\!\!/\rho + \left[2\omega\underline{\beta} + 2\hat{\underline{\chi}}\cdot\beta + {}^\star\nabla\!\!\!\!/\sigma - 3(\underline{\eta}\rho - {}^\star\underline{\eta}\sigma)\right]\\
\rho_3 &\equiv \mathbf{D}_3\rho + \frac{3}{2}\mathrm{tr}\underline{\chi}\rho = -\mathrm{d}\!\!\!/\mathrm{iv}\underline{\beta} - \left[\frac{1}{2}\hat{\chi}\cdot\underline{\alpha} - \zeta\cdot\underline{\beta} + 2\eta\cdot\underline{\beta}\right]\\
\rho_4 &\equiv \mathbf{D}_4\rho + \frac{3}{2}\mathrm{tr}\chi\rho = \mathrm{d}\!\!\!/\mathrm{iv}\beta - \left[\frac{1}{2}\hat{\underline{\chi}}\cdot\alpha - \zeta\cdot\beta - 2\underline{\eta}\cdot\beta\right]\\
\sigma_3 &\equiv \mathbf{D}_3\sigma + \frac{3}{2}\mathrm{tr}\underline{\chi}\sigma = -\mathrm{d}\!\!\!/\mathrm{iv}{}^\star\underline{\beta} + \left[\frac{1}{2}\hat{\chi}\cdot{}^\star\underline{\alpha} - \zeta\cdot{}^\star\underline{\beta} - 2\eta\cdot{}^\star\underline{\beta}\right]\\
\sigma_4 &\equiv \mathbf{D}_4\sigma + \frac{3}{2}\mathrm{tr}\chi\sigma = -\mathrm{d}\!\!\!/\mathrm{iv}{}^\star\beta + \left[\frac{1}{2}\hat{\underline{\chi}}\cdot{}^\star\alpha - \zeta\cdot{}^\star\beta - 2\underline{\eta}\cdot{}^\star\beta\right]\\
\beta_3 &\equiv \mathbf{D}\!\!\!/_3\beta + \mathrm{tr}\underline{\chi}\beta = \nabla\!\!\!\!/\rho + \left[2\underline{\omega}\beta + {}^\star\nabla\!\!\!\!/\sigma + 2\hat{\chi}\cdot\underline{\beta} + 3(\eta\rho + {}^\star\eta\sigma)\right]\\
\beta_4 &\equiv \mathbf{D}\!\!\!/_4\beta + 2\mathrm{tr}\chi\beta = \mathrm{d}\!\!\!/\mathrm{iv}\alpha - \left[2\omega\beta - (2\zeta + \underline{\eta})\alpha\right]\\
\alpha_3 &\equiv \mathbf{D}\!\!\!/_3\alpha + \frac{1}{2}\mathrm{tr}\underline{\chi}\alpha = \nabla\!\!\!\!/\widehat{\otimes}\beta + \left[4\underline{\omega}\alpha - 3(\hat{\chi}\rho + {}^\star\hat{\chi}\sigma) + (\zeta + 4\eta)\widehat{\otimes}\beta\right].
\end{aligned} \tag{3.2.8}$$

These equations are similar to the ones in Minkowski spacetime (see equations (2.2.13) of Chapter 2). The terms in square brackets, absent in the flat case, are products between the Weyl null components and the connection coefficients.[20]

---

[20] In the case of Schwarzschild spacetime the only terms in parentheses different from zero are those depending on ω, $\underline{\omega}$.

3.3 Canonical double null foliation of the spacetime

In this section we introduce the concept of a canonical double foliation which plays an important role in the proof of the Main Theorem.

We start by considering a bounded region of spacetime, denoted by $\mathcal{K}$, whose boundary is identified by the following:

- a finite region of a spacelike hypersurface Σ_0; $\mathcal{K}$ is in the future of Σ_0.
- a portion of an incoming null hypersurface $\underline{C}_*$; $S_*(\lambda_1) \equiv \underline{C}_* \cap \Sigma_0$ is diffeomorphic to S^2.
- a portion of an outgoing null hypersurface C_0; $S_{(0)}(\nu_0) \equiv C_0 \cap \Sigma_0$ is diffeomorphic to S^2. Also $C_0 \cap \underline{C}_*$ is a two-surface, diffeomorphic to S^2.

A *double null foliation of* $\mathcal{K}$ is given by two optical functions u and $\underline{u}$ such that

$$\underline{C}_* = \{p \in \mathcal{K} | \underline{u}(p) = \nu_*\}\ ,\ C_0 = \{p \in \mathcal{K} | u(p) = \lambda_0\}$$

with λ_0 and ν_* fixed constants.[21]

A canonical double null foliation of $\mathcal{K}$ is a double null foliation such that the restriction of u on $\underline{C}_*$ and of $\underline{u}$ on Σ_0 are "canonical" in a sense which will be clarified in this section.

Remark: We shall refer in the sequel to $\Sigma_0 \cap \mathcal{K}$ as the "initial slice" and to $\underline{C}_* \cap \mathcal{K}$ as the "last slice."

3.3.1 *Canonical foliation of the initial hypersurface*

We consider foliations on a region of the initial hypersurface Σ_0 specified by a "radial" function $r(p) = w(p)$. By this we mean a differentiable real function defined on all points of this region, which takes values in an interval (σ_0, ∞) and satisfies the following conditions:

1. w has no critical points.
2. The level surfaces $S_0(\sigma) \equiv \{p \in \Sigma_0 | w(p) = \sigma\}$ are diffeomorphic to the two-dimensional spheres S^2.

Let $K \subset \Sigma_0$ be a compact set[22] such that $\Sigma_0 \backslash K$ is diffeomorphic to the complement of the closed unit ball. Consider a radial foliation of $\Sigma_0 \backslash K$ given by the function $w(p)$. Its leaves are

$$S_0(\sigma) = \{p \in \Sigma_0 | w(p) = \sigma\}.$$

We assume that ∂K is a leaf of the foliation, $\partial K = S_0(\sigma_0)$.

We choose on Σ_0 a moving frame, $\{\tilde{N}, e_a\}$, adapted to this foliation where $\tilde{N}^i = \frac{1}{|\partial w|} g^{ij} \partial_j w$ is the unit vector field defined on Σ_0, and normal to each $S_0(\sigma)$. The metric on Σ_0 can be written

$$g(\cdot,\cdot) = a^2 dw^2 + \gamma_{ab} d\phi^a d\phi^b, \tag{3.3.1}$$

[21] The reason for the notation $S_*(\lambda_1)$ and $S_{(0)}(\nu_0)$ will be clear after Subsections 3.3.1 and 3.3.3.

[22] See Definition 3.6.2.

and, with this choice of coordinates, $\tilde{N} = \frac{1}{a}\frac{\partial}{\partial w}$ and $a^{-2} = |\partial w|^2$.

Using Gauss and Codazzi–Mainardi equations relative to the surfaces $S_0(\sigma)$ immersed in Σ_0, we obtain the following evolution equation for $\text{tr}\theta$:[23]

$$\nabla_{\tilde{N}}\text{tr}\theta + \frac{1}{2}(\text{tr}\theta)^2 + a^{-1}\Delta\!\!\!/ a = -|\hat{\theta}|^2 - R_{\tilde{N}\tilde{N}}, \tag{3.3.2}$$

which can be rewritten as

$$\nabla_{\tilde{N}}\text{tr}\theta + \frac{1}{2}(\text{tr}\theta)^2 = -(\Delta\!\!\!/ \log a + \rho) + \left[-|\nabla\!\!\!/ \log a|^2 - |\hat{\theta}|^2 + g(k)\right], \tag{3.3.3}$$

where ρ is the null component $-\frac{1}{4}\mathbf{R}_{3434}$ of the Riemann tensor relative to the null pair $\{e'_4 = \tilde{N} + T_0, e'_3 = \tilde{N} - T_0\}$,[24] and

$$g(k) \equiv k^2_{\tilde{N}\tilde{N}} + \sum_a |k_{e'_a\tilde{N}}|^2 .$$

Definition 3.3.1 *A foliation on $\Sigma_0 \setminus K$, defined by a radial function $\underline{u}_{(0)}(p)$, is said to be canonical if $\underline{u}_{(0)}(p)$ is a solution to the initial slice problem with initial condition on ∂K,*

$$|\nabla \underline{u}_{(0)}| = a^{-1}\ ,\ \underline{u}_{(0)}|_{\partial K} = \nu_0$$
$$\Delta\!\!\!/ \log a = -(\rho - \overline{\rho})\ ,\ \overline{\log a} = 0. \tag{3.3.4}$$

The leaves of the canonical foliation are denoted by

$$S_{(0)}(\nu) = \{p \in \Sigma_0 | \underline{u}_{(0)}(p) = \nu\}. \tag{3.3.5}$$

The next theorem assures the local and global existence of a canonical foliation on $\Sigma_0 \backslash K$.

Theorem 3.3.1 *Under appropriate smallness assumptions on $\Sigma_0 \backslash K$ there exists a canonical foliation on $\Sigma_0 \backslash K$.*

The precise statement of Theorem 3.3.1 is given in Section 3.7 and its proof is given in Chapter 7.

Remark: The canonical foliation on $\Sigma_0 \backslash K$ is required because we need to control θ up to third derivatives. Without the canonical foliation the control of the third derivatives of θ requires the control of g up to five derivatives and of k up to four derivatives. This would lead to stronger assumptions on the initial data than necessary.

3.3.2 Foliations on the last slice

A foliation on the last slice $\underline{C}_*$ is specified by giving a function u_* with the following properties:

1. u_* is a differentiable real function defined on all points of $\underline{C}_*$;
2. u_* has no critical points;
3. the level surfaces of $u_*(p)$, $S_*(\lambda) \equiv S(\lambda, \nu_*) = \{p \in \underline{C}_* | u_*(p) = \lambda\}$ are diffeomorphic to the two-dimensional spheres S^2.

---

[23] These are derived in Chapter 7, Subsection 7.1.1; see also [Ch-Kl] Chapter 5. R_{ij} is the Ricci tensor of Σ_0.

[24] See also footnote 7 of Subsection 7.1.3.

3.3.3 Canonical foliation of the last slice

The concept of the canonical foliation of the last slice is an important ingredient in the proof of the Main Theorem; see also the discussion in [Kl-Ni].

We start by defining the following functions which we call *mass aspect functions*:[25]

$$\mu = \mathbf{K} + \frac{1}{4}\mathrm{tr}\underline{\chi}\,\mathrm{tr}\chi - \mathrm{d}\!\!/\mathrm{iv}\,\eta$$
$$\underline{\mu} = \mathbf{K} + \frac{1}{4}\mathrm{tr}\chi\,\mathrm{tr}\underline{\chi} - \mathrm{d}\!\!/\mathrm{iv}\,\underline{\eta}. \tag{3.3.6}$$

We restrict our attention to $\underline{C}_*$ and its initial section $S_*(\lambda_1)$. Let S be an arbitrary section of $\underline{C}_*$. There is a unique null-outgoing normal L^* to S, conjugate to $\underline{L}$ such that $g(\underline{L}, L^*) = -2$. We recall (see (3.1.32), (3.1.33)) that in the normalized null frame $\{\hat{N} = 2\Omega^{-1}L^*,\ \hat{\underline{N}} = 2\Omega\underline{L}\}$, $\underline{\eta} = -\zeta + \nabla\!\!\!/ \log\Omega$. Therefore

$$\underline{\eta}(X) = -\frac{1}{2}\mathbf{g}(\mathbf{D}_X L^*, \underline{L}) \tag{3.3.7}$$

follows easily, with $X \in TS$.[26] Hence to obtain $\underline{\eta}$ knowledge of Ω is not required. Once $\underline{C}_*$ and its null geodesic vector field $\underline{L}$ are given, $\underline{\eta}$ is uniquely defined by the section S. Clearly $\mathrm{tr}\chi\,\mathrm{tr}\underline{\chi}$ and the Gaussian curvature $\mathbf{K}$ are also independent of Ω. Consequently the quantity (see (3.3.6))

$$\underline{\mu} = \mathbf{K} + \frac{1}{4}\mathrm{tr}\chi\,\mathrm{tr}\underline{\chi} - \mathrm{d}\!\!/\mathrm{iv}\,\underline{\eta}$$

is also independent of Ω.

Consider a given scalar function u_* on $\underline{C}_*$ and let u be the outgoing solution of the eikonal equation such that $u|_{\underline{C}_*} = u_*$. Let L be the null geodesic vector field, $L^\mu = -g^{\mu\nu}\partial_\nu u$. The relation between the affine function v of $\underline{L}$ and the function $u_* = u|_{\underline{C}_*}$ is

$$\frac{du_*}{dv} = \underline{L}(u_*) = -g^{\mu\nu}\partial_\mu\underline{u}\partial_\nu u|_{\underline{C}_*} = -\mathbf{g}(\underline{L}, L)|_{\underline{C}_*} = (2\Omega^2)^{-1}\ .$$

We want to choose u_* on $\underline{C}_*$ such that the mass aspect function μ is constant on the surfaces

$$S_*(\lambda) = \{p \in \underline{C}_* | u_*(p) = \lambda\}, \tag{3.3.8}$$

the leaves of the foliation induced by u_* on $\underline{C}_*$. In other words we require that μ satisfy the equation

$$\mu - \overline{\mu} = 0, \tag{3.3.9}$$

with $\overline{\mu}$ the average of μ on $S_*(\lambda)$.

---

[25] These were first introduced in [Ch-Kl], Chapter 13.

[26] $\frac{1}{2}\mathbf{g}(\mathbf{D}_X L^*, \underline{L}) = \frac{1}{2}\mathbf{g}(\mathbf{D}_X \hat{N}, \hat{\underline{N}}) + \frac{1}{2}\Omega^{-1}(\mathbf{D}_X\Omega)\mathbf{g}(\hat{N}, \hat{\underline{N}}) = \zeta(X) - \nabla\!\!\!/_X \log\Omega = -\underline{\eta}(X)$.

This can be viewed as an equation for Ω at each $S_*(\lambda)$. According to (3.3.6) and relation $\eta+\underline{\eta}=2\nabla\!\!\!/\log\Omega$ (see (3.1.33)) we have

$$\overline{\mu}+\underline{\mu}=\mu+\underline{\mu}=2\mathbf{K}+\frac{1}{2}\mathrm{tr}\chi\,\mathrm{tr}\underline{\chi}-2\Delta\!\!\!/\log\Omega. \tag{3.3.10}$$

Therefore

$$\begin{aligned}\Delta\!\!\!/\log\Omega &= \mathbf{K}+\frac{1}{4}\mathrm{tr}\chi\,\mathrm{tr}\underline{\chi}-\frac{1}{2}(\underline{\mu}+\overline{\mu}) \\ &= \frac{1}{2}\mathrm{d}\!\!\!/\mathrm{iv}\,\underline{\eta}+\frac{1}{2}\left(\mathbf{K}-\overline{\mathbf{K}}+\frac{1}{4}(\mathrm{tr}\chi\,\mathrm{tr}\underline{\chi}-\overline{\mathrm{tr}\chi\,\mathrm{tr}\underline{\chi}})\right).\end{aligned} \tag{3.3.11}$$

Observe that the right-hand side of (3.3.11) does not depend on Ω.[27]

Definition 3.3.2 *A foliation on the last slice given by the level sets of u_* is said to be canonical if the functions u_* and Ω satisfy the following system of equations:*

$$\begin{aligned}&\Delta\!\!\!/\log\Omega=\frac{1}{2}\mathrm{d}\!\!\!/\mathrm{iv}\,\underline{\eta}+\frac{1}{2}\left(\mathbf{K}-\overline{\mathbf{K}}+\frac{1}{4}(\mathrm{tr}\chi\,\mathrm{tr}\underline{\chi}-\overline{\mathrm{tr}\chi\,\mathrm{tr}\underline{\chi}})\right)\\ &\overline{\log 2\Omega}=0\\ &\frac{du_*}{dv}=(2\Omega^2)^{-1};\quad u_*|_{\underline{C}_*\cap\Sigma_0}=\lambda_1.\end{aligned} \tag{3.3.12}$$

Note that Ω is uniquely defined by the first two equations in 3.3.12.

The next theorem proves the existence of a canonical foliation on $\underline{C}_*$.

Theorem 3.3.2 *Assume given on $\underline{C}_*$ a background foliation whose connection coefficients and null curvature components satisfy appropriate smallness assumptions. Then it is possible to foliate the whole $\underline{C}_*$ with a canonical foliation close to the background one.*

Remark: As Theorem 3.3.2 plays an important role in the proof of the Main Theorem, we will state it again with all the details in Section 3.7 after we have introduced the appropriate families of norms for the connection coefficients and the Riemann curvature tensor. The proof of Theorem 3.3.2 is given in Chapter 7.

We can now define the *canonical double null foliation* of the spacetime, a property that will be used in the Bootstrap assumption **B1** of the Main Theorem.

Definition 3.3.3 *A double null foliation of $\mathcal{K}$ is called canonical if:*

---

[27] Observe that the mass aspect function $\underline{\mu}$ can be connected to the Hawking mass, defined by $2\frac{m_H}{r}=1+\frac{1}{16\pi}\int_S \mathrm{tr}\chi\mathrm{tr}\underline{\chi}d\mu_\gamma$, according to the following equation

$$\frac{8\pi m_H}{r}=\int_S\underline{\mu}d\mu_\gamma$$

Indeed integrating the first line of (3.3.10) we obtain $\overline{\mu}|S|+\int_S\underline{\mu}d\mu_\gamma=8\pi+\frac{1}{2}\int_S\mathrm{tr}\chi\mathrm{tr}\underline{\chi}d\mu_\gamma$, where $|S|=4\pi r^2$. On the other hand, from the second equation in (3.3.6), we have $\int_S\underline{\mu}d\mu_\gamma=4\pi+\frac{1}{4}\int_S\mathrm{tr}\chi\mathrm{tr}\underline{\chi}d\mu_\gamma$. Therefore the result follows.

i) the $C(\lambda)$ null hypersurfaces are defined by $u(p) = \lambda$, where $\lambda \in [\lambda_1, \lambda_0]$; u is the incoming solution of the eikonal equation with "final data" given by the canonical function u_* on the last slice;

ii) the $\underline{C}(\nu)$ null hypersurfaces are defined by $\underline{u}(p) = \nu$, where $\nu \in [\nu_0, \nu_*]$; $\underline{u}$ is the outgoing solution of the eikonal equation with initial data given by the canonical function" $\underline{u}_{(0)}$ on the initial hypersurface Σ_0.

Definition 3.3.4 *$\mathcal{K}$ is the causal past of $S(\lambda_0, \nu_*)$, in the future of Σ_0.*

The *canonical double null foliation* of $\mathcal{K}$ consists, therefore, of the $C(\lambda)$ null hypersurfaces, with $\lambda \in [\lambda_1, \lambda_0]$ and the $\underline{C}(\nu)$ null hypersurfaces with $\nu \in [\nu_0, \nu_*]$; each point $p \in \mathcal{K}$ belongs to one and only one pair of the hypersurfaces $C(\lambda)$ and $\underline{C}(\nu)$.[28] Given this canonical double null foliation the two-dimensional surfaces

$$S(\lambda, \nu) = C(\lambda) \cap \underline{C}(\nu)$$

define a codimension-two double null integrable S-foliation.

Remark: The global spacetime of our Main Theorem will be constructed, by a continuity argument, with the help of a continous family of spacetime regions $\mathcal{K}$ each endowed with a canonical foliation. While the canonical foliation plays an essential part in our construction it has one undesirable feature; the foliations on Σ_0 induced by the two families of null hypersurfaces $C(\lambda)$ and $\underline{C}(\nu)$ differ from each other. In particular the canonical surfaces on $\Sigma_0 \backslash K$ do not belong to the S-foliation associated to the double null foliation, $\{S(\lambda, \nu) = C(\lambda) \cap \underline{C}(\nu)\}$.

In order to correct for this we construct, in a small neighborhood of $\Sigma_0 \backslash K$, a different foliation which we denote the initial layer foliation. We shall discuss this in the next section.

3.3.4 Initial layer foliation

Starting with the canonical foliation on $\Sigma_0 \backslash K$, defined by the level surfaces of $\underline{u}_{(0)}$, we consider the null-incoming hypersurfaces $\underline{C}(\nu)$ and the null-outgoing hypersurfaces $C'(\lambda)$. More precisely,

1. the $C'(\lambda')$ null hypersurfaces are given by $u'(p) = \lambda'$, where $\lambda' \in [-\nu_0, -\nu_*]$; with u' the outgoing solution of the eikonal equation with initial condition $u' = -\underline{u}_{(0)}$ on Σ_0;

2. the $\underline{C}(\nu)$ null hypersurfaces are defined as before by $\underline{u}(p) = \nu$, where $\nu \in [\nu_0, \nu_*]$; with $\underline{u}$ the incoming solution of the eikonal equation with initial condition $\underline{u} = \underline{u}_{(0)}$ on the initial hypersurface Σ_0.

Consider the region $\mathcal{K}'_{\delta_0} \subset \mathcal{K}$ specified by the condition

$$\frac{1}{2}(u'(p) + \underline{u}(p)) \leq \delta_0. \tag{3.3.13}$$

[28] We sometimes use the more precise definition $C(\lambda, [\nu_a, \nu_b])$ and $\underline{C}(\nu, [\lambda_a, \lambda_b])$ where the interval where the functions $u(p)$ and $\underline{u}(p)$ vary is written explicitly.

Definition 3.3.5 *For a fixed δ_0, sufficiently small, we shall call ${\mathcal K}'_{\delta_0}$ the initial layer region of height δ_0. The double null foliation induced on ${\mathcal K}'_{\delta_0}$ by the optical functions u', $\underline{u}$ defined above is called the initial layer foliation. Its two-dimensional surfaces are denoted by*

$$S'(\lambda', \nu) = C'(\lambda') \cap \underline{C}(\nu). \tag{3.3.14}$$

Remarks:

1. The leaves of the canonical foliation of Σ_0, $S_{(0)}(\nu)$, belong to the initial layer foliation. More precisely,

$$S_{(0)}(\nu) = S'(-\nu, \nu). \tag{3.3.15}$$

2. Relative to the initial layer foliation we associate, as before, the normalized null pair $\{\hat{N}' = 2\Omega' L', \ \hat{\underline{N}}' = 2\Omega'\underline{L}\}$, with

$$2\Omega'^2 = -g(L', \underline{L})^{-1} = -(g^{\rho\sigma}\partial_\rho u' \partial_\sigma \underline{u})^{-1}, \tag{3.3.16}$$

exactly as in Definition 3.1.12 for Ω.

3. Hereafter when we say that the spacetime $\mathcal{K}$ is foliated by a double null canonical foliation, we intend above the initial layer region which, vice-versa, is foliated by the initial layer foliation.[29]

4. We shall also make use of the null equivariant pair $N' = 2\Omega'^2 L'$, $\underline{N}' = 2\Omega'^2 \underline{L}$.

The next proposition shows that, given a double null foliation, it is possible to introduce a global time function and prove that the associated three-dimensional spacelike hypersurfaces define a spacelike foliation.

Proposition 3.3.1 *Assume a double null foliation specified by the functions $u(p)$, $\underline{u}(p)$. Let us define the global time function $t(p) = \frac{1}{2}(u(p) + \underline{u}(p))$. Then the three-dimensional spacelike hypersurfaces*

$$\tilde{\Sigma}_t \equiv \{p \in \mathcal{K} | t(p) = t\}$$

define a three-dimensional spacelike foliation of $\mathcal{K}$. Each two-dimensional surface $S(\lambda, \nu)$ is immersed in the hypersurface $\tilde{\Sigma}_t$ with $t = \frac{1}{2}(\lambda + \nu)$. Moreover

$$dt = -\frac{1}{4\Omega^2}(n + \underline{n}) \ , \ \ \frac{\partial}{\partial t} = (N + \underline{N}), \tag{3.3.17}$$

where n, $\underline{n}$ are the 1-forms corresponding to N, $\underline{N}$.

Finally, given the hypersurfaces $\tilde{\Sigma}_t$, their second fundamental form k has the following expression in terms of the connection coefficients,[30]

$$k_{\tilde{N}\tilde{N}} = \omega + \underline{\omega} \ , \ \ k_{e_a\tilde{N}} = \zeta_a \ , \ \ k_{e_a e_b} = -\frac{1}{2}(\chi_{ab} + \underline{\chi}_{ab}) \ . \tag{3.3.18}$$

[29] This is discussed in detail in Chapter 4.

[30] In terms of the coordinate t and coordinates $\{x^i\}$ adapted to $\tilde{\Sigma}_t$, the second fundamental form has the following expression: $k_{ij} = -(4\Omega)^{-1}\partial_t g_{ij}$.

Because the two-dimensional surfaces $S_{(0)}(\nu)$, $\nu \in [\nu_0, \nu_*]$, which canonically foliate Σ_0 (see Definition 3.3.1) do not belong to the family $\{S(\lambda, \nu) = C(\lambda) \cap \underline{C}(\nu)\}$, this implies that $\Sigma_0 \neq \tilde{\Sigma}_{t=0}$. Moreover it is easy to prove that the $\tilde{\Sigma}_t$ hypersurfaces are not maximal since, by direct computation, it follows that $\text{tr}k \neq 0$.[31]

Using Proposition 3.3.1, it is also possible to introduce a different spacelike foliation, adapted to the initial layer foliation whose spacelike hypersurfaces are defined by the global time function $t'(p) = \frac{1}{2}(u'(p) + \underline{u}(p))$. Its three-dimensional spacelike hypersurfaces are

$$\Sigma'_{t'} \equiv \{p \in \mathcal{K} | t'(p) = t\}.$$

Observe that Σ_0 is a leaf of this foliation, $\Sigma_0 = \Sigma'_{t'=0}$.

Remark: As explained in a previous remark, we need the initial layer foliation to connect the initial hypersurface Σ_0 and its surfaces $\{S_{(0)}(\nu)\}$ with the canonical foliation of $\mathcal{K}$ and the surfaces $S(\lambda, \nu)$. This is discussed in detail in Chapter 4.

3.4 Deformation tensors

3.4.1 *Approximate Killing and conformal Killing vector fields*

The functions u, $\underline{u}$ of the double null foliation, along with the null pair $\{e_3 = \hat{N}, e_4 = \hat{\underline{N}}\}$, allow us to define the vector fields $T, S, K_0, \bar{K}$ analogous to the ones used earlier for the Minkowski spacetime, Chapter 2, Subsection 2.2.1,[32]

$$\begin{aligned} T &= \frac{1}{2}(e_3 + e_4)\,,\ S = \frac{1}{2}(ue_3 + \underline{u}e_4)\,, \\ K_0 &= \frac{1}{2}(u^2 e_3 + \underline{u}^2 e_4)\,,\ \bar{K} = \frac{1}{2}(\tau_+^2 e_4 + \tau_-^2 e_3), \end{aligned} \tag{3.4.1}$$

where[33]

$$\tau_+ = (1 + \underline{u}^2)^{\frac{1}{2}}\,,\ \tau_- = (1 + u^2)^{\frac{1}{2}}. \tag{3.4.2}$$

Unlike the case of Minkowski spacetime, these vector fields are not conformal Killing. We show, however, that their deformation tensors, or rather their traceless parts, are asymptotically vanishing in a sufficiently strong sense. We can also define approximate Killing rotation vector fields, ${}^{(i)}O$, $i \in \{1, 2, 3\}$, also known as angular momentum vector fields, which play the same role as the rotation vector fields of Minkowski spacetime. They are constructed, geometrically, as follows:

We start from the asymptotic region of the initial hypersurface Σ_0. There, in view of our strong asymptotic flatness assumptions (see Subsection 3.6.1) this manifold looks Euclidean. We can thus define the canonical angular momentum vector fields at infinity and pull them back with the help of the diffeomorphism generated by the flow normal to the S surfaces along Σ_0.[34] The vector fields can then be pushed forward in the same way along

[31] Nevertheless it will follow from the results of the next chapters that $\text{tr}k$ is small; see Theorem 3.7.3. Observe that the initial data hypersurface Σ_0 (see Theorem 3.7.1) is chosen maximal.

[32] See also [Ch-Kl1] and [Kl-Ni].

[33] In the sequel the vector fields K_0 and $\bar{K}$ can both be used since, in the Main Theorem, u is bounded from below.

[34] See [Kl-Ni] and Chapter 5 in[Ch-Kl].

the last slice $\underline{C}_*$ using the diffeomorphism $\underline{\phi}_\tau$ generated by $\underline{N}$. Finally we pull them back once more along the hypersurfaces $C(\lambda)$ with the help of the diffeomorphism generated by the null-outgoing equivariant vector field N. These steps define the vector fields ${}^{(i)}O$ at any point of our spacetime $\mathcal{K}$.[35] By definition they are tangent to the S-foliation and commute with N. Moreover they satisfy the canonical commutation relations. Thus, finally, the "extended" rotation generators — or angular vector fields — ${}^{(i)}O$ satisfy[36]

$$\begin{aligned}
&[{}^{(i)}O, {}^{(j)}O] = \epsilon_{ijk}\,{}^{(k)}O \\
&[N, {}^{(i)}O] = 0 \\
&\mathbf{g}({}^{(i)}O, e_4) = \mathbf{g}({}^{(i)}O, e_3) = 0.
\end{aligned} \tag{3.4.3}$$

All these steps are described in complete detail in Chapter 4, Section 4.6 and Chapter 7.

3.4.2 *Deformation tensors of the vector fields T, S, K_0*

We use the adapted null frame $\{\hat{N}, \hat{\underline{N}}, e_1, e_2\}$ associated to the canonical double null integral foliation introduced in the previous section.

Let X be a vector field on $\mathcal{K}$. If X were a Killing vector field, then

$${}^{(X)}\pi \equiv \mathcal{L}_X \mathbf{g} = 0$$

would hold and the diffeomorphism generated by the integral curves of X would be an isometry of $(\mathcal{K}, g)$. If X is not a Killing vector field the previous relation does not hold, but if the spacetime is not "too different" from the Minkowski spacetime, we expect to control the magnitude of some appropriate norms of the deformation tensor ${}^{(X)}\pi$. We recall that

$${}^{(X)}\pi_{\mu\nu} = D_\mu X_\nu + D_\nu X_\mu,$$

and its traceless part is

$${}^{(X)}\hat{\pi}_{\mu\nu} = {}^{(X)}\pi_{\mu\nu} - \frac{1}{4} g_{\mu\nu} \mathrm{tr}\pi.$$

In the null frame associated to the canonical foliation, therefore,

$$\begin{aligned}
{}^{(X)}\pi_{ab} &= \mathbf{g}(\mathbf{D}_{e_a} X, e_b) + \mathbf{g}(\mathbf{D}_{e_b} X, e_a) \\
{}^{(X)}\pi_{4a} &= \mathbf{g}(\mathbf{D}_{\hat{N}} X, e_a) + \mathbf{g}(\mathbf{D}_{e_a} X, \hat{N}) \\
{}^{(X)}\pi_{3a} &= \mathbf{g}(\mathbf{D}_{\hat{\underline{N}}} X, e_a) + \mathbf{g}(\mathbf{D}_{e_a} X, \hat{\underline{N}}) \\
{}^{(X)}\pi_{34} &= \mathbf{g}(\mathbf{D}_{\hat{\underline{N}}} X, \hat{N}) + \mathbf{g}(\mathbf{D}_{\hat{N}} X, \hat{\underline{N}}) \\
{}^{(X)}\pi_{33} &= \mathbf{g}(\mathbf{D}_{\hat{\underline{N}}} X, \hat{\underline{N}}) + \mathbf{g}(\mathbf{D}_{\hat{\underline{N}}} X, \hat{\underline{N}}) \\
{}^{(X)}\pi_{44} &= \mathbf{g}(\mathbf{D}_{\hat{N}} X, \hat{N}) + \mathbf{g}(\mathbf{D}_{\hat{N}} X, \hat{N}),
\end{aligned} \tag{3.4.4}$$

---

[35] Let $q \in S(\lambda, \nu)$ be an arbitrary point of $\mathcal{K}$. As $S(\lambda, \nu)$ is diffeomorphic via $\phi_{\Delta=(\nu_*-\nu)}$ to $S(\lambda, \nu_*) \subset \underline{C}_*$, $\exists p \in S(\lambda, \nu_*)$ such that $q = \phi_\Delta^{-1}(p)$. We define the element O of the rotation group operating over q as $(O, q) \equiv \phi_\Delta^{-1}(O_*, p = \phi_\Delta(q))$ where (O, q) is a point of $S(\lambda, \nu)$, while $(O_*, p = \phi_\Delta(q))$ is the point of $S(\lambda, \nu_*)$ obtained applying O_* to the point p.

[36] The commutator $[{}^{(i)}O, \underline{N}] \in TS$ is different from zero, see Subsection 4.6.1. This shows that one could have defined the rotation vector fields in a different way starting from the Σ_0 hypersurface and using the diffeomorphism $\underline{\phi}_t$ generated by $\underline{N}$.

and

$$\begin{aligned}
{}^{(X)}\hat{\pi}_{ab} &= {}^{(X)}\pi_{ab} - \tfrac{1}{4}\delta_{ab}\mathrm{tr}\,{}^{(X)}\pi \\
{}^{(X)}\hat{\pi}_{4a} &= {}^{(X)}\pi_{4a} \\
{}^{(X)}\hat{\pi}_{3a} &= {}^{(X)}\pi_{3a} \\
{}^{(X)}\hat{\pi}_{34} &= {}^{(X)}\pi_{34} + \tfrac{1}{2}\mathrm{tr}\,{}^{(X)}\pi \\
{}^{(X)}\hat{\pi}_{33} &= {}^{(X)}\pi_{33} \\
{}^{(X)}\hat{\pi}_{44} &= {}^{(X)}\pi_{44}.
\end{aligned} \tag{3.4.5}$$

We denote the various components of the deformation tensors, with respect to a null frame, as

$$\begin{aligned}
{}^{(X)}\mathbf{i}_{ab} &= {}^{(X)}\hat{\pi}_{ab} \;; \quad {}^{(X)}\mathbf{j} = {}^{(X)}\hat{\pi}_{34} \\
{}^{(X)}\mathbf{m}_a &= {}^{(X)}\hat{\pi}_{4a} \;; \quad {}^{(X)}\underline{\mathbf{m}}_a = {}^{(X)}\hat{\pi}_{3a} \\
{}^{(X)}\mathbf{n} &= {}^{(X)}\hat{\pi}_{44} \;; \quad {}^{(X)}\underline{\mathbf{n}} = {}^{(X)}\hat{\pi}_{33}.
\end{aligned} \tag{3.4.6}$$

Their explicit expressions relative to the vectors defined in (3.4.1) are

$$\begin{aligned}
{}^{(T)}\mathbf{i}_{ab} &= \hat{\chi}_{ab} + \hat{\underline{\chi}}_{ab} + \frac{1}{2}\delta_{ab}\big(\frac{1}{2}(\mathrm{tr}\chi + \mathrm{tr}\underline{\chi}) + (\omega + \underline{\omega})\big) \\
{}^{(T)}\mathbf{j} &= \frac{1}{2}(\mathrm{tr}\chi + \mathrm{tr}\underline{\chi}) + (\omega + \underline{\omega}) \\
{}^{(T)}\mathbf{m}_a &= 2\underline{\eta}_a - \nabla\!\!\!\!/_a \log\Omega = \underline{\eta}_a - \zeta_a \\
{}^{(T)}\underline{\mathbf{m}}_a &= 2\eta_a - \nabla\!\!\!\!/_a \log\Omega = \eta_a + \zeta_a \\
{}^{(T)}\mathbf{n} &= -4\omega = 2\mathbf{D}_4 \log\Omega \\
{}^{(T)}\underline{\mathbf{n}} &= -4\underline{\omega} = 2\mathbf{D}_3 \log\Omega
\end{aligned} \tag{3.4.7}$$

$$\begin{aligned}
{}^{(S)}\mathbf{i}_{ab} &= \underline{u}\hat{\chi}_{ab} + u\hat{\underline{\chi}}_{ab} + \frac{1}{2}\delta_{ab}\big(\frac{1}{2}(\underline{u}\mathrm{tr}\chi + u\mathrm{tr}\underline{\chi}) + (\underline{u}\omega + u\underline{\omega}) - \frac{1}{\Omega}\big) \\
{}^{(S)}\mathbf{j} &= \frac{1}{2}(\underline{u}\mathrm{tr}\chi + u\mathrm{tr}\underline{\chi}) + (\underline{u}\omega + u\underline{\omega}) - \frac{1}{\Omega} \\
{}^{(S)}\mathbf{m}_a &= u(2\underline{\eta}_a - \nabla\!\!\!\!/_a \log\Omega) = u(\underline{\eta}_a - \zeta_a) = u\,{}^{(T)}\mathbf{m}_a \\
{}^{(S)}\underline{\mathbf{m}}_a &= \underline{u}(2\eta_a - \nabla\!\!\!\!/_a \log\Omega) = \underline{u}(\eta_a + \zeta_a) = \underline{u}\,{}^{(T)}\underline{\mathbf{m}}_a \\
{}^{(S)}\mathbf{n} &= u(-4\omega) = 2u\mathbf{D}_4 \log\Omega = u\,{}^{(T)}\mathbf{n} \\
{}^{(S)}\underline{\mathbf{n}} &= \underline{u}(-4\underline{\omega}) = 2\underline{u}\mathbf{D}_3 \log\Omega = \underline{u}\,{}^{(T)}\underline{\mathbf{n}}
\end{aligned} \tag{3.4.8}$$

$$\begin{aligned}
{}^{(K_0)}\mathbf{i}_{ab} &= \underline{u}^2\hat{\chi}_{ab} + u^2\hat{\underline{\chi}}_{ab} + \frac{1}{2}\delta_{ab}\big(\frac{1}{2}(\underline{u}^2\mathrm{tr}\chi + u^2\mathrm{tr}\underline{\chi}) + (\underline{u}^2\omega + u^2\underline{\omega}) - \frac{\underline{u} + u}{\Omega}\big) \\
{}^{(K_0)}\mathbf{j} &= \frac{1}{2}(\underline{u}^2\mathrm{tr}\chi + u^2\mathrm{tr}\underline{\chi}) + (\underline{u}^2\omega + u^2\underline{\omega}) - \frac{\underline{u} + u}{\Omega} \\
{}^{(K_0)}\mathbf{m}_a &= u^2(2\underline{\eta}_a - \nabla\!\!\!\!/_a \log\Omega) = u^2(\underline{\eta}_a - \zeta_a) = u^2\,{}^{(T)}\mathbf{m}_a \\
{}^{(K_0)}\underline{\mathbf{m}}_a &= \underline{u}^2(2\eta_a - \nabla\!\!\!\!/_a \log\Omega) = \underline{u}^2(\eta_a + \zeta_a) = \underline{u}^2\,{}^{(T)}\underline{\mathbf{m}}_a \\
{}^{(K_0)}\mathbf{n} &= u^2(-4\omega) = 2u^2\mathbf{D}_4 \log\Omega = u^2\,{}^{(T)}\mathbf{n} \\
{}^{(K_0)}\underline{\mathbf{n}} &= \underline{u}^2(-4\underline{\omega}) = 2\underline{u}^2\mathbf{D}_3 \log\Omega = \underline{u}^2\,{}^{(T)}\underline{\mathbf{n}}.
\end{aligned} \tag{3.4.9}$$

3.4.3 Rotation deformation tensors

In what follows we display the form of the various components of the deformation tensors associated with the rotation vector fields ${}^{(i)}O$. Recalling (3.4.4) and $N = \Omega\hat{N}$ we derive the commutation relations

$$[{}^{(i)}O, \hat{N}] = [{}^{(i)}O, \Omega^{-1}N] = -{}^{(i)}O(\log\Omega)\hat{N} = {}^{(i)}F\hat{N}, \qquad (3.4.10)$$

where

$${}^{(i)}F \equiv -(\nabla\!\!\!/_c \log\Omega){}^{(i)}O_c\,.$$

One can also easily check the following expressions

$$\begin{aligned}
&\mathbf{g}(\mathbf{D}_a{}^{(i)}O, e_4) = -\chi_{ab}{}^{(i)}O_b\\
&\mathbf{g}(\mathbf{D}_4{}^{(i)}O, e_a) = \chi_{ab}{}^{(i)}O_b\\
&\mathbf{g}(\mathbf{D}_4{}^{(i)}O, e_4) = 0\\
&\mathbf{g}(\mathbf{D}_a{}^{(i)}O, e_3) = -\underline{\chi}_{ab}{}^{(i)}O_b\\
&\mathbf{g}(\mathbf{D}_3{}^{(i)}O, e_3) = 0\\
&\mathbf{g}(\mathbf{D}_4{}^{(i)}O, e_3) = -2\underline{\eta}_b{}^{(i)}O_b\\
&\mathbf{g}(\mathbf{D}_3{}^{(i)}O, e_4) = -2\eta_b{}^{(i)}O_b.
\end{aligned} \qquad (3.4.11)$$

Using these equations we compute explicitly the following components of the deformation tensor. Denoting ${}^{({}^{(i)}O)}\pi \equiv {}^{(i)}\pi$, we have

$$\begin{aligned}
{}^{(i)}\pi_{44} &= 2\mathbf{g}(\mathbf{D}_4{}^{(i)}O, e_4) = 0\\
{}^{(i)}\pi_{4a} &= \mathbf{g}(\mathbf{D}_4{}^{(i)}O, e_a) + \mathbf{g}(\mathbf{D}_a{}^{(i)}O, e_4) = 0\\
{}^{(i)}\pi_{33} &= 2\mathbf{g}(\mathbf{D}_3{}^{(i)}O, e_3) = 0.
\end{aligned} \qquad (3.4.12)$$

The remaining components are denoted by

$$\begin{aligned}
2{}^{(i)}H_{ab} &\equiv {}^{(i)}\pi_{ab} = \mathbf{g}(\mathbf{D}_a{}^{(i)}O, e_b) + \mathbf{g}(\mathbf{D}_b{}^{(i)}O, e_a)\\
4{}^{(i)}Z_a &\equiv {}^{(i)}\pi_{3a} = \mathbf{g}(\mathbf{D}_a{}^{(i)}O, e_3) + \mathbf{g}(\mathbf{D}_3{}^{(i)}O, e_a)\\
&= -{}^{(i)}O_b\underline{\chi}_{ab} + \underline{\hat{N}}({}^{(i)}O_a) + {}^{(i)}O_b\mathbf{g}(\mathbf{D}\!\!\!/_3 e_b, e_a)\\
4{}^{(i)}F &\equiv {}^{(i)}\pi_{34} = -2(\eta_b + \underline{\eta}_b){}^{(i)}O_b = -4(\nabla\!\!\!/_b \log\Omega){}^{(i)}O_b.
\end{aligned} \qquad (3.4.13)$$

In order to evaluate ${}^{(i)}F$, ${}^{(i)}Z_a$, ${}^{(i)}H_{ab}$ we use the evolution equations for these quantities. They are derived, together with those for their derivatives, in Sections 4.6 and 4.7 of Chapter 4.

3.5 The definitions of the fundamental norms

All the norms we introduce in the next subsections are simple generalizations of the following:

- *Pointwise $S(\lambda,\nu)$ norms:* Let f be a $\binom{n}{m}$ tensor, tangent at q to $S(\lambda,\nu)$, $f \in TS_q(\lambda,\nu)$. Its pointwise norm at q is

$$|f(q)| \equiv \left| f^{b_1,\dots,b_n}_{a_1,\dots,a_m} f^{b'_1,\dots,b'_n}_{a'_1,\dots,a'_m} \gamma_q^{a_1a'_1} \cdot\cdot \gamma_q^{a_ma'_m} \gamma_{q\,b_1b'_1} \cdot\cdot \gamma_{q\,b_mb'_m} \right|^{\frac{1}{2}}$$

  where $\gamma(q)(\cdot,\cdot)$ is the Riemannian metric induced on $S(\lambda,\nu)$ by the metric $\mathbf{g}$ of $\mathcal{K}$.

- *$L^p(S)$ norms:* Let V be an $S(\lambda,\nu)$-tangent vector field. Its $L^p(S(\lambda,\nu))$ norms are

$$|V|_{p,S(\lambda,\nu)} \equiv \left(\int_{S(\lambda,\nu)} |V(\omega)|^p d\mu_\gamma\right)^{\frac{1}{p}}, \quad p \in [2,\infty)$$
$$|V|_{\infty,S(\lambda,\nu)} \equiv \sup_{\omega\in S(\lambda,\nu)} |V(\omega)|, \quad p = \infty$$

  where $d\mu_\gamma$ is the measure associated to $\gamma(\cdot,\cdot)$.

- *$L^2(C(\lambda)\cap V(\lambda,\nu))$, $L^2(\underline{C}(\nu)\cap V(\lambda,\nu))$ norms:* These are L^2 norms relative to the null hypersurfaces $C(\lambda)$ and $\underline{C}(\nu)$, respectively, or to portions of these hypersurfaces. Let w be an $S(\lambda,\nu)$-tangent tensor field. Its $L^2(C(\lambda)\cap V(\lambda,\nu))$ and $L^2(\underline{C}(\nu)\cap V(\lambda,\nu))$ norms are defined as follows:

$$||w||_{2,C(\lambda)\cap V(\lambda,\nu)} \equiv \left(\int_{C(\lambda)} |w|^2\right)^{\frac{1}{2}} = \left(\int_{\nu_0(\lambda)}^{\nu} d\nu' \int_{S(\lambda,\nu')} |w|^2 d\mu_\gamma\right)^{\frac{1}{2}}$$
$$||w||_{2,\underline{C}(\nu)\cap V(\lambda,\nu)} \equiv \left(\int_{\underline{C}(\nu)} |w|^2\right)^{\frac{1}{2}} = \left(\int_{\lambda_0(\nu)}^{\lambda_0} d\lambda' \int_{S(\lambda',\nu)} |w|^2 d\mu_\gamma\right)^{\frac{1}{2}}$$

  where $\nu_0(\lambda) = \underline{u}|_{C(\lambda)\cap\Sigma_0}$, $\lambda_0(\nu) = u|_{\underline{C}(\nu)\cap\Sigma_0}$.

- *$L^2(\Sigma_0)$ norms:* These are L^2 norms relative to the initial hypersurface Σ_0 or to a portion of it. Let g be a vector field defined on Σ_0, tangent to the two-dimensional surfaces $S_{(0)}(\nu)$ of the initial canonical foliation; see subsection 3.3.1. Then

$$||g||_{2,\Sigma_0} \equiv \left(\int_{\underline{u}_{(0)}}^{\infty} d\nu \int_{S_{(0)}(\nu)} |g(\nu,\omega)|^2 d\mu_{\gamma_0}\right)^{\frac{1}{2}},$$

  where γ_0 is the metric induced on $S_{(0)}(\nu)$ by the metric assigned on Σ_0 with the initial data.

3.5.1 *$\mathcal{Q}$ integral norms*

Consider the spacetime $\mathcal{K}$ endowed with the canonical double null foliation $\{C(\lambda), \underline{C}(\nu)\}$ and the corresponding normalized null pair $\{e_4, e_3\}$. Denoting by $\mathcal{Q}(\mathbf{R})$ the Bell–Robinson tensor (see (3.2.5)) associated to the curvature tensor $\mathbf{R}$ and saturating it with the vector fields K_0, T (see (3.4.1)) e_4 and e_3 we obtain

$$
\begin{aligned}
Q(\mathbf{R})(K_0,K_0,T,e_4) &= \frac{1}{4}\underline{u}^4|\alpha|^2+\frac{1}{2}(\underline{u}^4+2\underline{u}^2u^2)|\beta|^2+\frac{1}{2}(u^4+2\underline{u}^2u^2)(\rho^2+\sigma^2)\\
&\quad+\frac{1}{2}u^4|\underline{\beta}|^2\\
Q(\mathbf{R})(K_0,K_0,T,e_3) &= \frac{1}{4}u^4|\underline{\alpha}|^2+\frac{1}{2}(u^4+2\underline{u}^2u^2)|\underline{\beta}|^2+\frac{1}{2}(\underline{u}^4+2\underline{u}^2u^2)(\rho^2+\sigma^2)\\
&\quad+\frac{1}{2}\underline{u}^4|\beta|^2
\end{aligned}
\tag{3.5.1}
$$

$$
\begin{aligned}
Q(\mathbf{R})(K_0,K_0,K_0,e_4) &= \frac{1}{4}\underline{u}^6|\alpha|^2+\frac{3}{2}\underline{u}^4u^2|\beta|^2+\frac{3}{2}u^4\underline{u}^2(\rho^2+\sigma^2)+\frac{1}{2}u^6|\underline{\beta}|^2\\
Q(\mathbf{R})(K_0,K_0,K_0,e_3) &= \frac{1}{4}u^6|\underline{\alpha}|^2+\frac{3}{2}u^4\underline{u}^2|\underline{\beta}|^2+\frac{3}{2}\underline{u}^4u^2(\rho^2+\sigma^2)+\frac{1}{2}\underline{u}^6|\beta|^2
\end{aligned}
\tag{3.5.2}
$$

We also have

$$
\begin{aligned}
Q(\mathbf{R})(K_0,K_0,T,T) &= \frac{1}{8}\underline{u}^4|\alpha|^2+\frac{1}{8}u^4|\underline{\alpha}|^2+\frac{1}{2}(\underline{u}^4+\frac{1}{2}\underline{u}^2u^2)|\beta|^2\\
&+\frac{1}{2}(u^4+4\underline{u}^2u^2+u^4)(\rho^2+\sigma^2)+\frac{1}{2}(u^4+\frac{1}{2}u^2\underline{u}^2)|\underline{\beta}|^2\\
Q(\mathbf{R})(K_0,K_0,K_0,T) &= \frac{1}{8}\underline{u}^6|\alpha|^2+\frac{1}{8}u^6|\underline{\alpha}|^2+\frac{1}{4}u^4(u^2+3\underline{u}^2)|\underline{\beta}|^2\\
&+\frac{3}{4}(\underline{u}^2+u^2)\underline{u}^2u^2(\rho^2+\sigma^2)+\frac{1}{4}\underline{u}^4(\underline{u}^2+3u^2)|\beta|^2.
\end{aligned}
\tag{3.5.3}
$$

Using the vector fields $\bar{K}$, S, T, and ${}^{(i)}O$ (see (3.4.1)) and denoting

$$
V(\lambda,\nu)=J^-(S(\lambda,\nu)),
\tag{3.5.4}
$$

we define the following energy-type norms

$$
\begin{aligned}
\mathcal{Q}(\lambda,\nu) &= \mathcal{Q}_1(\lambda,\nu)+\mathcal{Q}_2(\lambda,\nu)\\
\underline{\mathcal{Q}}(\lambda,\nu) &= \underline{\mathcal{Q}}_1(\lambda,\nu)+\underline{\mathcal{Q}}_2(\lambda,\nu),
\end{aligned}
\tag{3.5.5}
$$

where

$$
\begin{aligned}
\mathcal{Q}_1(\lambda,\nu) &\equiv \int_{C(\lambda)\cap V(\lambda,\nu)} Q(\hat{\mathcal{L}}_T\mathbf{R})(\bar{K},\bar{K},\bar{K},e_4)\\
&\quad+\int_{C(\lambda)\cap V(\lambda,\nu)} Q(\hat{\mathcal{L}}_O\mathbf{R})(\bar{K},\bar{K},T,e_4)\\
\mathcal{Q}_2(\lambda,\nu) &\equiv \int_{C(\lambda)\cap V(\lambda,\nu)} Q(\hat{\mathcal{L}}_O\hat{\mathcal{L}}_T\mathbf{R})(\bar{K},\bar{K},\bar{K},e_4)\\
&\quad+\int_{C(\lambda)\cap V(\lambda,\nu)} Q(\hat{\mathcal{L}}_O^2\mathbf{R})(\bar{K},\bar{K},T,e_4)\\
&\quad+\int_{C(\lambda)\cap V(\lambda,\nu)} Q(\hat{\mathcal{L}}_S\hat{\mathcal{L}}_T\mathbf{R})(\bar{K},\bar{K},\bar{K},e_4)
\end{aligned}
\tag{3.5.6}
$$

$$
\begin{aligned}
\underline{\mathcal{Q}}_1(\lambda,\nu) &\equiv \sup_{V(\lambda,\nu)\cap\Sigma_0}|r^3\overline{\rho}|^2+\int_{\underline{C}(\nu)\cap V(\lambda,\nu)}Q(\hat{\mathcal{L}}_T\mathbf{R})(\bar{K},\bar{K},\bar{K},e_3)\\
&\quad+\int_{\underline{C}(\nu)\cap V(\lambda,\nu)}Q(\hat{\mathcal{L}}_O\mathbf{R})(\bar{K},\bar{K},T,e_3)\\
\underline{\mathcal{Q}}_2(\lambda,\nu) &\equiv \int_{\underline{C}(\nu)\cap V(\lambda,\nu)}Q(\hat{\mathcal{L}}_O\hat{\mathcal{L}}_T\mathbf{R})(\bar{K},\bar{K},\bar{K},e_3)\\
&\quad+\int_{\underline{C}(\nu)\cap V(\lambda,\nu)}Q(\hat{\mathcal{L}}_O^2\mathbf{R})(\bar{K},\bar{K},T,e_3)\\
&\quad+\int_{\underline{C}(\nu)\cap V(\lambda,\nu)}Q(\hat{\mathcal{L}}_S\hat{\mathcal{L}}_T\mathbf{R})(\bar{K},\bar{K},\bar{K},e_3).
\end{aligned}
\tag{3.5.7}
$$

We also introduce the quantity

$$
\mathcal{Q}_{\mathcal{K}} \equiv \sup_{\{\lambda,\nu|S(\lambda,\nu)\subset\mathcal{K}\}}\{\mathcal{Q}(\lambda,\nu)+\underline{\mathcal{Q}}(\lambda,\nu)\}. \tag{3.5.8}
$$

Similarly on the initial hypersurface Σ_0 we define

$$
\mathcal{Q}_{\Sigma_0\cap\mathcal{K}} = \sup_{\{\lambda,\nu|S(\lambda,\nu)\subset\mathcal{K}\}}\{\mathcal{Q}_{1\,\Sigma_0\cap V(\lambda,\nu)}+\mathcal{Q}_{2\,\Sigma_0\cap V(\lambda,\nu)}\}, \tag{3.5.9}
$$

where

$$
\begin{aligned}
\mathcal{Q}_{1\,\Sigma_0\cap V(\lambda,\nu)} &\equiv \int_{\Sigma_0\cap V(\lambda,\nu)}Q(\hat{\mathcal{L}}_T\mathbf{R})(\bar{K},\bar{K},\bar{K},T)\\
&+\int_{\Sigma_0\cap V(\lambda,\nu)}Q(\hat{\mathcal{L}}_O\mathbf{R})(\bar{K},\bar{K},T,T)+\sup_{\Sigma_0\cap V(\lambda,\nu)}|r^3\overline{\rho}|^2
\end{aligned}
\tag{3.5.10}
$$

$$
\begin{aligned}
\mathcal{Q}_{2\,\Sigma_0\cap V(\lambda,\nu)} &\equiv \int_{\Sigma_0\cap V(\lambda,\nu)}Q(\hat{\mathcal{L}}_O\hat{\mathcal{L}}_T\mathbf{R})(\bar{K},\bar{K},\bar{K},T)\\
&\quad+\int_{\Sigma_0\cap V(\lambda,\nu)}Q(\hat{\mathcal{L}}_O^2\mathbf{R})(\bar{K},\bar{K},T,T)\\
&\quad+\int_{\Sigma_0\cap V(\lambda,\nu)}Q(\hat{\mathcal{L}}_S\hat{\mathcal{L}}_T\mathbf{R})(\bar{K},\bar{K},\bar{K},T).
\end{aligned}
\tag{3.5.11}
$$

Remark: Unlike the quantities $\mathcal{Q}$ and $\underline{\mathcal{Q}}$ defined in Chapter 2, see 2.2.14, 2.2.23, 2.2.24, these quantities include the integrals containing $Q(\hat{\mathcal{L}}_S\hat{\mathcal{L}}_T\mathbf{R})(\bar{K},\bar{K},\bar{K},T)$. These terms are needed for estimating α_{44} and $\underline{\alpha}_{33}$ which are generated in the error estimates of Chapter 6. One can view the quantities $\mathcal{Q}$, $\underline{\mathcal{Q}}$ defined in 3.5.7, 3.5.8, as those containing the smallest number of terms needed for the proof of Theorem 3.7.10 (Theorem **M8**).

3.5.2 $\mathcal{R}$ norms for the Riemann null components

In the proof of the Main Theorem we have to use a quantity $\mathcal{R}$ defined as a sum of weighted $L^2(C)$, $L^2(\underline{C})$ norms of the Riemann curvature tensor and their derivatives up to second order. These weighted $L^2(C)$ and $L^2(\underline{C})$ norms are intimately connected with the energy-type $\mathcal{Q}$ integrals, defined in Subsection 3.5.1.

The quantity $\mathcal{R}$ that enters in the statement of the Main Theorem is

$$\mathcal{R} \equiv \mathcal{R}_{[2]} + \underline{\mathcal{R}}_{[2]}, \tag{3.5.12}$$

where

$$\begin{aligned} &\mathcal{R}_{[2]} = \mathcal{R}_{[1]} + \mathcal{R}_2\,, \ \underline{\mathcal{R}}_{[2]} = \underline{\mathcal{R}}_{[1]} + \underline{\mathcal{R}}_2 \\ &\mathcal{R}_{[1]} = \mathcal{R}_{[0]} + \mathcal{R}_1\,, \ \underline{\mathcal{R}}_{[1]} = \underline{\mathcal{R}}_{[0]} + \underline{\mathcal{R}}_1 \\ &\mathcal{R}_{[0]} = \mathcal{R}_0\,, \ \underline{\mathcal{R}}_{[0]} = \underline{\mathcal{R}}_0 + \sup_{\mathcal{K}} r^3|\bar{\rho}| \end{aligned} \tag{3.5.13}$$

with

$$\begin{aligned} \mathcal{R}_0 &= \left(\mathcal{R}_0[\alpha]^2 + \mathcal{R}_0[\beta]^2 + \mathcal{R}_0[(\rho,\sigma)]^2 + \mathcal{R}_0[\underline{\beta}]^2\right)^{1/2} \\ \underline{\mathcal{R}}_0 &= \left(\underline{\mathcal{R}}_0[\beta]^2 + \underline{\mathcal{R}}_0[(\rho,\sigma)]^2 + \underline{\mathcal{R}}_0[\underline{\beta}]^2 + \underline{\mathcal{R}}_0[\underline{\alpha}]^2\right)^{1/2} \\ \mathcal{R}_1 &= \left(\mathcal{R}_1[\alpha]^2 + \mathcal{R}_1[\beta]^2 + \mathcal{R}_1[(\rho,\sigma)]^2 + \mathcal{R}_1[\underline{\beta}]^2\right)^{1/2} \\ \underline{\mathcal{R}}_1 &= \left(\underline{\mathcal{R}}_1[\beta]^2 + \underline{\mathcal{R}}_1[(\rho,\sigma)]^2 + \underline{\mathcal{R}}_1[\underline{\beta}]^2 + \underline{\mathcal{R}}_1[\underline{\alpha}]^2\right)^{1/2} \\ \mathcal{R}_2 &= \left(\mathcal{R}_2[\alpha]^2 + \mathcal{R}_2[\beta]^2 + \mathcal{R}_2[(\rho,\sigma)]^2 + \mathcal{R}_2[\underline{\beta}]^2\right)^{1/2} \\ \underline{\mathcal{R}}_2 &= \left(\underline{\mathcal{R}}_2[\beta]^2 + \underline{\mathcal{R}}_2[(\rho,\sigma)]^2 + \mathcal{R}_2[\underline{\beta}]^2 + \underline{\mathcal{R}}_2[\underline{\alpha}]^2\right)^{1/2} \end{aligned} \tag{3.5.14}$$

and

$$\begin{aligned} \mathcal{R}_{0,1,2}[w] &\equiv \sup_{\mathcal{K}} \mathcal{R}_{0,1,2}[w](\lambda,\nu) \\ \underline{\mathcal{R}}_{0,1,2}[w] &\equiv \sup_{\mathcal{K}} \underline{\mathcal{R}}_{0,1,2}[w](\lambda,\nu). \end{aligned}$$

The terms $\mathcal{R}_{0,1,2}[w](u,\underline{u})$ denote the L^2 norms of the zero, first and second derivatives of the null component w, along the portion of null hypersurface $C(\lambda)\cap V(\lambda,\nu)$. Recall that $V(\lambda,\nu)$ (see (3.5.4)) is the causal past of $S(\lambda,\nu)$ relative to $\mathcal{K}$, whose boundary is formed by the union of the portions of the null hypersurfaces $C(\lambda)$ and $\underline{C}(\nu)$ lying in $V(\lambda,\nu)$ and by $J^-(S(\lambda,\nu))\cap\Sigma_0$.

An analogous definition holds for the terms $\underline{\mathcal{R}}_{0,1,2}[w](\lambda,\nu)$ relative to the null hypersurface $\underline{C}(\nu)\cap V(\lambda,\nu)$. We write all of them explicitly below.

1.) L^2 norms for the zero derivatives of the Riemann components:

$$\begin{aligned} \mathcal{R}_0[\alpha](\lambda,\nu) &= \|r^2\alpha\|_{2,C(\lambda)\cap V(\lambda,\nu)} \\ \mathcal{R}_0[\beta](\lambda,\nu) &= \|r^2\beta\|_{2,C(\lambda)\cap V(\lambda,\nu)} \\ \mathcal{R}_0[(\rho,\sigma)](\lambda,\nu) &= \|\tau_- r(\rho-\overline{\rho},\sigma-\overline{\sigma})\|_{2,C(\lambda)\cap V(\lambda,\nu)} \\ \mathcal{R}_0[\underline{\beta}](\lambda,\nu) &= \|\tau_-^2\underline{\beta}\|_{2,C(\lambda)\cap V(\lambda,\nu)} \end{aligned}$$

$$
\begin{aligned}
\underline{\mathcal{R}}_0[\beta](\lambda,\nu) &= \|r^2\beta\|_{2,\underline{C}(\nu)\cap V(\lambda,\nu)} \\
\underline{\mathcal{R}}_0[(\rho,\sigma)](\lambda,\nu) &= \|r^2(\rho-\overline{\rho},\sigma-\overline{\sigma})\|_{2,\underline{C}(\nu)\cap V(\lambda,\nu)} \\
\underline{\mathcal{R}}_0[\underline{\beta}](\lambda,\nu) &= \|\tau_- r\underline{\beta}\|_{2,\underline{C}(\nu)\cap V(\lambda,\nu)} \\
\underline{\mathcal{R}}_0[\underline{\alpha}](\lambda,\nu) &= \|\tau_-^2\underline{\alpha}\|_{2,\underline{C}(\nu)\cap V(\lambda,\nu)}.
\end{aligned}
\tag{3.5.15}
$$

2. L^2 norms for the first derivatives of the Riemann components:

$$
\begin{aligned}
\mathcal{R}_1[\alpha](\lambda,\nu) &= \|r^3\not\nabla\alpha\|_{2,C(\lambda)\cap V(\lambda,\nu)} + \|r^3\alpha_3\|_{2,C(\lambda)\cap V(\lambda,\nu)} \\
&\quad + \|r^3\alpha_4\|_{2,C(\lambda)\cap V(\lambda,\nu)} \\
\mathcal{R}_1[\beta](\lambda,\nu) &= \|r^3\not\nabla\beta\|_{2,C(\lambda)\cap V(\lambda,\nu)} + \|\tau_- r^2\beta_3\|_{2,C(\lambda)\cap V(\lambda,\nu)} \\
&\quad + \|r^3\beta_4\|_{2,C(\lambda)\cap V(\lambda,\nu)} \\
\mathcal{R}_1[(\rho,\sigma)](\lambda,\nu) &= \|\tau_- r^2\not\nabla(\rho,\sigma)\|_{2,C(\lambda)\cap V(\lambda,\nu)} + \|r^3(\rho,\sigma)_4\|_{2,C(\lambda)\cap V(\lambda,\nu)} \\
&\quad + \|r\tau_-^2(\rho,\sigma)_3\|_{2,C(\lambda)\cap V(\lambda,\nu)} \\
\mathcal{R}_1[\underline{\beta}](\lambda,\nu) &= \|\tau_-^2 r\not\nabla\underline{\beta}\|_{2,C(\lambda)\cap V(\lambda,\nu)} + \|\tau_- r^2\underline{\beta}_4\|_{2,C(\lambda)\cap V(\lambda,\nu)} \\
\underline{\mathcal{R}}_1[\beta](\lambda,\nu) &= \|r^3\not\nabla\beta\|_{2,\underline{C}(\nu)\cap V(\lambda,\nu)} + \|r^3\beta_3\|_{2,\underline{C}(\nu)\cap V(\lambda,\nu)} \\
\underline{\mathcal{R}}_1[(\rho,\sigma)](\lambda,\nu) &= \|r^3\not\nabla(\rho,\sigma)\|_{2,\underline{C}(\nu)\cap V(\lambda,\nu)} + \|\tau_- r^2(\rho,\sigma)_3\|_{2,\underline{C}(\nu)\cap V(\lambda,\nu)} \\
\underline{\mathcal{R}}_1[\underline{\beta}](\lambda,\nu) &= \|\tau_- r^2\not\nabla\underline{\beta}\|_{2,\underline{C}(\nu)\cap V(\lambda,\nu)} + \|\tau_-^2 r\underline{\beta}_3\|_{2,\underline{C}(\nu)\cap V(\lambda,\nu)} \\
&\quad + \|r^3\underline{\beta}_4\|_{2,\underline{C}(\nu)\cap V(\lambda,\nu)} \\
\underline{\mathcal{R}}_1[\underline{\alpha}](\lambda,\nu) &= \|\tau_-^2 r\not\nabla\underline{\alpha}\|_{2,\underline{C}(\nu)\cap V(\lambda,\nu)} + \|\tau_-^3\underline{\alpha}_3\|_{2,\underline{C}(\nu)\cap V(\lambda,\nu)} \\
&\quad + \|\tau_- r^2\underline{\alpha}_4\|_{2,\underline{C}(\nu)\cap V(\lambda,\nu)}.
\end{aligned}
\tag{3.5.16}
$$

3. L^2 norms for the second derivatives of the Riemann components:

$$
\begin{aligned}
\mathcal{R}_2[\alpha](\lambda,\nu) &= \|r^4\not\nabla^2\alpha\|_{2,C(\lambda)\cap V(\lambda,\nu)} + \|r^4\not\nabla\alpha_3\|_{2,C(\lambda)\cap V(\lambda,\nu)} \\
&\quad + \|r^4\not\nabla\alpha_4\|_{2,C(\lambda)\cap V(\lambda,\nu)} + \|\tau_- r^3\alpha_{33}\|_{2,C(\lambda)\cap V(\lambda,\nu)} \\
&\quad + \|r^4\alpha_{34}\|_{2,C(\lambda)\cap V(\lambda,\nu)} + \|\tau_- r^3\alpha_{44}\|_{2,C(\lambda)\cap V(\lambda,\nu)} \\
\mathcal{R}_2[\beta](\lambda,\nu) &= \|r^4\not\nabla^2\beta\|_{2,C(\lambda)\cap V(\lambda,\nu)} + \|\tau_- r^3\not\nabla\beta_3\|_{2,C(\lambda)\cap V(\lambda,\nu)} \\
&\quad + \|r^4\not\nabla\beta_4\|_{2,C(\lambda)\cap V(\lambda,\nu)} + \|\tau_-^2 r^2\beta_{33}\|_{2,C(\lambda)\cap V(\lambda,\nu)} \\
&\quad + \|r^4\beta_{34}\|_{2,C(\lambda)\cap V(\lambda,\nu)} + \|r^4\beta_{44}\|_{2,C(\lambda)\cap V(\lambda,\nu)} \\
\mathcal{R}_2[(\rho,\sigma)](u,\underline{u}) &= \|\tau_- r^3\not\nabla^2(\rho,\sigma)\|_{2,C(\lambda)\cap V(\lambda,\nu)} + \|\tau_-^2 r^2\not\nabla(\rho,\sigma)_3\|_{2,C(\lambda)\cap V(\lambda,\nu)} \\
&\quad + \|r^4\not\nabla(\rho,\sigma)_4\|_{2,C(\lambda)\cap V(\lambda,\nu)} + \|\tau_- r^3(\rho,\sigma)_{34}\|_{2,C(\lambda)\cap V(\lambda,\nu)} \\
&\quad + \|r^4(\rho,\sigma)_{44}\|_{2,C(\lambda)\cap V(\lambda,\nu)} + \|\tau_-^3 r(\rho,\sigma)_{33}\|_{2,C(\lambda)\cap V(\lambda,\nu)} \\
\mathcal{R}_2[\underline{\beta}](\lambda,\nu) &= \|\tau_-^2 r^2\not\nabla^2\underline{\beta}\|_{2,C(\lambda)\cap V(\lambda,\nu)} + \|\tau_- r^3\not\nabla\underline{\beta}_4\|_{2,C(\lambda)\cap V(\lambda,\nu)} \\
&\quad + \|\tau_-^3 r\not\nabla\underline{\beta}_3\|_{2,C(\lambda)\cap V(\lambda,\nu)} + \|\tau_-^2 r^2\underline{\beta}_{34}\|_{2,C(\lambda)\cap V(\lambda,\nu)} \\
&\quad + \|r^4\underline{\beta}_{44}\|_{2,C(\lambda)\cap V(\lambda,\nu)} + \|\tau_-^3 r\underline{\beta}_{33}\|_{2,C(\lambda)\cap V(\lambda,\nu)}
\end{aligned}
\tag{3.5.17}
$$

$$
\underline{\mathcal{R}}_2[\beta](\lambda,\nu) = \|r^4\not\nabla^2\beta\|_{2,\underline{C}(\nu)\cap V(\lambda,\nu)} + \|r^4\not\nabla\beta_3\|_{2,\underline{C}(\nu)\cap V(\lambda,\nu)}
$$

$$
\begin{aligned}
&+\|r^4\nabla\beta_4\|_{2,\underline{C}(\nu)\cap V(\lambda,\nu)} + \|r^4\beta_{43}\|_{2,\underline{C}(\nu)\cap V(\lambda,\nu)}\\
&+\|\tau_-r^3\beta_{33}\|_{2,\underline{C}(\nu)\cap V(\lambda,\nu)} + \|\tau_-r^3\beta_{43}\|_{2,\underline{C}(\nu)\cap V(\lambda,\nu)}\\
\underline{\mathcal{R}}_2[(\rho,\sigma)](\lambda,\nu) &= \|r^4\nabla^2(\rho,\sigma)\|_{2,\underline{C}(\nu)\cap V(\lambda,\nu)} + \|\tau_-r^3\nabla(\rho,\sigma)_3\|_{2,\underline{C}(\nu)\cap V(\lambda,\nu)}\\
&+\|r^4\nabla(\rho,\sigma)_4\|_{2,\underline{C}(\nu)\cap V(\lambda,\nu)} + \|\tau_-r^3(\rho,\sigma)_{34}\|_{2,\underline{C}(\nu)\cap V(\lambda,\nu)}\\
&+\|\tau_-^2r^2(\rho,\sigma)_{33}\|_{2,\underline{C}(\nu)\cap V(\lambda,\nu)} + \|r^4(\rho,\sigma)_{44}\|_{2,\underline{C}(\nu)\cap V(\lambda,\nu)}\\
\underline{\mathcal{R}}_2[\underline{\beta}](\lambda,\nu) &= \|\tau_-r^3\nabla^2\underline{\beta}\|_{2,\underline{C}(\nu)\cap V(\lambda,\nu)} + \|\tau_-^2r^2\nabla\underline{\beta}_3\|_{2,\underline{C}(\nu)\cap V(\lambda,\nu)}\\
&+\|r^4\nabla\underline{\beta}_4\|_{2,\underline{C}(\nu)\cap V(\lambda,\nu)} + \|\tau_-^3r\underline{\beta}_{33}\|_{2,\underline{C}(\nu)\cap V(\lambda,\nu)}\\
&+\|\tau_-r^3\underline{\beta}_{34}\|_{2,\underline{C}(\nu)\cap V(\lambda,\nu)} + \|r^4\underline{\beta}_{44}\|_{2,\underline{C}(\nu)\cap V(\lambda,\nu)}\\
\underline{\mathcal{R}}_2[\underline{\alpha}](\lambda,\nu) &= \|\tau_-^2r^2\nabla^2\underline{\alpha}\|_{2,\underline{C}(\nu)\cap V(\lambda,\nu)} + \|\tau_-^3r\nabla\underline{\alpha}_3\|_{2,\underline{C}(\nu)\cap V(\lambda,\nu)}\\
&+\|\tau_-r^3\nabla\underline{\alpha}_4\|_{2,\underline{C}(\nu)\cap V(\lambda,\nu)} + \|\tau_-^4\underline{\alpha}_{33}\|_{2,\underline{C}(\nu)\cap V(\lambda,\nu)}\\
&+\|\tau_-^2r^2\underline{\alpha}_{34}\|_{2,\underline{C}(\nu)\cap V(\lambda,\nu)} + \|\tau_-r^3\underline{\alpha}_{44}\|_{2,\underline{C}(\nu)\cap V(\lambda,\nu)}
\end{aligned}
$$

Remark: The explicit definitions of $\alpha_3, \beta_4, \beta_3 \dots$ were given in (3.2.8). The explicit definitions of $\alpha_{33}, \alpha_{34}, \beta_{34} \dots$ are given in Chapter 5.

In addition to the basic norms defined above we shall need some other curvature norms,

$$
\begin{aligned}
\mathcal{R}_0^\infty &\equiv \sup_{\mathcal{K}}\left[|r^{7/2}\alpha| + |r^{7/2}\beta| + |r^3u^{\frac12}(\rho-\overline{\rho},\sigma)|\right]\\
\underline{\mathcal{R}}_0^\infty &\equiv \sup_{\mathcal{K}}\left[|r^3\rho| + |r^2u^{\frac32}\underline{\beta}| + |ru^{\frac52}\underline{\alpha}|\right]
\end{aligned}
\tag{3.5.18}
$$

$$
\mathcal{R}_1^S = \sup_{p\in[2,4]}\mathcal{R}_1^{p,S}\ ,\ \underline{\mathcal{R}}_1^S = \sup_{p\in[2,4]}\underline{\mathcal{R}}_1^{p,S},
\tag{3.5.19}
$$

where

$$
\begin{aligned}
\mathcal{R}_1^{p,S} &= \sup_{\mathcal{K}}\Big[|r^{\frac92-\frac2p}\nabla\alpha|_{p,S(\lambda,\nu)} + |r^{\frac92-\frac2p}\nabla\beta|_{p,S(u,\underline{u})} + |r^{4-\frac2p}u^{\frac12}\nabla(\rho,\sigma)|_{p,S(\lambda,\nu)}\\
&\quad+|r^{\frac92-\frac2p}\mathbf{D}_4\alpha|_{p,S(\lambda,\nu)}\Big] \qquad (3.5.20)\\
\underline{\mathcal{R}}_1^{p,S} &= \Big[|r^{4-\frac2p}u^{\frac12}\nabla\underline{\beta}|_{p,S(\lambda,\nu)} + |r^{1-\frac2p}u^{\frac52}\nabla\underline{\alpha}|_{p,S(\lambda,\nu)} + |r^{1-\frac2p}u^{\frac52}\mathbf{D}_3\underline{\alpha}|_{p,S(\lambda,\nu)}\Big].
\end{aligned}
$$

Estimates for these auxiliary norms can be obtained, using the global Sobolev inequalities of Chapter 4, in terms of the $\mathcal{R}$ norms introduced above.[37]

3.5.3 *$\mathcal{O}$ norms for the connection coefficients*

In addition to the quantities $\mathcal{Q}$ and $\mathcal{R}$, the other basic quantity that enters into the statement of the Main Theorem is

$$
\mathcal{O} \equiv \mathcal{O}_{[3]} + \underline{\mathcal{O}}_{[3]},
\tag{3.5.21}
$$

[37] With the exception of $\sup_{\mathcal{K}}|r^3\rho|$ in $\underline{\mathcal{R}}_0^\infty$.

where

$$
\begin{aligned}
\mathcal{O}_{[3]} &\equiv \left[\mathcal{O}_3 + \mathcal{O}_{[2]}\right] \\
\mathcal{O}_{[2]} &\equiv \left[\mathcal{O}_2 + \tilde{\mathcal{O}}_2(\underline{\omega})\right] + \mathcal{O}_{[1]} \\
\mathcal{O}_{[1]} &\equiv \left[\mathcal{O}_1 + \sup_{p\in[2,4]} \tilde{\mathcal{O}}_1(\underline{\omega})\right] + \mathcal{O}^{\infty}_{[0]} \\
\mathcal{O}^{\infty}_{[0]} &\equiv \mathcal{O}^{\infty}_0 + \sup_{\mathcal{K}} |r^2(\overline{\mathrm{tr}\chi} - \frac{2}{r})| + \sup_{\mathcal{K}} |r(\Omega - \frac{1}{2})|,
\end{aligned}
\tag{3.5.22}
$$

and for the underlined quantities[38]

$$
\begin{aligned}
\underline{\mathcal{O}}_{[3]} &\equiv \left[\underline{\mathcal{O}}_3 + \underline{\mathcal{O}}_{[2]}\right] \\
\underline{\mathcal{O}}_{[2]} &\equiv \left[\underline{\mathcal{O}}_2 + \sup_{p\in[2,4]} \mathcal{O}^{p,S}_2(\omega)\right] + \underline{\mathcal{O}}_{[1]} \\
\underline{\mathcal{O}}_{[1]} &\equiv \left[\underline{\mathcal{O}}_1 + \sup_{p\in[2,4]} \mathcal{O}^{p,S}_0(\mathbf{D}_4\omega)\right] + \underline{\mathcal{O}}^{\infty}_{[0]} \\
\underline{\mathcal{O}}^{\infty}_{[0]} &\equiv \underline{\mathcal{O}}^{\infty}_0 + \sup_{\mathcal{K}} |r\tau_-(\overline{\mathrm{tr}\underline{\chi}} + \frac{2}{r})|.
\end{aligned}
\tag{3.5.23}
$$

The explicit forms of the various quantities in (3.5.22), (3.5.23) can be derived from the following definitions:

$$
\begin{aligned}
&\mathcal{O}_{0,1,2} \equiv \sup_{p\in[2,4]} \mathcal{O}^p_{0,1,2} \ , \ \underline{\mathcal{O}}_{0,1,2} \equiv \sup_{p\in[2,4]} \underline{\mathcal{O}}^p_{0,1,2} \\
&\mathcal{O}_3 \equiv \mathcal{O}^{p=2}_3 \ , \ \underline{\mathcal{O}}_3 \equiv \underline{\mathcal{O}}^{p=2}_3.
\end{aligned}
\tag{3.5.24}
$$

The $\mathcal{O}^p_q$, $\underline{\mathcal{O}}^p_q$ norms have, for $q = 0, 1, 2$, the following expressions depending on different connection coefficients:

$$
\begin{aligned}
\mathcal{O}^p_q &= \mathcal{O}^{p,S}_q(\mathrm{tr}\chi) + \mathcal{O}^{p,S}_q(\hat{\chi}) + \mathcal{O}^{p,S}_q(\eta) + \mathcal{O}^{p,S}_q(\underline{\omega}) \\
\underline{\mathcal{O}}^p_q &= \mathcal{O}^{p,S}_q(\mathrm{tr}\underline{\chi}) + \mathcal{O}^{p,S}_q(\underline{\hat{\chi}}) + \mathcal{O}^{p,S}_q(\underline{\eta}) + \mathcal{O}^{p,S}_q(\omega).
\end{aligned}
\tag{3.5.25}
$$

In the $q = 3$ case we define:

$$
\begin{aligned}
\mathcal{O}^{p=2}_3 &= \mathcal{O}^{p=2}_3(\mathrm{tr}\chi) + \mathcal{O}^{p=2}_3(\hat{\chi}) + \mathcal{O}^{p=2}_3(\eta) + \mathcal{O}^{p=2}_3(\underline{\omega}) \\
\underline{\mathcal{O}}^{p=2}_3 &= \mathcal{O}^{p=2}_3(\mathrm{tr}\underline{\chi}) + \mathcal{O}^{p=2}_3(\underline{\hat{\chi}}) + \mathcal{O}^{p=2}_3(\underline{\eta}).
\end{aligned}
\tag{3.5.26}
$$

Remark: We have systematically assigned to the $\mathcal{O}$ norms those quantities that are estimated along the C null hypersurfaces and to the $\underline{\mathcal{O}}$ norms those that are estimated along the $\underline{C}$ null hypersurfaces.

---

[38]The reason the terms $\tilde{\mathcal{O}}_2(\omega)$ and $\tilde{\mathcal{O}}_1(\omega)$, analogous to $\tilde{\mathcal{O}}_2(\underline{\omega})$ and $\tilde{\mathcal{O}}_1(\underline{\omega})$, do not appear in the underlined norms (3.5.23) is discussed in Chapter 6; see Subsections 6.3.5 and 6.3.7.

For an arbitrary connection coefficient X, we have

$$\mathcal{O}_q^{p,S}(X) \equiv \sup_{\mathcal{K}} \mathcal{O}_q^{p,S}(X)(\lambda,\nu)\ ,\ \mathcal{O}_3^{p=2}(X) \equiv \sup_{\mathcal{K}} \mathcal{O}_3^{p=2}(X)(\lambda,\nu). \tag{3.5.27}$$

The explicit expressions of all these norms are given in the following definitions:

$q \leq 2$:

$$\begin{aligned}
&\mathcal{O}_q^{p,S}(tr\chi)(\lambda,\nu) = |r^{(2+q-\frac{2}{p})}\nabla\!\!\!\!/^q(\mathrm{tr}\chi - \overline{\mathrm{tr}\chi})|_{p,S(\lambda,\nu)}\\
&\mathcal{O}_q^{p,S}(tr\underline{\chi})(\lambda,\nu) = |r^{(2+q-\frac{2}{p})}\nabla\!\!\!\!/^q(\mathrm{tr}\underline{\chi} - \overline{\mathrm{tr}\underline{\chi}})|_{p,S(\lambda,\nu)}\\
&\mathcal{O}_q^{p,S}(\hat{\chi})(\lambda,\nu) = |r^{(2+q-\frac{2}{p})}\nabla\!\!\!\!/^q\hat{\chi}|_{p,S(\lambda,\nu)}\\
&\mathcal{O}_q^{p,S}(\underline{\hat{\chi}})(\lambda,\nu) = |r^{(1+q-\frac{2}{p})}\tau_-\nabla\!\!\!\!/^q\underline{\hat{\chi}}|_{p,S(\lambda,\nu)}\\
&\mathcal{O}_q^{p,S}(\eta)(\lambda,\nu) = |r^{(2+q-\frac{2}{p})}\nabla\!\!\!\!/^q\eta|_{p,S(\lambda,\nu)}\\
&\mathcal{O}_q^{p,S}(\underline{\eta})(\lambda,\nu) = |r^{(2+q-\frac{2}{p})}\nabla\!\!\!\!/^q\underline{\eta}|_{p,S(\lambda,\nu)}\\
&\mathcal{O}_q^{p,S}(\omega)(\lambda,\nu) = |r^{(2+q-\frac{2}{p})}\nabla\!\!\!\!/^q\omega|_{p,S(\lambda,\nu)}\\
&\mathcal{O}_q^{p,S}(\underline{\omega})(\lambda,\nu) = |r^{(1+q-\frac{2}{p})}\tau_-\nabla\!\!\!\!/^q\underline{\omega}|_{p,S(\lambda,\nu)}\\
&\mathcal{O}_q^{p,S}(\mathbf{D}_4\omega)(\lambda,\nu) = |r^{(3+q-\frac{2}{p})}\nabla\!\!\!\!/^q\mathbf{D}_4\omega|_{p,S(\lambda,\nu)}\\
&\mathcal{O}_q^{p,S}(\mathbf{D}_3\underline{\omega})(\lambda,\nu) = |r^{(1+q-\frac{2}{p})}\tau_-^2\nabla\!\!\!\!/^q\mathbf{D}_3\underline{\omega}|_{p,S(\lambda,\nu)};
\end{aligned} \tag{3.5.28}$$

$q = 3$:

$$\begin{aligned}
&\mathcal{O}_3^{p=2}(\hat{\chi})(\lambda,\nu) = r^{\frac{1}{2}}(\lambda,\nu)||r^3\nabla\!\!\!\!/^3\hat{\chi}||_{L^2(\underline{C}(\nu;[\lambda_0,\lambda])}\\
&\mathcal{O}_3^{p=2}(\mathrm{tr}\chi)(\lambda,\nu) = r^{\frac{1}{2}}(\lambda,\nu)||r^3\nabla\!\!\!\!/^3\mathrm{tr}\chi||_{L^2(\underline{C}(\nu;[\lambda_0,\lambda])}\\
&\mathcal{O}_3^{p=2}(\eta)(\lambda,\nu) = r^{\frac{1}{2}}(\lambda,\nu)||r^3\nabla\!\!\!\!/^3\eta||_{L^2(\underline{C}(\nu;[\lambda_0,\lambda])}\\
&\mathcal{O}_3^{p=2}(\underline{\omega})(\lambda,\nu) = r^{\frac{1}{2}}(\lambda,\nu)||r^3\nabla\!\!\!\!/^3\underline{\omega}||_{L^2(\underline{C}(\nu;[\lambda_0,\lambda])}\\
&\mathcal{O}_3^{p=2}(\underline{\hat{\chi}})(\lambda,\nu) = r^{\frac{1}{2}}(\lambda,\nu)||r^3\nabla\!\!\!\!/^3\underline{\hat{\chi}}||_{L^2(\underline{C}(\nu;[\lambda_0,\lambda])}\\
&\mathcal{O}_3^{p=2}(\mathrm{tr}\underline{\chi})(\lambda,\nu) = r^{\frac{1}{2}}(\lambda,\nu)||r^3\nabla\!\!\!\!/^3\mathrm{tr}\underline{\chi}||_{L^2(\underline{C}(\nu;[\lambda_0,\lambda])}\\
&\mathcal{O}_3^{p=2}(\underline{\eta})(\lambda,\nu) = r^{\frac{1}{2}}(\lambda,\nu)||r^3\nabla\!\!\!\!/^3\underline{\eta}||_{L^2(\underline{C}(\nu;[\lambda_0,\lambda])}.
\end{aligned} \tag{3.5.29}$$

We also define the norms $\tilde{\mathcal{O}}_1(\underline{\omega})$, $\tilde{\mathcal{O}}_2(\underline{\omega})$. They involve the second null and mixed derivatives of $\underline{\omega}$ and will be needed in the proof of the Main Theorem,

$$\begin{aligned}
&\tilde{\mathcal{O}}_1(\underline{\omega}) \equiv ||\frac{1}{\sqrt{\tau_+}}\tau_-^2\mathbf{D}_3\underline{\omega}||_{L_2(C\cap\mathcal{K})}\\
&\tilde{\mathcal{O}}_2(\underline{\omega}) \equiv ||\frac{1}{\sqrt{\tau_+}}r\tau_-^2\nabla\!\!\!\!/\mathbf{D}_3\underline{\omega}||_{L_2(C\cap\mathcal{K})} + ||\frac{1}{\sqrt{\tau_+}}\tau_-^3\mathbf{D}_3^2\underline{\omega}||_{L_2(C\cap\mathcal{K})}.
\end{aligned} \tag{3.5.30}$$

3.5.4 Norms on the initial layer region

In addition to the $\mathcal{O}$, $\mathcal{R}$ norms expressed relative to the canonical double null foliation[39] we shall also need similar norms defined in the initial layer region $\mathcal{K}'_{\delta_0}$, relative to the initial layer foliation. These norms have exactly the same expressions; to distinguish them from the main ones defined above we denote them by $\mathcal{O}'$, $\mathcal{R}'$.

3.5.5 $\mathcal{O}$ norms on the initial and final hypersurfaces

We consider now the previous $\mathcal{O}$ norms restricted to the initial hypersurface, Σ_0, and to the last slice, $\underline{C}_*$.[40] Observe that the $\mathcal{O}$ norms restricted to the initial hypersurface are tied to the initial layer foliation, while those restricted to the last slice $\underline{C}_*$ are tied to the canonical double null foliation . The only difference with respect to the previous definitions is that some of the previous norms are absent as their restrictions on $\underline{C}_*$ or on Σ_0 are not needed. They are:

$$
\begin{aligned}
&\mathcal{O}_{[1]}(\Sigma_0) \equiv \mathcal{O}_1(\Sigma_0) + \mathcal{O}^{\infty}_{[0]}(\Sigma_0)\\
&\underline{\mathcal{O}}_{[1]}(\Sigma_0) \equiv \left[\underline{\mathcal{O}}_1(\Sigma_0) + \sup_{p\in[2,4]} \mathcal{O}^{p,S}_0(\mathbf{D}_4\omega)(\Sigma_0)\right] + \underline{\mathcal{O}}^{\infty}_{[0]}(\Sigma_0)\\
&\mathcal{O}_{[2]}(\Sigma_0) = \mathcal{O}_2(\Sigma_0) + \mathcal{O}_{[1]}(\Sigma_0)\\
&\underline{\mathcal{O}}_{[2]}(\Sigma_0) = \left[\underline{\mathcal{O}}_2(\Sigma_0) + \sup_{p\in[2,4]} \left(\mathcal{O}^{p,S}_1(\mathbf{D}_4\omega) + \mathcal{O}^{p,S}_0(\mathbf{D}^2_4\omega)(\Sigma_0)\right)\right] + \underline{\mathcal{O}}_{[1]}(\Sigma_0)\\
&\mathcal{O}_{[1]}(\underline{C}_*) \equiv \left[\mathcal{O}_1(\underline{C}_*) + \sup_{p\in[2,4]} \mathcal{O}^{p,S}_0(\mathbf{D}_3\underline{\omega})(\underline{C}_*)\right] + \mathcal{O}^{\infty}_{[0]}(\underline{C}_*)\\
&\underline{\mathcal{O}}_{[1]}(\underline{C}_*) \equiv \underline{\mathcal{O}}_1(\underline{C}_*) + \underline{\mathcal{O}}^{\infty}_{[0]}(\underline{C}_*) \qquad (3.5.31)\\
&\mathcal{O}_{[2]}(\underline{C}_*) = \left[\mathcal{O}_2(\underline{C}_*) + \sup_{p\in[2,4]} \left(\mathcal{O}^{p,S}_1(\mathbf{D}_3\underline{\omega}) + \mathcal{O}^{p,S}_0(\mathbf{D}^2_3\underline{\omega})(\underline{C}_*)\right)\right] + \mathcal{O}_{[1]}(\underline{C}_*)\\
&\underline{\mathcal{O}}_{[2]}(\underline{C}_*) = \underline{\mathcal{O}}_2(\underline{C}_*) + \underline{\mathcal{O}}_{[1]}(\underline{C}_*),
\end{aligned}
$$

where, for $q \leq 2$,

$$
\begin{aligned}
\mathcal{O}^{p,S}_q(\underline{C}_*)(X) &\equiv \sup_{\underline{C}_*} \mathcal{O}^{p,S}_q(X)(\lambda,\nu)\\
\mathcal{O}^{p,S}_q(\Sigma_0)(\underline{X}) &\equiv \sup_{\Sigma_0} \mathcal{O}^{p,S}_q(\underline{X})(\lambda,\nu). \qquad (3.5.32)
\end{aligned}
$$

Finally we introduce the $\mathcal{O}_3(\underline{C}_*)$ and $\underline{\mathcal{O}}_3(\Sigma_0)$ norms on the initial and final slice. They are defined in the following way:

$$
\begin{aligned}
\mathcal{O}_3(\underline{C}_*) &= \mathcal{O}_3(\underline{C}_*)(\mathrm{tr}\chi) + \mathcal{O}_3(\underline{C}_*)(\underline{\omega})\\
\underline{\mathcal{O}}_3(\Sigma_0) &= \mathcal{O}_3(\Sigma_0)(\mathrm{tr}\underline{\chi}) + \mathcal{O}_3(\Sigma_0)(\omega), \qquad (3.5.33)
\end{aligned}
$$

[39] In Chapter 4, Subsection 4.1.3, there is an accurate discussion about the region where these norms are defined.

[40] Although we refer here to norms on Σ_0 we shall need these norms on $\Sigma_0 \backslash K$.

where

$$
\begin{aligned}
\mathcal{O}_3(\underline{C}_*)(\mathrm{tr}\chi) &= r^{\frac{1}{2}}(\lambda,\nu)||r^3\nabla\!\!\!/^3\mathrm{tr}\chi||_{L^2(\underline{C}_*\cap V(\lambda,\nu))}\\
\mathcal{O}_3(\underline{C}_*)(\underline{\omega}) &= r^{\frac{1}{2}}(\lambda,\nu)||r^3\nabla\!\!\!/^3\underline{\omega}||_{L^2(\underline{C}_*\cap V(\lambda,\nu))}\\
\mathcal{O}_3(\Sigma_0)(\mathrm{tr}\underline{\chi}) &= r^{\frac{1}{2}}(\lambda,\nu)||r^3\nabla\!\!\!/^3\mathrm{tr}\underline{\chi}||_{L^2(\Sigma_0\cap V(\lambda,\nu))}\\
\mathcal{O}_3(\Sigma_0)(\omega) &= r^{\frac{1}{2}}(\lambda,\nu)||r^3\nabla\!\!\!/^3\omega||_{L^2(\Sigma_0\cap V(\lambda,\nu))}.
\end{aligned}
\tag{3.5.34}
$$

3.5.6 $\mathcal{D}$ norms for the rotation deformation tensors

We introduce the quantity

$$
\mathcal{D}=\mathcal{D}_0+\mathcal{D}_1+\mathcal{D}_2,
\tag{3.5.35}
$$

where

$$
\mathcal{D}_{0,1}\equiv\sup_{p\in[2,4]}\mathcal{D}^p_{0,1}
\tag{3.5.36}
$$

and

$$
\mathcal{D}^p_0 = \mathcal{D}^{p,S}_0({}^{(i)}O)+\mathcal{D}^{p,S}_0({}^{(i)}F)+\mathcal{D}^{p,S}_0({}^{(i)}H)+\mathcal{D}^{p,S}_0({}^{(i)}Z),
\tag{3.5.37}
$$

$$
\begin{aligned}
\mathcal{D}^p_1 = {} & \mathcal{D}^{p,S}_1({}^{(i)}O)+\mathcal{D}^{p,S}_1({}^{(i)}F)+\mathcal{D}^{p,S}_1({}^{(i)}H)+\mathcal{D}^{p,S}_1({}^{(i)}Z)\\
&+\mathcal{D}^{p,S}_0(\mathbf{D}_3{}^{(i)}F)+\mathcal{D}^{p,S}_0(\mathbf{D}_3{}^{(i)}H)+\mathcal{D}^{p,S}_0(\mathbf{D}_3{}^{(i)}Z)\\
&+\mathcal{D}^{p,S}_0(\mathbf{D}_4{}^{(i)}F)+\mathcal{D}^{p,S}_0(\mathbf{D}_4{}^{(i)}H)+\mathcal{D}^{p,S}_0(\mathbf{D}_4{}^{(i)}Z).
\end{aligned}
\tag{3.5.38}
$$

Finally, denoting by X any of the quantities ${}^{(i)}O$, ${}^{(i)}F$, ${}^{(i)}H$ and ${}^{(i)}Z$ and their derivatives,

$$
\mathcal{D}^{p,S}_q(X)\equiv\sup_{\mathcal{K}}\mathcal{D}^{p,S}_q(X)(\lambda,\nu).
\tag{3.5.39}
$$

Explicitly we define:

$$
\begin{aligned}
\mathcal{D}^{p,S}_q({}^{(i)}O)(\lambda,\nu) &= |r^{(-1+q-\frac{2}{p})}\nabla\!\!\!/^q\,{}^{(i)}O|_{p,S}(\lambda,\nu)\\
\mathcal{D}^{p,S}_q({}^{(i)}F)(\lambda,\nu) &= |r^{(1+q-\frac{2}{p})}\nabla\!\!\!/^q\,{}^{(i)}F|_{p,S}(\lambda,\nu)\\
\mathcal{D}^{p,S}_q({}^{(i)}H)(\lambda,\nu) &= |r^{(1+q-\frac{2}{p})}\nabla\!\!\!/^q\,{}^{(i)}H|_{p,S}(\lambda,\nu)\\
\mathcal{D}^{p,S}_q({}^{(i)}Z)(\lambda,\nu) &= |r^{(1+q-\frac{2}{p})}\nabla\!\!\!/^q\,{}^{(i)}Z|_{p,S}(\lambda,\nu)
\end{aligned}
\tag{3.5.40}
$$

$$
\begin{aligned}
\mathcal{D}^{p,S}_q(\mathbf{D}_4{}^{(i)}F)(\lambda,\nu) &= |r^{(2+q-\frac{2}{p})}\nabla\!\!\!/^q\mathbf{D}_4{}^{(i)}F|_{p,S(\lambda,\nu)}\\
\mathcal{D}^{p,S}_q(\mathbf{D}_4{}^{(i)}H)(\lambda,\nu) &= |r^{(2+q-\frac{2}{p})}\nabla\!\!\!/^q\mathbf{D}_4{}^{(i)}H|_{p,S(\lambda,\nu)}\\
\mathcal{D}^{p,S}_q(\mathbf{D}_4{}^{(i)}Z)(\lambda,\nu) &= |r^{(2+q-\frac{2}{p})}\nabla\!\!\!/^q\mathbf{D}_4{}^{(i)}Z|_{p,S(\lambda,\nu)}
\end{aligned}
\tag{3.5.41}
$$

$$\mathcal{D}_q^{p,S}(\mathbf{D}_3{}^{(i)}F)(\lambda,\nu) = |r^{(1+q-\frac{2}{p})}\tau_-\nabla\!\!\!/^q\mathbf{D}_3{}^{(i)}F|_{p,S(\lambda,\nu)}$$
$$\mathcal{D}_q^{p,S}(\mathbf{D}_3{}^{(i)}H)(\lambda,\nu) = |r^{(1+q-\frac{2}{p})}\tau_-\nabla\!\!\!/^q\mathbf{D}_3{}^{(i)}H|_{p,S(\lambda,\nu)}$$
$$\mathcal{D}_q^{p,S}(\mathbf{D}_3{}^{(i)}Z)(\lambda,\nu) = |r^{(1+q-\frac{2}{p})}\tau_-\nabla\!\!\!/^q\mathbf{D}_3{}^{(i)}Z|_{p,S(\lambda,\nu)}. \tag{3.5.42}$$

Their restrictions on the last slice $\underline{C}_*$ will be denoted by[41]

$$\mathcal{D}_q^{p,S}(\underline{C}_*)(X) \equiv \sup_{\underline{C}_*}\mathcal{D}_q^{p,S}(X)(\lambda,\nu). \tag{3.5.43}$$

It remains to define the norms appearing in $\mathcal{D}_2$. They are the more delicate ones since they depend on the third derivatives of the connection coefficients

$$\mathcal{D}_2 = \sup_{\mathcal{K}}\mathcal{D}_2(\lambda,\nu), \tag{3.5.44}$$

where, with $\epsilon > 0$,

$$\begin{aligned}\mathcal{D}_2(\lambda,\nu) \quad = \quad & ||r\nabla\!\!\!/^2 H||_{L^2(\underline{C}(\nu)\cap V(\lambda,\nu))} + ||r\nabla\!\!\!/^2 Z||_{L^2(\underline{C}(\nu)\cap V(\lambda,\nu))}\\ + \quad & ||\frac{1}{\sqrt{r^{1-2\epsilon}}}r\nabla\!\!\!/\mathbf{D}\!\!\!\!/_3 Z||_{L^2(\underline{C}(\nu)\cap V(\lambda,\nu))}.\end{aligned} \tag{3.5.45}$$

Finally we denote $\mathcal{D}_2(\underline{C}_*) = \sup_{\underline{C}_*}\mathcal{D}_2(\lambda,\nu)$, its restriction on $\underline{C}_*$.

$\mathcal{D}$ norms relative to the initial layer foliation

In addition to the $\mathcal{D}$ norms expressed relative to the canonical double null foliation we shall also need similar norms defined, in the initial layer $\mathcal{K}'_{\delta_0}$, relative to the initial layer foliation. These norms have exactly the same expression, to distinguish them from the main ones defined above we denote them by $\mathcal{D}'$.

3.6 The initial data

3.6.1 *Global initial data conditions*

We restrict ourselves to initial data sets $\{\Sigma_0, g, k\}$ with Σ_0 diffeomorphic to R^3; moreover we assume they are strongly asymptotically flat in the following sense.[42]

Definition 3.6.1 *An initial data set $\{\Sigma_0, g, k\}$ is strongly asymptotically flat (see [Ch-Kl] equations (1.0.9a), (1.0.9b)) if there exists a compact set B, such that its complement*

---

[41] For the deformation tensor norms we need only the restriction on the last slice; for more details see Chapter 4 and Chapter 6.

[42] Observe that this definition is stronger than the usual definition of an asymptotically flat initial data set: *an initial data set $\{\Sigma_0, g, k\}$ is asymptotically flat if there exists a compact set B, such that its complement $\Sigma_0\backslash B$ is diffeomorphic to the complement of the closed unit ball in R^3. Moreover there exists a coordinate system (x^1, x^2, x^3) defined in a neighborhood of infinity such that, as $r = [\sum_{i=1}^3(x^i)^2]^{\frac{1}{2}} \to \infty$, we have*

$$g_{ij} = (1+2M/r)\delta_{ij} + o(r^{-1})\,.$$

$\Sigma_0 \backslash B$ is diffeomorphic to the complement of the closed unit ball in R^3. Moreover there exists a coordinate system (x^1, x^2, x^3) defined in a neighborhood of infinity such that, as $r = \sqrt{\sum_{i=1}^{3}(x^i)^2} \to \infty$, we have[43]

$$g_{ij} = (1 + 2M/r)\delta_{ij} + o_4(r^{-\frac{3}{2}})$$
$$k_{ij} = +o_3(r^{-\frac{5}{2}}). \tag{3.6.1}$$

Remark: Strong asymptotic flatness guarantees that the ADM energy, linear momentum P_i and angular momentum J_i are well defined. Moreover from (3.6.1) it follows that $P_i = 0, i \in \{1, 2, 3\}$, implying that we are placing ourselves in a center of mass frame. In this case $E = M$ and M is an invariant quantity finite and greater than zero due to the positive energy theorem ([Sc-Yau1], [Sc-Yau2], also [Ch-Kl] p. 11).

In [Ch-Kl] the global smallness condition was defined with the help of the quantity

$$J_0(\{\Sigma_0, g, k\}; b) = \sup_{\Sigma_0}\{b^{-2}(d_0^2 + b^2)^3|Ric|^2\} + b^{-3}\left\{\int_{\Sigma_0}\sum_{l=0}^{3}(d_0^2 + b^2)^{l+1}|\nabla^l k|^2 + \int_{\Sigma_0}\sum_{l=0}^{1}(d_0^2 + b^2)^{l+3}|\nabla^l B|^2\right\}, \tag{3.6.2}$$

where d_0 is the Riemann geodesic distance from a fixed point O on Σ_0, Ric is the Ricci tensor, B is the tensor $B_{ij} = (\text{curl}\hat{R})_{ij}$ and $\hat{R}$ is the traceless part of the Ricci tensor. ∇ is the covariant derivative with respect to the metric on Σ_0 and b is a positive constant with the dimensions of length. The components of the metric tensor g_{ij} have dimension zero in units of length, $[g_{ij}] = L^0$ and, therefore, $[(\text{Ric})_{ij}] = L^{-2}$, $[B_{ij}] = L^{-3}$ and $[k_{ij}] = L^{-1}$. Because b has the dimensions of length we can consider it the "natural" unit of length and so define new coordinates $\tilde{x}$ in these units as $x = \tilde{x}b$. In these coordinates the second and third terms of $J_0(\{\Sigma_0, g, k\}; b)$ do not change; that is, they are invariant under the rescaling $x \to \tilde{x} = \frac{x}{b}$.[44] In fact we have

$$b^{-3}\left\{\int_{\Sigma_0}\sum_{l=0}^{3}(d_0^2 + b^2)^{l+1}|\nabla^l k|^2 + \int_{\Sigma_0}\sum_{l=0}^{1}(d_0^2 + b^2)^{l+3}|\nabla^l B|^2\right\}$$
$$= \int_{\Sigma_0}\sum_{l=0}^{3}(\tilde{d}_0^2 + 1)^{l+1}|\tilde{\nabla}^l\tilde{k}|^2 + \int_{\Sigma_0}\sum_{l=0}^{1}(\tilde{d}_0^2 + 1)^{l+3}|\tilde{\nabla}^l\tilde{B}|^2. \tag{3.6.3}$$

The situation is different for the first term of (3.6.2) due to the presence in Ric of a part depending on the mass M; see (3.6.1). In fact the term $\{b^{-2}(d_0^2 + b^2)^3|\text{Ric}(M)|^2\}$ is invariant under the rescaling[45]

$$x \to \tilde{x} = \frac{x}{b}, \quad M \to \tilde{M} = \frac{M}{b}. \tag{3.6.4}$$

[43] A function f is $o_m(r^{-k})$ if $\partial^l f = o(r^{-k-l})$ for $l = 0, 1, \ldots, m$.

[44] They are invariant with respect to a rescaling $x \to \tilde{x} = \frac{x}{\lambda}$ for an arbitrary λ. Of course $b \to \tilde{b} = \frac{b}{\lambda}$ and, if $\lambda = b$, $\tilde{b} = 1$.

[45] The need for rescaling M also is obvious in the Schwarzschild case where a direct computation gives $\{b^{-2}(d_0^2 + b^2)^3|\text{Ric}(M)|^2\} = \{(\tilde{d}_0^2 + 1)^3|\tilde{\text{Ric}}(\tilde{M})|^2\}$.

This implies that, in view of the scale invariance property of the Einstein vacuum equations, the asymptotically flat initial data set specified by $\{\Sigma, g, k; M\}$ and the rescaled one specified by $\{\Sigma, \tilde{g}, \tilde{k}; \tilde{M}\}$, where $\tilde{g} = g$, $\tilde{k} = b^{-1}k$ and $\tilde{M} = b^{-1}M$, give rise to two equivalent developments $(\mathcal{M}, \mathbf{g})$ and $(\tilde{\mathcal{M}}, \tilde{\mathbf{g}})$.[46] Observe that by choosing b to be large we can make $\tilde{M}$ arbitrarily small.

Remark: In the case of Strong asymptotic flatness the rescaling of M follows automatically from the rescaling of g and k and the explicit expressions of the ADM energy. In fact, because $\widetilde{\partial g} = b^{-1}\partial g$, $\tilde{E} = b^{-1}E$, where

$$E \equiv \frac{1}{16\pi} \lim_{r\to\infty} \int_{S_r} \sum_{i,j} (\partial_i g_{ij} - \partial_j g_{ii}) N^j dA. \tag{3.6.5}$$

Therefore it is sufficient to define the global initial data smallness condition (see Definition 3.6.3) with the help of the quantity[47]

$$\begin{aligned} J_0(\Sigma_0, g, k) \quad &= \quad \sup_{\Sigma_0}\left((d_0^2+1)^3|Ric|^2\right) + \int_{\Sigma_0}\sum_{l=0}^{3}(d_0^2+1)^{l+1}|\nabla^l k|^2 \\ &+ \int_{\Sigma_0}\sum_{l=0}^{1}(d_0^2+1)^{l+3}|\nabla^l B|^2. \end{aligned} \tag{3.6.6}$$

Observe that by choosing b sufficiently large, the mass of $\{\Sigma_0, g, k\}$ can be chosen appropriately small.

Definition 3.6.2 *Given an initial data set $\{\Sigma_0, g, k\}$ and a compact set $K \subset \Sigma_0$ such that $\Sigma_0 \backslash K$ is diffeomorphic to the complement of the closed unit ball in R^3, we define $J_K(\Sigma_0, g, k)$ as follows:*

i) We denote by $\mathcal{G}$ the set of all smooth extensions $(\tilde{g}, \tilde{k})$ to the whole of Σ_0 of the data (g, k) restricted to $\Sigma_0 \backslash K$, with $\tilde{g}$ Riemannian and $\tilde{k}$ a symmetric 2-tensor;

ii) We denote by $\tilde{d}_0$ the geodesic distance from a fixed point O in K relative to the metric $\tilde{g}$;

iii) We denote

$$J_K(\Sigma_0, g, k) = \inf_{\mathcal{G}} J_0(\Sigma_0, \tilde{g}, \tilde{k}). \tag{3.6.7}$$

Definition 3.6.3 *We say that the initial data satisfy the global smallness initial data condition if for a sufficient small $\varepsilon > 0$*

$$J_0(\Sigma_0, g, k) \le \varepsilon^2.$$

Definition 3.6.4 *Consider an initial data set $\{\Sigma_0, g, k\}$ where K is a compact set such that $\Sigma_0 \backslash K$ is diffeomorphic to the complement of the closed unit ball in R^3. We say that the initial data set satisfies the exterior global smallness condition if, given $\varepsilon > 0$ sufficiently small,*

$$J_K(\Sigma_0, g, k) \le \varepsilon^2 .$$

[46] In other words if $(\tilde{g}_{ij}(\tilde{x}), \tilde{k}_{ij}(\tilde{x}))$ is a solution with initial data $\{\Sigma, \tilde{g}, \tilde{k}; \tilde{M}\}$, then $(g_{ij}(x), k_{ij}(x))$ is a solution with initial data $\{\Sigma, g, k; M\}$.

[47] The $\tilde{}$ have been suppressed.

Remarks:

1. An alternative definition of J_K could be given with the help of the geodesic distance function starting from the boundary of K.

2. Given an initial data set $\{\Sigma_0, g, k\}$ with $J_0(\Sigma_0, g, k) < \infty$ it should not be difficult to prove that for given $\varepsilon > 0$ sufficiently small we can find a sufficiently large compact set K such that $J_K(\Sigma_0, g, k) < \varepsilon^2$.

3. The same statement as in the previous remark should hold true for an arbitrary strongly asymptotically flat initial data set.[48]

3.7 The Main Theorem

Theorem 3.7.1 (Main Theorem) *Consider a strongly asymptotically flat, maximal initial data set $\{\Sigma_0, g, k\}$.[49] Assume that the initial data set satisfies the exterior global smallness condition, $J_K(\Sigma_0, g, k) < \varepsilon^2$, where K is a sufficiently large compact set $\subset \Sigma_0$ with $\Sigma_0 \backslash K$ diffeomorphic to $R^3 \backslash B$. The initial data set has a unique development $(\mathcal{M}, \mathbf{g})$, defined outside the domain of influence of K, with the following properties:*

i) $\mathcal{M} = \mathcal{M}^+ \cup \mathcal{M}^-$ where $\mathcal{M}^+$ consists of the part of $\mathcal{M}$ that is in the future of $\Sigma_0 \backslash K$, $\mathcal{M}^-$ the part in the past;

ii) $(\mathcal{M}^+, g)$ can be foliated by a canonical double null foliation $\{C(\lambda), \underline{C}(\nu)\}$ *whose outgoing leaves $C(\lambda)$ are complete for all $|\lambda| \geq |\lambda_0|$.[50] The boundary of K can be chosen to be the intersection of $C(\lambda_0) \cap \Sigma_0$;*

iii) The norms $\mathcal{O}$, $\mathcal{D}$ and $\mathcal{R}$ are bounded by a constant $\leq c\varepsilon$;

iv) In particular the null Riemann components have the following asymptotic behavior:

$$\begin{aligned}
&\sup_{\mathcal{K}} r^{7/2}|\alpha| \leq C_0\,, \quad \sup_{\mathcal{K}} r|u|^{\frac{5}{2}}|\underline{\alpha}| \leq C_0\\
&\sup_{\mathcal{K}} r^{7/2}|\beta| \leq C_0\,, \quad \sup_{\mathcal{K}} r^2|u|^{\frac{3}{2}}|\underline{\beta}| \leq C_0\\
&\sup_{\mathcal{K}} r^3|\rho| \leq C_0\,, \quad \sup_{\mathcal{K}} r^3|u|^{\frac{1}{2}}|(\rho - \overline{\rho}, \sigma)| \leq C_0
\end{aligned} \tag{3.7.1}$$

with C_0 a constant depending on the initial data;

v) $(\mathcal{M}^-, g)$ satisfies the same properties as $(\mathcal{M}^+, g)$;

vi) If $J_0(\Sigma_0, g, k)$ is sufficiently small we should be able to extend $(\mathcal{M}, \mathbf{g})$ to a smooth, complete solution compatible with the global stability of the Minkowski space.[51]

---

[48] Observe that the finiteness of $J_0(\Sigma_0, g, k)$ or $J_K(\Sigma_0, g, k)$ is consistent with the stronger version of the asymptotic flatness assumption introduced in definition 3.6.1, called strong asymptotic flatness.

[49] The requirement that Σ_0 be maximal is not essential. It can be avoided, as suggested in [Ch-Kl], by starting with a local solution of the Einstein equations and using the result of [Ba] concerning the existence of a maximal hypersurface.

[50] By this we mean that the null geodesics generating $C(\lambda)$ can be indefinitely extended toward the future.

[51] We do not address this issue here.

The proof of the Main Theorem which is given in Section 3.7.9, hinges on a sequence of basic theorems stated in Subsections 3.7.2–3.7.7, concerning estimates for the $\mathcal{O}$, $\mathcal{D}$, $\mathcal{R}$ and $\mathcal{Q}$ norms. Their proofs are lengthy and form the content of the next four chapters. In the statements of the theorems given below c refers systematically to a constant that is independent on all the main quantities appearing in the theorems.

3.7.1 Estimates for the initial layer foliation

Theorem 3.7.2 (Theorem M0) *Consider an initial data set that satisfies the exterior global smallness condition $J_K(\Sigma_0, g, k) \leq \varepsilon^2$, with ε sufficiently small. There exists an initial layer foliation on $\mathcal{K}'_{\delta_0} \subset \mathcal{K}$ of fixed height $\delta_0 < 1$, such that the following estimates hold:*[52]

$$\mathcal{O}_{[3]}{}' \leq c\varepsilon \ , \ \underline{\mathcal{O}}_{[3]}{}' \leq c\varepsilon$$
$$\mathcal{R}_{[2]}{}' \leq c\varepsilon \ , \ \underline{\mathcal{R}}_{[2]}{}' \leq c\varepsilon. \tag{3.7.2}$$

The proof of Theorem **M0** is discussed in Chapter 7.

Remark: Theorem **M0**, which describes the properties of the initial layer foliation, is totally independent of the global results stated here and proved in the next chapters. Nevertheless the structure of its proof follows, in a far simpler local situation, all the main steps needed in the proof of Theorems **M1**,...,**M9**. We shall have a short discussion of its proof at the end of Chapter 7, after the proof of Theorems **M1**,...,**M9** has been completely addressed.

3.7.2 Estimates for the $\mathcal{O}$ norms in $\mathcal{K}$

Theorem 3.7.3 (Theorem M1) *Assume that, relative to the double null canonical foliation of $\mathcal{K}$,*

$$\mathcal{R} \leq \Delta. \tag{3.7.3}$$

Moreover we assume that

$$\underline{\mathcal{O}}_{[3]}(\Sigma_0) \leq \mathcal{I}_0 \ , \ \mathcal{O}_{[3]}(\underline{C}_*) \leq \mathcal{I}_*. \tag{3.7.4}$$

Then, if Δ, $\mathcal{I}_0$, $\mathcal{I}_*$ are sufficiently small, the following estimate holds:

$$\mathcal{O} \leq c(\mathcal{I}_0 + \mathcal{I}_* + \Delta). \tag{3.7.5}$$

The proof of Theorem **M1** is in Section 4.2

3.7.3 Estimates for the $\mathcal{D}$ norms in $\mathcal{K}$

Theorem 3.7.4 (Theorem M2) *Assume that, relative to a double null foliation of $\mathcal{K}$,*

$$\mathcal{R} \leq \Delta.$$

[52] The initial layer region can in fact be extended to a height which is, at least, proportional to $1/\varepsilon_0$.

Moreover we assume

$$\mathcal{D}(\underline{C}_*) \leq \mathcal{I}_*, \tag{3.7.6}$$

and the results of Theorem 3.7.3.[53] *Then, if $\mathcal{I}_0, \mathcal{I}_*, \Delta$ are sufficiently small, the following estimate holds:*

$$\mathcal{D} \leq c(\mathcal{I}_0 + \mathcal{I}_* + \Delta). \tag{3.7.7}$$

The proof of Theorem **M2** is in Section 4.7.

3.7.4 Estimates for the $\mathcal{O}$ norms on the initial hypersurface

Theorem 3.3.1 (Theorem M3) *Consider an initial data set which satisfies the exterior global smallness condition $J_K(\Sigma_0, g, k) \leq \varepsilon^2$, with ε sufficiently small. There exists a canonical foliation on $\Sigma_0 \backslash K$, such that the following estimates hold:*

$$\mathcal{O}_{[3]}(\Sigma_0 \backslash K) \leq c\varepsilon \; , \; \underline{\mathcal{O}}_{[3]}(\Sigma_0 \backslash K) \leq c\varepsilon.$$

The proof of Theorem **M3** is in Subsection 7.1.3.

3.7.5 Estimates for the $\mathcal{O}$ norms and the $\mathcal{D}$ norms on the last slice

Theorem 3.7.5 (Theorem M4) *Consider a canonical foliation on $\underline{C}_*$ relative to which*

$$\mathcal{R} \leq \Delta\,.$$

Moreover we assume

$$\mathcal{O}_{[2]}(\underline{C}_* \cap \Sigma_0) + \mathcal{O}_3(\Sigma_0) + \underline{\mathcal{O}}_{[3]}(\Sigma_0) \leq \mathcal{I}_0.$$

If Δ, $\mathcal{I}_0$ are sufficiently small, then the following estimate holds:

$$\underline{\mathcal{O}}_{[2]}(\underline{C}_*) + \mathcal{O}_{[3]}(\underline{C}_*) \leq c(\mathcal{I}_0 + \Delta).$$

Remark: A stronger version of Theorem 3.7.5 will be proved in Chapter 7, Section 7.4. The proof of Theorem **M4** is in Subsection 3.5.5.

Theorem 3.7.6 (Theorem M5) *Consider a canonical foliation on $\underline{C}_*$ relative to which*

$$\mathcal{R} \leq \Delta\,.$$

Moreover we assume

$$\mathcal{O}_{[2]}(\underline{C}_* \cap \Sigma_0) + \mathcal{O}_3(\Sigma_0) + \underline{\mathcal{O}}_{[3]}(\Sigma_0) \leq \mathcal{I}_0.$$

If Δ, $\mathcal{I}_0$ are sufficiently small, then the following estimate holds:

$$\mathcal{D}(\underline{C}_*) \leq c(\mathcal{I}_0 + \Delta).$$

[53] The fact that we do not have to make assumptions for the $\mathcal{D}$ norms on Σ_0 follows from the construction of the rotation vector fields; see Chapter 4 Section 4.6.

The proof of Theorem **M5** is in Section 7.5

Corollary 3.7.7 *If the double null foliation is canonical and* Δ, $\mathcal{I}_0$ *are sufficiently small, we have*

$$\mathcal{O} + \mathcal{D} \leq c\,(\mathcal{I}_0 + \Delta)\,.$$

In addition we shall also need in the proof of the Main Theorem the following precise version of Theorem 3.3.2.

Theorem 3.3.2 (Theorem M6): *Assume given on* $\underline{C}_*$ *a foliation that we call a background foliation that is not necessarily canonical and whose connection coefficients and null curvature components satisfy the inequalities, with* $\epsilon_{0'}$ *sufficiently small,*[54]

$$\begin{aligned}\mathcal{R}'(\underline{C}_*) &\equiv \mathcal{R}_{[2]}{}'(\underline{C}_*) + \underline{\mathcal{R}}_{[2]}{}'(\underline{C}_*) \leq \epsilon_0' \\ \mathcal{O}'(\underline{C}_*) &\equiv \underline{\mathcal{O}}_{[2]}{}'(\underline{C}_*) + \mathcal{O}_{[2]}{}'(\underline{C}_*) \leq \epsilon_0',\end{aligned} \tag{3.7.8}$$

where $\mathcal{R}_{[2]}{}'(\underline{C}_*)$, $\underline{\mathcal{R}}_{[2]}{}'(\underline{C}_*)$, $\underline{\mathcal{O}}_{[2]}{}'(\underline{C}_*)$, $\mathcal{O}_{[2]}{}'(\underline{C}_*)$ *are the norms introduced in Section 3.5, restricted to* $\underline{C}_*$, *relative to the background foliation.*[55] *Then there exists a canonical foliation, on* $\underline{C}_*$ *relative to which we have*

$$\begin{aligned}\mathcal{R}(\underline{C}_*) &\equiv \mathcal{R}_{[2]}(\underline{C}_*) + \underline{\mathcal{R}}_{[2]}(\underline{C}_*) \leq c\epsilon_0' \\ \mathcal{O}(\underline{C}_*) &\equiv \underline{\mathcal{O}}_{[2]}(\underline{C}_*) + \mathcal{O}_{[3]}(\underline{C}_*) \leq c\epsilon_0'.\end{aligned} \tag{3.7.9}$$

In addition it can be shown that these two foliations remain close to each other in a sense that can be made precise.

The proof of Theorem **M6** is in Section 7.3.

3.7.6 Estimates for the $\mathcal{R}$ norms

Theorem 3.7.8 (Theorem M7) *Assume that relative to a double null foliation on* $\mathcal{K}$

$$\mathcal{O}_{[2]} + \underline{\mathcal{O}}_{[2]} \leq \Gamma. \tag{3.7.10}$$

Then, if Γ *is sufficiently small, we have*

$$\mathcal{R} \leq c\mathcal{Q}_{\mathcal{K}}^{\frac{1}{2}}. \tag{3.7.11}$$

The proof of Theorem **M7** is in Chapter 5.

Corollary 3.7.9 *Under the same assumptions of the previous theorem the following inequality holds:*

$$\mathcal{R}_0^{\infty} + \underline{\mathcal{R}}_0^{\infty} \leq c\mathcal{Q}_{\mathcal{K}}^{\frac{1}{2}}. \tag{3.7.12}$$

---

[54]The background foliation will be specified during the proof of the theorem.

[55]These are the appropriate smallness assumptions of the first version of the theorem.

3.7.7 Estimates for the $\mathcal{Q}$ integral norms

Lemma 3.7.1 *The exterior global smallness assumptions $J_K(\Sigma_0, g, k) \leq \varepsilon^2$, with ε sufficiently small, imply*

$$\mathcal{Q}_{\Sigma_0 \cap \mathcal{K}} \leq c\varepsilon^2.$$

Theorem 3.7.10 (Theorem M8) *Assume that relative to a double null foliation on $\mathcal{K}$,*

$$\mathcal{O} \leq \varepsilon_0 \,,\ \mathcal{R} \leq \varepsilon_0,$$

with a constant ε_0 sufficiently small.[56] *Then the following estimate holds:*

$$\mathcal{Q}_{\mathcal{K}} \leq c\mathcal{Q}_{\Sigma_0 \cap \mathcal{K}}, \tag{3.7.13}$$

where c is a constant, independent of ε_0.

The proof of Theorem **M8** is in Chapter 6.

3.7.8 Extension theorem

Theorem 3.7.11 (Theorem M9) *Consider the spacetime $\mathcal{K}(\lambda_0, \nu_*)$ together with its double null (canonical) foliation given by the functions u and $\underline{u}$ such that*

i) The norms $\mathcal{Q}$, $\mathcal{O}$, $\mathcal{R}$ are sufficiently small

$$\mathcal{Q} \leq \epsilon'_0 \,,\ \mathcal{O} \leq \epsilon'_0 \,,\ \mathcal{R} \leq \epsilon'_0 \,;$$

ii) The initial conditions on Σ_0 are such that

$$\mathcal{O}(\Sigma_0[\nu_*, \nu_* + \delta]) \leq \epsilon'_0 \ ,$$

where $\Sigma_0[\nu_*, \nu_* + \delta] \equiv \{p \in \Sigma_0 | \underline{u}_{(0)}(p) \in [\nu_*, \nu_* + \delta]\}$.

Then we can extend the spacetime $\mathcal{K}(\lambda_0, \nu_*)$ and the double null foliation $\{u, \underline{u}\}$ to a larger spacetime $\mathcal{K}(\lambda_0, \nu_* + \delta)$, with δ sufficiently small, such that the extended norms, denoted $\mathcal{O}'$, $\mathcal{R}'$ satisfy

$$\mathcal{O}' \leq c\epsilon'_0 \,,\ \mathcal{R}' \leq c\epsilon'_0 \,.$$

The proof of Theorem **M9** is in Section 7.6.

3.7.9 Proof of the Main Theorem

Step 1: Using the result stated in Theorem **M3** we can construct a canonical foliation on $\Sigma_0 \backslash K$ that satisfies

$$\mathcal{O}_{[3]}(\Sigma_0 \backslash K) \leq c\varepsilon \,,\ \underline{\mathcal{O}}_{[3]}(\Sigma_0 \backslash K) \leq c\varepsilon \,.$$

We use this foliation to extend the rotation vector fields from spacelike infinity to $\Sigma_0 \backslash K$.

[56]The assumption $\mathcal{R} \leq \varepsilon_0$ is needed to control the deformation tensors of the angular momentum vector fields. In fact the assumptions on $\mathcal{O}$ and on $\mathcal{R}$ imply, via Theorem 3.7.4, that $\mathcal{D} \leq c\varepsilon_0$.

Step 2: We define $\mathcal{U}$ as the set of values ν_1 such that there exists a spacetime $\mathcal{K} = \mathcal{K}(\lambda_0, \nu_1)$ with the following properties, called *Bootstrap assumptions*.

Bootstrap assumption B1:
The spacetime $\mathcal{K} = \mathcal{K}(\lambda_0, \nu_1)$ is foliated by a canonical double null foliation, as specified in Definition 3.3.2, made from the two families of null hypersurfaces $\{C(\lambda)\}$ and $\{\underline{C}(\nu)\}$, with λ and ν varying in the finite intervals $[\lambda_1, \lambda_0]$ and $[\nu_0, \nu_1]$ respectively.

Bootstrap assumption B2:

Relative to the canonical double null foliation of $\mathcal{K} = \mathcal{K}(\lambda_0, \nu_1)$ we have

$$\mathcal{O} \leq \varepsilon_0 \ , \quad \mathcal{R} \leq \varepsilon_0 .$$

Step 3: We show that the set $\mathcal{U}$ is not empty. To do this we first construct a local solution, starting from the initial data on Σ_0, and prove the existence on it of a canonical double null foliation, satisfying **B1**. Then using the initial data assumptions, the properties of the local solution and the canonical double null foliation it is easy to check that **B2** is also satisfied. The two nontrivial parts of this step are the actual local existence result, which has already been discussed in detail in [Ch-Kl], Chapter 10, and the construction of the canonical foliation of the last slice which has been discussed in Section 3.3 and proved in Chapter 7.[57] This completes the proof that $\mathcal{U}$ is not empty.

Step 4: This is the main step of the proof. Define ν_* to be the supremum of the set $\mathcal{U}$. If $\nu_* = \infty$ the result is achieved. If ν_* is finite we may assume $\nu_* \in \mathcal{U}$ and proceed in the following way.[58]

1. We consider the region $\mathcal{K} = \mathcal{K}(\lambda_0, \nu_1)$. Making use of the properties **B1** and **B2**, for sufficiently small ε_0, we find, using Theorem **M8**, that the main quantity $\mathcal{Q}_{\mathcal{K}}$ is bounded by $c\mathcal{Q}_{\Sigma_0 \cap \mathcal{K}}$. As $\mathcal{Q}_{\Sigma_0 \cap \mathcal{K}}$ is expressed in terms of initial data it follows that (see Lemma 3.7.1)

$$\mathcal{Q}_{\mathcal{K}} \leq c\varepsilon^2. \tag{3.7.14}$$

2. We use Theorems **M7**, **M8** to show that $\mathcal{R} \leq c\varepsilon$. Moreover recalling Theorems **M1, M4** and Corollary 3.7.7 we find that

$$\mathcal{O} \leq c(\mathcal{I}_0 + \Delta) \ .$$

In view of the fact that $\mathcal{R} \leq c\varepsilon$ we can choose $\Delta \leq c\varepsilon$. Recall that $\mathcal{I}_0$ is an upper bound for $\mathcal{O}(\Sigma_0 \backslash K)$. In view of Step 1 we can choose $\mathcal{I}_0 \leq c\varepsilon$, and therefore, we infer that $\mathcal{O} \leq c\varepsilon$.

3. From the previous steps we have that, under the bootstrap assumptions

$$\mathcal{O} \leq \varepsilon_0 \ , \mathcal{R} \leq \varepsilon_0 \ ,$$

---

[57] We remark that Theorem 3.3.2 concerning the existence of a canonical foliation on the last slice is used twice in the proof of the Main Theorem, the first time to prove that $\mathcal{U}$ is not empty, the second time in Step 7 to show that $\nu_* < \infty$ leads to a contradiction.

[58] In fact the argument below works for any fixed $\nu < \nu_*$ arbitrary close to ν_*.

with ε_0 sufficiently small, and assuming also the results of Step 1,

$$\mathcal{O} \leq c\varepsilon \ , \ \mathcal{R} \leq c\varepsilon. \tag{3.7.15}$$

Therefore if ε is sufficiently small we obtain the improved estimate

$$\mathcal{O} \leq \frac{1}{2}\varepsilon_0 \ , \ \mathcal{R} \leq \frac{1}{2}\varepsilon_0. \tag{3.7.16}$$

Step 5: With the help of Theorem **M9**, for $\epsilon_0' = c\varepsilon$, the value on the right-hand side of (3.7.15), we show that the spacetime $\mathcal{K}(\lambda_0, \nu_*)$ can be extended to a spacetime $\mathcal{K}(\lambda_0, \nu_* + \delta)$, for δ sufficiently small, foliated by a double null foliation which extends the canonical double null foliation of $\mathcal{K}(\lambda_0, \nu_*)$. Moreover, the norms $\mathcal{R}'$ and $\mathcal{O}'$, relative to the extended double null foliation, cannot become larger than $c^2\varepsilon$,

$$\mathcal{O}' \leq c^2\varepsilon \ , \ \mathcal{R}' \leq c^2\varepsilon. \tag{3.7.17}$$

We remark that the extended double null foliation fails to be canonical on $\mathcal{K}(\lambda_0, \nu_* + \delta)$. In fact $\underline{u}$ is canonical on Σ_0 but the extended u fails to be canonical on the new last slice, which we denote $\underline{C}_{**} \equiv \underline{C}(\nu_* + \delta)$.

Step 6: Finally we are able to show that the assumption $\nu_* < \infty$ leads to a contradiction. In fact the new spacetime $\mathcal{K}(\lambda_0, \nu_* + \delta)$ is a good candidate for our family of spacetimes satisfying the bootstrap assumptions **B1** and **B2**. The only property still missing is that the extended function u be canonical on $\underline{C}_{**}$.

Using Theorem **M6** with $\epsilon_0' = c^2\varepsilon$ where $c^2\varepsilon$ is the constant on the right-hand side of (3.7.17) we can construct a canonical foliation on $\underline{C}_{**}$ relative to which the new norms $\mathcal{R}(\underline{C}_{**})$ and $\mathcal{O}(\underline{C}_{**})$ satisfy

$$\mathcal{O}(\underline{C}_{**}) + \mathcal{R}(\underline{C}_{**}) \leq c^3\varepsilon. \tag{3.7.18}$$

Starting with this new canonical foliation, on the new last slice, we extend it to the interior of the spacetime and thus obtain a new extended canonical double null foliation near the previous one. In view of the continuity properties of the propagation equations of the double null foliation we can check that, for small δ, the new norms $\mathcal{O}$ and $\mathcal{R}$ remain arbitrarily close to the old ones; in fact we show that these new norms satisfy

$$\mathcal{O} < 100c^3\varepsilon \ , \ \mathcal{R} < 100c^3\varepsilon. \tag{3.7.19}$$

We shall prove this fact in the remark below. Therefore, for ε sufficiently small, we still have the inequalities

$$\mathcal{O} < \varepsilon_0 \ , \ \mathcal{R} < \varepsilon_0 \, .$$

We have, therefore, constructed the spacetime $\mathcal{K}(\lambda_0, \nu_* + \delta)$ satisfying all the bootstrap assumptions.

This proves that ν_* is not the supremum of the the set $\mathcal{U}$, which contradicts our assumption. Therefore the only way to avoid a contradiction is for $\nu_* = \infty$.

Remark: In what follows we show in more detail how the new norms $\mathcal{O}$ and $\mathcal{R}$ defined in the extended spacetime $\mathcal{K}(\lambda_0, \nu_* + \delta)$ satisfy the inequality (3.7.19).

1. We start with the inequality (3.7.18) on $\underline{C}_{**}$

$$\mathcal{O}(\underline{C}_{**}) + \mathcal{R}(\underline{C}_{**}) \leq c^3 \varepsilon. \tag{3.7.20}$$

We also know that relative to the old foliation we have, on the whole spacetime $\mathcal{K}(\nu_* + \delta)$,

$$\mathcal{O}' \leq c^2 \varepsilon \ , \ \mathcal{R}' \leq c^2 \varepsilon. \tag{3.7.21}$$

2. To prove our result we first observe that we can pass from the norms $\mathcal{R}$ to the norms $\mathcal{R}'$ with the help of the following estimate, provided $\mathcal{O}$, $\mathcal{O}'$ are sufficiently small,[59]

$$\mathcal{R} \leq \mathcal{R}' + c\mathcal{O}_{[2]} \cdot (1 + \mathcal{O}'_{[2]}) \cdot \mathcal{R}' + [\text{higher order terms}]. \tag{3.7.22}$$

This is proved in Chapter 7, Corollary 7.7.1. We then use the same bootstrap argument as in the proof of Theorem **M1**; see Chapter 4, Theorem 4.2.1. More precisely consider the region $\Delta(\lambda_2, \nu_2) \subset \mathcal{K}(\lambda_0, \nu_* + \delta)$ defined by

$$\Delta(\lambda_2, \nu_2) = \left\{ p \in \mathcal{K}(\lambda_0, \nu_* + \delta) | \ (u(p), \underline{u}(p)) \in [\overline{\lambda}_1, \lambda_2) \times [\nu_* + \delta, \nu_2) \right\}, \tag{3.7.23}$$

where $\overline{\lambda}_1 = u(p)|_{\underline{C}_{**} \cap \Sigma_0}$. Repeating the argument of Theorem 4.2.1 we obtain that, for any two-dimensional surface S contained in Δ, we have the inequality

$$\mathcal{O}|_\Delta \leq c \left(\mathcal{O}(\underline{C}_{**} \cap \Delta) + \mathcal{I}_0 + \mathcal{R}|_\Delta \right), \tag{3.7.24}$$

provided that $\mathcal{R}|_\Delta$ is sufficiently small. Therefore using (3.7.20), (3.7.22) and $\mathcal{I}_0 < c\varepsilon$, we conclude that

$$\mathcal{O}|_\Delta \leq c \left(c^3 \varepsilon + c\varepsilon + \mathcal{R}'|_\Delta + \mathcal{O}|_\Delta \cdot \mathcal{R}'|_\Delta + \mathcal{O}'|_\Delta \cdot \mathcal{R}'|_\Delta \right).$$

Now using the estimates (3.7.21) for $\mathcal{O}'$ and $\mathcal{R}'$ and taking ε sufficiently small,

$$\mathcal{O}|_\Delta \leq c \left(c^3 \varepsilon + c\varepsilon + \mathcal{R}'|_\Delta + \mathcal{O}'|_\Delta \cdot \mathcal{R}'|_\Delta \right) \leq 4c^3 \varepsilon. \tag{3.7.25}$$

Therefore in view of 3.7.22 we also have

$$\mathcal{R}|_\Delta \leq 4c^3 \varepsilon \tag{3.7.26}$$

By a standard continuity argument we can show that the region Δ can be chosen equal to the whole extended spacetime $\mathcal{K}(\lambda_0, \nu_* + \delta)$.

[59] Since we proceed by a continuity argument, starting from the last slice, where $\mathcal{O}$ is small, this assumption is justified.

3.8 Appendix

3.8.1 *Proof of Proposition 3.1.1*

We start by considering the Gauss equation which expresses the Riemann tensor of the submanifold S. We denoted $\not{\mathbf{R}}$ in terms of the Riemann tensor of the embedding manifold $(\mathcal{M}, \mathbf{g})$ and the null second fundamental forms χ, $\underline{\chi}$,[60]

$$\not{R}^{\nu}_{\mu\rho\sigma} = \Pi^{\tau}_{\mu}\Pi^{\nu}_{\delta}\Pi^{\lambda}_{\rho}\Pi^{\zeta}_{\sigma}R^{\delta}_{\tau\lambda\zeta} - \frac{1}{2}(\chi^{\nu}_{\rho}\underline{\chi}_{\mu\sigma} - \chi^{\nu}_{\ \sigma}\underline{\chi}_{\mu\rho}) - \frac{1}{2}(\underline{\chi}^{\nu}_{\ \rho}\chi_{\mu\sigma} - \underline{\chi}^{\nu}_{\ \sigma}\chi_{\mu\rho}), \tag{3.8.1}$$

where $\Pi^{\mu}_{\nu} = \delta^{\mu}_{\nu} + \frac{1}{2}(e_3{}^{\mu}e_{4\nu} + e_4{}^{\mu}e_{3\nu})$ projects on the tangent space TS. Contracting the indices ν and ρ with respect to the metric g of $\mathcal{M}$, we obtain an expression for the Ricci tensor relative to S

$$\not{R}_{\mu\sigma} = R^{\delta}_{\tau\lambda\zeta}\Pi^{\tau}_{\mu}\Pi^{\lambda}_{\delta}\Pi^{\zeta}_{\sigma} - \frac{1}{2}(\mathrm{tr}\chi\underline{\chi}_{\mu\sigma} + \mathrm{tr}\underline{\chi}\chi^{\mu}_{\sigma}) + \frac{1}{2}\left((\chi\cdot\underline{\chi})_{\mu\sigma} + (\underline{\chi}\cdot\chi)_{\mu\sigma}\right),$$

and contracting the indices μ, σ, we get

$$\begin{aligned} \not{R} &= R^{\delta}_{\tau\lambda\zeta}\Pi^{\lambda}_{\delta}\Pi^{\tau\zeta} - \mathrm{tr}\chi\,\mathrm{tr}\underline{\chi} + \chi\cdot\underline{\chi} \\ &= \mathbf{R} + \mathbf{Ricci}(e_4, e_3) - \frac{1}{2}\mathbf{R}(e_4, e_3, e_4, e_3) - \mathrm{tr}\chi\,\mathrm{tr}\underline{\chi} + \chi\cdot\underline{\chi}. \end{aligned}$$

In an Einstein vacuum manifold $\mathbf{R} = \mathbf{Ricci} = 0$ and, recalling the null decomposition of the Riemann tensor, the previous equation reduces to

$$\not{R} + \mathrm{tr}\chi\,\mathrm{tr}\underline{\chi} - (\chi\cdot\underline{\chi}) = -\frac{1}{2}\mathbf{R}(e_4, e_3, e_4, e_3) = -2\rho. \tag{3.8.2}$$

Written in terms of the scalar curvature of S, $K = \frac{1}{2}\not{R}$, (3.8.2) becomes

$$\mathbf{K} = -\frac{1}{4}\mathrm{tr}\chi\,\mathrm{tr}\underline{\chi} + \frac{1}{2}(\hat{\chi}\cdot\hat{\underline{\chi}}) - \rho. \tag{3.8.3}$$

Proceeding in a similar way we compute the Codazzi equations which connect the S tangential derivatives of χ and $\underline{\chi}$ to the Riemann tensor of $\mathcal{M}$ and to χ, $\underline{\chi}$, ζ,

$$\not\nabla_{\rho}\chi^{\rho}_{\mu} - \not\nabla_{\mu}\chi^{\rho}_{\rho} = \frac{1}{2}R_{\tau\nu\delta\sigma}e^{\tau}_{4}e^{\nu}_{3}e^{\delta}_{4}\Pi^{\sigma}_{\mu} + \mathrm{tr}\chi\,\zeta_a e_{a\mu} - \chi_{ab}\zeta_b e_{a\mu}\,.$$

Since

$$\not\nabla_{\rho}\chi^{\rho}_{\mu} = (D_{e_b}\chi)_{ba}e_{a\mu} = (\not{\mathrm{d}}\mathrm{iv}\,\chi)_a e_{a\mu},$$

---

[60]Let Y be a vector field defined on $\mathcal{M}$ and in TS. $\not\nabla Y$ its covariant derivative in S,

$$\not\nabla_{\mu}Y^{\rho} = \Pi^{\gamma}_{\mu}\Pi^{\rho}_{\beta}D_{\gamma}Y^{\beta}.$$

$\not{\mathbf{R}}$ is obtained by computing the right-hand side of the equation $\not{R}^{\nu}_{\mu\rho\sigma}Y^{\mu} = \not\nabla_{\rho}\not\nabla_{\sigma}Y^{\nu} - \not\nabla_{\sigma}\not\nabla_{\rho}Y^{\nu}$. Because $\not\nabla_{\rho}\not\nabla_{\sigma}Y^{\rho} = \Pi^{\lambda}_{\rho}\Pi^{\zeta}_{\sigma}\Pi^{\nu}_{\delta}D_{\lambda}D_{\zeta}Y^{\delta} + \Pi^{\lambda}_{\rho}\Pi^{\nu}_{\delta}\Pi^{\tau}_{\sigma}(D_{\lambda}\Pi^{\tau}_{\sigma})D_{\zeta}Y^{\delta} + \Pi^{\lambda}_{\rho}\Pi^{\zeta}_{\sigma}\Pi^{\nu}_{\gamma}(D_{\lambda}\Pi^{\gamma}_{\delta})D_{\zeta}Y^{\delta}$ and $\Pi^{\lambda}_{\rho}\Pi^{\zeta}_{\sigma}\Pi^{\nu}_{\gamma}(D_{\lambda}\Pi^{\gamma}_{\delta}) = \frac{1}{2}\Pi^{\zeta}_{\sigma}(\chi^{\nu}_{\ \rho}e_{3\delta} + \underline{\chi}^{\nu}_{\ \rho}e_{4\delta})$, the result follows.

we obtain

$$\not{d}\mathrm{iv}\,\chi_a + \chi_{ab}\zeta_b = \nabla\!\!\!/_a \mathrm{tr}\chi + \zeta_a \mathrm{tr}\chi - \beta_a, \tag{3.8.4}$$

and in the same way, with $\underline{\chi}$ instead of χ,

$$\not{d}\mathrm{iv}\,\chi_a - \underline{\chi}_{ab}\zeta_b = \nabla\!\!\!/_a \mathrm{tr}\underline{\chi} - \zeta_a \mathrm{tr}\underline{\chi} + \underline{\beta}_a. \tag{3.8.5}$$

3.8.2 *Derivation of the structure equations*

We give an example of the derivation of the explicit expressions of the structure equations (3.1.50), (3.1.51).

We start with

$$\begin{aligned}\mathbf{R}^{\delta}_{\gamma\alpha\beta} &= \Omega^{\delta}_{\gamma}(e_\alpha, e_\beta) = (d\omega^{\delta}_{\gamma} + \omega^{\delta}_{\sigma} \wedge \omega^{\sigma}_{\gamma})(e_\alpha, e_\beta) \\ &= e_\alpha(\Gamma^{\delta}_{\beta\gamma}) - e_\beta(\Gamma^{\delta}_{\alpha\gamma}) + \Gamma^{\lambda}_{\beta\gamma}\Gamma^{\delta}_{\alpha\lambda} - \Gamma^{\lambda}_{\alpha\gamma}\Gamma^{\delta}_{\beta\lambda} - \omega^{\delta}_{\gamma}([e_\alpha, e_\beta]),\end{aligned} \tag{3.8.6}$$

and observe that[61]

$$\mathbf{R}_{\delta\gamma\alpha\beta} = <\mathbf{R}(e_\alpha, e_\beta)e_\gamma, \tilde{e_\delta}> = \#(\delta)\mathbf{R}^{\tilde{\delta}}_{\gamma\alpha\beta}, \tag{3.8.7}$$

where

$$\#(\delta) = \begin{cases} -2 & if\ \delta \in \{3,4\} \\ 1 & if\ \delta \in \{1,2\}. \end{cases} \tag{3.8.8}$$

Choosing $\{(\delta,\gamma) = (a,3),\ (\alpha,\beta) = (3,b)\}$ we obtain

$$\mathbf{R}_{a33b} = (\mathbf{D\!\!\!/}_3\underline{\chi})_{ba} - 2(\nabla\!\!\!/\underline{\xi})_{ba} + 2\underline{\omega}\underline{\chi}_{ba} + (\underline{\chi}_{\cdot c}\underline{\chi}_{\cdot c})_{ba} + 4\zeta_b\underline{\xi}_a - 2\eta_b\underline{\xi}_a - 2\underline{\eta}_a\underline{\xi}_b.$$

Decomposing this equation into its trace and traceless part, we obtain

$$\begin{aligned}&(\mathbf{D\!\!\!/}_3\hat{\underline{\chi}})_{ba} + 2\underline{\omega}\hat{\underline{\chi}}_{ba} - (\nabla\!\!\!/\widehat{\otimes}\underline{\xi})_{ba} + ((2\zeta - \underline{\eta} - \eta)\widehat{\otimes}\underline{\xi})_{ba} + \hat{\underline{\chi}}_{ba}\mathrm{tr}\underline{\chi} = \hat{\mathbf{R}}_{a33b} \\ &\mathbf{D}_3\mathrm{tr}\underline{\chi} + 2\underline{\omega}\mathrm{tr}\underline{\chi} + |\hat{\underline{\chi}}|^2 + \frac{1}{2}(\mathrm{tr}\underline{\chi})^2 - 2\not{d}\mathrm{iv}\underline{\xi} - 2\underline{\xi}\cdot(\eta + \underline{\eta} - 2\zeta) = \delta_{ba}\mathbf{R}_{a33b}.\end{aligned}$$

Now we consider the indices $(\delta,\gamma) = (a,3)$, $(\alpha,\beta) = (4,b)$ and obtain

$$\mathbf{R}_{a34b} = (\mathbf{D\!\!\!/}_4\underline{\chi})_{ba} - 2(\nabla\!\!\!/\underline{\eta})_{ba} + 2\omega\underline{\chi}_{ba} + (\chi_{\cdot c}\underline{\chi}_{c\cdot})_{ba} - 2\underline{\eta}_b\underline{\eta}_a - 2\xi_b\underline{\xi}_a.$$

Proceeding exactly as in the previous case, we decompose this equation into its trace and traceless part to obtain

$$\begin{aligned}&(\mathbf{D\!\!\!/}_4\hat{\underline{\chi}})_{ba} - 2\omega\hat{\underline{\chi}}_{ba} + (\widehat{\chi_{bc}\underline{\chi}_{ca}}) - (\nabla\!\!\!/\widehat{\otimes}\underline{\eta})_{ba} - (\underline{\eta}\widehat{\otimes}\underline{\eta})_{ba} - (\xi\widehat{\otimes}\underline{\xi})_{ba} = \mathcal{S}(\widehat{\mathbf{R}}_{a34b}) \\ &\mathbf{D}_4\mathrm{tr}\underline{\chi} - 2\omega\mathrm{tr}\underline{\chi} + \delta_{ba}(\chi_{bc}\underline{\chi}_{ca}) - 2\not{d}\mathrm{iv}\underline{\eta} - 2|\underline{\eta}|^2 - 2\xi\cdot\underline{\xi} = \delta_{ba}\mathcal{S}(\mathbf{R}_{a34b}).\end{aligned}$$

Choosing $(\delta,\gamma) = (a,3)$, $(\alpha,\beta) = (b,c)$ it follows that

$$(\nabla\!\!\!/_b\underline{\chi})_{ca} - (\nabla\!\!\!/_c\underline{\chi})_{ba} - (\zeta_b\underline{\chi}_{ca} - \zeta_c\underline{\chi}_{ba}) = \mathbf{R}_{a3bc},$$

---

[61] In fact $\mathbf{R}_{\tilde{\delta}\gamma\alpha\beta} = R_{\mu\nu\rho\sigma}e^{\nu}_{\gamma}e^{\rho}_{\alpha}e^{\sigma}_{\beta}e^{\mu}_{\tilde{\delta}} = R^{\mu'}_{\nu\rho\sigma}e^{\nu}_{\gamma}e^{\rho}_{\alpha}e^{\sigma}_{\beta}(g_{\mu'\mu}e^{\mu}_{\tilde{\delta}}) = \#(\delta)\mathbf{R}^{\tilde{\delta}}_{\gamma\alpha\beta}.$

and the trace and traceless part with respect to (a, c) are

$$(\nabla\!\!\!/_b \hat{\underline{\chi}})_{ca} - \widehat{(\nabla\!\!\!/_c \underline{\chi})}_{ba} - (\zeta_b \hat{\underline{\chi}}_{ca} - \widehat{\zeta_c \underline{\chi}_{ba}}) = \mathbf{R}_{a3bc}$$
$$\nabla\!\!\!/_b(\mathrm{tr}\underline{\chi}) - (\mathrm{d}\!\!\!/\mathrm{iv}\,\underline{\chi})_b + (\zeta \cdot \underline{\chi})_b - \zeta_b \mathrm{tr}\underline{\chi} = \delta_{ac}\mathbf{R}_{a3bc}\,.$$

Choosing $(\delta, \gamma) = (a, 3)$, $(\alpha, \beta) = (4, 3)$ it follows that

$$(\mathbf{D}\!\!\!\!/_4 \underline{\xi})_a - (\mathbf{D}\!\!\!\!/_3 \underline{\eta})_a + ((\eta - \underline{\eta}) \cdot \underline{\chi})_a - 4\omega \underline{\xi}_a = \frac{1}{2}\mathbf{R}_{a343}. \tag{3.8.9}$$

This result completes all the computations with $(\delta, \gamma) = (a, 3)$.

Most of the equations for $(\delta, \gamma) = (a, 4)$ are not independent from the equations for $(\delta, \gamma) = (a, 3)$. The independent equations can be obtained by swapping indices 3 and 4. Therefore without additional computations for $(\delta, \gamma) = (a, 4)$, $(\alpha, \beta) = (3, b)$, we obtain

$$\mathbf{R}_{a43b} = (\mathbf{D}\!\!\!\!/_3 \chi)_{ba} - 2(\nabla\!\!\!/ \eta)_{ba} + 2\omega \chi_{ba} + (\underline{\chi}_{\cdot c} \chi_{c\cdot})_{ba} - 2\eta_b \eta_a - 2\underline{\xi}_b \xi_a.$$

Proceeding exactly as in the case of $(\delta, \gamma) = (a, 3)$, $(\alpha, \beta) = (4, b)$, we decompose this equation into its trace and traceless part to obtain

$$(\mathbf{D}\!\!\!\!/_3 \hat{\chi})_{ba} - 2\underline{\omega} \hat{\chi}_{ba} + (\widehat{\underline{\chi}_{bc} \chi_{ca}}) - (\nabla\!\!\!/ \widehat{\otimes} \eta)_{ba} - (\eta \widehat{\otimes} \eta)_{ba} - (\underline{\xi} \widehat{\otimes} \xi)_{ba} = \mathcal{S}(\hat{\mathbf{R}}_{a43b})$$
$$\mathbf{D}_3 \mathrm{tr}\chi - 2\underline{\omega}\mathrm{tr}\chi + \delta_{ba}(\underline{\chi}_{bc}\chi_{ca}) - 2\mathrm{d}\!\!\!/\mathrm{iv}\,\eta - 2|\eta|^2 - 2\underline{\xi} \cdot \xi = \delta_{ba}\mathcal{S}(\mathbf{R}_{a43b}).$$

Now we set $(\delta, \gamma) = (a, 4)$, $(\alpha, \beta) = (4, b)$ and apply the substitutions described in (3.1.34) to get

$$(\mathbf{D}\!\!\!\!/_4 \hat{\chi})_{ba} + 2\omega \hat{\chi}_{ba} - (\nabla\!\!\!/ \widehat{\otimes} \xi)_{ba} - ((2\zeta + \eta + \underline{\eta}) \widehat{\otimes} \xi)_{ba} + \hat{\chi}_{ba}\mathrm{tr}\chi = \mathcal{S}(\hat{\mathbf{R}}_{a44b})$$
$$\mathbf{D}_4 \mathrm{tr}\chi + 2\omega \mathrm{tr}\chi + |\hat{\chi}|^2 + \frac{1}{2}(\mathrm{tr}\chi)^2 - 2\mathrm{d}\!\!\!/\mathrm{iv}\,\xi - 2\xi \cdot (\underline{\eta} + \eta + 2\zeta) = \delta_{ba}\mathcal{S}(\mathbf{R}_{a44b}).$$

As in the case of a double null integrable foliation the relations (3.1.33) hold. It is easy to obtain, with this procedure, the structure equations (3.1.46), (3.1.47), (3.1.48).

3.8.3 Some remarks on the definition of the adapted null frame

It is possible to choose the null frame in such a way that it is transported along $C(\lambda)$, once defined on a generic $S(\lambda, \nu)$, such that it remains null orthonormal, that is, $\mathbf{g}(e_a, e_b) = \delta_{ab}$ on the whole $C(\lambda)$. The analogous situation can be obtained by starting again from $S(\lambda, \nu)$ and extending the null orthonormal frame along $\underline{C}(\nu)$. In fact according to the equations

$$\begin{aligned}
\mathbf{D}_4 e_a &= \mathbf{D}\!\!\!\!/_4 e_a + (-\zeta_a + \nabla\!\!\!/_a \log\Omega)e_4 = \mathbf{D}\!\!\!\!/_4 e_a + \eta_a e_4 \\
\mathbf{D}_4 e_4 &= (\mathbf{D}_4 \log\Omega)e_4 \\
\mathbf{D}_4 e_3 &= -(\mathbf{D}_4 \log\Omega)e_3 + 2(-\zeta_b + \nabla\!\!\!/_b \log\Omega)e_b,
\end{aligned} \tag{3.8.10}$$

if we impose $\mathbf{D}\!\!\!\!/_4 e_a = 0$, then the null orthonormal frame $\{e_a, \hat{N}, \hat{\underline{N}}\}$ can be extended along $C(\lambda)$, once defined on a generic $S(\lambda, \nu)$, remaining null orthonormal, that is $\mathbf{g}(e_a, e_b) = \delta_{ab}$, on the whole $C(\lambda)$. From equations (3.1.45) this implies

$$\mathcal{L}_N e_a + \Omega \chi_{ac} e_c = 0. \tag{3.8.11}$$

Starting again from $S(\lambda, \nu)$ we can also extend the null orthonormal frame along $\underline{C}(\nu)$ using the equations[62]

$$\begin{aligned}\mathbf{D}_3 e_a &= \not{\mathbf{D}}_3 e_a + (\zeta_a + \nabla\!\!\!/_a \log \Omega) e_3 = \not{\mathbf{D}}_3 e_a + \eta_a e_3 \\ \mathbf{D}_3 e_3 &= (\mathbf{D}_3 \log \Omega) e_3 \\ \mathbf{D}_3 e_4 &= (\mathbf{D}_3 \log \Omega) e_4 + 2(\zeta_b + \nabla\!\!\!/_b \log \Omega) e_b,\end{aligned} \tag{3.8.12}$$

and, again, $g(e_a, e_b) = \delta_{ab}$ on the whole $\underline{C}(\nu)$ if we impose

$$\mathcal{L}_{\underline{N}} e_a + \Omega \underline{\chi}_{ac} e_c = 0. \tag{3.8.13}$$

3.8.4 Proof of Proposition 3.3.1

We restate the proposition in a slightly more general way.

Proposition 3.3.1: *Let $\triangle_p$ be the three-dimensional subspace of $T\mathcal{K}_p$ spanned by $TS_p \oplus \tilde{N}_p$, where $\tilde{N} \equiv (N - \underline{N})$. Let us consider the three-dimensional distribution on $\mathcal{K}$, $p \to \triangle_p$. This distribution is integrable. Moreover $\mathcal{K}$ is foliated by the three-dimensional spacelike hypersurfaces*

$$\tilde{\Sigma}_t \equiv \{p \in \mathcal{K} | t(p) = t\},$$

where $t(p) = \frac{1}{2}(u + \underline{u})$ and each two-dimensional surface $S(\lambda, \nu)$ is immersed in the hypersurface $\tilde{\Sigma}_t$ with $t = \frac{1}{2}(\lambda + \nu)$. The global time function $t(p)$ satisfies

$$dt = -\frac{1}{4\Omega^2}(n + \underline{n}) \ , \ \frac{\partial}{\partial t} = (N + \underline{N}),$$

where $n, \underline{n}$ are the 1-forms corresponding to $N, \underline{N}$. Finally, given the hypersurfaces $\tilde{\Sigma}_t$, their second fundamental form k has the following expression in terms of the connection coefficients,

$$k_{\tilde{N}\tilde{N}} = \omega + \underline{\omega} \ , \ k_{e_a \tilde{N}} = \zeta_a \ , \ k_{e_a e_b} = -\frac{1}{2}(\chi_{ab} + \underline{\chi}_{ab}).$$

Proof: We observe that

$$\begin{aligned}[\tilde{N}, e_a] &= ([N, e_a] - [\underline{N}, e_a]) = (\not{\mathbf{D}}_4 e_a - \not{\mathbf{D}}_3 e_a) - (\chi_{ab} - \underline{\chi}_{ab}) e_b \\ [e_a, e_b] &= \nabla\!\!\!/_a e_b - \nabla\!\!\!/_b e_a.\end{aligned} \tag{3.8.14}$$

Therefore, as a result of Frobenius' theorem (see [Sp] Vol. I, Chapter 6) the distribution $p \to \triangle_p$ is locally integrable. This implies that for a fixed generic point $p \in \mathcal{K}$, there is a neighborhood U of p such that, given $q \in U$ it is possible to define a submanifold $\tilde{\Sigma} \subset \mathcal{K}$, containing q, whose tangent space is, at each point $p' \in \tilde{\Sigma}$, $\triangle_{p'}$.

[62] The null frames one builds extending the orthogonal vectors $\{e_a\}$ along the null hypersurfaces $\{C(\lambda)\}$ or $\{\underline{C}(\nu)\}$ are different and will be used in different situations.

Let $t(p)$ be the function whose $\tilde{\Sigma}$ is a level surface. In the neighborhood U we have

$$dt(\cdot) = (dt)_\mu dx^\mu(\cdot) = \alpha(N + \underline{N})_\mu dx^\mu(\cdot),$$

where α is a regular scalar function on U.[63] The result becomes valid in the whole $\mathcal{K}$ if we choose $\alpha = -\frac{1}{4\Omega^2}$. In fact, recalling (3.1.38),

$$-\frac{1}{4\Omega^2}(N + \underline{N})_\mu = -\frac{1}{2}(L + \underline{L})_\mu = \frac{1}{2}\partial_\mu(u + \underline{u}) = \partial_\mu t, \tag{3.8.15}$$

so that

$$dt(\cdot) = \frac{\partial t}{\partial x^\mu} dx^\mu(\cdot) = \frac{1}{2}\frac{\partial(u + \underline{u})}{\partial x^\mu} dx^\mu(\cdot)\,.$$

Defining

$$\tilde{\Sigma}_t \equiv \{p \in \mathcal{K} | t(p) = t\}, \tag{3.8.16}$$

we can choose t as a coordinate for $\mathcal{K}$ and define the vector field $\frac{\partial}{\partial t}$ as the vector field satisfying $dt(\frac{\partial}{\partial t}) = 1$. From

$$g^{\mu\nu}\frac{\partial t}{\partial x^\mu}\frac{\partial t}{\partial x^\nu} = -\frac{1}{4\Omega^2},$$

it follows that

$$\left(\frac{\partial}{\partial t}\right)^\mu = (N + \underline{N})^\mu. \tag{3.8.17}$$

This proves that a global time t and a global foliation using the spacelike hypersurfaces $\{\tilde{\Sigma}_t\}$ exists,[64] whereas from the Frobenius theorem, the result will be only local and not unique. The expression of the second fundamental form k in terms of the connection coefficients follows from a direct computation.

Remark: Observe that the spacetime foliation relative to the canonical double null foliation is not the one used in [Ch-Kl] because the $\tilde{\Sigma}_t$ hypersurfaces are not maximal. In fact in [Ch-Kl], p. 268, the condition tr$k = 0$ is written as[65]

$$\delta = -\mathrm{tr}\eta_{(C.K.)} = -\sum_{a=1}^{2} k_{aa}\,.$$

Observing that in the present notation,

$$\sum_{a=1}^{2} k_{aa} = -(\mathrm{tr}\chi + \mathrm{tr}\underline{\chi}) \quad \text{and} \quad \delta = \omega + \underline{\omega}\,,$$

[63] In principle one can define $\tilde{N} \equiv \frac{1}{2}(\hat{N} - \hat{\underline{N}})$ and given the vector field $T = \frac{1}{2}(\hat{N} + \hat{\underline{N}})$ build locally a time function t' just by considering the flow of T. This is a local result and one does not know apriori if it holds globally.

[64] Asymptotically, $\underline{u}$ is basically $r_* = r + m\log(r/2m-1)$ is the coordinate used in the Schwarzschild spacetime.

[65] $\eta_{(C.K.)}$ is not the η connection coefficient we use here.

the maximality condition becomes

$$\omega + \underline{\omega} = \frac{1}{2}(\mathrm{tr}\chi + \mathrm{tr}\underline{\chi}). \tag{3.8.18}$$

Equation (3.8.18) is not satisfied in the present approach. In fact it cannot be imposed in our foliation because Ω is already completely determined by the structure equations, the initial conditions on the Σ_0 hypersurface and those on the last slice $\underline{C}_*$.

4

Estimates for the Connection Coefficients

4.1 Preliminary results

4.1.1 *Elliptic estimates for Hodge systems*

We consider Hodge systems of equations defined on a compact two-dimensional Riemann surface. We recall Definition 3.1.4 of Chapter 3.

Definition 3.1.4 *Given the 1-form ξ on S we define its Hodge dual*[1]

$$ {}^\star\xi_a = \in_{ab} \xi^b . $$

Clearly ${}^\star({}^\star\xi) = -\xi$. *If ξ is a symmetric traceless 2-tensor, we define the following left, ${}^\star\xi$, and right, $\xi^\star$, Hodge duals*

$$ {}^\star\xi_{ab} = \in_{ac} \xi^c{}_b\, ,\; \xi^\star_{ab} = \xi_a{}^c \in_{cb} . $$

Observe that the tensors ${}^\star\xi$, $\xi^\star$ are also symmetric traceless and satisfy

$$ {}^\star\xi = -\xi^\star\, ,\; {}^\star({}^\star\xi) = -\xi. $$

We will need estimates for the following elliptic systems of equations.

$\mathbf{H}_0$: *Hodge system of type 0*

This refers to the scalar Poisson equation

$$ \Delta\phi = f, $$

where f is an arbitrary scalar function on S and Δ is the Laplacian relative to the induced metric on S.

---

[1]Here a, b are just coordinate indices.

$\mathbf{H}_1$: *Hodge system of type 1*

This concerns 1-forms ξ satisfying

$$\begin{aligned} \not{d}\mathrm{iv}\,\xi &= f \\ \mathrm{c}\not{u}\mathrm{rl}\,\xi &= f_*, \end{aligned}$$

where $f,\ f_*$ are given scalar functions on S and the operators $\not{d}\mathrm{iv}$, $\mathrm{c}\not{u}\mathrm{rl}$ are defined as

$$\not{d}\mathrm{iv}\,\xi = \not{\nabla}^a \xi_a \ ; \ \mathrm{c}\not{u}\mathrm{rl}\,\xi = \in^{ab} \not{\nabla}_a \xi_b .$$

$\mathbf{H}_2$: *Hodge system of type 2*

This concerns traceless symmetric 2-forms ξ satisfying

$$\not{d}\mathrm{iv}\,\xi = f,$$

where f is a given vector field and $\not{d}\mathrm{iv}\,\xi$ is defined by

$$\not{d}\mathrm{iv}\,\xi_a = \not{\nabla}^b \xi_{ab}.$$

For these three systems of equations we have the following $L^2(S)$ estimates, see [Ch-Kl], Chapter 2.

Proposition 4.1.1 *On an arbitrary compact Riemannian manifold (S,γ), with K the Gauss curvature of S,*

i) If ϕ is a solution of $\mathbf{H}_0$ the following estimate holds

$$\int_S \{|\not{\nabla}^2\phi|^2 + K|\not{\nabla}\phi|^2\} = \int_S |f|^2;$$

ii) If the vector field ξ is a solution of $\mathbf{H}_1$ then

$$\int_S \{|\not{\nabla}\xi|^2 + K|\xi|^2\} = \int_S \{|f|^2 + |f_*|^2\};$$

iii) If the symmetric traceless 2-tensor ξ is a solution of $\mathbf{H}_2$ then

$$\int_S \{|\not{\nabla}\xi|^2 + 2K|\xi|^2\} = 2\int_S |f|^2.$$

Definition 4.1.1 *Assume (S,γ) is an arbitrary compact Riemannian manifold with K its Gauss curvature. We introduce the following quantities:*[2]

$$k_m = \min_S r^2 K \ , \ k_M = \max_S r^2 K \ , \ k_1 \equiv \left(\int_S |\not{\nabla} K|^2\right)^{\frac{1}{2}} .$$

[2] r has been defined in (3.1.2).

Proposition 4.1.2 *Assume $k_m > 0$. If ξ is a solution of $\mathbf{H}_1$ or $\mathbf{H}_2$, then the following inequalities hold:*

$$\int_S \{|\nabla\xi|^2 + r^{-2}|\xi|^2\} = c_1 \int_S |f|^2$$

$$\int_S |\nabla^2\xi|^2 \;\leq\; c_2 \int_S \left(|\nabla f|^2 + r^{-2}|f|^2\right),$$

where c_1, c_2 are two constants depending on k_m, k_M. Assume, moreover, that k_1 is finite. Then there exists a constant c_3, depending on k_m, k_M and k_1, such that

$$\int_S |\nabla^3\xi|^2 \;\leq\; c \int_S \left(|\nabla^2 f|^2 + r^{-2}|\nabla f|^2 + r^{-4}|f|^2\right),$$

where, in the $\mathbf{H}_1$ case, $f = (f, f_*)$ and $|f|^2 = |f|^2 + |f_*|^2$.

We will also need some L^p estimates for the above systems, which we recall from [Ch-Kl], Chapter 2.

Proposition 4.1.3 *Assume that S is an arbitrary compact 2-surface satisfying $k_m > 0$ and $k_M < \infty$. Then the following statements hold.*

i) Let ϕ be a solution to the Poisson equation $\mathbf{H_0}$ on S. There exists a constant c that depends only on ${k_m}^{-1}$, k_M, p such that

$$|\nabla^2\phi|_{L^p} + r^{-1}|\nabla\phi|_{L^p} + r^{-2}|\phi - \bar{\phi}|_{L^p} \leq c|f|_{L^p}$$
$$|\nabla^3\phi|_{L^p} \leq c_p\left(|\nabla f|_{L^p} + r^{-1}|f|_{L^p}\right).$$

ii) Let ξ be a solution of either $\mathbf{H}_1$ or $\mathbf{H}_2$. Then we have:

First-derivative estimates in L^p. There exists a constant c that depends only on ${k_m}^{-1}$, k_M and p such that, for all $2 \leq p < \infty$,

$$\int_S (|\nabla\xi|^p + r^{-p}|\xi|^p) \leq c \int_S |f|^p.$$

Second-derivative estimates in L^p. There exists a constant c that depends only on ${k_m}^{-1}$, k_M and p such that, for all $2 \leq p < \infty$,

$$\int_S |\nabla^2\xi|^p \leq c \int_S (|\nabla f|^p + r^{-p}|f|^p).$$

4.1.2 Global Sobolev inequalities

In this subsection we assume that the spacetime $\mathcal{K}$ has a double null foliation[3] and that the following assumptions hold

1. $$\sup_{\mathcal{K}} |\mathrm{tr}\chi - \frac{2}{r}| \leq \delta\ ,\ \sup_{\mathcal{K}} |\mathrm{tr}\underline{\chi} + \frac{2}{r}| \leq \delta \quad \text{with } \delta \text{ small;}$$

[3]Note that in the subsequent applications the spacetime $\mathcal{K}$ is foliated by two double null foliations, the *double null canonical foliation* and the *initial layer foliation* in the layer region near Σ_0. In that case u_0 and $\underline{u}_0$ have different expressions; see discussion in Subsection 4.1.3.

2. $k_m > 0$ on any surface $S(u,\underline{u}) = C(u) \cap \underline{C}(\underline{u})$.

Moreover we use the notation

$$V(u,\underline{u}) = J^{-}(S(u,\underline{u}))\ ,\ u_0 = u|_{\underline{C}(\underline{u})\cap\Sigma_0}\ ,\ \underline{u}_0 = \underline{u}|_{C(u)\cap\Sigma_0}\ .$$

Proposition 4.1.4 *Let F be a smooth S-tangent tensor field.* [4] *The following nondegenerate version of the global Sobolev inequality along $C(u)$ holds true:*

$$\begin{aligned}\sup_{S(u,\underline{u})} (r^{\frac{3}{2}}|F|) \ \leq\ & c\Bigg[\Big(\int_{S(u,\underline{u}_0)} r^4|F|^4\Big)^{\frac{1}{4}} + \Big(\int_{S(u,\underline{u}_0)} r^4|r\nabla\!\!\!/ F|^4\Big)^{\frac{1}{4}}\\ &+\Bigg(\int_{C(u)\cap V(u,\underline{u})} |F|^2 + r^2|\nabla\!\!\!/ F|^2 + r^2|\mathbf{D}\!\!\!\!/_4 F|^2\\ &+r^4|\nabla\!\!\!/^2 F|^2 + r^4|\nabla\!\!\!/\mathbf{D}\!\!\!\!/_4 F|^2\Bigg)^{\frac{1}{2}}\ \Bigg].\end{aligned} \tag{4.1.1}$$

We also have the degenerate version:

$$\begin{aligned}\sup_{S(u,\underline{u})} (r\tau_-^{\frac{1}{2}}|F|) \ \leq\ & c\Bigg[\Big(\int_{S(u,\underline{u}_0)} r^2\tau_-^2|F|^4\Big)^{\frac{1}{4}} + \Big(\int_{S(u,\underline{u}_0)} r^2\tau_-^2|r\nabla\!\!\!/ F|^4\Big)^{\frac{1}{4}}\\ &+\Bigg(\int_{C(u)\cap V(u,\underline{u})} (|F|^2 + r^2|\nabla\!\!\!/ F|^2 + \tau_-^2|\mathbf{D}\!\!\!\!/_4 F|^2\\ &+r^4|\nabla\!\!\!/^2 F|^2 + r^2\tau_-^2|\nabla\!\!\!/\mathbf{D}\!\!\!\!/_4 F|^2\Bigg)^{\frac{1}{2}}\ \Bigg].\end{aligned} \tag{4.1.2}$$

Analogous estimates are obtained along the null-incoming hypersurfaces $\underline{C}(\underline{u})$:

$$\begin{aligned}\sup_{S(u,\underline{u})} (r^{\frac{3}{2}}|F|) \ \leq\ & c\Bigg[\Big(\int_{S(u_0,\underline{u})} r^4|F|^4\Big)^{\frac{1}{4}} + \Big(\int_{S(u_0,\underline{u})} r^4|r\nabla\!\!\!/ F|^4\Big)^{\frac{1}{4}}\\ &+\Bigg(\int_{\underline{C}(\underline{u})\cap V(u,\underline{u})} |F|^2 + r^2|\nabla\!\!\!/ F|^2 + r^2|\mathbf{D}\!\!\!\!/_3 F|^2\\ &+r^4|\nabla\!\!\!/^2 F|^2 + r^4|\nabla\!\!\!/\mathbf{D}\!\!\!\!/_3 F|^2\Bigg)^{\frac{1}{2}}\ \Bigg].\end{aligned} \tag{4.1.3}$$

and

$$\begin{aligned}\sup_{S(u,\underline{u})} (r\tau_-^{\frac{1}{2}}|F|) \ \leq\ & c\Bigg[\Big(\int_{S(u_0,\underline{u})} r^2\tau_-^2|F|^4\Big)^{\frac{1}{4}} + \Big(\int_{S(u_0,\underline{u})} r^2\tau_-^2|r\nabla\!\!\!/ F|^4\Big)^{\frac{1}{4}}\\ &+\Bigg(\int_{\underline{C}(\underline{u})\cap V(u,\underline{u})} |F|^2 + r^2|\nabla\!\!\!/ F|^2 + \tau_-^2|\mathbf{D}\!\!\!\!/_3 F|^2\\ &+r^4|\nabla\!\!\!/^2 F|^2 + r^2\tau_-^2|\nabla\!\!\!/\mathbf{D}\!\!\!\!/_3 F|^2\Bigg)^{\frac{1}{2}}\ \Bigg].\end{aligned} \tag{4.1.4}$$

---

[4] This means that at any point it is tangent to the 2-surface $S(u,\underline{u})$ passing through that point.

Proof: The proof of inequalities (4.1.1), (4.1.2) is based on the following lemma.

Lemma 4.1.1 *Let F be a smooth tensor field on $\mathcal{K}$, tangent to the two-dimensional surfaces $S(u,\underline{u})$ at every point. We introduce the quantities:*

$$
\begin{aligned}
A(F) &\equiv \sup_{C(u)\cap V(u,\underline{u})} \left(\int_{S(u,\underline{u})} r^4|F|^4\right)^{\frac{1}{4}}\\
B(F) &\equiv \left(\int_{C(u)\cap V(u,\underline{u})} r^6|F|^6\right)^{\frac{1}{6}}\\
E(F) &\equiv \left(\int_{C(u)\cap V(u,\underline{u})} |F|^2+r^2|\nabla\!\!\!/ F|^2+r^2|\mathbf{D}\!\!\!\!/_4 F|^2\right)^{\frac{1}{2}}
\end{aligned}
\tag{4.1.5}
$$

and

$$
\begin{aligned}
A_*(F) &\equiv \sup_{C(u)\cap V(u,\underline{u})} \left(\int_{S(u,\underline{u})} r^2\tau_-^2|F|^4\right)^{\frac{1}{4}}\\
B_*(F) &\equiv \left(\int_{C(u)\cap V(u,\underline{u})} r^4\tau_-^2|F|^6\right)^{\frac{1}{6}}\\
E_*(F) &\equiv \left(\int_{C(u)\cap V(u,\underline{u})} |F|^2+r^2|\nabla\!\!\!/ F|^2+\tau_-^2|\mathbf{D}\!\!\!\!/_4 F|^2\right)^{\frac{1}{2}}.
\end{aligned}
\tag{4.1.6}
$$

The following nondegenerate inequalities hold:

$$B \leq c(I)A^{2/3}E^{1/3} \tag{4.1.7}$$

$$A \leq A_0 + c(I)B^{3/4}E^{1/4}. \tag{4.1.8}$$

We also have the degenerate estimates

$$B_* \leq c(I)A_*^{2/3}E_*^{1/3} \tag{4.1.9}$$

$$A_* \leq A_{*0} + c(I)B_*^{3/4}E_*^{1/4}, \tag{4.1.10}$$

where

$$A_0 = \left(\int_{S(u,\underline{u}_0)} r^4|F|^4\right)^{\frac{1}{4}}, \quad A_{*0} = \left(\int_{S(u,\underline{u}_0)} r^2\tau_-^2|F|^4\right)^{\frac{1}{4}},$$

and $c(I)$ is a constant depending on $I=\sup_{C(u)} I(u,\underline{u})$, where $I(u,\underline{u})$ is the isoperimetric constant of $S(u,\underline{u})$.

Proof: The proof of inequalities 4.1.3, 4.1.4 is based on the analogue of Lemma 4.1.1.

Lemma 4.1.2 *Let G be a smooth tensor field on $\mathcal{K}$ tangent to the two-dimensional surfaces $S(u,\underline{u})$ at every point. We introduce the quantities*

$$\begin{aligned}
\underline{A}(F) &\equiv \sup_{\underline{C}(\underline{u})\cap V(u,\underline{u})} \left(\int_{S(u,\underline{u})} r^4|F|^4\right)^{\frac{1}{4}} \\
\underline{B}(F) &\equiv \left(\int_{\underline{C}(\underline{u})\cap V(u,\underline{u})} r^6|F|^6\right)^{\frac{1}{6}} \\
\underline{E}(F) &\equiv \left(\int_{\underline{C}(\underline{u})\cap V(u,\underline{u})} |F|^2 + r^2|\nabla\!\!\!/ F|^2 + r^2|\mathbf{D}\!\!\!\!/_3 F|^2\right)^{\frac{1}{2}}
\end{aligned} \tag{4.1.11}$$

and

$$\begin{aligned}
\underline{A}_{de.}(F) &\equiv \sup_{\underline{C}(\underline{u})\cap V(u,\underline{u})} \left(\int_{S(u,\underline{u})} r^2\tau_-^2|F|^4\right)^{\frac{1}{4}} \\
\underline{B}_{de.}(F) &\equiv \left(\int_{\underline{C}(\underline{u})\cap V(u,\underline{u})} r^4\tau_-^2|F|^6\right)^{\frac{1}{6}} \\
\underline{E}_{de.}(F) &\equiv \left(\int_{\underline{C}(\nu)\cap V(u,\underline{u})} |F|^2 + r^2|\nabla\!\!\!/ F|^2 + \tau_-^2|\mathbf{D}\!\!\!\!/_3 F|^2\right)^{\frac{1}{2}}.
\end{aligned} \tag{4.1.12}$$

Then the following inequalities hold:

$$\underline{B} \le c(I)\underline{A}^{2/3}\underline{E}^{1/3} \tag{4.1.13}$$

$$\underline{A} \le \underline{A}_0 + c(I)\underline{B}^{3/4}\underline{E}^{1/4} \tag{4.1.14}$$

and

$$\underline{B}_{de.} \le c(I)\underline{A}_{de.}^{2/3}\underline{E}_{de.}^{1/3} \tag{4.1.15}$$

$$\underline{A}_{de.} \le \underline{A}_{de.0} + c(I)\underline{B}_{de.}^{3/4}\underline{E}_{de.}^{1/4} \tag{4.1.16}$$

where

$$\underline{A}_0 = \left(\int_{S(u_0,\underline{u})} r^4|F|^4\right)^{\frac{1}{4}}, \quad \underline{A}_{de.0} = \left(\int_{S(u_0,\underline{u})} r^2\tau_-^2|F|^4\right)^{\frac{1}{4}}. \tag{4.1.17}$$

Corollary 4.1.1 *Under the assumptions of Lemma 4.1.1 and Lemma 4.1.2 the following estimates hold*

$$\begin{aligned}
\left(\int_{S(u,\underline{u})} r^4|F|^4\right)^{\frac{1}{4}} \le& \left(\int_{S(u,\underline{u}_0)} r^4|F|^4\right)^{\frac{1}{4}} \\
&+ c\left(\int_{C(u)\cap V(u,\underline{u})} |F|^2 + r^2|\nabla\!\!\!/ F|^2 + r^2|\mathbf{D}\!\!\!\!/_4 F|^2\right)^{\frac{1}{2}},
\end{aligned}$$

$$\left(\int_{S(u,\underline{u})} r^2\tau_-^2|F|^4\right)^{\frac{1}{4}} \leq \left(\int_{S(u,\underline{u}_0)} r^2\tau_-^2|F|^4\right)^{\frac{1}{4}} + c\left(\int_{C(u)\cap V(u,\underline{u})} |F|^2 + r^2|\nabla\!\!\!/ F|^2 + \tau_-^2|\mathbf{D}\!\!\!/_4 F|^2\right)^{\frac{1}{2}}. \tag{4.1.18}$$

Also,

$$\left(\int_{S(u,\underline{u})} r^4|F|^4\right)^{\frac{1}{4}} \leq \left(\int_{S(u_0,\underline{u})} r^4|F|^4\right)^{\frac{1}{4}} + c\left(\int_{\underline{C}(\underline{u})\cap V(u,\underline{u})} |F|^2 + r^2|\nabla\!\!\!/ F|^2 + r^2|\mathbf{D}\!\!\!/_3 F|^2\right)^{\frac{1}{2}}$$

$$\left(\int_{S(u,\underline{u})} r^2\tau_-^2|F|^4\right)^{\frac{1}{4}} \leq \left(\int_{S(u_0,\underline{u})} r^2\tau_-^2|F|^4\right)^{\frac{1}{4}} + c\left(\int_{\underline{C}(\underline{u})\cap V(u,\underline{u})} |F|^2 + r^2|\nabla\!\!\!/ F|^2 + \tau_-^2|\mathbf{D}\!\!\!/_3 F|^2\right)^{\frac{1}{2}}. \tag{4.1.19}$$

The proof of Proposition 4.1.4 follows immediately from this corollary combined with the following form of the standard Sobolev inequalities for the sphere.

Lemma 4.1.3 *Let G be a tensor field tangent to the spheres $S(u,\underline{u})$. Then*

$$\sup_{S(u,\underline{u})} |G| \leq cr^{-\frac{1}{2}}\left(\int_{S(u,\underline{u})} |G|^4 + r^4|\nabla\!\!\!/ G|^4\right)^{\frac{1}{4}}. \tag{4.1.20}$$

Indeed, it suffices to apply this lemma to $G = rF$ or $G = r^{\frac{1}{2}}\tau_-^{\frac{1}{2}}F$ and then to take Lemma 4.1.1 into account. We now present the main steps in the proof of the nondegenerate version of Lemma 4.1.1.

To prove (4.1.7) we recall the following version of the isoperimetric inequality,(see [Cha]) for a compact two-dimensional surface S of strictly positive Gauss curvature:

$$\int_S (\Phi - \bar{\Phi})^2 \leq I(S)\left(\int_S |\nabla\!\!\!/ \Phi|\right)^2, \tag{4.1.21}$$

where Φ is a scalar function on a sphere S in $\mathcal{K}$ and the isoperimetric constant $I(S)$ can be bounded by a constant that depends only on k_M.[5] Applying (4.1.21) to the surfaces $S(u,\underline{u}) \subset C(u) \cap V(u,\underline{u})$ with $\Phi = |F|^3$ and using the Hölder inequality we derive

$$\int_{S(u,\underline{u})} |F|^6 \leq c\left(r^{-2}\int_{S(u,\underline{u})} |F|^4\right)\left(\int_{S(u,\underline{u})} |F|^2 + r^2|\nabla\!\!\!/ F|^2\right). \tag{4.1.22}$$

---

[5] $I(S)^{-\frac{1}{2}} = \inf_\Gamma \left(L(\Gamma)/\min\{A(D_1), A(D_2)\}^{\frac{1}{2}}\right)$ where Γ is an arbitrary closed curve on $S(u,\underline{u})$, $L(\Gamma)$ its total length and $A(D_1)$, $A(D_2)$ the areas of the two components of S/Γ.

Multiplying equation (4.1.22) by r^6 and integrating with respect to $\underline{u}$ we easily derive (4.1.7). To obtain (4.1.8) we express, with the help of the divergence theorem and assuming that everywhere $\mathrm{tr}\chi$ is near to $\frac{2}{r}$, the integral $\int_{S(u,\underline{u})} r^4|F|^4$ in terms of an integral over $C(u) \cap V(u, \underline{u})$ and an integral over $S(u, \underline{u}_0)$. Applying also the Cauchy–Schwartz inequality gives

$$\int_{S(u,\underline{u})} r^4|F|^4 \leq \int_{S(u,\underline{u}_0)} r^4|F|^4 + c\left(\int_{C(u)\cap V(u,\underline{u})} r^6|F|^6\right)^{\frac{1}{2}}\left(\int_{C(u)\cap V(u,\underline{u})} r^2|\mathbf{D}\!\!\!/_4 F|^2\right)^{\frac{1}{2}},$$

which proves (4.1.8).

To prove the degenerate estimates (4.1.9), (4.1.10) of Lemma 4.1.1 we proceed precisely in the same way with the quantities A_*, B_* and E_*. In this case the inequality (4.1.15) follows by multiplying (4.1.22) by $r^4\tau_-^2$ and integrating in $\underline{u}$. The corresponding inequality (4.1.10) follows, as in the nondegenerate case, by applying the divergence theorem to $\int_{S(u,\underline{u})} r^2\tau_-^2|F|^4$.

We conclude this subsection by recalling the Gronwall inequality (see [Ho]) and the Evolution Lemma, which will be used, repeatedly in the following sections.

Lemma 4.1.4 (Gronwall inequality) *Let $f, g : [a, b) \to R$ be continuous and nonnegative. Assume*

$$f(t) \leq A + \int_a^t f(s)g(s)ds\ , \quad A \geq 0.$$

Then

$$f(t) \leq A \exp \int_a^t g(s)ds\ , \quad \textit{for } t \in [a, b).$$

Lemma 4.1.5 (Evolution Lemma) *Consider the spacetime $\mathcal{K}$ foliated by a double null foliation.*

i) Assume that for $\delta > 0$ sufficiently small,

$$|\Omega\mathrm{tr}\chi - \overline{\Omega\mathrm{tr}\chi}| \leq \delta r^{-2}. \tag{4.1.23}$$

Let U, F be k-covariant S-tangent tensor fields satisfying the outgoing evolution equation

$$\frac{dU_{a_1\ldots a_k}}{d\underline{u}} + \lambda_0\Omega\mathrm{tr}\chi\, U_{a_1\ldots a_k} = F_{a_1\ldots a_k}, \tag{4.1.24}$$

with λ_0 a nonnegative real number and

$$U_{a_1\ldots a_k} \equiv U(e_{a_1}, e_{a_2}, \ldots, e_{a_k})\ ,\ F_{a_1\ldots a_k} \equiv F(e_{a_1}, e_{a_2}, \ldots, e_{a_k}),$$

the components relative to an arbitrary orthonormal frame on S. Denoting $\lambda_1 = 2(\lambda_0 - \frac{1}{p})$, we have along $C(u)$

$$|r^{\lambda_1}U|_{p,S}(u, \underline{u}) \leq c_0\left(|r^{\lambda_1}U|_{p,S}(u, \underline{u}_*) + \int_{\underline{u}}^{\underline{u}_*} |r^{\lambda_1}F|_{p,S}(u, \underline{u}')d\underline{u}'\right). \tag{4.1.25}$$

Here $\underline{u}_*$ is the value that the function $\underline{u}(p)$ assumes on $\underline{C}_*$.

ii) Assume that for $\delta > 0$ sufficiently small,

$$|\Omega\mathrm{tr}\underline{\chi} - \overline{\Omega\mathrm{tr}\underline{\chi}}| \leq \delta r^{-1}\tau_-^{-1}. \tag{4.1.26}$$

Let V, $\underline{F}$ be k-covariant S-tangent tensor fields satisfying the incoming evolution equation

$$\frac{dV_{a_1\ldots a_k}}{du} + \lambda_0\Omega\mathrm{tr}\underline{\chi}V_{a_1\ldots a_k} = \underline{F}_{a_1\ldots a_k}. \tag{4.1.27}$$

Denoting $\lambda_1 = 2(\lambda_0 - \frac{1}{p})$, we have along $\underline{C}(\underline{u})$

$$|r^{\lambda_1}V|_{p,S}(u,\underline{u}) \leq c_0\left(|r^{\lambda_1}V|_{p,S(u_0(\underline{u}),\underline{u})} + \int_{u_0(\underline{u})}^{u}|r^{\lambda_1}\underline{F}|_{p,S}(u',\underline{u})du'\right) \tag{4.1.28}$$

where $S(u_0(\underline{u}),\underline{u}) \equiv C(u_0(\underline{u})) \cap \underline{C}(\underline{u}) \subset \mathcal{K}$.

Remark: *Here $u_0(\underline{u}) \neq u|_{\underline{C}(\underline{u})\cap\Sigma_0}$. When we apply part ii) of the Lemma we will choose the two-dimensional surface $S(u_0(\underline{u}),\underline{u})$ in a convenient way.*[6]

Proof: From Lemma 3.1.3 we have, for any scalar function f,

$$\frac{d}{d\underline{u}}\int_{S(u,\underline{u})} f d\mu_\gamma = \int_{S(u,\underline{u})}\left(\frac{df}{d\underline{u}} + \Omega\mathrm{tr}\chi f\right)d\mu_\gamma. \tag{4.1.29}$$

In particular, setting $f = 1$ and denoting $|S(u,\underline{u})|$ the area of $S(u,\underline{u})$ and $\overline{h}$ the average of h over $S(u,\underline{u})$, we obtain

$$\frac{d}{d\underline{u}}|S(u,\underline{u})| = |S(u,\underline{u})|\overline{\Omega\mathrm{tr}\chi}.$$

From Definition 3.1.2 and $r(u,\underline{u}) \equiv \sqrt{\frac{1}{4\pi}|S(u,\underline{u})|}$,

$$\frac{d}{d\underline{u}}r(u,\underline{u}) = \frac{r(u,\underline{u})}{2}\overline{\Omega\mathrm{tr}\chi}. \tag{4.1.30}$$

Hence, for any function f and any real number λ,

$$\begin{aligned}\frac{d}{d\underline{u}}\left(\int_{S(u,\underline{u})} r^\lambda f d\mu_\gamma\right) &= \int_{S(u,\underline{u})} r^\lambda\left(\frac{df}{d\underline{u}} + (1+\frac{\lambda}{2})\Omega\mathrm{tr}\chi f\right)\\ &- \frac{\lambda}{2}\int_{S(u,\underline{u})} r^\lambda f(\Omega\mathrm{tr}\chi - \overline{\Omega\mathrm{tr}\chi})\end{aligned} \tag{4.1.31}$$

The equation satisfied by the tensor field U implies

$$\begin{aligned}&\frac{d}{d\underline{u}}|U|^p + \lambda_0 p\Omega\mathrm{tr}\chi|U|^p \leq p|F||U|^{p-1},\\ &\frac{d}{d\underline{u}}|U|^p + \lambda_0 p\Omega\mathrm{tr}\chi|U|^p \geq -p|F||U|^{p-1}.\end{aligned} \tag{4.1.32}$$

[6] We will choose it such that it allows us to connect the norm on $S(u_0(\underline{u}),\underline{u})$ with the initial norms on $S_{(0)}(\underline{u}) \subset \Sigma_0$.

Therefore setting $f = |U|^p$ in equation (4.1.31) and $\lambda = \lambda_1 p$ where

$$\lambda_1 = \left(2\lambda_0 - \frac{2}{p}\right)$$

we obtain

$$\begin{aligned}\frac{d}{d\underline{u}}\int_{S(u,\underline{u})} r^{\lambda_1 p}|U|^p d\mu_\gamma &= \int_{S(u,\underline{u})} r^{\lambda_1 p}\left(\frac{d}{d\underline{u}}|U|^p + \lambda_0 p\Omega\mathrm{tr}\chi|U|^p\right)d\mu_\gamma \\ &- \frac{\lambda_1 p}{2}\int_{S(u,\underline{u})} r^{\lambda_1 p}|U|^p(\Omega\mathrm{tr}\chi - \overline{\Omega\mathrm{tr}\chi})d\mu_\gamma .\end{aligned}$$

Using the second inequality of (4.1.32) we find that

$$\begin{aligned}\frac{d}{d\underline{u}}\int_{S(u,\underline{u})} r^{\lambda_1 p}|U|^p d\mu_\gamma &\geq -p\int_{S(u,\underline{u})} r^{\lambda_1 p}|F||U|^{p-1}d\mu_\gamma \\ &-\frac{\lambda_1 p}{2}\int_{S(u,\underline{u})} r^{\lambda_1 p}|U|^p(\Omega\mathrm{tr}\chi - \overline{\Omega\mathrm{tr}\chi})d\mu_\gamma .\end{aligned} \tag{4.1.33}$$

Applying the Hölder inequality

$$\int_{S(u,\underline{u})} r^{\lambda_1 p}|F||U|^{p-1} \leq \left(\int_{S(u,\underline{u})} r^{\lambda_1 p}|F|^p\right)^{\frac{1}{p}}\left(\int_{S(u,\underline{u})} r^{\lambda_1 p}|U|^p\right)^{\frac{p-1}{p}},$$

we obtain

$$\begin{aligned}-\frac{d}{d\underline{u}}\int_{S(u,\underline{u})} r^{\lambda_1 p}|U|^p d\mu_\gamma &\leq p\left(\int_{S(u,\underline{u})} r^{\lambda_1 p}|F|^p\right)^{\frac{1}{p}}\left(\int_{S(u,\underline{u})} r^{\lambda_1 p}|U|^p\right)^{\frac{p-1}{p}} \\ &+ \frac{|\lambda_1|p}{2}\int_{S(u,\underline{u})} r^{\lambda_1 p}|U|^p|\Omega\mathrm{tr}\chi - \overline{\Omega\mathrm{tr}\chi}|d\mu_\gamma .\end{aligned}$$

We now make use of the first inequality (4.1.23) and derive the inequality

$$-\frac{d}{d\underline{u}}|r^{\lambda_1}U|_{p,S} \leq \left(|r^{\lambda_1}F|_{p,S} + c\delta r^{-2}|r^{\lambda_1}U|_{p,S}\right), \tag{4.1.34}$$

which, upon integration in the interval $[\underline{u}, \underline{u}_*]$, yields

$$\begin{aligned}|r^{\lambda_1}U|_{p,S}(u,\underline{u}) &\leq |r^{\lambda_1}U|_{p,S}(u,\underline{u}_*) + \int_{\underline{u}}^{\underline{u}_*}|r^{\lambda_1}F|_{p,S}(u,\underline{u}')d\underline{u}' \\ &+ (\sup_{[\underline{u},\underline{u}_*]}|r^{\lambda_1}U|_{p,S})\left(c\delta\int_{\underline{u}}^{\underline{u}_*} r^{-2}\right).\end{aligned} \tag{4.1.35}$$

Choosing δ such that $\left(c\delta\int_{\underline{u}}^{\underline{u}_*} r^{-2}\right) \leq \delta' < 1$ we have, for fixed $\underline{u}$ and any $\underline{u}' \in [\underline{u}, \underline{u}_*]$,

$$\begin{aligned}|r^{\lambda_1}U|_{p,S}(u,\underline{u}') &\leq |r^{\lambda_1}U|_{p,S}(u,\underline{u}_*) + \int_{\underline{u}}^{\underline{u}_*}|r^{\lambda_1}F|_{p,S}(u,\underline{u}'')d\underline{u}'' \\ &+ \delta'\sup_{[\underline{u},\underline{u}_*]}|r^{\lambda_1}U|_{p,S} .\end{aligned}$$

Taking the sup with respect to $\underline{u}'$ in $[\underline{u}, \underline{u}_*]$, we obtain

$$(1-\delta')\sup_{[\underline{u},\underline{u}_*]} |r^{\lambda_1}U|_{p,S} \leq |r^{\lambda_1}U|_{p,S}(u,\underline{u}_*) + \int_{\underline{u}}^{\underline{u}_*} |r^{\lambda_1}F|_{p,S}(u,\underline{u}')d\underline{u}' \tag{4.1.36}$$

from which (i) of the Evolution Lemma follows with $c_0 = \frac{1}{(1-\delta')}$.

To obtain (ii) we proceed in the same way, using the differential inequalities

$$\begin{aligned}&\frac{d}{du}|V|^p + \lambda_0 p\Omega \mathrm{tr}\underline{\chi}|V|^p \leq p|\underline{F}||V|^{p-1}\\ &\frac{d}{du}|V|^p + \lambda_0 p\Omega \mathrm{tr}\underline{\chi}|V|^p \geq -p|\underline{F}||V|^{p-1},\end{aligned} \tag{4.1.37}$$

and assumption (4.1.26).

4.1.3 The initial layer foliation

In the previous section we encountered a difficulty with part ii) of the Evolution Lemma 4.1.5 along the null-incoming hypersurfaces $\underline{C}(\nu)$. The double null canonical foliation of $\mathcal{K}$ does not allow us to connect the surfaces $S(\lambda,\nu)$ with the surfaces $S_{(0)}(\nu)$ on Σ_0; see also Proposition 3.3.1.

We show in this subsection how to overcome this difficulty by using, in a small neighborhood of Σ_0, the initial layer foliation introduced in Chapter 3; see Definition 3.3.4 which we recall below.

The initial layer foliation is the double null foliation defined by the null-incoming surfaces $\underline{C}(\underline{u})$ and the null-outgoing hypersurfaces $C'(u')$, both intersecting Σ_0 along the canonical foliation $S_{(0)}(\nu)$. More precisely, see Definition 3.3.5,

1. The $C'(\lambda')$ null hypersurfaces are given by $u'(p)=\lambda'$, where $\lambda' \in [-\nu_0, -\nu_*]$, with u' the outgoing solution of the eikonal equation with initial condition $u' = -\underline{u}_{(0)}$ on Σ_0.
2. The $\underline{C}(\nu)$ null hypersurfaces are defined as before by $\underline{u}(p) = \nu$, where $\nu \in [\nu_0, \nu_*]$, with $\underline{u}$ the incoming solution of the eikonal equation with initial condition $\underline{u} = \underline{u}_{(0)}$ on Σ_0.

Observe that $S_{(0)}(\nu) = C'(-\nu) \cap \underline{C}(\nu)$. The initial layer region $\mathcal{K}'_{\delta_0} \subset \mathcal{K}$ is specified by the condition

$$\frac{1}{2}(u'(p) + \underline{u}(p)) \leq \delta_0. \tag{4.1.38}$$

As discussed in Chapter 3 (see Proposition 3.3.1) the initial layer K'_{δ_0} also comes equipped with an adapted spacelike foliation, $\{\Sigma'_{t'}\}$ with $t' = \frac{1}{2}(u' + \underline{u})$. With this definition Σ_0 satisfies $\Sigma_0 = \Sigma'_{t'=0}$. Thus the height δ_0 of the layer K'_{δ_0} corresponds to the time interval $0 \leq t' \leq \delta_0$.

We are now ready to state our main result concerning the compatibility between the canonical and initial layer foliations.

Lemma 4.1.6 (Oscillation Lemma) *Consider a space time region $\mathcal{K}$ with the canonical double null foliation generated by $u(p), \underline{u}(p)$. Consider also an initial layer region ${\mathcal{K}'}_{\delta_0}$, of height δ_0, with the initial layer foliation generated by $u'(p), \underline{u}(p)$. We make the following assumptions:*

i) On the surface $S'_* = \Sigma'_{\delta_0} \cap \underline{C}_* = S'(2\delta_0 - \nu_*, \nu_*)$

$$\left(\sup_{(p,p')\in S'_*} |u(p) - u(p')|\right) \leq \epsilon_0, \tag{4.1.39}$$

and

$$|r^2 \tau_-^{\frac{1}{2}} \eta| \leq \epsilon_0 \ , \ |r'^2 \tau'_- \mathbf{g}(L', L)| \leq \epsilon_0 \ , \ |r'^3 \tau'_- \nabla\!\!\!\!/ \mathbf{g}(L', L)| \leq \epsilon_0. \tag{4.1.40}$$

ii) On the initial hypersurface Σ_0,

$$|r'^{\frac{5}{2}} \eta'| \leq \epsilon_0. \tag{4.1.41}$$

iii) On $\mathcal{K}/\mathcal{K}'_{\delta_0}$,

$$\mathcal{O}^{\infty}_{[1]} + \underline{\mathcal{O}}^{\infty}_{[1]} \leq \epsilon_0. \tag{4.1.42}$$

iv) On the initial layer $\mathcal{K}'_{\delta_0}$,

$$\mathcal{O}'^{\infty}_{[1]} + \underline{\mathcal{O}}'^{\infty}_{[1]} \leq \epsilon_0. \tag{4.1.43}$$

Then, if ϵ_0 is sufficiently small,

$$\mathrm{Osc}(u)(\Sigma'_{\delta_0}) \equiv \sup_{\nu\in[\nu_0,\nu_*]} \left(\sup_{(p,p')\in S'(2\delta_0-\nu,\nu)} |u(p) - u(p')|\right) \leq c\epsilon_0. \tag{4.1.44}$$

Remarks:

1. The norms appearing in (4.1.40), (4.1.41), (4.1.42), and (4.1.43) are pointwise.
2. The assumptions (4.1.39), (4.1.40) are satisfied in view of the canonicity of the foliation on the last slice $\underline{C}_*$; see Proposition 7.4.1 and Lemma 7.7.2.
3. The assumptions (4.1.41) are satisfied in view of the canonicity of the foliation on the initial slice Σ_0 and are used in Lemma 4.8.2.

Proof of the Oscillation Lemma: The detailed proof of the Oscillation Lemma is given in the appendix to this chapter.

Corollary 4.1.2 *Given an null-incoming hypersurface $\underline{C}(\nu)$ there exists a two-dimensional surface S relative to the double null canonical foliation belonging to $\underline{C}(\nu)$ and included in the initial layer region $\mathcal{K}'_{\delta_0}$ for any $\nu \in [\nu_0, \nu_*]$.*

Proof: We define

$$\tilde{\delta}_0 = \frac{1}{2} \inf_{\nu\in[\nu_0,\nu_*]} \left(\inf_{p\in S'(2\delta_0-\nu,\nu)} (u(p) + \nu)\right). \tag{4.1.45}$$

From the Oscillation Lemma, it follows that

$$|\tilde{\delta}_0 - \delta_0| \leq c\epsilon_0. \tag{4.1.46}$$

$S(\lambda_0(\nu), \nu)$, with $\lambda_0(\nu) = 2\tilde{\delta}_0 - \nu$, is a two-dimensional surface relative to the double null canonical foliation included in the initial layer region $\mathcal{K}'_{\tilde{\delta}_0}$ for any $\nu \in [\nu_0, \nu_*]$. These are the surfaces which we refer to, in part ii of the Evolution Lemma; see 4.1.28.

Remarks:

1. In Chapter 7, assuming tha initial and final slice are endowed with a canonical foliation, all the estimates in the assumptions of the Oscillation Lemma are proved with $\epsilon_0 \leq c\epsilon$.

2. Recalling the definition of the $\tilde{\Sigma}_{\tilde{t}}$ spacelike hypersurfaces associated with the double null canonical foliation (see Proposition 3.3.1) the previous lemma and its corollary imply that we can extend the double null canonical foliation to $\tilde{\Sigma}_{\tilde{\delta}_0}$ with $\tilde{\delta}_0 \geq \delta_0 - 2c\epsilon_0 > 0$ a little below Σ'_{δ_0}. In other words we can find a spacelike hypersurface foliated by the $S(\lambda, \nu)$ two-dimensional surfaces, relative to the double null canonical foliation contained in the initial layer region at a distance ϵ_0 from Σ'_{δ_0}.

We use Lemma 4.1.6 to express the estimates of part ii), equation (4.1.28) of the Evolution Lemma in terms of the initial data norms on $S_{(0)}$. We rewrite first the estimates of part ii) of the Evolution Lemma,[7]

$$|r^{\lambda_1} V|_{p,S}(u, \underline{u}) \leq c_0 \left(|r^{\lambda_1} V|_{p,S(u_0(\underline{u}),\underline{u})} + \int_{u_0(\underline{u})}^{u} |r^{\lambda_1} \underline{F}|_{p,S}(u', \underline{u}) du' \right). \tag{4.1.47}$$

In the next lemma we show, using the Oscillation Lemma, how to control the difference between the norm $|r^{\lambda_1} V|_{p,S(u_0(\nu),\nu)}$ and the norm $|r'^{\lambda_1} V'|_{p,S_{(0)}(\nu)}$, where V' indicates the S-tangent tensor field analogous to V, but relative to the *initial layer foliation*; see the Evolution Lemma.[8] Recall that the norm $|r^{\lambda_1} V|_{p,S(u_0(\nu),\nu)}$ refers to a surface $S(u_0(\nu), \nu)$ associated with the double null canonical foliation and $r = (\frac{1}{4\pi}|S(u_0(\nu), \nu)|)^{\frac{1}{2}}$, while $|r'^{\lambda_1} V'|_{p,S_{(0)}(\nu)}$ refers to a surface contained in Σ_0 associated with the initial layer foliation and, therefore $r' = (\frac{1}{4\pi}|S_{(0)}(\nu)|)^{\frac{1}{2}}$.

Lemma 4.1.7 *Let V be a tensor field satisfying the evolution equation*

$$\mathbf{D}_N V + \lambda_0 \Omega \mathrm{tr}\underline{\chi} V = \underline{F} .$$

Assume that on $\tilde{\Sigma}_{\tilde{\delta}_0}$ and on Σ'_{δ_0}, with ϵ_0 sufficiently small, the following inequalities hold; see 4.1.42, 4.1.43,

$$\begin{aligned} \mathcal{O}^{\infty}_{[1]} + \underline{\mathcal{O}}^{\infty}_{[1]} &\leq \epsilon_0 \\ \mathcal{O}'^{\infty}_{[1]} + \underline{\mathcal{O}}'^{\infty}_{[1]} &\leq \epsilon_0, \end{aligned} \tag{4.1.48}$$

---

[7] We use here the notation $u_0(\nu)$ instead of $\lambda_0(\nu)$ to avoid confusion with the exponents of the estimates like (4.1.47).

[8] V and V' are not the same tensor field since they are tangent to different two-dimensional surfaces. Nevertheless V can be expressed in terms of S'-tangent tensor fields and vice-versa. See also the proof of the next lemma.

and that, in the initial layer region,

$$\mathcal{O}_{[2]}{}' \leq \epsilon_0\,,\ \underline{\mathcal{O}}_{[2]}{}' \leq \epsilon_0$$
$$\mathcal{R}_{[2]}{}' \leq \epsilon_0\,,\ \underline{\mathcal{R}}_{[2]}{}' \leq \epsilon_0. \tag{4.1.49}$$

Then, with $\lambda_1 = 2\left(\lambda_0 - \frac{1}{p}\right)$,

$$|r^{\lambda_1} V|_{p,S(u_0(\underline{u}),\underline{u})} \leq c|r'^{\lambda_1} V'|_{p,S_{(0)}(\underline{u})} + c\epsilon_0. \tag{4.1.50}$$

Remark: Throughout Chapter 4, where this result is used, the tensor field V describes the various underlined connection coefficients and their derivatives.

Proof: The proof of this lemma is in the appendix to this chapter.

4.1.4 Comparison estimates for the function $r(u, \underline{u})$

In the proofs of this chapter and of Chapters 5 and 6 we often use some estimates that connect the function $r(u, \underline{u})$ with the functions $u(p)$ and $\underline{u}(p)$ and also with the functions $v(p)$ and $\underline{v}(p)$, the affine parameters of the null geodesics generating the null hypersurfaces $C(u)$ and $\underline{C}(\underline{u})$. We collect all these estimates here.

We recall that $u(p)$ and $\underline{u}(p)$ are solutions of the eikonal equation, the first one having as initial data the function $u_*(p)$ defined on the last slice $\underline{C}_*$ solution of the last slice problem, see Definition 3.3.2, and $\underline{u}(p)$ having as initial data the function $\underline{u}_{(0)}(p)$ defined on Σ_0, solution of the initial slice problem; see Definition 3.3.1. The first estimate for the function $r(u, \underline{u}) \equiv \sqrt{\frac{|S(u,\underline{u})|}{4\pi}}$ is provided by the following lemma.[9]

Lemma 4.1.8 *Assume in the spacetime $\mathcal{K}$ the estimate (see (4.2.9))*

$$|r^2(\overline{\Omega \mathrm{tr}\chi} - \frac{2}{r})| \leq c\,(\mathcal{I}_0 + \mathcal{I}_* + \Delta_0)$$
$$|r\tau_-(\overline{\mathrm{tr}\underline{\chi}} + \frac{2}{r})| \leq c\,(\mathcal{I}_0 + \mathcal{I}_* + \Delta_0)\,.$$

Assume that on $\underline{C}_*$

$$|r(\lambda, \nu_*) - \frac{1}{2}(\nu_* - \lambda)| \leq c\mathcal{I}_* \log r(\lambda, \nu_*).$$

Then if $(\mathcal{I}_0 + \mathcal{I}_* + \Delta_0)$ is sufficently small, $r(u, \underline{u})$ satisfies the inequality

$$|r(u, \underline{u}) - \frac{1}{2}(\underline{u} - u)| \leq c\,(\mathcal{I}_0 + \mathcal{I}_* + \Delta_0) \log r(u, \underline{u}). \tag{4.1.51}$$

Moreover, there exists a constant c such that

$$c^{-1}\tau_+ \leq r(u, \underline{u}) \leq c\tau_+. \tag{4.1.52}$$

[9] We discuss this lemma and the following one here, although they require, to be proved, the results of Theorem 4.2.1. Of course the proof of these results does not depend on these lemmas.

Proof: We integrate the equation (see 4.1.30):

$$\frac{d}{d\underline{u}}r(u,\underline{u})=\frac{1}{2}+\frac{r(u,\underline{u})}{2}\left(\overline{\Omega\mathrm{tr}\chi}-\frac{1}{r}\right),$$

along $C(u)$ from $\underline{u}=\nu_*$ to $\underline{u}$, to obtain

$$r(u,\nu_*)-r(u,\underline{u})=\frac{1}{2}(\nu_*-\underline{u})+\int_{\underline{u}}^{\nu_*}\frac{r}{2}\left(\overline{\Omega\mathrm{tr}\chi}-\frac{1}{r}\right),$$

which we rewrite as

$$r(u,\underline{u})-\frac{1}{2}(\underline{u}-u)=r(u,\nu_*)-\frac{1}{2}(\nu_*-u)-\int_{\underline{u}}^{\nu_*}\frac{r}{2}\left(\overline{\Omega\mathrm{tr}\chi}-\frac{1}{r}\right).$$

Using the assumptions we obtain

$$(r(u,\underline{u})-\frac{1}{2}(\underline{u}-u))+c\,(\mathcal{I}_0+\mathcal{I}_*+\Delta_0)\log r(u,\underline{u})\le c\mathcal{I}_*,$$

which implies that

$$|r(u,\underline{u})-\frac{1}{2}(\underline{u}-u)|\le c\,(\mathcal{I}_0+\mathcal{I}_*+\Delta_0)\log r(u,\underline{u}), \tag{4.1.53}$$

thus proving the first part of the lemma.[10] The second part follows immediately.

Remark: The result of Lemma 4.1.8

$$|r(u,\underline{u})-\frac{1}{2}(\underline{u}-u)|\le c\epsilon_0\log r(u,\underline{u}),$$

although sufficient for our purposes is not the optimal one. A better and more delicate result is stated in the following lemma, whose proof we do not present here.[11]

Lemma 4.1.9 *Under appropriate assumptions, consistent with the bootstrap assumptions for $\mathcal{O}$ of Theorem* **M1**, $r(u,\underline{u})$ *satisfies the following inequalities*

$$\begin{aligned}|r(u,\underline{u})-(\underline{v}-\frac{u}{2})|&\le c(\mathcal{I}_*+\Delta_0)\\ |r(u,\underline{u})+(v-\frac{u}{2})|&\le c(\mathcal{I}_0+\Delta_0)\end{aligned} \tag{4.1.54}$$

where v, $\underline{v}$ are the affine parameters of the null geodesics generating $C(u)$ and $\underline{C}(\underline{u})$ respectively.

---

[10] The assumption relative to the last slice is proved in Chapter 7.

[11] The proof is, anyway, an adapted version of Proposition 9.1.3 of [Ch-Kl].

4.2 Proof of Theorem **M1**

We are now ready to start proving Theorem **M1**. While we have structured the proof in a way that we believe is optimal for the comprehension of the reader, we omit detailed motivations for various important technical steps. For these we refer the reader to our review paper [Kl-Ni].

We divide Theorem **M1** into three theorems, the first referring to zero and first derivatives, the second referring to second derivatives and the third to third derivatives.

Remarks:

1. It is important to note that to prove Theorem **M1** relative to a double null foliation we need to introduce as an auxiliary assumption the result of the Oscillation Lemma. To remove this auxiliary assumption we need that $\mathcal{K}$ be endowed with a double null canonical foliation.

2. It is also important to realize that all the norm assumptions relative to the initial hypersurface Σ_0 are relative to the connection coefficients relative to the initial layer foliation. These norms are connected to those relative to the double null canonical foliation through the Oscillation Lemma and Lemma 4.1.7.

Theorem 4.2.1 *Assume that*

$$\begin{aligned}\mathcal{R}_0^\infty + \underline{\mathcal{R}}_0^\infty &\leq \Delta_0\\ \mathcal{R}_1^S + \underline{\mathcal{R}}_1^S &\leq \Delta_1\end{aligned} \tag{4.2.1}$$

and

$$\mathcal{O}_{[1]}(\underline{C}_*) \leq \mathcal{I}_*\,,\ \underline{\mathcal{O}}_{[1]}(\Sigma_0) \leq \mathcal{I}_0. \tag{4.2.2}$$

Assume further that in the initial layer region

$$\begin{aligned}\mathcal{O}_{[1]}{}' \leq \mathcal{I}_0\,,\ \underline{\mathcal{O}}_{[1]}{}' \leq \mathcal{I}_0\\ \mathcal{R}_{[1]}{}' \leq \Delta_0\,,\ \underline{\mathcal{R}}_{[1]}{}' \leq \Delta_0.\end{aligned} \tag{4.2.3}$$

Assume finally that Δ_0, Δ_1, $\mathcal{I}_0$, $\mathcal{I}_*$ are sufficiently small; then there exists a constant such that the following estimate holds:[12]

$$\mathcal{O}_{[1]} + \underline{\mathcal{O}}_{[1]} \leq c(\mathcal{I}_0 + \mathcal{I}_* + \Delta_0 + \Delta_1). \tag{4.2.4}$$

Proof: We present here the strategy of the proof. All the details are given in Section 4.3. The proof is divided into the following steps.

1. We make the following additional bootstrap assumptions:

$$\mathcal{O}_{[0]}^\infty + \underline{\mathcal{O}}_{[0]}^\infty \leq \Gamma_0, \tag{4.2.5}$$

[12] We denote by c a constant that does not depend on the relevant parameters. It can be different in different estimates.

$$\mathrm{Osc}(u)(\Sigma'_{\delta_0}) \leq \Gamma_0, \tag{4.2.6}$$

with $\Gamma_0 > 0$ sufficiently small.[13] Then we prove that the following inequalities hold:

$$\begin{aligned} \mathcal{O}_0 &\leq c\,(\mathcal{I}_* + \Delta_0) \\ \underline{\mathcal{O}}_0 &\leq c\,(\mathcal{I}_0 + \Delta_0) \\ \mathcal{O}_1 &\leq c\,(\mathcal{I}_0 + \mathcal{I}_* + \Delta_0) \\ \underline{\mathcal{O}}_1 &\leq c\,(\mathcal{I}_0 + \mathcal{I}_* + \Delta_0) \end{aligned} \tag{4.2.7}$$

$$\underline{\tilde{\mathcal{O}}}_1(\omega) \leq c\,(\mathcal{I}_0 + \mathcal{I}_* + \Delta_0 + \Delta_1) \tag{4.2.8}$$

and finally,

$$\begin{aligned} |r(\Omega - \frac{1}{2})| &\leq c\,(\mathcal{I}_0 + \mathcal{I}_* + \Delta_0) \\ |r^2(\overline{\mathrm{tr}\chi} - \frac{2}{r})| &\leq c\,(\mathcal{I}_0 + \mathcal{I}_* + \Delta_0) \\ |r\tau_-(\overline{\mathrm{tr}\underline{\chi}} + \frac{2}{r})| &\leq c\,(\mathcal{I}_0 + \mathcal{I}_* + \Delta_0)\,. \end{aligned} \tag{4.2.9}$$

This is the main part of the proof. The details will be given in the subsequent sections. The next steps (2) and (3) are standard.

2. Using the estimates (4.2.7), (4.2.8) and the Sobolev Lemma 4.1.3, we infer that

$$\mathcal{O}^{\infty}_{[0]} + \underline{\mathcal{O}}^{\infty}_{[0]} \leq c\,(\mathcal{I}_0 + \mathcal{I}_* + \Delta_0 + \Delta_1) \leq \frac{\Gamma_0}{2}, \tag{4.2.10}$$

provided we choose $(\mathcal{I}_0 + \mathcal{I}_* + \Delta_0 + \Delta_1)$ sufficiently small.

3. To remove the additional assumption (4.2.5) we consider the region $\mathcal{S}(\Gamma_0)$ contained in $\mathcal{K}$ defined by the properties[14]

a) $\mathcal{S}(\Gamma_0) = \{p \in \mathcal{K}|\ (u(p), \underline{u}(p)) \in [\lambda_1, \lambda_2) \times (\nu_2, \nu_*]\}$;

b) The inequality holds in $\mathcal{S}(\Gamma_0)$

$$\mathcal{O}^{\infty}_{[0]} + \underline{\mathcal{O}}^{\infty}_{[0]} < \Gamma_0\,.$$

Using the result in (2) we infer that this region is, simultaneously, open and closed and, therefore, must coincide with the whole $\mathcal{K}$. From this result the estimates (4.2.7), (4.2.8) and (4.2.9) hold in $\mathcal{K}$ and the theorem follows.

Details of the implementation of step (1) are given in Section 4.3.

---

[13] Γ_0 must be such that $\Gamma_0^2 < (\mathcal{I}_0 + \mathcal{I}_* + \Delta_0) < \Gamma_0$.

[14] Recall that λ_1 is the value of $u(p)$ on $\underline{C}_* \cap \Sigma_0$; see Chapter 3.

Remarks:

1. Instead of the bootstrap assumption (4.2.5) we could have used a stronger bootstrap assumption involving the full norms $\mathcal{O}_{[3]}$ and $\underline{\mathcal{O}}_{[3]}$. Because of the importance of this result we prefer, however, this proof which emphasizes the fact that only the norms $\mathcal{O}_{[1]}$ and $\underline{\mathcal{O}}_{[1]}$ are needed to break the nonlinear structure of the null structure equations.

2. It is easy to check that the bootstrap assumption 4.2.5 implies all the assumptions needed in the proofs of all the preliminary results of previous section.

Theorem 4.2.2 *Assume that*

$$\begin{aligned} &\mathcal{R}_0^\infty + \underline{\mathcal{R}}_0^\infty \le \Delta_0 \\ &\mathcal{R}_1^S + \underline{\mathcal{R}}_1^S \le \Delta_1 \\ &\mathcal{R}_2 + \underline{\mathcal{R}}_2 \le \Delta_2 \end{aligned} \tag{4.2.11}$$

and also

$$\mathcal{O}_{[2]}(\underline{C}_*) \le \mathcal{I}_* \,,\ \underline{\mathcal{O}}_{[2]}(\Sigma_0) \le \mathcal{I}_0 \tag{4.2.12}$$

Assume further that in the initial layer region

$$\begin{aligned} &\mathcal{O}_{[2]}{}' \le \mathcal{I}_0 \,,\ \underline{\mathcal{O}}_{[2]}{}' \le \mathcal{I}_0 \\ &\mathcal{R}_{[2]}{}' \le \Delta_0 \,,\ \underline{\mathcal{R}}_{[2]}{}' \le \Delta_0 \end{aligned} \tag{4.2.13}$$

Assume finally that $\Delta_0,\ \Delta_1,\ \Delta_2,\ \mathcal{I}_0,\ \mathcal{I}_*$ are sufficiently small. Then there exists a generic constant c such that

$$\mathcal{O}_{[2]} + \underline{\mathcal{O}}_{[2]} \le c(\mathcal{I}_0 + \mathcal{I}_* + \Delta_0 + \Delta_1 + \Delta_2). \tag{4.2.14}$$

Proof: We present here the strategy of the proof. All the details are given in Section 4.4. We divide the proof into four steps.

1. We assume (4.2.6) and the auxiliary bootstrap assumption

$$\mathcal{O}_1^\infty + \underline{\mathcal{O}}_1^\infty \le \Gamma_1, \tag{4.2.15}$$

with Γ_1 sufficiently small. Then the following inequalities hold:[15]

$$\begin{aligned} &\mathcal{O}_2 \le c\,(\mathcal{I}_0 + \mathcal{I}_* + \Delta_0 + \Delta_1) \\ &\underline{\mathcal{O}}_2 \le c\,(\mathcal{I}_0 + \mathcal{I}_* + \Delta_0 + \Delta_1) \end{aligned} \tag{4.2.16}$$

$$\tilde{\mathcal{O}}_2(\omega) \le c\,(\mathcal{I}_0 + \mathcal{I}_* + \Delta_0 + \Delta_1 + \Delta_2)\,. \tag{4.2.17}$$

2. These inequalities, together with the estimates

$$\begin{aligned} &\mathcal{O}_{[1]} \le c\,(\mathcal{I}_0 + \mathcal{I}_* + \Delta_0 + \Delta_1) \\ &\underline{\mathcal{O}}_{[1]} \le c\,(\mathcal{I}_0 + \mathcal{I}_* + \Delta_0 + \Delta_1) \end{aligned}$$

[15] To prove these inequalities we also use the results of the previous theorem.

proved in the previous theorem allow us, by applying Lemma 4.1.3, to estimate $\mathcal{O}_1^\infty + \underline{\mathcal{O}}_1^\infty$ in terms of $\mathcal{O}_{[1]} + \underline{\mathcal{O}}_{[1]}$, $\mathcal{O}_2 + \underline{\mathcal{O}}_2$. Therefore we obtain

$$\mathcal{O}_1^\infty + \underline{\mathcal{O}}_1^\infty \leq c\,(\mathcal{I}_0 + \mathcal{I}_* + \Delta_0 + \Delta_1)\,, \tag{4.2.18}$$

so that by choosing $\mathcal{I}_0 + \mathcal{I}_* + \Delta_0 + \Delta_1$ sufficiently small we may infer that

$$\mathcal{O}_1^\infty + \underline{\mathcal{O}}_1^\infty \leq \frac{\Gamma_1}{2}. \tag{4.2.19}$$

3. Introducing again, as in Theorem 4.2.1, a region $\mathcal{S}(\Gamma_1)$ contained in $\mathcal{K}$ we repeat the previous argument and check that

$$\mathcal{O}_1^\infty + \underline{\mathcal{O}}_1^\infty \leq \Gamma_1 \tag{4.2.20}$$

holds in the entire spacetime $\mathcal{K}$. In view of this result, the inequalities (4.2.16), (4.2.17), (4.2.18) hold in $\mathcal{K}$.

Details of the implementation of steps (1), (2), (3) are given in Section 4.4.

Theorem 4.2.3 *Assume that*

$$\begin{aligned} &\mathcal{R}_0^\infty + \underline{\mathcal{R}}_0^\infty \leq \Delta_0\\ &\mathcal{R}_1^S + \underline{\mathcal{R}}_1^S \leq \Delta_1\\ &\mathcal{R}_2 + \underline{\mathcal{R}}_2 \leq \Delta_2 \end{aligned} \tag{4.2.21}$$

and also

$$\mathcal{O}_{[3]}(\underline{C}_*) \leq \mathcal{I}_*\,,\ \ \underline{\mathcal{O}}_{[3]}(\Sigma_0) \leq \mathcal{I}_0. \tag{4.2.22}$$

Assume further that Δ_0, Δ_1, Δ_2, $\mathcal{I}_0$, $\mathcal{I}_*$ are sufficiently small; then there exists a generic constant c such that

$$\mathcal{O}_3 + \underline{\mathcal{O}}_3 \leq c(\mathcal{I}_0 + \mathcal{I}_* + \Delta_0 + \Delta_1 + \Delta_2). \tag{4.2.23}$$

The proof of this theorem is discussed in Section 4.5.

To complete the proof of Theorem **M1**, we now eliminate the assumptions on the oscillation of u.

Corollary 4.2.4 *Under the assumptions of Theorem* **M1** *relative to a double null canonical foliation, the following inequality holds:* [16]

$$\text{Osc}(u)(\Sigma'_{\delta_0}) \leq (\mathcal{I}_0 + \mathcal{I}_* + \Delta_0) \leq \frac{\Gamma_0}{2}, \tag{4.2.24}$$

provided we choose $\mathcal{I}_0, \mathcal{I}_*, \Delta_0$ sufficiently small. Therefore this allows us to conclude that $\text{Osc}(u)(\Sigma'_{\delta_0}) \leq \epsilon_0$ *on the whole* Σ'_{δ_0}.

[16] This implies stronger assumptions on the initial and last slice (satisfied with their canonical foliations). See, in particular, Proposition 4.3.17 and Corollary 4.4.1.

4.3 Proof of Theorem 4.2.1 and estimates for the zero and first derivatives of the connection coefficents

We concentrate on the proof of part (1) of the theorem and divide the proof into many steps.

4.3.1 *Estimate for* $\mathcal{O}_{0,1}^{p,S}(\mathrm{tr}\chi)$ *and* $\mathcal{O}_{0,1}^{p,S}(\hat{\chi})$ *with* $p \in [2,4]$

Proposition 4.3.1 *Assuming (4.2.1), (4.2.2) and the bootstrap assumption (4.2.5) the following estimates hold:*

$$
\begin{aligned}
&|r^{3-2/p}\nabla\!\!\!/\hat{\chi}|_{p,S}(u,\underline{u}) \leq c(\mathcal{I}_0+\mathcal{I}_*+\Delta_0)\\
&|r^{2-2/p}\hat{\chi}|_{p,S}(u,\underline{u}) \leq c(\mathcal{I}_0+\mathcal{I}_*+\Delta_0).
\end{aligned}
\tag{4.3.1}
$$

Proof: We derive first an evolution equation for $\nabla\!\!\!/\mathrm{tr}\chi$ by differentiating the evolution equation of $\mathrm{tr}\chi$ along the outgoing direction (see 3.1.46),

$$
\nabla\!\!\!/\mathbf{D}_4\mathrm{tr}\chi - \nabla\!\!\!/((\mathbf{D}_4\log\Omega)\mathrm{tr}\chi) = -\mathrm{tr}\chi\nabla\!\!\!/\mathrm{tr}\chi - \nabla\!\!\!/|\hat{\chi}|^2. \tag{4.3.2}
$$

Using the commutation relation (see 4.8.1) in the appendix to this chapter,

$$
[\nabla\!\!\!/, \mathbf{D}\!\!\!/_4]\mathrm{tr}\chi = \chi\cdot\nabla\!\!\!/\mathrm{tr}\chi - (\zeta+\underline{\eta})\mathbf{D}_4\mathrm{tr}\chi,
$$

equation (4.3.2) becomes

$$
\mathbf{D}\!\!\!/_4\nabla\!\!\!/\mathrm{tr}\chi + \chi\cdot\nabla\!\!\!/\mathrm{tr}\chi = (\nabla\!\!\!/\log\Omega)\mathbf{D}_4\mathrm{tr}\chi + \nabla\!\!\!/((\mathbf{D}_4\log\Omega)\mathrm{tr}\chi) - \mathrm{tr}\chi\nabla\!\!\!/\mathrm{tr}\chi - \nabla\!\!\!/|\hat{\chi}|^2,
$$

which we rewrite as

$$
\begin{aligned}
&\mathbf{D}\!\!\!/_4\nabla\!\!\!/\mathrm{tr}\chi - (\mathbf{D}_4\log\Omega)\nabla\!\!\!/\mathrm{tr}\chi + \frac{1}{2}\mathrm{tr}\chi\nabla\!\!\!/\mathrm{tr}\chi = -\hat{\chi}\cdot\nabla\!\!\!/\mathrm{tr}\chi + (\nabla\!\!\!/\log\Omega)\mathbf{D}_4\mathrm{tr}\chi\\
&\quad + (\nabla\!\!\!/\mathbf{D}_4\log\Omega)\mathrm{tr}\chi - \mathrm{tr}\chi\nabla\!\!\!/\mathrm{tr}\chi - \nabla\!\!\!/|\hat{\chi}|^2.
\end{aligned}
\tag{4.3.3}
$$

Defining

$$
U \equiv \Omega^{-1}\nabla\!\!\!/\mathrm{tr}\chi,
$$

and choosing a null frame such that the vector fields e_a satisfy $\mathbf{D}\!\!\!/_4 e_a = 0$, the previous equation becomes

$$
\frac{d}{d\underline{u}}U_a + \frac{3}{2}\Omega\mathrm{tr}\chi U_a = F_a, \tag{4.3.4}
$$

where

$$
F = -\Omega\hat{\chi}\cdot U - \nabla\!\!\!/|\hat{\chi}|^2 + (\nabla\!\!\!/\mathbf{D}_4\log\Omega)\mathrm{tr}\chi - (\nabla\!\!\!/\log\Omega)\left(\frac{1}{2}(\mathrm{tr}\chi)^2 - (\mathbf{D}_4\log\Omega)\mathrm{tr}\chi + |\hat{\chi}|^2\right).
$$

Equation (4.3.4) is not quite suited for our purposes because it leads to logarithmic divergences when we try to apply the Evolution Lemma to it with $\lambda_1 = 3 - \frac{2}{p}$.[17] To avoid this difficulty we introduce the tensor

$$\not{\Psi} = \Omega^{-1}\not{\nabla}\mathrm{tr}\chi + \Omega^{-1}\mathrm{tr}\chi\zeta = U + \Omega^{-1}\mathrm{tr}\chi\zeta \tag{4.3.5}$$

Recalling (see 3.1.46) that ζ satisfies the equation

$$\not{\mathbf{D}}_4\zeta = -\not{\nabla}\mathbf{D}_4\log\Omega + \chi(\underline{\eta} - \zeta) - (\mathbf{D}_4\log\Omega)\not{\nabla}\log\Omega - \beta,$$

it is easy to see that $\not{\Psi}$ satisfies

$$\frac{d}{d\underline{u}}\not{\Psi}_a + \frac{3}{2}\Omega\mathrm{tr}\chi\not{\Psi}_a = \not{F}_a, \tag{4.3.6}$$

where

$$\not{F} \equiv -\Omega\hat{\chi}\cdot\not{\Psi} - \not{\nabla}|\hat{\chi}|^2 - \eta|\hat{\chi}|^2 + \mathrm{tr}\chi\ \hat{\chi}\cdot\underline{\eta} - \mathrm{tr}\chi\beta. \tag{4.3.7}$$

Applying the Evolution Lemma to the evolution equation (4.3.6) we obtain

$$|r^{3-\frac{2}{p}}\not{\Psi}|_{p,S}(u,\underline{u}) \le c_0\left(|r^{3-\frac{2}{p}}\not{\Psi}|_{p,S}(u,\underline{u}_*) + \int_{\underline{u}}^{\underline{u}_*}|r^{3-\frac{2}{p}}\not{F}|_{p,S}\right). \tag{4.3.8}$$

It is easy to show that the integral on the right-hand side is bounded for $\underline{u}_* \to \infty$. In fact,

$$\begin{aligned} |r^{3-\frac{2}{p}}\not{F}|_{p,S} \ &\le\ |r^{3-\frac{2}{p}}\Omega\hat{\chi}\cdot\not{\Psi}|_{p,S} + |r^{3-\frac{2}{p}}\not{\nabla}|\hat{\chi}|^2|_{p,S} + \Big[|r^{3-\frac{2}{p}}\eta|\hat{\chi}|^2|_{p,S} \\ &+\ |r^{3-\frac{2}{p}}\underline{\eta}\cdot\hat{\chi}\mathrm{tr}\chi|_{p,S} + |r^{3-\frac{2}{p}}\mathrm{tr}\chi\beta|_{p,S}\Big]. \end{aligned} \tag{4.3.9}$$

Using the notation $\sup = \sup_{\mathcal{K}}$, the bootstrap assumption (4.2.5), and assumption (4.2.1) for β, we have

$$|r^{3-\frac{2}{p}}\Omega\hat{\chi}\not{\Psi}|_{p,S} \le \sup|\Omega|\sup|r^2\hat{\chi}||r^{3-\frac{2}{p}}\not{\Psi}|_{p,S}\frac{1}{r^2} \le c(\Gamma_0)\Gamma_0\frac{1}{r^2}|r^{3-\frac{2}{p}}\not{\Psi}|_{p,S}$$

$$|r^{3-\frac{2}{p}}\not{\nabla}|\hat{\chi}|^2|_{p,S} \le 2\sup|r^2\hat{\chi}||r^{3-\frac{2}{p}}\not{\nabla}\hat{\chi}|_{p,S}\frac{1}{r^2} \le c(\Gamma_0)\Gamma_0\frac{1}{r^2}|r^{3-\frac{2}{p}}\not{\nabla}\hat{\chi}|_{p,S}$$

---

[17] Choosing $\lambda_1 = 3 - \frac{2}{p}$ we obtain, using the Evolution Lemma,

$$|r^{3-\frac{2}{p}}U|_{p,S}(u,\underline{u}) \le c_0(|r^{3-\frac{2}{p}}U|_{p,S}(u,\underline{u}_*) + \int_{\underline{u}}^{\underline{u}_*}|r^{3-\frac{2}{p}}F|_{p,S})$$

where
$$\begin{aligned} |r^{3-\frac{2}{p}}F|_{p,S} &\le |r^{3-\frac{2}{p}}\Omega\hat{\chi}\cdot U|_{p,S} + |r^{3-\frac{2}{p}}\not{\nabla}|\hat{\chi}|^2|_{p,S} + |r^{3-\frac{2}{p}}(\zeta+\underline{\eta})|\hat{\chi}|^2|_{p,S} \\ &+ |r^{3-\frac{2}{p}}(\not{\nabla}\mathbf{D}_4\log\Omega)\mathrm{tr}\chi|_{p,S} + \frac{1}{2}|r^{3-\frac{2}{p}}(\zeta+\underline{\eta})(\mathrm{tr}\chi)^2)|_{p,S} \end{aligned}$$

and from the previous assumptions on $\mathrm{tr}\chi$, ζ, $\underline{\eta}$, it follows that the last term of $|r^{3-\frac{2}{p}}F|_{p,S}$ decays too slowly for the integral $\int_{\underline{u}}^{\underline{u}_*}|r^{3-\frac{2}{p}}F|_{p,S}$ to converge when $\underline{u}_* \to \infty$. This problem was already discussed and solved in [Ch-Kl]. Moreover if we assume for $\not{\nabla}\mathbf{D}_4\log\Omega$ the expected asymptotic behavior, which will be proved later on, one realizes that the term $|r^{3-\frac{2}{p}}(\not{\nabla}\mathbf{D}_4\log\Omega)\mathrm{tr}\chi|_{p,S}$ also has a bad asymptotic behavior.

$$|r^{3-\frac{2}{p}}\eta|\hat{\chi}|^2|_{p,S} \le \sup|r^2\hat{\chi}|^2 \sup|r^2\eta|_{p,S}\frac{1}{r^3} \le \Gamma_0^3\frac{1}{r^3} \tag{4.3.10}$$
$$|r^{3-\frac{2}{p}}\underline{\eta}\cdot\hat{\chi}\mathrm{tr}\chi|_{p,S} \le \sup|r^2\hat{\chi}|\sup|r\mathrm{tr}\chi|\sup|r^2\underline{\eta}|\frac{1}{r^2} \le c(\Gamma_0)\Gamma_0^2\frac{1}{r^2}$$
$$|r^{3-\frac{2}{p}}\mathrm{tr}\chi\beta|_{p,S} \le \sup|r\mathrm{tr}\chi|\sup|r^{\frac{7}{2}}\beta|\frac{1}{r^{\frac{3}{2}}} \le c(\Gamma_0)\Delta_0\frac{1}{r^{\frac{3}{2}}}$$

where $c(\Gamma_0)$ is a constant depending on Γ_0 that can be bounded by $c(1+\Gamma_0)$.

From these estimates the following inequality holds:

$$\begin{aligned}\int_{\underline{u}}^{\underline{u}_*}|r^{3-\frac{2}{p}}F|_{p,S} &\le c(\Gamma_0)\Gamma_0\int_{\underline{u}}^{\underline{u}_*}\frac{1}{r^2}\left(|r^{3-\frac{2}{p}}\Psi|_{p,S}+|r^{3-\frac{2}{p}}\nabla\hat{\chi}|_{p,S}\right)\\ &+ c(\Gamma_0)(\Delta_0+\Gamma_0^2+\Gamma_0^3)\frac{1}{r^{\frac{1}{2}}}.\end{aligned} \tag{4.3.11}$$

Using the final slice assumption $\mathcal{O}_{[1]}(\underline{C}_*) \le \mathcal{I}_*$ to control $|r^{3-\frac{2}{p}}\Psi|_{p,S}(u,\underline{u}_*)$ we obtain, for $p\in[2,4]$,

$$\begin{aligned}|r^{3-\frac{2}{p}}\Psi|_{p,S}(u,\underline{u}) &\le c(\Gamma_0)\left(\mathcal{I}_*+\Delta_0+\Gamma_0^2+\Gamma_0^3\right)\\ &+ c(\Gamma_0)\Gamma_0\int_{\underline{u}}^{\underline{u}_*}\frac{1}{r^2}\left(|r^{3-\frac{2}{p}}\Psi|_{p,S}+|r^{3-\frac{2}{p}}\nabla\hat{\chi}|_{p,S}\right).\end{aligned} \tag{4.3.12}$$

We then apply the elliptic L^p estimates of Proposition 4.1.3 to the Codazzi equation (see (3.1.47)) expressed relative to the tensor Ψ,

$$\mathrm{d}\!\!/\mathrm{iv}\,\hat{\chi}+\zeta\cdot\hat{\chi} = \frac{1}{2}\Omega\Psi-\beta, \tag{4.3.13}$$

and derive

$$\begin{aligned}|r^{3-2/p}\nabla\hat{\chi}|_{p,S} &\le c\left(|r^{3-\frac{2}{p}}\Omega\Psi|_{p,S}+r^{-\frac{1}{2}}|r^{\frac{7}{2}-\frac{2}{p}}\beta|_{p,S}+r^{-1}|r^{4-\frac{2}{p}}\zeta\cdot\hat{\chi}|_{p,S}\right)\\ &\le c\left(|r^{3-\frac{2}{p}}\Psi|_{p,S}+r^{-\frac{1}{2}}\Delta_0+r^{-1}\Gamma_0^2\right).\end{aligned} \tag{4.3.14}$$

Substituting this estimate into the inequality (4.3.12) we obtain

$$\begin{aligned}|r^{3-\frac{2}{p}}\Psi|_{p,S}(u,\underline{u}) &\le c(\Gamma_0)\left(\mathcal{I}_*+\Delta_0+\Gamma_0^2+\Gamma_0^3\right)\\ &+c(\Gamma_0)\Gamma_0\int_{\underline{u}}^{\underline{u}_*}\frac{1}{r^2}|r^{3-\frac{2}{p}}\Psi|_{p,S}.\end{aligned} \tag{4.3.15}$$

Finally we apply the Gronwall inequality to (4.3.15) and, assuming Γ_0 sufficiently small, we obtain

$$|r^{3-2/p}\Psi|_{p,S}(u,\underline{u}) \le c(\Gamma_0)\left(\mathcal{I}_*+\Delta_0+\Gamma_0^2+\Gamma_0^3\right). \tag{4.3.16}$$

This estimate together with the elliptic L^p estimates of Proposition 4.1.3 applied to (4.3.13) implies, for $p\in[2,4]$,

$$|r^{3-2/p}\nabla\hat{\chi}|_{p,S}(u,\underline{u}) \le c(\Gamma_0)\left(\mathcal{I}_*+\Delta_0+\Gamma_0^2+\Gamma_0^3\right) \le c(\mathcal{I}_0+\mathcal{I}_*+\Delta_0)$$
$$|r^{2-2/p}\hat{\chi}|_{p,S}(u,\underline{u}) \le c(\Gamma_0)\left(\mathcal{I}_*+\Delta_0+\Gamma_0^2+\Gamma_0^3\right) \le c(\mathcal{I}_0+\mathcal{I}_*+\Delta_0)$$

completing the proof.

4.3.2 Estimates for $|r^{2-\frac{2}{p}}(\mathrm{tr}\chi - \overline{\mathrm{tr}\chi})|_{p,S}$ and $|r^{3-2/p}\nabla\!\!\!/ \mathrm{tr}\chi|_{p,S}$, with $p \in [2,4]$

The estimates are

$$|r^{3-2/p}\nabla\!\!\!/ \mathrm{tr}\chi|_{p,S} \leq c(\Gamma_0)\left(\mathcal{I}_* + \Delta_0 + \Gamma_0^2 + \Gamma_0^3\right) \leq c(\mathcal{I}_0 + \mathcal{I}_* + \Delta_0)$$
$$|r^{2-\frac{2}{p}}(\mathrm{tr}\chi - \overline{\mathrm{tr}\chi})|_{p,S} \leq c(\Gamma_0)\big(\mathcal{I}_* + \Delta_0 + \Gamma_0^2 + \Gamma_0^3\big) \leq c(\mathcal{I}_0 + \mathcal{I}_* + \Delta_0). \quad (4.3.17)$$

Remark: To prove these estimates we need, in view of Definition 4.3.5, an estimate of ζ which will be proved later on; see Subsection 4.3.9. We do not present the proof of (4.3.17).[18] As in the estimates in Subsection 4.3.3, we are going to prove a slightly stronger version which will be used in the Main Theorem.

4.3.3 Estimates for $|r^{2-\frac{2}{p}}\tau_-^{\frac{1}{2}}(\mathrm{tr}\chi - \overline{\mathrm{tr}\chi})|_{p,S}$ with $p \in [2,4]$

Remarks:

1. Observe that this estimate is slightly stronger than the one suggested by the bootstrap assumption (4.2.5) and discussed in Subsection 4.3.2.

2. The proof of this estimate, given in Proposition 4.3.14, implies a stronger assumption for $(\mathrm{tr}\chi - \overline{\mathrm{tr}\chi})$ on the last slice and an appropriate estimate for $(\Omega\mathbf{D}_4\log\Omega - \overline{\Omega\mathbf{D}_4\log\Omega})$. Therefore we delay its proof until this last estimate has been proved in Proposition 4.3.13.

4.3.4 Estimate for $|r^{2-\frac{2}{p}}\left(\overline{\Omega\mathrm{tr}\chi} - \frac{1}{r}\right)|_{p,S}$ with $p \in [2,4]$

Remark: As for the previous estimate, the estimate for $\left(\overline{\Omega\mathrm{tr}\chi} - \frac{1}{r}\right)$ requires, preliminary, an estimate for $\Omega\mathbf{D}_4\log\Omega$. Therefore we delay the proof of this result until this last estimate has been proved in Proposition 4.3.4.

4.3.5 Estimates for $\mathcal{O}_{0,1}^{p,S}(\mathrm{tr}\underline{\chi})$ and $\mathcal{O}_{0,1}^{p,S}(\hat{\underline{\chi}})$ with $p \in [2,4]$

Proposition 4.3.2 *Assuming (4.2.1), (4.2.2), (4.2.3) and the bootstrap assumptions (4.2.5), (4.2.6) the following estimates hold:*

$$|r^{2-2/p}\tau_-\nabla\!\!\!/\hat{\underline{\chi}}|_{p,S}(u,\underline{u}) \leq c(\mathcal{I}_0 + \mathcal{I}_* + \Delta_0)$$
$$|r^{1-2/p}\tau_-\hat{\underline{\chi}}|_{p,S}(u,\underline{u}) \leq c(\mathcal{I}_0 + \mathcal{I}_* + \Delta_0). \quad (4.3.18)$$

Proof: We proceed, basically, as in Proposition 4.3.1. We introduce the tensor field

$$\underline{\psi\!\!\!/} = \Omega^{-1}\nabla\!\!\!/ \mathrm{tr}\underline{\chi} - \Omega^{-1}\mathrm{tr}\underline{\chi}\zeta, \quad (4.3.19)$$

---

[18] Given an estimate for $|r^{2-2/p}\zeta|_{p,S}$ we deduce an estimate for $|r^{3-2/p}\nabla\!\!\!/ \mathrm{tr}\chi|_{p,S}$. Differentiating (4.3.5), we obtain $\Delta\!\!\!/ \mathrm{tr}\chi = \mathrm{d}\!\!\!/\mathrm{iv}(\Omega\psi\!\!\!/) - \mathrm{d}\!\!\!/\mathrm{iv}(\mathrm{tr}\chi\zeta)$. $\nabla\!\!\!/\psi\!\!\!/$ can be easily estimated from its evolution equation, obtained differentiating (4.3.6). The result is

$$|r^{4-2/p}\nabla\!\!\!/\psi\!\!\!/|_{p,S}(u,\underline{u}) \leq c(\Gamma_0)\left(\mathcal{I}_0 + \mathcal{I}_* + \Delta_0 + \Gamma_0^2 + \Gamma_0^3\right).$$

Using the estimate already obtained for $\nabla\!\!\!/ \mathrm{tr}\chi$ and the estimate for $\nabla\!\!\!/\zeta$ obtained from Subsection 4.3.9 and applying Lemma 4.1.3 we obtain the result.

and show that $\underline{\Psi}$ satisfies the evolution equation

$$\frac{d}{d\underline{u}}\underline{\Psi}_a + \frac{3}{2}\Omega tr\underline{\chi}\underline{\Psi}_a = \underline{F}_a, \tag{4.3.20}$$

where

$$\underline{F} \equiv -\Omega\hat{\underline{\chi}}\cdot\underline{\Psi} - \underline{\nabla}|\hat{\underline{\chi}}|^2 - \underline{\eta}|\hat{\underline{\chi}}|^2 + \mathrm{tr}\underline{\chi}\;\hat{\underline{\chi}}\cdot\eta - \mathrm{tr}\underline{\chi}\;\underline{\beta}, \tag{4.3.21}$$

satisfies

$$\begin{aligned}|r^{3-\frac{2}{p}}\underline{F}|_{p,S} \;&\le\; |r^{3-\frac{2}{p}}\Omega\hat{\underline{\chi}}\cdot\underline{\Psi}|_{p,S} + |r^{3-\frac{2}{p}}\underline{\nabla}|\hat{\underline{\chi}}|^2|_{p,S} + \Big[|r^{3-\frac{2}{p}}\underline{\eta}|\hat{\underline{\chi}}|^2|_{p,S}\\ &+ |r^{3-\frac{2}{p}}\eta\cdot\hat{\underline{\chi}}\mathrm{tr}\underline{\chi}|_{p,S} + |r^{3-\frac{2}{p}}\mathrm{tr}\underline{\chi}\;\underline{\beta}|_{p,S}\Big].\end{aligned}$$

The various terms on the right-hand side are estimated, with the help of the bootstrap assumption (4.2.5) and assumption (4.2.1) for $\underline{\beta}$,

$$\begin{aligned}
|r^{3-\frac{2}{p}}\Omega\hat{\underline{\chi}}\,\underline{\Psi}|_{p,S} &\le \sup|\Omega|\sup|r\tau_-^{\frac{3}{2}}\hat{\underline{\chi}}||r^{3-\frac{2}{p}}\underline{\Psi}|_{p,S}\frac{1}{r\tau_-^{\frac{3}{2}}} \le c(\Gamma_0)\Gamma_0|r^{3-\frac{2}{p}}\underline{\Psi}|_{p,S}\frac{1}{r\tau_-^{\frac{3}{2}}}\\
|r^{3-\frac{2}{p}}\underline{\nabla}|\hat{\underline{\chi}}|^2|_{p,S} &\le 2\sup|r\tau_-^{\frac{3}{2}}\hat{\underline{\chi}}||r^{2-\frac{2}{p}}\tau_-\underline{\nabla}\hat{\underline{\chi}}|_{p,S}\frac{1}{\tau_-^{\frac{5}{2}}} \le c(\Gamma_0)\Gamma_0\frac{1}{\tau_-^{\frac{5}{2}}}|r^{2-\frac{2}{p}}\tau_-\underline{\nabla}\hat{\underline{\chi}}|_{p,S}\\
|r^{3-\frac{2}{p}}\underline{\eta}|\hat{\underline{\chi}}|^2|_{p,S} &\le \sup|r\tau_-^{\frac{3}{2}}\hat{\underline{\chi}}|^2\sup|r^2\underline{\eta}|_{p,S}\frac{1}{r\tau_-^3} \le \Gamma_0^3\frac{1}{r\tau_-^3}\\
|r^{3-\frac{2}{p}}\eta\hat{\underline{\chi}}\mathrm{tr}\underline{\chi}|_{p,S} &\le \sup|r\tau_-^{\frac{3}{2}}\hat{\underline{\chi}}|\sup|r\mathrm{tr}\underline{\chi}|\sup|r^2\eta|\frac{1}{r\tau_-^{\frac{3}{2}}} \le c(\Gamma_0)\Gamma_0^2\frac{1}{r\tau_-^{\frac{3}{2}}}\\
|r^{3-\frac{2}{p}}\mathrm{tr}\underline{\chi}\;\underline{\beta}|_{p,S} &\le \sup|r\mathrm{tr}\underline{\chi}|\sup|r^2\tau_-^{\frac{3}{2}}\underline{\beta}|\frac{1}{\tau_-^{\frac{3}{2}}} \le c(\Gamma_0)\Delta_0\frac{1}{\tau_-^{\frac{3}{2}}}.
\end{aligned} \tag{4.3.22}$$

From these estimates we infer that

$$\begin{aligned}\int_{u_0(\underline{u})}^{u}|r^{3-\frac{2}{p}}\underline{F}|_{p,S} \;&\le\; c(\Gamma_0)\Gamma_0\int_{u_0(\underline{u})}^{u}\left(\frac{1}{r\tau_-^{\frac{3}{2}}}|r^{3-\frac{2}{p}}\underline{\Psi}|_{p,S} + \frac{1}{\tau_-^{\frac{5}{2}}}|r^{2-\frac{2}{p}}\tau_-\underline{\nabla}\hat{\underline{\chi}}|_{p,S}\right)\\ &+\; c(\Gamma_0)(\Delta_0+\Gamma_0^2+\Gamma_0^3)\left(\frac{1}{r\tau_-^2}+\frac{1}{r\tau_-^{\frac{1}{2}}}+\frac{1}{\tau_-^{\frac{1}{2}}}\right),\end{aligned} \tag{4.3.23}$$

where $u_0 = u_0(\underline{u}) = 2\tilde{\delta}_0 - \underline{u}$. Using the initial assumption $\mathcal{O}_{[1]}(\Sigma_0) \le \mathcal{I}_0$ to control $|r^{3-\frac{2}{p}}\underline{\Psi}|_{p,S_{(0)}}(\underline{u})$ and the results of Lemma 4.1.7 applied to $V = \underline{\Psi}$ we derive

$$\begin{aligned}|r^{3-\frac{2}{p}}\underline{\Psi}|_{p,S}(u,\underline{u}) \;&\le\; c(\Gamma_0)\left(\mathcal{I}_0+\Delta_0+\Gamma_0^2+\Gamma_0^3\right)\\ &+\; c(\Gamma_0)\Gamma_0\int_{u_0}^{u}\left(\frac{1}{r\tau_-^{\frac{3}{2}}}|r^{3-\frac{2}{p}}\underline{\Psi}|_{p,S} + \frac{1}{\tau_-^{\frac{5}{2}}}|r^{2-\frac{2}{p}}\tau_-\underline{\nabla}\hat{\underline{\chi}}|_{p,S}\right),\end{aligned} \tag{4.3.24}$$

for $p \in [2,4]$. The main difference with the previous case is that the integration is made here along the null-incoming hypersurfaces $\underline{C}$.

We are now ready to apply the elliptic L^p estimates of Proposition 4.1.3 to the Hodge system (see (3.1.47))

$$\mathrm{d}\!\!/\mathrm{iv}\,\hat{\underline{\chi}} = \zeta\hat{\underline{\chi}} + \frac{1}{2}\Omega\underline{\Psi\!\!\!/} - \underline{\beta}. \tag{4.3.25}$$

We thus obtain

$$\begin{aligned} |r^{2-\frac{2}{p}}\tau_-\nabla\!\!\!\!/\hat{\underline{\chi}}|_{p,S} &\leq c(\Gamma_0)\Bigg(|r^{3-\frac{2}{p}}\underline{\Psi\!\!\!/}|_{p,S}\frac{\tau_-}{r} + \sup|r^{2-\frac{2}{p}}\tau_-^{3/2}\underline{\beta}|\frac{1}{\tau_-^{\frac{1}{2}}} \\ &\quad + \sup|r^2\zeta|\sup|r\tau_-^{\frac{3}{2}}\hat{\underline{\chi}}|\frac{1}{r\tau_-^{\frac{1}{2}}}\Bigg) \\ &\leq c(\Gamma_0)\Bigg(|r^{3-\frac{2}{p}}\underline{\Psi\!\!\!/}|_{p,S}\frac{\tau_-}{r} + \frac{\Delta_0}{\tau_-^{\frac{1}{2}}} + \frac{\Gamma_0^2}{r\tau_-^{\frac{1}{2}}}\Bigg), \end{aligned} \tag{4.3.26}$$

which, substituted in (4.3.24) gives

$$\begin{aligned} |r^{3-\frac{2}{p}}\underline{\Psi\!\!\!/}|_{p,S}(u,\underline{u}) &\leq c(\Gamma_0)\big(\mathcal{I}_0 + \Delta_0 + \Gamma_0^2 + \Gamma_0^3\big) + c(\Gamma_0)\Gamma_0\Bigg[\int_{u_0}^{u}\frac{1}{r\tau_-^{\frac{3}{2}}}|r^{3-\frac{2}{p}}\underline{\Psi\!\!\!/}|_{p,S} \\ &\quad + \int_{u_0}^{u}\frac{1}{\tau_-^3}\left(\frac{\Gamma_0^2}{r} + \Delta_0\right)\Bigg], \end{aligned} \tag{4.3.27}$$

and by the Gronwall inequality,

$$|r^{3-2/p}\underline{\Psi\!\!\!/}|_{p,S}(u,\underline{u}) \leq c(\Gamma_0)\big(\mathcal{I}_0 + \Delta_0 + \Gamma_0^2 + \Gamma_0^3\big). \tag{4.3.28}$$

From (4.3.28), going back to (4.3.26), we deduce that

$$\begin{aligned} &|r^{1-2/p}\tau_-\hat{\underline{\chi}}|_{p,S}(u,\underline{u}) \leq c(\Gamma_0)\left(\mathcal{I}_0 + \Delta_0 + \Gamma_0^2 + \Gamma_0^3\right) \leq c(\mathcal{I}_0 + \Delta_0) \\ &|r^{2-2/p}\tau_-\nabla\!\!\!\!/\hat{\underline{\chi}}|_{p,S}(u,\underline{u}) \leq c(\Gamma_0)\left(\mathcal{I}_0 + \Delta_0 + \Gamma_0^2 + \Gamma_0^3\right) \leq c(\mathcal{I}_0 + \Delta_0), \end{aligned}$$

proving the proposition.

4.3.6 Estimates for $|r^{2-\frac{2}{p}}(\mathrm{tr}\underline{\chi} - \overline{\mathrm{tr}\underline{\chi}})|_{p,S}$ and $|r^{3-2/p}\nabla\!\!\!\!/\mathrm{tr}\underline{\chi}|_{p,S}$, with $p \in [2,4]$

$$\begin{aligned} &|r^{3-2/p}\nabla\!\!\!\!/\mathrm{tr}\underline{\chi}|_{p,S} \leq c(\Gamma_0)\left(\mathcal{I}_* + \Delta_0 + \Gamma_0^2 + \Gamma_0^3\right) \leq c(\mathcal{I}_0 + \mathcal{I}_* + \Delta_0) \\ &|r^{2-\frac{2}{p}}(\mathrm{tr}\underline{\chi} - \overline{\mathrm{tr}\underline{\chi}})|_{p,S} \leq c(\Gamma_0)\big(\mathcal{I}_0 + \Delta_0 + \Gamma_0^2 + \Gamma_0^3\big) \leq c(\mathcal{I}_0 + \mathcal{I}_* + \Delta_0). \end{aligned} \tag{4.3.29}$$

Remark: To prove these estimates we need an estimate of ζ which will be proved later on; see Subsection 4.3.9. We do not present the proof of (4.3.29).[19] In the estimates in Subsection 4.3.7 we are going to prove a slightly stronger version which will be used in the Main Theorem.

---

[19]Given an estimate for $|r^{2-2/p}\zeta|_{p,S}$ we deduce an estimate for $|r^{3-2/p}\nabla\!\!\!\!/\mathrm{tr}\underline{\chi}|_{p,S}$. Differentiating (4.3.19), we obtain $\Delta\!\!\!\!/\mathrm{tr}\underline{\chi} = \mathrm{d}\!\!/\mathrm{iv}\,(\Omega\underline{\Psi\!\!\!/}) + \mathrm{d}\!\!/\mathrm{iv}\,(\mathrm{tr}\underline{\chi}\zeta)$. $\nabla\!\!\!\!/\underline{\Psi\!\!\!/}$ can be easily estimated from its evolution equation, obtained by differentiating (4.3.20).

4.3.7 Estimate for $|r^{2-\frac{2}{p}}\tau_-^{\frac{1}{2}}(\mathrm{tr}\underline{\chi}-\overline{\mathrm{tr}\underline{\chi}})|_{p,S}$ with $p\in[2,4]$

Remarks:

1. Observe that, as in the case of $(\mathrm{tr}\chi-\overline{\mathrm{tr}\chi})$, the estimate we are going to prove is slightly stronger than the one suggested by the bootstrap assumption (4.2.5) and in fact implies it.[20]

2. The proof of this estimate, given in Proposition 4.3.14, implies a stronger assumption for $(\mathrm{tr}\underline{\chi}-\overline{\mathrm{tr}\underline{\chi}})$ on the initial hypersurface and some estimate for $(\Omega\mathbf{D}_3\log\Omega-\overline{\Omega\mathbf{D}_3\log\Omega})$. Therefore we delay its proof until this last estimate has been proved in Proposition 4.3.13.

4.3.8 Estimate for $|r^{1-\frac{2}{p}}\tau_-\left(\overline{\Omega\mathrm{tr}\underline{\chi}}+\frac{1}{r}\right)|_{p,S}$ with $p\in[2,4]$

Remark: As for the previous estimate, the estimate for $\left(\overline{\Omega\mathrm{tr}\underline{\chi}}+\frac{1}{r}\right)$ requires, preliminarily, an estimate for $\Omega\mathbf{D}_3\log\Omega$. Therefore we delay the proof of this result until this last estimate has been proved in Proposition 4.3.4.

4.3.9 Estimates for $\mathcal{O}_{0,1}^{p,S}(\eta)$ and $\mathcal{O}_{0,1}^{p,S}(\underline{\eta})$ with $p\in[2,4]$

Proposition 4.3.3 *Assuming (4.2.1), (4.2.2), (4.2.3) and (4.2.5), (4.2.6), then the following estimates hold:*

$$\begin{aligned}
|r^{2-2/p}\eta|_{p,S}(u,\underline{u}) &\le c\,(\mathcal{I}_*+\mathcal{I}_0+\Delta_0)\\
|r^{3-2/p}\nabla\!\!\!/\,\eta|_{p,S}(u,\underline{u}) &\le c\,(\mathcal{I}_*+\mathcal{I}_0+\Delta_0)\\
|r^{2-2/p}\underline{\eta}|_{p,S}(u,\underline{u}) &\le c\,(\mathcal{I}_*+\mathcal{I}_0+\Delta_0)\\
|r^{3-2/p}\nabla\!\!\!/\,\underline{\eta}|_{p,S}(u,\underline{u}) &\le c\,(\mathcal{I}_*+\mathcal{I}_0+\Delta_0)\,.
\end{aligned}\tag{4.3.30}$$

Proof: To obtain the norm estimates for η, $\underline{\eta}$ and their first tangential derivatives we recall the equations (see (3.1.46))

$$\begin{aligned}
\mathbf{D}\!\!\!\!/_4\eta &= -\chi\cdot\eta+\chi\cdot\underline{\eta}-\beta\\
\mathbf{D}\!\!\!\!/_3\underline{\eta} &= -\underline{\chi}\cdot\underline{\eta}+\underline{\chi}\cdot\eta+\underline{\beta}\,.
\end{aligned}$$

If we differentiate these equations to obtain estimates for $\nabla\!\!\!/\,\eta$ or $\nabla\!\!\!/\,\underline{\eta}$ their orders of differentiability will be the same as those of $\nabla\!\!\!/\,\beta$, $\nabla\!\!\!/\,\underline{\beta}$. To avoid this loss of derivatives we proceed as in [Ch-Kl], with the help of the mass aspect functions introduced in Chapter 2, equations (3.3.6)

$$\begin{aligned}
\mu &= -\mathrm{d}\!\!\!/\mathrm{iv}\,\eta+\frac{1}{2}\hat{\chi}\cdot\underline{\hat{\chi}}-\rho\\
\underline{\mu} &= -\mathrm{d}\!\!\!/\mathrm{iv}\,\underline{\eta}+\frac{1}{2}\underline{\hat{\chi}}\cdot\hat{\chi}-\rho
\end{aligned}\tag{4.3.31}$$

The result is

$$|r^{4-2/p}\nabla\!\!\!/\,\underline{\psi\!\!\!/}|_{p,S}(u,\underline{u})\le c(\Gamma_0)\left(\mathcal{I}_0+\mathcal{I}_*+\Delta_0+\Gamma_0^2+\Gamma_0^3\right).$$

Using the estimate already obtained for $\nabla\!\!\!/\,\mathrm{tr}\chi$ and the estimate for $\nabla\!\!\!/\,\zeta$ obtained from Subsection 4.3.9 and applying Lemma 4.1.3 we obtain the result.

[20] Recall that $\mathcal{O}^{p,S}_{q=0}(\mathrm{tr}\underline{\chi})(\lambda,\nu)=|r^2(\mathrm{tr}\underline{\chi}-\overline{\mathrm{tr}\underline{\chi}})|_{p,S}$.

which satisfy the following lemma, proved by a direct computation:

Lemma 4.3.1 *The scalar functions μ, $\underline{\mu}$ satisfy the following evolution equations*

$$\begin{aligned} \frac{d}{d\underline{u}}\mu + (\Omega\mathrm{tr}\chi)\mu &= G + \frac{1}{2}(\Omega\mathrm{tr}\chi)\underline{\mu} \\ \frac{d}{du}\underline{\mu} + (\Omega\mathrm{tr}\underline{\chi})\underline{\mu} &= \underline{G} + \frac{1}{2}(\Omega\mathrm{tr}\underline{\chi})\mu, \end{aligned} \tag{4.3.32}$$

where

$$\begin{aligned} G &\equiv \Omega\hat{\chi}\cdot(\not\nabla\widehat{\otimes}\eta) + \Omega^2(\eta-\underline{\eta})\cdot\not\Psi - \frac{1}{4}\Omega\mathrm{tr}\underline{\chi}|\hat{\chi}|^2 \\ &\quad + \frac{1}{2}\Omega\mathrm{tr}\chi\left(-\hat{\chi}\cdot\hat{\underline{\chi}} + 2\rho - |\underline{\eta}|^2\right) + 2\Omega(\eta\cdot\hat{\chi}\cdot\underline{\eta} - \eta\cdot\beta) \\ \underline{G} &\equiv \Omega\hat{\underline{\chi}}\cdot(\not\nabla\widehat{\otimes}\underline{\eta}) + \Omega^2(\underline{\eta}-\eta)\cdot\underline{\not\Psi} - \frac{1}{4}\Omega\mathrm{tr}\chi|\hat{\underline{\chi}}|^2 \\ &\quad + \frac{1}{2}\Omega\mathrm{tr}\underline{\chi}\left(-\hat{\underline{\chi}}\cdot\hat{\chi} + 2\rho - |\eta|^2\right) + 2\Omega(\underline{\eta}\cdot\hat{\underline{\chi}}\cdot\eta + \underline{\eta}\cdot\underline{\beta}) \end{aligned} \tag{4.3.33}$$

Using equations (4.3.32) it is possible to obtain estimates for μ, $\underline{\mu}$ which, together with the Hodge systems

$$\begin{aligned} \not{\mathrm{d}}\mathrm{iv}\,\eta &= -\mu + \frac{1}{2}\hat{\chi}\cdot\hat{\underline{\chi}} - \rho \\ \mathrm{c}\not{\mathrm{u}}\mathrm{rl}\,\eta &= \sigma - \frac{1}{2}\hat{\underline{\chi}}\wedge\hat{\chi} \end{aligned} \tag{4.3.34}$$

$$\begin{aligned} \not{\mathrm{d}}\mathrm{iv}\,\underline{\eta} &= -\underline{\mu} + \frac{1}{2}\hat{\underline{\chi}}\cdot\hat{\chi} - \rho \\ \mathrm{c}\not{\mathrm{u}}\mathrm{rl}\,\underline{\eta} &= -\sigma - \frac{1}{2}\hat{\chi}\wedge\hat{\underline{\chi}} \end{aligned} \tag{4.3.35}$$

enables one to control the norms $\mathcal{O}^{p,S}_{0,1}(\eta)$, $\mathcal{O}^{p,S}_{0,1}(\underline{\eta})$. We choose, nevertheless, a slightly different method that allows one to obtain the result in a easier way and, moreover, to obtain a somewhat improved estimate if the assumptions on the initial and final slices are stronger.

We start by introducing two different functions $\tilde{\mu}$ and $\underline{\tilde{\mu}}$:

$$\begin{aligned} \tilde{\mu} &= (\mu - \overline{\mu}) + \frac{1}{4}(\mathrm{tr}\chi\mathrm{tr}\underline{\chi} - \overline{\mathrm{tr}\chi\mathrm{tr}\underline{\chi}}) = -\not{\mathrm{d}}\mathrm{iv}\,\eta + \frac{1}{2}(\chi\cdot\underline{\chi} - \overline{\chi\cdot\underline{\chi}}) - (\rho - \overline{\rho}) \\ \underline{\tilde{\mu}} &= (\underline{\mu} - \overline{\underline{\mu}}) + \frac{1}{4}(\mathrm{tr}\underline{\chi}\mathrm{tr}\chi - \overline{\mathrm{tr}\underline{\chi}\mathrm{tr}\chi}) = -\not{\mathrm{d}}\mathrm{iv}\,\underline{\eta} + \frac{1}{2}(\underline{\chi}\cdot\chi - \overline{\underline{\chi}\cdot\chi}) - (\rho - \overline{\rho}). \end{aligned} \tag{4.3.36}$$

Relative to these functions η and $\underline{\eta}$ satisfy the following Hodge systems:

$$\begin{aligned} \not{\mathrm{d}}\mathrm{iv}\,\eta &= -\tilde{\mu} + \frac{1}{2}(\chi\cdot\underline{\chi} - \overline{\chi\cdot\underline{\chi}}) - (\rho - \overline{\rho}) \\ \mathrm{c}\not{\mathrm{u}}\mathrm{rl}\,\eta &= \sigma - \frac{1}{2}\hat{\underline{\chi}}\wedge\hat{\chi} \end{aligned} \tag{4.3.37}$$

$$\not{d}\text{iv}\,\underline{\eta} = -\underline{\tilde{\mu}} + \frac{1}{2}(\underline{\chi}\cdot\chi - \overline{\underline{\chi}\cdot\chi}) - (\rho - \overline{\rho})$$
$$\not{c}\text{url}\,\underline{\eta} = -\sigma - \frac{1}{2}\hat{\chi}\wedge\underline{\hat{\chi}}. \tag{4.3.38}$$

The functions $\tilde{\mu}$ and $\underline{\tilde{\mu}}$ satisfy the following lemma,

Lemma 4.3.2 *$\tilde{\mu}$ and $\underline{\tilde{\mu}}$ satisfy the following evolution equations:*

$$\frac{d}{d\underline{u}}\tilde{\mu} + (\Omega\text{tr}\chi)\tilde{\mu} = (\Omega\tilde{F} - \overline{\Omega\tilde{F}}) + (\Omega\tilde{H} - \overline{\Omega\tilde{H}}) \tag{4.3.39}$$
$$\frac{d}{du}\underline{\tilde{\mu}} + (\Omega\text{tr}\underline{\chi})\underline{\tilde{\mu}} = (\Omega\underline{\tilde{F}} - \overline{\Omega\underline{\tilde{F}}}) + (\Omega\underline{\tilde{H}} - \overline{\Omega\underline{\tilde{H}}})$$

where

$$\tilde{F} = \hat{\chi}\cdot(\not{\nabla}\widehat{\otimes}\eta) + \Omega(\eta - \underline{\eta})\cdot\not{\Psi} + \frac{1}{2}\text{tr}\chi(|\eta|^2 - |\underline{\eta}|^2) \tag{4.3.40}$$
$$-\frac{1}{2}\left(\text{tr}\underline{\chi}|\hat{\chi}|^2 + \text{tr}\chi(\hat{\chi}\cdot\underline{\hat{\chi}})\right) + 2\eta\cdot\hat{\chi}\cdot\underline{\eta}$$
$$\tilde{H} = \left(-\text{tr}\chi\frac{1}{2}\overline{\chi\cdot\underline{\chi}} - 2\eta\cdot\beta + \text{tr}\chi(\rho + \overline{\rho})\right)$$
$$\underline{\tilde{F}} = \underline{\hat{\chi}}\cdot(\not{\nabla}\widehat{\otimes}\underline{\eta}) + \Omega(\underline{\eta} - \eta)\cdot\underline{\not{\Psi}} + \frac{1}{2}\Omega\text{tr}\underline{\chi}(|\underline{\eta}|^2 - |\eta|^2)$$
$$-\frac{1}{2}\left(\text{tr}\chi|\underline{\hat{\chi}}|^2 + \text{tr}\underline{\chi}(\underline{\hat{\chi}}\cdot\hat{\chi})\right) + 2\underline{\eta}\cdot\underline{\hat{\chi}}\cdot\eta$$
$$\underline{\tilde{H}} = \left(-\text{tr}\chi\frac{1}{2}\overline{\underline{\chi}\cdot\chi} + 2\underline{\eta}\cdot\underline{\beta} + \text{tr}\underline{\chi}(\rho + \overline{\rho})\right). \tag{4.3.41}$$

Proof: Recalling the definition

$$\tilde{\mu} = (\mu - \overline{\mu}) + \frac{1}{4}(\text{tr}\chi\,\text{tr}\underline{\chi} - \overline{\text{tr}\chi\,\text{tr}\underline{\chi}})\,,$$

we observe that

$$\mathbf{D}_4\left(\mu + \frac{1}{4}\text{tr}\chi\,\text{tr}\underline{\chi}\right) = \mathbf{D}_4\mu + \frac{1}{4}\left(\text{tr}\underline{\chi}\mathbf{D}_4\text{tr}\chi + \text{tr}\chi\mathbf{D}_4\text{tr}\underline{\chi}\right)$$
$$= \left[-\text{tr}\chi\mu + \Omega^{-1}G + \frac{1}{2}\text{tr}\chi\left(\frac{1}{2}\underline{\hat{\chi}}\cdot\hat{\chi} - \rho - \not{d}\text{iv}\,\underline{\eta}\right)\right]$$
$$+\frac{1}{4}\text{tr}\underline{\chi}\left[(\mathbf{D}_4\log\Omega)\text{tr}\chi - \frac{1}{2}(\text{tr}\chi)^2 - |\hat{\chi}|^2\right] \tag{4.3.42}$$
$$+\frac{1}{4}\text{tr}\chi\left[-(\mathbf{D}_4\log\Omega)\text{tr}\underline{\chi} - \frac{1}{2}\text{tr}\chi\,\text{tr}\underline{\chi} - \hat{\chi}\cdot\underline{\hat{\chi}} + 2\not{d}\text{iv}\,\underline{\eta} + 2|\eta|^2 + 2\rho\right],$$

which can be rewritten as

$$\frac{d}{d\underline{u}}\left(\mu + \frac{1}{4}\text{tr}\chi\,\text{tr}\underline{\chi}\right) + (\Omega\text{tr}\chi)\left(\mu + \frac{1}{4}\text{tr}\chi\,\text{tr}\underline{\chi}\right) = \Omega\tilde{F} + \Omega(\text{tr}\chi\rho - 2\eta\cdot\beta). \tag{4.3.43}$$

Posing $[\mu] \equiv \left(\mu + \frac{1}{4}\text{tr}\chi\,\text{tr}\underline{\chi}\right)$ we have

$$\begin{aligned}
\frac{d}{d\underline{u}}\overline{[\mu]} &= \frac{1}{|S|}\int_S \frac{d}{d\underline{u}}[\mu] - \frac{1}{|S|}\frac{\partial|S|}{\partial\underline{u}}\overline{[\mu]} + \frac{1}{|S|}\int_S \Omega\text{tr}\chi[\mu] \\
&= \frac{1}{|S|}\int_S \Omega\left(-\text{tr}\chi[\mu] + \tilde{F} + (\text{tr}\chi\rho - 2\eta\cdot\beta)\right) - (\overline{\Omega\text{tr}\chi})\overline{[\mu]} + \overline{\Omega\text{tr}\chi[\mu]} \\
&= \overline{\Omega\tilde{F}} + \overline{\Omega(\text{tr}\chi\rho - 2\eta\cdot\beta)} - (\overline{\Omega\text{tr}\chi})\overline{[\mu]} \qquad (4.3.44)
\end{aligned}$$

which implies

$$\frac{d}{d\underline{u}}\overline{[\mu]} + (\overline{\Omega\text{tr}\chi})\overline{[\mu]} = \overline{\Omega\tilde{F}} + \overline{\Omega(\text{tr}\chi\rho - 2\eta\cdot\beta)} + \overline{(\Omega\text{tr}\chi - \overline{\Omega\text{tr}\chi})[\mu]} \qquad (4.3.45)$$

$$= \overline{\Omega\tilde{F}} - 2\overline{\Omega\eta\cdot\beta} + \overline{(\Omega\text{tr}\chi - \overline{\Omega\text{tr}\chi})\frac{1}{2}\chi\cdot\underline{\chi}} \quad + \overline{\Omega\text{tr}\chi\rho} - (\overline{\Omega\text{tr}\chi} - \overline{\Omega\text{tr}\chi})\overline{\rho}.$$

Combining (4.3.42) and (4.3.45) we obtain

$$\begin{aligned}
\frac{d}{d\underline{u}}\tilde{\mu} + (\Omega\text{tr}\chi)\tilde{\mu} &= (\Omega\tilde{F} - \overline{\Omega\tilde{F}}) - \overline{(\Omega\text{tr}\chi - \overline{\Omega\text{tr}\chi})\frac{1}{2}\chi\cdot\underline{\chi}} - 2(\Omega\eta\cdot\beta - \overline{\Omega\eta\cdot\beta}) \\
&\quad + (\Omega\text{tr}\chi\rho - \overline{\Omega\text{tr}\chi\rho}) + (\Omega\text{tr}\chi - \overline{\Omega\text{tr}\chi})\overline{\rho}, \qquad (4.3.46)
\end{aligned}$$

which completes the proof of the first part of Lemma 4.3.2. The proof for $\underline{\tilde{\mu}}$ is exactly analogous to the one for $\tilde{\mu}$ and we do not present it here.

Using the evolution equations (4.3.39), the Evolution Lemma with $\lambda_1 = 2 - \frac{2}{p}$, and the elliptic estimates of Proposition 4.1.3 we prove the following lemma.

Lemma 4.3.3 *Under the bootstrap assumption (4.2.5), if $\Gamma_0 > 0$ is sufficiently small, then the following inequality holds:*

$$|r^{3-\frac{2}{p}}\tilde{\mu}|_{p,S}(u,\underline{u}) \leq c\left[|r^{3-\frac{2}{p}}\tilde{\mu}|_{p,S}(u,\underline{u}_*) + c(\Gamma_0)(\mathcal{I}_* + \Delta_0)\right], \qquad (4.3.47)$$

and the corresponding one for $\underline{\mu}$,

$$|r^{3-\frac{2}{p}}\underline{\tilde{\mu}}|_{p,S}(u,\underline{u}) \leq c\left[|r^{3-\frac{2}{p}}\underline{\tilde{\mu}}|_{p,S}(u_0,\underline{u}) + c(\Gamma_0)(\mathcal{I}_0 + \Delta_0)\right]. \qquad (4.3.48)$$

Proof of Lemma 4.3.3: We apply the Evolution Lemma with $\lambda_1 = 2 - \frac{2}{p}$ to equations (4.3.39) and derive

$$\begin{aligned}
|r^{2-\frac{2}{p}}\tilde{\mu}|_{p,S}(u,\underline{u}) &\leq c\bigg(|r^{2-\frac{2}{p}}\tilde{\mu}|_{p,S}(u,\underline{u}_*) + \int_{\underline{u}}^{\underline{u}_*} |r^{2-\frac{2}{p}}(\Omega\tilde{F} - \overline{\Omega\tilde{F}})|_{p,S} \\
&\quad + \int_{\underline{u}}^{\underline{u}_*} |r^{2-\frac{2}{p}}(\Omega\tilde{H} - \overline{\Omega\tilde{H}})|_{p,S}\bigg), \qquad (4.3.49) \\
|r^{2-\frac{2}{p}}\underline{\tilde{\mu}}|_{p,S}(u,\underline{u}) &\leq c\bigg(|r^{2-\frac{2}{p}}\underline{\tilde{\mu}}|_{p,S}(u_0,\underline{u}) + \int_{u_0}^{u} |r^{2-\frac{2}{p}}(\Omega\underline{\tilde{F}} - \overline{\Omega\underline{\tilde{F}}})|_{p,S}) \\
&\quad + \int_{u_0}^{u} |r^{2-\frac{2}{p}}(\Omega\underline{\tilde{H}} - \overline{\Omega\underline{\tilde{H}}})|_{p,S}\bigg). \qquad (4.3.50)
\end{aligned}$$

The various terms in $|r^{2-\frac{2}{p}}(\Omega\tilde{F}-\overline{\Omega\tilde{F}})|_{p,S}$, $|r^{2-\frac{2}{p}}(\Omega\underline{\tilde{F}}-\overline{\Omega\underline{\tilde{F}}})|_{p,S}$ are estimated using the bootstrap assumption (4.2.5), the estimates (4.3.16), (4.3.28) for $\not{\Psi}$ and $\underline{\not{\Psi}}$ and the assumption (4.2.1) for ρ, β and $\underline{\beta}$. We obtain

$$
\begin{aligned}
&|r^{2-\frac{2}{p}}\Omega\hat{\chi}\cdot(\not{\nabla}\widehat{\otimes}\eta)|_{p,S}\leq c(\Gamma_0)\Gamma_0|r^{2-\frac{2}{p}}\not{\nabla}\eta|_{p,S}\frac{1}{r^2}\\
&|r^{2-\frac{2}{p}}\Omega^2(\eta-\underline{\eta})\cdot\not{\Psi}|_{p,S}\leq c(\Gamma_0)\Gamma_0(\mathcal{I}_*+\Delta_0)\frac{1}{r^3}\\
&|r^{2-\frac{2}{p}}\Omega\mathrm{tr}\chi(|\underline{\eta}|^2-|\eta|^2)|_{p,S}\leq c(\Gamma_0)\Gamma_0^2\frac{1}{r^3}\\
&|r^{2-\frac{2}{p}}\Omega\eta\cdot\hat{\chi}\cdot\underline{\eta}|_{p,S}\leq c(\Gamma_0)\Gamma_0^3\frac{1}{r^4}\\
&|r^{2-\frac{2}{p}}\left(\mathrm{tr}\underline{\chi}|\hat{\chi}|^2+\mathrm{tr}\chi(\hat{\chi}\cdot\underline{\hat{\chi}})\right)|_{p,S}\leq c(\Gamma_0)\Gamma_0^2\frac{1}{r^2|u|}
\end{aligned}
\tag{4.3.51}
$$

$$
\begin{aligned}
&|r^{2-\frac{2}{p}}\Omega\underline{\hat{\chi}}\cdot(\not{\nabla}\widehat{\otimes}\underline{\eta})|_{p,S}\leq c(\Gamma_0)\Gamma_0|r^{2-\frac{2}{p}}\not{\nabla}\underline{\eta}|_{p,S}\frac{1}{r|u|}\\
&|r^{2-\frac{2}{p}}\Omega^2(\underline{\eta}-\eta)\cdot\underline{\not{\Psi}}|_{p,S}\leq c(\Gamma_0)\Gamma_0(\mathcal{I}_0+\Delta_0)\frac{1}{r^3}\\
&|r^{2-\frac{2}{p}}\Omega\mathrm{tr}\underline{\chi}(|\underline{\eta}|^2-|\eta|^2)|_{p,S}\leq c(\Gamma_0)\Gamma_0^2\frac{1}{r^3}\\
&|r^{2-\frac{2}{p}}\Omega\underline{\eta}\cdot\underline{\hat{\chi}}\cdot\eta|_{p,S}\leq c(\Gamma_0)\Gamma_0^3\frac{1}{r^3u}\\
&|r^{2-\frac{2}{p}}\left(\mathrm{tr}\chi|\underline{\hat{\chi}}|^2+\mathrm{tr}\underline{\chi}(\underline{\hat{\chi}}\cdot\hat{\chi})\right)|_{p,S}\leq c(\Gamma_0)\Gamma_0^2\frac{1}{ru^2}.
\end{aligned}
\tag{4.3.52}
$$

To estimate the terms appearing in the explicit expressions of $|r^{2-\frac{2}{p}}(\Omega\tilde{H}-\overline{\Omega\tilde{H}})|_{p,S}$ and $|r^{2-\frac{2}{p}}(\Omega\underline{\tilde{H}}-\overline{\Omega\underline{\tilde{H}}})|_{p,S}$ we observe that

$$
\begin{aligned}
(\Omega\tilde{H}-\overline{\Omega\tilde{H}}) &= (\Omega\mathrm{tr}\chi-\overline{\Omega\mathrm{tr}\chi})\frac{1}{2}\overline{\chi\cdot\underline{\chi}}-2(\Omega\eta\cdot\beta-\overline{\Omega\eta\cdot\beta})\\
&+(\Omega\mathrm{tr}\chi\rho-\Omega\mathrm{tr}\chi\overline{\rho})+2(\Omega\mathrm{tr}\chi-\overline{\Omega\mathrm{tr}\chi})\overline{\rho}-\overline{(\Omega\mathrm{tr}\chi\rho-\Omega\mathrm{tr}\chi\overline{\rho})}\\
(\Omega\underline{\tilde{H}}-\overline{\Omega\underline{\tilde{H}}}) &= (\Omega\mathrm{tr}\underline{\chi}-\overline{\Omega\mathrm{tr}\underline{\chi}})\frac{1}{2}\overline{\underline{\chi}\cdot\chi}+2(\Omega\underline{\eta}\cdot\underline{\beta}-\overline{\Omega\underline{\eta}\cdot\underline{\beta}})\\
&+(\Omega\mathrm{tr}\underline{\chi}\rho-\Omega\mathrm{tr}\underline{\chi}\overline{\rho})+2(\Omega\mathrm{tr}\underline{\chi}-\overline{\Omega\mathrm{tr}\underline{\chi}})\overline{\rho}-\overline{(\Omega\mathrm{tr}\underline{\chi}\rho-\Omega\mathrm{tr}\underline{\chi}\overline{\rho})}.
\end{aligned}
$$

Using the bootstrap assumption (4.2.5) and assumptions (4.2.1) for ρ, β and $\underline{\beta}$ we have

$$
\begin{aligned}
&|r^{2-\frac{2}{p}}(\Omega\mathrm{tr}\chi-\overline{\Omega\mathrm{tr}\chi})\overline{\chi\cdot\underline{\chi}}|_{p,S}\leq c(\Gamma_0)\Gamma_0^2\frac{1}{r^3|u|}\\
&|r^{2-\frac{2}{p}}(\Omega\mathrm{tr}\underline{\chi}-\overline{\Omega\mathrm{tr}\underline{\chi}})\overline{\underline{\chi}\cdot\chi}|_{p,S}\leq c(\Gamma_0)\Gamma_0^2\frac{1}{r^3|u|}\\
&|r^{2-\frac{2}{p}}(\Omega\eta\cdot\beta-\overline{\Omega\eta\cdot\beta})|_{p,S}\leq c(\Gamma_0)\Gamma_0\Delta_0\frac{1}{r^{\frac{7}{2}}}
\end{aligned}
$$

$$
\begin{aligned}
&|r^{2-\frac{2}{p}}(\Omega\underline{\eta}\cdot\underline{\beta}-\overline{\Omega\underline{\eta}\cdot\underline{\beta}})|_{p,S}\leq c(\Gamma_0)\Gamma_0\Delta_0\frac{1}{r^2|u|^{\frac{3}{2}}} \qquad (4.3.53)\\
&|r^{2-\frac{2}{p}}(\Omega\mathrm{tr}\chi\rho-\Omega\mathrm{tr}\chi\overline{\rho})|_{p,S}\leq c(\Gamma_0)\Delta_0\frac{1}{r^2|u|^{\frac{1}{2}}}\\
&|r^{2-\frac{2}{p}}(\Omega\mathrm{tr}\underline{\chi}\rho-\Omega\mathrm{tr}\underline{\chi}\overline{\rho})|_{p,S}\leq c(\Gamma_0)\Delta_0\frac{1}{r^2|u|^{\frac{1}{2}}}\\
&|r^{2-\frac{2}{p}}(\Omega\mathrm{tr}\chi-\overline{\Omega\mathrm{tr}\chi})\overline{\rho}|_{p,S}\leq c(\Gamma_0)\Delta_0\frac{1}{r^3}\\
&|r^{2-\frac{2}{p}}(\Omega\mathrm{tr}\underline{\chi}-\overline{\Omega\mathrm{tr}\underline{\chi}})\overline{\rho}|_{p,S}\leq c(\Gamma_0)\Delta_0\frac{1}{r^3}.
\end{aligned}
$$

Using the inequalities (4.3.51), (4.3.52) we derive the estimates

$$
\begin{aligned}
&|r^{2-\frac{2}{p}}(\Omega\tilde{F}-\overline{\Omega\tilde{F}})|_{p,S}\leq c(\Gamma_0)\left[\Gamma_0\frac{1}{r^2}|r^{2-\frac{2}{p}}\nabla\!\!\!/\eta|_{p,S}+\frac{1}{r^3}\left((\mathcal{I}_*+\Delta_0)+\Gamma_0^2\right)\right]\\
&|r^{2-\frac{2}{p}}(\Omega\underline{\tilde{F}}-\overline{\Omega\underline{\tilde{F}}})|_{p,S}\leq c(\Gamma_0)\left[\Gamma_0\frac{1}{ru}|r^{2-\frac{2}{p}}\nabla\!\!\!/\underline{\eta}|_{p,S}+\frac{1}{r^2u}\left((\mathcal{I}_0+\Delta_0)+\Gamma_0^2\right)\right]. \qquad (4.3.54)
\end{aligned}
$$

We estimate $|r^{2-\frac{2}{p}}\nabla\!\!\!/\eta|_{p,S}$, $|r^{2-\frac{2}{p}}\nabla\!\!\!/\underline{\eta}|_{p,S}$ with the help of the Hodge systems (4.3.37), (4.3.38). In view of Proposition 4.1.3 we derive

$$
\begin{aligned}
|r^{2-\frac{2}{p}}\nabla\!\!\!/\eta|_{p,S} &\leq c|r^{2-\frac{2}{p}}\tilde{\mu}|_{p,S}+c\frac{1}{r}|r^{3-\frac{2}{p}}(\rho-\overline{\rho},\sigma)|_{p,S}+\frac{1}{r|u|}\Gamma_0^2\\
&\quad +c\frac{1}{r}\left(\mathcal{I}_0+\mathcal{I}_*+\Delta_0+\Gamma_0^2\right)\\
&\leq c|r^{2-\frac{2}{p}}\tilde{\mu}|_{p,S}+c\frac{1}{r}\left(\mathcal{I}_0+\mathcal{I}_*+\Delta_0+\Gamma_0^2\right)+\frac{1}{r|u|}\Gamma_0^2\\
|r^{2-\frac{2}{p}}\nabla\!\!\!/\underline{\eta}|_{p,S} &\leq c|r^{2-\frac{2}{p}}\underline{\tilde{\mu}}|_{p,S}+c\frac{1}{r}|r^{3-\frac{2}{p}}(\rho,\sigma)|_{p,S}+\frac{1}{ru}\Gamma_0^2+c\frac{1}{r}\left(\mathcal{I}_0+\mathcal{I}_*+\Delta_0+\Gamma_0^2\right)\\
&\leq c|r^{2-\frac{2}{p}}\underline{\tilde{\mu}}|_{p,S}+c\frac{1}{r}\left(\mathcal{I}_0+\mathcal{I}_*+\Delta_0+\Gamma_0^2\right)+\frac{1}{r|u|}\Gamma_0^2. \qquad (4.3.55)
\end{aligned}
$$

Therefore inequalities (4.3.54) become

$$
\begin{aligned}
&|r^{2-\frac{2}{p}}(\Omega\tilde{F}-\overline{\Omega\tilde{F}})|_{p,S}\leq c(\Gamma_0)\left[\Gamma_0\frac{1}{r^2}|r^{2-\frac{2}{p}}\tilde{\mu}|_{p,S}+\frac{1}{r^3}\left((\mathcal{I}_0+\mathcal{I}_*+\Delta_0)+\Gamma_0\Delta_0+\Gamma_0^2\right)\right]\\
&|r^{2-\frac{2}{p}}(\Omega\underline{\tilde{F}}-\overline{\Omega\underline{\tilde{F}}})|_{p,S}\leq c(\Gamma_0)\left[\Gamma_0\frac{1}{r|u|}|r^{2-\frac{2}{p}}\underline{\tilde{\mu}}|_{p,S}+\frac{1}{r^2|u|}\left((\mathcal{I}_0+\mathcal{I}_*+\Delta_0)+\Gamma_0\Delta_0+\Gamma_0^2\right)\right].
\end{aligned}
\qquad (4.3.56)
$$

Using the inequalities (4.3.53) we derive the estimates

$$
\begin{aligned}
&|r^{2-\frac{2}{p}}(\Omega\tilde{H}-\overline{\Omega\tilde{H}})|_{p,S}\leq c(\Gamma_0)\left(\Gamma_0^2\frac{1}{r^2|u|}+\Gamma_0\Delta_0\frac{1}{r^{\frac{7}{2}}}+\Delta_0\frac{1}{r^2}\right)\\
&|r^{2-\frac{2}{p}}(\Omega\underline{\tilde{H}}-\overline{\Omega\underline{\tilde{H}}})|_{p,S}\leq c(\Gamma_0)\left(\Gamma_0^2\frac{1}{r^2|u|}+\Gamma_0\Delta_0\frac{1}{r^2|u|^{\frac{3}{2}}}+\Delta_0\frac{1}{r^2}\right). \qquad (4.3.57)
\end{aligned}
$$

Using these estimates in (4.3.49), (4.3.50), Lemma 4.3.3 follows immediately from an application of the Gronwall inequality.

Once the estimates for $\tilde{\mu}$ and $\underline{\tilde{\mu}}$, (4.3.47), (4.3.48) have been proved, Proposition 4.3.3 follows applying again the estimates of Proposition 4.1.2 to the Hodge systems (4.3.37), (4.3.38).

The following subsections are dedicated to the estimates of $\log\Omega$ and its derivatives up to fourth order. Specifically we control ω, $\underline{\omega}$ and the following derivatives

$$\nabla\!\!\!\!/\,\omega,\ \nabla\!\!\!\!/\,\underline{\omega},\ \mathbf{D}_3\underline{\omega},\ \mathbf{D}_4\omega\ ,\ \mathbf{D}_3^2\underline{\omega},\ \nabla\!\!\!\!/\,\mathbf{D}_3\underline{\omega},\ \nabla\!\!\!\!/\,\mathbf{D}_4\omega\ ,\ \ \nabla\!\!\!\!/^2\omega,\ \nabla\!\!\!\!/^2\underline{\omega},\ \nabla\!\!\!\!/^3\underline{\omega}\ .$$

4.3.10 *Estimates for $\mathcal{O}_0^{p,S}(\omega)$ and $\mathcal{O}_0^{p,S}(\underline{\omega})$ with $p \in [2,4]$*

Proposition 4.3.4 *Assuming (4.2.1), (4.2.2), (4.2.3), the bootstrap assumptions (4.2.5) and (4.2.6), then the following estimates hold:*

$$\begin{aligned}|r^{2-\frac{2}{p}}\Omega\mathbf{D}_4\log\Omega|_{p,S}(u,\underline{u}) &\le c\,(\mathcal{I}_0+\mathcal{I}_*+\Delta_0)\\ |r^{1-\frac{2}{p}}\tau_-\Omega\mathbf{D}_3\log\Omega|_{p,S}(u,\underline{u}) &\le c\,(\mathcal{I}_0+\mathcal{I}_*+\Delta_0)\,.\end{aligned} \tag{4.3.58}$$

Proof: To control ω and $\underline{\omega}$ we start from the equation (3.1.48) which we rewrite as

$$\mathbf{D}_3\mathbf{D}_4\log\Omega+\mathbf{D}_4\mathbf{D}_3\log\Omega=-2(\mathbf{D}_3\log\Omega)(\mathbf{D}_4\log\Omega)+2(\underline{\eta}\cdot\eta-2\zeta^2-\rho).$$

From, see Proposition 4.8.1 in the appendix to this chapter,

$$\mathbf{D}_4\mathbf{D}_3\log\Omega-\mathbf{D}_3\mathbf{D}_4\log\Omega=-4\zeta\cdot\nabla\!\!\!\!/\log\Omega\ ,$$

we derive

$$\begin{aligned}\mathbf{D}_3(\Omega\mathbf{D}_4\log\Omega)&=\hat{F}-\Omega\rho\\ \mathbf{D}_4(\Omega\mathbf{D}_3\log\Omega)&=\underline{\hat{F}}-\Omega\rho,\end{aligned} \tag{4.3.59}$$

where $\hat{F}$, $\underline{\hat{F}}$ are given by

$$\begin{aligned}\hat{F}&\equiv 2\Omega\zeta\cdot\nabla\!\!\!\!/\log\Omega+\Omega(\underline{\eta}\cdot\eta-2\zeta^2)\\ \underline{\hat{F}}&\equiv -2\Omega\zeta\cdot\nabla\!\!\!\!/\log\Omega+\Omega(\underline{\eta}\cdot\eta-2\zeta^2).\end{aligned} \tag{4.3.60}$$

Applying the Evolution Lemma to these equations we obtain

$$\begin{aligned}|r^{-\frac{2}{p}}\Omega\mathbf{D}_3\log\Omega|_{p,S}(u,\underline{u}) &\le c\left(|r^{-\frac{2}{p}}\Omega\mathbf{D}_3\log\Omega|_{p,S}(u,\underline{u}_*)+\int_{\underline{u}}^{\underline{u}_*}|r^{-\frac{2}{p}}\hat{F}|_{p,S}\right.\\ &\qquad\left.+\int_{\underline{u}}^{\underline{u}_*}|r^{-\frac{2}{p}}\Omega\rho|_{p,S}\right)\\ |r^{-\frac{2}{p}}\Omega\mathbf{D}_4\log\Omega|_{p,S}(u,\underline{u}) &\le c\left(|r^{-\frac{2}{p}}\Omega\mathbf{D}_4\log\Omega|_{p,S}(u_0,\underline{u})+\int_{u_0}^{u}|r^{-\frac{2}{p}}\underline{\hat{F}}|_{p,S}\right.\\ &\qquad\left.+\int_{u_0}^{u}|r^{-\frac{2}{p}}\Omega\rho|_{p,S}\right).\end{aligned} \tag{4.3.61}$$

Using assumptions (4.2.1) and the bootstrap assumption (4.2.5) we easily check that

$$|r^{-\frac{2}{p}}(\hat{F},\underline{\hat{F}})|_{p,S} \leq c(\Gamma_0)\Gamma_0^2\frac{1}{r^4}\ ,\quad |r^{-\frac{2}{p}}\rho|_{p,S} \leq cr^{-3}|r^{3-\frac{2}{p}}\rho|_{p,S} \leq c\frac{\Delta_0}{r^3}. \tag{4.3.62}$$

Multiplying both sides of the first inequality in (4.3.61) by r^2 and those of the second inequality by $r\tau_-$, and using the results of Lemma 4.1.7 applied to $V = \mathbf{D}_4\log\Omega$, we obtain

$$|r^{2-\frac{2}{p}}\Omega\mathbf{D}_4\log\Omega|_{p,S}(u,\underline{u}) \leq c\left(|r^{2-\frac{2}{p}}\Omega\mathbf{D}_4\log\Omega|_{p,S_{(0)}}(\underline{u}) + c(\Gamma_0)(\Delta_0+\Gamma_0^2)\right)$$
$$|r^{1-\frac{2}{p}}\tau_-\Omega\mathbf{D}_3\log\Omega|_{p,S}(u,\underline{u}) \leq c\left(|r^{1-\frac{2}{p}}\tau_-\Omega\mathbf{D}_3\log\Omega|_{p,S}(u,\underline{u}_*) + c(\Gamma_0)(\Delta_0+\Gamma_0^2)\right).$$

The exponents $(2-\frac{2}{p})$ and $(1-\frac{2}{p})$ are chosen in such a way that the norms

$$|r^{2-\frac{2}{p}}\Omega\mathbf{D}_4\log\Omega|_{p,S_{(0)}}(\underline{u})\ ,\quad |r^{1-\frac{2}{p}}\tau_-\Omega\mathbf{D}_3\log\Omega|_{p,S}(u,\underline{u}_*)$$

are controlled by assumptions (4.2.2).[21] The final result gives the following estimates:

$$|r^{2-\frac{2}{p}}\Omega\mathbf{D}_4\log\Omega|_{p,S}(u,\underline{u}) \leq c\left(\mathcal{I}_0+\mathcal{I}_*+\Delta_0+\Gamma_0^2\right)$$
$$|r^{1-\frac{2}{p}}\tau_-\Omega\mathbf{D}_3\log\Omega|_{p,S}(u,\underline{u}) \leq c\left(\mathcal{I}_0+\mathcal{I}_*+\Delta_0+\Gamma_0^2\right)$$

for any $p \in [2,4]$. Choosing $\Gamma_0^2 < \mathcal{I}_0+\mathcal{I}_*+\Delta_0$ we infer

$$\sup_{p\in[2,4]}\left(\mathcal{O}_0^{p,S}(\omega)+\mathcal{O}_0^{p,S}(\underline{\omega})\right) \leq c\left(\mathcal{I}_0+\mathcal{I}_*+\Delta_0\right),$$

proving the proposition.

4.3.11 Estimate for $\sup|r(\Omega-\frac{1}{2})|$

Proposition 4.3.5 *Assuming (4.2.1), (4.2.2), (4.2.3) and (4.2.5), (4.2.6), then the following estimate holds:*

$$\sup|r(\Omega-\frac{1}{2})| \leq c\left(\mathcal{I}_0+\mathcal{I}_*+\Delta_0\right). \tag{4.3.63}$$

Proof: We start from the inequalities (4.3.30). We recall that since $\eta+\underline{\eta} = 2\nabla\!\!\!/\log\Omega$,

$$\Delta\!\!\!/\log\Omega = \frac{1}{2}\nabla\!\!\!/(\eta+\underline{\eta})$$

and use the elliptic estimates of Proposition 4.1.3 to obtain

$$|r^{1-2/p}(\log\Omega-\overline{\log\Omega})|_{p,S}(u,\underline{u}) \leq c|r^{3-2/p}\nabla\!\!\!/(\eta+\underline{\eta})|_{p,S} \leq c(\mathcal{I}_*+\mathcal{I}_0+\Delta_0)$$
$$|r^{2-2/p}\nabla\!\!\!/\log\Omega|_{p,S}(u,\underline{u}) \leq c|r^{3-2/p}\nabla\!\!\!/(\eta+\underline{\eta})|_{p,S} \leq c(\mathcal{I}_*+\mathcal{I}_0+\Delta_0). \tag{4.3.64}$$

From the first inequality, to control $\sup_{\mathcal{K}}|r(\Omega-\frac{1}{2})|$ we have to control $r\overline{\log\Omega}$. To do so we use the following lemma.

[21] Note that, assuming $\mathcal{K}$ is endowed with a double null canonical foliation, we control on $\underline{C}_*$ the stronger norm $|r^{1-\frac{2}{p}}\tau_-^{\frac{3}{2}}\Omega\mathbf{D}_3\log\Omega|_{p,S_*}$; see Definition 7.4.3 in Chapter 7. Nevertheless this does not allow us to obtain a stronger estimate for $\Omega\mathbf{D}_3\log\Omega$ in the whole $\mathcal{K}$ due to the presence of the term $\Omega\rho$ in (4.3.59). An analogous observation holds for $\Omega\mathbf{D}_4\log\Omega$.

Lemma 4.3.4 *If the last slice is endowed with the canonical foliation, then $\overline{\log\Omega}$ satisfies, along the null hypersurfaces $C(u)$, the following equation:*

$$\overline{\log 2\Omega}(u,\underline{u}) = -\int_{\underline{u}}^{\underline{u}_*}\frac{1}{|S|}\int_S \Omega\mathrm{tr}\chi\,(\log\Omega-\overline{\log\Omega}) - \int_{\underline{u}}^{\underline{u}_*}\overline{\Omega\mathbf{D}_4\log\Omega} \tag{4.3.65}$$

Proof:

$$\begin{aligned}\frac{\partial}{\partial\underline{u}}\left(\frac{1}{|S|}\int_S\log\Omega\right) &= \frac{1}{|S|}\int_S\frac{\partial}{\partial\underline{u}}\log\Omega - \frac{1}{|S|}\left(\frac{\partial}{\partial\underline{u}}|S|\right)\overline{\log\Omega}+\frac{1}{|S|}\int_S\Omega\mathrm{tr}\chi\log\Omega\\ &= \overline{\Omega\mathbf{D}_4\log\Omega}+\frac{1}{|S|}\int_S\Omega\mathrm{tr}\chi\,(\log\Omega-\overline{\log\Omega}),\end{aligned} \tag{4.3.66}$$

where $\frac{\partial}{\partial\underline{u}}|S| = |S|\overline{\Omega\mathrm{tr}\chi}$. Since the canonical foliation implies that, on the last slice, $\overline{\log 2\Omega}=0$, the result is achieved.

Using Lemma 4.3.4 as well as (4.3.64) we infer that

$$\begin{aligned}|\overline{\log 2\Omega}(u,\underline{u})| &\leq \int_{\underline{u}}^{\underline{u}_*}\frac{1}{|S|}\int_S\Omega|\mathrm{tr}\chi||(\log\Omega-\overline{\log\Omega})|+\int_{\underline{u}}^{\underline{u}_*}\overline{|\Omega\mathbf{D}_4\log\Omega|}\\ &\leq \int_{\underline{u}}^{\underline{u}_*}\frac{1}{|S|}\left(\int_S|\Omega\mathrm{tr}\chi|^2\right)^{\frac12}\left(\int_S|(\log\Omega-\overline{\log\Omega})|^2\right)^{\frac12}+\int_{\underline{u}}^{\underline{u}_*}\overline{|\Omega\mathbf{D}_4\log\Omega|}\\ &\leq c(\Gamma_0)\int_{\underline{u}}^{\underline{u}_*}\frac{1}{r^2}|(\log\Omega-\overline{\log\Omega})|_{p=2,S}+\int_{\underline{u}}^{\underline{u}_*}\overline{|\Omega\mathbf{D}_4\log\Omega|}\\ &\leq c(\Gamma_0)(\mathcal{I}_*+\mathcal{I}_0+\Delta_0)\int_{\underline{u}}^{\underline{u}_*}\frac{1}{r^2}+c(\Gamma_0)\frac{1}{r}\left(\mathcal{I}_0+\mathcal{I}_*+\Delta_0\right)\\ &\leq c(\Gamma_0)\frac{1}{r}\left(\mathcal{I}_0+\mathcal{I}_*+\Delta_0\right).\end{aligned} \tag{4.3.67}$$

Plugging this estimate back into (4.3.64) we obtain

$$\begin{aligned}|r^{1-2/p}\log 2\Omega|_{p,S}(u,\underline{u}) &\leq |r^{1-2/p}\overline{\log 2\Omega}|_{p,S}(u,\underline{u})+c(\Gamma_0)(\mathcal{I}_*+\mathcal{I}_0+\Delta_0)\\ &\leq c\left(\mathcal{I}_0+\mathcal{I}_*+\Delta_0\right).\end{aligned}$$

This result together with the second inequality of (4.3.64)

$$|r^{2-2/p}\nabla\!\!\!/\log 2\Omega|_{p,S}(u,\underline{u}) \leq \frac{1}{2}|r^{2-2/p}(\eta+\underline{\eta})|_{p,S}(u,\underline{u})\leq c\left(\mathcal{I}_*+\mathcal{I}_0+\Delta_0\right)$$

allows one to conclude, using Lemma 4.1.3, that

$$\sup|r\log 2\Omega|\leq c\left(\mathcal{I}_0+\mathcal{I}_*+\Delta_0\right). \tag{4.3.68}$$

From

$$2(\Omega-\frac{1}{2}) = \exp(\log 2\Omega)-1 = \Sigma_{k=1}^{\infty}\frac{(\log 2\Omega)^k}{k!},$$

it follows that the inequality (4.3.68) implies the inequality

$$\sup|r(\Omega-\frac{1}{2})|\leq c\left(\mathcal{I}_0+\mathcal{I}_*+\Delta_0\right),$$

thus completing the proof.

4.3.12 *Completion of the estimates for* $\mathrm{tr}\chi$ *and* $\mathrm{tr}\underline{\chi}$

Proposition 4.3.6 *Assume (4.2.1), (4.2.2) and the bootstrap assumption (4.2.5). With the help of the first result of Proposition 4.3.4,*

$$|r^{2-\frac{2}{p}}\Omega\mathbf{D}_4\log\Omega|_{p,S}(u,\underline{u})\leq c(\mathcal{I}_0+\mathcal{I}_*+\Delta_0), \tag{4.3.69}$$

we prove

$$\sup|r^2(\overline{\Omega\mathrm{tr}\chi}-\frac{1}{r})|\leq c\,(\mathcal{I}_0+\mathcal{I}_*+\Delta_0)\,. \tag{4.3.70}$$

Proof: The evolution equation for $\overline{\Omega\mathrm{tr}\chi}$ is the following:[22]:

$$\frac{d}{d\underline{u}}(\overline{\Omega\mathrm{tr}\chi})+\frac{1}{2}\Omega\mathrm{tr}\chi\,(\overline{\Omega\mathrm{tr}\chi})=\frac{1}{2}(\overline{\Omega\mathrm{tr}\chi})V+\frac{1}{2}\overline{V^2}+\overline{E}, \tag{4.3.71}$$

where

$$V\equiv\left(\Omega\mathrm{tr}\chi-\overline{\Omega\mathrm{tr}\chi}\right)\;,\;E\equiv\left[2\Omega\mathrm{tr}\chi\,(\Omega\mathbf{D}_4\log\Omega)-\Omega^2|\hat{\chi}|^2\right]\,.$$

Moreover (see 4.1.30))

$$\frac{d}{d\underline{u}}\frac{1}{r}=-\frac{1}{r^2}\frac{\partial r}{\partial\underline{u}}=-\frac{1}{2r}\overline{\Omega\mathrm{tr}\chi}=-\frac{1}{2}\Omega\mathrm{tr}\chi\frac{1}{r}+\frac{1}{2r}V, \tag{4.3.72}$$

and denoting $W\equiv\left(\overline{\Omega\mathrm{tr}\chi}-\frac{1}{r}\right)$,

$$\frac{d}{d\underline{u}}W+\frac{1}{2}\Omega\mathrm{tr}\chi\,W=\frac{1}{2}WV+\frac{1}{2}\overline{V^2}+\overline{E}. \tag{4.3.73}$$

We control $\overline{E}=[2\Omega\mathrm{tr}\chi\,(\Omega\mathbf{D}_4\log\Omega)-\Omega^2|\hat{\chi}|^2]$ using the bootstrap assumption (4.2.5) and assumption (4.3.69,

$$|r^{1-\frac{2}{p}}E|_{p,S}(u,\underline{u})\leq c(\Gamma_0)\left((\mathcal{I}_0+\mathcal{I}_*+\Delta_0)+\Gamma_0^2\right)\frac{1}{r^2}\,.$$

We now apply the Evolution Lemma to (4.3.73) with $\lambda_1=1-\frac{2}{p}$. Applying estimate (4.3.2) to V, applying the Gronwall inequality and multiplying both sides by r, we obtain for $p\in[2,4]$

$$\left|r^{2-\frac{2}{p}}\left(\overline{\Omega\mathrm{tr}\chi}-\frac{1}{r}\right)\right|_{p,S}\leq c\left(\mathcal{I}_0+\mathcal{I}_*+\Delta_0+\Gamma_0^2\right)\leq c(\mathcal{I}_0+\mathcal{I}_*+\Delta_0)\,.$$

Using Lemma 4.1.3, since $\nabla\!\!\!/\left(\overline{\Omega\mathrm{tr}\chi}-\frac{1}{r}\right)=0$, we obtain

$$\sup|r^2(\overline{\Omega\mathrm{tr}\chi}-\frac{1}{r})|\leq c\,(\mathcal{I}_0+\mathcal{I}_*+\Delta_0)\,,$$

concluding the proof.

22

$$\begin{aligned}\frac{d}{d\underline{u}}(\overline{\Omega\mathrm{tr}\chi})+\frac{1}{2}\Omega\mathrm{tr}\chi(\overline{\Omega\mathrm{tr}\chi})&=\frac{1}{2}\overline{(\Omega\mathrm{tr}\chi)^2}+\frac{1}{2}\Omega\mathrm{tr}\chi(\overline{\Omega\mathrm{tr}\chi})-(\overline{\Omega\mathrm{tr}\chi})^2+\overline{\left[2\Omega\mathrm{tr}\chi(\Omega\mathbf{D}_4\log\Omega)-\Omega^2|\hat{\chi}|^2\right]}\\&=\frac{1}{2}(\overline{\Omega\mathrm{tr}\chi})\left(\Omega\mathrm{tr}\chi-\overline{\Omega\mathrm{tr}\chi}\right)+\frac{1}{2}\left(\overline{(\Omega\mathrm{tr}\chi)^2}-(\overline{\Omega\mathrm{tr}\chi})^2\right)+\overline{\left[2\Omega\mathrm{tr}\chi(\Omega\mathbf{D}_4\log\Omega)-\Omega^2|\hat{\chi}|^2\right]}.\end{aligned}$$

Proposition 4.3.7 *Assume (4.2.1), (4.2.2), (4.2.3), the bootstrap assumptions (4.2.5) and (4.2.6). With the help of the second result of Proposition 4.3.4,*

$$|r^{1-\frac{2}{p}}\tau_-\Omega\mathbf{D}_3\log\Omega|_{p,S}(u,\underline{u})\leq c(\mathcal{I}_0+\mathcal{I}_*+\Delta_0), \tag{4.3.74}$$

we prove,

$$\sup|r\tau_-(\overline{\Omega\mathrm{tr}\underline{\chi}}+\frac{1}{r})|\leq c\,(\mathcal{I}_0+\mathcal{I}_*+\Delta_0)\,. \tag{4.3.75}$$

Proof: We proceed as for $(\overline{\Omega\mathrm{tr}\chi}-\frac{1}{r})$. Defining $\underline{W}\equiv\left(\overline{\Omega\mathrm{tr}\underline{\chi}}+\frac{1}{r}\right)$, we obtain for the evolution equation

$$\frac{d}{du}\underline{W}+\frac{1}{2}\overline{\Omega\mathrm{tr}\underline{\chi}}\,\underline{W}=\frac{1}{2}\underline{W}\,\underline{V}+\frac{1}{2}\overline{\underline{V}^2}+\overline{\underline{E}}. \tag{4.3.76}$$

To control $\underline{E}\equiv[2\Omega\mathrm{tr}\underline{\chi}(\Omega\mathbf{D}_3\log\Omega)-\Omega^2|\hat{\underline{\chi}}|^2]$ we use the bootstrap assumption (4.2.5) and inequality (4.3.74)

$$|r^{1-\frac{2}{p}}\underline{E}|_{p,S}(u,\underline{u})\leq c(\Gamma_0)\left((\mathcal{I}_0+\mathcal{I}_*+\Delta_0)+\Gamma_0^2\right)\frac{1}{r\tau_-}\,.$$

We apply the Evolution Lemma to equation (4.3.76) with $\lambda_1=1-\frac{2}{p}$. Using estimate (4.3.6) to control $\underline{V}$, the Gronwall inequality and the results of Lemma 4.1.7 applied to $\underline{W}$, and multiplying both sides by τ_-, we obtain, for $p\in[2,4]$,

$$\left|r^{1-\frac{2}{p}}\tau_-\left(\overline{\Omega\mathrm{tr}\underline{\chi}}+\frac{1}{r}\right)\right|_{p,S}\leq c\,(\mathcal{I}_0+\mathcal{I}_*+\Delta_0)\,. \tag{4.3.77}$$

In view of Lemma 4.1.3, we conclude that[23]

$$\sup|r\tau_-(\overline{\Omega\mathrm{tr}\underline{\chi}}+\frac{1}{r})|\leq c\,(\mathcal{I}_0+\mathcal{I}_*+\Delta_0)\,.$$

The next corollary, which we state without proof, is an elementary consequence of previous results.

Corollary 4.3.1 *From Propositions 4.3.6, 4.3.7 and estimates (4.3.2), (4.3.6) the following inequality holds:*

$$\sup|r\tau_-(\mathrm{tr}\chi+\mathrm{tr}\underline{\chi})|\leq c(\mathcal{I}_0+\mathcal{I}_*+\Delta_0). \tag{4.3.78}$$

[23] Observe that the decay cannot be improved due to the term $2\Omega\mathrm{tr}\underline{\chi}(\Omega\mathbf{D}_3\log\Omega)$ in $\underline{E}$; see (4.3.76).

4.3.13 Estimates for $\mathcal{O}_1^{p,S}(\omega)$ and $\mathcal{O}_1^{p,S}(\underline{\omega})$ with $p \in [2,4]$

Proposition 4.3.8 *Under assumptions (4.2.1), (4.2.2) and the bootstrap assumptions (4.2.5), (4.2.6), the following estimates hold, for any $p \in [2,4]$:*[24]

$$
\begin{aligned}
&|r^{3-\frac{2}{p}}\nabla\!\!\!/\Omega\mathbf{D}_4\log\Omega|_{p,S}(u,\underline{u}) \leq c\,(\mathcal{I}_0+\mathcal{I}_*+\Delta_0+\Delta_1)\\
&|r^{2-\frac{2}{p}}\tau_-\nabla\!\!\!/\Omega\mathbf{D}_3\log\Omega|_{p,S}(u,\underline{u}) \leq c\,(\mathcal{I}_0+\mathcal{I}_*+\Delta_0+\Delta_1)\,.
\end{aligned}
\tag{4.3.79}
$$

Proof: $\nabla\!\!\!/\omega=-\frac{1}{2}\nabla\!\!\!/\mathbf{D}_4\log\Omega$ and $\nabla\!\!\!/\underline{\omega}=-\frac{1}{2}\nabla\!\!\!/\mathbf{D}_3\log\Omega$ satisfy, along the $\underline{C}(\underline{u})$ and the $C(u)$ null hypersurfaces, the following evolution equations, obtained by deriving tangentially equations (4.3.59) and applying the commutation relations (4.8.2)[25]

$$
\begin{aligned}
\mathbf{D}\!\!\!\!/_3(\Omega\nabla\!\!\!/\mathbf{D}_4\log\Omega)+\frac{1}{2}\mathrm{tr}\underline{\chi}(\Omega\nabla\!\!\!/\mathbf{D}_4\log\Omega) &= -\hat{\underline{\chi}}(\Omega\nabla\!\!\!/\mathbf{D}_4\log\Omega)+\nabla\!\!\!/\rho\\
&-(\mathbf{D}_4\log\Omega)(\Omega\nabla\!\!\!/\mathbf{D}_3\log\Omega)-\hat{H}\\
\mathbf{D}\!\!\!\!/_4(\Omega\nabla\!\!\!/\mathbf{D}_3\log\Omega)+\frac{1}{2}\mathrm{tr}\chi(\Omega\nabla\!\!\!/\mathbf{D}_3\log\Omega) &= -\hat{\chi}(\Omega\nabla\!\!\!/\mathbf{D}_3\log\Omega)+\nabla\!\!\!/\rho\\
&-(\mathbf{D}_3\log\Omega)(\Omega\nabla\!\!\!/\mathbf{D}_4\log\Omega)-\underline{\hat{H}}
\end{aligned}
\tag{4.3.80}
$$

where

$$
\begin{aligned}
\hat{H} &= (\nabla\!\!\!/\log\Omega)\Big[-(\mathbf{D}_3\log\Omega)(\mathbf{D}_4\log\Omega)+2\zeta\cdot\nabla\!\!\!/\log\Omega+(\underline{\eta}\cdot\eta-2\zeta^2-\rho)\Big]\\
&\quad+\nabla\!\!\!/\Big[2\zeta\cdot\nabla\!\!\!/\log\Omega+(\underline{\eta}\cdot\eta-2\zeta^2)\Big]\\
\underline{\hat{H}} &= (\nabla\!\!\!/\log\Omega)\Big[-(\mathbf{D}_4\log\Omega)(\mathbf{D}_3\log\Omega)-2\zeta\cdot\nabla\!\!\!/\log\Omega+(\eta\cdot\underline{\eta}-2\zeta^2-\rho)\Big]\\
&\quad+\nabla\!\!\!/\Big[-2\zeta\cdot\nabla\!\!\!/\log\Omega+(\eta\cdot\underline{\eta}-2\zeta^2)\Big].
\end{aligned}
$$

These evolution equations have to be estimated simultaneously. Using the bootstrap assumptions (4.2.5), (4.2.6) and assumptions 4.2.1, 4.2.2 we easily check [26] that, for $p \in [2,4]$,

$$
\begin{aligned}
&|r^{-\frac{2}{p}}(\hat{H},\underline{\hat{H}})|_{p,S} \leq cr^{-5}\Gamma_0\left(\Delta_0+(\mathcal{I}_0+\mathcal{I}_*+\Delta_0)+\Gamma_0^2\right) \leq cr^{-5}\Gamma_0(\mathcal{I}_0+\mathcal{I}_*+\Delta_0)\\
&|r^{4-\frac{2}{p}}\tau_-^{\frac{1}{2}}\nabla\!\!\!/\rho|_{p,S} \leq \Delta_1.
\end{aligned}
$$

Also,

$$
\begin{aligned}
&|(\Omega\nabla\!\!\!/\mathbf{D}_3\log\Omega)(\mathbf{D}_4\log\Omega)|_{p,S} \leq r^{-4}\tau_-^{-1}\Gamma_0|r^{2-\frac{2}{p}}\tau_-(\Omega\nabla\!\!\!/\mathbf{D}_3\log\Omega)|_{p,S}\\
&|(\Omega\nabla\!\!\!/\mathbf{D}_4\log\Omega)(\mathbf{D}_3\log\Omega)|_{p,S} \leq r^{-4}\tau_-^{-1}\Gamma_0|r^{3-\frac{2}{p}}(\Omega\nabla\!\!\!/\mathbf{D}_4\log\Omega)|_{p,S}\,.
\end{aligned}
$$

---

[24] The estimates of $\mathcal{O}_1^{p,S}(\omega)$ and $\mathcal{O}_1^{p,S}(\underline{\omega})$ discussed in this subsection can also be obtained in a different way together with the estimates of $\mathcal{O}_2^{p,S}(\omega)$ and $\mathcal{O}_2^{p,S}(\underline{\omega})$; see Proposition 4.4.1.

[25] We choose the null frame such that $\mathbf{D}\!\!\!\!/_3 e_a=0$ or $\mathbf{D}\!\!\!\!/_4 e_a=0$.

[26] Observe that the generic term of $\nabla\!\!\!/[2\zeta\cdot\nabla\!\!\!/\log\Omega+(\underline{\eta}\cdot\eta-2\zeta^2)]$ has to be estimated as a product $|A|_\infty|\nabla\!\!\!/B|_{L^p(S)}$ which is bounded by $\Gamma_0(\mathcal{I}_0+\mathcal{I}_*+\Delta_0)$.

Applying the Evolution Lemma to the first evolution equation (4.3.80) with $\lambda_1 = 1 - \frac{2}{p}$ and then multiplying both sides by r^2 we obtain[27]

$$|r^{3-\frac{2}{p}}(\Omega \nabla\!\!\!/ \mathbf{D}_4 \log\Omega)|_{p,S}(u,\underline{u}) \le c\Bigg(|r^{3-\frac{2}{p}}(\Omega \nabla\!\!\!/ \mathbf{D}_4 \log\Omega)|_{p,S_{(0)}}(\underline{u})$$
$$+\Gamma_0 \int_{u_0}^{u} \frac{1}{r\tau_-}|r^{3-\frac{2}{p}}(\Omega \nabla\!\!\!/ \mathbf{D}_4 \log\Omega)|_{p,S} + \Gamma_0 \int_{u_0}^{u} \frac{1}{r\tau_-}|r^{2-\frac{2}{p}}\tau_-(\Omega \nabla\!\!\!/ \mathbf{D}_3 \log\Omega)|_{p,S}$$
$$+\frac{1}{r^{\frac{1}{2}}}\left(\Gamma_0(\mathcal{I}_0 + \mathcal{I}_* + \Delta_0) + \Delta_1\right)\Bigg) \tag{4.3.81}$$

Applying the Evolution Lemma once again with $\lambda_1 = 1 - \frac{2}{p}$ to the second evolution equation (4.3.80) and then multiplying both sides by $r\tau_-$, we obtain

$$|r^{2-\frac{2}{p}}\tau_-(\Omega \nabla\!\!\!/ \mathbf{D}_3 \log\Omega)|_{p,S}(u,\underline{u}) \le c\Bigg(|r^{2-\frac{2}{p}}\tau_-(\Omega \nabla\!\!\!/ \mathbf{D}_3 \log\Omega)|_{p,S}(u,\underline{u}_*)$$
$$+\Gamma_0 \int_{\underline{u}}^{\underline{u}_*} \frac{1}{r^2}|r^{2-\frac{2}{p}}\tau_-(\Omega \nabla\!\!\!/ \mathbf{D}_3 \log\Omega)|_{p,S} + \Gamma_0 \int_{\underline{u}}^{\underline{u}_*} \frac{1}{r^2}|r^{3-\frac{2}{p}}(\Omega \nabla\!\!\!/ \mathbf{D}_4 \log\Omega)|_{p,S}$$
$$+\frac{1}{r^{\frac{1}{2}}}\left(\Gamma_0(\mathcal{I}_0 + \mathcal{I}_* + \Delta_0) + \Delta_1\right)\Bigg). \tag{4.3.82}$$

Applying the Gronwall Lemma to the inequalities (4.3.81) and (4.3.82) gives

$$|r^{3-\frac{2}{p}}(\Omega \nabla\!\!\!/ \mathbf{D}_4 \log\Omega)|_{p,S}(u,\underline{u}) \le \Bigg(|r^{3-\frac{2}{p}}(\Omega \nabla\!\!\!/ \mathbf{D}_4 \log\Omega)|_{p,S_{(0)}}(\underline{u})$$
$$+\Gamma_0 \int_{u_0}^{u} \frac{1}{r\tau_-}|r^{2-\frac{2}{p}}\tau_-(\Omega \nabla\!\!\!/ \mathbf{D}_3 \log\Omega)|_{p,S} + \frac{1}{r^{\frac{1}{2}}}\left(\Gamma_0(\mathcal{I}_0 + \mathcal{I}_* + \Delta_0) + \Delta_1\right)\Bigg)$$
$$|r^{2-\frac{2}{p}}\tau_-(\Omega \nabla\!\!\!/ \mathbf{D}_3 \log\Omega)|_{p,S}(u,\underline{u}) \le \Bigg(|r^{2-\frac{2}{p}}\tau_-(\Omega \nabla\!\!\!/ \mathbf{D}_3 \log\Omega)|_{p,S}(u,\underline{u}_*)$$
$$+\Gamma_0 \int_{\underline{u}}^{\underline{u}_*} \frac{1}{r^2}|r^{3-\frac{2}{p}}(\Omega \nabla\!\!\!/ \mathbf{D}_4 \log\Omega)|_{p,S} + \frac{1}{r^{\frac{1}{2}}}\left(\Gamma_0(\mathcal{I}_0 + \mathcal{I}_* + \Delta_0) + \Delta_1\right)\Bigg),$$

and by combining the two, we easily derive

$$|r^{3-\frac{2}{p}}(\Omega \nabla\!\!\!/ \mathbf{D}_4 \log\Omega)|_{p,S}(u,\underline{u}) \le c\Bigg(\sup_{\Sigma_0\cap\mathcal{K}} |r^{3-\frac{2}{p}}(\Omega \nabla\!\!\!/ \mathbf{D}_4 \log\Omega)|_{p,S_{(0)}}(\underline{u})$$
$$+\Gamma_0 \sup_{\underline{C}_*\cap\mathcal{K}} |r^{2-\frac{2}{p}}\tau_-(\Omega \nabla\!\!\!/ \mathbf{D}_3 \log\Omega)|_{p,S}(u,\underline{u}_*) + \frac{1}{r^{\frac{1}{2}}}(1+\Gamma_0)\left(\Gamma_0(\mathcal{I}_0 + \mathcal{I}_* + \Delta_0) + \Delta_1\right)\Bigg)$$
$$+\Gamma_0^2 \int_{u_0}^{u} \frac{1}{r\tau_-}\int_{\underline{u}}^{\underline{u}_*} \frac{1}{r^2}|r^{3-\frac{2}{p}}(\Omega \nabla\!\!\!/ \mathbf{D}_4 \log\Omega)|_{p,S} \tag{4.3.83}$$

[27] Each time we use the evolution equations along the null-incoming hypersurfaces we also use Lemma 4.1.7. Because this is always done in the same way we will not repeat it anymore.

and its counterpart. Choosing Γ_0 sufficiently small and taking the sup of $|r^{3-\frac{2}{p}}(\Omega \nabla\!\!\!/ \mathbf{D}_4 \log \Omega)|_{p,S}$, we obtain

$$|r^{3-\frac{2}{p}}(\Omega \nabla\!\!\!/ \mathbf{D}_4 \log \Omega)|_{p,S}(u,\underline{u}) \leq c\left(\sup_{\Sigma_0 \cap \mathcal{K}} |r^{3-\frac{2}{p}}(\Omega \nabla\!\!\!/ \mathbf{D}_4 \log \Omega)|p, S_{(0)}(\underline{u})\right.$$
$$\left. + \Gamma_0 \sup_{\underline{C}_* \cap \mathcal{K}} |r^{2-\frac{2}{p}}\tau_-(\Omega \nabla\!\!\!/ \mathbf{D}_3 \log \Omega)|_{p,S}(u,\underline{u}_*) + (1+\Gamma_0)\left(\Gamma_0(\mathcal{I}_0 + \mathcal{I}_* + \Delta_0) + \Delta_1\right)\right). \tag{4.3.84}$$

and its counterpart,

$$|r^{2-\frac{2}{p}}\tau_- \Omega \nabla\!\!\!/ \mathbf{D}_3 \log \Omega|_{p,S}(u,\underline{u}) \leq c\left(\sup_{\underline{C}_* \cap \mathcal{K}} |r^{2-\frac{2}{p}}\tau_-(\Omega \nabla\!\!\!/ \mathbf{D}_3 \log \Omega)|_{p,S}(u,\underline{u}_*)\right.$$
$$\left. + \Gamma_0 \sup_{\Sigma_0 \cap \mathcal{K}} |r^{3-\frac{2}{p}}(\Omega \nabla\!\!\!/ \mathbf{D}_4 \log \Omega)|p, S_{(0)}(\underline{u}) + (1+\Gamma_0)\left(\Gamma_0(\mathcal{I}_0 + \mathcal{I}_* + \Delta_0) + \Delta_1\right)\right). \tag{4.3.85}$$

Finally, from the initial and final slice assumptions (4.2.2)

$$\mathcal{O}_{[1]}(\underline{C}_*) \leq \mathcal{I}_*\ ,\ \underline{\mathcal{O}}_{[1]}(\Sigma_0) \leq \mathcal{I}_0\ ,$$

and choosing Γ_0 sufficiently small, we derive for any $p \in [2,4]$

$$|r^{3-\frac{2}{p}} \nabla\!\!\!/ \Omega \mathbf{D}_4 \log \Omega|_{p,S}(u,\underline{u}) \leq c(\Gamma_0)\left(\mathcal{I}_0 + \mathcal{I}_* + \Delta_0 + \Delta_1\right)$$
$$|r^{2-\frac{2}{p}}\tau_- \nabla\!\!\!/ \Omega \mathbf{D}_3 \log \Omega|_{p,S}(u,\underline{u}) \leq c(\Gamma_0)\left(\mathcal{I}_0 + \mathcal{I}_* + \Delta_0 + \Delta_1\right),$$

which proves the proposition.

The estimates (4.3.79) of Proposition 4.3.8 complete the control of the norms $\mathcal{O}_1^{p,S}(\omega)+\mathcal{O}_1^{p,S}(\underline{\omega})$, so that we obtain

$$\sup_{p\in[2,4]} \left(\mathcal{O}_1^{p,S}(\omega) + \mathcal{O}_1^{p,S}(\underline{\omega})\right) \leq c\left(\mathcal{I}_0 + \mathcal{I}_* + \Delta_0 + \Delta_1\right).$$

4.3.14 *Estimates for $\mathcal{O}_0^{p,S}(\mathbf{D}_4\omega)$ and $\mathcal{O}_0^{p,S}(\mathbf{D}_3\underline{\omega})$ with $p \in [2,4]$*

Proposition 4.3.9 *Under the assumptions of Theorem 4.2.1 and the bootstrap assumptions (4.2.5), (4.2.6), the following estimates hold with $p \in [2,4]$:*

$$|r^{3-\frac{2}{p}}\mathbf{D}_4^2 \log \Omega|_{p,S} \leq c\left(\mathcal{I}_0 + \mathcal{I}_* + \Delta_0 + \Delta_1\right)$$
$$|r^{1-\frac{2}{p}}\tau_-^2 \mathbf{D}_3^2 \log \Omega|_{p,S} \leq c\left(\mathcal{I}_0 + \mathcal{I}_* + \Delta_0 + \Delta_1\right). \tag{4.3.86}$$

Proof: To control the norms of $\mathbf{D}_3^2 \log \Omega$ and $\mathbf{D}_4^2 \log \Omega$ we derive, in the next lemma, their evolution equations along $C(u)$ and $\underline{C}(\underline{u})$.

Lemma 4.3.5 $(\Omega\mathbf{D}_4)^2\log\Omega$ *and* $(\Omega\mathbf{D}_3)^2\log\Omega$ *satisfy the equations*

$$\mathbf{D}_3(\Omega\mathbf{D}_4)^2\log\Omega = M - \Omega^2\mathbf{D}_4\rho$$
$$\mathbf{D}_4(\Omega\mathbf{D}_3)^2\log\Omega = \underline{M} - \Omega^2\mathbf{D}_3\rho \tag{4.3.87}$$

where

$$\begin{aligned} M &= 2\Omega\Big[2\zeta\cdot\nabla\!\!\!/\,\Omega\mathbf{D}_4\log\Omega + 2(\Omega\mathbf{D}_4\log\Omega)\zeta\cdot\nabla\!\!\!/\log\Omega + (\Omega\mathbf{D}_4\log\Omega)(\underline{\eta}\cdot\eta - 2\zeta^2)\\ &\quad +\Omega\eta\cdot\mathbf{D}_4\nabla\!\!\!/\log\Omega - (-3\zeta+\nabla\!\!\!/\log\Omega)\left(\nabla\!\!\!/\,\Omega\mathbf{D}_4\log\Omega - \Omega\chi\cdot(\underline{\eta}-\zeta)\right)\Big]\\ &\quad +2\Omega\Big(-\Omega(-3\zeta+\nabla\!\!\!/\log\Omega)\beta - (\Omega\mathbf{D}_4\log\Omega)\rho\Big) \end{aligned} \tag{4.3.88}$$

and $\underline{M}$ *is obtained from* M *by making the obvious changes (see (3.1.34)).*

The proof of Lemma 4.3.5 is in the appendix to this chapter, Subsection 4.8.2.

In view of the assumptions of Theorem 4.2.1 and the bootstrap assumptions (4.2.5), it is easy to check that

$$|r^{5-\frac{2}{p}}M|_{p,S}(u,\underline{u}) \le c(\Gamma_0)\Gamma_0\left((\mathcal{I}_0+\mathcal{I}_*+\Delta_0)+\Gamma_0^2\right). \tag{4.3.89}$$

Moreover we control $|r^{4-\frac{2}{p}}\mathbf{D}_4\rho|_{p,S}$ for $p\in[2,4]$ with

$$|r^{4-\frac{2}{p}}\mathbf{D}_4\rho|_{p,S} \le c(\Gamma_0)(\Delta_0+\Delta_1). \tag{4.3.90}$$

We apply the Evolution Lemma to the first equation of (4.3.87) and obtain, using initial and final slice assumptions (4.2.2), choosing Γ_0 and $(\mathcal{I}_0+\mathcal{I}_*+\Delta_0)$ sufficiently small and multiplying both sides by r^3,

$$\begin{aligned} &|r^{3-\frac{2}{p}}\mathbf{D}_4^2\log\Omega|_{p,S}(u,\underline{u})\\ &\le |r^{3-\frac{2}{p}}\mathbf{D}_4^2\log\Omega|_{p,S_{(0)}}(\underline{u}) + c\left[(\Delta_0+\Delta_1)+\frac{1}{r}\Gamma_0\left((\mathcal{I}_0+\mathcal{I}_*+\Delta_0)+\Gamma_0^2\right)\right]\\ &\le c\,(\mathcal{I}_0+\mathcal{I}_*+\Delta_0+\Delta_1). \end{aligned} \tag{4.3.91}$$

The same calculation can be repeated for $\mathbf{D}_3^2\log\Omega$ by interchanging the underlined quantities with those not underlined and viceversa; see (3.1.34). We obtain, for $p\in[2,4]$,

$$|r^{4-\frac{2}{p}}\tau_-\underline{M}|_{p,S}(u,\underline{u}) \le c(\Gamma_0)\Gamma_0\left((\mathcal{I}_0+\mathcal{I}_*+\Delta_0)+\Gamma_0^2\right), \tag{4.3.92}$$

and observe that

$$|r^{3-\frac{2}{p}}\tau_-\mathbf{D}_3\rho|_{p,S} \le c(\Gamma_0)(\Delta_0+\Delta_1).$$

We apply, finally, the Evolution Lemma to the second equation of (4.3.87) and obtain, using the initial and final slice assumptions (4.2.2) and multiplying both sides by $r\tau_-^2$,

$$\begin{aligned} |r^{1-\frac{2}{p}}\tau_-^2\mathbf{D}_3^2\log\Omega|_{p,S}(u,\underline{u}) &\le |r^{1-\frac{2}{p}}\tau_-^2\mathbf{D}_3^2\log\Omega|_{p,S_{(0)}}(\underline{u})\\ &+ c\left[(\Delta_0+\Delta_1)+\frac{\tau_-}{r^2}\Gamma_0\left((\mathcal{I}_0+\mathcal{I}_*+\Delta_0)+\Gamma_0^2\right)\right]\\ &\le c\,(\mathcal{I}_0+\mathcal{I}_*+\Delta_0+\Delta_1) \end{aligned} \tag{4.3.93}$$

for $p \in [2,4]$, thus completing Proposition 4.3.9.

The estimates (4.3.86) of Proposition 4.3.9 allow control of $\mathcal{O}_0^{p,S}(\mathbf{D}_4\omega)$ and $\underline{\mathcal{O}}_0^{p,S}(\mathbf{D}_3\underline{\omega})$, and allow one to obtain

$$\sup_{p\in[2,4]}\left(\mathcal{O}_0^{p,S}(\mathbf{D}_4\omega)+\underline{\mathcal{O}}_0^{p,S}(\mathbf{D}_3\underline{\omega})\right)\leq c\,(\mathcal{I}_0+\mathcal{I}_*+\Delta_0+\Delta_1)\,. \tag{4.3.94}$$

Remark: The control of the norm $\mathcal{O}_0^{p,S}(\mathbf{D}_3\underline{\omega})$ obtained in the previous estimate is not yet sufficient to control the error terms, discussed in Chapter 6. In fact it would produce a logarithmic divergence. The control of the following norm (see (3.5.30))

$$\underline{\tilde{\mathcal{O}}}_1(\underline{\omega})\equiv||\frac{1}{\sqrt{\tau_+}}\tau_-^2\mathbf{D}_3\underline{\omega}||_{L_2(C\cap\mathcal{K})},$$

avoids the problem.

4.3.15 Estimate for $\tilde{\mathcal{O}}_1(\underline{\omega})$ with $p \in [2,4]$

Proposition 4.3.10 *Under the assumptions of Theorem 4.2.1 and the bootstrap assumptions (4.2.5) the following estimate holds:*

$$||\frac{1}{\sqrt{\tau_+}}\tau_-^2\mathbf{D}_3\underline{\omega}||_{L_2(C\cap\mathcal{K})}\leq c\,(\mathcal{I}_0+\mathcal{I}_*+\Delta_0+\Delta_1)\,. \tag{4.3.95}$$

Proof: It follows immediately from the following inequality that

$$\begin{aligned}|r^{1-\frac{2}{p}}\tau_-^2\Omega^2\mathbf{D}_3^2\log\Omega|_{p,S}(u,\underline{u}) &\leq c\left[\frac{r(u,\underline{u})}{r(u,\underline{u}_*)}|r^{1-\frac{2}{p}}\tau_-^2\Omega^2\mathbf{D}_3^2\log\Omega|_{p,S}(u,\underline{u}_*)\right.\\ &+\left. c\left((\Delta_0+\Delta_1)+\frac{\tau_-}{r^2}\Gamma_0\left((\mathcal{I}_0+\mathcal{I}_*+\Delta_0)+\Gamma_0^2\right)\right)\right],\end{aligned}$$

an elementary improvement of (4.3.93).

Proposition 4.3.10 completes the proof of part (1) of Theorem 4.2.1. Part (2) of Theorem 4.2.1 now follows immediately. In fact, once we have proved inequalities (4.2.7), (4.2.8), we will be able to prove inequality (4.2.10) using Lemma 4.1.3 and conclude immediately, recalling Definitions 3.5.22 and 3.5.23, that

$$\begin{aligned}&\mathcal{O}_0^{\infty}+\mathcal{O}_0^{\infty,S}(\omega)\leq c\,(\mathcal{I}_0+\mathcal{I}_*+\Delta_0+\Delta_1)\\ &\underline{\mathcal{O}}_0^{\infty}+\underline{\mathcal{O}}_0^{\infty,S}(\underline{\omega})\leq c\,(\mathcal{I}_0+\mathcal{I}_*+\Delta_0+\Delta_1)\,.\end{aligned} \tag{4.3.96}$$

This concludes the proof of part (2) of Theorem 4.2.1.

4.3.16 Improved estimates under stronger assumptions on Σ_0 and $\underline{C}_*$

The estimates of the following propositions are proved making stronger assumptions for the various quantities on the initial hypersurface Σ_0 and on the last slice $\underline{C}_*$. These stronger assumptions are proved in Chapter 7 relative to a double null canonical foliation.

Proposition 4.3.11 *Assuming (4.2.1), the bootstrap assumption (4.2.5) and, on the last slice,*

$$|r^{3-2/p}\tau_-^{\frac{1}{2}}\underline{\Psi}|_{p,S}(u,\underline{u}_*) \leq \mathcal{I}_*, \tag{4.3.97}$$

then the following estimates hold:

$$\begin{aligned}
&|r^{3-2/p}\tau_-^{\frac{1}{2}}\nabla\!\!\!/\hat{\chi}|_{p,S}(u,\underline{u}) \leq c(\mathcal{I}_0+\mathcal{I}_*+\Delta_0)\\
&|r^{2-2/p}\tau_-^{\frac{1}{2}}\hat{\chi}|_{p,S}(u,\underline{u}) \leq c(\mathcal{I}_0+\mathcal{I}_*+\Delta_0).
\end{aligned} \tag{4.3.98}$$

Proof: The result is obtained as in Proposition 4.3.1, but using the stronger assumption (4.3.96) on the last slice.

Proposition 4.3.12 *Assuming (4.2.1), (4.2.3), the bootstrap assumptions (4.2.5), (4.2.6) and, on the initial slice,*

$$|r^{\frac{7}{2}-\frac{2}{p}}\underline{\Psi}|_{p,S_{(0)}}(\underline{u}) \leq \mathcal{I}_0, \tag{4.3.99}$$

then the following estimates hold:

$$\begin{aligned}
&|r^{2-2/p}\tau_-^{\frac{3}{2}}\nabla\!\!\!/\underline{\hat{\chi}}|_{p,S}(u,\underline{u}) \leq c(\mathcal{I}_0+\mathcal{I}_*+\Delta_0)\\
&|r^{1-2/p}\tau_-^{\frac{3}{2}}\underline{\hat{\chi}}|_{p,S}(u,\underline{u}) \leq c(\mathcal{I}_0+\mathcal{I}_*+\Delta_0).
\end{aligned} \tag{4.3.100}$$

Proof: The result is obtained as in Proposition 4.3.2, but using the stronger assumption, (4.3.99), on the initial slice.

The next result will be used in most of the subsequent propositions.

Proposition 4.3.13 *Under the assumptions of Theorem 4.2.1 and the bootstrap assumptions (4.2.5), (4.2.6), and assuming, moreover, on Σ_0 and on the last slice $\underline{C}_*$ the following inequalities:*

$$|r^{\frac{5}{2}-\frac{2}{p}}(\Omega\mathbf{D}_4\log\Omega-\overline{\Omega\mathbf{D}_4\log\Omega})|_{p,S_{(0)}}(\underline{u}) \leq \mathcal{I}_0 \tag{4.3.101}$$

$$|r^{1-\frac{2}{p}}\tau_-^{\frac{3}{2}}(\Omega\mathbf{D}_3\log\Omega-\overline{\Omega\mathbf{D}_3\log\Omega})|_{p,S}(u,\underline{u}_*) \leq \mathcal{I}_*, \tag{4.3.102}$$

we have that in $\mathcal{K}$,

$$\begin{aligned}
&|r^{2-\frac{2}{p}}-\tau_-^{\frac{1}{2}}\left(\Omega\mathbf{D}_4\log\Omega-\overline{\Omega\mathbf{D}_4\log\Omega}\right)|_{p,S}(u,\underline{u}) \leq c(\mathcal{I}_0+\mathcal{I}_*+\Delta_0)\\
&|r^{1-\frac{2}{p}}\tau_-^{\frac{3}{2}}(\Omega\mathbf{D}_3\log\Omega-\overline{\Omega\mathbf{D}_3\log\Omega})|_{p,S}(u,\underline{u}) \leq c(\mathcal{I}_0+\mathcal{I}_*+\Delta_0).
\end{aligned} \tag{4.3.103}$$

Proof: It is easy to prove from the first of equation of (4.3.59) that

$$\mathbf{D}_3\overline{\Omega\mathbf{D}_4\log\Omega} = \overline{(\Omega\mathrm{tr}\underline{\chi}-\overline{\Omega\mathrm{tr}\underline{\chi}})\Omega\mathbf{D}_4\log\Omega} + \overline{\hat{F}} - \overline{\Omega\rho}. \tag{4.3.104}$$

Therefore, if $Y \equiv \Omega\mathbf{D}_4 \log\Omega$, then the evolution equation is

$$\frac{d}{du}(Y-\overline{Y}) = \Omega\Big[(\hat{F}-\overline{\hat{F}}) + (\Omega\rho - \overline{\Omega\rho}) + \overline{(\Omega\mathrm{tr}\underline{\chi} - \overline{\Omega\mathrm{tr}\underline{\chi}})Y}\,\Big]. \tag{4.3.105}$$

From assumptions (4.2.5) and (4.2.1), we have

$$|r^{4-\frac{2}{p}}\hat{F}|_{p,S} \leq c(\Gamma_0)\Gamma_0^2\ ,\ |r\tau_-(\Omega\mathrm{tr}\underline{\chi} - \overline{\Omega\mathrm{tr}\underline{\chi}})| \leq \Gamma_0\ ,\ |r^3\tau_-^{\frac{1}{2}}(\Omega\rho - \overline{\Omega\rho})| \leq c\Delta_0,$$

which enables us to obtain

$$\begin{aligned}|r^{-\frac{2}{p}}(Y-\overline{Y})|_{p,S}(u,\underline{u}) &\leq |r^{-\frac{2}{p}}(Y-\overline{Y})|_{p,S_{(0)}}(\underline{u})\\ &+ \frac{1}{r^2\tau_-^{\frac{1}{2}}}c(\Gamma_0)\big(\Delta_0 + \Gamma_0(\mathcal{I}_0+\mathcal{I}_*+\Delta_0)+\Gamma_0^2\big).\end{aligned}$$

From that inequality, using assumption (4.3.101) and the results of Lemma 4.1.7 applied to $V = \mathbf{D}_4\log\Omega - \overline{\Omega\mathbf{D}_4\log\Omega}$, we can derive

$$|r^{\frac{5}{2}-\frac{2}{p}}\big(\Omega\mathbf{D}_4\log\Omega - \overline{\Omega\mathbf{D}_4\log\Omega}\big)|_{p,S}(u,\underline{u}) \leq c(\mathcal{I}_0+\mathcal{I}_*+\Delta_0)\ .$$

To prove the second part of the proposition we observe that, from the second equation in (4.3.59),

$$\mathbf{D}_4\overline{\Omega\mathbf{D}_3\log\Omega} = \overline{(\Omega\mathrm{tr}\chi\ - \overline{\Omega\mathrm{tr}\chi})\Omega\mathbf{D}_3\log\Omega} + \overline{\underline{\hat{F}}} - \overline{\Omega\rho}. \tag{4.3.106}$$

Denoting $\underline{Y} \equiv \Omega\mathbf{D}_3\log\Omega$, we obtain

$$\frac{d}{d\underline{u}}(\underline{Y}-\overline{\underline{Y}}) = \Omega\Big[(\underline{\hat{F}}-\overline{\underline{\hat{F}}}) + (\Omega\rho - \overline{\Omega\rho}) + \overline{(\Omega\mathrm{tr}\chi\ - \overline{\Omega\mathrm{tr}\chi})\underline{Y}}\Big]. \tag{4.3.107}$$

Since, again,

$$|r^{4-\frac{2}{p}}\underline{\hat{F}}|_{p,S} \leq c(\Gamma_0)\Gamma_0^2\ ,\ |r^3\tau_-^{\frac{1}{2}}(\Omega\rho - \overline{\Omega\rho})| \leq c\Delta_0\ , \text{and}\ |r^2(\Omega\mathrm{tr}\chi\ - \overline{\Omega\mathrm{tr}\chi})| \leq \Gamma_0\ ,$$

we conclude that

$$\begin{aligned}|r^{-\frac{2}{p}}(\underline{Y}-\overline{\underline{Y}})|_{p,S}(u,\underline{u}) &\leq c|r^{-\frac{2}{p}}(\underline{Y}-\overline{\underline{Y}})|_{p,S}(u,\underline{u}_*)\\ &+ \frac{1}{r^2\tau_-^{\frac{1}{2}}}c(\Gamma_0)\big(\Delta_0 + \Gamma_0(\mathcal{I}_0+\mathcal{I}_*+\Delta_0)+\Gamma_0^2\big).\end{aligned}$$

From that inequality the second inequality of the proposition,

$$|r^{1-\frac{2}{p}}\tau_-^{\frac{3}{2}}(\Omega\mathbf{D}_3\log\Omega - \overline{\Omega\mathbf{D}_3\log\Omega})|_{p,S}(u,\underline{u}) \leq c(\mathcal{I}_0+\mathcal{I}_*+\Delta_0)\ ,$$

follows, provided that, on the last slice, assumption (4.3.102) is satisfied.

Proposition 4.3.14 *Assume (4.2.1), the bootstrap assumption (4.2.5) and, on the last slice* $\underline{C}_*$,

$$|r^{2-\frac{2}{p}}\tau_-^{\frac{1}{2}}\left(\Omega\mathrm{tr}\chi-\overline{\Omega\mathrm{tr}\chi}\right)|_{p,S}(u,\underline{u}_*).\leq\mathcal{I}_* \tag{4.3.108}$$

Then, with the help of the first inequality of (4.3.103) in Proposition 4.3.13,

$$|r^{2-\frac{2}{p}}\tau_-^{\frac{1}{2}}\left(\Omega\mathrm{tr}\chi-\overline{\Omega\mathrm{tr}\chi}\right)|_{p,S}(u,\underline{u})\leq c\left(\mathcal{I}_0+\mathcal{I}_*+\Delta_0\right). \tag{4.3.109}$$

Proof: The evolution equation for $\mathrm{tr}\chi$ (see 3.1.46) can be rewritten as

$$\frac{d}{d\underline{u}}(\Omega\mathrm{tr}\chi)+\frac{1}{2}\Omega\mathrm{tr}\chi(\Omega\mathrm{tr}\chi)=\left[2\Omega\mathrm{tr}\chi(\Omega\mathbf{D}_4\log\Omega)-\Omega^2|\hat{\chi}|^2\right].$$

To control $(\mathrm{tr}\chi-\overline{\mathrm{tr}\chi})$ we derive the evolution equation for $\overline{\Omega\mathrm{tr}\chi}$,

$$\frac{d}{d\underline{u}}\overline{\Omega\mathrm{tr}\chi}=\frac{1}{2}\overline{(\Omega\mathrm{tr}\chi)^2}-(\overline{\Omega\mathrm{tr}\chi})^2+\overline{\left[2\Omega\mathrm{tr}\chi(\Omega\mathbf{D}_4\log\Omega)-\Omega^2|\hat{\chi}|^2\right]},$$

and from it,

$$\begin{aligned}\frac{d}{d\underline{u}}(\Omega\mathrm{tr}\chi-\overline{\Omega\mathrm{tr}\chi})&=-\left(\frac{1}{2}(\Omega\mathrm{tr}\chi)^2+\frac{1}{2}\overline{(\Omega\mathrm{tr}\chi)^2}-(\overline{\Omega\mathrm{tr}\chi})^2\right)\\&+\left[2\Omega\mathrm{tr}\chi(\Omega\mathbf{D}_4\log\Omega)-\Omega^2|\hat{\chi}|^2\right]-\overline{\left[2\Omega\mathrm{tr}\chi(\Omega\mathbf{D}_4\log\Omega)-\Omega^2|\hat{\chi}|^2\right]}.\end{aligned}$$

Denoting $V\equiv\left(\Omega\mathrm{tr}\chi-\overline{\Omega\mathrm{tr}\chi}\right)$, this equation can be written as

$$\begin{aligned}\frac{d}{d\underline{u}}V+\Omega\mathrm{tr}\chi\, V&=\frac{1}{2}V^2-\overline{V^2}+2(\Omega\mathbf{D}_4\log\Omega)V-\left[\Omega^2|\hat{\chi}|^2-\overline{\Omega^2|\hat{\chi}|^2}\right] \qquad (4.3.110)\\&+2(\overline{\Omega\mathrm{tr}\chi})\left(\Omega\mathbf{D}_4\log\Omega-\overline{\Omega\mathbf{D}_4\log\Omega}\right)-2\overline{(\Omega\mathrm{tr}\chi)\left(\Omega\mathbf{D}_4\log\Omega-\overline{\Omega\mathbf{D}_4\log\Omega}\right)}.\end{aligned}$$

We now use the first inequality of (4.3.101) and the inequalities

$$|\Omega\mathrm{tr}\chi|\leq c(\Gamma_0)\frac{1}{r}\;,\;\;|r^{4-\frac{2}{p}}V^2|_{p,S}\leq\Gamma_0^2,$$

which follow from the bootstrap assumptions (4.2.5) of Theorem 4.2.1. Applying the Evolution Lemma to equation (4.3.110) with $\lambda_1=2-\frac{2}{p}$ we obtain, using assumption (4.3.108),

$$\begin{aligned}|r^{2-\frac{2}{p}}V|_{p,S}(u,\underline{u})&\leq c|r^{2-\frac{2}{p}}V|_{p,S}(u,\underline{u}_*)+c(\Gamma_0)\left(\Gamma_0^2\int_{\underline{u}}^{\underline{u}_*}\frac{1}{r^2}+(\mathcal{I}_0+\mathcal{I}_*+\Delta_0)\int_{\underline{u}}^{\underline{u}_*}\frac{1}{r^{\frac{3}{2}}}\right)\\&\leq c\left(|r^{2-\frac{2}{p}}V|_{p,S}(u,\underline{u}_*)+\frac{1}{r}\Gamma_0^2+\frac{1}{r^{\frac{1}{2}}}(\mathcal{I}_0+\mathcal{I}_*+\Delta_0)\right)\\&\leq c\tau_-^{-\frac{1}{2}}(\mathcal{I}_0+\mathcal{I}_*+\Delta_0)\,,\end{aligned} \tag{4.3.111}$$

if $(\mathcal{I}_0+\mathcal{I}_*+\Delta_0)$ and Γ_0 are sufficiently small.

Proposition 4.3.15 *Assume (4.2.1), (4.2.2), the bootstrap assumptions (4.2.5) and (4.2.6), and, on the initial slice Σ_0,*

$$|r^{\frac{5}{2}-\frac{2}{p}}\left(\Omega\mathrm{tr}\underline{\chi}-\overline{\Omega\mathrm{tr}\underline{\chi}}\right)|_{p,S_{(0)}}(\underline{u})\leq\mathcal{I}_0. \tag{4.3.112}$$

Then, with the help of the second inequality of (4.3.103) in Proposition 4.3.13,

$$|r^{2-\frac{2}{p}}\tau_-^{\frac{1}{2}}\left(\Omega\mathrm{tr}\underline{\chi}-\overline{\Omega\mathrm{tr}\underline{\chi}}\right)|_{p,S}(u,\underline{u})\leq c\left(\mathcal{I}_0+\mathcal{I}_*+\Delta_0\right). \tag{4.3.113}$$

Proof: The evolution equation for $\mathrm{tr}\underline{\chi}$, see 3.1.46, can be rewritten as

$$\frac{d}{du}(\Omega\mathrm{tr}\underline{\chi})+\frac{1}{2}\Omega\mathrm{tr}\underline{\chi}(\Omega\mathrm{tr}\underline{\chi})=\left[2\Omega\mathrm{tr}\underline{\chi}(\Omega\mathbf{D}_3\log\Omega)-\Omega^2|\hat{\underline{\chi}}|^2\right].$$

To control $(\mathrm{tr}\underline{\chi}-\overline{\mathrm{tr}\underline{\chi}})$ we derive the evolution equation for $\overline{\Omega\mathrm{tr}\underline{\chi}}$,

$$\frac{d}{du}(\overline{\Omega\mathrm{tr}\underline{\chi}})=\frac{1}{2}\overline{(\Omega\mathrm{tr}\underline{\chi})^2}-(\overline{\Omega\mathrm{tr}\underline{\chi}})^2+\overline{\left[2\Omega\mathrm{tr}\underline{\chi}(\Omega\mathbf{D}_3\log\Omega)-\Omega^2|\hat{\underline{\chi}}|^2\right]},$$

so that, defining $\underline{V}=\left(\Omega\mathrm{tr}\underline{\chi}-\overline{\Omega\mathrm{tr}\underline{\chi}}\right)$, we obtain

$$\frac{d}{du}\underline{V}+\Omega\mathrm{tr}\underline{\chi}\,\underline{V}=\frac{1}{2}\underline{V}^2-\overline{\underline{V}^2}+2(\Omega\mathbf{D}_3\log\Omega)\underline{V}-\left[\Omega^2|\hat{\underline{\chi}}|^2-\overline{\Omega^2|\hat{\underline{\chi}}|^2}\right] \tag{4.3.114}$$
$$+2(\overline{\Omega\mathrm{tr}\underline{\chi}})\left(\Omega\mathbf{D}_3\log\Omega-\overline{\Omega\mathbf{D}_3\log\Omega}\right)-2\overline{(\Omega\mathrm{tr}\underline{\chi})\left(\Omega\mathbf{D}_3\log\Omega-\overline{\Omega\mathbf{D}_3\log\Omega}\right)}.$$

From the bootstrap assumptions (4.2.5) of Theorem 4.2.1, it follows that

$$|\Omega\mathrm{tr}\underline{\chi}|\leq c(\Gamma_0)\frac{1}{r}\ ,\quad|\Omega(\mathrm{tr}\underline{\chi}-\overline{\mathrm{tr}\underline{\chi}})|\leq c(\Gamma_0)\Gamma_0\frac{1}{r^2}\,.$$

We use the second inequality of (4.3.103) and proceed as in the case of $(\mathrm{tr}\chi-\overline{\mathrm{tr}\chi})$. Applying the Evolution Lemma to equation (4.3.114) with $\lambda_1=1-\frac{2}{p}$, and using assumption (4.3.112) and the results of Lemma 4.1.7 applied to $V=\mathrm{tr}\underline{\chi}-\overline{\mathrm{tr}\underline{\chi}}$, we conclude that for $p\in[2,4]$

$$|r^{2-\frac{2}{p}}\tau_-^{\frac{1}{2}}(\mathrm{tr}\underline{\chi}-\overline{\mathrm{tr}\underline{\chi}})|_{p,S}\leq c\left(\mathcal{I}_0+\mathcal{I}_*+\Delta_0\right),$$

proving the proposition.

Proposition 4.3.16 *Assume (4.2.1), (4.2.2), the bootstrap assumptions (4.2.5) and (4.2.6), and that, on the initial and final slices,*

$$|r^{\frac{5}{2}-\frac{2}{p}}\left(\Omega\mathrm{tr}\underline{\chi}-\overline{\Omega\mathrm{tr}\underline{\chi}}\right)|_{p,S_{(0)}}(\underline{u})\leq\mathcal{I}_0$$
$$|r^{2-\frac{2}{p}}\tau_-^{\frac{1}{2}}\left(\Omega\mathrm{tr}\chi-\overline{\Omega\mathrm{tr}\chi}\right)|_{p,S}(u,\underline{u}_*)\leq\mathcal{I}_*$$
$$|r^{\frac{7}{2}-\frac{2}{p}}(\underline{\mu}-\overline{\underline{\mu}})|_{p,S_{(0)}}(\underline{u})\leq c\mathcal{I}_0$$
$$|r^{2-\frac{2}{p}}\tau_-^{\frac{1}{2}}(\underline{\mu}-\overline{\underline{\mu}})|_{p,S}(u,\underline{u}_*)\leq c\mathcal{I}_*.$$

Then,

$$
\begin{aligned}
|r^{2-2/p}\tau_-^{\frac{1}{2}}\eta|_{p,S}(u,\underline{u}) &\leq c\,(\mathcal{I}_*+\mathcal{I}_0+\Delta_0)\\
|r^{3-2/p}\tau_-^{\frac{1}{2}}\nabla\!\!\!\!/\,\eta|_{p,S}(u,\underline{u}) &\leq c\,(\mathcal{I}_*+\mathcal{I}_0+\Delta_0)\\
|r^{2-2/p}\tau_-^{\frac{1}{2}}\underline{\eta}|_{p,S}(u,\underline{u}) &\leq c\,(\mathcal{I}_*+\mathcal{I}_0+\Delta_0)\\
|r^{3-2/p}\tau_-^{\frac{1}{2}}\nabla\!\!\!\!/\,\underline{\eta}|_{p,S}(u,\underline{u}) &\leq c\,(\mathcal{I}_*+\mathcal{I}_0+\Delta_0)\,.
\end{aligned}
\tag{4.3.115}
$$

Proof: Recall that

$$
\begin{aligned}
\tilde{\mu} &= \left[(\mu-\overline{\mu})+\frac{1}{4}(\mathrm{tr}\chi\,\mathrm{tr}\underline{\chi}-\overline{\mathrm{tr}\chi\,\mathrm{tr}\underline{\chi}})\right]\\
\underline{\tilde{\mu}} &= \left[(\underline{\mu}-\overline{\underline{\mu}})+\frac{1}{4}(\mathrm{tr}\underline{\chi}\,\mathrm{tr}\chi-\overline{\mathrm{tr}\underline{\chi}\,\mathrm{tr}\chi})\right].
\end{aligned}
\tag{4.3.116}
$$

It follows from the assumptions that on the initial and final slice we have

$$
\begin{aligned}
&|r^{2-\frac{2}{p}}\tau_-^{\frac{1}{2}}\tilde{\mu}|_{p,S}(u,\underline{u}_*)\leq c\mathcal{I}_*\\
&|r^{2-\frac{2}{p}}\tau_-^{\frac{1}{2}}\underline{\tilde{\mu}}|_{p,S_{(0)}}(\underline{u})\leq c\mathcal{I}_0.
\end{aligned}
\tag{4.3.117}
$$

Using these estimates in Lemma 4.3.3 and the results of Propositions 4.3.14, 4.3.15 we obtain the inequalities

$$
\begin{aligned}
&|r^{3-\frac{2}{p}}\tau_-^{\frac{1}{2}}\tilde{\mu}|_{p,S}(u,\underline{u})\leq c(\mathcal{I}_0+\mathcal{I}_*+\Delta_0)\\
&|r^{3-\frac{2}{p}}\tau_-^{\frac{1}{2}}\underline{\tilde{\mu}}|_{p,S}(u,\underline{u})\leq c(\mathcal{I}_0+\mathcal{I}_*+\Delta_0).
\end{aligned}
\tag{4.3.118}
$$

This result also follows from propositions 4.3.11, 4.3.12 and from Proposition 4.1.2 applied to the Hodge systems (4.3.37), (4.3.38). From the last proposition the next corollary immediately follows.

Corollary 4.3.2 *Under the same assumptions as in Proposition 4.3.16,*

$$
\begin{aligned}
&|r^{3-2/p}\tau_-^{\frac{1}{2}}\nabla\!\!\!\!/\,\mathrm{tr}\chi|_{p,S}(u,\underline{u})\leq c(\mathcal{I}_0+\mathcal{I}_*+\Delta_0)\\
&|r^{3-2/p}\tau_-^{\frac{1}{2}}\nabla\!\!\!\!/\,\mathrm{tr}\underline{\chi}|_{p,S}(u,\underline{u})\leq c(\mathcal{I}_0+\mathcal{I}_*+\Delta_0).
\end{aligned}
\tag{4.3.119}
$$

Proof: This follows immediately from the relation $2\zeta=(\eta-\underline{\eta})$ and the definitions of $\Psi\!\!\!\!/$ and $\underline{\Psi\!\!\!\!/}$.

Proposition 4.3.17 *Assume (4.2.1), (4.2.2), (4.2.3), the bootstrap assumptions (4.2.5), (4.2.6). Assume, moreover, that on the last slice $\underline{C}_*$ and on the initial slice Σ_0 the following estimates hold:*

$$
\begin{aligned}
&|r^{2-\frac{2}{p}}\tau_-^{\frac{3}{2}}\nabla\!\!\!\!/\,\Omega\mathbf{D}_3\log\Omega|_{p,S}(u,\underline{u}_*)\leq\mathcal{I}_*\\
&|r^{\frac{7}{2}-\frac{2}{p}}\nabla\!\!\!\!/\,\Omega\mathbf{D}_4\log\Omega|_{p,S_{(0)}}(\underline{u})\leq\mathcal{I}_0.
\end{aligned}
\tag{4.3.120}
$$

Then we derive, for any $p \in [2, 4]$,

$$|r^{3-\frac{2}{p}}\tau_-^{\frac{1}{2}}\nabla\!\!\!/\,\Omega\mathbf{D}_4 \log\Omega|_{p,S}(u,\underline{u}) \leq c(\Gamma_0)\,(\mathcal{I}_0 + \mathcal{I}_* + \Delta_0 + \Delta_1)$$
$$|r^{2-\frac{2}{p}}\tau_-^{\frac{3}{2}}\nabla\!\!\!/\,\Omega\mathbf{D}_3 \log\Omega|_{p,S}(u,\underline{u}) \leq c(\Gamma_0)\,(\mathcal{I}_0 + \mathcal{I}_* + \Delta_0 + \Delta_1)\,. \qquad (4.3.121)$$

Proof: The result is obtained by proceeding as in Proposition 4.3.8, but using the stronger assumptions (4.3.120) on initial and last slices.

4.4 Proof of Theorem 4.2.2 and estimates for the second derivatives of the connection coefficients

Let us examine the various second derivatives whose norm estimates are provided by Theorem 4.2.2.

1. Estimates for the second derivatives $\nabla\!\!\!/^2\mathrm{tr}\chi$, $\nabla\!\!\!/^2\hat{\chi}$ are obtained by exactly the same procedure as that used for the corresponding first derivatives with the help of the equation for $\nabla\!\!\!/$, (4.3.6), and the Codazzi equation, (4.3.13), which have to be differentiated once more in the angular direction. Because there are no new ideas required, just technical drudgery, we shall omit the proof.

2. Estimates for the second derivatives $\nabla\!\!\!/^2\mathrm{tr}\underline{\chi}$, $\nabla\!\!\!/^2\hat{\underline{\chi}}$ are obtained via the same considerations as for (1). One uses the basic transport equation (4.3.19) and the Codazzi equation (4.3.25), which have to be differentiated to provide the appropriate equations.

3. Estimates for the second derivatives $\nabla\!\!\!/^2\eta$, $\nabla\!\!\!/^2\underline{\eta}$ are again obtained by the same procedure starting from the Hodge systems (4.3.37), (4.3.38) coupled with the transport equations (4.3.39) for $\tilde{\mu}$ and $\underline{\tilde{\mu}}$. We have again to take angular derivatives of these equations and proceed as before.

Remark: Observe that for the second derivatives $\nabla\!\!\!/^2\mathrm{tr}\chi$, $\nabla\!\!\!/^2\hat{\chi}$, $\nabla\!\!\!/^2\mathrm{tr}\underline{\chi}$, $\nabla\!\!\!/^2\hat{\underline{\chi}}$, $\nabla\!\!\!/^2\eta$, $\nabla\!\!\!/^2\underline{\eta}$, better estimates can be obtained, provided stronger assumptions hold on the initial and final slices. In this case their $|\cdot|_{p,S}$ norms gain a factor of $\tau_-^{\frac{1}{2}}$ exactly as proved for their zero and first derivatives in Subsection 4.3.16.

4. Estimates for the angular derivatives and the null-directions derivatives of ω and $\underline{\omega}$ are more delicate. They require some new ideas and will be examined in full detail.

4.4.1 *Estimates for* $\mathcal{O}_2^{p,S}(\omega)$ *and* $\mathcal{O}_2^{p,S}(\underline{\omega})$ *with* $p \in [2, 4]$

Remark: This subsection is devoted to the control of $\nabla\!\!\!/^2\mathbf{D}_3\log\Omega$, $\nabla\!\!\!/^2\mathbf{D}_4\log\Omega$. Nevertheless, by proving these estimates one also obtains an estimate for $\nabla\!\!\!/\mathbf{D}_3\log\Omega$ and $\nabla\!\!\!/\mathbf{D}_4\log\Omega$, which have already been estimated in a more direct way in Subsection 4.3.13.

Proposition 4.4.1 *Under the assumptions of Theorem 4.2.2 and the bootstrap assumptions (4.2.5) and (4.2.15), for any* $p \in [2, 4]$,

$$|r^{2-\frac{2}{p}}\tau_-\nabla\!\!\!/\mathbf{D}_3 \log\Omega|_{p,S} \leq c(\mathcal{I}_0 + \mathcal{I}_* + \Delta_0 + \Delta_1)$$

$$
\begin{aligned}
&|r^{3-\frac{2}{p}} \nabla\!\!\!/ \mathbf{D}_4 \log\Omega|_{p,S} \le c(\mathcal{I}_0 + \mathcal{I}_* + \Delta_0 + \Delta_1)\\
&|r^{3-\frac{2}{p}} \tau_- \nabla\!\!\!/^2 \mathbf{D}_3 \log\Omega|_{p,S} \le c(\mathcal{I}_0 + \mathcal{I}_* + \Delta_0 + \Delta_1)\\
&|r^{4-\frac{2}{p}} \nabla\!\!\!/^2 \mathbf{D}_4 \log\Omega|_{p,S} \le c(\mathcal{I}_0 + \mathcal{I}_* + \Delta_0 + \Delta_1).
\end{aligned}
\tag{4.4.1}
$$

Remark: In the proof of Theorem 4.2.2 we can, of course, rely on the results of Theorem 4.2.1. However since the terms to which the results of the previous theorem apply appear in nonlinear expressions, it suffices to use the weaker estimate $\mathcal{O}^{\infty}_{[0]} + \underline{\mathcal{O}}^{\infty}_{[0]} \le \Gamma_0$ which is, in fact, the bootstrap assumption of Theorem 4.2.1. Since, moreover, we also make the new bootstrap assumption (4.2.15), $\mathcal{O}^{\infty}_1 + \underline{\mathcal{O}}^{\infty}_1 \le \Gamma_1$, this means that we can rely on the assumption $\mathcal{O}^{\infty}_{[1]} + \underline{\mathcal{O}}^{\infty}_{[1]} \le \Gamma_0 + \Gamma_1$.

Proof: We only discuss the estimates for $\nabla\!\!\!/ \mathbf{D}_3 \log\Omega$ and $\nabla\!\!\!/^2 \mathbf{D}_3 \log\Omega$ since the estimates for $\nabla\!\!\!/ \mathbf{D}_4 \log\Omega$ and $\nabla\!\!\!/^2 \mathbf{D}_4 \log\Omega$ are found in the same way. The evolution equations for $\nabla\!\!\!/^2 \mathbf{D}_3 \log\Omega$ and $\nabla\!\!\!/^2 \mathbf{D}_4 \log\Omega$ are found in the following lemma, whose proof is obtained by deriving tangentially the evolution equations for $\nabla\!\!\!/ \mathbf{D}_3 \log\Omega$ and $\nabla\!\!\!/ \mathbf{D}_4 \log\Omega$; see 4.3.80. The details are in the appendix to this chapter.

Lemma 4.4.1 *$\nabla\!\!\!/^2 \mathbf{D}_3 \log\Omega$ and $\nabla\!\!\!/^2 \mathbf{D}_4 \log\Omega$ satisfy the following evolution equations. Denoting $V = \nabla\!\!\!/ \mathbf{D}_3 \log\Omega$, $\underline{V} = \nabla\!\!\!/ \mathbf{D}_4 \log\Omega$,*

$$
\mathbf{D}\!\!\!\!/_4 \nabla\!\!\!/_a V_b + \mathrm{tr}\chi \nabla\!\!\!/_a V_b = -\left(\hat{\chi}_{ac} \nabla\!\!\!/_c V_b + \hat{\chi}_{bc} \nabla\!\!\!/_a V_c\right) + \Omega^{-1} \nabla\!\!\!/_a \nabla\!\!\!/_b (\Omega \mathbf{D}_4 \mathbf{D}_3 \log\Omega) + \underline{Q}_{ab}
$$

$$
\mathbf{D}\!\!\!\!/_3 \nabla\!\!\!/_a V_b + \mathrm{tr}\underline{\chi} \nabla\!\!\!/_a V_b = -\left(\hat{\underline{\chi}}_{ac} \nabla\!\!\!/_c V_b + \hat{\underline{\chi}}_{bc} \nabla\!\!\!/_a V_c\right) + \Omega^{-1} \nabla\!\!\!/_a \nabla\!\!\!/_b (\Omega \mathbf{D}_3 \mathbf{D}_4 \log\Omega) + Q_{ab}, \tag{4.4.2}
$$

where

$$
\underline{Q}_{ab} = -\left[(\nabla\!\!\!/_a \log\Omega)\chi_{bc} V_c + (\nabla\!\!\!/_a \chi_{bc}) V_c + \underline{\eta}_b \chi_{ac} V_c - \chi_{ab} \underline{\eta}_c V_c - e_4^{\tau} e_a^{\rho} \left([D_\tau, D_\rho] V_\sigma\right) e_b^{\sigma}\right]
$$

$$
Q_{ab} = -\left[(\nabla\!\!\!/_a \log\Omega)\underline{\chi}_{bc} \underline{V}_c + (\nabla\!\!\!/_a \underline{\chi}_{bc}) \underline{V}_c + \eta_b \underline{\chi}_{ac} \underline{V}_c - \underline{\chi}_{ab} \eta_c \underline{V}_c - e_3^{\tau} e_a^{\rho} \left([D_\tau, D_\rho] \underline{V}_\sigma\right) e_b^{\sigma}\right].
$$

Using this equation directly is not efficient. From the explicit expression of $\Omega \mathbf{D}_4 \mathbf{D}_3 \log\Omega$, (see 4.3.59),

$$
\Omega \mathbf{D}_4 \mathbf{D}_3 \log\Omega = \Omega\left[-2\zeta \cdot \nabla\!\!\!/ \log\Omega - (\mathbf{D}_3 \log\Omega)(\mathbf{D}_4 \log\Omega) + (\eta \cdot \underline{\eta} - 2\zeta^2)\right] - \Omega\rho\,,
$$

we have

$$
\Omega^{-1} \nabla\!\!\!/_a \nabla\!\!\!/_b (\Omega \mathbf{D}_4 \mathbf{D}_3 \log\Omega) = -\nabla\!\!\!/_a \nabla\!\!\!/_b \rho - (\mathbf{D}_4 \log\Omega)(\nabla\!\!\!/_a V_b) + \underline{L}_{ab} \tag{4.4.3}
$$

where

$$
\begin{aligned}
\underline{L}_{ab} \;=\; & \frac{1}{\Omega}\Big[- (\nabla\!\!\!/_a \nabla\!\!\!/_b \Omega)\rho - 2(\nabla\!\!\!/_a \Omega)\nabla\!\!\!/_b \rho \Big]\\
& - \frac{1}{\Omega}\Big[\nabla\!\!\!/_a \nabla\!\!\!/_b \Big(\Omega(\mathbf{D}_3 \log\Omega)(\mathbf{D}_4 \log\Omega)\Big) - \Omega(\mathbf{D}_4 \log\Omega) \nabla\!\!\!/_a V_b\Big]\\
& + \frac{1}{\Omega}\Big[\nabla\!\!\!/_a \nabla\!\!\!/_b \Big(\Omega(-2\zeta \cdot \nabla\!\!\!/ \log\Omega + \eta \cdot \underline{\eta} - 2\zeta^2)\Big)\Big]
\end{aligned}
$$

depends only on the first derivatives of the Riemann components[28] and the second derivatives of the connection coefficients. Plugging this expression in the previous equation we obtain

$$\begin{aligned}\mathbf{D}_4(\Omega \nabla\!\!\!\!/_a V_b) + \mathrm{tr}\chi(\Omega \nabla\!\!\!\!/_a V_b) &= -\Big(\hat{\chi}_{ac}(\Omega \nabla\!\!\!\!/_c V_b) + \hat{\chi}_{bc}(\Omega \nabla\!\!\!\!/_a V_c)\Big)\\ &+ \Omega\left[\underline{Q}+\underline{L}\right]_{ab} - \Omega \nabla\!\!\!\!/_a \nabla\!\!\!\!/_b \rho. \end{aligned} \tag{4.4.4}$$

The term $\Omega \nabla\!\!\!\!/_a \nabla\!\!\!\!/_b \rho$ makes it impossible to obtain an estimate of $\Omega \nabla\!\!\!\!/_a V_b$ in the $|\cdot|_{p,S}$ norm.[29] We overcome this problem using the Bianchi equations to transform a tangential derivative into a null derivative, a procedure which has been repeatedly used in [Ch-Kl]. Using the equation (see (3.2.8)),[30]

$$\underline{\beta}_4 \equiv \mathbf{D}\!\!\!\!/_4 \underline{\beta} + \mathrm{tr}\chi \underline{\beta} = -\nabla\!\!\!\!/ \rho + {}^*\nabla\!\!\!\!/ \sigma + 2\hat{\underline{\chi}} \cdot \beta + 2\omega \underline{\beta} - 3(\underline{\eta}\rho - {}^*\underline{\eta}\sigma),$$

we write $\nabla\!\!\!\!/_a \nabla\!\!\!\!/_b \rho$ in terms of $\mathbf{D}_4 \nabla\!\!\!\!/_a \underline{\beta}_b$ plus lower order terms[31]

$$\mathbf{D}_4(\Omega \nabla\!\!\!\!/_a \underline{\beta}_b) + \frac{3}{2}\mathrm{tr}\chi\,\Omega \nabla\!\!\!\!/_a \underline{\beta}_b = -\Omega \nabla\!\!\!\!/_a \nabla\!\!\!\!/_b \rho + \Omega \nabla\!\!\!\!/_a {}^*\nabla\!\!\!\!/_b \sigma - \hat{\underline{\chi}}_{ac}\Omega \nabla\!\!\!\!/_c \underline{\beta}_b + \underline{H}_{ab},$$

and the evolution equation for $\nabla\!\!\!\!/_a (V_b - \underline{\beta}_b)$ is

$$\begin{aligned}\mathbf{D}_4(\Omega \nabla\!\!\!\!/_a (V_b - \underline{\beta}_b)) + \mathrm{tr}\chi(\Omega \nabla\!\!\!\!/_a (V_b - \underline{\beta}_b)) &= -\Big(\hat{\chi}_{ac}(\Omega \nabla\!\!\!\!/_c V_b) + \hat{\chi}_{bc}\Omega \nabla\!\!\!\!/_a V_c\Big)\\ &+\frac{1}{2}\mathrm{tr}\chi\,\Omega \nabla\!\!\!\!/_a \underline{\beta}_b + \Omega\left[\underline{Q}+\underline{L}+\underline{H}\right]_{ab} - \Omega \nabla\!\!\!\!/_a {}^*\nabla\!\!\!\!/_b \sigma \end{aligned}\tag{4.4.5}$$

To achieve the result of avoiding, in the right hand side, the second derivatives of the Riemann tensor we have to get rid of the term $\nabla\!\!\!\!/ {}^*\nabla\!\!\!\!/ \sigma$. This is obtained considering instead of $\Omega \nabla\!\!\!\!/_a (V_b - \underline{\beta}_b)$ the quantity

$$\underline{\omega\!\!\!/} \equiv \Omega(\mathrm{d\!\!\!/iv}\, V - \mathrm{d\!\!\!/iv}\underline{\beta}) = -\Omega(2\Delta\!\!\!\!/\,\underline{\omega} + \mathrm{d\!\!\!/iv}\underline{\beta}) \tag{4.4.6}$$

which satisfies the evolution equation

$$\begin{aligned}\mathbf{D}_4\underline{\omega\!\!\!/} + \mathrm{tr}\chi\underline{\omega\!\!\!/} &= -2\hat{\chi}\cdot(\Omega \nabla\!\!\!\!/ V) + \frac{1}{2}\mathrm{tr}\chi\,\Omega\mathrm{d\!\!\!/iv}\underline{\beta} + \Omega\mathrm{tr}(\underline{Q}+\underline{L}+\underline{H})\\ &= -2\hat{\chi}\cdot\Omega(\nabla\!\!\!\!/ V - \nabla\!\!\!\!/ \underline{\beta}) - 2\hat{\chi}\cdot\Omega \nabla\!\!\!\!/ \underline{\beta} + \frac{1}{2}\mathrm{tr}\chi\,\Omega\mathrm{d\!\!\!/iv}\underline{\beta}\\ &+ \Omega\mathrm{tr}(\underline{Q}+\underline{L}+\underline{H}). \end{aligned}\tag{4.4.7}$$

The evolution equation for $\underline{\omega\!\!\!/}$ does not contain any second tangential derivatives of ρ or σ and, therefore, can be estimated with the $|\cdot|_{p,S}$ norms, with $p \in [2,4]$. We obtain the following inequality

$$|r^{3-\frac{2}{p}}\tau_-\underline{\omega\!\!\!/}\,|_{p,S}(u,\underline{u}) \le c|r^{3-\frac{2}{p}}\tau_-\underline{\omega\!\!\!/}\,|_{p,S}(u,\underline{u}_*) + c\tau_-^{-\frac{1}{2}}(\mathcal{I}_0 + \mathcal{I}_* + \Delta_0 + \Delta_1) \tag{4.4.8}$$

---

[28] This is true, notwithstanding the presence of $\nabla\!\!\!\!/_a \nabla\!\!\!\!/_b \mathbf{D}_4 \log\Omega$ in $\underline{L}_{ab}$, since the same argument we are developing here holds for it.

[29] We recall that, in this norm, we do not control the second derivatives of the Riemann components.

[30] Recall that ${}^*\nabla\!\!\!\!/_a \sigma = \epsilon_{ac} \nabla\!\!\!\!/^c \sigma$.

[31] Here "lower order" is in the sense of order of derivatives.

We control $\Omega\nabla\!\!\!\!/_a V_b = \Omega\nabla\!\!\!\!/_a\nabla\!\!\!\!/_b \mathbf{D}_3 \log\Omega$ using the estimate of $\underline{\wp}$, (4.4.8), the elliptic estimates for the equation

$$\Delta\!\!\!\!/(\mathbf{D}_3 \log\Omega) = \Omega^{-1}\underline{\wp} + \mathrm{d}\!\!\!/\mathrm{iv}\,\underline{\beta}\,,$$

the estimate of $\mathrm{d}\!\!\!/\mathrm{iv}\,\underline{\beta}$ and the last slice estimate derived from (4.2.12),

$$|r^{3-\frac{2}{p}}\tau_-(\Omega\nabla\!\!\!\!/^2\mathbf{D}_3\log\Omega)|_{p,S}(u,\underline{u}_*) \leq \mathcal{I}_*. \tag{4.4.9}$$

We obtain in this way the estimates that prove Proposition 4.4.1

$$|r^{2-\frac{2}{p}}\tau_-(\Omega\nabla\!\!\!\!/\mathbf{D}_3\log\Omega)|_{p,S} \leq c(\mathcal{I}_0+\mathcal{I}_*+\Delta_0+\Delta_1)$$
$$|r^{3-\frac{2}{p}}\tau_-(\Omega\nabla\!\!\!\!/^2\mathbf{D}_3\log\Omega)|_{p,S} \leq c(\mathcal{I}_0+\mathcal{I}_*+\Delta_0+\Delta_1)\,.$$

Corollary 4.4.1 *Assume inequalities (4.2.1), (4.2.5) and, on the last slice $\underline{C}_*$ and the initial slice Σ_0,*

$$|r^{3-\frac{2}{p}}\tau_-^{\frac{3}{2}}\nabla\!\!\!\!/^2\Omega\mathbf{D}_3\log\Omega|_{p,S}(u,\underline{u}_*) \leq \mathcal{I}_*$$
$$|r^{\frac{9}{2}-\frac{2}{p}}\nabla\!\!\!\!/^2\Omega\mathbf{D}_4\log\Omega|_{p,S_{(0)}}(\underline{u}) \leq \mathcal{I}_0. \tag{4.4.10}$$

Then we derive for any $p \in [2,4]$

$$|r^{2-\frac{2}{p}}\tau_-^{\frac{3}{2}}(\Omega\nabla\!\!\!\!/\mathbf{D}_3\log\Omega)|_{p,S} \leq c(\mathcal{I}_0+\mathcal{I}_*+\Delta_0+\Delta_1)$$
$$|r^{3-\frac{2}{p}}\tau_-^{\frac{3}{2}}(\Omega\nabla\!\!\!\!/^2\mathbf{D}_3\log\Omega)|_{p,S} \leq c(\mathcal{I}_0+\mathcal{I}_*+\Delta_0+\Delta_1)$$
$$|r^{3-\frac{2}{p}}\tau_-^{\frac{1}{2}}(\Omega\nabla\!\!\!\!/\mathbf{D}_4\log\Omega)|_{p,S} \leq c(\mathcal{I}_0+\mathcal{I}_*+\Delta_0+\Delta_1) \tag{4.4.11}$$
$$|r^{4-\frac{2}{p}}\tau_-^{\frac{1}{2}}(\Omega\nabla\!\!\!\!/^2\mathbf{D}_4\log\Omega)|_{p,S} \leq c(\mathcal{I}_0+\mathcal{I}_*+\Delta_0+\Delta_1).$$

Proof: The result is obtained as in Proposition 4.4.1, but using the stronger assumption (4.4.10) on the last and initial slices.

4.4.2 *Estimate for $\tilde{\mathcal{O}}_2(\underline{\omega})$ with $p \in [2,4]$*

We recall the definition (see 3.5.30)

$$\tilde{\mathcal{O}}_2(\underline{\omega}) \equiv ||\frac{1}{\sqrt{\tau_+}}\tau_-^3\mathbf{D}_3^2\underline{\omega}||_{L_2(C\cap\mathcal{K})} + ||\frac{1}{\sqrt{\tau_+}}r\tau_-^2\nabla\!\!\!\!/\mathbf{D}_3\underline{\omega}||_{L_2(C\cap\mathcal{K})}.$$

The terms on the right-hand side are estimated in the following proposition.

Proposition 4.4.2 *Under the assumptions of Theorem 4.2.2 and the bootstrap assumptions (4.2.5) and (4.2.15), and choosing Γ_0 and Γ_1 sufficiently small,*

$$||\frac{1}{\sqrt{\tau_+}}r\tau_-^2\nabla\!\!\!\!/\mathbf{D}_3\underline{\omega}||_{L_2(C\cap\mathcal{K})} \leq c\,(\mathcal{I}_0+\mathcal{I}_*+\Delta_0+\Delta_1+\Delta_2) \tag{4.4.12}$$
$$||\frac{1}{\sqrt{\tau_+}}\tau_-^3\mathbf{D}_3^2\underline{\omega}||_{L_2(C\cap\mathcal{K})} \leq c(\mathcal{I}_0+\mathcal{I}_*+\Delta_0+\Delta_1+\Delta_2). \tag{4.4.13}$$

Proof: The result is a direct consequence of the following estimates:

$$|r^{2-\frac{2}{p}}\tau_-^2 \nabla\!\!\!/ \mathbf{D}_3^2 \log\Omega|_{p=2,S}(u,\underline{u}) \le c\left[\frac{r(u,\underline{u})}{r(u,\underline{u}_*)}|r^{2-\frac{2}{p}}\tau_-^2 \nabla\!\!\!/ \mathbf{D}_3^2 \log\Omega|_{p=2,S}(u,\underline{u}_*) + \frac{1}{r^{\frac{1}{2}}}(\mathcal{I}_0+\mathcal{I}_*+\Delta_0+\Delta_1+\Delta_2)\right] \tag{4.4.14}$$

$$|r^{1-\frac{2}{p}}\tau_-^3 (\Omega\mathbf{D}_3)^3 \log\Omega|_{p=2,S}(u,\underline{u}) \le c\left[\frac{r(u,\underline{u})}{r(u,\underline{u}_*)}|r^{1-\frac{2}{p}}\tau_-^3 (\Omega\mathbf{D}_3)^3 \log\Omega|_{p=2,S}(u,\underline{u}_*) + \sqrt{\frac{\tau_-}{r}}\Delta_2 + \frac{\tau_-}{r}(\mathcal{I}_0+\mathcal{I}_*+\Delta_0+\Delta_1+\Delta_2)\right]. \tag{4.4.15}$$

To prove inequality (4.4.14) we need the evolution equation for $\nabla\!\!\!/ \mathbf{D}_3^2 \log\Omega$. This is obtained by deriving tangentially the evolution equation for $(\Omega\mathbf{D}_3)^2\log\Omega$ (see 4.3.87) and using the commutation relation

$$[\nabla\!\!\!/_a, \mathbf{D}\!\!\!\!/_4]f = -(\nabla\!\!\!/ \log\Omega)\mathbf{D}_4 f + \chi_{ac}\nabla\!\!\!/_c f\ ,$$

proved in Proposition 4.8.1 in the appendix of this chapter. The result is

$$\mathbf{D}_4\big(\nabla\!\!\!/ (\Omega\mathbf{D}_3)^2\log\Omega\big) + \frac{1}{2}\mathrm{tr}\chi\big(\nabla\!\!\!/ (\Omega\mathbf{D}_3)^2\log\Omega\big) = \hat{\chi}\cdot\nabla\!\!\!/ (\Omega\mathbf{D}_3)^2\log\Omega - \Omega^2\nabla\!\!\!/ \mathbf{D}_3\rho + \underline{W},$$

where, with $\underline{M}$ defined in the previous subsection, (see 4.3.88)

$$\underline{W} = (\nabla\!\!\!/ \log\Omega)\underline{M} - 3\Omega^2(\nabla\!\!\!/ \log\Omega)\mathbf{D}_3\rho + \nabla\!\!\!/ \underline{M}\ .$$

Applying the Evolution Lemma, we derive

$$|r^{1-\frac{2}{p}}\tau_-^2(\nabla\!\!\!/ \Omega^2\mathbf{D}_3^2\log\Omega)|_{p,S}(u,\underline{u}) \le c\left(|r^{1-\frac{2}{p}}\tau_-^2(\nabla\!\!\!/ \Omega^2\mathbf{D}_3^2\log\Omega)|_{p,S}(u,\underline{u}_*) + \int_{\underline{u}}^{\underline{u}_*}|r^{1-\frac{2}{p}}\tau_-^2(\nabla\!\!\!/ \log\Omega)\left(\underline{M}-3\Omega^2\mathbf{D}_3\rho\right)|_{p,S} + \int_{\underline{u}}^{\underline{u}_*}|r^{1-\frac{2}{p}}\tau_-^2\nabla\!\!\!/ \underline{M}|_{p,S} + \int_{\underline{u}}^{\underline{u}_*}|r^{1-\frac{2}{p}}\tau_-^2\Omega^2\nabla\!\!\!/ \mathbf{D}_3\rho|_{p,S}\right). \tag{4.4.16}$$

The right-hand side of (4.4.16) depends on $\nabla\!\!\!/ \mathbf{D}_3\rho$. Because the second derivatives of the Riemann tensor are not bounded in the $|\cdot|_{p,S}$ norms, but only in the $L^2(C)$, $L^2(\underline{C})$ norms, we consider inequality (4.4.16) with $p=2$ and obtain

$$\begin{aligned}
&\int_{\underline{u}}^{\underline{u}_*}|r^{1-\frac{2}{p}}\tau_-^2(\nabla\!\!\!/ \log\Omega)\underline{M}|_{p=2,S} \le c\frac{\tau_-}{r^4}\Gamma_0^2\left((\mathcal{I}_0+\mathcal{I}_*+\Delta_0)+\Gamma_0+\Gamma_1\right)\\
&\int_{\underline{u}}^{\underline{u}_*}|r^{1-\frac{2}{p}}\tau_-^2\nabla\!\!\!/ \underline{M}|_{p=2,S} \le c\frac{\tau_-}{r^3}\left((\mathcal{I}_0+\mathcal{I}_*+\Delta_0+\Delta_1)+\Gamma_0^2+\Gamma_0\Gamma_1\right)\\
&\int_{\underline{u}}^{\underline{u}_*}|r^{1-\frac{2}{p}}\tau_-^2(\nabla\!\!\!/ \log\Omega)\Omega^2\mathbf{D}_3\rho|_{p=2,S} \le c\frac{\tau_-}{r^3}\Gamma_0(\Delta_0+\Delta_1)\\
&\int_{\underline{u}}^{\underline{u}_*}|r^{1-\frac{2}{p}}\tau_-^2\nabla\!\!\!/ \mathbf{D}_3\rho|_{p=2,S} \le c\frac{1}{r^{\frac{3}{2}}}(\Delta_1+\Delta_2)
\end{aligned} \tag{4.4.17}$$

where the second inequality uses the estimate for $\nabla\!\!\!/ \underline{M}$[32]

$$|r^{5-\frac{2}{p}}\tau_-\nabla\!\!\!/ \underline{M}|_{p,S} \leq c\left((\mathcal{I}_0+\mathcal{I}_*+\Delta_0+\Delta_1)+\Gamma_0^2+\Gamma_0\Gamma_1\right) \tag{4.4.18}$$

proved in Proposition 4.4.1. The last inequality in (4.4.17) is obtained as follows:

$$\begin{aligned}
\int_{\underline{u}}^{\underline{u}_*}|r^{1-\frac{2}{p}}\tau_-^2\nabla\!\!\!/ \mathbf{D}_3\rho|_{p=2,S} &= \int_{\underline{u}}^{\underline{u}_*}\frac{1}{r^2}\left(\int_{S(u,\underline{u}')}|r^2\tau_-^2\nabla\!\!\!/ \mathbf{D}_3\rho|^2\right)^{\frac{1}{2}}\\
&\leq \left(\int_{\underline{u}}^{\underline{u}_*}\frac{1}{r^4}\right)^{\frac{1}{2}}\left(\int_{\underline{u}}^{\underline{u}_*}\int_{S(u,\underline{u}')}|r^2\tau_-^2\nabla\!\!\!/ \mathbf{D}_3\rho|^2\right)^{\frac{1}{2}}\\
&\leq c\frac{1}{r^{\frac{3}{2}}}||r^2\tau_-^2\nabla\!\!\!/ \mathbf{D}_3\rho||_{L^2(C(u))}\leq c\frac{1}{r^{\frac{3}{2}}}\Delta_2,
\end{aligned}$$

where the last line uses the estimate for $||r^2\tau_-^2\nabla\!\!\!/ \mathbf{D}_3\rho||_{L^2(C(u))}$ contained in the assumptions (4.2.11) of Theorem 4.2.2. From these estimates, observing that assumptions (4.2.22) imply, on the last slice, for $p \in [2,4]$,

$$|r^{2-2/p}\tau_-^2\nabla\!\!\!/ \mathbf{D}_3^2\log\Omega|_{p,S}(u,\underline{u}_*) \leq \mathcal{I}_*\ ,$$

the estimate (4.4.14) follows.

To prove inequality (4.4.15) we use the previous results and the commutation relation (see Proposition 4.8.1),

$$[\mathbf{D}_3,\mathbf{D}_4]f = (\mathbf{D}_4\log\Omega)\mathbf{D}_3 f - (\mathbf{D}_3\log\Omega)\mathbf{D}_4 f + 4\zeta\cdot\nabla\!\!\!/ f\ ,$$

with f a scalar function. The evolution equation for $(\Omega\mathbf{D}_3)^3\log\Omega$ turns out to be

$$\mathbf{D}_4\left((\Omega\mathbf{D}_3)^3\log\Omega\right) = \underline{P} - \Omega^3\mathbf{D}_3^2\rho, \tag{4.4.19}$$

where

$$\underline{P} = \left[\mathbf{D}_3(\Omega\underline{M}) - (\mathbf{D}_3\Omega^3)\mathbf{D}_3\rho - 4\Omega\zeta\cdot\nabla\!\!\!/ (\Omega\mathbf{D}_3)^2\log\Omega\right]. \tag{4.4.20}$$

Because the right-hand side of (4.4.19) depends on the second derivatives of the Riemann tensor, we can only obtain an estimate for $p=2$. Proceeding as in the previous proposition (see (4.4.16)) we derive

$$\begin{aligned}
|r^{-1}\tau_-^2(\Omega\mathbf{D}_3)^3\log\Omega|_{p=2,S}(u,\underline{u}) &\leq c\left(|r^{-1}\tau_-^2(\Omega\mathbf{D}_3)^3\log\Omega|_{p=2,S}(u,\underline{u}_*)\right.\\
&\quad\left.+\int_{\underline{u}}^{\underline{u}_*}|r^{-1}\tau_-^2\underline{P}|_{p=2,S}+\int_{\underline{u}}^{\underline{u}_*}|r^{-1}\tau_-^2\mathbf{D}_3^2\rho|_{p=2,S}\right).
\end{aligned} \tag{4.4.21}$$

[32] Apparently $\nabla\!\!\!/ \underline{M}$ also depends on the second derivatives of the Riemann tensor due to the term $\nabla\!\!\!/^2\mathbf{D}_3\log\Omega$ in its expression. Nevertheless, as discussed in Proposition 4.4.1 (see also [Ch-Kl] page 373), we have a better estimate (4.4.18).

The more delicate terms in $\underline{P}$ are $\zeta \cdot \nabla\!\!\!/(\Omega\mathbf{D}_3)^2 \log\Omega$ and $(\mathbf{D}_3 \log\Omega)\mathbf{D}_3\rho$. Considering the first contribution, we have

$$\int_{\underline{u}}^{\underline{u}_*} |r^{-1}\tau_-^2\zeta \cdot \nabla\!\!\!/(\Omega\mathbf{D}_3)^2 \log\Omega|_{p=2,S} \le c\Gamma_0 \int_{\underline{u}}^{\underline{u}_*} \frac{1}{r^2}|r^{-1}\tau_-^2\nabla\!\!\!/(\Omega\mathbf{D}_3)^2 \log\Omega|_{p=2,S} \,.$$

Recalling that, from inequality (4.4.16) and the subsequent estimates (4.4.17),

$$|r^{-1}\tau_-^2\nabla\!\!\!/(\Omega\mathbf{D}_3)^2 \log\Omega|_{p=2,S} \le c\frac{1}{r^2}(\mathcal{I}_0 + \mathcal{I}_* + \Delta_0 + \Delta_1 + \Delta_2)\,,$$

we derive

$$\int_{\underline{u}}^{\underline{u}_*} |r^{-1}\tau_-^2\zeta \cdot \nabla\!\!\!/(\Omega\mathbf{D}_3)^2 \log\Omega|_{p=2,S} \quad\le\quad c\frac{1}{r^3}\Gamma_0(\mathcal{I}_0 + \mathcal{I}_* + \Delta_0 + \Delta_1 + \Delta_2).$$

To estimate the contribution due to the term $(\mathbf{D}_3 \log\Omega)\mathbf{D}_3\rho$, we recall the estimate

$$\sup|\mathbf{D}_3 \log\Omega| \le c\frac{1}{r\tau_-}\Gamma_0,$$

which follows from the bootstrap assumptions, and obtain

$$\int_{\underline{u}}^{\underline{u}_*} |r^{-1}\tau_-^2(\mathbf{D}_3 \log\Omega)\mathbf{D}_3\rho|_{p=2,S} \quad\le\quad c\Gamma_0\frac{1}{\tau_-}\int_{\underline{u}}^{\underline{u}_*} \frac{1}{r}|r^{-1}\tau_-^2\mathbf{D}_3\rho|_{p=2,S}.$$

The estimate

$$|r^{-\frac{2}{p}}\tau_-^2\mathbf{D}_3\rho|_{p=2,S} \le c\frac{\tau_-}{r^3}(\Delta_0 + \Delta_1)$$

follows from assumptions (4.2.11),[33] and so

$$\int_{\underline{u}}^{\underline{u}_*} |r^{-1}\tau_-^2(\mathbf{D}_3 \log\Omega)\mathbf{D}_3\rho|_{p=2,S} \le c\frac{1}{r^3}\Gamma_0(\Delta_0 + \Delta_1)\,.$$

Collecting these estimates we obtain

$$\int_{\underline{u}}^{\underline{u}_*} |r^{-1}\tau_-^2\underline{P}|_{p=2,S} \le c\frac{1}{r^3}\Gamma_0\,(\mathcal{I}_* + \Delta_0 + \Delta_1 + \Delta_2)\,. \tag{4.4.22}$$

The second integral in the right-hand side of (4.4.21) is estimated as

$$\begin{aligned}\int_{\underline{u}}^{\underline{u}_*} |r^{-1}\tau_-^2\mathbf{D}_3^2\rho|_{p=2,S} \quad&=\quad \frac{1}{\sqrt{\tau_-}}\int_{\underline{u}}^{\underline{u}_*} \frac{1}{r^2}\left(\int_S |r\tau_-^{\frac{5}{2}}\mathbf{D}_3^2\rho|^2\right)^{\frac{1}{2}} \qquad (4.4.23)\\ &\le c\frac{1}{\sqrt{\tau_-}}\left(\int_{\underline{u}}^{\underline{u}_*} \frac{1}{r^4}\right)^{\frac{1}{2}}\left(\int_{\underline{u}}^{\underline{u}_*}\int_S |r\tau_-^{\frac{5}{2}}\mathbf{D}_3^2\rho|^2\right)^{\frac{1}{2}}\\ &\le c\frac{1}{\tau_-^{\frac{1}{2}}r^{\frac{3}{2}}}||r\tau_-^{\frac{5}{2}}\mathbf{D}_3^2\rho||_{L_2(C(u))} \le c\frac{1}{\tau_-^{\frac{1}{2}}r^{\frac{3}{2}}}\Delta_2,\end{aligned}$$

---

[33]In fact from the Bianchi equations we have, for $p \in [2,4]$, $|r^{-\frac{2}{p}}\tau_-^2\mathbf{D}_3\rho|_{p=2,S} \le \frac{\tau_-^2}{r^4}|r^{4-\frac{2}{p}}tr\underline{\chi}\rho|_{p=2,S} + \frac{\sqrt{\tau_-}}{r^3}|r^{3-\frac{2}{p}}\tau_-^{\frac{3}{2}}\mathrm{d}\!\!/\mathrm{iv}\,\underline{\beta}|_{p=2,S} \le c\frac{\tau_-}{r^3}(\Delta_0 + \Delta_1)$.

where the inequality in the last line follows from assumptions (4.2.11).

This completes the proof of inequality (4.4.15). Inequality (4.4.13) also follows immediately recalling that, from assumptions (4.2.22), on the last slice we have

$$|r^{1-2/p}\tau_-^3(\Omega \mathbf{D}_3)^3 \log \Omega|_{p=2,S}(u, \underline{u}_*) \leq \mathcal{I}_* \, .$$

Remark: The estimates of Propositions 4.4.1, 4.4.2 complete the more delicate part of the control of the second derivatives of the connection coefficients and, therefore, of Theorem 4.2.2. In the next section we provide the estimates concerning the third derivatives of the connection coefficients. These estimates are needed to control the $\mathcal{D}$ norms introduced in Subsection 3.5.6 and estimated in Section 4.7.

4.5 Proof of Theorem 4.2.3 and control of third derivatives of the connection coefficients

The proof of Theorem 4.2.3 is achieved once we prove the following proposition.

Proposition 4.5.1 *Under the assumptions of Theorem 4.2.3 and using the results of Theorems 4.2.1, 4.2.2, and denoting* $\Delta \equiv \Delta_0 + \Delta_1 + \Delta_2$, *we obtain*

$$\begin{aligned}
r^{\frac{1}{2}}(u, \underline{u})\|r^3 \nabla\!\!\!/^3 \hat{\chi}\|_{L^2(\underline{C}(\underline{u}) \cap V(u,\underline{u}))} &\leq c(\mathcal{I}_0 + \mathcal{I}_* + \Delta) \\
r^{\frac{1}{2}}(u, \underline{u})\|r^3 \nabla\!\!\!/^3 \mathrm{tr}\chi\|_{L^2(\underline{C}(\underline{u}) \cap V(u,\underline{u}))} &\leq c(\mathcal{I}_0 + \mathcal{I}_* + \Delta) \\
r^{\frac{1}{2}}(u, \underline{u})\|r^3 \nabla\!\!\!/^3 \eta\|_{L^2(\underline{C}(\underline{u}) \cap V(u,\underline{u}))} &\leq c(\mathcal{I}_0 + \mathcal{I}_* + \Delta) \\
r^{\frac{1}{2}}(u, \underline{u})\|r^3 \nabla\!\!\!/^3 \underline{\omega}\|_{L^2(\underline{C}(\underline{u}) \cap V(u,\underline{u}))} &\leq c(\mathcal{I}_0 + \mathcal{I}_* + \Delta) \\
r^{\frac{1}{2}}(u, \underline{u})\|r^3 \nabla\!\!\!/^3 \hat{\underline{\chi}}\|_{L^2(\underline{C}(\underline{u}) \cap V(u,\underline{u}))} &\leq c(\mathcal{I}_0 + \mathcal{I}_* + \Delta) \\
r^{\frac{1}{2}}(u, \underline{u})\|r^3 \nabla\!\!\!/^3 \mathrm{tr}\underline{\chi}\|_{L^2(\underline{C}(\underline{u}) \cap V(u,\underline{u}))} &\leq c(\mathcal{I}_0 + \mathcal{I}_* + \Delta).
\end{aligned}$$

Remark: Proposition 4.5.1 does not require that the foliations on the initial and final slices be canonical. In this case, as discussed in Chapter 7, one can prove the boundedness of slightly stronger norms.

Proof: All these norms are estimated in essentially the same way. We give only a detailed account of the first estimate.

Estimate of $r^{\frac{1}{2}}(u, \underline{u})\|r^3 \nabla\!\!\!/^3 \hat{\chi}\|_{L^2(\underline{C}(\underline{u}) \cap V(u,\underline{u}))}$

From the definition, $\psi\!\!\!/ = \Omega^{-1}(\nabla\!\!\!/ \mathrm{tr}\chi + \mathrm{tr}\chi\zeta)$ (see (4.3.5)) and the Codazzi equation, see (4.3.13),

$$\mathrm{d}\!\!\!/\mathrm{iv}\,\hat{\chi} + \zeta \cdot \hat{\chi} = \frac{1}{2}\Omega\psi\!\!\!/ - \beta,$$

it follows that $\nabla\!\!\!/^3 \hat{\chi}$ and $\nabla\!\!\!/^3 \mathrm{tr}\chi$ are controlled in terms of $\nabla\!\!\!/^2 \psi\!\!\!/$.

Let us consider first $\nabla\!\!\!/^3\hat{\chi}$. Applying Proposition 4.1.2 to the Codazzi equation, we derive

$$|r^{3-\frac{2}{p}}\nabla\!\!\!/^3\hat{\chi}|_{p=2,S} \leq c\left(|r^{3-\frac{2}{p}}\nabla\!\!\!/^2\Psi\!\!\!/|_{p=2,S} + |r^{3-\frac{2}{p}}\nabla\!\!\!/^2\beta|_{p=2,S}\right) + [\cdots], \tag{4.5.1}$$

where $[\cdots]$ does not contain second derivatives of the Riemann tensor and behaves as $O(r^{-\frac{5}{2}})$. To estimate $\nabla\!\!\!/^2\Psi\!\!\!/$ we introduce the scalar quantity[34]

$$\Psi\!\!\!/\!\!\!/ \equiv \mathrm{d}\!\!\!/\mathrm{iv}\,\Psi\!\!\!/ + \Omega^{-1}\mathrm{tr}\chi\rho \tag{4.5.2}$$

and write its evolution equation along the null-outgoing hypersurfaces[35]

$$\frac{d}{d\underline{u}}\Psi\!\!\!/\!\!\!/ + 2\Omega\mathrm{tr}\chi\Psi\!\!\!/\!\!\!/ = -2\Omega\hat{\chi}\Psi\!\!\!/\!\!\!/ - 4\hat{\chi}\cdot\nabla\!\!\!/\beta + \mathcal{U}_1 \tag{4.5.3}$$

where $\mathcal{U}_1$ denotes all the terms that do not depend on the first derivatives of the Riemann tensor and have an appropriate asymptotic behavior.[36] The one-form $\Psi\!\!\!/$ satisfies the Hodge system,

$$\begin{aligned}\mathrm{d}\!\!\!/\mathrm{iv}\,\Psi\!\!\!/ &= \Psi\!\!\!/\!\!\!/ - \Omega^{-1}\mathrm{tr}\chi\rho\\ \mathrm{c}\!\!\!/\mathrm{url}\,\Psi\!\!\!/ &= \Omega^{-1}\mathrm{tr}\chi\,\mathrm{c}\!\!\!/\mathrm{url}\,\zeta + \zeta\cdot{}^*\Psi\!\!\!/.\end{aligned} \tag{4.5.4}$$

Applying Proposition 4.1.2 to the Hodge system (4.5.4) we obtain

$$|r^{3-\frac{2}{p}}\nabla\!\!\!/^2\Psi\!\!\!/|_{p=2,S} \leq c\left(|r^{3-\frac{2}{p}}\nabla\!\!\!/\Psi\!\!\!/\!\!\!/|_{p=2,S} + |\mathrm{tr}\chi|_{\infty,S}|r^{3-\frac{2}{p}}\nabla\!\!\!/\rho|_{p=2,S}\right) + [\cdots], \tag{4.5.5}$$

where, again, $[\cdots]$ indicates terms with a better asymptotic behavior and that can be estimated in the $L^p(S)$ norms. Combining (4.5.5) and (4.5.1) we derive

$$\begin{aligned}|r^{4-\frac{2}{p}}\nabla\!\!\!/^3\hat{\chi}|_{p=2,S} &\leq c\left(|r^{4-\frac{2}{p}}\nabla\!\!\!/^2\Psi\!\!\!/|_{p=2,S} + |r^{4-\frac{2}{p}}\nabla\!\!\!/^2\beta|_{p=2,S}\right) + [\cdots]\\ &\leq c\left(|r^{4-\frac{2}{p}}\nabla\!\!\!/\Psi\!\!\!/\!\!\!/|_{p=2,S} + |r^{4-\frac{2}{p}}\nabla\!\!\!/^2\beta|_{p=2,S}\right) + [\cdots]\end{aligned} \tag{4.5.6}$$

where $[\cdots] = O(r^{-\frac{3}{2}})$ and does not depend on second Riemann derivatives. These terms will be, hereafter, neglected. Using this estimate we write,

$$\int_{\underline{C}(\underline{u},[u_0,u])}|r^3\nabla\!\!\!/^3\hat{\chi}|^2 \leq c\int_{u_0}^{u}\frac{1}{r^2}|r^{5-\frac{2}{p}}\nabla\!\!\!/\Psi\!\!\!/\!\!\!/|^2_{p=2,S} + c\int_{u_0}^{u}|r^{4-\frac{2}{p}}\nabla\!\!\!/^2\beta|^2_{p=2,S}. \tag{4.5.7}$$

---

[34] We follow here the discussion in Chapter 13 of [Ch-Kl].

[35] The evolution equation is obtained deriving tangentially the evolution equation of $\Psi\!\!\!/$ and using the commutation relations in Proposition 4.8.1. The term $\Omega^{-1}\mathrm{tr}\chi\rho$, added in the definition of $\Psi\!\!\!/\!\!\!/$, is necessary to cancel the term $-\mathrm{tr}\chi\,\mathrm{d}\!\!\!/\mathrm{iv}\,\beta$ which appears on the right-hand side of the evolution equation for $\mathrm{d}\!\!\!/\mathrm{iv}\,\Psi\!\!\!/$ and which prevents the integrability of the evolution equation for $\mathrm{d}\!\!\!/\mathrm{iv}\,\Psi\!\!\!/$, $\frac{d}{d\underline{u}}(\mathrm{d}\!\!\!/\mathrm{iv}\,\Psi\!\!\!/) + 2\Omega\mathrm{tr}\chi(\mathrm{d}\!\!\!/\mathrm{iv}\,\Psi\!\!\!/) = -\mathrm{tr}\chi\,\mathrm{d}\!\!\!/\mathrm{iv}\,\beta - 2\Omega\hat{\chi}\cdot(\nabla\!\!\!/\Psi\!\!\!/) - 2\chi_{cd}\triangle\!\!\!/\chi_{cd} + \mathcal{U}_2$, where $\mathcal{U}_2$ depends on $\nabla\!\!\!/\chi$, $\Psi\!\!\!/$, $\underline{\eta}$, β and decays as $O(r^{-6})$.

[36] The term $-4\Omega\hat{\chi}\cdot\nabla\!\!\!/\beta$, although decaying properly, as $O(r^{-\frac{13}{2}})$, has been written explicitly because it is the only term depending on the first derivative of the curvature tensor. It arises from the term $-2\chi_{cd}\triangle\!\!\!/\chi_{cd}$ appearing in the evolution equation of $\mathrm{d}\!\!\!/\mathrm{iv}\,\Psi\!\!\!/$, deriving the Codazzi equation; see 3.1.47.

The second integral of (4.5.7) satisfies, using assumptions (4.2.11) of Theorem 4.2.2 (see also (3.5.17))

$$\int_{u_0}^{u}|r^{4-\frac{2}{p}}\nabla\!\!\!/^2\beta|^2_{p=2,S}\leq\int_{\underline{C}(\underline{u},[u_0,u])}r^2|r^2\nabla\!\!\!/^2\beta|^2\leq c\frac{1}{r^2}\underline{\mathcal{R}}_2[\beta]^2\leq c\frac{1}{r^2}\Delta_2^2. \tag{4.5.8}$$

To estimate the first integral on the right-hand side of (4.5.7) we write the evolution of $\nabla\!\!\!/\Psi\!\!\!/$ along the null-outgoing direction. A long but straightforward computation, where the evolution equation for $\Psi\!\!\!/$ has been derived tangentially, gives

$$\frac{d}{d\underline{u}}\nabla\!\!\!/\Psi\!\!\!/+\frac{5}{2}\Omega\mathrm{tr}\chi\nabla\!\!\!/\Psi\!\!\!/=-4\hat{\chi}\nabla\!\!\!/\Psi\!\!\!/-4\hat{\chi}\nabla\!\!\!/^2\beta+\mathcal{U}_3, \tag{4.5.9}$$

where $\mathcal{U}_3$ denotes all the terms that do not depend on the second derivatives of the Riemann tensor and, therefore, can be estimated in the $L^p(S)$ norms. These terms are easier to treat, have the appropriate asymptotic behavior, and, hereafter, will be omitted.

Applying the Evolution Lemma and the Gronwall inequality to (4.5.9) we obtain, using the previous results on $\mathcal{O}_{[2]}$,

$$\begin{aligned}|r^{5-\frac{2}{p}}\nabla\!\!\!/\Psi\!\!\!/|_{p=2,S}(u,\underline{u}')&\leq c|r^{5-\frac{2}{p}}\nabla\!\!\!/\Psi\!\!\!/|_{p=2,S}(u,\underline{u}_*)+c\int_{\underline{u}'}^{\underline{u}_*}|r^{5-\frac{2}{p}}\hat{\chi}\nabla\!\!\!/^2\beta|_{p=2,S}\\&\leq c|r^{5-\frac{2}{p}}\nabla\!\!\!/\Psi\!\!\!/|_{p=2,S}(u,\underline{u}_*)+c\Gamma_0\int_{\underline{u}'}^{\underline{u}_*}|r^{3-\frac{2}{p}}\nabla\!\!\!/^2\beta|_{p=2,S}\\&\leq c|r^{5-\frac{2}{p}}\nabla\!\!\!/\Psi\!\!\!/|_{p=2,S}(u,\underline{u}_*)+c\Gamma_0\Delta_2\frac{1}{r^{\frac{3}{2}}},\end{aligned} \tag{4.5.10}$$

where the last integral of (4.5.10) has been estimated with the help of (3.5.17) as

$$\int_{\underline{u}'}^{\underline{u}_*}|r^{3-\frac{2}{p}}\nabla\!\!\!/^2\beta|_{p=2,S}\leq\left(\int_{\underline{u}'}^{\underline{u}_*}\frac{1}{r^2}\right)^{\frac{1}{2}}\left(\int_{C(u)\cap V(u,\underline{u}')}r^2|r^2\nabla\!\!\!/^2\beta|^2\right)^{\frac{1}{2}}\leq c\Delta_2\frac{1}{r^{\frac{3}{2}}}.$$

In conclusion we have

$$\begin{aligned}||r^3\nabla\!\!\!/^3\hat{\chi}||_{L^2(\underline{C}(\underline{u})\cap V(u,\underline{u}))}&\leq\left(\int_{u_0}^{u}du'\frac{1}{r^2}|r^{5-\frac{2}{p}}\nabla\!\!\!/\Psi\!\!\!/|^2_{p=2,S}(u',\underline{u}_*)\right)^{\frac{1}{2}}\\&+c\Delta_2\frac{1}{r(u,\underline{u})}(1+\Gamma_0\frac{1}{r(u,\underline{u})})\\&\leq||r^3\nabla\!\!\!/\Psi\!\!\!/||_{L^2(\underline{C}_*\cap V(u,\underline{u}))}+c\Delta_2\frac{1}{r(u,\underline{u})}.\end{aligned} \tag{4.5.11}$$

The estimate is achieved once we observe that the final slice assumption of Theorem **M1**, $\mathcal{O}_3(\underline{C}_*)\leq\mathcal{I}_*$, implies the inequality

$$r^{\frac{1}{2}}(u,\underline{u})||r^3\nabla\!\!\!/\Psi\!\!\!/||_{L^2(\underline{C}_*\cap V(u,\underline{u}))}\leq\mathcal{I}_*\,.$$

The estimate of the norm $||r^3\nabla\!\!\!/^3\mathrm{tr}\chi||_{L^2(\underline{C}(\underline{u})\cap V(u,\underline{u}))}$ is immediately reduced to the previous one using the identity

$$\triangle\!\!\!/\mathrm{tr}\chi=\Omega\mathrm{d}\!\!\!/\mathrm{iv}\Psi\!\!\!/-\mathrm{tr}\chi\,\mathrm{d}\!\!\!/\mathrm{iv}\zeta-\zeta\nabla\!\!\!/\mathrm{tr}\chi+(\nabla\!\!\!/\log\Omega)\mathrm{tr}\chi\zeta$$

and Proposition 4.1.3.

Estimate of $r^{\frac{1}{2}}(u,\underline{u})||r^3\nabla\!\!\!/^3\eta||_{L^2(\underline{C}(\underline{u})\cap V(u,\underline{u}))}$

We apply Proposition 4.1.2 to the Hodge system (see 4.3.37))

$$
\begin{aligned}
\mathrm{d\!\!\!/iv}\,\eta &= -\tilde{\mu} + \frac{1}{2}(\chi\cdot\underline{\chi} - \overline{\chi\cdot\underline{\chi}}) - (\rho - \overline{\rho})\\
\mathrm{cu\!\!\!/rl}\,\eta &= \sigma - \frac{1}{2}\underline{\hat{\chi}}\wedge\hat{\chi},
\end{aligned}
$$

and obtain

$$
\begin{aligned}
|r^{4-\frac{2}{p}}\nabla\!\!\!/^3\eta|_{p=2,S} &\le c\Big(|r^{4-\frac{2}{p}}\nabla\!\!\!/^2\tilde{\mu}|_{p=2,S} + |r^{4-\frac{2}{p}}\nabla\!\!\!/^2(\rho,\sigma)|_{p=2,S}\Big) &(4.5.12)\\
&+c\frac{1}{r}(\mathcal{I}_0 + \mathcal{I}_* + \Delta)(1 + \Gamma_0\frac{1}{\tau_-})\,. &(4.5.13)
\end{aligned}
$$

Using this inequality it follows that

$$
\begin{aligned}
\int_{u_0}^{u}|r^{4-\frac{2}{p}}\nabla\!\!\!/^3\eta|^2_{p=2,S}(u',\underline{u}) \;&\le\; c\int_{u_0}^{u}\frac{1}{r^2}|r^{5-\frac{2}{p}}\nabla\!\!\!/^2\tilde{\mu}|^2_{p=2,S} + c\int_{u_0}^{u}|r^{4-\frac{2}{p}}\nabla\!\!\!/^2(\rho,\sigma)|^2_{p=2,S}\\
&+\; c\frac{1}{r}(\mathcal{I}_0 + \mathcal{I}_* + \Delta)^2. \qquad (4.5.14)
\end{aligned}
$$

The last integral is estimated using assumptions (4.2.11) of Theorem 4.2.2 (see also (3.5.17))

$$
\begin{aligned}
&\int_{u_0}^{u}|r^{4-\frac{2}{p}}\nabla\!\!\!/^2(\rho,\sigma)|^2_{p=2,S} = \frac{1}{r^2(u,\underline{u})}\int_{\underline{C}(\underline{u};[u_0,u])} r^4|r^2\nabla\!\!\!/^2(\rho,\sigma)|^2\\
&\le c\frac{1}{r^2(u,\underline{u})}\underline{\mathcal{R}}_2[(\rho,\sigma)](u,\underline{u}) \le c\frac{1}{r^2}\Delta_2^2. \qquad (4.5.15)
\end{aligned}
$$

To estimate the first integral on the right-hand side of (4.5.14) we have to control $|r^{5-\frac{2}{p}}\nabla\!\!\!/^2\tilde{\mu}|_{p=2,S}$. The estimates for $|r^{5-\frac{2}{p}}\nabla\!\!\!/^2\tilde{\mu}|_{p=2,S}$ are obtained in a way similar to Lemma 4.3.3 for $\tilde{\mu}$ and $\underline{\tilde{\mu}}$. The evolution equation for $\nabla\!\!\!/^2\tilde{\mu}$ can be written as

$$
\frac{d}{d\underline{u}}(\nabla\!\!\!/^2\tilde{\mu}) + 2(\Omega\mathrm{tr}\chi)(\nabla\!\!\!/^2\tilde{\mu}) = \tilde{F}_2 - 2\Omega\eta\cdot\nabla\!\!\!/^2\beta, \qquad (4.5.16)
$$

where $\tilde{F}_2$ does not depend on the second derivatives of the Riemann tensor and, therefore, can be estimated with the $|\cdot|_{p,S}$ norms. Applying to this evolution equation the Evolution Lemma and Gronwall's inequality we obtain,

$$
\begin{aligned}
|r^{4-\frac{2}{p}}\nabla\!\!\!/^2\tilde{\mu}|_{p=2,S}(u,\underline{u}) \;&\le\; c\Bigg(|r^{4-\frac{2}{p}}\nabla\!\!\!/^2\tilde{\mu}|_{p=2,S}(u,\underline{u}_*) + \int_{\underline{u}}^{\underline{u}_*}|r^3\tilde{F}_2|_{p=2,S}\\
&+\; \sup|2\Omega\eta|\int_{\underline{u}}^{\underline{u}_*}|r^3\nabla\!\!\!/^2\beta|_{p=2,S}\Bigg)\,. \qquad (4.5.17)
\end{aligned}
$$

Moreover on the last slice the canonical foliation of $\underline{C}_*$ implies (see 3.3.9))

$$|r^{4-\frac{2}{p}}\nabla\!\!\!/^2\tilde{\mu}|_{p=2,S}(u,\underline{u}_*) \leq |r^{4-\frac{2}{p}}\nabla\!\!\!/^2(\mathrm{tr}\chi\,\mathrm{tr}\underline{\chi})|_{p=2,S}(u,\underline{u}_*) \leq c\frac{1}{r}\mathcal{I}_*. \tag{4.5.18}$$

The estimates (4.5.17) and (4.5.18) allow control of the integral of $|r^{4-\frac{2}{p}}\nabla\!\!\!/^2\tilde{\mu}|^2_{p=2,S}$ and, therefore, $||r^3\nabla\!\!\!/^2\tilde{\mu}||_{L^2(\underline{C}(\underline{u})\cap V(u,\underline{u}))}$, completing the estimate of $\nabla\!\!\!/^3\eta$.
The estimates for the remaining $\mathcal{O}_3$ norms proceed in the same way and we do not discuss them here.

4.6 Rotation tensor estimates

The rotation vector fields ${}^{(i)}O$ form the Lie algebra of the rotation group $SO(3)$. They satisfy the commutation relations

$$[{}^{(i)}O, {}^{(j)}O] = \epsilon_{ijk}{}^{(k)}O \ , \ i,j,k \in \{1,2,3\}$$

and are defined on the tangent space TS_p, for any $p \in \mathcal{M}$.
Since the spacetime $\mathcal{K}$ is not flat but, in a appropriate sense, "nearly" flat we expect that the rotation vector fields will produce a set of diffeomorphisms which are "nearly" isometries and, therefore, that the deformation tensors associated with these fields have small norms. This will be crucial in the error estimates of Chapter 6 where we need to control the norms of the various components of the Riemann curvature tensor in terms of the initial data.

To define these vector fields we start by transporting the canonical generators of the rotation group defined at spacelike infinity of Σ_0, backward along Σ_0 up to the surface

$$S_{(0)}(\underline{u}_*) \equiv \underline{C}_* \cap \Sigma_0,$$

using the diffeomorphism induced by the flow normal to our canonical foliation. We then continue to transport them along the null hypersurface $\underline{C}_*$ using the diffeomorphism $\underline{\phi}_t$, restricted to $\underline{C}_*$, generated by the equivariant null vector field $\underline{N} = \frac{d}{du_*}$. On $\underline{C}_*$ we denote them ${}^{(i)}O_*, i = 1,2,3$.

Finally, starting from any surface $S(u,\underline{u}_*)$ foliating $\underline{C}_*$, we use the diffeomorphism ϕ_t generated by N to transport the rotation vector fields to any surface $S(u,\underline{u})$ of $\mathcal{K}$.

The discussion about the construction of the ${}^{(i)}O$ vector fields on the initial hypersurface, Σ_0, and on the last slice, $\underline{C}_*$, is in Chapter 7. Here we show how to extend the vector fields ${}^{(i)}O$ from the last slice to the whole $\mathcal{K}$.

4.6.1 *Technical aspects*

The rotation fields ${}^{(i)}O_*$ defined on $\underline{C}_*$ satisfy, on any $S(u,\underline{u}_*)$ surface,

$$[{}^{(i)}O_*, {}^{(j)}O_*] = \epsilon_{ijk}{}^{(k)}O_* \ , \ ijk \in \{1,2,3\} \ .$$

Let $q \in S(u,\underline{u})$ be a generic point of $\mathcal{K}$. Since $S(u,\underline{u})$ is diffeomorphic to $S(u,\underline{u}_*)$ via ϕ_Δ, $\Delta = \underline{u}_* - \underline{u}$, there exists a $p \in S(u,\underline{u}_*)$ such that $q = \phi_\Delta^{-1}(p)$. We define the element O

of the rotation group operating over q in the following way:[37]

$$(O;q) \equiv \phi_{\Delta}^{-1}(O_*;p)$$

where $(O;q)$ is a point of $S(u,\underline{u})$ and $(O_*;p)$ is the point of $S(u,\underline{u}_*)$ obtained by applying O_* to the point p. This extension of the action of the rotation group to the whole of $\mathcal{K}$ satisfies

$$O = \phi_t^{-1} O \phi_t$$

This implies that the generators ${}^{(i)}O$ satisfy

$$[N, {}^{(i)}O] = 0$$

where ${}^{(i)}O$ is the extension of ${}^{(i)}O_*$ to $\mathcal{K}$. From the previous definitions we can easily check that

$$[{}^{(i)}O, {}^{(j)}O] = \epsilon_{ijk}{}^{(k)}O \ , \ ijk \in \{1,2,3\} \, .$$

Since ${}^{(i)}O \in TS_q$ we have that $g({}^{(i)}O, e_4) = g({}^{(i)}O, e_3) = 0$. In conclusion the generators ${}^{(i)}O$, defined on the whole $\mathcal{K}$, satisfy

$$\begin{aligned} &[{}^{(i)}O, {}^{(j)}O] = \epsilon_{ijk}{}^{(k)}O \\ &[N, {}^{(i)}O] = 0 \\ &g({}^{(i)}O, e_4) = g({}^{(i)}O, e_3) = 0. \end{aligned} \tag{4.6.1}$$

Moreover, since $N = \Omega\hat{N} = \Omega e_4$, it follows that

$$[{}^{(i)}O, e_4] = {}^{(i)}F e_4, \tag{4.6.2}$$

where

$${}^{(i)}F \equiv -{}^{(i)}O_c(\nabla\!\!\!/_c \log\Omega) \ , \ {}^{(i)}O_c \equiv g({}^{(i)}O, e_c). \tag{4.6.3}$$

Proposition 4.6.1 *The quantities ${}^{(i)}O_a$ and $\nabla\!\!\!/_b {}^{(i)}O_a \equiv (\nabla\!\!\!/ {}^{(i)}O)_{ab}$ satisfy the following evolution equations:*

$$\frac{d}{d\underline{u}}{}^{(i)}O_b = \Omega\chi_{bc}{}^{(i)}O_c \tag{4.6.4}$$

$$\frac{d}{d\underline{u}}(\nabla\!\!\!/ {}^{(i)}O)_{ab} = \Omega\left(\hat{\chi}_{bc}(\nabla\!\!\!/_a {}^{(i)}O)_c - \hat{\chi}_{ac}(\nabla\!\!\!/_c {}^{(i)}O)_b\right) + (\mathcal{F}_1)_{ab} \tag{4.6.5}$$

where

$$\begin{aligned} (\mathcal{F}_1)_{ab} \ \equiv \ & \Omega\Big[{}^{(i)}O_c(\chi_{cb}\underline{\eta}_a - \chi_{ca}\underline{\eta}_b) + {}^{(i)}O_c R_{4acb} + {}^{(i)}O_c\chi_{cb}\zeta_a \\ & +\chi_{ab}(\underline{\eta}_c {}^{(i)}O_c) + {}^{(i)}O_c(\nabla\!\!\!/_a\chi)_{cb}\Big]. \end{aligned} \tag{4.6.6}$$

[37] At the differential level we can define the extension as

$${}^{(i)}O \equiv \phi_{*-\Delta}{}^{(i)}O_* \, .$$

Proof: From $[N, {}^{(i)}O] = 0$ we infer that

$$\Omega \mathbf{D}_4 {}^{(i)}O = {}^{(i)}O_c (\nabla\!\!\!/_c \log \Omega) N + \Omega \mathbf{D}_{{}^{(i)}O} e_4, \tag{4.6.7}$$

and choosing a moving frame satisfying $\mathbf{D}\!\!\!/_4 e_b = 0$,

$$\frac{d}{d\underline{u}} {}^{(i)}O_b = \Omega \chi_{bc} {}^{(i)}O_c.$$

To obtain an evolution equation for $\nabla\!\!\!/_b {}^{(i)}O_a \equiv (\nabla\!\!\!/ {}^{(i)}O)_{ab}$ we start from equation (4.6.7), which we rewrite as

$$\mathbf{D}_4 {}^{(i)}O = {}^{(i)}O_c \chi_{cb} e_b + {}^{(i)}O_c \underline{\eta}_c e_4. \tag{4.6.8}$$

Using the commutation relations in the appendix to this chapter, see Proposition 4.8.1, we derive

$$\begin{aligned} \frac{d}{d\underline{u}} (\nabla\!\!\!/ {}^{(i)}O)_{ab} &= \Omega \Big[\hat{\chi}_{bc} (\nabla\!\!\!/_a {}^{(i)}O)_c - \hat{\chi}_{ac} (\nabla\!\!\!/_c {}^{(i)}O)_b + {}^{(i)}O_c (\chi_{cb} \underline{\eta}_a - \chi_{ca} \underline{\eta}_b) \\ &\quad + {}^{(i)}O_c R_{4acb} + {}^{(i)}O_c \chi_{cb} \zeta_a + \chi_{ab} (\underline{\eta}_c {}^{(i)}O_c) + {}^{(i)}O_c (\nabla\!\!\!/_a \chi)_{cb} \Big]. \end{aligned}$$

In view of (4.6.1) and (4.6.8) we have

$$\begin{aligned} &g(\mathbf{D}_a {}^{(i)}O, e_4) = -\chi_{ab} {}^{(i)}O_b \\ &g(\mathbf{D}_4 {}^{(i)}O, e_a) = \chi_{ab} {}^{(i)}O_b \\ &g(\mathbf{D}_4 {}^{(i)}O, e_4) = 0 \\ &g(\mathbf{D}_a {}^{(i)}O, e_3) = -\underline{\chi}_{ab} {}^{(i)}O_b \\ &g(\mathbf{D}_3 {}^{(i)}O, e_3) = 0 \\ &g(\mathbf{D}_4 {}^{(i)}O, e_3) = -2\underline{\eta}_b {}^{(i)}O_b \\ &g(\mathbf{D}_3 {}^{(i)}O, e_4) = -2\eta_b {}^{(i)}O_b. \end{aligned} \tag{4.6.9}$$

Using this we compute some components of the deformation tensor relative to the rotation vector fields. Denoting ${}^{({}^{(i)}O)}\pi \equiv {}^{(i)}\pi$, we obtain

$$\begin{aligned} &{}^{(i)}\pi_{44} = 2g(\mathbf{D}_4 {}^{(i)}O, e_4) = 0 \\ &{}^{(i)}\pi_{33} = 2g(\mathbf{D}_3 {}^{(i)}O, e_3) = 0 \\ &{}^{(i)}\pi_{4a} = g(\mathbf{D}_4 {}^{(i)}O, e_a) + g(\mathbf{D}_a {}^{(i)}O, e_4) = 0 \\ &{}^{(i)}\pi_{34} = -2(\eta_b + \underline{\eta}_b) {}^{(i)}O_b = -4(\nabla\!\!\!/_b \log \Omega) {}^{(i)}O_b = 4 {}^{(i)}F, \end{aligned} \tag{4.6.10}$$

where the remaining components are

$$\begin{aligned} &{}^{(i)}\pi_{ab} = g(\mathbf{D}_a {}^{(i)}O, e_b) + g(\mathbf{D}_b {}^{(i)}O, e_a) \equiv 2 {}^{(i)}H_{ab} \\ &{}^{(i)}\pi_{3a} = g(\mathbf{D}_a {}^{(i)}O, e_3) + g(\mathbf{D}_3 {}^{(i)}O, e_a) \equiv 4 {}^{(i)}Z_a. \end{aligned} \tag{4.6.11}$$

Observe that ${}^{(i)}Z_a$ can be written

$${}^{(i)}Z_a = \frac{1}{4} \Big(-{}^{(i)}O_b \underline{\chi}_{ab} + e_3({}^{(i)}O_a) + {}^{(i)}O_b g(\mathbf{D}\!\!\!/_3 e_b, e_a) \Big)$$

and in view of $[{}^{(i)}O, e_3] = -4{}^{(i)}Z_b e_b + {}^{(i)}F e_3$, we have

$$[{}^{(i)}O, \underline{N}] = -4\Omega{}^{(i)}Z_b e_b. \tag{4.6.12}$$

To control the quantities ${}^{(i)}H_{ab}$, ${}^{(i)}Z_b$ we derive their evolution equations along the $C(u)$ hypersurfaces.

Proposition 4.6.2 *The quantities ${}^{(i)}H_{ab}$ and ${}^{(i)}Z_b$ satisfy the following evolution equations*

$$\begin{aligned}
\frac{d}{d\underline{u}}{}^{(i)}H_{ab} &= -\Omega\left(\hat{\chi}_{ac}{}^{(i)}H_{cb} + \hat{\chi}_{bc}{}^{(i)}H_{ca}\right) + \Omega\chi_{ab}(\nabla\!\!\!/_c \log\Omega){}^{(i)}O_c \\
&\quad +\Omega\left(\hat{\chi}_{bc}\nabla\!\!\!/_a{}^{(i)}O_c + \hat{\chi}_{ac}\nabla\!\!\!/_b{}^{(i)}O_c\right) + \Omega{}^{(i)}O_c(\nabla\!\!\!/_c\chi)_{ab} \\
\frac{d}{d\underline{u}}{}^{(i)}Z_a &= \frac{1}{2}\Omega\mathrm{tr}\chi{}^{(i)}Z_a + \Omega\hat{\chi}_{ab}{}^{(i)}Z_b + 2\Omega\omega{}^{(i)}Z_a + \Omega\bigg[\frac{1}{2}(\mathcal{L}_{{}^{(i)}O} - {}^{(i)}F)(\zeta+\eta))_a \\
&\quad +\frac{1}{2}\nabla\!\!\!/_a{}^{(i)}F - 2\zeta_b{}^{(i)}H_{ab} - \frac{1}{2}(\eta - \underline{\eta})_a{}^{(i)}F)\bigg].
\end{aligned} \tag{4.6.13}$$

Proof: The proof is a long but direct computation and is reported in the appendix to this chapter.

4.6.2 *Derivatives of the rotation deformation tensors*

The following relations hold

$$\begin{aligned}
&(\mathbf{D}_4{}^{(i)}\pi)_{44} = 0\,,\ (\mathbf{D}_a{}^{(i)}\pi)_{44} = 0 \\
&(\mathbf{D}_3{}^{(i)}\pi)_{44} = 0\,,\ (\mathbf{D}_4{}^{(i)}\pi)_{a4} = 0 \\
&(\mathbf{D}_b{}^{(i)}\pi)_{a4} = -2(\chi_{ba}{}^{(i)}F + \chi_{bc}{}^{(i)}H_{ca}) \\
&(\mathbf{D}_3{}^{(i)}\pi)_{a4} = -4(\eta_a{}^{(i)}F + \eta_b{}^{(i)}H_{ab}) \\
&(\mathbf{D}_4{}^{(i)}\pi)_{34} = 4\mathbf{D}_4{}^{(i)}F \\
&(\mathbf{D}_a{}^{(i)}\pi)_{34} = 4(\nabla\!\!\!/_a{}^{(i)}F - \chi_{ab}{}^{(i)}Z_b) \\
&(\mathbf{D}_3{}^{(i)}\pi)_{34} = 4(\mathbf{D}_3{}^{(i)}F - 2\eta_b{}^{(i)}Z_b) \\
&(\mathbf{D}_4{}^{(i)}\pi)_{33} = -16\underline{\eta}_b{}^{(i)}Z_b \\
&(\mathbf{D}_a{}^{(i)}\pi)_{33} = -8\underline{\chi}_{ab}{}^{(i)}Z_b \\
&(\mathbf{D}_3{}^{(i)}\pi)_{33} = 0 \\
&(\mathbf{D}_4{}^{(i)}\pi)_{ab} = 2\mathbf{D}_4{}^{(i)}H_{ab} = 2\Omega^{-1}\frac{d}{d\underline{u}}{}^{(i)}H_{ab} \\
&(\mathbf{D}_c{}^{(i)}\pi)_{ab} = 2(\nabla\!\!\!/_c{}^{(i)}H)_{ab} - 2\chi_{ca}{}^{(i)}Z_b - 2\chi_{cb}{}^{(i)}Z_a \\
&(\mathbf{D}_3{}^{(i)}\pi)_{ab} = 2(\mathbf{D}_3{}^{(i)}H)_{ab} - 4(\eta_a{}^{(i)}Z_b + \eta_b{}^{(i)}Z_a) \\
&(\mathbf{D}_4{}^{(i)}\pi)_{a3} = 4\Omega^{-2}\frac{d}{d\underline{u}}(\Omega{}^{(i)}Z_a) - 4(\underline{\eta}_a{}^{(i)}F + \underline{\eta}_b{}^{(i)}H_{ba}) \\
&(\mathbf{D}_b{}^{(i)}\pi)_{a3} = 4(\nabla\!\!\!/_b{}^{(i)}Z)_a - 2\underline{\chi}_{ba}{}^{(i)}F - 2\underline{\chi}_{bc}{}^{(i)}H_{ca} - 4\zeta_b{}^{(i)}Z_a \\
&(\mathbf{D}_3{}^{(i)}\pi)_{a3} = 4(\mathbf{D}_3{}^{(i)}Z)_a - 4(\mathbf{D}_3\log\Omega){}^{(i)}Z_a.
\end{aligned} \tag{4.6.14}$$

Most of the terms on the right-hand side have already been estimated or their estimates follow immediately from the previous results. The derivative terms we still need to control are

$$\mathbf{D}_3{}^{(i)}H\ ,\ \nabla\!\!\!/_c{}^{(i)}H\ ,\ \mathbf{D}_3{}^{(i)}Z\ ,\ \nabla\!\!\!/_b{}^{(i)}Z\ .$$

Not all of them are independent; in fact a long but straightforward computation gives

$$\begin{aligned}(\mathbf{D}_3{}^{(i)}\pi)_{ab} &= (\mathbf{D}_a{}^{(i)}\pi)_{3b} + (\mathbf{D}_b{}^{(i)}\pi)_{3a} + {}^{(i)}O_c(R_{3abc} + R_{3bac})\\ &+ 2((\nabla\!\!\!/_a{}^{(i)}O)_c\hat{\chi}_{cb} + (\nabla\!\!\!/_b{}^{(i)}O)_c\underline{\hat{\chi}}_{ca}) + \mathrm{tr}\underline{\chi}((\nabla\!\!\!/_a{}^{(i)}O)_b + (\nabla\!\!\!/_b{}^{(i)}O)_a)\\ &+ {}^{(i)}O_c((\nabla\!\!\!/_a\underline{\chi})_{bc} + (\nabla\!\!\!/_b\underline{\chi})_{ac}) - 2\underline{\chi}_{ab}\underline{\eta}_c{}^{(i)}O_c - {}^{(i)}O_c(\zeta_a\underline{\chi}_{bc} + \zeta_b\underline{\chi}_{ac}).\end{aligned}$$

This equation allows one to write $\mathbf{D}_3{}^{(i)}H_{ab}$ in terms of $\nabla\!\!\!/_a{}^{(i)}Z_b$ and of $\nabla\!\!\!/_b{}^{(i)}Z_a$ so that we have only to control $\nabla\!\!\!/_c{}^{(i)}H,\ \mathbf{D}_3{}^{(i)}Z,\ \nabla\!\!\!/_b{}^{(i)}Z$.

Proposition 4.6.3 *The quantities $\nabla\!\!\!/_c{}^{(i)}H_{ab}$, $\nabla\!\!\!/_c{}^{(i)}Z_b$ and $\mathbf{D}_3{}^{(i)}Z_b$ satisfy the following evolution equations:*

$$\begin{aligned}\frac{d}{d\underline{u}}(\nabla\!\!\!/_c{}^{(i)}H_{ab}) + \frac{1}{2}\Omega\mathrm{tr}\chi(\nabla\!\!\!/_c{}^{(i)}H_{ab}) &= -\Omega\hat{\chi}_{ad}(\nabla\!\!\!/^{(i)}H)_{cdb}\\ &+ \hat{\chi}_{ad}\left[(\nabla\!\!\!/^{(i)}H)_{bdc} - (\nabla\!\!\!/^{(i)}H)_{dbc}\right]\\ &+ \hat{\chi}_{bd}\left[(\nabla\!\!\!/^{(i)}H)_{adc} - (\nabla\!\!\!/^{(i)}H)_{dac}\right] + \mathcal{H}_1\end{aligned} \tag{4.6.15}$$

$$\frac{d}{d\underline{u}}(\nabla\!\!\!/_a{}^{(i)}Z_b) = -\Omega\hat{\chi}_{ac}\nabla\!\!\!/_c{}^{(i)}Z_b + \Omega\hat{\chi}_{bc}\nabla\!\!\!/_a{}^{(i)}Z_c + \mathcal{Z}_1 \tag{4.6.16}$$

$$\begin{aligned}\frac{d}{d\underline{u}}(\mathbf{D}_3{}^{(i)}Z_b) &= 4\Omega\omega(\mathbf{D}_3{}^{(i)}Z_b) - 4\Omega(\zeta\cdot\nabla\!\!\!/)^{(i)}Z_b + \Omega\mathbf{R}_{bc43}{}^{(i)}Z_c\\ &+ \frac{1}{2}\Omega\mathrm{tr}\chi(\mathbf{D}_3{}^{(i)}Z_b) + \Omega\hat{\chi}_{bc}(\mathbf{D}_3{}^{(i)}Z_c) + \mathcal{Z}_{11}\end{aligned} \tag{4.6.17}$$

where

$$\begin{aligned}\mathcal{H}_1 &= \Omega\Big\{(\nabla\!\!\!/\log\Omega)_c\mathbf{D}_{e_4}({}^{(i)}H_{ab}) + \left[(\mathbf{R}_{ad4c} - \underline{\eta}_a\chi_{dc})^{(i)}H_{db} + (\mathbf{R}_{bd4c} - \underline{\eta}_b\chi_{cd})^{(i)}H_{da}\right]\\ &- \left[(\nabla\!\!\!/\hat{\chi})_{cad}{}^{(i)}H_{db} + (\nabla\!\!\!/\hat{\chi})_{cbd}{}^{(i)}H_{da}\right] + (\mathcal{L}_O\nabla\!\!\!/\hat{\chi})_{cab} + \frac{1}{2}\delta_{ab}\nabla\!\!\!/_c(\mathcal{L}_O\mathrm{tr}\chi) - (\nabla\!\!\!/^{(i)}F\chi)_{cab}\Big\}\end{aligned} \tag{4.6.18}$$

$$\begin{aligned}\mathcal{Z}_1 &= \Omega\Big\{(\nabla\!\!\!/_a\log\Omega)\mathbf{D}\!\!\!/_N{}^{(i)}Z_b + \Big[(\nabla\!\!\!/_a\chi_{bc})^{(i)}Z_c + \frac{1}{2}L_O\nabla\!\!\!/_a(\zeta+\eta)_b\\ &+ \frac{1}{2}\nabla\!\!\!/_a\nabla\!\!\!/_b{}^{(i)}F - \frac{1}{2}\left(\nabla\!\!\!/_a{}^{(i)}F(\zeta+\eta)\right) - 2(\nabla\!\!\!/_a\zeta_c)^{(i)}H_{cb} - 2\zeta_c(\nabla\!\!\!/_a{}^{(i)}H_{cb})\\ &- \frac{1}{2}\nabla\!\!\!/_a({}^{(i)}F(\zeta-\underline{\eta})_b)\Big] + \left[R_{bca4}{}^{(i)}Z_c - \Omega\underline{\eta}_b\chi_{ac}{}^{(i)}Z_c + \Omega\chi_{ab}\underline{\eta}_c{}^{(i)}Z_c\right]\\ &+ \frac{1}{2}\left(\nabla\!\!\!/_a{}^{(i)}H_{cb} + \nabla\!\!\!/_b{}^{(i)}H_{ca} - \nabla\!\!\!/_c{}^{(i)}H_{ab}\right)(\zeta+\eta)_c\Big\}\end{aligned} \tag{4.6.19}$$

$$\mathcal{Z}_{11} = \Big\{2(\mathbf{D}_3\Omega\omega)^{(i)}Z_b + \frac{1}{2}(\mathbf{D}_3\Omega\mathrm{tr}\chi)^{(i)}Z_b + (\mathbf{D}_3\Omega\hat{\chi}_{bc})^{(i)}Z_c \tag{4.6.20}$$
$$+\mathbf{D}_3\left(\Omega\left[\frac{1}{2}(\mathcal{L}_{^{(i)}O} - {}^{(i)}F)(\zeta+\eta))_a + \frac{1}{2}\nabla\!\!\!/_a{}^{(i)}F - 2\zeta_b{}^{(i)}H_{ab} - \frac{1}{2}(\eta-\underline{\eta})_a{}^{(i)}F)\right]\right)\Big\}.$$

Proof: The proof is a long but direct computation and is reported in the appendix to this chapter.

4.7 Proof of Theorem M2 and estimates for the $\mathcal{D}$ norms of the rotation deformation tensors

We discuss how to control the $\mathcal{D}_0$ and $\mathcal{D}_1$ norms relative to the zero- and first-order derivatives and the $\mathcal{D}_2$ norms for the second-order derivatives in two different propositions, as their estimates are somewhat different.

Proposition 4.7.1 *Assume on $\underline{C}_*$ the following estimates for $p \in [2,4]$:*

$$|r^{-1(i)}O|_{p,S}(u,\underline{u}_*) \leq c\mathcal{I}_* \ , \quad |r^{-\frac{2}{p}}\nabla\!\!\!/^{(i)}O|_{p,S}(u,\underline{u}_*) \leq c\mathcal{I}_*$$
$$|r^{(i)}H_{ab}|_{p,S}(u,\underline{u}_*) \leq c\mathcal{I}_* \ , \quad |r^{2-\frac{2}{p}}\nabla\!\!\!/^{(i)}H_{ab}|_{p,S}(u,\underline{u}_*) \leq c\mathcal{I}_*$$
$$|r^{(i)}Z_a|_{p,S}(u,\underline{u}_*) \leq c\mathcal{I}_* \ , \quad |r^{2-\frac{2}{p}}\nabla\!\!\!/^{(i)}Z_{ab}|_{p,S}(u,\underline{u}_*) \leq c\mathcal{I}_*$$
$$|r^{1-\frac{2}{p}}\tau_-\mathbf{D}_3{}^{(i)}H_{ab}|_{p,S} \leq c\mathcal{I}_* \ , \quad |r^{1-\frac{2}{p}}\tau_-\mathbf{D}_3{}^{(i)}Z_a|_{p,S} \leq c\mathcal{I}_*$$
$$|r^{1-\frac{2}{p}}\tau_-\mathbf{D}_3{}^{(i)}F|_{p,S} \leq c\mathcal{I}_*.$$

Then, in view of the results of Theorem 4.2.1, Theorem 4.2.2 and Theorem 4.2.3, we prove the following inequalities for $p \in [2,4]$:

$$\begin{aligned}
&|r^{(-1-\frac{2}{p})(i)}O|_{p,S} \leq c(\mathcal{I}_* + \Delta_0)\\
&|r^{(1-\frac{2}{p})(i)}F|_{p,S} \leq c(\mathcal{I}_0 + \mathcal{I}_* + \Delta_0)\\
&|r^{(1-\frac{2}{p})(i)}H|_{p,S} \leq c(\mathcal{I}_0 + \mathcal{I}_* + \Delta_0)\\
&|r^{(1-\frac{2}{p})(i)}Z|_{p,S} \leq c(\mathcal{I}_0 + \mathcal{I}_* + \Delta_0)
\end{aligned} \tag{4.7.1}$$

$$\begin{aligned}
&|r^{-\frac{2}{p}}\nabla\!\!\!/^{(i)}O|_{p,S} \leq c(\mathcal{I}_0 + \mathcal{I}_* + \Delta_0)\\
&|r^{(2-\frac{2}{p})}\nabla\!\!\!/^{(i)}F|_{p,S} \leq c(\mathcal{I}_0 + \mathcal{I}_* + \Delta_0)\\
&|r^{(2-\frac{2}{p})}\nabla\!\!\!/^{(i)}H|_{p,S} \leq c(\mathcal{I}_0 + \mathcal{I}_* + \Delta_0 + \Delta_1)\\
&|r^{(2-\frac{2}{p})}\nabla\!\!\!/^{(i)}Z|_{p,S} \leq c(\mathcal{I}_0 + \mathcal{I}_* + \Delta_0 + \Delta_1)
\end{aligned} \tag{4.7.2}$$

$$\begin{aligned}
&|r^{2-\frac{2}{p}}\mathbf{D}_4{}^{(i)}H_{ab}|_{p,S} \leq c(\mathcal{I}_0 + \mathcal{I}_* + \Delta_0 + \Delta_1)\\
&|r^{2-\frac{2}{p}}\mathbf{D}_4{}^{(i)}Z_a|_{p,S} \leq c(\mathcal{I}_0 + \mathcal{I}_* + \Delta_0 + \Delta_1)\\
&|r^{2-\frac{2}{p}}\mathbf{D}_4{}^{(i)}F|_{p,S} \leq c(\mathcal{I}_0 + \mathcal{I}_* + \Delta_0 + \Delta_1)
\end{aligned} \tag{4.7.3}$$

$$
\begin{aligned}
&|r^{1-\frac{2}{p}}\tau_-\mathbf{D}_3{}^{(i)}H_{ab}|_{p,S} \le c(\mathcal{I}_0+\mathcal{I}_*+\Delta_0+\Delta_1)\\
&|r^{1-\frac{2}{p}}\tau_-\mathbf{D}_3{}^{(i)}Z_a|_{p,S} \le c(\mathcal{I}_0+\mathcal{I}_*+\Delta_0+\Delta_1)\\
&|r^{1-\frac{2}{p}}\tau_-\mathbf{D}_3{}^{(i)}F|_{p,S} \le c(\mathcal{I}_0+\mathcal{I}_*+\Delta_0+\Delta_1).
\end{aligned}
\tag{4.7.4}
$$

Moreover from the inequalities (4.7.1), (4.7.2) the following estimates hold:

$$
\begin{aligned}
&|r^{-1(i)}O|_{\infty,S} \le c(\mathcal{I}_*+\Delta_0)\\
&|r^{(i)}F|_{\infty,S} \le c(\mathcal{I}_0+\mathcal{I}_*+\Delta_0)\\
&|r^{(i)}H_{ab}|_{\infty,S} \le c(\mathcal{I}_0+\mathcal{I}_*+\Delta_0+\Delta_1)\\
&|r^{(i)}Z_a|_{\infty,S} \le c(\mathcal{I}_0+\mathcal{I}_*+\Delta_0+\Delta_1).
\end{aligned}
\tag{4.7.5}
$$

Proof: Applying the Evolution Lemma to equations (4.6.4) and (4.6.5), we obtain the following inequalities

$$
\begin{aligned}
|r^{(-1-\frac{2}{p})(i)}O|_{p,S}(u,\underline{u}) &\le c\left(|r^{(-1-\frac{2}{p})(i)}O|_{p,S}(u,\underline{u}_*)+\int_{\underline{u}}^{\underline{u}_*}\Omega|\hat{\chi}|_{\infty,S}|r^{(-1-\frac{2}{p})(i)}O|_{p,S}\right)\\
|r^{-\frac{2}{p}}\nabla\!\!\!/^{(i)}O|_{p,S}(u,\underline{u}) &\le c\left(|r^{-\frac{2}{p}}\nabla\!\!\!/^{(i)}O|_{p,S}(u,\underline{u}_*)+\int_{\underline{u}}^{\underline{u}_*}\Omega|\hat{\chi}|_{\infty,S}|r^{-\frac{2}{p}}\nabla\!\!\!/^{(i)}O|_{p,S}\right.\\
&\quad\left.+\int_{\underline{u}}^{\underline{u}_*}|r^{-\frac{2}{p}}\mathcal{F}_1|_{p,S}\right),
\end{aligned}
\tag{4.7.6}
$$

where (see (4.6.9))

$$
|r^{-\frac{2}{p}}\mathcal{F}_1|_{p,S} \le c\frac{(\mathcal{I}_0+\mathcal{I}_*+\Delta_0)}{r^3}|^{(i)}O|_{p,S} \le c\frac{(\mathcal{I}_0+\mathcal{I}_*+\Delta_0)}{r^2}\mathcal{I}_*\,.
$$

Using the Gronwall inequality, we obtain

$$
\begin{aligned}
&|r^{(-1-\frac{2}{p})(i)}O|_{p,S}(u,\underline{u}) \le c|r^{(-1-\frac{2}{p})(i)}O|_{p,S}(u,\underline{u}_*) \le c\mathcal{I}_*\\
&|r^{-\frac{2}{p}}\nabla\!\!\!/^{(i)}O|_{p,S}(u,\underline{u}) \le |r^{-\frac{2}{p}}\nabla\!\!\!/^{(i)}O|_{p,S}(u,\underline{u}_*)+c\frac{(\mathcal{I}_0+\mathcal{I}_*+\Delta_0)}{r}\mathcal{I}_*,
\end{aligned}
\tag{4.7.7}
$$

and applying Lemma 4.1.3,

$$
|r^{-1(i)}O|_{\infty,S} \le c(\mathcal{I}_0+\mathcal{I}_*+\Delta_0)\,.
$$

Using the explicit expression of the Lie derivatives with respect to ${}^{(i)}O$ and the previous results for the connection coefficients we can prove the following inequalities:[38]

$$
\begin{aligned}
&|r^2\mathcal{L}_{{}^{(i)}O}\mathrm{tr}\chi|_{\infty,S} \le c(\mathcal{I}_0+\mathcal{I}_*+\Delta_0)\\
&|r^2\mathcal{L}_{{}^{(i)}O}\hat{\chi}|_{\infty,S} \le c(\mathcal{I}_0+\mathcal{I}_*+\Delta_0),
\end{aligned}
$$

[38] In fact these require control of the norm $|\nabla\!\!\!/^{(i)}O|_{\infty,S}$, which, in turn, requires control of $|r^{1-\frac{2}{p}}\nabla\!\!\!/^{2(i)}O|_{p,S}$. The latter can be proved in the same way as $|r^{-\frac{2}{p}}\nabla\!\!\!/^{(i)}O|_{p,S}$ by deriving once again (4.6.5).

and in view of (4.6.10), we also obtain

$$|r\,{}^{(i)}F|_{\infty,S} = |r\,{}^{(i)}O\nabla\!\!\!/ \log\Omega|_{\infty,S} \leq c(\mathcal{I}_0 + \mathcal{I}_* + \Delta_0)\ .$$

To derive the estimates (4.7.1), (4.7.2) for ${}^{(i)}H$ and ${}^{(i)}Z$ we use their evolution equations (4.6.13). Thus, in the case of ${}^{(i)}H$, we obtain, using the Evolution Lemma,

$$\begin{aligned}\frac{d}{d\underline{u}}|r^{-\frac{2}{p}}\,{}^{(i)}H|_{p,S} &\leq c\frac{(\mathcal{I}_0 + \mathcal{I}_* + \Delta_0)}{r^2}\Big(|r^{-\frac{2}{p}}\,{}^{(i)}H|_{p,S} + (|r^3\nabla\!\!\!/\hat{\chi}|_{\infty,S} + |r^2\hat{\chi}|_{\infty,S}\\ &\qquad + |r^3\nabla\!\!\!/\mathrm{tr}\chi|_{\infty,S})\Big)\\ &\leq \frac{(\mathcal{I}_0 + \mathcal{I}_* + \Delta_0)}{r^2}\Big(|r^{-\frac{2}{p}}\,{}^{(i)}H|_{p,S} + c(\mathcal{I}_0 + \mathcal{I}_* + \Delta_0)\Big)\ .\end{aligned}$$

Integrating from $\underline{u}$ to $\underline{u}_*$ and applying the Gronwall inequality gives[39]

$$|r^{1-\frac{2}{p}}\,{}^{(i)}H|_{p,S} \leq c(\mathcal{I}_0 + \mathcal{I}_* + \Delta_0). \tag{4.7.8}$$

Proceeding in the same way, starting from (4.6.16), we obtain

$$\frac{d}{d\underline{u}}|r^{1-\frac{2}{p}}\nabla\!\!\!/{}^{(i)}H|_{p,S} \leq c\frac{(\mathcal{I}_0 + \mathcal{I}_* + \Delta_0)}{r^2}\left(|r^{1-\frac{2}{p}}\nabla\!\!\!/{}^{(i)}H|_{p,S} + \frac{1}{r}(\mathcal{I}_0 + \mathcal{I}_* + \Delta_0 + \Delta_1)\right),$$

which, again, integrating from $\underline{u}$ to $\underline{u}_*$ and applying the Gronwall inequality, gives

$$|r^{2-\frac{2}{p}}\nabla\!\!\!/{}^{(i)}H|_{p,S} \leq c(\mathcal{I}_0 + \mathcal{I}_* + \Delta_0 + \Delta_1)\ .$$

Combining these results and using Lemma 4.1.3 we conclude

$$|r^{(i)}H|_{\infty,S} \leq c(\mathcal{I}_0 + \mathcal{I}_* + \Delta_0 + \Delta_1). \tag{4.7.9}$$

Proceeding exactly in the same way for ${}^{(i)}Z$ we obtain

$$\begin{aligned}&|r^{1-\frac{2}{p}}\,{}^{(i)}Z|_{p,S} \leq c(\mathcal{I}_0 + \mathcal{I}_* + \Delta_0)\\ &|r^{2-\frac{2}{p}}\nabla\!\!\!/{}^{(i)}Z|_{p,S} \leq c(\mathcal{I}_0 + \mathcal{I}_* + \Delta_0 + \Delta_1),\end{aligned}$$

and therefore,

$$|r^{(i)}Z|_{\infty,S} \leq c(\mathcal{I}_0 + \mathcal{I}_* + \Delta_0 + \Delta_1)\ .$$

The estimates (4.7.3) are obtained writing their explicit expressions and estimating them using the previous results. The estimates (4.7.4) require to use the evolution equation for $\mathbf{D}_3{}^{(i)}Z$ (see (4.6.16)) and then proceed as before.

[39] Observe that the different asymptotic behavior of ${}^{(i)}H$ and $\nabla\!\!\!/{}^{(i)}O$ is due to the different estimates they satisfy on $\underline{C}_*$.

Proposition 4.7.1 allows one to estimate the components of the traceless part of the rotation deformation tensor[40],

$$
\begin{aligned}
{}^{(O)}\mathbf{i}_{ab} &\equiv {}^{(i)}\hat{\pi}_{ab} = 2{}^{(i)}H_{ab} - 2^{-1}\delta_{ab}({}^{(i)}H_{aa} - 2{}^{(i)}F) \\
{}^{(O)}\mathbf{j} &\equiv {}^{(i)}\hat{\pi}_{34} = (2{}^{(i)}F + {}^{(i)}H_{aa}) \\
{}^{(O)}\underline{\mathbf{m}}_a &\equiv {}^{(i)}\hat{\pi}_{3a} = 4{}^{(i)}Z_a \\
{}^{(O)}\mathbf{m}_a &\equiv {}^{(i)}\hat{\pi}_{4a} = 0 \\
{}^{(O)}\underline{\mathbf{n}} &\equiv {}^{(i)}\hat{\pi}_{33} = 2g(\mathbf{D}_3{}^{(i)}O, e_3) = 0 \\
{}^{(O)}\mathbf{n} &\equiv {}^{(i)}\hat{\pi}_{44} = 2g(\mathbf{D}_4{}^{(i)}O, e_4) = 0,
\end{aligned}
\tag{4.7.10}
$$

which proves the following result.

Corollary 4.7.1 *In $\mathcal{K}$, the following inequalities hold:*

$$
\begin{aligned}
&|r^{1-\frac{2}{p}}({}^{(O)}\mathbf{i}, {}^{(O)}\mathbf{j}, {}^{(O)}\underline{\mathbf{m}})|_{p,S} \le c\,(\mathcal{I}_0 + \mathcal{I}_* + \Delta_0) \\
&|r^{2-\frac{2}{p}}\nabla\!\!\!\!/\,({}^{(O)}\mathbf{i}, {}^{(O)}\mathbf{j}, {}^{(O)}\underline{\mathbf{m}})|_{p,S} \le c\,(\mathcal{I}_0 + \mathcal{I}_* + \Delta_0 + \Delta_1) \\
&|r^{2-\frac{2}{p}}\mathbf{D}\!\!\!\!/_4({}^{(O)}\mathbf{i}, {}^{(O)}\mathbf{j}, {}^{(O)}\underline{\mathbf{m}})|_{p,S} \le c\,(\mathcal{I}_0 + \mathcal{I}_* + \Delta_0 + \Delta_1) \\
&|r^{1-\frac{2}{p}}\tau_-\mathbf{D}\!\!\!\!/_3({}^{(O)}\mathbf{i}, {}^{(O)}\mathbf{j}, {}^{(O)}\underline{\mathbf{m}})|_{p,S} \le c\,(\mathcal{I}_0 + \mathcal{I}_* + \Delta_0 + \Delta_1)\,.
\end{aligned}
\tag{4.7.11}
$$

The first line is for $p \in [2, \infty)$,[41] the others for $p \in [2, 4]$.

The next proposition provides us with the estimates for the $\mathcal{D}_2$ norm; see Definition 3.5.45. The $\mathcal{D}_2$ norm collects the norms of the second derivatives of the rotation deformation tensors which will be used in Chapter 6 for the estimates of the error terms.

Proposition 4.7.2 *Assume that on $\underline{C}_*$ the following inequality holds:*

$$
||r\nabla\!\!\!\!/^2 H||_{L^2(\underline{C}_*\cap V(u,\underline{u}_*))} = \left(\int_{u_{(0)}(\underline{u}_*)}^{u} du'|r^{2-\frac{2}{p}}\nabla\!\!\!\!/^2 H|^2_{p=2,S}(u', \underline{u}_*)\right)^{\frac{1}{2}} \le c\mathcal{I}_*.
$$

Then, in view of the results of Theorems 4.2.1, 4.2.2, 4.2.3 and of Proposition 4.7.1, the following estimates hold, the last one for $\delta > \epsilon > 0$:

$$
||r\nabla\!\!\!\!/^2 H||_{L^2(\underline{C}(\underline{u}')\cap V(u,\underline{u}'))} \le c\,(\mathcal{I}_0 + \mathcal{I}_* + \Delta) \tag{4.7.12}
$$

$$
||r\nabla\!\!\!\!/^2 Z||_{L^2(\underline{C}(\underline{u}')\cap V(u,\underline{u}'))} \le c\,(\mathcal{I}_0 + \mathcal{I}_* + \Delta) \tag{4.7.13}
$$

$$
||\frac{r}{\sqrt{r^{1-2\epsilon}}}\nabla\!\!\!\!/\mathbf{D}\!\!\!\!/_3 Z||_{L^2(\underline{C}(\underline{u}')\cap V(u,\underline{u}'))} \le c\frac{1}{|u|\sqrt{r^{1-2\delta}}}(\mathcal{I}_0 + \mathcal{I}_* + \Delta)\,, \tag{4.7.14}
$$

where we have defined $\Delta \equiv \Delta_0 + \Delta_1 + \Delta_2$.

---

[40] Recall that ${}^{(i)}\hat{\pi}_{ab} = {}^{(i)}\pi_{ab} - \frac{1}{4}\delta_{ab}\mathrm{tr}\pi$ and $\mathrm{tr}{}^{(i)}\pi = {}^{(i)}\pi_{aa} - {}^{(i)}\pi_{34}$.

[41] The estimates hold also for $p = \infty$, but in this case, their bound also depends on Δ_1 (see (4.7.8), (4.7.9)).

Proof of inequality 4.7.12:

To estimate $\nabla\!\!\!/^{2\,(i)}H$ we use its evolution equation, obtained by deriving tangentially the evolution equation for $\nabla\!\!\!/^{(i)}H$; see 4.6.16. It can be written as

$$\frac{d}{d\underline{u}}(\nabla\!\!\!/^2 H) + \Omega\mathrm{tr}\chi(\nabla\!\!\!/^2 H) = \hat{\chi}(\nabla\!\!\!/^2 H) + (\mathcal{L}_O\nabla\!\!\!/^2\chi) + \mathcal{H}_2, \tag{4.7.15}$$

where $\mathcal{H}_2$ collects all the terms that do not depend on third-order derivatives of the connection coefficients and, therefore, can be estimated in the $L^p(S)$ norms with $p \in [2,4]$.[42] Applying the Evolution Lemma and the Gronwall inequality to (4.7.15), we obtain

$$\begin{aligned}|r^{2-\frac{2}{p}}\nabla\!\!\!/^2 H|_{p=2,S}(u,\underline{u}) \le c\Bigg(&|r^{2-\frac{2}{p}}\nabla\!\!\!/^2 H|_{p=2,S}(u,\underline{u}_*)\\ &+\int_{\underline{u}}^{\underline{u}_*} d\underline{u}'\Big[|r^{2-\frac{2}{p}}\mathcal{L}_O\nabla\!\!\!/^2\chi|_{p=2,S} + |r^{2-\frac{2}{p}}\mathcal{H}_2|_{p=2,S}\Big](u,\underline{u}')\Bigg),\end{aligned} \tag{4.7.16}$$

where the integral with $\mathcal{H}_2$, which depends on lower derivatives, does not give problems and, hereafter, will be neglected. Substituting (4.7.16) on the left-hand side of (4.7.12) we have[43]

$$\begin{aligned}
||r\nabla\!\!\!/^2 H||^2_{L^2(\underline{C}(\underline{u}')\cap V(u,\underline{u}'))} &= \int_{u_0(\underline{u}')}^{u} du'|r^{2-\frac{2}{p}}\nabla\!\!\!/^2 H|^2_{p=2,S}(u',\underline{u})\\
&\le c\int_{u_0(\underline{u}')}^{u} du'\left[|r^{2-\frac{2}{p}}\nabla\!\!\!/^2 H|^2_{p=2,S}(u',\underline{u}_*) + \left(\int_{\underline{u}'}^{\underline{u}_*} d\underline{u}''|r^{3-\frac{2}{p}}\nabla\!\!\!/^3\chi|_{p=2,S}(u',\underline{u}'')\right)^2\right]\\
&\le c\mathcal{I}_*^2 + c\int_{u_0(\underline{u}')}^{u} du'\left(\int_{\underline{u}'}^{\underline{u}_*}\frac{1}{r(u',\underline{u}')}\right)\left(\int_{\underline{u}'}^{\underline{u}_*} d\underline{u}''|r^{4-\frac{2}{p}}\nabla\!\!\!/^3\chi|^2_{p=2,S}(u',\underline{u}'')\right)\\
&\le c\mathcal{I}_*^2 + c\int_{u_0(\underline{u}')}^{u} du'\frac{1}{r(u',\underline{u}')}||r^3\nabla\!\!\!/^3\chi||^2_{L^2(C(u',[\underline{u}',\underline{u}_*]))}\\
&\le c\mathcal{I}_*^2 + c\left(\sup_u r^{\frac{1}{2}}(u,\underline{u}')||r^3\nabla\!\!\!/^3\chi||_{L^2(C(u,[\underline{u}',\underline{u}_*]))}\right)^2\left(\int_{u_0(\underline{u}')}^{u} du'\frac{1}{r^2(u',\underline{u}')}\right) \qquad (4.7.17)\\
&\le c\mathcal{I}_*^2 + c\left(\sup_u r^{\frac{1}{2}}(u,\underline{u}')||r^3\nabla\!\!\!/^3\chi||_{L^2(C(u,[\underline{u}',\underline{u}_*]))}\right)^2 \le c\mathcal{I}_*^2 + (\mathcal{I}_0+\mathcal{I}_*+\Delta)^2,
\end{aligned}$$

using the result of Proposition 4.5.1.

Proof of inequality 4.7.13:

To estimate $\nabla\!\!\!/^{2\,(i)}Z$ we look at its evolution equation, obtained by deriving tangentially the evolution equation for $\nabla\!\!\!/^{(i)}Z$ (see (4.6.16)),

$$\frac{d}{d\underline{u}}(\nabla\!\!\!/^2 Z)+\frac{1}{2}\Omega\mathrm{tr}\chi(\nabla\!\!\!/^2 Z)=\Omega\hat{\chi}(\nabla\!\!\!/^2 Z)+\mathcal{L}_O\nabla\!\!\!/^2(\zeta+\eta)+\nabla\!\!\!/^3 F+(\zeta+\eta)\nabla\!\!\!/^2 H+\mathcal{Z}_2,$$

[42] Moreover they also have the appropriate asymptotic behavior to control their integration.

[43] Here ϵ_0 is a small constant that bounds $c\mathcal{I}_*$. In general by ϵ_0 we denote a small constant satisfying $c(\mathcal{I}_0+\mathcal{I}_*+\Delta) \le \epsilon_0$.

where $\mathcal{Z}_2$ collects all the terms that do not depend on third-order derivatives of the connection coefficients and, therefore, can be estimated in the $L^p(S)$ norms with $p \in [2,4]$. Therefore we will neglect it in the sequel. Applying the Evolution Lemma and the Gronwall inequality to this evolution equation we obtain,

$$
\begin{aligned}
|r^{2-\frac{2}{p}}\nabla\!\!\!/^2 Z|_{p=2,S}(u,\underline{u}) &\leq c\Bigg(|r^{2-\frac{2}{p}}\nabla\!\!\!/^2 Z|_{p=2,S}(u,\underline{u}_*)\\
&+\int_{\underline{u}}^{\underline{u}_*}\Big[|r^{2-\frac{2}{p}}\mathcal{L}_O\nabla\!\!\!/^2\eta|_{p=2,S}+|r^{2-\frac{2}{p}}\mathcal{L}_O\nabla\!\!\!/^2\underline{\eta}|_{p=2,S}+|r^{3-\frac{2}{p}}\nabla\!\!\!/^3(\eta+\underline{\eta})|_{p=2,S}\Big]\\
&+\int_{\underline{u}}^{\underline{u}_*}\frac{1}{r^2}|r^{2-\frac{2}{p}}\nabla\!\!\!/^2 H|_{p=2,S}\Bigg)\\
&\leq c\Bigg(|r^{2-\frac{2}{p}}\nabla\!\!\!/^2 H|_{p=2,S}(u,\underline{u}_*)+\int_{\underline{u}}^{\underline{u}_*}|r^{3-\frac{2}{p}}\nabla\!\!\!/^3(\eta+\underline{\eta})|_{p=2,S}+\int_{\underline{u}}^{\underline{u}_*}\frac{1}{r^2}|r^{3-\frac{2}{p}}\nabla\!\!\!/^3\chi|_{p=2,S}\Bigg).
\end{aligned}
$$

The last integral in the last line can be treated as in the previous estimate and gives a better contribution due to the factor r^{-2}. Therefore we are left with the inequality

$$
\begin{aligned}
||r\nabla\!\!\!/^2 Z||^2_{L^2(\underline{C}(\underline{u}')\cap V(u,\underline{u}'))} &\leq \int_{u_0(\underline{u}')}^{u} du'|r^{2-\frac{2}{p}}\nabla\!\!\!/^2 Z|^2_{p=2,S}(u',\underline{u}_*) \qquad (4.7.18)\\
&+\int_{u_0(\underline{u}')}^{u} du'\left(\int_{\underline{u}'}^{\underline{u}_*} d\underline{u}''|r^{3-\frac{2}{p}}\nabla\!\!\!/^3(\eta+\underline{\eta})|_{p=2,S}(u',\underline{u}'')\right)^2 + c\,(\mathcal{I}_0+\mathcal{I}_*+\Delta)^2.
\end{aligned}
$$

The estimate of the last integral on the right-hand side is obtained by a straightforward application of the Schwartz inequality and Proposition 4.5.1.

Proof of inequality 4.7.14:

To estimate $\nabla\!\!\!/\mathbf{D}\!\!\!/_3{}^{(i)}Z$ we have to look at its evolution equation, obtained by applying $\mathbf{D}\!\!\!/_3$ to the evolution equation for $\nabla\!\!\!/^{(i)}Z$ (see (4.6.16)),

$$
\begin{aligned}
\frac{d}{d\underline{u}}(\nabla\!\!\!/\mathbf{D}\!\!\!/_3 Z) &= \Omega\Big[\hat{\chi}(\nabla\!\!\!/\mathbf{D}\!\!\!/_3 Z)+\frac{1}{2}\mathcal{L}_O\nabla\!\!\!/\mathbf{D}\!\!\!/_3(\zeta+\eta)+\frac{1}{2}\nabla\!\!\!/^2\mathbf{D}\!\!\!/_3 F \qquad (4.7.19)\\
&- 2(\zeta+\eta)\nabla\!\!\!/\mathbf{D}\!\!\!/_3 H\Big]+\mathcal{S}_2,
\end{aligned}
$$

where $\mathcal{S}_2$ collects all the terms that do not depend on third-order derivatives of the connection coefficients and, therefore, can be estimated in the $L^p(S)$ norms with $p \in [2,4]$. Among the terms in brackets the more delicate are $r\nabla\!\!\!/^3\underline{\omega}$ and $r\nabla\!\!\!/^2\underline{\beta}$, which are present in the explicit expressions of $\mathcal{L}_O\nabla\!\!\!/\mathbf{D}\!\!\!/_3(\zeta+\eta)$ and $\nabla\!\!\!/^2\mathbf{D}\!\!\!/_3 F$.[44] Considering only these terms and applying the Evolution Lemma and the Gronwall inequality to the evolution equation we obtain[45]

$$
|r^{(\frac{3}{2}+\epsilon)-\frac{2}{p}}\nabla\!\!\!/\mathbf{D}\!\!\!/_3 Z|_{p=2,S}(u',\underline{u}') \leq c\int_{\underline{u}'}^{\underline{u}_*}\Big[|r^{(\frac{5}{2}+\epsilon)-\frac{2}{p}}\nabla\!\!\!/^3\underline{\omega}|_{p=2,S}+\frac{1}{r^2}|r^{(\frac{5}{2}+\epsilon)-\frac{2}{p}}\nabla\!\!\!/^2\underline{\beta}|_{p=2,S}\Big].
$$

---

[44] All the terms in $[\cdots]$ of (4.7.19) have to be estimated in the L^2-norms, but we focus our attention on these since they have the slowest decay.

[45] The term $|r^{(\frac{3}{2}+\epsilon)-\frac{2}{p}}\nabla\!\!\!/\mathbf{D}\!\!\!/_3 Z|_{p=2,S}(u,\underline{u}_*)$ is missing since in Chapter 7 we prove (see Proposition 7.5.1) that $Z=0$ on $\underline{C}_*$.

Therefore

$$
\begin{aligned}
|r^{(\frac{3}{2}+\epsilon)-\frac{2}{p}}\nabla\!\!\!\!/ \mathbf{D}\!\!\!\!/_3 Z|^2_{p=2,S}(u',\underline{u}') \;&\leq\; 2\left(\int_{\underline{u}'}^{\underline{u}_*}|r^{(\frac{5}{2}+\epsilon)-\frac{2}{p}}\nabla\!\!\!\!/^3\underline{\omega}|_{p=2,S}(u'\underline{u}'')\right)^2\\
&+\;2\left(\int_{\underline{u}'}^{\underline{u}_*}|r^{(\frac{5}{2}+\epsilon)-\frac{2}{p}}\nabla\!\!\!\!/^2\underline{\beta}|_{p=2,S}(u'\underline{u}'')\right)^2, \qquad (4.7.20)
\end{aligned}
$$

and from it,

$$
\begin{aligned}
&||\frac{r}{\sqrt{r^{1-2\epsilon}}}\nabla\!\!\!\!/ \mathbf{D}\!\!\!\!/_3 Z||^2_{L^2(\underline{C}(\underline{u}')\cap V(u,\underline{u}'))}\leq c\int_{u_0(\underline{u}')}^{u}du'\left(\int_{\underline{u}'}^{\underline{u}_*}d\underline{u}''|r^{(\frac{5}{2}+\epsilon)-\frac{2}{p}}\nabla\!\!\!\!/^3\underline{\omega}|_{p=2,S}(u'\underline{u}'')\right)^2\\
&+c\int_{u_0(\underline{u}')}^{u}du'\left(\int_{\underline{u}'}^{\underline{u}_*}d\underline{u}''|r^{(\frac{5}{2}+\epsilon)-\frac{2}{p}}\nabla\!\!\!\!/^2\underline{\beta}|_{p=2,S}(u'\underline{u}'')\right)^2\equiv c\,(\mathcal{I}_3+\mathcal{I}_4)\,. \qquad (4.7.21)
\end{aligned}
$$

To estimate the $\mathcal{I}_4$ integral we apply the Schwartz inequality and, chosing $\eta>0$,

$$
\begin{aligned}
\mathcal{I}_4 \;&\leq\; c\int_{\underline{u}'}^{\underline{u}_*}d\underline{u}''\left(\int_{\underline{u}'}^{\underline{u}_*}d\underline{u}''\frac{1}{r^{1+2\eta}}\right)\left(\int_{u_0(\underline{u}')}^{u}du'|r^{(3+(\epsilon+\eta))-\frac{2}{p}}\nabla\!\!\!\!/^2\underline{\beta}|^2_{p=2,S}(u'\underline{u}'')\right)\\
&\leq\; c\frac{1}{u^2}\int_{\underline{u}'}^{\underline{u}_*}d\underline{u}''\frac{1}{r^{2-2\epsilon}(u'\underline{u}'')}\int_{\underline{C}(\underline{u}'';[u_0,u])}r^2u'^2|\underline{\beta}(\hat{\mathcal{L}}_O^2W)|^2 \qquad (4.7.22)\\
&\leq\; c\frac{1}{u^2r^{(1-2\epsilon)}(u,\underline{u}')}||\tau_-r^3\nabla\!\!\!\!/^2\underline{\beta}||^2_{L^2}\leq c\frac{\Delta_2^2}{u^2r^{(1-2\epsilon)}}\leq c\,(\mathcal{I}_0+\mathcal{I}_*+\Delta)^2\,\frac{1}{u^2r^{(1-2\epsilon)}},
\end{aligned}
$$

where in the last line we used assumptions (4.2.11). Proceeding in the same way for $\mathcal{I}_3$, we obtain

$$
\begin{aligned}
\mathcal{I}_3 \;&\leq\; \int_{u_0(\underline{u}')}^{u}du'\left(\int_{\underline{u}'}^{\underline{u}_*}d\underline{u}''\frac{1}{r^{1+2\eta}}\right)\left(\int_{\underline{u}'}^{\underline{u}_*}d\underline{u}''|r^{(3+(\epsilon+\eta))-\frac{2}{p}}\nabla\!\!\!\!/^3\underline{\omega}|^2_{p=2,S}(u'\underline{u}'')\right)\\
&\leq\; c\int_{u_0(\underline{u}')}^{u}du'\int_{\underline{u}'}^{\underline{u}_*}d\underline{u}''|r^{(3+(\epsilon+\eta))-\frac{2}{p}}\nabla\!\!\!\!/^3\underline{\omega}|^2_{p=2,S}(u'\underline{u}'')\\
&\leq\; c\int_{\underline{u}'}^{\underline{u}_*}d\underline{u}''\int_{u_0(\underline{u}')}^{u}du'|r^{(3+(\epsilon+\eta))-\frac{2}{p}}\nabla\!\!\!\!/^3\underline{\omega}|^2_{p=2,S}(u'\underline{u}'')\\
&\leq\; c\int_{\underline{u}'}^{\underline{u}_*}d\underline{u}''\frac{1}{r^{2(1-(\epsilon+\eta))}}\int_{u_0(\underline{u}')}^{u}du'|r^{4-\frac{2}{p}}\nabla\!\!\!\!/^3\underline{\omega}|^2_{p=2,S}(u'\underline{u}'')\\
&\leq\; c\left(\sup_{\underline{u}}r^{\frac{1}{2}}(u,\underline{u})||r^3\nabla\!\!\!\!/^3\underline{\omega}||_{L^2(\underline{C}(\underline{u})\cap V(u,\underline{u}))}\right)^2\leq(\mathcal{I}_0+\mathcal{I}_*+\Delta)^2\,, \qquad (4.7.23)
\end{aligned}
$$

using Proposition 4.5.1.

4.8 Appendix

4.8.1 Some commutation relations

We collect in the following proposition the proof of some commutation relations which will be repeatedly used in various estimates.

Proposition 4.8.1

i) Let f be a scalar function. Then[46]

$$
\begin{aligned}
[\mathbf{D}_3, \mathbf{D}_4]f &= (\mathbf{D}_4 \log\Omega)\mathbf{D}_3 f - (\mathbf{D}_3 \log\Omega)\mathbf{D}_4 f + 4\zeta\cdot\slashed{\nabla} f\\
[\slashed{\mathbf{D}}_4, \slashed{\nabla}]f &= (\zeta + \underline{\eta})\mathbf{D}_4 f - \chi\cdot\slashed{\nabla} f.
\end{aligned}
\tag{4.8.1}
$$

ii) Let V be a vector field tangent to S. Then[47]

$$
\begin{aligned}
[\slashed{\mathbf{D}}_4, \slashed{\nabla}_a]V_b &= -\chi_{ac}\slashed{\nabla}_c V_b - \underline{\eta}_b\chi_{ac}V_c + \chi_{ab}(\underline{\eta}\cdot V)\\
&\quad + (\slashed{\nabla}_a \log\Omega)\slashed{\mathbf{D}}_4 V_b + \big([D_\tau, D_\rho]V_\sigma\big)e_4^\tau e_a^\rho e_b^\sigma\\
[\slashed{\mathbf{D}}_3, \slashed{\nabla}_a]V_b &= -\underline{\chi}_{ac}\slashed{\nabla}_c V_b - \eta_b\underline{\chi}_{ac}V_c + \underline{\chi}_{ab}(\eta\cdot V)\\
&\quad + (\slashed{\nabla}_a \log\Omega)\slashed{\mathbf{D}}_3 V_b + \big([D_\tau, D_\rho]V_\sigma\big)e_3^\tau e_a^\rho e_b^\sigma.
\end{aligned}
\tag{4.8.2}
$$

iii) Let f be a scalar function. Then[48]

$$
\begin{aligned}
[\slashed{\Delta}, \mathbf{D}_3]f = &-\Big[-\eta_a\underline{\chi}_{ab}\slashed{\nabla}_b f + \mathrm{tr}\underline{\chi}\eta_b\slashed{\nabla}_b f - 2\underline{\chi}_{ab}(\slashed{\nabla}\slashed{\nabla} f)_{ab}\\
&\quad - \zeta_a\underline{\chi}_{ab}\slashed{\nabla}_b f - (\slashed{\mathrm{div}}\,\underline{\chi})_b\slashed{\nabla}_b f - \underline{\beta}_b\slashed{\nabla}_b f\Big]\\
&-\Big[(\slashed{\nabla}_a\log\Omega)(\mathbf{D}_3\slashed{\nabla} f)_a - \zeta_a(\mathbf{D}_3\slashed{\nabla} f)_a + \eta_a\slashed{\nabla}_a\mathbf{D}_3 f\\
&\quad +(\slashed{\Delta}\log\Omega)\mathbf{D}_3 f + \zeta_a(\slashed{\nabla}_a\log\Omega)\mathbf{D}_3 f\Big].
\end{aligned}
\tag{4.8.3}
$$

iv) Let X be a vector field tangent to S. Then

$$
\begin{aligned}
[\slashed{\mathbf{D}}_3, \slashed{\mathrm{div}}]X &= -\underline{\chi}_{ac}\slashed{\nabla}_c X_a + \slashed{\nabla}_a \log\Omega\slashed{\mathbf{D}}_3 X_a - \eta_a\underline{\chi}_{ac}X_c + \mathrm{tr}\underline{\chi}(\eta\cdot X)\\
&\quad + e_a^\rho e_a^\sigma e_3^\tau[D_\tau, D_\rho]X_\sigma\\
[\slashed{\mathbf{D}}_4, \slashed{\mathrm{div}}]X &= -\chi_{ac}\slashed{\nabla}_c X_a + \slashed{\nabla}_a \log\Omega\slashed{\mathbf{D}}_4 X_a - \underline{\eta}_a\chi_{ac}X_c + \mathrm{tr}\chi(\underline{\eta}\cdot X)\\
&\quad + e_a^\rho e_a^\sigma e_4^\tau[D_\tau, D_\rho]X_\sigma.
\end{aligned}
\tag{4.8.4}
$$

Proof: To prove the second line of (4.8.1) we recall the definitions[49]

$$
\begin{aligned}
\slashed{\nabla}_a\mathbf{D}_4 f &= e_a^\mu D_\mu\mathbf{D}_4 f = (\mathbf{D}_{e_a}e_4)^\sigma D_\sigma f + e_a^\mu e_4^\sigma D_\mu D_\sigma f\\
\slashed{\mathbf{D}}_4\slashed{\nabla}_a f &\equiv (\slashed{\mathbf{D}}_4\slashed{\nabla}_\mu f)e_a^\mu = \mathbf{D}_4(\Pi_\mu^\rho D_\rho f)e_a^\mu\\
&= e_a^\mu(\mathbf{D}_4\Pi)_\mu^\rho D_\rho f + e_a^\mu e_4^\sigma D_\sigma D_\mu f.
\end{aligned}
\tag{4.8.5}
$$

---

[46] Of course $[\slashed{\mathbf{D}}_4, \slashed{\nabla}]f = \slashed{\mathbf{D}}_4\slashed{\nabla} f - \slashed{\nabla}\mathbf{D}_4 f$.

[47] Where $[\slashed{\mathbf{D}}_4, \slashed{\nabla}_a]V_b \equiv (\slashed{\mathbf{D}}_4\slashed{\nabla} V - \slashed{\nabla}\slashed{\mathbf{D}}_4 V)_{ab}$.

[48] $\slashed{\Delta} f \equiv e_a^\mu e_a^\nu\slashed{\nabla}_\mu\slashed{\nabla}_\nu f$. On a scalar function f, $\slashed{\nabla}_\mu f = \Pi_\mu^\sigma D_\sigma f$ and, on a vector field X tangent to S, $\slashed{\nabla}_\mu X^\nu = \Pi_\mu^\sigma\Pi_\rho^\nu D_\sigma X^\rho$ where Π is the tensor projecting on TS. Finally, assuming that the vector fields $\{e_a\}$ satisfy $\slashed{\mathbf{D}}_3 e_a = 0$ we have $\mathbf{D}_3\slashed{\Delta} f = e_a^\mu e_a^\nu\mathbf{D}_3\slashed{\nabla}_\mu\slashed{\nabla}_\nu f$.

[49] We always use the notation $\mathbf{D}W_a \equiv (\mathbf{D}W)_a$, $\slashed{\nabla} W_a \equiv (\slashed{\nabla} W)_a$, unless it causes confusion.

Therefore

$$
\begin{aligned}
(\nabla\!\!\!/_a \mathbf{D}_4 - \mathbf{D}\!\!\!/_4 \nabla\!\!\!/_a) f &= (\mathbf{D}_{e_a} e_4)^\sigma D_\sigma f - e_a^\mu (\mathbf{D}_4 \Pi)_\mu^\rho D_\rho f + e_a^\mu e_4^\sigma [D_\sigma, D_\mu] f \\
&= -\frac{1}{2} (\mathbf{D}_{e_a} e_4)^\nu e_{3\nu} e_4^\sigma D_\sigma f + (\mathbf{D}_{e_a} e_4)^\nu e_{c\nu} e_c^\sigma D_\sigma f - e_a^\mu (\mathbf{D}_4 \Pi)_\mu^\rho D_\rho f \\
&= -\frac{1}{2} g(\mathbf{D}_{e_a} e_4, e_3) \mathbf{D}_4 f + g(\mathbf{D}_{e_a} e_4, e_c) \mathbf{D}_c f - e_a^\mu (\mathbf{D}_4 \Pi)_\mu^\rho D_\rho f \\
&= -\zeta_a \mathbf{D}_4 f + \chi_{ac} \nabla\!\!\!/_c f - e_a^\mu (\mathbf{D}_4 \Pi)_\mu^\rho D_\rho f.
\end{aligned} \tag{4.8.6}
$$

The proof is completed by observing that $e_a^\mu \mathbf{D}_4 \Pi_\mu^\rho = \underline{\eta}_a e_4^\rho$. To prove (4.8.2) observe that

$$
\begin{aligned}
\mathbf{D}\!\!\!/_4 \nabla\!\!\!/_a V_b &\equiv e_a^\mu e_b^\nu \mathbf{D}_4 \Pi_\mu^\rho \Pi_\nu^\sigma D_\rho V_\sigma = e_a^\mu (\mathbf{D}_4 \Pi_\mu^\rho) e_b^\sigma D_\rho V_\sigma \\
&+ e_a^\rho e_b^\nu (\mathbf{D}_4 \Pi_\nu^\sigma) D_\rho V_\sigma + e_a^\rho e_b^\sigma \mathbf{D}_4 D_\rho V_\sigma,
\end{aligned} \tag{4.8.7}
$$

which with

$$
e_a^\mu (\mathbf{D}_4 \Pi_\mu^\rho) = \underline{\eta}_a e_4^\rho \ , \ e_b^\nu (\mathbf{D}_4 \Pi_\nu^\sigma) = \underline{\eta}_b e_4^\sigma \tag{4.8.8}
$$

can be written as

$$
\mathbf{D}\!\!\!/_4 \nabla\!\!\!/_a V_b = \underline{\eta}_a \mathbf{D}\!\!\!/_4 V_b - \underline{\eta}_b \chi_{ac} V_c + e_a^\rho e_b^\sigma e_4^\tau D_\rho D_\tau V_\sigma + e_a^\rho e_b^\sigma e_4^\tau [D_\tau, D_\rho] V_\sigma . \tag{4.8.9}
$$

On the other side we have

$$
\begin{aligned}
\nabla\!\!\!/_a \mathbf{D}\!\!\!/_4 V_b &\equiv e_a^\mu e_b^\nu \left(\Pi_\mu^\rho \Pi_\nu^\sigma D_\rho \Pi_\sigma^\tau \mathbf{D}_4 V_\tau \right) = e_a^\rho e_b^\sigma \left(D_\rho \Pi_\sigma^\tau \mathbf{D}_4 V_\tau \right) \\
&= e_a^\rho (D_\rho \Pi_\sigma^\tau) e_b^\sigma \mathbf{D}_4 V_\tau + e_a^\rho e_b^\sigma D_\rho e_4^\tau D_\tau V_\sigma ,
\end{aligned} \tag{4.8.10}
$$

and since

$$
e_a^\rho (D_\rho \Pi_\sigma^\tau) e_b^\sigma = \frac{1}{2} \underline{\chi}_{ab} e_4^\tau + \frac{1}{2} \chi_{ab} e_3^\tau , \tag{4.8.11}
$$

we obtain

$$
\nabla\!\!\!/_a \mathbf{D}\!\!\!/_4 V_b = -\chi_{ab} (\underline{\eta} \cdot V) + e_a^\rho e_b^\sigma (D_\rho e_4^\tau) D_\tau V_\sigma + e_a^\rho e_b^\sigma e_4^\tau D_\rho D_\tau V_\sigma . \tag{4.8.12}
$$

Moreover, because

$$
e_a^\rho e_b^\sigma (D_\rho e_4^\tau) D_\tau V_\sigma = \chi_{ac} \nabla\!\!\!/_c V_b - \zeta_a \mathbf{D}\!\!\!/_4 V_b , \tag{4.8.13}
$$

(4.8.12) can be rewritten as

$$
\nabla\!\!\!/_a \mathbf{D}\!\!\!/_4 V_b = -\chi_{ab} (\underline{\eta} \cdot V) + \chi_{ac} \nabla\!\!\!/_c V_b - \zeta_a \mathbf{D}\!\!\!/_4 V_b + e_a^\rho e_b^\sigma e_4^\tau D_\rho D_\tau V_\sigma , \tag{4.8.14}
$$

which together with (4.8.9) proves relation (4.8.2).

To prove (4.8.3) observe that

$$
\begin{aligned}
[\mathbf{D}_3, \Delta\!\!\!/] f &= \big\{ e_a^\mu (\mathbf{D}_3 \Pi)_\mu^\sigma e_a^\tau D_\sigma \Pi_\tau^\lambda D_\lambda f + e_a^\sigma e_a^\nu (\mathbf{D}_3 \Pi)_\nu^\tau D_\sigma \Pi_\tau^\lambda D_\lambda f \\
&+ e_a^\sigma e_a^\tau D_\sigma (\mathbf{D}_3 \Pi)_\tau^\lambda D_\lambda f - e_a^\sigma e_a^\tau (D_\sigma e_3)^\delta D_\delta \Pi_\tau^\lambda D_\lambda f \\
&- e_a^\sigma e_a^\tau D_\sigma \Pi_\tau^\lambda (D_\lambda e_3)^\delta D_\delta f - e_a^\sigma e_a^\tau e_3^\delta R^\gamma{}_{\tau\delta\sigma} \nabla\!\!\!/_\gamma f \big\} ,
\end{aligned} \tag{4.8.15}
$$

where the six terms in brackets have the following expressions:

$$
\begin{aligned}
e_a^\mu(\mathbf{D}_3\Pi)_\mu^\sigma e_a^\tau D_\sigma \Pi_\tau^\lambda D_\lambda f &= \eta_a \mathbf{D}_3(\nabla\!\!\!/ f)_a\\
e_a^\sigma e_a^\nu(\mathbf{D}_3\Pi)_\nu^\tau D_\sigma \Pi_\tau^\lambda D_\lambda f &= -\eta_a \underline{\chi}_{ab}\nabla\!\!\!/_b f\\
e_a^\sigma e_a^\tau D_\sigma(\mathbf{D}_3\Pi)_\tau^\lambda D_\lambda f &= \mathrm{tr}\underline{\chi}\eta_b\nabla\!\!\!/_b f + \eta_a\nabla\!\!\!/_a\mathbf{D}_3 f + (\mathrm{d}\!\!\!/\mathrm{iv}\,\eta)\mathbf{D}_3 f\\
-e_a^\sigma e_a^\tau(D_\sigma e_3)^\delta D_\delta \Pi_\tau^\lambda D_\lambda f &= -\underline{\chi}_{ab}(\nabla\!\!\!/\nabla\!\!\!/ f)_{ab} - \zeta_a(\mathbf{D}_3\nabla\!\!\!/ f)_a\\
-e_a^\sigma e_a^\tau D_\sigma \Pi_\tau^\lambda(D_\lambda e_3)^\delta D_\delta f &= -\underline{\chi}_{ab}(\nabla\!\!\!/\nabla\!\!\!/ f)_{ab} - (\mathrm{d}\!\!\!/\mathrm{iv}\,\underline{\chi})_b\nabla\!\!\!/_b f\\
&\quad -(\mathrm{d}\!\!\!/\mathrm{iv}\,\zeta)\mathbf{D}_3 f - \zeta_a(\mathbf{D}_3\nabla\!\!\!/ f)_a\\
&\quad -\zeta_a\underline{\chi}_{ab}\nabla\!\!\!/_b f + \zeta_a(\nabla\!\!\!/_a\log\Omega)\mathbf{D}_3 f\\
-e_a^\sigma e_a^\tau e_3^\delta R^\gamma{}_{\tau\delta\sigma}\nabla\!\!\!/_\gamma f &= -\underline{\beta}_a\nabla\!\!\!/_a f.
\end{aligned}
\tag{4.8.16}
$$

Collecting all these expressions we obtain the result. To prove the first line of (4.8.4), observe that

$$
\mathbf{D}\!\!\!/_3\mathrm{d}\!\!\!/\mathrm{iv}\,X = e_a^\mu(\mathbf{D}_3\Pi_\mu^\rho)e_a^\sigma D_\rho X_\sigma + e_a^\rho e_a^\nu(\mathbf{D}_3\Pi_\nu^\sigma)D_\rho X_\sigma + e_a^\rho e_b^\sigma\mathbf{D}_3 D_\rho X_\sigma,
$$

and since

$$
e_a^\mu(\mathbf{D}_3\Pi_\mu^\rho) = \eta_a e_3^\rho\ ,\ \ e_a^\nu(\mathbf{D}_3\Pi_\nu^\sigma) = \eta_a e_3^\sigma,
\tag{4.8.17}
$$

the result follows.

4.8.2 *Proof of Lemma 4.3.5*

We start with

$$
\mathbf{D}_3(\Omega\mathbf{D}_4)^2\log\Omega = \Omega\mathbf{D}_4\mathbf{D}_3(\Omega\mathbf{D}_4\log\Omega) + [\mathbf{D}_3,\Omega\mathbf{D}_4](\Omega\mathbf{D}_4\log\Omega).
\tag{4.8.18}
$$

Since $[\mathbf{D}_3,\Omega\mathbf{D}_4]f = (\mathbf{D}_3\Omega)\mathbf{D}_4 f + \Omega[\mathbf{D}_3,\mathbf{D}_4]f$, using (4.8.1), we have that

$$
\begin{aligned}
[\mathbf{D}_3,\Omega\mathbf{D}_4]f &= (\mathbf{D}_3\Omega)\mathbf{D}_4 f + \Omega\Big((\mathbf{D}_4\log\Omega)\mathbf{D}_3 f - (\mathbf{D}_3\log\Omega)\mathbf{D}_4 f + 4\zeta\cdot\nabla\!\!\!/ f\Big)\\
&= (\mathbf{D}_4\log\Omega)\Omega\mathbf{D}_3 f + 4\Omega\zeta\cdot\nabla\!\!\!/ f,
\end{aligned}
$$

and if $f = \Omega\mathbf{D}_4\log\Omega$, then

$$
\begin{aligned}
[\mathbf{D}_3,\Omega\mathbf{D}_4](\Omega\mathbf{D}_4\log\Omega) &= 4\Omega\zeta\cdot\nabla\!\!\!/(\Omega\mathbf{D}_4\log\Omega) - (\Omega\mathbf{D}_3\log\Omega)\mathbf{D}_4(\Omega\mathbf{D}_4\log\Omega)\\
&\quad + (\Omega\mathbf{D}_4\log\Omega)\mathbf{D}_3(\Omega\mathbf{D}_4\log\Omega) + (\mathbf{D}_3\Omega)\mathbf{D}_4(\Omega\mathbf{D}_4\log\Omega)\\
&= (\Omega\mathbf{D}_4\log\Omega)\mathbf{D}_3(\Omega\mathbf{D}_4\log\Omega) + 4\Omega\zeta\cdot\nabla\!\!\!/(\Omega\mathbf{D}_4\log\Omega)\ .
\end{aligned}
$$

Substituting in (4.8.18) gives

$$
\begin{aligned}
\mathbf{D}_3(\Omega\mathbf{D}_4)^2\log\Omega &= \Omega\mathbf{D}_4\mathbf{D}_3(\Omega\mathbf{D}_4\log\Omega) + (\Omega\mathbf{D}_4\log\Omega)\mathbf{D}_3(\Omega\mathbf{D}_4\log\Omega)\\
&\quad + 4\Omega\zeta\cdot\nabla\!\!\!/(\Omega\mathbf{D}_4\log\Omega).
\end{aligned}
\tag{4.8.19}
$$

Since (see (4.3.59))

$$\mathbf{D}_3(\Omega\mathbf{D}_4 \log\Omega) = 2\Omega\zeta\cdot\nabla\!\!\!/ \log\Omega + \Omega(\underline{\eta}\cdot\eta - 2\zeta^2 - \rho)$$

(4.8.19) becomes

$$\begin{aligned}
\mathbf{D}_3\left((\Omega\mathbf{D}_4)^2\log\Omega\right) &= \Omega\mathbf{D}_4\left[2\Omega\zeta\cdot\nabla\!\!\!/\log\Omega + \Omega(\underline{\eta}\cdot\eta - 2\zeta^2-\rho)\right]\\
&\quad+(\Omega\mathbf{D}_4\log\Omega)\left[2\Omega\zeta\cdot\nabla\!\!\!/\log\Omega+\Omega(\underline{\eta}\cdot\eta-2\zeta^2-\rho)\right]+4\Omega\zeta\cdot\nabla\!\!\!/(\Omega\mathbf{D}_4\log\Omega)\\
&=\Omega^2\left[2\mathbf{D}_4(\zeta\cdot\nabla\!\!\!/\log\Omega)+\mathbf{D}_4(\underline{\eta}\cdot\eta-2\zeta^2-\rho)\right]\\
&\quad+\ 2(\Omega\mathbf{D}_4\log\Omega)\left[2\Omega\zeta\cdot\nabla\!\!\!/\log\Omega+\Omega(\underline{\eta}\cdot\eta-2\zeta^2-\rho)\right]+4\Omega\zeta\cdot\nabla\!\!\!/(\Omega\mathbf{D}_4\log\Omega)\\
&=\left[4\Omega\zeta\cdot\nabla\!\!\!/(\Omega\mathbf{D}_4\log\Omega)+2(\Omega\mathbf{D}_4\log\Omega)\left(2\Omega\zeta\cdot\nabla\!\!\!/\log\Omega+\Omega(\underline{\eta}\cdot\eta-2\zeta^2-\rho)\right)\right]\\
&\quad+\ \left[\Omega^2\left(2\mathbf{D}_4\zeta\cdot\nabla\!\!\!/\log\Omega+\mathbf{D}_4(\underline{\eta}\cdot\eta-2\zeta^2)\right)\right]-\Omega^2\mathbf{D}_4\rho\\
&=[1]+[2]-\Omega^2\mathbf{D}_4\rho
\end{aligned}\tag{4.8.20}$$

Using the evolution equation for ζ (see (3.1.46))

$$\mathbf{D}\!\!\!/_4\zeta = -2\chi\cdot\zeta - \mathbf{D}\!\!\!/_4\nabla\!\!\!/\log\Omega - \beta,$$

and the commutation relation (see (4.8.1))

$$[\mathbf{D}\!\!\!/_4,\nabla\!\!\!/]\log\Omega = (\zeta+\underline{\eta})\mathbf{D}_4\log\Omega - \chi\cdot\nabla\!\!\!/\log\Omega,$$

it is easy to show that [2] can be rewritten as

$$[2]=\Omega^2\left[(6\zeta-2\nabla\!\!\!/\log\Omega)\Omega^{-1}\left(\nabla\!\!\!/\Omega\mathbf{D}_4\log\Omega+\Omega\chi\cdot(\zeta-\underline{\eta})+\Omega\beta\right)+2\eta\mathbf{D}_4\nabla\!\!\!/\log\Omega\right].$$

So finally, we have

$$\mathbf{D}_3\left((\Omega\mathbf{D}_4)^2\log\Omega\right) = M - \Omega^2\mathbf{D}_4\rho,$$

and

$$\begin{aligned}
M \;=\; 2\Omega\Bigg\{&\bigg[2\zeta\cdot\nabla\!\!\!/\Omega\mathbf{D}_4\log\Omega + 2(\Omega\mathbf{D}_4\log\Omega)\zeta\cdot\nabla\!\!\!/\log\Omega\\
&+(\Omega\mathbf{D}_4\log\Omega)(\underline{\eta}\cdot\eta-2\zeta^2)+\Omega\eta\cdot\mathbf{D}_4\nabla\!\!\!/\log\Omega\\
&-(-3\zeta+\nabla\!\!\!/\log\Omega)\left(\nabla\!\!\!/\Omega\mathbf{D}_4\log\Omega-\Omega\chi\cdot(\underline{\eta}-\zeta)\right)\bigg]\\
&+\bigg[-\Omega(-3\zeta+\nabla\!\!\!/\log\Omega)\beta-(\Omega\mathbf{D}_4\log\Omega)\rho\bigg]\Bigg\}.
\end{aligned}\tag{4.8.21}$$

4.8.3 Proof of Lemma 4.4.1

Starting from

$$\not{D}_4(\not{\nabla} V) = \not{\nabla}\not{D}_4 V + [\not{D}_4, \not{\nabla}]V, \tag{4.8.22}$$

with $V = \not{\nabla}\mathbf{D}_3 \log\Omega$ and using (4.8.2) to compute $[\not{D}_4, \not{\nabla}]V$, we obtain

$$\begin{aligned}
\not{D}_4\not{\nabla}_a V_b &= -\chi_{ac}\not{\nabla}_c V_b + \not{\nabla}_a\not{D}_4 V_b + (\not{\nabla}_a \log\Omega)\not{D}_4 V_b \\
&\quad + \left\{-\underline{\eta}_b(\chi\cdot V)_a + \chi_{ab}(\underline{\eta}\cdot V) + e_4^\tau e_a^\rho\left([D_\tau, D_\rho]V_\sigma\right)e_b^\sigma\right\} \\
&= -\chi_{ac}\not{\nabla}_c V_b + \frac{1}{\Omega}\not{\nabla}_a\Omega\not{D}_4 V_b \\
&\quad + \left\{-\underline{\eta}_b(\chi\cdot V)_a + \chi_{ab}(\underline{\eta}\cdot V) + e_4^\tau e_a^\rho\left([D_\tau, D_\rho]V_\sigma\right)e_b^\sigma\right\}.
\end{aligned} \tag{4.8.23}$$

Recalling that

$$\begin{aligned}
\not{D}_4 V_b &= \not{\nabla}_b\mathbf{D}_4\mathbf{D}_3\log\Omega + (\not{\nabla}_b\log\Omega)\mathbf{D}_4\mathbf{D}_3\log\Omega - \chi_{bc}V_c \\
&= \frac{1}{\Omega}\not{\nabla}_b\Omega\mathbf{D}_4\mathbf{D}_3\log\Omega - \chi_{bc}V_c,
\end{aligned} \tag{4.8.24}$$

and plugging it into the previous equation we obtain

$$\begin{aligned}
\not{D}_4\not{\nabla}_a V_b &= -\chi_{ac}\not{\nabla}_c V_b + \frac{1}{\Omega}\not{\nabla}_a(\not{\nabla}_b\Omega\mathbf{D}_4\mathbf{D}_3\log\Omega - \Omega\chi_{bc}V_c) \\
&\quad + \left\{-\underline{\eta}_b(\chi\cdot V)_a + \chi_{ab}(\underline{\eta}\cdot V) + e_4^\tau e_a^\rho\left([D_\tau, D_\rho]V_\sigma\right)e_b^\sigma\right\} \\
&= -\chi_{ac}\not{\nabla}_c V_b - \chi_{bc}\not{\nabla}_a V_c + \frac{1}{\Omega}\not{\nabla}_a\not{\nabla}_b\Omega\mathbf{D}_4\mathbf{D}_3\log\Omega \\
&\quad - \left[(\not{\nabla}_a\log\Omega)\chi_{bc}V_c + (\not{\nabla}_a\chi_{bc})V_c\right] \\
&\quad + \left\{-\underline{\eta}_b(\chi\cdot V)_a + \chi_{ab}(\underline{\eta}\cdot V) + e_4^\tau e_a^\rho\left([D_\tau, D_\rho]V_\sigma\right)e_b^\sigma\right\}.
\end{aligned} \tag{4.8.25}$$

Finally

$$\begin{aligned}
\not{D}_4\not{\nabla}_a V_b + \mathrm{tr}\chi\not{\nabla}_a V_b &= \frac{1}{\Omega}\not{\nabla}_a\not{\nabla}_b\Omega\mathbf{D}_4\mathbf{D}_3\log\Omega - \left(\hat{\chi}_{ac}\not{\nabla}_c V_b + \hat{\chi}_{bc}\not{\nabla}_a V_c\right) \\
&\quad - \Big[(\not{\nabla}_a\log\Omega)\chi_{bc}V_c + (\not{\nabla}_a\chi_{bc})V_c + \underline{\eta}_b\chi_{ac}V_c \\
&\quad - \chi_{ab}(\underline{\eta}_c V_c) - e_4^\tau e_a^\rho\left([D_\tau, D_\rho]V_\sigma\right)e_b^\sigma\Big].
\end{aligned} \tag{4.8.26}$$

4.8.4 Proof of Proposition 4.6.2

To prove the first line of Proposition 4.6.2 we observe (see (4.6.11)) that

$${}^{(i)}H_{ab} = \frac{1}{2}\left((\not{\nabla}_a{}^{(i)}O)_b + (\not{\nabla}_b{}^{(i)}O)_a\right).$$

Therefore, using the evolution equation for ${}^{(i)}O$, we obtain

$$\begin{aligned}\frac{d}{d\underline{u}}{}^{(i)}H_{ab} &= \frac{\Omega}{2}\Big[\big(\hat{\chi}_{bc}(\nabla\!\!\!/_a{}^{(i)}O)_c + \hat{\chi}_{ac}(\nabla\!\!\!/_b{}^{(i)}O)_c\big) - \big(\hat{\chi}_{ac}(\nabla\!\!\!/_c{}^{(i)}O)_b + \hat{\chi}_{bc}(\nabla\!\!\!/_c{}^{(i)}O)_a\big) \\ &\quad + {}^{(i)}O_c\left((\nabla\!\!\!/_a\chi)_{cb} + (\nabla\!\!\!/_b\chi)_{ca}\right) + {}^{(i)}O_c\left(\chi_{cb}\zeta_a + \chi_{ca}\zeta_b\right) + 2\chi_{ab}(\underline{\eta}_c{}^{(i)}O_c) \\ &\quad + {}^{(i)}O_c\left(R_{4acb} + R_{4bca}\right)\Big]. \end{aligned} \tag{4.8.27}$$

A simple computation gives the explicit expression of $(R_{4acb} + R_{4bca})$,

$$R_{4acb} + R_{4bca} = -(\nabla\!\!\!/_a\chi)_{cb} - (\nabla\!\!\!/_b\chi)_{ca} + 2(\nabla\!\!\!/_c\chi)_{ab} + 2\zeta_c\chi_{ba} - (\chi_{cb}\zeta_a + \chi_{ca}\zeta_b),$$

which, when substituted in (4.8.27) gives

$$\begin{aligned}\frac{d}{d\underline{u}}{}^{(i)}H_{ab} &= -\Omega\left(\hat{\chi}_{ac}{}^{(i)}H_{cb} + \hat{\chi}_{bc}{}^{(i)}H_{ca}\right) + \Omega\chi_{ab}(\nabla\!\!\!/_c\log\Omega){}^{(i)}O_c \\ &\quad +\Omega\left(\hat{\chi}_{bc}\nabla\!\!\!/_a{}^{(i)}O_c + \hat{\chi}_{ac}\nabla\!\!\!/_b{}^{(i)}O_c\right) + \Omega{}^{(i)}O_c(\nabla\!\!\!/_c\chi)_{ab}.\end{aligned}$$

To prove the second line of Proposition 4.6.2 we denote $W = \mathcal{L}_O g$ and write, omitting the indices ${}^{(i)}$,

$$\begin{aligned}\frac{d}{d\underline{u}}W(e_a, e_3) &= \Omega\mathcal{L}_{e_4}W(e_a, e_3) \\ &= \Omega\left[(\mathcal{L}_{e_4}W)(e_a, e_3) + W([e_4, e_a], e_3) + W(e_a, [e_4, e_3])\right] \\ &= \Omega\left[(\mathcal{L}_{e_4}W)(e_a, e_3) - \chi_{ab}W(e_b, e_3) + (\nabla\!\!\!/_a\log\Omega)W(e_4, e_3)\right. \\ &\quad \left. -(\mathbf{D}_4\log\Omega)W(e_a, e_3) - 4\zeta_b W(e_a, e_b)\right], \end{aligned} \tag{4.8.28}$$

which can be rewritten as

$$\frac{d}{d\underline{u}}Z_a = \Omega\left[\frac{1}{4}(\mathcal{L}_{e_4}W)(e_a, e_3) - \chi_{ab}Z_b + (\nabla\!\!\!/_a\log\Omega)F - (\mathbf{D}_4\log\Omega)Z_a - 2\zeta_b H_{ab}\right]. \tag{4.8.29}$$

The first term on the right-hand side of (4.8.29) can be rewritten as[50]

$$\begin{aligned}\mathcal{L}_{e_4}W(e_a, e_3) &= \mathcal{L}_{e_4}\mathcal{L}_O g(e_a, e_3) = \mathcal{L}_O\mathcal{L}_{e_4}g(e_a, e_3) + \mathcal{L}_{[e_4, O]}g(e_a, e_3) \\ &= \mathcal{L}_O\mathcal{L}_{e_4}g(e_a, e_3) - \mathcal{L}_{[O, e_4]}g(e_a, e_3) = \mathcal{L}_O\mathcal{L}_{e_4}g(e_a, e_3) - \mathcal{L}_{Fe_4}g(e_a, e_3) \\ &= \mathcal{L}_O\mathcal{L}_{e_4}g(e_a, e_3) - F(\mathcal{L}_{e_4}g)(e_a, e_3) - 2\nabla\!\!\!/_a F \\ &= \mathcal{L}_O\mathcal{L}_{e_4}g(e_a, e_3) - 2(\zeta + \eta)_a F - 2\nabla\!\!\!/_a F. \end{aligned} \tag{4.8.30}$$

Plugging (4.8.30) into (4.8.29) we obtain

$$\begin{aligned}\frac{d}{d\underline{u}}Z_a + \Omega(\mathbf{D}_4\log\Omega)Z_a + \Omega\chi_{ab}Z_b &= \frac{\Omega}{4}\left[(\mathcal{L}_O\mathcal{L}_{e_4}g)(e_a, e_3) - 2(\zeta + \eta)_a F - 2\nabla\!\!\!/_a F\right] \\ &+ \Omega\left[(\nabla\!\!\!/_a\log\Omega)F - 2\zeta_b H_{ab}\right]. \end{aligned} \tag{4.8.31}$$

---

[50] We use the equation

$$([\mathcal{L}_{e_4}, \mathcal{L}_{{}^{(i)}O}]g)_{\mu\nu} = (\mathcal{L}_{-({}^{(i)}Fe_4)}g)_{\mu\nu} = -{}^{(i)}F(\mathcal{L}_{e_4}g)_{\mu\nu} - (D_\mu{}^{(i)}F)g_{4\nu} - (D_\nu{}^{(i)}F)g_{\mu 4},$$

which follows from the relation $\mathcal{L}_X\mathcal{L}_Y U - \mathcal{L}_Y\mathcal{L}_X U = \mathcal{L}_{[X,Y]}U$ and the commutation relation $[{}^{(i)}O, e_4] = {}^{(i)}Fe_4$.

The first term on the right-hand side of (4.8.31) can be written as [51]

$$(\mathcal{L}_O \mathcal{L}_{e_4} g)(e_a, e_3) = 2\,(\mathcal{L}_O(\zeta + \eta))_a - (\mathcal{L}_{e_4} g)_{a\nu}[O, e_3]^\nu \tag{4.8.32}$$

and since (see 4.6.12))

$$[O, e_3] = -4 Z_b e_b - O_c(\nabla\!\!\!/_c \log \Omega) e_3,$$

we obtain

$$(\mathcal{L}_O \mathcal{L}_{e_4} g)(e_a, e_3) = 2\,((\mathcal{L}_O - F)(\zeta + \eta))_a + 8\chi_{ab} Z_b. \tag{4.8.33}$$

Inserting this last expression into (4.8.31) we obtain the expected result:

$$\begin{aligned}\frac{d}{d\underline{u}}(\Omega^{(i)} Z_a) \quad &= \quad \Omega\chi_{ab}(\Omega^{(i)} Z_b) + \Omega^2 \Big[\frac{1}{2}(\mathcal{L}_{{}^{(i)}O} - {}^{(i)}F)(\zeta + \eta))_a \\ &+ \quad \frac{1}{2}\nabla\!\!\!/_a {}^{(i)}F - 2\zeta_b {}^{(i)}H_{ab} - \frac{1}{2}(\eta - \underline{\eta})_a {}^{(i)}F)\Big].\end{aligned} \tag{4.8.34}$$

4.8.5 *Proof of Proposition 4.6.3*

We start with the equality

$$\frac{d}{d\underline{u}}(\nabla\!\!\!/_c H_{ab}) = \Omega\left[(\nabla\!\!\!/ \mathbf{D}\!\!\!\!/_4 H)_{cab} + ([\mathbf{D}\!\!\!\!/_4, \nabla\!\!\!/] H)_{cab}\right], \tag{4.8.35}$$

where we have chosen the null frame such that $\mathbf{D}\!\!\!\!/_4 e_a = 0$. A standard computation gives

$$\begin{aligned}([\mathbf{D}\!\!\!\!/_4, \nabla\!\!\!/] H)_{cab} \quad &= \quad [\cdots]^{\mu\nu\rho}_{cab} D_\mu H_{\nu\rho} - (\mathbf{D}_c e_4)^\sigma (D_\sigma H_{\nu\rho}) e^\nu_a e^\rho_b \\ &\quad + e^\mu_c e^\nu_a e^\rho_b e_4{}^\sigma (D_\sigma D_\mu - D_\mu D_\sigma) H_{\nu\rho},\end{aligned} \tag{4.8.36}$$

where

$$[\cdots]^{\mu\nu\rho}_{cab} = (\mathbf{D}_4\Pi)^\mu_c e^\nu_a e^\rho_b + e^\mu_c (\mathbf{D}_4\Pi)^\nu_a e^\rho_b + e^\mu_c e^\nu_a (\mathbf{D}_4\Pi)^\rho_b.$$

Since

$$(\mathbf{D}_4\Pi)^\mu_c = \underline{\eta}_c e^\mu_4,$$

it follows immediately that

$$[\cdots]^{\mu\nu\rho}_{cab} D_\mu H_{\nu\rho} = \underline{\eta}_c \mathbf{D}_4(H_{ab}) - \underline{\eta}_a \chi_{cd} H_{db} - \underline{\eta}_b \chi_{cd} H_{ad}\,.$$

Moreover,

$$e^\mu_c e^\nu_a e^\rho_b e_4{}^\sigma (D_\sigma D_\mu - D_\mu D_\sigma) H_{\nu\rho} = (R_{ad4c} H_{db} + R_{bd4c} H_{da})$$

---

[51] Note that

$$\begin{aligned}(\mathcal{L}_O \mathcal{L}_{e_4} g)(e_a, e_3) \quad &= \quad ((\mathcal{L}_O \mathcal{L}_{e_4} g)_{\mu\nu} e^\nu_3 \Pi^\mu_\sigma) e^\sigma_a = \left[\mathcal{L}_O((\mathcal{L}_{e_4} g)_{\mu\nu} e^\nu_3 \Pi^\mu_\sigma)\right] e^\sigma_a \\ &- \quad \left[(\mathcal{L}_{e_4} g)_{\mu\nu}(\mathcal{L}_O \Pi)^\mu_\sigma e^\nu_3 + (\mathcal{L}_{e_4} g)_{\mu\nu} \Pi^\mu_\sigma (\mathcal{L}_O e_3)^\nu\right] e^\sigma_a\,.\end{aligned}$$

and

$$-(\mathbf{D}_c e_4)^{\sigma}(D_{\sigma} H)_{ab} = -\chi_{cd} \nabla\!\!\!\!/_d H_{ab} + \zeta_c \mathbf{D}_4(H_{ab}) \,.$$

Therefore the left-hand side of (4.8.36) becomes

$$\begin{aligned} \left([\mathbf{D}\!\!\!\!/_4, \nabla\!\!\!\!/]H\right)_{cab} &= -\chi_{cd}(\nabla\!\!\!\!/_d H_{ab}) + (\nabla\!\!\!\!/ \log\Omega)\mathbf{D}_4(H_{ab}) \\ &+ \left[(R_{ad4c} - \underline{\eta}_a \chi_{dc})H_{db} + (R_{bd4c} - \underline{\eta}_b \chi_{cd})H_{da}\right] \end{aligned} \tag{4.8.37}$$

which, when substituted into (4.8.35), gives

$$\begin{aligned} \frac{d}{d\underline{u}}(\nabla\!\!\!\!/_c H_{ab}) + \Omega\chi_{cd}(\nabla\!\!\!\!/_d H_{ab}) &= \Omega(\nabla\!\!\!\!/ \mathbf{D}\!\!\!\!/_4 H)_{cab} + \Omega(\nabla\!\!\!\!/_c \log\Omega)(\mathbf{D}_4 H)_{ab} \\ &+ \Omega\left[(R_{ad4c} - \underline{\eta}_a \chi_{dc})H_{db} + (R_{bd4c} - \underline{\eta}_b \chi_{cd})H_{da}\right]. \end{aligned} \tag{4.8.38}$$

Starting from[52]

$$\mathbf{D}_4 H_{\mu\nu} = -(\hat{\chi}_{\mu d} H_{d\nu} + \hat{\chi}_{\nu d} H_{d\mu}) + (\mathcal{L}_O - F)\hat{\chi}_{\mu\nu} + \frac{1}{2}\Pi_{\mu\nu}(\mathcal{L}_O - F)\mathrm{tr}\chi \,,$$

we compute explicitly

$$\begin{aligned} (\nabla\!\!\!\!/_c \mathbf{D}\!\!\!\!/_4 H)_{ab} &= -\left((\nabla\!\!\!\!/_c \hat{\chi}_{ad})H_{db} + (\nabla\!\!\!\!/_c \hat{\chi}_{bd})H_{da}\right) - (\hat{\chi}_{ad}\nabla\!\!\!\!/_c H_{db} + \hat{\chi}_{bd}\nabla\!\!\!\!/_c H_{da}) \\ &+ (\nabla\!\!\!\!/_c(\mathcal{L}_O - F)\hat{\chi})_{ab} + \frac{1}{2}\Pi_{\mu\nu}(\nabla\!\!\!\!/_c(\mathcal{L}_O - F)\mathrm{tr}\chi). \end{aligned} \tag{4.8.39}$$

Inserting this last expression into (4.8.38) we obtain

$$\begin{aligned} \frac{d}{d\underline{u}}(\nabla\!\!\!\!/_c H_{ab}) + \frac{1}{2}\Omega\mathrm{tr}\chi(\nabla\!\!\!\!/_c H_{ab}) &= -\Omega\left(\hat{\chi}_{ad}\nabla\!\!\!\!/_c H_{db} + \hat{\chi}_{bd}\nabla\!\!\!\!/_c H_{da} + \hat{\chi}_{cd}\nabla\!\!\!\!/_d H_{ab}\right) \\ &+\Omega\left\{(\nabla\!\!\!\!/ \log\Omega)_c \mathbf{D}_4(H_{ab})\right. \\ &+\left[(R_{ad4c} - \underline{\eta}_a \chi_{dc})H_{db} + (R_{bd4c} - \underline{\eta}_b \chi_{cd})H_{da}\right] \\ &-\left((\nabla\!\!\!\!/_c \hat{\chi}_{ad})H_{db} + (\nabla\!\!\!\!/_c \hat{\chi}_{bd})H_{da}\right) \\ &\left.+(\nabla\!\!\!\!/(\mathcal{L}_O - F)\hat{\chi})_{ab} + \frac{1}{2}\delta_{ab}(\nabla\!\!\!\!/(\mathcal{L}_O - F)\mathrm{tr}\chi)\right\}. \end{aligned} \tag{4.8.40}$$

---

[52]Starting from (4.6.13)

$$(\mathbf{D}_4 H)_{ab} = -\left(\hat{\chi}_{ad} H_{db} + \hat{\chi}_{bd} H_{da}\right) + \left((\mathcal{L}_O - F)\hat{\chi}\right)_{ab} + \frac{1}{2}\delta_{ab}(\mathcal{L}_O - F)\mathrm{tr}\chi.$$

We rewrite (4.8.42) in a more formal way as

$$\begin{aligned}
\frac{d}{d\underline{u}}(\nabla\!\!\!/_c H_{ab}) + \frac{1}{2}\Omega \mathrm{tr}\chi(\nabla\!\!\!/_c H_{ab}) &= -\Omega\hat{\chi}_{ad}(\nabla\!\!\!/ H)_{cdb} + \Omega\left\{(\nabla\!\!\!/ \log\Omega)_c \mathbf{D}_4(H_{ab})\right.\\
&+\left[(R_{ad4c} - \underline{\eta}_a\chi_{dc})H_{db} + (R_{bd4c} - \underline{\eta}_b\chi_{cd})H_{da}\right]\\
&-\left((\nabla\!\!\!/\hat{\chi})_{cad}H_{db} + (\nabla\!\!\!/\hat{\chi})_{cbd}H_{da}\right)\\
&-\left(\hat{\chi}_{bd}(\nabla\!\!\!/ H)_{cda} + \hat{\chi}_{cd}(\nabla\!\!\!/ H)_{dab}\right)\\
&\left.+(\mathcal{L}_O\nabla\!\!\!/\hat{\chi})_{cab} + \frac{1}{2}\delta_{ab}\nabla\!\!\!/_c(\mathcal{L}_O\mathrm{tr}\chi) - (\nabla\!\!\!/ F\chi)_{cab}\right\}.\\
&+\Omega([\nabla\!\!\!/, \mathcal{L}_O]\hat{\chi})_{cab}. \qquad (4.8.41)
\end{aligned}$$

Substituting the commutation relation

$$\begin{aligned}
\left([\nabla\!\!\!/, \mathcal{L}_O]\hat{\chi}\right)_{cab} &= \hat{\chi}_{bd}\left(\nabla\!\!\!/_a H_{dc} + \nabla\!\!\!/_c H_{da} - \nabla\!\!\!/_d H_{ac}\right)\\
&+ \hat{\chi}_{ad}\left(\nabla\!\!\!/_b H_{dc} + \nabla\!\!\!/_c H_{db} - \nabla\!\!\!/_d H_{bc}\right), \qquad (4.8.42)
\end{aligned}$$

into the previous expression we obtain the result of Proposition 4.6.3

$$\begin{aligned}
\frac{d}{d\underline{u}}(\nabla\!\!\!/_c H_{ab}) &+ \frac{1}{2}\Omega\mathrm{tr}\chi(\nabla\!\!\!/_c H_{ab}) = -\Omega\hat{\chi}_{ad}(\nabla\!\!\!/ H)_{cdb}\\
&+\hat{\chi}_{ad}\left[(\nabla\!\!\!/ H)_{bdc} - (\nabla\!\!\!/ H)_{dbc}\right] + \hat{\chi}_{bd}\left[(\nabla\!\!\!/ H)_{adc} - (\nabla\!\!\!/ H)_{dac}\right]\\
&+\Omega\left\{(\nabla\!\!\!/ \log\Omega)_c\mathbf{D}_4(H_{ab})\right.\\
&+\left[(R_{ad4c} - \underline{\eta}_a\chi_{dc})H_{db} + (R_{bd4c} - \underline{\eta}_b\chi_{cd})H_{da}\right]\\
&-\left[(\nabla\!\!\!/\hat{\chi})_{cad}H_{db} + (\nabla\!\!\!/\hat{\chi})_{cbd}H_{da}\right]\\
&\left.+(\mathcal{L}_O\nabla\!\!\!/\hat{\chi})_{cab} + \frac{1}{2}\delta_{ab}\nabla\!\!\!/_c(\mathcal{L}_O\mathrm{tr}\chi) - (\nabla\!\!\!/ F\chi)_{cab}\right\}. \qquad (4.8.43)
\end{aligned}$$

The proofs of the second and third lines of Proposition 4.6.3 are similar and we do not present them here.

4.8.6 Proof of the Oscillation Lemma

We repeat here the statement of the lemma.

Lemma 4.1.6 (Oscillation Lemma) *Consider a space time region $\mathcal{K}$ with the canonical double null foliation generated by $u(p)$, $\underline{u}(p)$. Consider also an initial layer region $\mathcal{K}'_{\delta_0}$, of height δ_0, with the initial layer foliation generated by $u'(p)$, $\underline{u}(p)$. We make the following assumptions:*

i) On the surface $S'_* = \Sigma'_{\delta_0} \cap \underline{C}_* = S'(2\delta_0 - \nu_*, \nu_*)$

$$\left(\sup_{(p,p')\in S'_*} |u(p) - u(p')|\right) \leq \epsilon_0. \qquad (4.8.44)$$

Also,

$$|r^2\tau_-^{\frac{1}{2}}\eta| \le \epsilon_0 \ , \ |r'^2\tau'_-\mathbf{g}(L',L)| \le \epsilon_0 \ , \ |r'^3\tau'_-\nabla\!\!\!/\mathbf{g}(L',L)| \le \epsilon_0. \tag{4.8.45}$$

ii) On the initial hypersurface Σ_0,

$$|r'^{\frac{5}{2}}\eta'| \le \epsilon_0. \tag{4.8.46}$$

iii) On $\mathcal{K}/\mathcal{K}'_{\delta_0}$,

$$\mathcal{O}^\infty_{[1]} + \underline{\mathcal{O}}^\infty_{[1]} \le \epsilon_0. \tag{4.8.47}$$

iv) On the initial layer $\mathcal{K}'_{\delta_0}$,

$$\mathcal{O}'^\infty_{[1]} + \underline{\mathcal{O}}'^\infty_{[1]} \le \epsilon_0. \tag{4.8.48}$$

Then, if ϵ_0 *is sufficiently small,*

$$\mathrm{Osc}(u)(\Sigma'_{\delta_0}) \equiv \sup_{\nu\in[\nu_0,\nu_*]}\left(\sup_{(p,p')\in S'(2\delta_0-\nu,\nu)}|u(p)-u(p')|\right) \le c\epsilon_0. \tag{4.8.49}$$

Remarks:

1. The norms appearing in (4.8.45), (4.8.46), (4.8.47), and (4.8.48) are pointwise.
2. The assumptions (4.8.46), (4.8.47) are satisfied in view of the canonicity of the foliation on the last slice $\underline{C}_*$; see Proposition 7.4.1 and Lemma 7.7.2.
3. The assumptions (4.8.46) are verified in view of the canonicity of the foliation on the initial slice Σ_0 and are used in Lemma 4.8.2.

Proof: The proof requires a bootstrap mechanism that has been used many times in this chapter. We assume that the oscillation of u is bounded by a small quantity, Γ'_0,

$$\mathrm{Osc}(u)(\Sigma'_{\delta_0}) \le \Gamma'_0 \tag{4.8.50}$$

and prove the better inequality

$$\mathrm{Osc}(u)(\Sigma'_{\delta_0}) \le c\epsilon_0 \le \frac{\Gamma'_0}{2}. \tag{4.8.51}$$

Denoting by $\Sigma'_{\delta_0}[r'_*, r'_1]$ the portion of Σ'_{δ_0} where (4.8.50) is satisfied, the inequality (4.8.51) allows us to conclude that $\Sigma'_{\delta_0}[r'_*, r'_1]$ coincides with the whole Σ'_{δ_0} and therefore (4.8.50) holds everywhere.

Let $\{e'_4, e'_3, e'_a\}$ be the normalized null frame adapted to the initial layer foliation with

$$e'_4 = 2\Omega' L' \ , \ e'_3 = 2\Omega'\underline{L}, \tag{4.8.52}$$

where Ω' is given by

$$(2\Omega'^2)^{-1} = -\mathbf{g}(L', \underline{L}) = -(g^{\rho\sigma}\partial_\rho u'\partial_\sigma \underline{u}),$$

as in Definition 3.1.12 (see also (3.3.16)). Let us define

$$\tilde{N}' \equiv \frac{1}{2}(e_4' - e_3') \ , \ T' \equiv \frac{1}{2}(e_4' + e_3'). \tag{4.8.53}$$

$\tilde{N}'$ is a unit vector field tangent to Σ'_{δ_0}, orthogonal to the leaves $S'(2\delta_0 - \nu, \nu)$ and contained in Σ'_{δ_0}, and T' is the unit vector field normal to Σ'_{δ_0}. Introducing the functions

$$t' = \frac{1}{2}(\underline{u} + u') \ , \ r' = \frac{1}{2}(\underline{u} - u'), \tag{4.8.54}$$

we observe that (see Proposition 3.3.1),

$$T' = \frac{1}{2\Omega'}\frac{\partial}{\partial t'} \ , \ \tilde{N}' = \frac{1}{2\Omega'}\frac{\partial}{\partial r'}. \tag{4.8.55}$$

Therefore, denoting $r'_* = r'|_{\underline{C}_* \cap \Sigma'_{\delta_0}}$ and introducing the angular coordinates $\phi = \{\phi^1, \phi^2\}$ on Σ'_{δ_0},[53]

$$\begin{aligned} u(r', \phi) &= u(r'_*, \phi) - \int_{r'}^{r'_*} \frac{\partial u}{\partial r'}(\gamma(r'', \phi)) = u(r'_*, \phi) - \int_{r'}^{r'_*} 2\Omega' \tilde{N}'(u)(r'', \phi) \\ &= u(r'_*, \phi) - \int_{r'}^{r'_*} \Omega' \left(e_4'(u) - e_3'(u)\right), \end{aligned} \tag{4.8.56}$$

where $\gamma(r'', \phi)$ is the integral curve starting from $p = (r'_*, \phi) \in S'(2\delta_0 - \nu_*, \nu_*)$, with tangent vector $\tilde{N}'$.

To estimate the right-hand side we express the null frame $\{e_4', e_3', e_a'\}$ adapted to the initial layer foliation in terms of the null frame relative to the double null canonical foliation, $\{e_4, e_3, e_a\}$, where

$$e_4 = \hat{N} = 2\Omega L \ , \ e_3 = \underline{\hat{N}} = 2\Omega \underline{L}. \tag{4.8.57}$$

It is easy to show, since both frames are null frames, that

$$\begin{aligned} e_4' &= \frac{\Omega}{\Omega'} e_4 + \Omega'\Omega(-2\mathbf{g}(L', L))e_3 - 2\Omega(-2\mathbf{g}(L', L))^{\frac{1}{2}}\hat{\sigma}_a e_a \\ e_3' &= \frac{\Omega'}{\Omega} e_3 \\ e_a' &= e_a - \Omega'(-2\mathbf{g}(L', L))^{\frac{1}{2}}\hat{\sigma}_a e_3 \end{aligned} \tag{4.8.58}$$

and

$$\begin{aligned} e_4 &= \frac{\Omega'}{\Omega} e_4' + \Omega\Omega'(-2\mathbf{g}(L', L))e_3' + 2\Omega'(-2\mathbf{g}(L', L))^{\frac{1}{2}}\hat{\sigma}_a e_a' \\ e_3 &= \frac{\Omega}{\Omega'} e_3' \\ e_a &= e_a' + \Omega(-2\mathbf{g}(L', L))^{\frac{1}{2}}\hat{\sigma}_a e_3', \end{aligned} \tag{4.8.59}$$

---

[53] The angular coordinates ϕ are introduced in the standard way starting from the surface $S'_* = \underline{C}_* \cap \Sigma'_{\delta_0}$ and moving along the radial curves with tangent vector field $\tilde{N}'$.

where the vectors $\hat{\sigma}_a$ satisfy $|\hat{\sigma}|^2 = \sum_a \hat{\sigma}_a^2 = 1$. Recall also that

$$e_4(u) = \Omega^{-1}N(u) = 0\ ,\ e_a(u) = 0\ ,\ e_3(u) = \Omega^{-1}\underline{N}(u) = \Omega^{-1}, \tag{4.8.60}$$

where N is the outgoing equivariant vector field relative to the double null canonical foliation. Therefore the last line of (4.8.56) can be rewritten as

$$u(r',\phi) = u(r'_*,\phi) - \int_{r'}^{r'_*} \Omega'\left(e'_4(u) - \frac{\Omega'}{\Omega^2}\right)(r'',\phi). \tag{4.8.61}$$

The oscillation of $u(r',\phi)$ on the surface $S'(2\delta_0-\nu,\nu)$ satisfies

$$\begin{aligned}
&\sup_{p,p'\in S'(2\delta_0-\nu,\nu)} |u(p)-u(p')| = \sup_{\phi,\phi'\in S^2} |u(r'(\nu),\phi)-u(r'(\nu),\phi')|\\
&\leq \sup_{\phi,\phi'\in S^2}\left[|u(r'_*,\phi)-u(r'_*,\phi')| + \int_{s(\phi)}^{s(\phi')}\frac{d}{ds}\left(\int_{r'}^{r'_*}\Omega'\left(e'_4(u)-\frac{\Omega'}{\Omega^2}\right)(r'',\sigma(s))\right)\right]\\
&\leq \sup_{\phi,\phi'\in S^2}|u(r'_*,\phi)-u(r'_*,\phi')| + c\left(\sup_{(a,\phi\in S^2)}\left|\int_{r'}^{r'_*}\frac{\partial}{\partial\phi^a}\Omega'\left(e'_4(u)-\frac{\Omega'}{\Omega^2}\right)(r'',\phi)\right|\right)\\
&\leq \epsilon_0 + c\left(\sup_{(a,\phi\in S^2)}\left|\int_{r'}^{r'_*} r''\nabla\!\!\!\!/'_a\left[\Omega'\left(e'_4(u)-\frac{\Omega'}{\Omega^2}\right)(r'',\phi)\right]dr''\right|\right),
\end{aligned} \tag{4.8.62}$$

where in the first inequality $\sigma(s)$ is a regular curve on S^2 from ϕ to ϕ'. We are left with estimating the integrals on the right-hand side of (4.8.62),

$$\int_{r'}^{r'_*} r''\nabla\!\!\!\!/'_a\left(\Omega' e'_4(u)\right) - \int_{r'}^{r'_*} r''\nabla\!\!\!\!/'_a\left(\frac{\Omega'^2}{\Omega^2}\right) \equiv \textbf{(I)}+\textbf{(II)}, \tag{4.8.63}$$

which in the next lemma are estimated in terms of the connection coefficients relative to the canonical and the initial layer foliations and of the quantities $g(L,L')$ and $\nabla\!\!\!\!/' g(L,L')$.

Lemma 4.8.1 *The integrals in (4.8.63) satisfy the following inequalities:*

$$\begin{aligned}
\textbf{(I)} &= \int_{r'}^{r'_*} r''\nabla\!\!\!\!/'_a(\Omega' e'_4(u)) \leq 2\int_{r'}^{r'_*} r''|\Omega'^2(\eta'+\underline{\eta}')||\mathbf{g}(L',L)|\\
&+ 2\int_{r'}^{r'_*} r''\Omega'^2|\nabla\!\!\!\!/'_a\mathbf{g}(L',L)|\\
\textbf{(II)} &= \int_{r'}^{r'_*} r''\nabla\!\!\!\!/'_a\frac{\Omega'^2}{\Omega^2}\\
&\leq \int_{r'}^{r'_*}\frac{\Omega'^2}{\Omega^2}r''(-2\mathbf{g}(L',L))^{\frac{1}{2}}\left(\Omega|\underline{\chi}'|+|\eta'-\eta|+4|\Omega'\underline{\omega}+\Omega\underline{\omega}'|\right).
\end{aligned} \tag{4.8.64}$$

Proof: To estimate **(I)** observe that

$$\begin{aligned}
\textbf{(I)} &= \int_{r'}^{r'_*} r''\nabla\!\!\!\!/'_a(\Omega' e'_4(u)) = \int_{r'}^{r'_*} r''\nabla\!\!\!\!/'_a\left[\Omega'\left(\frac{\Omega}{\Omega'}e_4(u) + \Omega'\Omega(-2\mathbf{g}(L',L))e_3(u)\right.\right.\\
&\left.\left. - 2\Omega(-2\mathbf{g}(L',L))^{\frac{1}{2}}\hat{\sigma}_a e_a(u)\right)\right] = \int_{r'}^{r'_*} r''\nabla\!\!\!\!/'_a\left(\Omega'^2(-2\mathbf{g}(L',L))\right),
\end{aligned} \tag{4.8.65}$$

where the last equality follows from (4.8.60). Finally

$$\begin{aligned}
\int_{r'}^{r'_*} r''\nabla\!\!\!/'_a \left(\Omega'^2(-2\mathbf{g}(L',L))\right) &= 2\int_{r'}^{r'_*} r''\Omega'^2\nabla\!\!\!/'_a \log\Omega'(-2\mathbf{g}(L',L)) \\
&\quad -2\int_{r'}^{r'_*} r''\Omega'^2(\nabla\!\!\!/'_a\mathbf{g}(L',L)) \\
&= \int_{r'}^{r'_*} r''\nabla\!\!\!/'_a \left(\Omega'^2(-2\mathbf{g}(L',L))\right) - 2\int_{r'}^{r'_*} r''\Omega'^2(\nabla\!\!\!/'_a\mathbf{g}(L',L)) \\
&\le 2\int_{r'}^{r'_*} r''|\Omega'^2(\eta'+\underline{\eta}')||\mathbf{g}(L',L)| + 2\int_{r'}^{r'_*} r''\Omega'^2|\nabla\!\!\!/'_a\mathbf{g}(L',L)|. \qquad (4.8.66)
\end{aligned}$$

To estimate **(II)** we write

$$\begin{aligned}
\int_{r'}^{r'_*} r''\nabla\!\!\!/'_a\frac{\Omega'^2}{\Omega^2} &= 2\int_{r'}^{r'_*} r''\left(\frac{\Omega'^2}{\Omega^2}\nabla\!\!\!/'_a\log\Omega' - \frac{\Omega'^2}{\Omega^2}\nabla\!\!\!/'_a\log\Omega\right) \\
&= 2\int_{r'}^{r'_*} r''\left(\frac{\Omega'^2}{\Omega^2}\frac{1}{2}(\eta'+\underline{\eta}')_a - \frac{\Omega'^2}{\Omega^2}\left(\nabla\!\!\!/_a\log\Omega - \Omega'(-2\mathbf{g}(L',L))^{\frac{1}{2}}\hat{\sigma}_a\mathbf{D}_{e_3}\log\Omega\right)\right) \\
&= \int_{r'}^{r'_*}\frac{\Omega'^2}{\Omega^2}r''\left(\left[(\eta'+\underline{\eta}')_a - (\eta+\underline{\eta})_a\right] - 4\Omega'(-2\mathbf{g}(L',L))^{\frac{1}{2}}\hat{\sigma}_a\underline{\omega}\right). \qquad (4.8.67)
\end{aligned}$$

We write the term $\left[(\eta'+\underline{\eta}')_a - (\eta+\underline{\eta})_a\right]$, expressing $(\eta+\underline{\eta})$ in terms of the primed quantities. A long but easy computation gives

$$\begin{aligned}
\underline{\eta}_a &= \underline{\eta}'_a + \Omega(-2\mathbf{g}(L,L'))^{\frac{1}{2}}\hat{\sigma}_b\underline{\chi}'_{ba} \qquad (4.8.68)\\
\eta_a &= \eta'_a + (-2\mathbf{g}(L,L'))^{\frac{1}{2}}\hat{\sigma}_a(\Omega'\underline{\omega}+\Omega\underline{\omega}') - \Omega\hat{\sigma}_a\mathbf{D}_{e'_3}(-2\mathbf{g}(L,L'))^{\frac{1}{2}}.
\end{aligned}$$

Finally the explicit computation of $\mathbf{D}_{e'_3}(-2\mathbf{g}(L,L'))^{\frac{1}{2}}$ gives, as discussed in Chapter 7 (see Lemma 7.7.2)),

$$\mathbf{D}_{e'_3}(-2\mathbf{g}(L,L'))^{\frac{1}{2}} = \Omega^{-1}(-2\mathbf{g}(L,L'))^{\frac{1}{2}}\left[2(\Omega'\underline{\omega}+\Omega\underline{\omega}') - \hat{\sigma}\cdot(\eta'-\eta)\right]. \quad (4.8.69)$$

Using (4.8.68) and (4.8.69) on the right-hand side of (4.8.67) we estimate **(II)** as

$$\begin{aligned}
\int_{r'}^{r'_*} r''\nabla\!\!\!/'_a\frac{\Omega'^2}{\Omega^2} &= \int_{r'}^{r'_*}\frac{\Omega'^2}{\Omega^2}r''\left(\left[(\eta'+\underline{\eta}')_a - (\eta+\underline{\eta})_a\right] - 4\Omega'(-2\mathbf{g}(L',L))^{\frac{1}{2}}\hat{\sigma}_a\underline{\omega}\right) \\
&= \int_{r'}^{r'_*}\frac{\Omega'^2}{\Omega^2}r''(-2\mathbf{g}(L',L))^{\frac{1}{2}}\hat{\sigma}_a\left(-\Omega\underline{\chi}'_{ba}\hat{\sigma}_b + \hat{\sigma}\cdot(\eta'-\eta) - 4(\Omega'\underline{\omega}+\Omega\underline{\omega}')\right) \\
&\le \int_{r'}^{r'_*}\frac{\Omega'^2}{\Omega^2}r''(-2\mathbf{g}(L',L))^{\frac{1}{2}}\left(\Omega|\underline{\chi}'| + |\eta'-\eta| + 4|\Omega'\underline{\omega}+\Omega\underline{\omega}'|\right). \qquad (4.8.70)
\end{aligned}$$

Therefore the estimates for the Oscillation Lemma require the estimates for $\mathbf{g}(L',L)$ and $\nabla\!\!\!/'\mathbf{g}(L',L)$ along Σ'_{δ_0}. These estimates are obtained writing the evolution equations for these quantities along Σ'_{δ_0}. The result is in the following lemma.

Lemma 4.8.2 *Under the assumptions of the Oscillation Lemma, the following estimates hold on Σ'_{δ_0}:*

$$|r'^3\mathbf{g}(L',L)| \le c\epsilon_0\ ,\ |r'^4\nabla\!\!\!\!/'\mathbf{g}(L',L)| \le c\epsilon_0. \tag{4.8.71}$$

Proof: Using equations (4.8.58) and (4.8.59) a long computation gives

$$\begin{aligned}\frac{d}{dr'}\mathbf{g}(L',L) &= \Omega(-2\mathbf{g}(L,L'))\frac{\mathrm{tr}\chi}{2} + \Big\{\Omega(-2\mathbf{g}(L,L'))\hat{\sigma}_a\hat{\sigma}_b\hat{\chi}_{ab}\\ &+2\frac{\Omega}{\Omega'}(-2\mathbf{g}(L,L'))(\Omega'\underline{\omega}+\Omega\underline{\omega}') - 2\Omega\Omega'(-2\mathbf{g}(L,L'))^{\frac{3}{2}}\hat{\sigma}\cdot\eta\\ &+2\Omega'^2\Omega(-2\mathbf{g}(L,L'))^2\underline{\omega} + \frac{\Omega'}{\Omega}(-2\mathbf{g}(L,L'))^{\frac{1}{2}}\hat{\sigma}\cdot(\eta-\eta')\Big\}.\end{aligned} \tag{4.8.72}$$

The first term on the right-hand side can be written as

$$\begin{aligned}\Omega(-2\mathbf{g}(L,L'))\frac{\mathrm{tr}\chi}{2} &= -\Omega\mathbf{g}(L,L')\left(\frac{1}{2}(\mathrm{tr}\chi-\mathrm{tr}\underline{\chi})+\frac{1}{2}(\mathrm{tr}\chi+\mathrm{tr}\underline{\chi})\right) \qquad (4.8.73)\\ = -\frac{1}{2}\mathrm{tr}\theta\mathbf{g}(L,L')&+\left(\Omega-\frac{1}{2}\right)\frac{\mathrm{tr}\theta}{2}(-2\mathbf{g}(L,L'))+\frac{\Omega}{4}(\mathrm{tr}\chi+\mathrm{tr}\underline{\chi})(-2\mathbf{g}(L,L')),\end{aligned}$$

where

$$\mathrm{tr}\theta = \frac{1}{2}(\mathrm{tr}\chi - \mathrm{tr}\underline{\chi}) = \delta^{ab}\mathbf{g}(\mathbf{D}_{e_a}\tilde{N},e_b), \tag{4.8.74}$$

and the evolution equation (4.8.72) becomes

$$\begin{aligned}\frac{d}{dr'}\mathbf{g}(L',L) &+ \frac{1}{2}\mathrm{tr}\theta\mathbf{g}(L,L')\\ &= \left[\left(\Omega-\frac{1}{2}\right)\frac{\mathrm{tr}\theta}{2}(-2\mathbf{g}(L,L'))+\frac{\Omega}{4}(\mathrm{tr}\chi+\mathrm{tr}\underline{\chi})(-2\mathbf{g}(L,L'))\right]\\ &+ \Big\{\Omega(-2\mathbf{g}(L,L'))\hat{\sigma}_a\hat{\sigma}_b\hat{\chi}_{ab}+2\frac{\Omega}{\Omega'}(-2\mathbf{g}(L,L'))(\Omega'\underline{\omega}+\Omega\underline{\omega}')\\ &- 2\Omega\Omega'(-2\mathbf{g}(L,L'))^{\frac{3}{2}}\hat{\sigma}\cdot\eta+2\Omega'^2\Omega(-2\mathbf{g}(L,L'))^2\underline{\omega}\\ &+ \frac{\Omega'}{\Omega}(-2\mathbf{g}(L,L'))^{\frac{1}{2}}\hat{\sigma}\cdot(\eta-\eta')\Big\}.\end{aligned} \tag{4.8.75}$$

To apply the analogous of the Evolution Lemma to (4.8.75) we have still to replace $\mathrm{tr}\theta$, defined in (4.8.74), with

$$\mathrm{tr}\theta' = \frac{1}{2}(\mathrm{tr}\chi'-\mathrm{tr}\underline{\chi}') = \delta^{ab}\mathbf{g}(\mathbf{D}_{e'_a}\tilde{N}',e'_b), \tag{4.8.76}$$

as $\mathrm{tr}\theta'$ is the second fundamental form associated to the $S'(2\delta_0-\nu,\nu)$ surfaces foliating Σ'_{δ_0}, while $\mathrm{tr}\theta$ refers to the $S(\lambda,\nu)$ two-dimensional surfaces of the double null canonical foliation. Making this substitution, (4.8.75) becomes

$$
\begin{aligned}
&\frac{d}{dr'}\mathbf{g}(L',L)+\frac{\mathrm{tr}\theta'}{2}\mathbf{g}(L,L') = \left[\frac{1}{4}(\mathrm{tr}\theta-\mathrm{tr}\theta')(-2\mathbf{g}(L,L'))\right.\\
&+\left.\left(\Omega-\frac{1}{2}\right)\frac{\mathrm{tr}\theta}{2}(-2\mathbf{g}(L,L'))+\frac{\Omega}{4}(\mathrm{tr}\chi+\mathrm{tr}\underline{\chi})(-2\mathbf{g}(L,L'))\right]\\
&+\left\{\Omega(-2\mathbf{g}(L,L'))\left(\hat{\sigma}_a\hat{\sigma}_b\hat{\chi}_{ab}+2\frac{\Omega}{\Omega'}(\Omega'\underline{\omega}+\Omega\underline{\omega}')\right)-2\Omega\Omega'(-2\mathbf{g}(L,L'))^{\frac{3}{2}}\hat{\sigma}\cdot\eta\right.\\
&+\left.2\,\Omega'^2\Omega(-2\mathbf{g}(L,L'))^2\underline{\omega}\right\}+\frac{\Omega'}{\Omega}(-2\mathbf{g}(L,L'))^{\frac{1}{2}}\hat{\sigma}\cdot(\eta-\eta'), \qquad (4.8.77)
\end{aligned}
$$

which we rewrite as

$$
\begin{aligned}
\frac{d}{dr'}\mathbf{g}(L',L)+\frac{\mathrm{tr}\theta'}{2}\mathbf{g}(L',L) &= (-2\mathbf{g}(L',L))^{\frac{1}{2}}F_{\frac{1}{2}}+(-2\mathbf{g}(L',L))F_1\\
&+(-2\mathbf{g}(L',L))^{\frac{3}{2}}F_{\frac{3}{2}}+(-2\mathbf{g}(L',L))^2F_2.
\end{aligned}
$$

where

$$
\begin{aligned}
F_{\frac{1}{2}} &= \frac{\Omega'}{\Omega}\hat{\sigma}\cdot(\eta-\eta')\\
F_1 &= \left[\frac{1}{4}\left(\mathrm{tr}\theta-\frac{2}{r}\right)+\frac{1}{2}\left(\frac{1}{r}-\frac{1}{r'}\right)+\frac{1}{4}\left(\frac{2}{r'}-\mathrm{tr}\theta'\right)+\left(\Omega-\frac{1}{2}\right)\frac{\mathrm{tr}\theta}{2}\right.\\
&\left.+\frac{\Omega}{4}(\mathrm{tr}\chi+\mathrm{tr}\underline{\chi})+\left(\Omega\hat{\sigma}_a\hat{\sigma}_b\hat{\chi}_{ab}+2\frac{\Omega}{\Omega'}(\Omega'\underline{\omega}+\Omega\underline{\omega}')\right)\right]\\
F_{\frac{3}{2}} &= -2\Omega\Omega'\hat{\sigma}\cdot\eta\\
F_2 &= +2\Omega'^2\Omega\underline{\omega}. \qquad (4.8.78)
\end{aligned}
$$

Using assumptions (4.1.42), (4.1.43) for the Oscillation Lemma and the auxiliary assumption (4.8.50) it follows that[54]

$$
\begin{aligned}
|F_1|(r',\phi) &\le c\frac{(\Gamma_0'+c\epsilon_0)}{r'^2}\log r'\\
|F_{\frac{3}{2}}|(r',\phi) &\le c\frac{\epsilon_0}{r'^{\frac{5}{2}}}\\
|F_2|(r',\phi) &\le c\frac{\epsilon_0}{r'^2}\\
|F_{\frac{1}{2}}|(r',\phi) &\le c\frac{\epsilon_0}{r'^{\frac{5}{2}}}.
\end{aligned} \qquad (4.8.79)
$$

Observe that to estimate $|F_{\frac{1}{2}}|(r',\phi)$ and $|F_{\frac{3}{2}}|(r',\phi)$ we used the following estimates for η and η' on Σ'_{δ_0}:

$$
|r^{\frac{5}{2}}\eta|\le c\epsilon_0\,,\ |r'^{\frac{5}{2}}\eta'|\le c\epsilon_0. \qquad (4.8.80)
$$

---

[54]While r' is defined in (4.8.54), the variable r introduced in (4.8.78) is defined analogously by $r=\frac{1}{2}(\underline{u}-u)$. They differ logarithmically from the standard definition of r given in (3.1.2).

These estimates follow from the assumptions (4.1.40), (4.1.41) and Proposition 4.3.16.[55] Using assumptions (4.8.79) the following inequality holds:

$$\begin{aligned}\frac{d}{dr'}\mathbf{g}(L',L)+\frac{\mathrm{tr}\theta'}{2}\mathbf{g}(L',L) &\le \frac{(\Gamma_0'+c\epsilon_0)}{r'^2}(-\mathbf{g}(L',L))+\frac{c\epsilon_0}{r'^{\frac{5}{2}}}(-\mathbf{g}(L',L))^{\frac{3}{2}}\\ &+\frac{c\epsilon_0}{r'^2}(-\mathbf{g}(L',L))^2+\frac{c\epsilon_0}{r'^{\frac{5}{2}}}(-\mathbf{g}(L',L))^{\frac{1}{2}}. \qquad (4.8.81)\end{aligned}$$

Applying the Evolution Lemma to the evolution inequality on Σ'_{δ_0} we obtain

$$\begin{aligned}|r'^{1-\frac{2}{p}}\mathbf{g}(L,L')|_{p,S'}(\overline{r}') &\le |r'^{1-\frac{2}{p}}\mathbf{g}(L,L')|_{p,S'}(r'_*)\\ &+\int_{\overline{r}'}^{r'_*}\frac{(\Gamma_0'+c\epsilon_0)}{r'^2}|r'^{1-\frac{2}{p}}\mathbf{g}(L,L')|_{p,S'}\\ &+\int_{\overline{r}'}^{r'_*}\frac{c\epsilon_0}{r'^{\frac{5}{2}}}|r'^{1-\frac{2}{p}}\mathbf{g}(L,L')|_{p,S'}|\mathbf{g}^{\frac{1}{2}}(L,L')|_{\infty,S'}\\ &+\int_{\overline{r}'}^{r'_*}\frac{c\epsilon_0}{r'^2}|r'^{1-\frac{2}{p}}\mathbf{g}(L,L')|_{p,S'}|\mathbf{g}(L,L')|_{\infty,S'} \qquad (4.8.82)\\ &+\int_{\overline{r}'}^{r'_*}\frac{c\epsilon_0}{r'^{\frac{5}{2}}}r'|\mathbf{g}^{\frac{1}{2}}(L,L')|_{\mathrm{inf}ty,S'}\end{aligned}$$

We again use a bootstrap mechanism. We assume that, on Σ'_{δ_0}, $|\mathbf{g}(L,L')|_{\infty,S'}$ satisfies

$$|\mathbf{g}(L,L')|_{\infty,S'}\le\frac{\tilde{\Gamma}}{r'^3} \qquad (4.8.83)$$

with $\epsilon_0<\tilde{\Gamma}<c\epsilon_0$, and we prove that

$$|\mathbf{g}(L,L')|_{\infty,S'}\le\frac{1}{2}\frac{\tilde{\Gamma}}{r'^3}\,. \qquad (4.8.84)$$

This allows to conclude that

$$|\mathbf{g}(L,L')|_{\infty,S'}\le\frac{c\epsilon_0}{r'^3} \qquad (4.8.85)$$

on the whole Σ'_{δ_0}. To prove inequality (4.8.84) we start applying Gronwall inequality to (4.8.82) obtaining

$$\begin{aligned}|r'^{1-\frac{2}{p}}\mathbf{g}(L,L')|_{p,S'}(\overline{r}') &\le \left(|r'^{1-\frac{2}{p}}\mathbf{g}(L,L')|_{p,S'}(r'_*)+\int_{\overline{r}'}^{r'_*}\frac{c\epsilon_0}{r'^4}r'\tilde{\Gamma}\right)\\ &\le \left(|r'^{1-\frac{2}{p}}\mathbf{g}(L,L')|_{p,S'}(r'_*)+\frac{c\epsilon_0\tilde{\Gamma}^{\frac{1}{2}}}{r'^2}\right)\\ &\le \left(\frac{\epsilon_0}{r'^2}+\frac{c\epsilon_0\tilde{\Gamma}^{\frac{1}{2}}}{r'^2}\right)\le\frac{c\epsilon_0}{r'^2}. \qquad (4.8.86)\end{aligned}$$

[55] To estimate η' we use a slightly modified version of Proposition 4.3.16.

To complete the proof we have only to obtain the analogous estimate for $|r'^{2-\frac{2}{p}}\nabla\!\!\!/'\mathbf{g}(L,L')|_{p,S'}(\overline{r}')$, deriving tangentially the previous evolution equation, and, finally, apply Lemma 4.1.3.

Once Lemma 4.8.2 is proved we complete the proof of the Oscillation Lemma using the result of this lemma to estimate the right-hand sides of (4.8.64).

4.8.7 Proof of Lemma 4.1.7

Lemma 4.1.7 controls the difference between the norm $|r^{\lambda_1}V|_{p,S(u_0(\nu),\nu)}$ and the norm $|r'^{\lambda_1}V'|_{p,S_{(0)}(\nu)}$. Observe that the norm $|r^{\lambda_1}V|_{p,S(u_0(\nu),\nu)}$ refers to a surface $S(u_0(\nu),\nu)$ associated to the double null canonical foliation with $r=(\frac{1}{4\pi}|S(u_0(\nu),\nu)|)^{\frac{1}{2}}$, while $|r'^{\lambda_1}V'|_{p,S_{(0)}(\nu)}$ refers to a two-dimensional surface contained in Σ_0 associated to the initial layer foliation, and therefore, $r'=(\frac{1}{4\pi}|S_{(0)}(\nu)|)^{\frac{1}{2}}$.

We recall also that this result, completing part II of the Evolution Lemma, is applied, in this chapter, to the estimates of the underlined connection coefficients.

Let, therefore, V be a connection coefficient which satisfies the evolution equation

$$\mathbf{D}_{\underline{N}}V+\lambda_0\Omega\mathrm{tr}\underline{\chi}\,V=\underline{F}.$$

Using the relations (4.8.58), (4.8.59), between the normalized null frame adapted to the double null canonical foliation and the one relative to the initial layer foliation, it is possible to express the connection coefficient V in terms of a linear combination of the connection coefficients relative to the initial layer foliation. The explicit expressions for $\underline{\chi}$, $\underline{\eta}$ and ω are

$$\underline{\chi}_{ab}=\frac{\Omega}{\Omega'}\underline{\chi}'_{ab}\ ,\ \ \underline{\eta}_a=\underline{\eta}'_a+\Omega(-2\mathbf{g}(L',L))^{\frac{1}{2}}\hat{\sigma}_b\underline{\chi}'_{ab} \tag{4.8.87}$$
$$\omega=\frac{\Omega'}{\Omega}\omega'-\Omega'(-2\mathbf{g}(L',L))^{\frac{1}{2}}\hat{\sigma}_a\underline{\eta}'-\frac{\Omega\Omega'}{2}(-2\mathbf{g}(L',L))\hat{\sigma}_a\hat{\sigma}_b\underline{\chi}'_{ab}.$$

Let us prove the lemma in the case $V=\omega$; the proof in the other cases proceeds exactly along the same lines. We observe that, using the results of the Oscillation Lemma and Lemma 4.8.2,

$$|r^{\lambda_1}\omega|^p_{p,S(u_0(\nu),\nu)}=\int_{S(u_0(\nu),\nu)}|r^{\lambda_1}\omega|^p\le c\int_{S(u_0(\nu),\nu)}|r'^{\lambda_1}\omega'|^p$$
$$+c\left[\epsilon_0^{\frac{1}{2}}\int_{S(u_0(\nu),\nu)}|r'^{(\lambda_1-\frac{3}{2})}\underline{\eta}'|^p+\epsilon_0\int_{S(u_0(\nu),\nu)}|r'^{(\lambda_1-3)}\underline{\chi}'|^p\right]. \tag{4.8.88}$$

It is easy to realize that the terms in the $[\cdot\cdot]$ brackets can be treated as corrections. Therefore we fix the attention on the first term, $\int_{S(u_0(\nu),\nu)}|r'^{\lambda_1}\omega'|^p$, which we rewrite in the following way

$$\int_{S(u_0(\nu),\nu)}|r'^{\lambda_1}\omega'|^p\le c\int_{S^2}d\phi r'^2|r'^{\lambda_1}\omega'|^p(u'(\phi),\phi), \tag{4.8.89}$$

choosing $u'(\phi)$ in such a way that

$$S(u_0(\nu),\nu)=\{p\in\underline{C}(\nu)|u'(p)=u'(\phi)\}\,.$$

The connection coefficient $\omega' = -\frac{1}{2}\mathbf{D}_{e'_4} \log \Omega'$ satisfies in the initial layer region the following evolution equation (see Subsection 4.3.10),

$$\mathbf{D}'_3(\Omega' \mathbf{D}'_4 \log \Omega') = \hat{F}' - \Omega' \rho' = F', \tag{4.8.90}$$

where $\hat{F}'$ is given by

$$\hat{F}' \equiv 2\Omega' \zeta' \cdot \nabla\!\!\!\!/' \log \Omega' + \Omega'(\underline{\eta}' \cdot \eta' - 2\zeta'^2). \tag{4.8.91}$$

The scalar function $|\omega'|^p(u', \phi)$ satisfies the evolution equation

$$\frac{d|\omega'|^p}{du'} = \frac{p}{2}|\omega'|^{p-1}\frac{\omega'}{|\omega'|}F',$$

and therefore,

$$\frac{d|\omega'|}{du'}(u', \phi) \le \frac{1}{2}|F'|(u', \phi). \tag{4.8.92}$$

Integrating with respect to u', at fixed ϕ, starting from

$$S_{(0)}(\underline{u}) = \{p \in \underline{C}(\underline{u}) | u'(p) = u'_0\},$$

where $u'_0 = -\underline{u}_{(0)}$, we obtain, multiplying both sides with r'^{λ_1},

$$|r'^{\lambda_1}\omega'|(u'(\phi), \phi) \le c\left(|r'^{\lambda_1}\omega'|(u'_0, \phi) + \int_{u'_0}^{u'(\phi)} du''|r'^{\lambda_1}F'|(u'', \phi)\right).$$

Substituting this last expression in (4.8.89) we obtain

$$\begin{aligned}
\int_{S(u_0(\nu),\nu)} |r'^{\lambda_1}\omega'|^p &= \int_{S^2} d\phi r'^2 |r'^{\lambda_1}\omega'|^p(u'(\phi), \phi) \\
&\le c\int_{S^2} d\phi r'^2 |r'^{\lambda_1}\omega'|^p(u'_0, \phi) + c\int_{S^2} d\phi r'^2 \left(\int_{u'_0}^{u'(\phi)} du''|r'^{\lambda_1}F'|(u'', \phi)\right)^p \\
&\le c|r'^{\lambda_1}\omega'|^p_{p,S_{(0)}}(\nu) + c\left\|\left(\int_{u'_0}^{u'_S} du'' r'^2 |r'^{\lambda_1}F'|(u'', \phi)\right)\right\|^p_{L^p(S^2)}
\end{aligned} \tag{4.8.93}$$

where in the last inequality we have chosen

$$u'_S(\nu) = 2\delta_0 - \nu, \tag{4.8.94}$$

and used that δ_0 is small to bring r' inside the integral. Applying the Minkowski inequality to the norm in the last line of (4.8.93), for $p \in [2, 4]$, we obtain

$$|r'^{\lambda_1}\omega'|_{p,S(u_0(\underline{u}),\underline{u})} \le c|r'^{\lambda_1}\omega'|_{p,S_{(0)}(\underline{u})} + c\int_{u'_0}^{u'_S} \left|r'^{\lambda_1}F'\right|_{p,S'}(u'', \underline{u})du''. \tag{4.8.95}$$

The choice of λ_1 is dictated by the explicit expression of F' and by the estimate for ω' on the initial slice Σ_0, it is $\lambda_1 = 2 - \frac{2}{p}$, as expected. The estimate for integral

$\int_{u'_0}^{u'_S} \left|r^{\lambda_1} F'\right|_{p,S}(u'', \underline{u})du''$ is exactly of the same type as all the estimates done in the rest of this chapter, with the only difference that all the quantities are here relative to the initial layer foliation. Its estimate is, therefore, a repetition of the previous ones.

The estimates for the remaining terms of (4.8.88) proceeds in the same way, and the final result is, therefore,

$$|r^{\lambda_1}\omega|_{p,S(u_0(\underline{u}),\underline{u})} \leq c|r'^{\lambda_1}\omega'|_{p,S_{(0)}(\underline{u})} + c\epsilon_0, \tag{4.8.96}$$

proving the lemma.

5

Estimates for the Riemann Curvature Tensor

This chapter is devoted to the proof of Theorem **M7** in terms of the fundamental quantities $\mathcal{Q}$, $\underline{\mathcal{Q}}$. These quantities can be expressed, according to (3.5.1), as weighted integrals of the null components of $\hat{\mathcal{L}}_O\mathbf{R}$, $\hat{\mathcal{L}}_T\mathbf{R}$, $\hat{\mathcal{L}}_O^2\mathbf{R}$, $\hat{\mathcal{L}}_O\hat{\mathcal{L}}_T\mathbf{R}$ and $\hat{\mathcal{L}}_S\hat{\mathcal{L}}_T\mathbf{R}$ along the null hypersurfaces $C(\lambda)$ and $\underline{C}(\nu)$. We recall their explicit expressions:

$$\mathcal{Q}(\lambda,\nu)=\mathcal{Q}_1(\lambda,\nu)+\mathcal{Q}_2(\lambda,\nu)$$
$$\underline{\mathcal{Q}}(\lambda,\nu)=\underline{\mathcal{Q}}_1(\lambda,\nu)+\underline{\mathcal{Q}}_2(\lambda,\nu)$$

$$\begin{aligned}
\mathcal{Q}_1(\lambda,\nu) &\equiv \int_{C(\lambda)\cap V(\lambda,\nu)} Q(\hat{\mathcal{L}}_T\mathbf{R})(\bar{K},\bar{K},\bar{K},e_4)\\
&\quad+\int_{C(\lambda)\cap V(\lambda,\nu)} Q(\hat{\mathcal{L}}_O\mathbf{R})(\bar{K},\bar{K},T,e_4)\\
\mathcal{Q}_2(\lambda,\nu) &\equiv \int_{C(\lambda)\cap V(\lambda,\nu)} Q(\hat{\mathcal{L}}_O\hat{\mathcal{L}}_T\mathbf{R})(\bar{K},\bar{K},\bar{K},e_4)\\
&\quad+\int_{C(\lambda)\cap V(\lambda,\nu)} Q(\hat{\mathcal{L}}_O^2\mathbf{R})(\bar{K},\bar{K},T,e_4)\\
&\quad+\int_{C(\lambda)\cap V(\lambda,\nu)} Q(\hat{\mathcal{L}}_S\hat{\mathcal{L}}_T\mathbf{R})(\bar{K},\bar{K},\bar{K},e_4)
\end{aligned}$$

$$\begin{aligned}
\underline{\mathcal{Q}}_1(\lambda,\nu) &\equiv \sup_{V(\lambda,\nu)\cap\Sigma_0}|r^3\overline{\rho}|^2+\int_{\underline{C}(\nu)\cap V(\lambda,\nu)} Q(\hat{\mathcal{L}}_T\mathbf{R})(\bar{K},\bar{K},\bar{K},e_3)\\
&\quad+\int_{\underline{C}(\nu)\cap V(\lambda,\nu)} Q(\hat{\mathcal{L}}_O\mathbf{R})(\bar{K},\bar{K},T,e_3)\\
\underline{\mathcal{Q}}_2(\lambda,\nu) &\equiv \int_{\underline{C}(\nu)\cap V(\lambda,\nu)} Q(\hat{\mathcal{L}}_O\hat{\mathcal{L}}_T\mathbf{R})(\bar{K},\bar{K},\bar{K},e_3)
\end{aligned}$$

$$+\int_{\underline{C}(\nu)\cap V(\lambda,\nu)} Q(\hat{\mathcal{L}}_O^2\mathbf{R})(\bar{K},\bar{K},T,e_3)$$
$$+\int_{\underline{C}(\nu)\cap V(\lambda,\nu)} Q(\hat{\mathcal{L}}_S\hat{\mathcal{L}}_T\mathbf{R})(\bar{K},\bar{K},\bar{K},e_3),$$

where T, S, O, $\bar{K}$ are the vector fields defined in Chapter 3; see (3.4.1).

Theorem M7 *Assume that, relative to a double null foliation on $\mathcal{K}$*

$$\mathcal{O}_{[2]}+\underline{\mathcal{O}}_{[2]}\le\epsilon_0\,.$$

Then, if ϵ_0 is sufficiently small, we have, with c a positive constant,

$$\mathcal{R}\le c\mathcal{Q}_{\mathcal{K}}^{\frac{1}{2}}. \tag{5.0.1}$$

Proof: Let us introduce some definitions that we will use in the sequel,

$$\begin{aligned}
&\mathcal{Q}_{1,2}\equiv\sup_{\{\lambda,\nu|S(\lambda,\nu)\subset\mathcal{K}\}}\mathcal{Q}_{1,2}(\lambda,\nu)\,,\ \underline{\mathcal{Q}}_{1,2}\equiv\sup_{\{\lambda,\nu|S(\lambda,\nu)\subset\mathcal{K}\}}\underline{\mathcal{Q}}_{1,2}(\lambda,\nu)\\
&[\mathcal{Q}]_{1,2}(\lambda,\nu)\equiv\mathcal{Q}_{1,2}(\lambda,\nu)+\underline{\mathcal{Q}}_{1,2}(\lambda,\nu)\,,\ [\mathcal{Q}]_{1,2}\equiv\sup_{\{\lambda,\nu|S(\lambda,\nu)\subset\mathcal{K}\}}[\mathcal{Q}]_{1,2}(\lambda,\nu)\\
&\mathcal{Q}_{1,2\,\Sigma_0\cap\mathcal{K}}\equiv\sup_{\{\lambda,\nu|S(\lambda,\nu)\subset\mathcal{K}\}}\mathcal{Q}_{1,2\,\Sigma_0\cap V(\lambda,\nu)}.
\end{aligned} \tag{5.0.2}$$

Then

$$\mathcal{Q}_{\mathcal{K}}=[\mathcal{Q}]_1+[\mathcal{Q}]_2\ ,\quad \mathcal{Q}_{\Sigma_0\cap\mathcal{K}}=\mathcal{Q}_{1\,\Sigma_0\cap\mathcal{K}}+\mathcal{Q}_{2\,\Sigma_0\cap\mathcal{K}}$$

where $\mathcal{Q}_{1,2}(\lambda,\nu)$, $\underline{\mathcal{Q}}_{1,2}(\lambda,\nu)$, $\mathcal{Q}_{1,2\,\Sigma_0\cap V(\lambda,\nu)}$, $\mathcal{Q}_{\mathcal{K}}$ and $\mathcal{Q}_{\Sigma_0\cap\mathcal{K}}$ were defined in Chapter 3, equations (3.5.7)–(3.5.12). According to the statement of Theorem **M7**, we make use of the assumption $\mathcal{O}_{[2]}+\underline{\mathcal{O}}_{[2]}\le\epsilon_0$ and prove that

$$\begin{aligned}
&\mathcal{R}_{[1]}\le c[\mathcal{Q}]_1+c\epsilon_0\left[(\mathcal{R}_{[0]}+\underline{\mathcal{R}}_{[0]})+\mathcal{Q}_{1\ \Sigma_0\cap\mathcal{K}}^{\frac{1}{2}}\right]\\
&\underline{\mathcal{R}}_{[1]}\le c[\mathcal{Q}]_1+c\epsilon_0\left[(\mathcal{R}_{[0]}+\underline{\mathcal{R}}_{[0]})+\mathcal{Q}_{1\ \Sigma_0\cap\mathcal{K}}^{\frac{1}{2}}\right]\\
&\mathcal{R}_2\le c\,([\mathcal{Q}]_1+[\mathcal{Q}]_2)+c\epsilon_0\left[(\mathcal{R}_{[1]}+\underline{\mathcal{R}}_{[1]})+\mathcal{Q}_{1\ \Sigma_0\cap\mathcal{K}}^{\frac{1}{2}}\right]\\
&\underline{\mathcal{R}}_2\le c\,([\mathcal{Q}]_1+[\mathcal{Q}]_2)+c\epsilon_0\left[(\mathcal{R}_{[1]}+\underline{\mathcal{R}}_{[1]})+\mathcal{Q}_{1\ \Sigma_0\cap\mathcal{K}}^{\frac{1}{2}}\right].
\end{aligned} \tag{5.0.3}$$

From these estimates we conclude that, for sufficiently small ϵ_0, $\mathcal{R}_{[2]}+\underline{\mathcal{R}}_{[2]}$ is bounded by $c\mathcal{Q}_{\mathcal{K}}$.[1] We then use these results, together with the global Sobolev inequalities (see Subsection 4.1.2) to derive the estimates for $\mathcal{R}_0^\infty$, $\underline{\mathcal{R}}_0^\infty$, $\mathcal{R}_1^S$ and $\underline{\mathcal{R}}_1^{p,S}$.

---

[1] We also assume that ϵ_0 is such that $\epsilon_0\mathcal{Q}_{\Sigma_0\cap\mathcal{K}}\le c\mathcal{Q}_{\mathcal{K}}$. Observe also that the estimate $\mathcal{R}\le c(\mathcal{Q}_{\mathcal{K}}^{\frac{1}{2}}+\epsilon_0\mathcal{Q}_{\Sigma_0\cap\mathcal{K}}^{\frac{1}{2}})$ is sufficient for our purposes.

The proof of the first two inequalities in (5.0.3) is the content of Propositions 5.1.2 and 5.1.3 (which also use the results of Proposition 5.1.4). The proof of the last two inequalities which uses the previous ones is the content of Proposition 5.1.5.

Remark: All the main ideas in the proof of Theorem **M7** are already present in the flat case;[2] see Theorem 2.2.2. We view the proof of the flat case as a prerequisite to understand Theorem **M7**. The nontrivial character of the background spacetime, controlled by the smallness assumption $\mathcal{O} \leq \epsilon_0$, introduces quadratic or higher order correction terms which have, roughly, the following structure:[3]

1. linear relative to the null components of the curvature tensor;

2. linear relative to the connection coefficients $\text{tr}\chi - \frac{2}{r}$, $\text{tr}\underline{\chi} + \frac{2}{r}$, $\hat{\chi}$, $\hat{\underline{\chi}}$, η, $\underline{\eta}$, ω, $\underline{\omega}$;

3. linear relative to the deformation tensors and the Lie coefficients of the vector fields ${}^{(i)}O, S, T$.[4]

These correction terms have two different sources:

1. The corrections to the null Bianchi equations, (3.2.8), due to the nonflat character of the spacetime, which is reflected in the presence of the terms $\text{tr}\chi - \frac{2}{r}$, $\text{tr}\underline{\chi} + \frac{2}{r}$, $\hat{\chi}$, $\hat{\underline{\chi}}$, η, $\underline{\eta}$, ω, $\underline{\omega}$.

2. The terms generated by commuting $\hat{\mathcal{L}}_{{}^{(i)}O}$, $\hat{\mathcal{L}}_S$, $\hat{\mathcal{L}}_T$ with the null decomposition of the Riemann curvature tensor.

It is because of this structure that these corrections contribute to the terms of the form $\epsilon_0\mathcal{R}$ on the right-hand side of inequalities (5.0.3). These correction terms, unlike the error terms discussed in the next chapter, are very easy to treat. We shall show in detail how to handle them in some examples and later simply ignore them.

5.1 Preliminary tools

We collect in this section a large number of definitions and a family of propositions and lemmas, without proofs, which will be used in the rest of the chapter.

Definition 5.1.1 *Let X be a vector field in the family $\{T, S, {}^{(i)}O\}$. The Lie coefficients of X: ${}^{(X)}P$, ${}^{(X)}\underline{P}$, ${}^{(X)}Q$, ${}^{(X)}\underline{Q}$, ${}^{(X)}M$, ${}^{(X)}\underline{M}$, ${}^{(X)}N$, ${}^{(X)}\underline{N}$ are defined through the following commutation relations (see [Ch-Kl], Proposition 7.3.1):*[5]

$$\begin{aligned}
[X, e_3] &= {}^{(X)}\underline{P}_b e_b + {}^{(X)}\underline{M} e_3 + {}^{(X)}\underline{N} e_4 \\
[X, e_4] &= {}^{(X)}P_b e_b + {}^{(X)}N e_3 + {}^{(X)}M e_4 \\
[X, e_a] &= \Pi[X, e_a] + \frac{1}{2}{}^{(X)}Q_a e_3 + {}^{(X)}\underline{Q}_a e_4.
\end{aligned} \tag{5.1.1}$$

[2] The only exceptions are the estimates for $\overline{\rho}$ and $\overline{\sigma}$ which are trivial in the flat case.

[3] These estimates are the generalizations of those in Proposition 2.2.4.

[4] See Definition 5.1.1.

[5] $\Pi[X, e_a]$ is the projection on $TS(\lambda, \nu)$ of the vector field $[X, e_a]$.

Their explicit expressions are, see also [Ch-Kl], equation (7.3.6b),

$$\begin{aligned}
&\underline{P}_a = \mathbf{g}(\mathbf{D}_X e_3, e_a) - \mathbf{D}_3 X_a\ ,\ \ P_a = \mathbf{g}(\mathbf{D}_X e_4, e_a) - \mathbf{D}_4 X_a\\
&\underline{Q}_a = \mathbf{g}(\mathbf{D}_X e_3, e_a) + \mathbf{D}_a X_3\ ,\ \ Q_a = \mathbf{g}(\mathbf{D}_X e_4, e_a) + \mathbf{D}_a X_4\\
&\underline{M}_a = -\frac{1}{2}\mathbf{g}(\mathbf{D}_X e_3, e_4) + \frac{1}{2}\mathbf{D}_3 X_4\ ,\ \ M_a = \frac{1}{2}\mathbf{g}(\mathbf{D}_X e_4, e_3) + \frac{1}{2}\mathbf{D}_4 X_3\\
&\underline{N}_a = \frac{1}{2}\mathbf{D}_3 X_3\ ,\ \ N_a = \frac{1}{2}\mathbf{D}_4 X_4.
\end{aligned} \tag{5.1.2}$$

The Lie coefficients of the vector field X originate when we commute $\hat{\mathcal{L}}_X$ with the null decomposition of the Riemann curvature tensor. The result of this commutation is expressed in the following proposition, see also [Ch-Kl], Proposition 7.3.1.

Proposition 5.1.1 *Let W be an arbitrary Weyl tensor. Consider the null components $\alpha(W), \ldots, \underline{\alpha}(W)$ as well as the null components $\alpha(\hat{\mathcal{L}}_X W), \ldots, \underline{\alpha}(\hat{\mathcal{L}}_X W)$. Let $\not{\mathcal{L}}_X$ be the projection on $S(\lambda, \nu)$ of the Lie derivative $\mathcal{L}_X$, and let $\hat{\not{\mathcal{L}}}_X\alpha$, $\hat{\not{\mathcal{L}}}_X\underline{\alpha}$ be the traceless parts of the tensors $\hat{\mathcal{L}}\alpha$, $\hat{\mathcal{L}}\underline{\alpha}$. Then the following relations hold:*

$$\begin{aligned}
\alpha(\hat{\mathcal{L}}_X W)_{ab} &= \hat{\not{\mathcal{L}}}_X\alpha(W)_{ab} + \left(-({}^{(X)}M + {}^{(X)}\underline{M}) + \frac{1}{8}\mathrm{tr}{}^{(X)}\pi\right)\alpha(W)_{ab}\\
&- \left({}^{(X)}P_a + {}^{(X)}Q_a\right)\beta(W)_b - \left({}^{(X)}P_b + {}^{(X)}Q_b\right)\beta(W)_a\\
&+ \delta_{ab}({}^{(X)}P + {}^{(X)}Q)\cdot\beta(W)\\
\beta(\hat{\mathcal{L}}_X W)_a &= \not{\mathcal{L}}_X\beta(W)_a - \frac{1}{2}{}^{(X)}\hat{\pi}_{ab}\beta(W)_b + \left(-{}^{(X)}M - \frac{1}{8}\mathrm{tr}{}^{(X)}\pi\right)\beta(W)_a\\
&- \frac{3}{4}\left({}^{(X)}P_a + {}^{(X)}Q_a\right)\rho(W) - \frac{3}{4}\,\epsilon_{ab}\left({}^{(X)}P_b + {}^{(X)}Q_b\right)\sigma(W)\\
&- \frac{1}{4}\left({}^{(X)}\underline{P}_b + {}^{(X)}\underline{Q}_b\right)\alpha(W)_{ab}\\
\rho(\hat{\mathcal{L}}_X W) &= \mathcal{L}_X\rho(W) - \frac{1}{8}\mathrm{tr}{}^{(X)}\pi\rho(W)\\
&- \frac{1}{2}\left({}^{(X)}\underline{P}_a + {}^{(X)}\underline{Q}_a\right)\beta(W)_a + \frac{1}{2}\left({}^{(X)}P_a + {}^{(X)}Q_a\right)\underline{\beta}(W)_a\\
\sigma(\hat{\mathcal{L}}_X W) &= \mathcal{L}_X\sigma(W) - \frac{1}{8}\mathrm{tr}{}^{(X)}\pi\sigma(W)\\
&+ \frac{1}{2}\left({}^{(X)}\underline{P}_a + {}^{(X)}\underline{Q}_a\right){}^*\beta(W)_a + \frac{1}{2}\left({}^{(X)}P_a + {}^{(X)}Q_a\right){}^*\underline{\beta}(W)_a\\
\underline{\beta}(\hat{\mathcal{L}}_X W)_a &= \not{\mathcal{L}}_X\underline{\beta}(W)_a - \frac{1}{2}{}^{(X)}\hat{\pi}_{ab}\underline{\beta}(W)_b + \left(-{}^{(X)}\underline{M} - \frac{1}{8}\mathrm{tr}{}^{(X)}\pi\right)\underline{\beta}(W)_a\\
&+ \frac{3}{4}\left({}^{(X)}\underline{P}_a + {}^{(X)}\underline{Q}_a\right)\rho(W) - \frac{3}{4}\,\epsilon_{ab}\left({}^{(X)}\underline{P}_b + {}^{(X)}\underline{Q}_b\right)\sigma(W)\\
&+ \frac{1}{4}\left({}^{(X)}P_b + {}^{(X)}Q_b\right)\underline{\alpha}(W)_{ab}\\
\underline{\alpha}(\hat{\mathcal{L}}_X W)_{ab} &= \hat{\not{\mathcal{L}}}_X\underline{\alpha}(W)_{ab} + \left(-({}^{(X)}\underline{M} + {}^{(X)}M) + \frac{1}{8}\mathrm{tr}{}^{(X)}\pi\right)\underline{\alpha}(W)_{ab}\\
&+ \left({}^{(X)}\underline{P}_a + {}^{(X)}\underline{Q}_a\right)\underline{\beta}(W)_b + \left({}^{(X)}\underline{P}_b + {}^{(X)}\underline{Q}_b\right)\underline{\beta}(W)_a
\end{aligned}$$

$$- \quad \delta_{ab}({}^{(X)}\underline{P} + {}^{(X)}\underline{Q}) \cdot \underline{\beta}(W). \tag{5.1.3}$$

Proof: We sketch the proof for the first equation of (5.1.3), a detailed discussion is in [Ch-Kl], Proposition 7.3.1. We compute $\alpha(\mathcal{L}_X W)_{ab} = (\mathcal{L}_X W)_{4a4b}$ and obtain

$$\begin{aligned}\mathcal{L}_X\alpha(W)_{ab} &= \alpha(\mathcal{L}_X W)_{ab} + 2^{(X)}M\alpha(W)_{ab} + \left({}^{(X)}P_a + {}^{(X)}Q_a\right)\beta(W)_b \\ &+\left({}^{(X)}P_b + {}^{(X)}Q_b\right)\beta(W)_a - 2\delta_{ab}P \cdot \beta(W) - 2^{(X)}N\rho(W)\delta_{ab}.\end{aligned} \tag{5.1.4}$$

Recalling the definition of $\hat{\mathcal{L}}_X$, (3.2.2), it follows that

$$\alpha(\mathcal{L}_X W)_{ab} = \alpha(\hat{\mathcal{L}}_X W)_{ab} - \frac{3}{8}\mathrm{tr}^{(X)}\pi\alpha(W)_{ab} + \frac{1}{2}{}^{(X)}[W]_{a4b4}, \tag{5.1.5}$$

where (see (3.2.3))

$$\begin{aligned}{}^{(X)}[W]_{a4b4} &= \left({}^{(X)}\hat{\pi}_{ac}\alpha(W)_{cb} + {}^{(X)}\hat{\pi}_{bc}\alpha(W)_{ca}\right) + \mathrm{tr}^{(X)}\pi\alpha(W)_{ab} \\ &- {}^{(X)}\hat{\pi}_{34}\alpha(W)_{ab} + {}^{(X)}\hat{\pi}_{44}\rho(W)\delta_{ab} - 2\delta_{ab}{}^{(X)}\hat{\pi}_{4c}\beta(W)_c.\end{aligned} \tag{5.1.6}$$

Substituting (5.1.6) into (5.1.5) and using the expressions of the Lie coeeficients, (5.1.2), we obtain the result for $\hat{\not{\mathcal{L}}}_X\alpha(W)_{ab} = \mathcal{L}_X\alpha(W)_{ab} - \frac{1}{2}\mathrm{tr}(\mathcal{L}_X\alpha(W))$.

In the course of the various estimates of this chapter we will use systematically the fact that the rotation vector fields, defined in Chapter 3, Section 3.4, satisfy the following lemma; see also [Ch-Kl], Proposition 7.5.3,

Lemma 5.1.1 *The rotation vector fields ${}^{(i)}O$ defined in Chapter 3, Section 3.4 satisfy the following properties:*

Property 1: *Given an S-tangent tensor field f on $\mathcal{M}$ there exists a constant c_0 such that*

$$c_0^{-1}\int_{S(\lambda,\nu)} r^2|\nabla\!\!\!/ f|^2 \leq \int_{S(\lambda,\nu)} |\mathcal{L}_O f|^2 \leq c_0\int_{S(\lambda,\nu)} (|f|^2 + r^2|\nabla\!\!\!/ f|^2), \tag{5.1.7}$$

where $|\mathcal{L}_O f|^2 \equiv \sum_i |\mathcal{L}_{{}^{(i)}O} f|^2$.

Property 2: *The Lie coefficients ${}^{({}^{(i)}O)}P$, ${}^{({}^{(i)}O)}Q$, ${}^{({}^{(i)}O)}\underline{Q}$, ${}^{({}^{(i)}O)}N$ of the rotation vector fields ${}^{(i)}O$ are identically zero in $\mathcal{K}$. If f is a 1-form or a traceless symmetric 2-covariant tensor tangent to the surfaces $S(\lambda,\nu)$ the following inequality holds:*

$$c_0^{-1}\int_{S(\lambda,\nu)} |f|^2 \leq \int_{S(\lambda,\nu)} |\mathcal{L}_O f|^2. \tag{5.1.8}$$

In the L^2 estimates for the first and second derivatives of the Riemann null components we will often use the following relationship:[6]

$$\begin{aligned}\not{\mathcal{L}}_T\alpha_{ab} &= \not{\mathbf{D}}_T\alpha_{ab} + \left(\alpha_{ac}(\chi + \underline{\chi})_{cb} + \alpha_{bc}(\chi + \underline{\chi})_{ca}\right) + (\omega + \underline{\omega})\alpha_{ab} \\ &= \not{\mathbf{D}}_T\alpha_{ab} + \delta_{ab}\alpha \cdot (\hat{\chi} + \hat{\underline{\chi}}) + \left((\mathrm{tr}\chi + \mathrm{tr}\underline{\chi}) + (\omega + \underline{\omega})\right)\alpha_{ab}.\end{aligned} \tag{5.1.9}$$

[6]In the second equality we use the relation $u_{ac}v_{cb} + u_{bc}v_{ca} = \delta_{ab}u \cdot v$ valid for any two traceless symmetric form tangent to $S(\lambda,\nu)$.

In this chapter we give only the main ideas of the various estimates and do not discuss all the technical details.[7] They can be easily recovered using Chapter 7 of [Ch-Kl]. The estimates of the various Lie coefficients of the X vector fields and their derivatives can be obtained from their explicit expressions in terms of the null connection coefficients and the estimates proved in Chapter 4.

5.1.1 L^2 *estimates for the zero derivatives*

We recall the definitions of the $\mathcal{R}$ norms (see (3.5.13)).

$$\mathcal{R}_{[0]}=\mathcal{R}_0\ ,\ \underline{\mathcal{R}}_{[0]}=\underline{\mathcal{R}}_0+\sup_{\mathcal{K}} r^3|\bar{\rho}|$$

with

$$\mathcal{R}_0=\left(\mathcal{R}_0[\alpha]^2+\mathcal{R}_0[\beta]^2+\mathcal{R}_0[(\rho,\sigma)]^2+\mathcal{R}_0[\underline{\beta}]^2\right)^{1/2}$$
$$\underline{\mathcal{R}}_0=\left(\underline{\mathcal{R}}_0[\beta]^2+\underline{\mathcal{R}}_0[(\rho,\sigma)]^2+\underline{\mathcal{R}}_0[\underline{\beta}]^2+\underline{\mathcal{R}}_0[\underline{\alpha}]^2\right)^{1/2}$$

and

$$\mathcal{R}_{0,1,2}[w]\equiv\sup_{\mathcal{K}}\mathcal{R}_{0,1,2}[w](\lambda,\nu)\ ,\ \underline{\mathcal{R}}_{0,1,2}[w]\equiv\sup_{\mathcal{K}}\underline{\mathcal{R}}_{0,1,2}[w](\lambda,\nu).$$

Proposition 5.1.2 *Under the assumptions of Theorem* **M7** *the following inequalities hold:*[8]

$$\mathcal{R}_0[\alpha]^2(\lambda,\nu)\leq c\int_{C(\lambda)\cap V(\lambda,\nu)}Q(\hat{\mathcal{L}}_O W)(\bar{K},\bar{K},T,e_4)+c\epsilon_0^2{\mathcal{R}_{[0]}}^2$$
$$\mathcal{R}_0[\beta]^2(\lambda,\nu)\leq c\int_{C(\lambda)\cap V(\lambda,\nu)}Q(\hat{\mathcal{L}}_O W)(\bar{K},\bar{K},T,e_4)+c\epsilon_0^2{\mathcal{R}_{[0]}}^2$$
$$\mathcal{R}_0[(\rho,\sigma)]^2(\lambda,\nu)\leq c\int_{C(\lambda)\cap V(\lambda,\nu)}Q(\hat{\mathcal{L}}_O W)(\bar{K},\bar{K},T,e_4)+c\epsilon_0^2{\mathcal{R}_{[0]}}^2$$
$$\mathcal{R}_0[\underline{\beta}]^2(\lambda,\nu)\leq c\int_{C(\lambda)\cap V(\lambda,\nu)}Q(\hat{\mathcal{L}}_O W)(\bar{K},\bar{K},T,e_4)+c\epsilon_0^2{\mathcal{R}_{[0]}}^2 \tag{5.1.10}$$

$$\underline{\mathcal{R}}_0[\beta]^2(\lambda,\nu)\leq c\int_{\underline{C}(\nu)\cap V(\lambda,\nu)}Q(\hat{\mathcal{L}}_O W)(\bar{K},\bar{K},T,e_3)$$
$$+c\epsilon_0\Bigg[\int_{C(\lambda)\cap V(\lambda,\nu)}Q(\hat{\mathcal{L}}_O W)(\bar{K},\bar{K},T,e_4)+\int_{C(\lambda)\cap V(\lambda,\nu)}Q(\hat{\mathcal{L}}_T W)(\bar{K},\bar{K},T,e_4)$$
$$+\mathcal{Q}_{1\Sigma_0\cap V(\lambda,\nu)}+c\epsilon_0^2{\mathcal{R}_{[0]}}^2\Bigg]+c\epsilon_0^2{\underline{\mathcal{R}}_{[0]}}^2$$

---

[7]Some of them are examined in the appendix to this chapter.

[8]In all the estimates of this chapter W is an arbitrary Weyl field. In the proof of the Main Theorem, W is the Riemann curvature tensor of $(\mathcal{M},\mathbf{g})$.

$$\underline{\mathcal{R}}_0[(\rho,\sigma)]^2(\lambda,\nu) \leq c\int_{\underline{C}(\nu)\cap V(\lambda,\nu)} Q(\hat{\mathcal{L}}_O W)(\bar{K},\bar{K},T,e_3) + c\epsilon_0^2 \underline{\mathcal{R}}_{[0]}{}^2$$

$$\underline{\mathcal{R}}_0[\underline{\beta}]^2(\lambda,\nu) \leq c\int_{\underline{C}(\nu)\cap V(\lambda,\nu)} Q(\hat{\mathcal{L}}_O W)(\bar{K},\bar{K},T,e_3) + c\epsilon_0^2 \underline{\mathcal{R}}_{[0]}{}^2$$

$$\underline{\mathcal{R}}_0[\underline{\alpha}]^2(\lambda,\nu) \leq c\int_{\underline{C}(\nu)\cap V(\lambda,\nu)} Q(\hat{\mathcal{L}}_O W)(\bar{K},\bar{K},T,e_3) + c\epsilon_0^2 \underline{\mathcal{R}}_{[0]}{}^2. \qquad (5.1.11)$$

Proof:

1. $\mathcal{R}_0[\alpha](\lambda,\nu) = \|r^2\alpha\|_{2,C(\lambda)\cap V(\lambda,\nu)}$ To control $\|r^2\alpha\|_{2,C(\lambda)\cap V(\lambda,\nu)}$ we first use the inequality (5.1.8), Property 2 of Lemma 5.1.1, to infer

$$\int_{C(\lambda)\cap V(\lambda,\nu)} r^4|\alpha(W)|^2 \leq c\int_{C(\lambda)\cap V(\lambda,\nu)} r^4|\mathcal{L}_O\alpha(W)|^2. \qquad (5.1.12)$$

Then we express $\mathcal{L}_O\alpha(W)$ in terms of $\alpha(\mathcal{L}_O W)$ plus correction terms; see (5.1.4). Finally we write $\alpha(\mathcal{L}_O W)$ in terms of $\alpha(\hat{\mathcal{L}}_O W)$ using the relationship (5.1.5) with $X = O$. From these equations we obtain

$$\begin{aligned}(\mathcal{L}_X\alpha(W))_{ab} &= \alpha(\hat{\mathcal{L}}_X W)_{ab} + \left(\frac{1}{8}\mathrm{tr}^{(X)}\pi + 2^{(X)}M\right)\alpha(W)_{ab} \qquad (5.1.13)\\ &+ \frac{1}{2}\delta_{ab}{}^{(X)}\hat{\pi}\cdot\alpha(W) - \frac{1}{2}{}^{(X)}\hat{\pi}_{34}\alpha(W)_{ab} - \delta_{ab}(P+Q)\cdot\beta(W)\\ &+ \left(({}^{(X)}P_a + {}^{(X)}Q_a)\beta(W)_b + ({}^{(X)}P_b + {}^{(X)}Q_b)\beta(W)_a\right)\end{aligned}$$

In the case $X = O$, we have, recalling equations (5.1.2) and Property 2 of Lemma 5.1.1,[9]

$$\begin{aligned}&{}^{({}^iO)}M = {}^{({}^iO)}\underline{M} = -{}^{(i)}O_b \nabla\!\!\!/_b \log\Omega\\ &{}^{({}^iO)}P_a = {}^{({}^iO)}\underline{Q}_a = {}^{({}^iO)}Q_a = {}^{({}^iO)}\underline{N} = {}^{({}^iO)}N = 0,\end{aligned} \qquad (5.1.14)$$

and the previous expression (5.1.13) can be written as

$$\begin{aligned}(\mathcal{L}_O\alpha(W))_{ab} &= \alpha(\hat{\mathcal{L}}_O W)_{ab} + \left(\frac{1}{8}\mathrm{tr}^{(O)}\pi + 2^{(O)}M\right)\alpha(W)_{ab}\\ &+ \frac{1}{2}\delta_{ab}{}^{(O)}\hat{\pi}\cdot\alpha(W) - \frac{1}{2}{}^{(O)}\hat{\pi}_{34}\alpha(W)_{ab} \qquad (5.1.15)\\ &= \alpha(\hat{\mathcal{L}}_O W)_{ab} - \frac{1}{8}\mathrm{tr}^{(O)}\pi\alpha(W)_{ab} + \frac{1}{2}\delta_{ab}{}^{(O)}\hat{\pi}\cdot\alpha(W).\end{aligned}$$

Using (5.1.15), inequality (5.1.12) can be rewritten as

$$\begin{aligned}\textstyle\int_{C(\lambda)\cap V(\lambda,\nu)} r^4|\alpha(W)|^2 &\leq c\textstyle\int_{C(\lambda)\cap V(\lambda,\nu)} r^4|\alpha(\hat{\mathcal{L}}_O W)|^2 \qquad (5.1.16)\\ &+ \textstyle\int_{C(\lambda)\cap V(\lambda,\nu)} r^4 Qr[({}^{(O)}M, \mathrm{tr}^{(O)}\pi, {}^{(O)}\hat{\pi}); \alpha(W)]^2,\end{aligned}$$

[9] See also the definitions of the deformation tensors in Chapter 3, Subsection 3.4.3.

where $Qr[A, B]$ denotes a quadratic term linear in A and B. From the estimates for the sup norms of ${}^{(O)}M$, $\mathrm{tr}{}^{(O)}\pi$ and ${}^{(O)}\hat{\pi}$ discussed in Chapter 4, Subsection 4.7,[10]

$$\sup_{\mathcal{K}} |r \mathrm{tr}{}^{(O)}\pi| \,,\ \sup_{\mathcal{K}} |r{}^{(O)}\hat{\pi}| \,,\ \sup_{\mathcal{K}} |r{}^{(O)}M| \leq c\epsilon_0 \,,$$

we obtain

$$\int_{C(\lambda)\cap V(\lambda,\nu)} r^4 Qr[({}^{(O)}M, \mathrm{tr}{}^{(O)}\pi, {}^{(O)}\hat{\pi}); \alpha(W)]^2 \leq c\frac{\epsilon_0^2}{r(\lambda,\nu)^2}\int_{C(\lambda)\cap V(\lambda,\nu)} r^4|\alpha(W)|^2. \tag{5.1.17}$$

Substituting into (5.1.16) gives

$$\int_{C(\lambda)\cap V(\lambda,\nu)} r^4|\alpha(W)|^2 \leq c\int_{C(\lambda)\cap V(\lambda,\nu)} r^4|\alpha(\hat{\mathcal{L}}_O W)|^2 + [Correction]^2,$$

where the $[Correction]$ term satisfies the inequality

$$[Correction] \leq c\epsilon_0 \left(\int_{C(\lambda)\cap V(\lambda,\nu)} r^4|\alpha(W)|^2\right)^{\frac{1}{2}} \leq c\epsilon_0 \mathcal{R}_{[0]}. \tag{5.1.18}$$

Collecting these results we obtain

$$\int_{C(\lambda)\cap V(\lambda,\nu)} r^4|\hat{\mathcal{L}}_O\alpha(W)|^2 \leq c\int_{C(\lambda)\cap V(\lambda,\nu)} r^4|\alpha(\hat{\mathcal{L}}_O W)|^2 + c\epsilon_0^2 \mathcal{R}_{[0]}{}^2$$

where the first term on the right-hand side is controlled by the $\mathcal{Q}$ integral $\int_{C(\lambda)\cap V(\lambda,\nu)} \mathcal{Q}(\hat{\mathcal{L}}_O W)(\bar{K}, \bar{K}, T, e_4)$.

2. $\mathcal{R}_0[\beta](\lambda,\nu) = \|r^2\beta\|_{2,C(\lambda)\cap V(\lambda,\nu)}$

We use Lemma 5.1.1 to infer

$$\int_{C(\lambda)\cap V(\lambda,\nu)} r^4|\beta(W)|^2 \leq c\int_{C(\lambda)\cap V(\lambda,\nu)} r^4|\mathcal{L}_O\beta(W)|^2.$$

Then, using Proposition 5.1.1, we express $\not{\mathcal{L}}_O\beta(W)$ in terms of $\beta(\mathcal{L}_O W)$ plus correction terms and $\beta(\mathcal{L}_O W)$ in terms of $\beta(\hat{\mathcal{L}}_O W)$ and correction terms, obtaining

$$\not{\mathcal{L}}_O\beta(W) = \beta(\hat{\mathcal{L}}_O W) + \frac{1}{2}{}^{(O)}\hat{\pi}\cdot\beta(W) + ({}^{(O)}M + \frac{1}{8}\mathrm{tr}{}^{(O)}\pi)\beta(W) + \frac{1}{4}{}^{(O)}\underline{P}\cdot\alpha(W) \,.$$

Proceeding as in the previous case we prove that the correction terms satisfy the inequality

$$[Correction] \leq c\epsilon_0 \left[\left(\int_{C(\lambda)\cap V(\lambda,\nu)} r^4|\beta(W)|^2\right)^{\frac{1}{2}} + \left(\int_{C(\lambda)\cap V(\lambda,\nu)} r^4|\alpha(W)|^2\right)^{\frac{1}{2}}\right] \leq c\epsilon_0\mathcal{R}_{[0]}$$

so that, finally,

$$\int_{C(\lambda)\cap V(\lambda,\nu)} r^4|\beta(W)|^2 \leq c\int_{C(\lambda)\cap V(\lambda,\nu)} r^4|\beta(\hat{\mathcal{L}}_O W)|^2 + c\epsilon_0^2\mathcal{R}_{[0]}{}^2$$

and the right-hand side integral is bounded by $\int_{C(\lambda)\cap V(\lambda,\nu)} \mathcal{Q}(\hat{\mathcal{L}}_O W)(\bar{K}, \bar{K}, T, e_4)$.

---

[10] The estimate for ${}^{(O)}M$ can be easily obtained from the results of Subsection 4.7.

3. $\mathcal{R}_0[(\rho,\sigma)](\lambda,\nu) = \|\tau_- r(\rho-\overline{\rho},\sigma-\overline{\sigma})\|_{2,C(\lambda)\cap V(\lambda,\nu)}$

The control of $\rho-\overline{\rho}$ and of $\sigma-\overline{\sigma}$ is obtained in a similar way. From the Poincaré inequality,

$$\int_S (\Phi-\bar{\Phi})^2 \leq c\int_S |r\nabla\!\!\!/\Phi|^2 ,$$

the following inequality holds:

$$\int_{C(\lambda)\cap V(\lambda,\nu)} \tau_-^2 r^2|\rho-\overline{\rho}|^2 \leq c\int_{C(\lambda)\cap V(\lambda,\nu)} \tau_-^2 r^2|r\nabla\!\!\!/\rho|^2. \tag{5.1.19}$$

We estimate $|r\nabla\!\!\!/\rho(W)|$ in terms of $|\mathcal{L}_O\rho(W)|$ using Lemma 5.1.1, and repeating the same steps as for $\beta(W)$, we obtain

$$\mathcal{L}_O\rho(W) = \rho(\hat{\mathcal{L}}_O W) + \frac{1}{8}\mathrm{tr}^{(O)}\pi\,\rho(W) + \frac{1}{2}{}^{(O)}\underline{P}\cdot\beta(W)\,.$$

Therefore

$$\int_{C(\lambda)\cap V(\lambda,\nu)} \tau_-^2 r^2|(\rho-\overline{\rho})(W)|^2 \leq c\int_{C(\lambda)\cap V(\lambda,\nu)} \tau_-^2 r^2|\rho(\hat{\mathcal{L}}_O W)|^2 + [Correction]^2$$

and

$$\begin{aligned}[Correction] \;&\leq\; c\Bigg[\left(\int_{C(\lambda)\cap V(\lambda,\nu)} \tau_-^2 r^2|\mathrm{tr}^{(O)}\pi|^2|\rho(W)|^2\right)^{\frac{1}{2}} \\ &+ \left(\int_{C(\lambda)\cap V(\lambda,\nu)} \tau_-^2 r^2|{}^{(O)}\underline{P}|^2|\beta(W)|^2\right)^{\frac{1}{2}}\Bigg] \leq c\epsilon_0\mathcal{R}_{[0]}.\end{aligned} \tag{5.1.20}$$

So in conclusion,

$$\int_{C(\lambda)\cap V(\lambda,\nu)} \tau_-^2 r^2|(\rho-\overline{\rho})(W)|^2 \leq c\int_{C(\lambda)\cap V(\lambda,\nu)} \tau_-^2 r^2|\rho(\hat{\mathcal{L}}_O W)|^2 + c\epsilon_0^2{\mathcal{R}_{[0]}}^2,$$

where the right-hand side integral is bounded by $\int_{C(\lambda)\cap V(\lambda,\nu)} Q(\hat{\mathcal{L}}_O W)(\bar{K},\bar{K},T,e_4)$.

4. $\mathcal{R}_0[\underline{\beta}](\lambda,\nu) = \|\tau_-^2\underline{\beta}\|_{2,C(\lambda)\cap V(\lambda,\nu)}$

Using Lemma 5.1.1 again and controlling the correction terms exactly as before, it follows that

$$\int_{C(\lambda)\cap V(\lambda,\nu)} \tau_-^4|\underline{\beta}(W)|^2 \leq c\int_{C(\lambda)\cap V(\lambda,\nu)} \tau_-^4|\underline{\beta}(\hat{\mathcal{L}}_O W)|^2 + c\epsilon_0^2{\mathcal{R}_{[0]}}^2,$$

where the integral on the right-hand side is controlled by the Q integral $\int_{C(\lambda)\cap V(\lambda,\nu)} Q(\hat{\mathcal{L}}_O W)(\bar{K},\bar{K},T,e_4)$.

5. $\underline{\mathcal{R}}_0[\beta](\lambda,\nu)=\|r^2\beta\|_{2,\underline{C}(\nu)\cap V(\lambda,\nu)}$

Using Property 2 of Lemma 5.1.1, it follows that

$$\int_{\underline{C}(\nu)\cap V(\lambda,\nu)} r^4|\beta(W)|^2 \le c\int_{\underline{C}(\nu)\cap V(\lambda,\nu)} r^4|\beta(\mathcal{L}_O W)|^2.$$

Applying Proposition 5.1.1, we express $\not{\mathcal{L}}_O\beta(W)$ in terms of $\beta(\mathcal{L}_O W)$ plus correction terms and $\beta(\mathcal{L}_O W)$ in terms of $\beta(\hat{\mathcal{L}}_O W)$ and correction terms to obtain

$$\not{\mathcal{L}}_O\beta(W)=\beta(\hat{\mathcal{L}}_O W)+\frac{1}{2}{}^{(O)}\hat{\pi}\cdot\beta(W)+({}^{(O)}M+\frac{1}{8}\mathrm{tr}{}^{(O)}\pi)\beta(W)+\frac{1}{4}{}^{(O)}\underline{P}\cdot\alpha(W),$$

so that, finally,

$$\int_{\underline{C}(\nu)\cap V(\lambda,\nu)} r^4|\beta(W)|^2 \le c\int_{\underline{C}(\nu)\cap V(\lambda,\nu)} r^4|\beta(\mathcal{L}_O W)|^2+[Correction]^2.$$

As in the previous case the correction terms satisfy the inequality

$$[Correction]\le c\epsilon_0\left[\left(\int_{\underline{C}(\nu)\cap V(\lambda,\nu)} r^2|\beta(W)|^2\right)^{\frac{1}{2}}+\left(\int_{\underline{C}(\nu)\cap V(\lambda,\nu)} r^2|\alpha(W)|^2\right)^{\frac{1}{2}}\right].$$

Observe that the estimate of the second integral of the previous inequality has to be performed differently. Its bound is provided in Proposition 5.1.4, and from this result, we obtain, with $\epsilon<1$,

$$\int_{\underline{C}(\nu)\cap V(\lambda,\nu)} r^2|\alpha(W)|^2\le\frac{1}{r^{1+\epsilon}}\left[\int_{C(\lambda)\cap V(\lambda,\nu)} Q(\hat{\mathcal{L}}_O W)(\bar{K},\bar{K},T,e_4)\right.$$
$$\left.+\int_{C(\lambda)\cap V(\lambda,\nu)} Q(\hat{\mathcal{L}}_T W)(\bar{K},\bar{K},T,e_4)+\mathcal{Q}_{1\,\Sigma_0\cap V(\lambda,\nu)}+c\epsilon_0^2\mathcal{R}_{[0]}{}^2\right].$$

Finally, we have

$$\begin{aligned}\int_{\underline{C}(\nu)\cap V(\lambda,\nu)} r^4|\beta(W)|^2 &\le c\int_{\underline{C}(\nu)\cap V(\lambda,\nu)} Q(\hat{\mathcal{L}}_O W)(\bar{K},\bar{K},T,e_3)+c\epsilon_0^2\underline{\mathcal{R}}_{[0]}{}^2\\ &+c\epsilon_0^2\Big[\int_{C(\lambda)\cap V(\lambda,\nu)} Q(\hat{\mathcal{L}}_O W)(\bar{K},\bar{K},T,e_4)+\int_{C(\lambda)\cap V(\lambda,\nu)} Q(\hat{\mathcal{L}}_T W)(\bar{K},\bar{K},T,e_4)\\ &+\mathcal{Q}_{1\,\Sigma_0\cap V(\lambda,\nu)}+c\epsilon_0^2\mathcal{R}_{[0]}{}^2\Big].\end{aligned}$$

6. $\underline{\mathcal{R}}_0[(\rho,\sigma)](\lambda,\nu)=\|r^2(\rho-\overline{\rho},\sigma-\overline{\sigma})\|_{2,\underline{C}(\nu)\cap V(\lambda,\nu)}$

The estimate proceeds exactly as that for $\mathcal{R}_0[(\rho,\sigma)](\lambda,\nu)$ but with a better weight due to the expression of $Q(\hat{\mathcal{L}}_O W)(\bar{K},\bar{K},T,e_3)$; see (3.5.1). The $[Correction]$ term is controlled as in the previous cases. The final result for $\rho-\overline{\rho}$ is

$$\int_{\underline{C}(\nu)\cap V(\lambda,\nu)} r^4|(\rho-\overline{\rho})(W)|^2\le c\int_{\underline{C}(\nu)\cap V(u,\underline{u})} r^4|\rho(\hat{\mathcal{L}}_O W)|^2+c\epsilon_0^2\underline{\mathcal{R}}_{[0]}{}^2,$$

where the integral on the right-hand side is controlled by the $\mathcal{Q}$ integral $\int_{\underline{C}(\nu)\cap V(\lambda,\nu)} Q(\hat{\mathcal{L}}_O W)(\bar{K},\bar{K},T,e_3)$. An analogous result is obtained for $\sigma-\overline{\sigma}$.

7. $\underline{\mathcal{R}}_0[\underline{\beta}](\lambda,\nu) = \|\tau_- r \underline{\beta}\|_{2,\underline{C}(\nu)\cap V(\lambda,\nu)}$

Proceeding as in the estimate of $\mathcal{R}_0[\underline{\beta}](\lambda,\nu)$ we obtain

$$\int_{\underline{C}(\nu)\cap V(\lambda,\nu)} \tau_-^2 r^2 |\underline{\beta}(W)|^2 \leq c \int_{\underline{C}(\nu)\cap V(\lambda,\nu)} \tau_-^2 r^2 |\underline{\beta}(\hat{\mathcal{L}}_O W)|^2 + c\epsilon_0^2 \underline{\mathcal{R}}_{[0]}{}^2, \tag{5.1.21}$$

where the the integral on the right-hand side is controlled by the $\mathcal{Q}$ integral $\int_{\underline{C}(\nu)\cap V(\lambda,\nu)} Q(\hat{\mathcal{L}}_O W)(\bar{K},\bar{K},T,e_3)$.

8. $\underline{\mathcal{R}}_0[\underline{\alpha}](\lambda,\nu) = \|\tau_-^2 \underline{\alpha}\|_{2,\underline{C}(\nu)\cap V(\lambda,\nu)}$

The estimate proceeds as for $\mathcal{R}_0[\alpha](\lambda,\nu)$ but with the underlined and not underlined quantities interchanged and with e_3 replacing e_4; the estimate of the correction terms also proceeds in the same way. The final result is

$$\int_{\underline{C}(\nu)\cap V(\lambda,\nu)} \tau_-^4 |\underline{\alpha}(W)|^2 \leq c \int_{\underline{C}(\nu)\cap V(\lambda,\nu)} \tau_-^4 |\underline{\alpha}(\hat{\mathcal{L}}_O W)|^2 + c\epsilon_0^2 \underline{\mathcal{R}}_{[0]}{}^2,$$

where the right-hand side integral is bounded by $\int_{\underline{C}(\nu)\cap V(\lambda,\nu)} Q(\hat{\mathcal{L}}_O W)(\bar{K},\bar{K},T,e_3)$.

5.1.2 L^2 estimates for the first derivatives

Proposition 5.1.3 *Under the assumptions of Theorem* **M7** *the following inequalities hold:*

$$\begin{aligned}
\mathcal{R}_1[\alpha]^2(\lambda,\nu) &\leq c\left(\int_{C(\lambda)\cap V(\lambda,\nu)} Q(\hat{\mathcal{L}}_O W)(\bar{K},\bar{K},T,e_4)\right.\\
&+ \left.\int_{C(\lambda)\cap V(\lambda,\nu)} Q(\hat{\mathcal{L}}_T W)(\bar{K},\bar{K},\bar{K},e_4)\right) + c\epsilon_0^2 \mathcal{R}_{[0]}{}^2\\
\mathcal{R}_1[\beta]^2(\lambda,\nu) &\leq c\int_{C(\lambda)\cap V(\lambda,\nu)} Q(\hat{\mathcal{L}}_O W)(\bar{K},\bar{K},T,e_4) + c\epsilon_0^2 \mathcal{R}_{[0]}{}^2\\
\mathcal{R}_1[(\rho,\sigma)]^2(\lambda,\nu) &\leq c\left(\int_{C(\lambda)\cap V(\lambda,\nu)} Q(\hat{\mathcal{L}}_O W)(\bar{K},\bar{K},T,e_4)\right.\\
&+ \left.\int_{C(\lambda)\cap V(\lambda,\nu)} Q(\hat{\mathcal{L}}_T W)(\bar{K},\bar{K},\bar{K},e_4)\right) + c\epsilon_0^2 \mathcal{R}_{[0]}{}^2\\
\mathcal{R}_1[\underline{\beta}]^2(\lambda,\nu) &\leq c\int_{C(\lambda)\cap V(\lambda,\nu)} Q(\hat{\mathcal{L}}_O W)(\bar{K},\bar{K},T,e_4) + c\epsilon_0^2 \mathcal{R}_{[0]}{}^2,
\end{aligned}$$

and

$$\begin{aligned}
&\underline{\mathcal{R}}_1[\beta]^2(\lambda,\nu) \leq c\int_{\underline{C}(\nu)\cap V(\lambda,\nu)} Q(\hat{\mathcal{L}}_O W)(\bar{K},\bar{K},T,e_3)\\
&\quad + c\epsilon_0^2\Big[\int_{C(\lambda)\cap V(\lambda,\nu)} Q(\hat{\mathcal{L}}_O W)(\bar{K},\bar{K},T,e_4) + \int_{C(\lambda)\cap V(\lambda,\nu)} Q(\hat{\mathcal{L}}_O W)(\bar{K},\bar{K},T,e_4)\\
&\qquad + \mathcal{Q}_{1\,\Sigma_0\cap V(\lambda,\nu)} + c\epsilon_0^2 \mathcal{R}_{[0]}{}^2\Big] + c\epsilon_0^2 \underline{\mathcal{R}}_{[0]}{}^2\\
&\underline{\mathcal{R}}_1[(\rho,\sigma)]^2(\lambda,\nu) \leq c\int_{\underline{C}(\nu)\cap V(\lambda,\nu)} Q(\hat{\mathcal{L}}_O W)(\bar{K},\bar{K},T,e_3)
\end{aligned} \tag{5.1.22}$$

$$\underline{\mathcal{R}}_1[\underline{\beta}]^2(\lambda,\nu) \le c\int_{\underline{C}(\nu)\cap V(\lambda,\nu)} Q(\hat{\mathcal{L}}_O W)(\bar{K},\bar{K},T,e_3)$$

$$\underline{\mathcal{R}}_1[\underline{\alpha}]^2(\lambda,\nu) \le c\Big(\int_{\underline{C}(\nu)\cap V(\lambda,\nu)} Q(\hat{\mathcal{L}}_O W)(\bar{K},\bar{K},T,e_3) + \int_{\underline{C}(\nu)\cap V(\lambda,\nu)} Q(\hat{\mathcal{L}}_T W)(\bar{K},\bar{K},\bar{K},e_3)\Big).$$

Proof:

1. $\mathcal{R}_1[\alpha](\lambda,\nu) = \|r^3\nabla\!\!\!/\alpha\|_{2,C(\lambda)\cap V(\lambda,\nu)} + \|r^3\alpha_3\|_{2,C(\lambda)\cap V(\lambda,\nu)} + \|r^3\alpha_4\|_{2,C(\lambda)\cap V(\lambda,\nu)}$

As in Subsection 5.1.1, the various terms of $\mathcal{R}_1[\alpha](\lambda,\nu)$ are controlled in the following way:

(a) Using Lemma 5.1.1 $\|r^3\nabla\!\!\!/\alpha\|^2_{2,C(\lambda)\cap V(\lambda,\nu)}$ is bounded by

$$\int_{C(\lambda)\cap V(\lambda,\nu)} r^6|\nabla\!\!\!/\alpha(W)|^2 \le c\int_{C(\lambda)\cap V(\lambda,\nu)} r^4|\mathcal{L}_O\alpha(W)|^2.$$

Then, as in the previous estimate of $\|r^2\alpha\|^2_{2,C(\lambda)\cap V(\lambda,\nu)}$, we have

$$\begin{aligned}\int_{C(\lambda)\cap V(\lambda,\nu)} r^4|\mathcal{L}_O\alpha(W)|^2 &\le \int_{C(\lambda)\cap V(\lambda,\nu)} r^4|\alpha(\hat{\mathcal{L}}_O W)|^2 + [Correction]\\ &\le \int_{C(\lambda)\cap V(\lambda,\nu)} r^4|\alpha(\hat{\mathcal{L}}_O W)|^2 + c\epsilon_0^2\mathcal{R}_{[0]}{}^2\end{aligned}$$

where the right-hand side integral is bounded by $\int_{C(\lambda)\cap V(\lambda,\nu)} Q(\hat{\mathcal{L}}_O W)(\bar{K},\bar{K},T,e_4)$.

(b) $\|r^3\alpha_3\|^2_{2,C(\lambda)\cap V(\lambda,\nu)}$ is controlled using the Bianchi equation (see (3.2.8))

$$\alpha_3 \equiv \mathbf{D}\!\!\!\!/_3\alpha + \frac{1}{2}\mathrm{tr}\underline{\chi}\alpha = \nabla\!\!\!/\widehat{\otimes}\beta + \Big[4\underline{\omega}\alpha - 3(\hat{\chi}\rho + {}^*\hat{\chi}\sigma) + (\zeta+4\eta)\widehat{\otimes}\beta\Big]. \qquad (5.1.23)$$

Neglecting the terms in square brackets we are left to estimate $\int_{C(\lambda)\cap V(\lambda,\nu)} r^6|\nabla\!\!\!/\beta(W)|^2$. This term, using Lemma 5.1.1, is bounded by

$$\begin{aligned}\int_{C(\lambda)\cap V(\lambda,\nu)} r^6|\nabla\!\!\!/\beta(W)|^2 &\le c\int_{C(\lambda)\cap V(\lambda,\nu)} r^4|\mathcal{L}_O\beta(W)|^2\\ &\le c\int_{C(\lambda)\cap V(\lambda,\nu)} r^4|\beta(\hat{\mathcal{L}}_O W)|^2 + c\epsilon_0^2\mathcal{R}_{[0]}{}^2\end{aligned}$$

where the second inequality has already been obtained in the estimate of $\mathcal{R}_0[\beta](\lambda,\nu)$. To control the terms in square brackets in the Bianchi null equation (5.1.23), we observe that, since $\mathcal{O} \le \epsilon_0$ on the Ricci coefficients (see Theorem **M7**), these terms are small. Moreover, since they are quadratic, linear in the connection coefficients and linear in the null Riemann components, they have the same structure as the correction terms discussed in Subsection 5.1.1.[11] Therefore we write

$$\begin{aligned}\int_{C(\lambda)\cap V(\lambda,\nu)} r^6|\alpha_3|^2 &\le c\int_{C(\lambda)\cap V(\lambda,\nu)} r^6|\nabla\!\!\!/\beta(W)|^2 + [Correction]^2\\ &\le c\int_{C(\lambda)\cap V(\lambda,\nu)} r^6|\nabla\!\!\!/\beta(W)|^2 + c\epsilon_0^2\mathcal{R}_{[0]}{}^2,\end{aligned}$$

[11] Moreover they have better asymptotic behavior.

where the right-hand side integral is controlled by $\int_{C(\lambda)\cap V(\lambda,\nu)} Q(\hat{\mathcal{L}}_O W)(\bar{K},\bar{K},T,e_4)$.

(c) $\|r^3\alpha_4\|^2_{2,C(\lambda)\cap V(\lambda,\nu)}$ is estimated differently because the Bianchi equations do not provide an evolution equation for α along the null outgoing hypersurfaces. We write, therefore, $\not{\mathbf{D}}_4\alpha = -\not{\mathbf{D}}_3\alpha + 2\not{\mathbf{D}}_T\alpha$ and use the Bianchi equation along the $C(\lambda)$ null hypersurfaces, as we did before, to control the term $\|r^3\alpha_3\|_{2,C(\lambda)\cap V(\lambda,\nu)}$.

To control $\|r^3\not{\mathbf{D}}_T\alpha(W)\|_{2,C(\lambda)\cap V(\lambda,\nu)}$, we express first $\not{\mathbf{D}}_T\alpha(W)$ in terms of $\mathcal{L}_T\alpha(W)$ plus corrections using the relation (5.1.9),

$$\not{\mathbf{D}}_T\alpha(W) = \mathcal{L}_T\alpha(W) + {}^{(T)}H\cdot\alpha(W), \tag{5.1.24}$$

where

$$\left({}^{(T)}H\cdot\alpha(W)\right)_{ab} = \delta_{ab}\alpha\cdot(\hat{\chi}+\underline{\hat{\chi}}) + \Big((\mathrm{tr}\chi+\mathrm{tr}\underline{\chi}) + (\omega+\underline{\omega})\Big)\alpha_{ab}.$$

Moreover, using the estimates of Subsection 6.1.1, we find that ${}^{(T)}H$ satisfies

$$\sup|r\,{}^{(T)}H| \le c\epsilon_0\,.$$

We apply Proposition 5.1.1 (see (5.1.13)) with $X = T$, and using the estimates of Subsection 6.1.1 we obtain

$$\sup\left\{|r\,{}^{(T)}P|, |r\,{}^{(T)}Q|, |r\,{}^{(T)}M|, |r\,{}^{(T)}\pi|\right\} \le c\epsilon_0. \tag{5.1.25}$$

We arrive then at the inequality

$$\int_{C(\lambda)\cap V(\lambda,\nu)} r^6|\not{\mathbf{D}}_T\alpha(W)|^2 \le \int_{C(\lambda)\cap V(\lambda,\nu)} r^6|\alpha(\hat{\mathcal{L}}_T W)|^2 + [Correction]^2,$$

where the first integral on the right-hand side is bounded by the Q integral $\int_{C(\lambda)\cap V(\lambda,\nu)} Q(\hat{\mathcal{L}}_T W)(\bar{K},\bar{K},\bar{K},e_4)$ and the $[Correction]$ term has the same structure as discussed before and, therefore, satisfies

$$[Correction] \le c\epsilon_0\mathcal{R}_{[0]}\,.$$

All these estimates imply

$$\begin{aligned}\mathcal{R}_1[\alpha]^2(\lambda,\nu) \;\le\; & c\Bigg(\int_{C(\lambda)\cap V(\lambda,\nu)} Q(\hat{\mathcal{L}}_O W)(\bar{K},\bar{K},T,e_4) \\ & + \int_{C(\lambda)\cap V(\lambda,\nu)} Q(\hat{\mathcal{L}}_T W)(\bar{K},\bar{K},\bar{K},e_4)\Bigg) + c\epsilon_0^2{\mathcal{R}_{[0]}}^2.\end{aligned} \tag{5.1.26}$$

2. $\mathcal{R}_1[\beta](\lambda,\nu) = \|r^3\not{\nabla}\beta\|_{2,C(\lambda)\cap V(\lambda,\nu)} + \|\tau_- r^2\beta_3\|_{2,C(\lambda)\cap V(\lambda,\nu)} + \|r^3\beta_4\|_{2,C(\lambda)\cap V(\lambda,\nu)}$

The control of these norms proceeds as in the previous case. The main difference is that for β we have the evolution equations along both the forward and backward null directions (see (3.2.8))

$$\beta_3 \equiv \not{\mathbf{D}}_3\beta + \mathrm{tr}\underline{\chi}\beta = \not{\nabla}\rho + \left[2\underline{\omega}\beta + {}^\star\not{\nabla}\sigma + 2\hat{\chi}\cdot\underline{\beta} + 3(\eta\rho + {}^\star\eta\sigma)\right]$$

$$\beta_4 \equiv \not{\mathbf{D}}_4\beta + 2\mathrm{tr}\chi\beta = \not{d}\mathrm{iv}\,\alpha - \left[2\omega\beta - (2\zeta + \underline{\eta})\alpha\right].$$

Using these equations, taking into account the main terms and controlling the corrections in the square brackets as previously discussed, we obtain for $\mathcal{R}_1[\beta](\lambda, \nu)$ the bound[12]

$$\mathcal{R}_1[\beta]^2(\lambda, \nu) \leq c \int_{C(\lambda)\cap V(\lambda,\nu)} Q(\hat{\mathcal{L}}_O W)(\bar{K}, \bar{K}, T, e_4) + c\epsilon_0^2 {\mathcal{R}_{[0]}}^2 \tag{5.1.27}$$

3. $\mathcal{R}_1[(\rho, \sigma)](\lambda, \nu) =$

$$\|\tau_- r^2 \nabla\!\!\!\!/(\rho, \sigma)\|_{2,C(\lambda)\cap V(\lambda,\nu)} + \|r^3(\rho, \sigma)_4\|_{2,C(\lambda)\cap V(\lambda,\nu)} + \|r\tau_-^2(\rho, \sigma)_3\|_{2,C(\lambda)\cap V(\lambda,\nu)}$$

The norm $\|\tau_- r^2 \nabla\!\!\!\!/(\rho, \sigma)\|_{2,C(\lambda)\cap V(\lambda,\nu)}$ was estimated in the estimate of $\mathcal{R}_0[(\rho, \sigma)]$ and it is controlled, apart from corrections, by the $\mathcal{Q}$ integral $\int_{C(\lambda)\cap V(\lambda,\nu)} Q(\hat{\mathcal{L}}_O W)(\bar{K}, \bar{K}, T, e_4)$.

The norm $\|r^3(\rho, \sigma)_4\|_{2,C(\lambda)\cap V(\lambda,\nu)}$ is estimated using the Bianchi equations

$$\rho_4 \equiv \mathbf{D}_4\rho + \frac{3}{2}\mathrm{tr}\chi\rho = \mathrm{d}\!\!\!/\mathrm{iv}\,\beta - \left[\frac{1}{2}\underline{\hat{\chi}} \cdot \alpha - \zeta \cdot \beta - 2\underline{\eta} \cdot \beta\right]$$

$$\sigma_4 \equiv \mathbf{D}_4\sigma + \frac{3}{2}\mathrm{tr}\chi\sigma = -\mathrm{d}\!\!\!/\mathrm{iv}\,{}^\star\beta + \left[\frac{1}{2}\underline{\hat{\chi}} \cdot {}^\star\alpha - \zeta \cdot {}^\star\beta - 2\underline{\eta} \cdot {}^\star\beta\right]$$

and apart from the estimates of the terms in square brackets which produce standard correction terms controlled by $c\epsilon_0^2{\mathcal{R}_{[0]}}^2$, we must control the integral $\int_{C(\lambda)\cap V(\lambda,\nu)} r^6|\nabla\!\!\!\!/\beta|^2$ which, as already discussed in the previous estimate, is bounded by $\int_{C(\lambda)\cap V(\lambda,\nu)} Q(\hat{\mathcal{L}}_O W)(\bar{K}, \bar{K}, T, e_4)$.

The norm $\|r\tau_-^2(\rho, \sigma)_3\|_{2,C(\lambda)\cap V(\lambda,\nu)}$ has to be estimated in a slightly different way.[13] Using the decomposition

$$\rho_3 = 2\mathbf{D}_T\rho - \rho_4 + \frac{3}{2}\left(\mathrm{tr}\chi + \mathrm{tr}\underline{\chi}\right)\rho,$$

we must estimate the norms

$$||r\tau_-^2\mathbf{D}_T\rho||_{2,C(\lambda)\cap V(\lambda,\nu)},\ ||r\tau_-^2\rho_4||_{2,C(\lambda)\cap V(\lambda,\nu)},\ ||r\tau_-^2(\mathrm{tr}\chi + \mathrm{tr}\underline{\chi})\rho||_{2,C(\lambda)\cap V(\lambda,\nu)}\ .$$

The second and the third norms are immediately bounded by

$$\int_{C(\lambda)\cap V(\lambda,\nu)} Q(\hat{\mathcal{L}}_O W)(\bar{K}, \bar{K}, T, e_4) + c\epsilon_0^2{\mathcal{R}_{[0]}}^2\ .$$

This follows from the previous estimate for $\|r^3(\rho, \sigma)_4\|_{2,C(\lambda)\cap V(\lambda,\nu)}$, from the estimates of $\mathcal{R}_0[(\rho, \sigma)](\lambda, \nu)$ and from the fact that, as proved in Corollary 4.3.1, $(\mathrm{tr}\chi + \mathrm{tr}\underline{\chi}) = O(r^{-1}\tau_-^{-1})$.[14]

---

[12]The weight $\tau_- r^2$ for the $L^2(C)$ norm of β_3 is due to the presence of $\nabla\!\!\!\!/\rho$ in the Bianchi equation for β_3.

[13]In fact if we use the Bianchi equations

$$\rho_3 \equiv \mathbf{D}_3\rho + \frac{3}{2}\mathrm{tr}\underline{\chi}\rho = -\mathrm{d}\!\!\!/\mathrm{iv}\,\underline{\beta} - \left[\frac{1}{2}\hat{\chi} \cdot \underline{\alpha} - \zeta \cdot \underline{\beta} + 2\eta \cdot \underline{\beta}\right]$$

$$\sigma_3 \equiv \mathbf{D}_3\sigma + \frac{3}{2}\mathrm{tr}\underline{\chi}\sigma = -\mathrm{d}\!\!\!/\mathrm{iv}\,{}^\star\underline{\beta} + \left[\frac{1}{2}\hat{\chi} \cdot {}^\star\underline{\alpha} - \zeta \cdot {}^\star\underline{\beta} - 2\eta \cdot {}^\star\underline{\beta}\right]$$

the component $\underline{\alpha}$ appears in the bracket terms. Since $\underline{\alpha}$ never appears in the $\mathcal{Q}$ integrals along $C(\lambda)$, this requires one to estimate $\int_{C(\lambda)\cap V(\lambda,\nu)} r^{-2}\tau_-^4|\underline{\alpha}|^2$ differently. This is, nevertheless, possible and is discussed in Proposition 5.1.4.

[14]More precisely we need the norm estimate for $\overline{\rho}$, which is discussed later on in this chapter.

The norm $||r\tau_-^2\mathbf{D}_T\rho||_{2,C(\lambda)\cap V(\lambda,\nu)}$ is estimated similar to the estimate for the norm $\|r^3\mathbf{D}\!\!\!/_T\alpha(W)\|_{2,C(\lambda)\cap V(\lambda,\nu)}$. First we recall that

$$||r\tau_-^2\mathbf{D}_T\rho||_{2,C(\lambda)\cap V(\lambda,\nu)} = ||r\tau_-^2\mathcal{L}_T\rho||_{2,C(\lambda)\cap V(\lambda,\nu)}\ .$$

Then we apply Proposition 5.1.1 for $X = T$ and observe that, using the estimates of Subsection 6.1.1 (see also (5.1.25))

$$\sup\left\{|r^{(T)}P|, |r^{(T)}\underline{P}|, |r^{(T)}Q|, |r^{(T)}\underline{Q}|, |r^{(T)}M|, |r^{(T)}\pi|\right\} \leq c\epsilon_0\ .$$

In conclusion we obtain the inequality

$$\int_{C(\lambda)\cap V(\lambda,\nu)} r^2\tau_-^4|\mathbf{D}\!\!\!/_T\rho(W)|^2 \leq c\int_{C(\lambda)\cap V(\lambda,\nu)} r^2\tau_-^4|\rho(\hat{\mathcal{L}}_T(W))|^2 + [Correction]^2\ ,$$

where the first integral on the right-hand side is bounded by the $\mathcal{Q}$ integral $\int_{C(\lambda)\cap V(\lambda,\nu)} Q(\hat{\mathcal{L}}_T W)(\bar{K}, \bar{K}, \bar{K}, e_4)$ and the correction terms have the same structure as discussed before and satisfy $[Correction] \leq c\epsilon_0\mathcal{R}_{[0]}$. Finally

$$\begin{aligned}\mathcal{R}_1[(\rho,\sigma)]^2(\lambda,\nu) \ &\leq\ c\left(\int_{C(\lambda)\cap V(\lambda,\nu)} Q(\hat{\mathcal{L}}_O W)(\bar{K}, \bar{K}, T, e_4)\right.\\ &+\left.\int_{C(\lambda)\cap V(\lambda,\nu)} Q(\hat{\mathcal{L}}_T W)(\bar{K}, \bar{K}, \bar{K}, e_4)\right) + c\epsilon_0^2{\mathcal{R}_{[0]}}^2.\end{aligned} \tag{5.1.28}$$

4. $\mathcal{R}_1[\underline{\beta}](\lambda,\nu) = \|\tau_-^2 r\nabla\!\!\!\!/\,\underline{\beta}\|_{2,C(\lambda)\cap V(\lambda,\nu)} + \|\tau_- r^2\underline{\beta}_4\|_{2,C(\lambda)\cap V(\lambda,\nu)}$

To control the norm $\|\tau_-^2 r\nabla\!\!\!\!/\,\underline{\beta}\|_{2,C(\lambda)\cap V(\lambda,\nu)}$ we use Lemma 5.1.1 to obtain

$$\int_{C(\lambda)\cap V(\lambda,\nu)} \tau_-^4 r^2|\nabla\!\!\!\!/\,\underline{\beta}(W)|^2 \leq c\int_{C(\lambda)\cap V(\lambda,\nu)} \tau_-^4|\mathcal{L}_O\underline{\beta}(W)|^2.$$

The integral on the right-hand side has been already estimated in the control of $\mathcal{R}_0[\underline{\beta}](\lambda,\nu)$ so that, finally,

$$\int_{C(\lambda)\cap V(\lambda,\nu)} \tau_-^4 r^2|\nabla\!\!\!\!/\,\underline{\beta}(W)|^2 \leq c\int_{C(\lambda)\cap V(u,\underline{u})} Q(\hat{\mathcal{L}}_O W)(\bar{K}, \bar{K}, T, e_4) + c\epsilon_0^2{\mathcal{R}_{[0]}}^2\ .$$

The norm $\|\tau_- r^2\underline{\beta}_4\|_{2,C(\lambda)\cap V(\lambda,\nu)}$ is controlled using the Bianchi equation

$$\underline{\beta}_4 \equiv \mathbf{D}\!\!\!/_4\underline{\beta} + \mathrm{tr}\chi\,\underline{\beta} = -\nabla\!\!\!\!/\,\rho + {}^\star\nabla\!\!\!\!/\,\sigma + \left[2\omega\underline{\beta} + 2\underline{\hat{\chi}}\cdot\beta + -3(\underline{\eta}\rho - {}^\star\underline{\eta}\sigma)\right]\ .$$

Aside from terms in square brackets, this means controlling $\int_{C(\lambda)\cap V(\lambda,\nu)} \tau_-^2 r^4|\nabla\!\!\!\!/(\rho,\sigma)|^2$. These integrals are bounded, as in the estimate of $\mathcal{R}_0[(\rho,\sigma)](\lambda,\underline{\nu})$, by $\int_{C(\lambda)\cap V(\lambda,\nu)} Q(\hat{\mathcal{L}}_O W)(\bar{K}, \bar{K}, T, e_4) + [Correction]^2$.

The terms in square brackets appearing in the Bianchi equation produce, as already discussed in the case of $\|r^3\alpha_3\|^2_{2,C(\lambda)\cap V(\lambda,\nu)}$, terms with the same structure as the $[Correction]$ term[15] and, therefore, can be included in it. In conclusion,

$$\mathcal{R}_1[\underline{\beta}]^2(\lambda,\nu) \leq c\int_{C(\lambda)\cap V(\lambda,\nu)} Q(\hat{\mathcal{L}}_O W)(\bar{K}, \bar{K}, T, e_4) + c\epsilon_0^2{\mathcal{R}_{[0]}}^2. \tag{5.1.29}$$

[15]Moreover they have a better asymptotic behavior.

5. $\underline{\mathcal{R}}_1[\beta](\lambda,\nu) = \|r^3 \nabla\!\!\!/ \beta\|_{2,\underline{C}(\nu)\cap V(\lambda,\nu)} + \|r^3\beta_3\|_{2,\underline{C}(\nu)\cap V(\lambda,\nu)}$

The norm $\|r^3\nabla\!\!\!/\beta\|_{2,\underline{C}(\nu)\cap V(\lambda,\nu)}$ has already been bounded for the estimate of $\underline{\mathcal{R}}_0[\beta](u,\underline{u})$. The result is

$$\begin{aligned}\int_{\underline{C}(\nu)\cap V(\lambda,\nu)} r^6|\nabla\!\!\!/\beta|^2 \;&\leq\; c\int_{\underline{C}(\nu)\cap V(\lambda,\nu)} Q(\hat{\mathcal{L}}_O W)(\bar{K},\bar{K},T,e_3) + c\epsilon_0^2\underline{\mathcal{R}}_{[0]}{}^2\\ &+\; c\epsilon_0\Bigg[\int_{C(\lambda)\cap V(\lambda,\nu)} Q(\hat{\mathcal{L}}_O W)(\bar{K},\bar{K},T,e_4)\\ &+\; \int_{C(\lambda)\cap V(\lambda,\nu)} Q(\hat{\mathcal{L}}_T W)(\bar{K},\bar{K},T,e_4) + c\epsilon_0^2\mathcal{R}_{[0]}{}^2\Bigg].\end{aligned}$$

The norm $\|r^3\beta_3\|_{2,\underline{C}(\nu)\cap V(\lambda,\nu)}$ is controlled using the Bianchi equation

$$\beta_3 \equiv \mathbf{D}\!\!\!\!/_3\beta + \mathrm{tr}\underline{\chi}\,\beta = \nabla\!\!\!/\rho + \left[2\underline{\omega}\beta + {}^\star\nabla\!\!\!/\sigma + 2\hat{\chi}\cdot\underline{\beta} + 3(\eta\rho + {}^\star\eta\sigma)\right].$$

As before, apart from corrections, we have to control the integral $\int_{\underline{C}(\nu)\cap V(\lambda,\nu)} r^6|\nabla\!\!\!/\rho|^2$ which is bounded (see the estimate of $\underline{\mathcal{R}}_0[\rho]$) by $\int_{\underline{C}(\nu)\cap V(\lambda,\nu)} Q(\hat{\mathcal{L}}_O W)(\bar{K},\bar{K},T,e_3)$. Therefore, in conclusion,

$$\begin{aligned}\underline{\mathcal{R}}_1[\beta]^2(\lambda,\nu) \;&\leq\; c\int_{\underline{C}(\nu)\cap V(\lambda,\nu)} Q(\hat{\mathcal{L}}_O W)(\bar{K},\bar{K},T,e_3) + c\epsilon_0^2\underline{\mathcal{R}}_{[0]}{}^2\\ &+\; c\epsilon_0\Bigg[\int_{C(\lambda)\cap V(\lambda,\nu)} Q(\hat{\mathcal{L}}_O W)(\bar{K},\bar{K},T,e_4)\\ &+\; \int_{C(\lambda)\cap V(\lambda,\nu)} Q(\hat{\mathcal{L}}_T W)(\bar{K},\bar{K},T,e_4) + c\epsilon_0^2\mathcal{R}_{[0]}{}^2\Bigg].\end{aligned} \tag{5.1.30}$$

6. $\underline{\mathcal{R}}_1[(\rho,\sigma)](\lambda,\nu) = \|r^3\nabla\!\!\!/(\rho,\sigma)\|_{2,\underline{C}(\nu)\cap V(\lambda,\nu)} + \|\tau_- r^2(\rho,\sigma)_3\|_{2,\underline{C}(\nu)\cap V(\lambda,\nu)}$

The estimate of $\underline{\mathcal{R}}_1[(\rho,\sigma)](\lambda,\nu)$ proceeds exactly as for $\mathcal{R}_1[(\rho,\sigma)](\lambda,\nu)$ with the obvious substitutions. The norm $\|r^3\nabla\!\!\!/(\rho,\sigma)\|_{2,\underline{C}(\nu)\cap V(\lambda,\nu)}$ has already been bounded in the estimate of $\underline{\mathcal{R}}_0[(\rho,\sigma)]$; it is controlled, apart from correction terms, by $\int_{\underline{C}(\nu)\cap V(\lambda,\nu)} Q(\hat{\mathcal{L}}_O W)(\bar{K},\bar{K},T,e_3)$. The norm $\|\tau_- r^2(\rho,\sigma)_3\|_{2,\underline{C}(\nu)\cap V(\lambda,\nu)}$ is estimated using the Bianchi equations

$$\begin{aligned}\rho_3 \equiv \mathbf{D}_3\rho + \frac{3}{2}\mathrm{tr}\underline{\chi}\,\rho &= -\mathrm{d}\!\!\!/\mathrm{iv}\,\underline{\beta} - \left[\frac{1}{2}\hat{\chi}\cdot\underline{\alpha} - \zeta\cdot\underline{\beta} + 2\eta\cdot\underline{\beta}\right]\\ \sigma_3 \equiv \mathbf{D}_3\sigma + \frac{3}{2}\mathrm{tr}\underline{\chi}\,\sigma &= -\mathrm{d}\!\!\!/\mathrm{iv}\,{}^\star\underline{\beta} + \left[\frac{1}{2}\hat{\chi}\cdot{}^\star\underline{\alpha} - \zeta\cdot{}^\star\underline{\beta} - 2\eta\cdot{}^\star\underline{\beta}\right]\end{aligned}$$

and, again, this implies that, apart from the corrections arising from the terms in the square brackets, which we have already discussed, we have to control $\int_{\underline{C}(\nu)\cap V(\lambda,\nu)} \tau_-^2 r^4|\nabla\!\!\!/\underline{\beta}|^2$, which is bounded by $\int_{\underline{C}(\nu)\cap V(\lambda,\nu)} Q(\hat{\mathcal{L}}_O W)(\bar{K},\bar{K},T,e_3)$ plus corrections. Therefore, finally,

$$\underline{\mathcal{R}}_1[(\rho,\sigma)]^2(\lambda,\nu) \leq c\int_{\underline{C}(\nu)\cap V(\lambda,\nu)} Q(\hat{\mathcal{L}}_O W)(\bar{K},\bar{K},T,e_3) + c\epsilon_0^2\underline{\mathcal{R}}_{[0]}{}^2. \tag{5.1.31}$$

7. $\underline{\mathcal{R}}_1[\underline{\beta}](\lambda,\nu)=\|\tau_-r^2\nabla\!\!\!/\underline{\beta}\|_{2,\underline{C}(\nu)\cap V(\lambda,\nu)}+\|\tau_-^2r\underline{\beta}_3\|_{2,\underline{C}(\nu)\cap V(\lambda,\nu)}+\|r^3\underline{\beta}_4\|_{2,\underline{C}(\nu)\cap V(\lambda,\nu)}$

The estimate of the norm $\underline{\mathcal{R}}_1[\underline{\beta}](\lambda,\nu)$ proceeds exactly as for $\mathcal{R}_1[\beta](\lambda,\nu)$ with the obvious changes. The final result is

$$\underline{\mathcal{R}}_1[\underline{\beta}]^2(\lambda,\nu)\le c\int_{\underline{C}(\nu)\cap V(\lambda,\nu)} Q(\hat{\mathcal{L}}_O W)(\bar{K},\bar{K},T,e_3)+c\epsilon_0^2\underline{\mathcal{R}}_{[0]}^{\ 2}. \tag{5.1.32}$$

8. $\underline{\mathcal{R}}_1[\underline{\alpha}](\lambda,\nu)=\|\tau_-^2r\nabla\!\!\!/\underline{\alpha}\|_{2,\underline{C}(\nu)\cap V(\lambda,\nu)}+\|\tau_-^3\underline{\alpha}_3\|_{2,\underline{C}(\nu)\cap V(\lambda,\nu)}+\|\tau_-r^2\underline{\alpha}_4\|_{2,\underline{C}(\nu)\cap V(\lambda,\nu)}$.

The estimate of the norm $\underline{\mathcal{R}}_1[\underline{\alpha}](\lambda,\nu)$ proceeds exactly as for $\mathcal{R}_1[\alpha](\lambda,\nu)$ with the obvious changes. The final result is

$$\begin{aligned}\underline{\mathcal{R}}_1[\underline{\alpha}]^2(\lambda,\nu)\ &\le\ c\left(\int_{\underline{C}(\nu)\cap V(\lambda,\nu)} Q(\hat{\mathcal{L}}_O W)(\bar{K},\bar{K},T,e_3)\right.\\ &\quad\left.+\int_{\underline{C}(\nu)\cap V(\lambda,\nu)} Q(\hat{\mathcal{L}}_T W)(\bar{K},\bar{K},\bar{K},e_3)\right)+c\epsilon_0^2\underline{\mathcal{R}}_{[0]}^{\ 2}.\end{aligned} \tag{5.1.33}$$

Remark: Some observations relative to the previous proposition are now appropriate.

1. The quantity $\underline{\mathcal{R}}_1[\underline{\beta}]$ does not contain an L^2 norm for $\underline{\beta}_3$. Nevertheless, proceeding as in the case of $(\rho,\sigma)_3$, one can easily bound $\|\tau_-^3\underline{\beta}_3\|_{2,C(\lambda)\cap V(\lambda,\nu)}$.[16] A better result can be obtained using the Bianchi equation

$$\underline{\beta}_3\equiv \mathbf{D}\!\!\!\!/_3\underline{\beta}+2\mathrm{tr}\underline{\chi}\ \underline{\beta}=-\mathrm{d}\!\!\!/\mathrm{iv}\underline{\alpha}-\left[2\underline{\omega}\underline{\beta}+(-2\zeta+\eta)\cdot\underline{\alpha}\right].$$

The problem here is the presence of $\mathrm{d}\!\!\!/\mathrm{iv}\underline{\alpha}$ on the right-hand side. Since $\underline{\alpha}$ never appears in the $\mathcal{Q}$ integrals along $C(\lambda)$, we have to estimate integrals like $\int_{C(\lambda)\cap V(\lambda,\nu)} r^\sigma\tau_-^\delta|\nabla\!\!\!/\underline{\alpha}|^2$ differently. The final result, stated in Proposition 5.1.4, is the control of $\|r^\lambda\tau_-^{\frac{5}{2}}\underline{\beta}_3\|_{2,C(\lambda)\cap V(\lambda,\nu)}$, with $\lambda<\frac{1}{2}$.

2. In $\mathcal{R}_1[\beta]$ we would like to control the norm $\|r^3\beta_4\|_{2,\underline{C}(\nu)\cap V(\lambda,\nu)}$. The estimate we obtain is slightly weaker. In fact we control $\|r^\lambda\beta_4\|_{2,\underline{C}(\nu)\cap V(\lambda,\nu)}$ with $\lambda<3$. This result can be obtained either by writing

$$\beta_4=2\mathbf{D}\!\!\!\!/_T\beta-\beta_3+(2\mathrm{tr}\chi+\mathrm{tr}\underline{\chi})\beta,$$

and then estimating the corresponding norms[17] or by using the Bianchi equation[18]

$$\beta_4\equiv\mathbf{D}\!\!\!\!/_4\beta+2\mathrm{tr}\chi\beta=\mathrm{d}\!\!\!/\mathrm{iv}\alpha+\left[-2\omega\beta+(2\zeta+\underline{\eta})\cdot\alpha\right].$$

This result is discussed in Proposition 5.1.4.

---

[16]One obtains a weaker estimate than the one in [Ch-K].

[17]In this case we cannot obtain control for $\lambda=3$ due to the corrections that arise when we express the norm of $\mathbf{D}\!\!\!\!/_T\beta(W)$ in terms of the norm of $\beta(\hat{\mathcal{L}}_T W)$.

[18]In this case we cannot obtain control for $\lambda=3$ due to the presence of the component $\mathrm{d}\!\!\!/\mathrm{iv}\alpha$.

3. Under the assumptions of Theorem **M7** it is possible to control the norm $\|r^{\lambda}(\rho,\sigma)_4\|_{2,\underline{C}(\nu)\cap V(\lambda,\nu)}$ with $\lambda < 3$; see Proposition 5.1.4. To obtain it we must use the Bianchi equations[19]

$$\rho_4 \equiv \mathbf{D}_4\rho + \frac{3}{2}\mathrm{tr}\chi\rho = \not{d}\mathrm{iv}\,\beta - \left[\frac{1}{2}\hat{\underline{\chi}}\cdot\alpha - \zeta\cdot\beta - 2\underline{\eta}\cdot\beta\right]$$

$$\sigma_4 \equiv \mathbf{D}_4\sigma + \frac{3}{2}\mathrm{tr}\chi\sigma = -\not{d}\mathrm{iv}\,{}^{\star}\beta + \left[\frac{1}{2}\hat{\underline{\chi}}\cdot{}^{\star}\alpha - \zeta\cdot{}^{\star}\beta - 2\underline{\eta}\cdot{}^{\star}\beta\right].$$

We cannot obtain control for the value $\lambda = 3$ due to the presence of α on the right- hand side.

5.1.3 Auxiliary L^2 norms for the zero and first derivatives of the Riemann components

To prove Propositions 5.1.2 and 5.1.3 we need the estimates of some other L^2 norms of zero and first derivatives of Riemann components. These estimates are collected in Proposition 5.1.4. The idea is to use the estimates for the $|\cdot|_{p,S}$ norms expressed in terms of appropriate L^2 norms; see Corollary 4.1.1.

Proposition 5.1.4 *Under the assumptions of Theorem* **M7**, *we have the following L^2 norm estimates,*

$$\begin{aligned}
&\|r^{\delta}\tau_-^{\frac{5}{2}}\alpha\|_{2,C(\lambda)\cap V(\lambda,\nu)} \leq \\
&\qquad c\left([\mathcal{Q}]_1^{\frac{1}{2}}(\lambda,\nu) + \mathcal{Q}_1^{\frac{1}{2}}{}_{\Sigma_0\cap V(\lambda,\nu)} + \epsilon_0(\mathcal{R}_{[0]} + \underline{\mathcal{R}}_{[0]})\right)\ \delta < -\frac{1}{2}\\
&\|r^{\delta}\tau_-^{\frac{5}{2}}\not{\nabla}\alpha\|_{2,C(\lambda)\cap V(\lambda,\nu)} \leq \\
&\qquad c\left(\left([\mathcal{Q}]_1^{\frac{1}{2}} + [\mathcal{Q}]_2^{\frac{1}{2}}\right)(\lambda,\nu) + \mathcal{Q}_1^{\frac{1}{2}}{}_{\Sigma_0\cap V(\lambda,\nu)} + \epsilon_0(\mathcal{R}_{[0]} + \underline{\mathcal{R}}_{[0]})\right),\ \delta < \frac{1}{2}\\
&\|r^{\delta}\tau_-^{2}\alpha_4\|_{2,C(\lambda)\cap V(\lambda,\nu)} \leq \\
&\qquad c\left(\left([\mathcal{Q}]_1^{\frac{1}{2}} + [\mathcal{Q}]_2^{\frac{1}{2}}\right)(\lambda,\nu) + \mathcal{Q}_1^{\frac{1}{2}}{}_{\Sigma_0\cap V(\lambda,\nu)} + \epsilon_0(\mathcal{R}_{[0]} + \underline{\mathcal{R}}_{[0]})\right),\ \delta \leq 1\\
&\|r^{\delta}\tau_-^{\frac{7}{2}}\alpha_3\|_{2,C(\lambda)\cap V(\lambda,\nu)} \leq \\
&\qquad c\left(\left([\mathcal{Q}]_1^{\frac{1}{2}} + [\mathcal{Q}]_2^{\frac{1}{2}}\right)(\lambda,\nu) + \mathcal{Q}_1^{\frac{1}{2}}{}_{\Sigma_0\cap V(\lambda,\nu)} + \epsilon_0(\mathcal{R}_{[0]} + \underline{\mathcal{R}}_{[0]})\right),\ \delta < -\frac{1}{2}\\
&\|r^{\delta}\tau_-^{\frac{5}{2}}\beta_3\|_{2,C(\lambda)\cap V(\lambda,\nu)} \leq \qquad\qquad (5.1.34)\\
&\qquad c\left(\left([\mathcal{Q}]_1^{\frac{1}{2}} + [\mathcal{Q}]_2^{\frac{1}{2}}\right)(\lambda,\nu) + \mathcal{Q}_1^{\frac{1}{2}}{}_{\Sigma_0\cap V(\lambda,\nu)} + \epsilon_0(\mathcal{R}_{[0]} + \underline{\mathcal{R}}_{[0]})\right),\ \delta < \frac{1}{2}
\end{aligned}$$

$$\|r^{\delta}\alpha\|_{2,\underline{C}(\nu)\cap V(\lambda,\nu)} \leq$$

[19] In this case we cannot proceed as we did for the norm $\|r^3(\rho,\sigma)_3\|_{2,C(\lambda)\cap V(\lambda,\nu)}$.

$$c\left([\mathcal{Q}]_1^{\frac{1}{2}}(\lambda,\nu)+\mathcal{Q}_1^{\frac{1}{2}}{}_{\Sigma_0\cap V(\lambda,\nu)}+\epsilon_0(\mathcal{R}_{[0]}+\underline{\mathcal{R}}_{[0]})\right)\ ,\ \delta<2$$

$$\|r^\delta \nabla\!\!\!/\alpha\|_{2,\underline{C}(\nu)\cap V(\lambda,\nu)}\leq$$
$$c\left(\left([\mathcal{Q}]_1^{\frac{1}{2}}+[\mathcal{Q}]_2^{\frac{1}{2}}\right)(\lambda,\nu)+\mathcal{Q}_1^{\frac{1}{2}}{}_{\Sigma_0\cap V(\lambda,\nu)}+\epsilon_0(\mathcal{R}_{[0]}+\underline{\mathcal{R}}_{[0]})\right)\ ,\ \delta<3$$

$$\|r^\delta \alpha_3\|_{2,\underline{C}(\nu)\cap V(\lambda,\nu)}\leq$$
$$c\left(\left([\mathcal{Q}]_1^{\frac{1}{2}}+[\mathcal{Q}]_2^{\frac{1}{2}}\right)(\lambda,\nu)+\mathcal{Q}_1^{\frac{1}{2}}{}_{\Sigma_0\cap V(\lambda,\nu)}+\epsilon_0(\mathcal{R}_{[0]}+\underline{\mathcal{R}}_{[0]})\right)\ ,\ \delta<3$$

$$||r^\delta \alpha_4||_{2,\underline{C}(\nu)\cap V(\lambda,\nu)}\leq$$
$$c\left(\left([\mathcal{Q}]_1^{\frac{1}{2}}+[\mathcal{Q}]_2^{\frac{1}{2}}\right)(\lambda,\nu)+\mathcal{Q}_1^{\frac{1}{2}}{}_{\Sigma_0\cap V(\lambda,\nu)}+\epsilon_0(\mathcal{R}_{[0]}+\underline{\mathcal{R}}_{[0]})\right)\ ,\ \delta<3$$

$$\|r^\delta \beta_4\|_{2,\underline{C}(\nu)\cap V(\lambda,\nu)}\leq$$
$$c\left(\left([\mathcal{Q}]_1^{\frac{1}{2}}+[\mathcal{Q}]_2^{\frac{1}{2}}\right)(\lambda,\nu)+\mathcal{Q}_1^{\frac{1}{2}}{}_{\Sigma_0\cap V(\lambda,\nu)}+\epsilon_0(\mathcal{R}_{[0]}+\underline{\mathcal{R}}_{[0]})\right)\ ,\ \delta<3$$

$$\|r^\delta (\rho,\sigma)_4\|_{2,\underline{C}(\nu)\cap V(\lambda,\nu)}\leq \tag{5.1.35}$$
$$c\left(\left([\mathcal{Q}]_1^{\frac{1}{2}}+[\mathcal{Q}]_2^{\frac{1}{2}}\right)(\lambda,\nu)+\mathcal{Q}_1^{\frac{1}{2}}{}_{\Sigma_0\cap V(\lambda,\nu)}+\epsilon_0(\mathcal{R}_{[0]}+\underline{\mathcal{R}}_{[0]})\right)\ ,\ \delta<3.$$

Proof: See the appendix to this chapter.

Control of the L^2 norms of the second derivatives of the Riemann components

Proposition 5.1.5 *Under the assumptions of Theorem* **M7** *the following inequalities hold:*

$$\begin{aligned}
\mathcal{R}_2[\alpha]^2(\lambda,\nu) &\leq c\Bigg(\int_{C(\lambda)\cap V(\lambda,\nu)} Q(\hat{\mathcal{L}}_O^2 W)(\bar{K},\bar{K},T,e_4)\\
&\quad+\int_{C(\lambda)\cap V(\lambda,\nu)} Q(\hat{\mathcal{L}}_O\hat{\mathcal{L}}_T W)(\bar{K},\bar{K},\bar{K},e_4)\\
&\quad+\int_{C(\lambda)\cap V(\lambda,\nu)} Q(\hat{\mathcal{L}}_S\hat{\mathcal{L}}_T W)(\bar{K},\bar{K},\bar{K},e_4)\Bigg)+c(1+\epsilon_0^2)\mathcal{R}_{[1]}{}^2\\
\mathcal{R}_2[\beta]^2(\lambda,\nu) &\leq c\int_{C(\lambda)\cap V(\lambda,\nu)} Q(\hat{\mathcal{L}}_O^2 W)(\bar{K},\bar{K},T,e_4)+c(1+\epsilon_0^2)\mathcal{R}_{[1]}{}^2\\
\mathcal{R}_2[(\rho,\sigma)]^2(\lambda,\nu) &\leq c\Bigg(\int_{C(\lambda)\cap V(\lambda,\nu)} Q(\hat{\mathcal{L}}_O^2 W)(\bar{K},\bar{K},T,e_4)\\
&\quad+\int_{C(\lambda)\cap V(\lambda,\nu)} Q(\hat{\mathcal{L}}_O\hat{\mathcal{L}}_T W)(\bar{K},\bar{K},\bar{K},e_4)\Bigg)+c(1+\epsilon_0^2)\mathcal{R}_{[1]}{}^2\\
\mathcal{R}_2[\underline{\beta}]^2(\lambda,\nu) &\leq \Bigg(\int_{C(\lambda)\cap V(\lambda,\nu)} Q(\hat{\mathcal{L}}_O^2 W)(\bar{K},\bar{K},T,e_4)\\
&\quad+\int_{C(\lambda)\cap V(\lambda,\nu)} Q(\hat{\mathcal{L}}_O\hat{\mathcal{L}}_T W)(\bar{K},\bar{K},\bar{K},e_4)\\
&\quad+\int_{C(\lambda)\cap V(\lambda,\nu)} Q(\hat{\mathcal{L}}_S\hat{\mathcal{L}}_T W)(\bar{K},\bar{K},\bar{K},e_4)\Bigg)+c(1+\epsilon_0^2)\mathcal{R}_{[1]}{}^2
\end{aligned}$$

and

$$\begin{aligned}
\underline{\mathcal{R}}_2[\beta]^2(\lambda,\nu) \ &\leq\ c\left(\int_{\underline{C}(\nu)\cap V(\lambda,\nu)} Q(\hat{\mathcal{L}}_O^2 W)(\bar{K},\bar{K},T,e_3)\right.\\
&\quad\left.+\int_{\underline{C}(\nu)\cap V(\lambda,\nu)} Q(\hat{\mathcal{L}}_O\hat{\mathcal{L}}_T W)(\bar{K},\bar{K},\bar{K},e_3)\right)+c(1+\epsilon_0^2)\underline{\mathcal{R}}_{[1]}{}^2\\
\underline{\mathcal{R}}_2[(\rho,\sigma)]^2(\lambda,\nu) \ &\leq\ c\left(\int_{\underline{C}(\nu)\cap V(\lambda,\nu)} Q(\hat{\mathcal{L}}_O^2 W)(\bar{K},\bar{K},T,e_3)\right.\\
&\quad\left.+\int_{\underline{C}(\nu)\cap V(\lambda,\nu)} Q(\hat{\mathcal{L}}_O\hat{\mathcal{L}}_T W)(\bar{K},\bar{K},\bar{K},e_3)\right)+c(1+\epsilon_0^2)\underline{\mathcal{R}}_{[1]}{}^2\\
\underline{\mathcal{R}}_2[\underline{\beta}]^2(\lambda,\nu) \ &\leq\ c\left(\int_{\underline{C}(\nu)\cap V(\lambda,\nu)} Q(\hat{\mathcal{L}}_O^2 W)(\bar{K},\bar{K},T,e_3)\right.\\
&\quad\left.+\int_{\underline{C}(\nu)\cap V(\lambda,\nu)} Q(\hat{\mathcal{L}}_O\hat{\mathcal{L}}_T W)(\bar{K},\bar{K},\bar{K},e_3)\right)+c(1+\epsilon_0^2)\underline{\mathcal{R}}_{[1]}{}^2\\
\underline{\mathcal{R}}_2[\underline{\alpha}]^2(\lambda,\nu) \ &\leq\ c\left(\int_{\underline{C}(\nu)\cap V(\lambda,\nu)} Q(\hat{\mathcal{L}}_O^2 W)(\bar{K},\bar{K},T,e_3)\right. \qquad (5.1.36)\\
&\quad+\int_{\underline{C}(\nu)\cap V(\lambda,\nu)} Q(\hat{\mathcal{L}}_O\hat{\mathcal{L}}_T W)(\bar{K},\bar{K},\bar{K},e_3)\\
&\quad\left.+\int_{\underline{C}(\nu)\cap V(\lambda,\nu)} Q(\hat{\mathcal{L}}_S\hat{\mathcal{L}}_T W)(\bar{K},\bar{K},\bar{K},e_3)\right)+c(1+\epsilon_0^2)\underline{\mathcal{R}}_{[1]}{}^2.
\end{aligned}$$

Proof: The proof is in the appendix to this chapter.

Remarks: To understand in detail where the various integrals composing $\mathcal{Q}_2$ and $\underline{\mathcal{Q}}_2$ are needed, we make the following observations.

1. $\int_{C(\lambda)} Q(\hat{\mathcal{L}}_O^2 W)(\bar{K},\bar{K},T,e_4)$ and $\int_{\underline{C}(\nu)} Q(\hat{\mathcal{L}}_O^2 W)(\bar{K},\bar{K},T,e_3)$ are used to control the integrals

$$\begin{aligned}
&\int_{C(\lambda)} r^4\tau_+^\gamma\tau_-^\delta|\nabla\!\!\!/^2(\alpha,\beta,\rho,\sigma,\underline{\beta})|^2\\
&\int_{\underline{C}(\nu)} r^4\tau_+^\gamma\tau_-^\delta|\nabla\!\!\!/^2(\beta,\rho,\sigma,\underline{\beta},\underline{\alpha})|^2 \qquad (5.1.37)
\end{aligned}$$

with $\gamma+\delta=4$.

2. The presence of the integral norms

$$\int_{C(\lambda)} Q(\hat{\mathcal{L}}_O\hat{\mathcal{L}}_T W)(\bar{K},\bar{K},\bar{K},e_4)\ ,\ \int_{\underline{C}(\nu)} Q(\hat{\mathcal{L}}_O\hat{\mathcal{L}}_T W)(\bar{K},\bar{K},\bar{K},e_3)$$

and of the integral norms

$$\int_{C(\lambda)} Q(\hat{\mathcal{L}}_S\hat{\mathcal{L}}_T W)(\bar{K},\bar{K},\bar{K},e_4)\ ,\ \int_{\underline{C}(\nu)} Q(\hat{\mathcal{L}}_S\hat{\mathcal{L}}_T W)(\bar{K},\bar{K},\bar{K},e_3)$$

is due to two different reasons. The first one is that the Bianchi evolution equations for the Riemann components α and $\underline{\alpha}$ exist only relative to the e_3 and e_4 directions, respectively.

This implies that the terms $\mathbf{D}_4\alpha$ and $\mathbf{D}_3\underline{\alpha}$ in the integrand have to be transformed into $\mathbf{D}_3\alpha + 2\mathbf{D}_T\alpha$ and $\mathbf{D}_4\underline{\alpha} - 2\mathbf{D}_T\alpha$ and that the Bianchi equations can be applied only to the first part of them.

The second reason is that the norm integrals along $C(\lambda)$ do not contain terms with $\underline{\alpha}(W)$ and the norm integrals along $\underline{C}(\nu)$ do not contain terms with $\alpha(W)$.

Recalling the definition of signature given in Chapter 3 (see (3.1.24)), it follows that when we consider norms along $C(\lambda)$, each time the signature of the integrand is $+3$, giving signature $+1$ to $\mathbf{D}_4$ and -1 to $\mathbf{D}_3$, the integrand is of type $\nabla\!\!\!/\mathbf{D}_4\alpha$ or, using the Bianchi equations, contains a term of this type. This implies that, to estimate it, we have to estimate an integral norm of $\nabla\!\!\!/\mathbf{D}_T\alpha$ which requires $\int_{C(\lambda)} Q(\hat{\mathcal{L}}_O\hat{\mathcal{L}}_T W)(\bar{K}, \bar{K}, \bar{K}, e_4)$ to be bounded. If, on the other hand, the signature is $+4$, the integrand must contain $\mathbf{D}_4^2\alpha$ and, to estimate this term, we have to express $\mathbf{D}_4$ as $\mathbf{D}_3 + 2\mathbf{D}_T$. This implies that we have to consider integrals of $\mathbf{D}_T^2\alpha$ which require the norms $\int_{C(\lambda)} Q(\hat{\mathcal{L}}_S\hat{\mathcal{L}}_T W)(\bar{K}, \bar{K}, \bar{K}, e_4)$ to be controlled.[20]

The second reason shows up in the following way: when the integrand of a $C(\lambda)$ integral norm has signature -2, it contains a $\mathbf{D}_3\underline{\beta}$ or a $\mathbf{D}_3^2(\rho, \sigma)$ term which produces, using the Bianchi equations, a $\nabla\!\!\!/\underline{\alpha}(W)$ term. This term cannot be estimated in a straightforward way using the $\mathcal{Q}_2$ norms.[21] Therefore, in this case, the strategy is to substitute $\mathbf{D}_3$ with $-\mathbf{D}_4 + 2\mathbf{D}_T$ which, again, implies that we have to use the $\int_{C(\lambda)} Q(\hat{\mathcal{L}}_O\hat{\mathcal{L}}_T W)(\bar{K}, \bar{K}, \bar{K}, e_4)$ norm to control the terms containing $\mathbf{D}_T\underline{\beta}$ or $\nabla\!\!\!/\mathbf{D}_T\underline{\beta}$. If the integrand signature is -3, repeating the previous argument, it follows that we also have to control terms differentiated twice with respect to T, as for instance, $\mathbf{D}_T^2\underline{\beta}$, which again requires the norms $\int_{C(\lambda)} Q(\hat{\mathcal{L}}_S\hat{\mathcal{L}}_T W)(\bar{K}, \bar{K}, \bar{K}, e_4)$ to be controlled.

Exactly the same discussion holds for the integrals along $\underline{C}(\nu)$ with all the signatures interchanged. These integral norms are needed, in this case, for the integrands of signature $+2$, $+3$ and -3, -4.

5.1.4 The asymptotic behavior of $\overline{\rho}$ and $\overline{\sigma}$

The first lemma refers to the asymptotic behavior of $\overline{\rho}$.

Lemma 5.1.2 *Assume that the connection coefficients satisfy the inequality*

$$\mathcal{O} \leq \epsilon_0 \, .$$

Then the average of ρ on the two-dimensional surfaces $S(\lambda, \nu)$,

$$\overline{\rho}(\lambda, \nu) = \frac{1}{|S(\lambda, \nu)|} \int_{S(\lambda,\nu)} \rho \, ,$$

satisfies the following estimate:

$$\sup_{\mathcal{K}} |r^3 \overline{\rho}| \leq \sup_{\mathcal{K}\cap\Sigma_0} |r^3 \overline{\rho}| + c\epsilon_0 \left([\mathcal{Q}]_1 + \epsilon_0 \underline{\mathcal{R}}_{[0]} \right) .$$

[20]The reason the integral norm $\int_{C(\lambda)} Q(\hat{\mathcal{L}}_T^2 W)(\bar{K}, \bar{K}, \bar{K}, e_4)$ is not sufficient is connected to the weight of the integrand.

[21]Nevertheless see Proposition 5.1.4.

Proof: Proceeding as in Lemma 4.3.4 we obtain that $\overline{\rho}$ satisfies the evolution equation

$$\frac{d}{d\lambda}\overline{\rho} = -\overline{\Omega \mathrm{tr}\underline{\chi}}\,\overline{\rho} + \frac{1}{|S(\lambda,\nu)|}\int_{S(\lambda,\nu)}\left(\frac{d\rho}{d\lambda} + \Omega\mathrm{tr}\underline{\chi}\rho\right).$$

Using the Bianchi equation (3.2.4) for ρ along the $\underline{C}(\nu)$ null hypersurfaces

$$\frac{d}{d\lambda}\rho + \frac{3}{2}\Omega\mathrm{tr}\underline{\chi}\rho = \Omega(-\mathrm{d\!\!/iv}\,\underline{\beta} - 2\eta\cdot\underline{\beta} - \frac{1}{2}\hat{\chi}\cdot\underline{\alpha} + \zeta\cdot\underline{\beta})$$

and denoting $f \equiv r^3\overline{\rho}$, we obtain

$$\begin{aligned}\frac{d}{d\lambda}f &= \frac{d}{d\lambda}r^3\overline{\rho} = r^3\left(\frac{3}{2}\overline{\Omega\mathrm{tr}\underline{\chi}}\,\overline{\rho} + \frac{d}{d\lambda}\overline{\rho}\right)\\ &= \frac{1}{2}\frac{r^3}{|S(\lambda,\nu)|}\int_{S(\lambda,\nu)}(\overline{\Omega\mathrm{tr}\underline{\chi}} - \Omega\mathrm{tr}\underline{\chi})(\rho-\overline{\rho}) \qquad (5.1.38)\\ &+ \frac{r^3}{|S(\lambda,\nu)|}\int_{S(\lambda,\nu)}\Omega\left(-\mathrm{d\!\!/iv}\,\underline{\beta} - 2\eta\cdot\underline{\beta} - \frac{1}{2}\hat{\chi}\cdot\underline{\alpha} + \zeta\cdot\underline{\beta}\right).\end{aligned}$$

From the assumptions on the Ricci coefficients

$$\left|\frac{d}{d\lambda}f\right| \le c\epsilon_0\left(\int_{S(\lambda,\nu)}\frac{1}{r}|\rho-\overline{\rho}| + \int_{S(\lambda,\nu)}r|\mathrm{d\!\!/iv}\,\underline{\beta}| + \int_{S(\lambda,\nu)}\frac{1}{r}|\underline{\beta}| + \int_{S(\lambda,\nu)}\frac{1}{r}|\underline{\alpha}|\right),$$

so that, integrating along $\underline{C}(\nu)\cap V(\lambda,\nu)$,

$$\begin{aligned}\int_{\lambda_0}^{\lambda}d\lambda'|\frac{d}{d\lambda}f|(\lambda',\nu) \le c\epsilon_0\Bigg(&\int_{\underline{C}(\nu)\cap V(\lambda,\nu)}O(\frac{1}{r^3})|r^2(\rho-\overline{\rho})| + \int_{\underline{C}(\nu)\cap V(\lambda,\nu)}O(\frac{1}{r\lambda'})|r^2\lambda'\mathrm{d\!\!/iv}\,\underline{\beta}|\\ &+\int_{\underline{C}(\nu)\cap V(\lambda,\nu)}O(\frac{1}{r\lambda'^2})|\lambda'^2\underline{\alpha}| + \int_{\underline{C}(\nu)\cap V(\lambda,\nu)}O(\frac{1}{r^2\lambda'})|r\lambda'\underline{\beta}|\Bigg).\end{aligned}$$

Applying the Schwartz inequality, we obtain

$$\begin{aligned}\int_{\lambda_0}^{\lambda}d\lambda'|\frac{d}{d\lambda}f|(\lambda',\nu) \le c\epsilon_0\Bigg[&\left(\int_{\underline{C}(\nu)\cap V(\lambda,\nu)}r^4|(\rho-\overline{\rho})|^2\right)^{\frac{1}{2}} + \left(\int_{\underline{C}(\nu)\cap V(\lambda,\nu)}r^4\lambda'^2|\mathrm{d\!\!/iv}\,\underline{\beta}|^2\right)^{\frac{1}{2}}\\ &+\left(\int_{\underline{C}(\nu)\cap V(\lambda,\nu)}\lambda'^4|\underline{\alpha}|^2\right)^{\frac{1}{2}} + \left(\int_{\underline{C}(\nu)\cap V(\lambda,\nu)}r^2\lambda'^2|\underline{\beta}|^2\right)\Bigg]^{\frac{1}{2}}.\end{aligned}$$

Using the estimates for the L^2 weighted norms of the Riemann null components, proved in Propositions 5.1.2 and 5.1.3,

$$\begin{aligned}\int_{\underline{C}(\nu)\cap V(\lambda,\nu)}r^4|(\rho-\overline{\rho})|^2 &\le c\int_{\underline{C}(\nu)\cap V(\lambda,\nu)}Q(\hat{\mathcal{L}}_O W)(\bar{K},\bar{K},T,e_3)\\ \int_{\underline{C}(\nu)\cap V(\lambda,\nu)}r^4\lambda'^2|\mathrm{d\!\!/iv}\,\underline{\beta}|^2 &\le c\int_{\underline{C}(\nu)\cap V(\lambda,\nu)}Q(\hat{\mathcal{L}}_O W)(\bar{K},\bar{K},T,e_3)\\ \int_{\underline{C}(\nu)\cap V(\lambda,\nu)}r^2\lambda'^2|\underline{\beta}|^2 &\le c\int_{\underline{C}(\nu)\cap V(\lambda,\nu)}Q(\hat{\mathcal{L}}_O W)(\bar{K},\bar{K},T,e_3)\\ \int_{\underline{C}(\nu)\cap V(\lambda,\nu)}\lambda'^4|\underline{\alpha}|^2 &\le c\int_{\underline{C}(\nu)\cap V(\lambda,\nu)}Q(\hat{\mathcal{L}}_O W)(\bar{K},\bar{K},T,e_3)\,,\end{aligned}$$

taking the sup over $\mathcal{K}$ and recalling Definitions 5.0.2 the proof is completed. Using the previous results, Corollary 4.1.1 and Lemma 4.1.3, we have

$$\begin{aligned}\sup_{\mathcal{K}} |r^3 \tau_-^{\frac{1}{2}} (\rho - \overline{\rho})| \leq\ & c \sup_{\mathcal{K}} \left[\left(\int_{\underline{C}(\nu)\cap V(\lambda,\nu)} Q(\hat{\mathcal{L}}_O W)(\bar{K}, \bar{K}, T, e_3) \right)^{\frac{1}{2}} \right. \\ & \left. + \left(\int_{\underline{C}(\nu)\cap V(\lambda,\nu)} Q(\hat{\mathcal{L}}_O^2 W)(\bar{K}, \bar{K}, T, e_3) \right)^{\frac{1}{2}} \right] \qquad (5.1.39)\end{aligned}$$

and we conclude that

$$\begin{aligned}\sup_{\mathcal{K}} |r^3 \rho| \leq\ & \sup_{\mathcal{K}} |r^3 (\rho - \overline{\rho})| + \sup_{\mathcal{K}} |r^3 \overline{\rho}| \\ \leq\ & \sup_{r \in \mathcal{K}\cap\Sigma_0} |r^3 \overline{\rho}| + c \sup_{\mathcal{K}} \left[\left(\int_{\underline{C}(\nu)\cap V(\lambda,\nu)} Q(\hat{\mathcal{L}}_O W)(\bar{K}, \bar{K}, T, e_3) \right)^{\frac{1}{2}} \right. \\ & \left. + \left(\int_{\underline{C}(\nu)\cap V(\lambda,\nu)} Q(\hat{\mathcal{L}}_O^2 W)(\bar{K}, \bar{K}, T, e_3) \right)^{\frac{1}{2}} \right] . \qquad (5.1.40)\end{aligned}$$

The asymptotic behavior of $\overline{\sigma}$ is obtained in a different and easier way than the one for $\overline{\rho}$.

Lemma 5.1.3 *Assume that the connection coefficients satisfy the inequality*

$$\mathcal{O} \leq \epsilon_0 .$$

Then the average of σ on the two-dimensional surfaces $S(\lambda, \nu)$

$$\overline{\sigma}(\lambda, \nu) = \frac{1}{|S(\lambda, \nu)|} \int_{S(\lambda,\nu)} \sigma ,$$

satisfies the following estimate:

$$\sup_{\mathcal{K}} |r^3 \overline{\sigma}| \leq c\epsilon_0 .$$

Proof: Lemma 5.1.3 follows immediately by taking the average over $S(\lambda, \nu)$ of the structure equation (see (3.1.47))

$$\sigma = \mathrm{c}\not{\mathrm{u}}\mathrm{rl}\, \zeta - \frac{1}{2} \hat{\underline{\chi}} \wedge \hat{\chi} ,$$

and using the assumption of the lemma.

Remark: Assuming $\mathcal{K}$ foliated by a double null canonical foliation, see Subsection 4.3.16, we have the stronger estimate $\sup_{\mathcal{K}} |r^3 \tau_-^{\frac{1}{2}} \overline{\sigma}| \leq c\epsilon_0$. In this case the result is stronger due to the presence of a factor $\tau_-^{\frac{1}{2}}$. The reason is that, different from the ρ component of the Riemann tensor, σ is not connected to the ADM mass relative to the initial hypersurface Σ_0.

5.1.5 Asymptotic behavior of the null Riemann components

The sup norms in the inequalities (3.7.1) are bounded by $\mathcal{Q}_{\mathcal{K}}$,

$$\sup_{\mathcal{K}} r^{7/2}|\alpha| \leq c\mathcal{Q}_{\mathcal{K}}\ , \ \ \sup_{\mathcal{K}} r|u|^{\frac{5}{2}}|\underline{\alpha}| \leq c\mathcal{Q}_{\mathcal{K}}$$

$$\sup_{\mathcal{K}} r^{7/2}|\beta| \leq c\mathcal{Q}_{\mathcal{K}}\ , \ \ \sup_{\mathcal{K}} r^2|u|^{\frac{3}{2}}|\underline{\beta}| \leq c\mathcal{Q}_{\mathcal{K}}$$

$$\sup_{\mathcal{K}} r^3|\rho| \leq c\mathcal{Q}_{\mathcal{K}}\ , \ \ \sup_{\mathcal{K}} r^3|u|^{\frac{1}{2}}|(\rho-\overline{\rho},\sigma)| \leq c\mathcal{Q}_{\mathcal{K}}.$$

This is an immediate consequence of Corollary 4.1.1, Propositions 5.1.2, 5.1.3, 5.1.5 and the last lines of Lemma 5.1.2 and Lemma 5.1.3. Theorem **M8**, proved in the next chapter, completes the proof of inequalities (3.7.1).

5.2 Appendix

5.2.1 Proof of Proposition 5.1.4

We discuss the proof of the more relevant inequalities in (5.1.34). The other ones can be easily deduced using the same technique. First of all we recall Corollary 4.1.1.
Corollary 4.1.1: *Under the assumptions of Lemma 4.1.1 and Lemma 4.1.2 the following estimates hold*

$$\begin{aligned}
\left(\int_{S(\lambda,\nu)} r^4|F|^4\right)^{\frac{1}{4}} &\leq \left(\int_{C(\lambda)\cap\Sigma_0} r^4|F|^4\right)^{\frac{1}{4}} \\
&\quad + c\left(\int_{C(\lambda)\cap V(\lambda,\nu)} |F|^2 + r^2|\nabla\!\!\!/ F|^2 + r^2|\mathbf{D}\!\!\!\!/_4 F|^2\right)^{\frac{1}{2}} \\
\left(\int_{S(\lambda,\nu)} r^2\tau_-^2|F|^4\right)^{\frac{1}{4}} &\leq \left(\int_{C(\lambda)\cap\Sigma_0} r^2\tau_-^2|F|^4\right)^{\frac{1}{4}} \\
&\quad + c\left(\int_{C(\lambda)\cap V(\lambda,\nu)} |F|^2 + r^2|\nabla\!\!\!/ F|^2 + \tau_-^2|\mathbf{D}\!\!\!\!/_4 F|^2\right)^{\frac{1}{2}}
\end{aligned} \tag{5.2.1}$$

and

$$\begin{aligned}
\left(\int_{S(\lambda,\nu)} r^4|F|^4\right)^{\frac{1}{4}} &\leq \left(\int_{\underline{C}(\nu)\cap\Sigma_0} r^4|F|^4\right)^{\frac{1}{4}} \\
&\quad + c\left(\int_{\underline{C}(\nu)\cap V(\lambda,\nu)} |F|^2 + r^2|\nabla\!\!\!/ F|^2 + r^2|\mathbf{D}\!\!\!\!/_3 F|^2\right)^{\frac{1}{2}} \\
\left(\int_{S(\lambda,\nu)} r^2\tau_-^2|F|^4\right)^{\frac{1}{4}} &\leq \left(\int_{\underline{C}(\nu)\cap\Sigma_0} r^2\tau_-^2|F|^4\right)^{\frac{1}{4}} \\
&\quad + c\left(\int_{\underline{C}(\nu)\cap V(\lambda,\nu)} |F|^2 + r^2|\nabla\!\!\!/ F|^2 + \tau_-^2|\mathbf{D}\!\!\!\!/_3 F|^2\right)^{\frac{1}{2}}.
\end{aligned} \tag{5.2.2}$$

1. $\|r^{\delta}\tau_-^{\frac{5}{2}}\underline{\alpha}\|_{2,C(\lambda)\cap V(\lambda,\nu)}$

$$
\begin{aligned}
\|r^{\delta}\tau_-^{\frac{5}{2}}\underline{\alpha}\|^2_{2,C(\lambda)\cap V(\lambda,\nu)} &= \int_{C(\lambda)\cap V(\lambda,\nu)} r^{2\delta}\tau_-^5|\underline{\alpha}|^2 = \int_{\nu_0}^{\nu}\int_{S(\lambda,\nu')} r^{2\delta}\tau_-^5|\underline{\alpha}|^2\\
&\leq c\int_{\nu_0}^{\nu}\left(\int_{S(\lambda,\nu')}\right)^{\frac{1}{2}}\left[\left(\int_{S(\lambda,\nu')} r^{4\delta}\tau_-^{10}|\underline{\alpha}|^4\right)^{\frac{1}{4}}\right]^2\\
&\leq c\int_{\nu_0}^{\nu} r(\lambda,\nu')^{2\delta}\left[\left(\int_{S(\lambda,\nu')} r^2\tau_-^{10}|\underline{\alpha}|^4\right)^{\frac{1}{4}}\right]^2.
\end{aligned}
$$

The integrand on the right-hand side is estimated with Corollary 4.1.1.

$$
\begin{aligned}
\left(\int_{S(\lambda,\nu)} r^2\tau_-^{10}|\underline{\alpha}|^4\right)^{\frac{1}{4}} &\leq \left(\int_{\underline{C}(\nu)\cap\Sigma_0} r^2\tau_-^{10}|\underline{\alpha}|^4\right)^{\frac{1}{4}}\\
&+c\left(\int_{\underline{C}(\nu)\cap V(\lambda,\nu)} \tau_-^4|\underline{\alpha}|^2 + r^2\tau_-^4|\nabla\!\!\!/\underline{\alpha}|^2 + \tau_-^6|\mathbf{D}\!\!\!\!/_3\underline{\alpha}|^2\right)^{\frac{1}{2}}
\end{aligned}
$$

obtaining

$$
\begin{aligned}
\|r^{\delta}\tau_-^{\frac{5}{2}}\underline{\alpha}\|^2_{2,C(\lambda)\cap V(\lambda,\nu)} &\leq c\int_{\nu_0}^{\nu} r(\lambda,\nu')^{2\delta}\left(\int_{\underline{C}(\nu')\cap\Sigma_0} r^2\tau_-^{10}|\underline{\alpha}|^4\right)^{\frac{1}{2}} \qquad (5.2.3)\\
+c\int_{\nu_0}^{\nu} r(\lambda,\nu')^{2\delta}&\left(\int_{\underline{C}(\nu')\cap V(\lambda,\nu')} \tau_-^4|\underline{\alpha}|^2 + r^2\tau_-^4|\nabla\!\!\!/\underline{\alpha}|^2 + \tau_-^6|\mathbf{D}\!\!\!\!/_3\underline{\alpha}|^2\right)
\end{aligned}
$$

The integrand of the second term on the right-hand side of (5.2.3) is estimated as follows. From the explicit expressions of the $\mathcal{Q}$ integrals and from the discussion in Section 5.1,

$$
\begin{aligned}
&\int_{\underline{C}(\nu')\cap V(\lambda,\nu')} (\tau_-^4|\underline{\alpha}|^2 + r^2\tau_-^4|\nabla\!\!\!/\underline{\alpha}|^2)\\
&\qquad\leq c\int_{\underline{C}(\nu')\cap V(\lambda,\nu')} \mathcal{Q}(\hat{\mathcal{L}}_O W)(\bar{K},\bar{K},T,e_3) + [correction]^2.
\end{aligned}
$$

To estimate $\int_{\underline{C}(\nu)\cap V(\lambda,\nu)} \tau_-^6|\mathbf{D}\!\!\!\!/_3\underline{\alpha}|^2$ we express $\underline{\alpha}_3$ in terms of $\underline{\alpha}_4$ and $\mathbf{D}\!\!\!\!/_T\underline{\alpha}$:

$$\underline{\alpha}_3 = 2\mathbf{D}\!\!\!\!/_T\underline{\alpha} - \underline{\alpha}_4 + \left(\frac{5}{2}\mathrm{tr}\underline{\chi} + \frac{1}{2}\mathrm{tr}\chi\right)\underline{\alpha}\,.$$

Since $\mathrm{tr}\underline{\chi}, \mathrm{tr}\chi = O(\frac{1}{r})$ we are left with

$$||\tau_-^3\mathbf{D}\!\!\!\!/_T\underline{\alpha}||_{2,\underline{C}(\nu)\cap V(\lambda,\nu)},\ ||\tau_-^3\underline{\alpha}_4||_{2,\underline{C}(\nu)\cap V(\lambda,\nu)},\ ||\tau_-^2\underline{\alpha}||_{2,\underline{C}(\nu)\cap V(\lambda,\nu)}\,.$$

The only term we have to estimate is the first one. In fact the second one can be bounded by $||\tau_- r^2\underline{\alpha}_4||_{2,\underline{C}(\nu)\cap V(\lambda,\nu)}$ which has been estimated in Proposition 5.1.3 and the third one is controlled in Proposition 5.1.2. Apart from correction terms, this amounts to estimating

$\int_{\underline{C}(\nu)\cap V(\lambda,\nu)} \tau_-^6|\underline{\alpha}(\hat{\mathcal{L}}_T W)|^2$ which is controlled by $\int_{\underline{C}(\nu)\cap V(\lambda,\nu)} Q(\hat{\mathcal{L}}_T W)(\bar{K},\bar{K},\bar{K},e_3)$. In conclusion

$$\begin{aligned}&\int_{\underline{C}(\nu)\cap V(\lambda,\nu)}(\tau_-^4|\underline{\alpha}|^2+r^2\tau_-^4|\nabla\!\!\!/\underline{\alpha}|^2+\tau_-^6|\mathbf{D}\!\!\!\!/_3\underline{\alpha}|^2)\\ &\quad\le c\int_{\underline{C}(\nu)\cap V(\lambda,\nu)} Q(\hat{\mathcal{L}}_O W)(\bar{K},\bar{K},T,e_3)+\int_{\underline{C}(\nu)\cap V(\lambda,\nu)} Q(\hat{\mathcal{L}}_T W)(\bar{K},\bar{K},\bar{K},e_3)\\ &\quad+[correction]^2\le c[\mathcal{Q}]_1(\lambda,\nu)+c\epsilon_0^2\underline{\mathcal{R}}_{[0]}{}^2.\end{aligned}$$

Therefore, if $\delta<-\frac{1}{2}$, we have

$$||r^\delta\tau_-^{\frac{5}{2}}\underline{\alpha}||^2_{2,C(\lambda)\cap V(\lambda,\nu)}\le c\left[\sup_{\nu'\in V(\lambda,\nu)\cap\Sigma_0}|r^{1-\frac{2}{p}}\tau_-^{\frac{5}{2}}\underline{\alpha}|^2_{p=4,S_{(0)}(\nu')}+[\mathcal{Q}]_1(\lambda,\nu)+\epsilon_0^2\underline{\mathcal{R}}_{[0]}{}^2\right],$$

where $S_{(0)}(\nu')=\underline{C}(\nu')\cap\Sigma_0$.

The norm $|r^{1-\frac{2}{p}}\tau_-^{\frac{5}{2}}\underline{\alpha}|_{p=4,S_{(0)}(\nu')}=\left(\int_{S_{(0)}(\nu)}r^2\tau_-^{10}|\underline{\alpha}|^4\right)^{\frac{1}{4}}$ can be estimated with an L^2 norm integral on Σ_0 in a way similar to what we did for $\left(\int_{S(\lambda,\nu)}r^2\tau_-^{10}|\underline{\alpha}|^4\right)^{\frac{1}{4}}$. We use the analogue of Corollary 4.1.1 relative to a spacelike hypersurface (see [Ch-Kl] and also Proposition 2.2.4). The estimate one obtains is

$$\sup_{\nu'\in V(\lambda,\nu)\cap\Sigma_0}|r^{1-\frac{2}{p}}\tau_-^{\frac{5}{2}}\underline{\alpha}|^2_{p=4,S_{(0)}(\nu')}\le\mathcal{Q}_{1\,\Sigma_0\cap V(\lambda,\nu)},\tag{5.2.4}$$

which proves Proposition 5.1.4 for $\underline{\alpha}$.

All the remaining estimates are made in the same spirit and we point out only their main aspects.

2. $\|r^\delta\tau_-^\sigma\underline{\beta}_3\|_{2,C(\lambda)\cap V(\lambda,\nu)}$

To estimate this term we use the Bianchi equation

$$\underline{\beta}_3\equiv\mathbf{D}\!\!\!\!/_3\underline{\beta}+2\mathrm{tr}\underline{\chi}\;\underline{\beta}=-\mathrm{d}\!\!/\mathrm{iv}\underline{\alpha}-\left[2\underline{\omega}\underline{\beta}+(-2\zeta+\eta)\cdot\underline{\alpha}\right].$$

It turns out that the main term to estimate is $\|r^\delta\tau_-^\sigma\nabla\!\!\!/\underline{\alpha}\|_{2,C(\lambda)\cap V(\lambda,\nu)}$. From its estimate, which can be patterned upon the estimate of $||r^\delta\tau_-^{\frac{5}{2}}\underline{\alpha}||_{2,C(\lambda)\cap V(\lambda,\nu)}$, we conclude that this term is bounded for $\delta<\frac{1}{2}$, $\sigma\le\frac{5}{2}$. Therefore

$$\begin{aligned}\|r^\delta\tau_-^\sigma\underline{\beta}_3\|_{2,C(\lambda)\cap V(\lambda,\nu)}&\le c\sup_{\nu'\in V(\lambda,\nu)\cap\Sigma_0}\left[\left(\int_{\underline{C}(\nu')\cap V(\lambda,\nu')}Q(\hat{\mathcal{L}}_O W)(\bar{K},\bar{K},T,e_3)\right.\right.\\ &+\int_{\underline{C}(\nu')\cap V(\lambda,\nu')}Q(\hat{\mathcal{L}}_O^2 W)(\bar{K},\bar{K},T,e_3)+\left.\int_{\underline{C}(\nu')\cap V(\lambda,\nu')}Q(\hat{\mathcal{L}}_O\hat{\mathcal{L}}_T W)(\bar{K},\bar{K},\bar{K},e_3)\right)^{\frac{1}{2}}\\ &+\left.\mathcal{Q}^{\frac{1}{2}}_{\Sigma_0\cap V(\lambda,\nu')}+\epsilon_0\underline{\mathcal{R}}_{[0]}\right]\end{aligned}$$

for $\delta<\frac{1}{2}$, $\sigma\le\frac{5}{2}$.

3. $\|r^\delta \tau_-^\sigma \alpha\|_{2,\underline{C}(\nu)\cap V(\lambda,\nu)}$

This term is estimated in the same way as the term $\|r^\delta \tau_-^\sigma \underline{\alpha}\|_{2,C(\lambda)\cap V(\lambda,\nu)}$ with the appropriate weights. Therefore we have, for $\delta < 2$,

$$||r^\delta \alpha||_{2,\underline{C}(\nu)\cap V(\lambda,\nu)} \leq c \sup_{\lambda' \in V(\lambda,\nu)\cap\Sigma_0} \left[\left(\int_{C(\lambda')\cap V(\lambda',\nu)} Q(\hat{\mathcal{L}}_O W)(\bar{K},\bar{K},T,e_4)\right.\right.$$
$$\left.\left. + \int_{C(\lambda')\cap V(\lambda',\nu)} Q(\hat{\mathcal{L}}_T W)(\bar{K},\bar{K},\bar{K},e_4)\right)^{\frac{1}{2}} + \mathcal{Q}_1^{\frac{1}{2}}{}_{\Sigma_0\cap V(\lambda',\nu)} + \epsilon_0 \underline{\mathcal{R}}_{[0]}\right].$$

4. $\|r^\delta \tau_-^\sigma \beta_4\|_{2,\underline{C}(\nu)\cap V(\lambda,\nu)}$

This term is estimated in the same way as the term $\|r^\delta \tau_-^\sigma \underline{\beta}_3\|_{2,C(\lambda)\cap V(\lambda,\nu)}$. We conclude that

$$\|r^\lambda \beta_4\|_{2,\underline{C}(\nu)\cap V(\lambda,\nu)} \leq c \sup_{\lambda' \in V(\lambda,\nu)\cap\Sigma_0} \left[\left(\int_{C(\lambda)\cap V(\lambda,\nu)} Q(\hat{\mathcal{L}}_O W)(\bar{K},\bar{K},T,e_4)\right.\right.$$
$$\left. + \int_{C(\lambda)\cap V(\lambda,\nu)} Q(\hat{\mathcal{L}}_O^2 W)(\bar{K},\bar{K},T,e_4) + \int_{C(\lambda)\cap V(\lambda,\nu)} Q(\hat{\mathcal{L}}_O \hat{\mathcal{L}}_T W)(\bar{K},\bar{K},\bar{K},e_4)\right)^{\frac{1}{2}}$$
$$\left. + \mathcal{Q}_1^{\frac{1}{2}}{}_{\Sigma_0\cap V(\lambda',\nu)} + \epsilon_0 \underline{\mathcal{R}}_{[0]}\right]$$

with $\lambda < 3$.

5. $\|r^\delta \tau_-^\sigma (\rho,\sigma)_4\|_{2,\underline{C}(\nu)\cap V(\lambda,\nu)}$

This term is estimated differently from $\|r^\delta \tau_-^\sigma (\rho,\sigma)_3\|_{2,C(\lambda)\cap V(\lambda,\nu)}$.[22] From the Bianchi equation

$$\rho_4 \equiv \mathbf{D}_4\rho + \frac{3}{2}\mathrm{tr}\chi\rho = \mathrm{d}\!\!/\mathrm{iv}\,\beta - \left[\frac{1}{2}\underline{\hat{\chi}}\cdot\alpha - \zeta\cdot\beta - 2\underline{\eta}\cdot\beta\right],$$

we observe that the main term to control is $\|r^\delta \tau_-^\sigma \nabla\!\!\!/\beta\|_{2,C(\lambda)\cap V(\lambda,\nu)}$ which was bounded in the estimate of the norm $\underline{\mathcal{R}}_1[\beta](u,\underline{u})$ by the integral $\int_{\underline{C}(\nu)\cap V(\lambda,\nu)} Q(\hat{\mathcal{L}}_O W)(\bar{K},\bar{K},T,e_3)$ for $\lambda \leq 3$, $\sigma \leq 0$. For the main correction term we have to estimate

$$\|r^{(\delta-1)}\tau_-^{(\sigma-\frac{3}{2})}\alpha\|_{2,\underline{C}(\nu)\cap V(\lambda,\nu)}.$$

From the previous estimate of $\|r^\delta \tau_-^\sigma \alpha\|_{2,\underline{C}(\nu)\cap V(\lambda,\nu)}$ it follows that

$$||r^{\delta-1}\tau_-^{\sigma-\frac{3}{2}}\alpha||_{2,\underline{C}(\nu)\cap V(\lambda,\nu)} \leq c \sup_{\lambda' \in V(\lambda,\nu)\cap\Sigma_0} \left[\left(\int_{C(\lambda')\cap V(\lambda',\nu)} Q(\hat{\mathcal{L}}_O W)(\bar{K},\bar{K},T,e_4)\right.\right.$$
$$\left.\left. + \int_{C(\lambda')\cap V(\lambda',\nu)} Q(\hat{\mathcal{L}}_T W)(\bar{K},\bar{K},\bar{K},e_4)\right)^{\frac{1}{2}} + \mathcal{Q}_1^{\frac{1}{2}}{}_{\Sigma_0\cap V(\lambda',\nu)} + \epsilon_0 \underline{\mathcal{R}}_{[0]}\right],$$

[22] If one proceeds as for $\|r^\delta \tau_-^\sigma (\rho,\sigma)_3\|_{2,C(\lambda)\cap V(\lambda,\nu)}$ one obtains a weaker result.

with $\delta < 3$ and $\sigma \leq \frac{3}{2}$. So finally we have

$$\|r^{\delta}\tau_{-}^{\sigma}(\rho,\sigma)_4\|_{2,\underline{C}(\nu)\cap V(\lambda,\nu)} \leq c \sup_{\lambda'\in V(\lambda,\nu)\cap\Sigma_0}\Bigg[\Bigg(\int_{C(\lambda)\cap V(\lambda,\nu)} Q(\hat{\mathcal{L}}_O W)(\bar{K},\bar{K},T,e_4)$$
$$+\int_{C(\lambda)\cap V(\lambda,\nu)} Q(\hat{\mathcal{L}}_T W)(\bar{K},\bar{K},\bar{K},e_4)+\int_{C(\lambda)\cap V(\lambda,\nu)} Q(\hat{\mathcal{L}}_O W)(\bar{K},\bar{K},T,e_4)\Bigg)^{\frac{1}{2}}$$
$$+\mathcal{Q}_{1\,\Sigma_0\cap V(\lambda',\nu)}+\epsilon_0\underline{\mathcal{R}}_{[0]}\Bigg]$$

for $\delta < 3$ and $\sigma \leq 0$.

5.2.2 *Proof of Proposition 5.1.5*

We provide the main ideas of the proof, neglecting all the correction terms coming from the commutation relations, the relation between the covariant derivatives and the $\mathcal{L}_X$ derivatives and between the $\mathcal{L}_X$ and the modified $\hat{\mathcal{L}}_X$ derivatives. At the end we discuss, in some specific cases, the structure of the correction terms for the second derivatives. Therefore the equality signs appearing in the proof must always be interpreted as signifying "equal apart from correction terms."[23]

1. $\mathcal{R}_2[\alpha]$**:**

$$\begin{aligned}\mathcal{R}_2[\alpha](\lambda,\nu) = {} & \|r^4\nabla\!\!\!\!/^2\alpha\|_{2,C(\lambda)\cap V(\lambda,\nu)}+\|r^4\nabla\!\!\!\!/\alpha_3\|_{2,C(\lambda)\cap V(\lambda,\nu)}\\ &+\|r^4\nabla\!\!\!\!/\alpha_4\|_{2,C(\lambda)\cap V(\lambda,\nu)}+\|\tau_- r^3\alpha_{33}\|_{2,C(\lambda)\cap V(\lambda,\nu)}\\ &+\|r^4\alpha_{34}\|_{2,C(\lambda)\cap V(\lambda,\nu)}+\|\tau_- r^3\alpha_{44}\|_{2,C(\lambda)\cap V(\lambda,\nu)}.\end{aligned}$$

Estimate for $\|r^4\nabla\!\!\!\!/\alpha_4\|_{2,C(\lambda)\cap V(\lambda,\nu)}$

$$\begin{aligned}\|r^4\nabla\!\!\!\!/\alpha_4\|^2_{2,C(\lambda)\cap V(\lambda,\nu)} &= \int_{C(\lambda)\cap V(\lambda,\nu)} r^8|\nabla\!\!\!\!/\alpha_4(W)|^2\\ &= \int_{C(\lambda)\cap V(\lambda,\nu)} r^8|\nabla\!\!\!\!/\alpha_3(W)|^2+\int_{C(\lambda)\cap V(\lambda,\nu)} r^8|\nabla\!\!\!\!/\mathbf{D}\!\!\!\!/_T\alpha(W)|^2\\ &= \int_{C(\lambda)\cap V(\lambda,\nu)} r^8|\nabla\!\!\!\!/\alpha_3(W)|^2+\int_{C(\lambda)\cap V(\lambda,\nu)} r^6|\alpha(\hat{\mathcal{L}}_O\hat{\mathcal{L}}_T W)|^2\\ &\leq \int_{C(\lambda)\cap V(\lambda,\nu)} Q(\hat{\mathcal{L}}_O^2 W)(\bar{K},\bar{K},T,e_4)\\ &+\int_{C(\lambda)\cap V(\lambda,\nu)} Q(\hat{\mathcal{L}}_O\hat{\mathcal{L}}_T W)(\bar{K},\bar{K},\bar{K},e_4).\end{aligned} \tag{5.2.5}$$

Estimate for $\|\tau_- r^3\alpha_{44}\|_{2,C(\lambda)\cap V(\lambda,\nu)}$

$$\|\tau_- r^3\alpha_{44}\|^2_{2,C(\lambda)\cap V(\lambda,\nu)} = \int_{C(\lambda)\cap V(\lambda,\nu)} \tau_-^2 r^6|\alpha_{44}(W)|^2$$

[23] The results here are exact if we interpret the Riemann components as the components of a Weyl tensor with Minkowski space as the background space. We also omit the numerical coefficients in front of the various integrals.

$$
\begin{aligned}
&= \int_{C(\lambda)\cap V(\lambda,\nu)} \tau_-^2 r^6 |\alpha_{43}(W)|^2 + \int_{C(\lambda)\cap V(\lambda,\nu)} \tau_-^2 r^6 |(\not{\mathbf{D}}_T\alpha)_4(W)|^2\\
&= \int_{C(\lambda)\cap V(\lambda,\nu)} \tau_-^2 r^6 |\alpha_{43}(W)|^2 + \int_{C(\lambda)\cap V(\lambda,\nu)} \tau_-^2 r^6 |\alpha_4(\hat{\mathcal{L}}_T W)|^2\\
&= \int_{C(\lambda)\cap V(\lambda,\nu)} \tau_-^2 r^6 |\alpha_{43}(W)|^2 + \int_{C(\lambda)\cap V(\lambda,\nu)} \tau_-^2 r^6 |\alpha_3(\hat{\mathcal{L}}_T W)|^2\\
&\quad + \int_{C(\lambda)\cap V(\lambda,\nu)} \tau_-^2 r^6 |\not{\mathbf{D}}_T\alpha(\hat{\mathcal{L}}_T W)|^2\\
&= \int_{C(\lambda)\cap V(\lambda,\nu)} \tau_-^2 r^6 |\alpha_{43}(W)|^2 + \int_{C(\lambda)\cap V(\lambda,\nu)} \tau_-^2 r^6 |\not\nabla\beta(\hat{\mathcal{L}}_T W)|^2\\
&\quad + \int_{C(\lambda)\cap V(\lambda,\nu)} \tau_-^2 r^6 \frac{1}{\tau_+^2} |\not{\mathbf{D}}_S\alpha(\hat{\mathcal{L}}_T W)|^2\\
&= \int_{C(\lambda)\cap V(\lambda,\nu)} \tau_-^2 r^6 |\alpha_{43}(W)|^2 + \int_{C(\lambda)\cap V(\lambda,\nu)} \tau_-^2 r^4 |\beta(\hat{\mathcal{L}}_O\hat{\mathcal{L}}_T W)|^2\\
&\quad + \int_{C(\lambda)\cap V(\lambda,\nu)} \tau_-^2 r^4 |\alpha(\hat{\mathcal{L}}_S\hat{\mathcal{L}}_T W)|^2\\
&\le \int_{C(\lambda)\cap V(\lambda,\nu)} Q(\hat{\mathcal{L}}_O^2 W)(\bar{K},\bar{K},T,e_4) + \int_{C(\lambda)\cap V(\lambda,\nu)} Q(\hat{\mathcal{L}}_O\hat{\mathcal{L}}_T W)(\bar{K},\bar{K},\bar{K},e_4)\\
&\quad + \int_{C(\lambda)\cap V(\lambda,\nu)} Q(\hat{\mathcal{L}}_S\hat{\mathcal{L}}_T W)(\bar{K},\bar{K},\bar{K},e_4). \qquad (5.2.6)
\end{aligned}
$$

Remark: To estimate $\|\tau_- r^3\alpha_{44}\|_{2,C(\lambda)\cap V(\lambda,\nu)}$ we are obliged to use the $\mathcal{Q}$ norm $\int_{C(\lambda)\cap V(\lambda,\nu)} Q(\hat{\mathcal{L}}_S\hat{\mathcal{L}}_T W)(\bar{K},\bar{K},\bar{K},e_4)$.

2. $\mathcal{R}_2[\beta]$**:**

$$
\begin{aligned}
\mathcal{R}_2[\beta](\lambda,\nu) &= \|r^4\not\nabla^2\beta\|_{2,C(\lambda)\cap V(\lambda,\nu)} + \|\tau_- r^3\not\nabla\beta_3\|_{2,C(\lambda)\cap V(\lambda,\nu)}\\
&\quad + \|r^4\not\nabla\beta_4\|_{2,C(\lambda)\cap V(\lambda,\nu)} + \|\tau_-^2 r^2\beta_{33}\|_{2,C(\lambda)\cap V(\lambda,\nu)}\\
&\quad + \|r^4\beta_{34}\|_{2,C(\lambda)\cap V(\lambda,\nu)} + \|r^4\beta_{44}\|_{2,C(\lambda)\cap V(\lambda,\nu)}.
\end{aligned}
$$

Estimate for $\|r^4\beta_{44}\|_{2,C(\lambda)\cap V(\lambda,\nu)}$

$$
\begin{aligned}
\|r^4\beta_{44}\|^2_{2,C(\lambda)\cap V(\lambda,\nu)} &= \int_{C(\lambda)\cap V(\lambda,\nu)} r^8|\beta_{44}(W)|^2 = \int_{C(\lambda)\cap V(\lambda,\nu)} r^8|\not\nabla\alpha_4(W)|^2\\
&= \int_{C(\lambda)\cap V(\lambda,\nu)} r^8|\not\nabla\alpha_3(W)|^2 + \int_{C(\lambda)\cap V(\lambda,\nu)} r^8|\not\nabla\not{\mathbf{D}}_T\alpha(W)|^2\\
&= \int_{C(\lambda)\cap V(\lambda,\nu)} r^8|\not\nabla\alpha_3(W)|^2 + \int_{C(\lambda)\cap V(\lambda,\nu)} r^6|\alpha(\hat{\mathcal{L}}_O\hat{\mathcal{L}}_T W)|^2\\
&\le \int_{C(\lambda)\cap V(\lambda,\nu)} Q(\hat{\mathcal{L}}_O^2 W)(\bar{K},\bar{K},T,e_4)\\
&\quad + \int_{C(\lambda)\cap V(\lambda,\nu)} Q(\hat{\mathcal{L}}_O\hat{\mathcal{L}}_T W)(\bar{K},\bar{K},\bar{K},e_4). \qquad (5.2.7)
\end{aligned}
$$

3. $\mathcal{R}_2[(\rho,\sigma)]$**:**

$$\begin{aligned}\mathcal{R}_2[(\rho,\sigma)](u,\underline{u}) &= \|\tau_- r^3 \nabla\!\!\!/^2(\rho,\sigma)\|_{2,C(\lambda)\cap V(\lambda,\nu)} + \|\tau_-^2 r^2 \nabla\!\!\!/(\rho,\sigma)_3\|_{2,C(\lambda)\cap V(\lambda,\nu)}\\ &+ \|r^4 \nabla\!\!\!/(\rho,\sigma)_4\|_{2,C(\lambda)\cap V(\lambda,\nu)} + \|\tau_- r^3 (\rho,\sigma)_{34}\|_{2,C(\lambda)\cap V(\lambda,\nu)}\\ &+ \|r^4(\rho,\sigma)_{44}\|_{2,C(\lambda)\cap V(\lambda,\nu)} + \|\tau_-^3 r(\rho,\sigma)_{33}\|_{2,C(\lambda)\cap V(\lambda,\nu)}.\end{aligned}$$

Estimate for $\|\tau_-^3 r(\rho,\sigma)_{33}\|_{2,C(\lambda)\cap V(\lambda,\nu)}$

$$\begin{aligned}&\|\tau_-^3 r(\rho,\sigma)_{33}\|^2_{2,C(\lambda)\cap V(\lambda,\nu)}\\ &= \int_{C(\lambda)\cap V(\lambda,\nu)} \tau_-^6 r^2 |(\rho,\sigma)_{33}(W)|^2 = \int_{C(\lambda)\cap V(\lambda,\nu)} \tau_-^6 r^2 |\nabla\!\!\!/ \underline{\beta}_3(W)|^2\\ &= \int_{C(\lambda)\cap V(\lambda,\nu)} \tau_-^6 r^2 |\nabla\!\!\!/ \underline{\beta}_4(W)|^2 + \int_{C(\lambda)\cap V(\lambda,\nu)} \tau_-^6 r^2 |\nabla\!\!\!/ \mathbf{D}_T \underline{\beta}(W)|^2\\ &= \int_{C(\lambda)\cap V(\lambda,\nu)} \tau_-^6 r^2 |\nabla\!\!\!/^2(\rho,\sigma)(W)|^2 + \int_{C(\lambda)\cap V(\lambda,\nu)} \tau_-^6 r^2 |\nabla\!\!\!/ \underline{\beta}(\hat{\mathcal{L}}_T W)|^2\\ &= \int_{C(\lambda)\cap V(\lambda,\nu)} \frac{\tau_-^6}{r^2} |(\rho,\sigma)(\hat{\mathcal{L}}_O^2 W)|^2 + \int_{C(\lambda)\cap V(\lambda,\nu)} \tau_-^6 |\underline{\beta}(\hat{\mathcal{L}}_O \hat{\mathcal{L}}_T W)|^2\\ &\le \int_{C(\lambda)\cap V(\lambda,\nu)} Q(\hat{\mathcal{L}}_O^2 W)(\bar{K},\bar{K},T,e_4) + \int_{C(\lambda)\cap V(\lambda,\nu)} Q(\hat{\mathcal{L}}_O \hat{\mathcal{L}}_T W)(\bar{K},\bar{K},\bar{K},e_4).\end{aligned} \tag{5.2.8}$$

4. $\mathcal{R}_2[\underline{\beta}]$**:**

$$\begin{aligned}\mathcal{R}_2[\underline{\beta}](\lambda,\nu) &= \|\tau_-^2 r^2 \nabla\!\!\!/^2 \underline{\beta}\|_{2,C(\lambda)\cap V(\lambda,\nu)} + \|\tau_- r^3 \nabla\!\!\!/ \underline{\beta}_4\|_{2,C(\lambda)\cap V(\lambda,\nu)}\\ &+ \|\tau_-^3 r \nabla\!\!\!/ \underline{\beta}_3\|_{2,C(\lambda)\cap V(\lambda,\nu)} + \|\tau_-^2 r^2 \underline{\beta}_{34}\|_{2,C(\lambda)\cap V(\lambda,\nu)}\\ &+ \|r^4 \underline{\beta}_{44}\|_{2,C(\lambda)\cap V(\lambda,\nu)} + \|\tau_-^3 r \underline{\beta}_{33}\|_{2,C(\lambda)\cap V(\lambda,\nu)}.\end{aligned}$$

Estimate for $\|\tau_- r^3 \nabla\!\!\!/ \underline{\beta}_3\|_{2,C(\lambda)\cap V(\lambda,\nu)}$

$$\begin{aligned}\|\tau_-^3 r \nabla\!\!\!/ \underline{\beta}_3\|^2_{2,C(\lambda)\cap V(\lambda,\nu)} &= \int_{C(\lambda)\cap V(\lambda,\nu)} \tau_-^6 r^2 |\nabla\!\!\!/ \underline{\beta}_3|^2\\ &= \int_{C(\lambda)\cap V(\lambda,\nu)} \tau_-^6 r^2 |\nabla\!\!\!/ \underline{\beta}_4|^2 + \int_{C(\lambda)\cap V(\lambda,\nu)} \tau_-^6 r^2 |\nabla\!\!\!/ \mathbf{D}_T \underline{\beta}|^2\\ &= \int_{C(\lambda)\cap V(\lambda,\nu)} \tau_-^6 r^2 |\nabla\!\!\!/^2 \rho(W)|^2 + \int_{C(\lambda)\cap V(\lambda,\nu)} \tau_-^6 r^2 |\nabla\!\!\!/ \underline{\beta}(\hat{\mathcal{L}}_T W)|^2\\ &= \int_{C(\lambda)\cap V(\lambda,\nu)} \frac{\tau_-^6}{r^2} |\rho(\hat{\mathcal{L}}_O^2 W)|^2 + \int_{C(\lambda)\cap V(\lambda,\nu)} \tau_-^6 |\underline{\beta}(\hat{\mathcal{L}}_O \hat{\mathcal{L}}_T W)|^2\\ &\le \int_{C(\lambda)\cap V(\lambda,\nu)} Q(\hat{\mathcal{L}}_O^2 W)(\bar{K},\bar{K},T,e_4) + \int_{C(\lambda)\cap V(\lambda,\nu)} Q(\hat{\mathcal{L}}_O \hat{\mathcal{L}}_T W)(\bar{K},\bar{K},\bar{K},e_4).\end{aligned} \tag{5.2.9}$$

Estimate for $\|\tau_-^3 r\underline{\beta}_{33}\|_{2,C(\lambda)\cap V(\lambda,\nu)}$

$$\|\tau_-^3 r\underline{\beta}_{33}\|^2_{2,C(\lambda)\cap V(\lambda,\nu)} = \int_{C(\lambda)\cap V(\lambda,\nu)} \tau_-^6 r^2|\underline{\beta}_{33}|^2 = \int_{C(\lambda)\cap V(\lambda,\nu)} \tau_-^6 r^2|\underline{\beta}_{44}|^2$$
$$+\int_{C(\lambda)\cap V(\lambda,\nu)} \tau_-^6 r^2|\mathbf{D\!\!\!/}_T\underline{\beta}_4|^2+\int_{C(\lambda)\cap V(\lambda,\nu)} \tau_-^6 r^2|\mathbf{D\!\!\!/}_T^2\underline{\beta}|^2$$
$$=\int_{C(\lambda)\cap V(\lambda,\nu)} \tau_-^6 r^2|\nabla\!\!\!\!/^2\beta|^2+\int_{C(\lambda)\cap V(\lambda,\nu)} \tau_-^6 r^2|\mathbf{D\!\!\!/}_T\nabla\!\!\!\!/\rho(W)|^2$$
$$+\int_{C(\lambda)\cap V(\lambda,\nu)} \tau_-^6 r^2|\underline{\beta}(\hat{\mathcal{L}}_T^2 W)|^2 = \int_{C(\lambda)\cap V(\lambda,\nu)} \frac{\tau_-^6}{r^2}|\beta(\hat{\mathcal{L}}_O^2 W)|^2$$
$$+\int_{C(\lambda)\cap V(\lambda,\nu)} \tau_-^6|\rho(\hat{\mathcal{L}}_O\hat{\mathcal{L}}_T W)|^2+\int_{C(\lambda)\cap V(\lambda,\nu)} \tau_-^6|\underline{\beta}(\hat{\mathcal{L}}_S\hat{\mathcal{L}}_T W)|^2$$
$$\leq\int_{C(\lambda)\cap V(\lambda,\nu)} Q(\hat{\mathcal{L}}_O^2 W)(\bar{K},\bar{K},T,e_4)+\int_{C(\lambda)\cap V(\lambda,\nu)} Q(\hat{\mathcal{L}}_O\hat{\mathcal{L}}_T W)(\bar{K},\bar{K},\bar{K},e_4)$$
$$+\int_{C(\lambda)\cap V(\lambda,\nu)} Q(\hat{\mathcal{L}}_S\hat{\mathcal{L}}_T W)(\bar{K},\bar{K},\bar{K},e_4). \tag{5.2.10}$$

5. $\underline{\mathcal{R}}_2[\beta]$**:**

$$\underline{\mathcal{R}}_2[\beta](\lambda,\nu) = \|r^4\nabla\!\!\!\!/^2\beta\|_{2,\underline{C}(\nu)\cap V(\lambda,\nu)} + \|r^4\nabla\!\!\!\!/\beta_3\|_{2,\underline{C}(\nu)\cap V(\lambda,\nu)} + \|r^4\nabla\!\!\!\!/\beta_4\|_{2,\underline{C}(\nu)\cap V(\lambda,\nu)}$$
$$+ \|r^4\beta_{43}\|_{2,\underline{C}(\nu)\cap V(\lambda,\nu)} + \|\tau_- r^3\beta_{33}\|_{2,\underline{C}(\nu)\cap V(\lambda,\nu)} + \|\tau_- r^3\beta_{43}\|_{2,\underline{C}(\nu)\cap V(\lambda,\nu)}.$$

Estimate for $\|r^4\nabla\!\!\!\!/\beta_4\|_{2,\underline{C}(\nu)\cap V(\lambda,\nu)}$

$$\|r^4\nabla\!\!\!\!/\beta_4\|^2_{2,\underline{C}(\nu)\cap V(\lambda,\nu)} = \int_{\underline{C}(\nu)\cap V(\lambda,\nu)} r^8|\nabla\!\!\!\!/\beta_4(W)|^2$$
$$=\int_{\underline{C}(\nu)\cap V(\lambda,\nu)} r^8|\nabla\!\!\!\!/\beta_3(W)|^2+\int_{\underline{C}(\nu)\cap V(\lambda,\nu)} r^8|\nabla\!\!\!\!/\mathbf{D\!\!\!/}_T\beta(W)|^2$$
$$=\int_{\underline{C}(\nu)\cap V(\lambda,\nu)} r^8|\nabla\!\!\!\!/\beta_3(W)|^2+\int_{\underline{C}(\nu)\cap V(\lambda,\nu)} r^8|\nabla\!\!\!\!/\beta(\hat{\mathcal{L}}_T W)|^2$$
$$=\int_{\underline{C}(\nu)\cap V(\lambda,\nu)} r^8|\nabla\!\!\!\!/\beta_3(W)|^2+\int_{\underline{C}(\nu)\cap V(\lambda,\nu)} r^6|\beta(\hat{\mathcal{L}}_O\hat{\mathcal{L}}_T W)|^2 \tag{5.2.11}$$
$$\leq\int_{\underline{C}(\nu)\cap V(\lambda,\nu)} Q(\hat{\mathcal{L}}_O^2 W)(\bar{K},\bar{K},T,e_3)+\int_{\underline{C}(\nu)\cap V(\lambda,\nu)} Q(\hat{\mathcal{L}}_O\hat{\mathcal{L}}_T W)(\bar{K},\bar{K},\bar{K},e_3).$$

Estimate for $\|\tau_- r^3\beta_{44}\|_{2,\underline{C}(\nu)\cap V(\lambda,\nu)}$

$$\|\tau_- r^3\beta_{44}\|^2_{2,\underline{C}(\nu)\cap V(\lambda,\nu)} = \int_{\underline{C}(\nu)\cap V(\lambda,\nu)} \tau_-^2 r^6|\beta_{44}(W)|^2$$
$$=\int_{\underline{C}(\nu)\cap V(\lambda,\nu)} \tau_-^2 r^6|\beta_{33}(W)|^2+\int_{\underline{C}(\nu)\cap V(\lambda,\nu)} \tau_-^2 r^6|\mathbf{D\!\!\!/}_T\beta_3(W)|^2$$
$$+\int_{\underline{C}(\nu)\cap V(\lambda,\nu)} \tau_-^2 r^6|\mathbf{D\!\!\!/}_T^2\beta(W)|^2$$

$$= \int_{\underline{C}(\nu)\cap V(\lambda,\nu)} \tau_-^2 r^6 |\beta_{33}(W)|^2 + \int_{\underline{C}(\nu)\cap V(\lambda,\nu)} \tau_-^2 r^4 |\rho(\hat{\mathcal{L}}_O \hat{\mathcal{L}}_T W)|^2$$
$$+ \int_{\underline{C}(\nu)\cap V(\lambda,\nu)} \tau_-^2 r^4 |\beta(\hat{\mathcal{L}}_S \hat{\mathcal{L}}_T W)|^2$$
$$\leq \int_{\underline{C}(\nu)\cap V(\lambda,\nu)} Q(\hat{\mathcal{L}}_O^2 W)(\bar{K},\bar{K},T,e_3) + \int_{\underline{C}(\nu)\cap V(\lambda,\nu)} Q(\hat{\mathcal{L}}_O \hat{\mathcal{L}}_T W)(\bar{K},\bar{K},\bar{K},e_3)$$
$$\int_{\underline{C}(\nu)\cap V(\lambda,\nu)} Q(\hat{\mathcal{L}}_S \hat{\mathcal{L}}_T W)(\bar{K},\bar{K},\bar{K},e_3). \tag{5.2.12}$$

6. $\underline{\mathcal{R}}_2[(\rho,\sigma)]$:

$$\begin{aligned}\underline{\mathcal{R}}_2[(\rho,\sigma)](\lambda,\nu) &= \|r^4 \nabla\!\!\!/^2(\rho,\sigma)\|_{2,\underline{C}(\nu)\cap V(\lambda,\nu)} + \|\tau_- r^3 \nabla\!\!\!/(\rho,\sigma)_3\|_{2,\underline{C}(\nu)\cap V(\lambda,\nu)}\\ &+ \|r^4 \nabla\!\!\!/(\rho,\sigma)_4\|_{2,\underline{C}(\nu)\cap V(\lambda,\nu)} + \|\tau_- r^3 (\rho,\sigma)_{34}\|_{2,\underline{C}(\nu)\cap V(\lambda,\nu)}\\ &+ \|\tau_-^2 r^2 (\rho,\sigma)_{33}\|_{2,\underline{C}(\nu)\cap V(\lambda,\nu)} + \|r^4 (\rho,\sigma)_{44}\|_{2,\underline{C}(\nu)\cap V(\lambda,\nu)}.\end{aligned}$$

Estimate for $\|r^4(\rho,\sigma)_{44}\|_{2,\underline{C}(\nu)\cap V(\lambda,\nu)}$

$$\|r^4(\rho,\sigma)_{44}\|^2_{2,\underline{C}(\nu)\cap V(\lambda,\nu)} = \int_{\underline{C}(\nu)\cap V(\lambda,\nu)} r^8 |(\rho,\sigma)_{44}(W)|^2$$
$$= \int_{\underline{C}(\nu)\cap V(\lambda,\nu)} r^8 |(\rho,\sigma)_{34}(W)|^2 + \int_{\underline{C}(\nu)\cap V(\lambda,\nu)} r^8 |\left(\mathbf{D}_T(\rho,\sigma)\right)_4(W)|^2$$
$$= \int_{\underline{C}(\nu)\cap V(\lambda,\nu)} r^8 |(\rho,\sigma)_{34}(W)|^2 + \int_{\underline{C}(\nu)\cap V(\lambda,\nu)} r^8 |(\rho,\sigma)_4(\hat{\mathcal{L}}_T W)|^2$$
$$= \int_{\underline{C}(\nu)\cap V(\lambda,\nu)} r^8 |(\rho,\sigma)_{34}(W)|^2 + \int_{\underline{C}(\nu)\cap V(\lambda,\nu)} r^8 |\nabla\!\!\!/\beta(\hat{\mathcal{L}}_T W)|^2$$
$$= \int_{\underline{C}(\nu)\cap V(\lambda,\nu)} r^8 |(\rho,\sigma)_{34}(W)|^2 + \int_{\underline{C}(\nu)\cap V(\lambda,\nu)} r^6 |\beta(\hat{\mathcal{L}}_O \hat{\mathcal{L}}_T W)|^2 \tag{5.2.13}$$
$$\leq \int_{\underline{C}(\nu)\cap V(\lambda,\nu)} Q(\hat{\mathcal{L}}_O^2 W)(\bar{K},\bar{K},T,e_3) + \int_{\underline{C}(\nu)\cap V(\lambda,\nu)} Q(\hat{\mathcal{L}}_O \hat{\mathcal{L}}_T W)(\bar{K},\bar{K},\bar{K},e_3).$$

7. $\underline{\mathcal{R}}_2[\underline{\beta}]$:

$$\begin{aligned}\underline{\mathcal{R}}_2[\underline{\beta}](\lambda,\nu) &= \|\tau_- r^3 \nabla\!\!\!/^2 \underline{\beta}\|_{2,\underline{C}(\nu)\cap V(\lambda,\nu)} + \|\tau_-^2 r^2 \nabla\!\!\!/\underline{\beta}_3\|_{2,\underline{C}(\nu)\cap V(\lambda,\nu)}\\ &+ \|r^4 \nabla\!\!\!/\underline{\beta}_4\|_{2,\underline{C}(\nu)\cap V(\lambda,\nu)} + \|\tau_-^3 r \underline{\beta}_{33}\|_{2,\underline{C}(\nu)\cap V(\lambda,\nu)}\\ &+ \|\tau_- r^3 \underline{\beta}_{34}\|_{2,\underline{C}(\nu)\cap V(\lambda,\nu)} + \|r^4 \underline{\beta}_{44}\|_{2,\underline{C}(\nu)\cap V(\lambda,\nu)}.\end{aligned}$$

Estimate for $\|\tau_-^3 r \underline{\beta}_{33}\|_{2,\underline{C}(\nu)\cap V(\lambda,\nu)}$

$$\|\tau_-^3 r \underline{\beta}_{33}\|_{2,\underline{C}(\nu)\cap V(\lambda,\nu)} = \int_{\underline{C}(\nu)\cap V(\lambda,\nu)} \tau_-^6 r^2 |\underline{\beta}_{33}(W)|^2$$

$$
\begin{aligned}
&= \int_{\underline{C}(\nu)\cap V(\lambda,\nu)} \tau_-^6 r^2 |\nabla\!\!\!/ \underline{\alpha}_4(W)|^2 + \int_{\underline{C}(\nu)\cap V(\lambda,\nu)} \tau_-^6 r^2 |\nabla\!\!\!/ \mathbf{D}\!\!\!/_T \underline{\alpha}(W)|^2 \\
&= \int_{\underline{C}(\nu)\cap V(\lambda,\nu)} \tau_-^6 r^2 |\nabla\!\!\!/^2 \underline{\beta}(W)|^2 + \int_{\underline{C}(\nu)\cap V(\lambda,\nu)} \tau_-^6 r^2 |\nabla\!\!\!/ \underline{\alpha}(\hat{\mathcal{L}}_T W)|^2 \\
&= \int_{\underline{C}(\nu)\cap V(\lambda,\nu)} \frac{\tau_-^6}{r^2} |\underline{\beta}(\hat{\mathcal{L}}_O^2 W)|^2 + \int_{\underline{C}(\nu)\cap V(\lambda,\nu)} \tau_-^6 |\underline{\beta}(\hat{\mathcal{L}}_O \hat{\mathcal{L}}_T W)|^2 \qquad (5.2.14)\\
&\leq \int_{\underline{C}(\nu)\cap V(\lambda,\nu)} Q(\hat{\mathcal{L}}_O^2 W)(\bar{K},\bar{K},T,e_3) + \int_{\underline{C}(\nu)\cap V(\lambda,\nu)} Q(\hat{\mathcal{L}}_O \hat{\mathcal{L}}_T W)(\bar{K},\bar{K},T,e_3).
\end{aligned}
$$

8. $\underline{\mathcal{R}}_2[\underline{\alpha}]$**:**

$$
\begin{aligned}
\underline{\mathcal{R}}_2[\underline{\alpha}](\lambda,\nu) &= \|\tau_-^2 r^2 \nabla\!\!\!/^2 \underline{\alpha}\|_{2,\underline{C}(\nu)\cap V(\lambda,\nu)} + \|\tau_-^3 r \nabla\!\!\!/ \underline{\alpha}_3\|_{2,\underline{C}(\nu)\cap V(\lambda,\nu)} \\
&+ \|\tau_- r^3 \nabla\!\!\!/ \underline{\alpha}_4\|_{2,\underline{C}(\nu)\cap V(\lambda,\nu)} + \|\tau_-^4 \underline{\alpha}_{33}\|_{2,\underline{C}(\nu)\cap V(\lambda,\nu)} \\
&+ \|\tau_-^2 r^2 \underline{\alpha}_{34}\|_{2,\underline{C}(\nu)\cap V(\lambda,\nu)} + \|\tau_- r^3 \underline{\alpha}_{44}\|_{2,\underline{C}(\nu)\cap V(\lambda,\nu)}.
\end{aligned}
$$

Estimate for $\|\tau_-^3 r \nabla\!\!\!/ \underline{\alpha}_3\|_{2,\underline{C}(\nu)\cap V(\lambda,\nu)}$

$$
\begin{aligned}
\|\tau_-^3 r \nabla\!\!\!/ \underline{\alpha}_3\|_{2,\underline{C}(\nu)\cap V(\lambda,\nu)} &= \int_{\underline{C}(\nu)\cap V(\lambda,\nu)} \tau_-^6 r^2 |\nabla\!\!\!/ \underline{\alpha}_3(W)|^2 \\
&= \int_{\underline{C}(\nu)\cap V(\lambda,\nu)} \tau_-^6 r^2 |\nabla\!\!\!/ \mathbf{D}\!\!\!/_T \underline{\alpha}(W)|^2 + \int_{\underline{C}(\nu)\cap V(\lambda,\nu)} \tau_-^6 r^2 |\nabla\!\!\!/ \underline{\alpha}_4(W)|^2 \\
&= \int_{\underline{C}(\nu)\cap V(\lambda,\nu)} \tau_-^6 r^2 |\nabla\!\!\!/ \underline{\alpha}(\hat{\mathcal{L}}_T W)|^2 + \int_{\underline{C}(\nu)\cap V(\lambda,\nu)} \tau_-^6 r^2 |\nabla\!\!\!/^2 \underline{\beta}(W)|^2 \\
&= \int_{\underline{C}(\nu)\cap V(\lambda,\nu)} \tau_-^6 |\underline{\alpha}(\hat{\mathcal{L}}_O \hat{\mathcal{L}}_T W)|^2 + \int_{\underline{C}(\nu)\cap V(\lambda,\nu)} \frac{\tau_-^6}{r^2} |\underline{\beta}(\hat{\mathcal{L}}_O^2 W)|^2 \qquad (5.2.15)\\
&\leq \int_{\underline{C}(\nu)\cap V(\lambda,\nu)} Q(\hat{\mathcal{L}}_O^2 W)(\bar{K},\bar{K},T,e_3) + \int_{\underline{C}(\nu)\cap V(\lambda,\nu)} Q(\hat{\mathcal{L}}_O \hat{\mathcal{L}}_T W)(\bar{K},\bar{K},\bar{K},e_3).
\end{aligned}
$$

Estimate for $\|\tau_-^2 r^2 \underline{\alpha}_{34}\|_{2,\underline{C}(\nu)\cap V(\lambda,\nu)}$

$$
\begin{aligned}
\|\tau_-^2 r^2 \underline{\alpha}_{34}\|_{2,\underline{C}(\nu)\cap V(\lambda,\nu)} &= \int_{\underline{C}(\nu)\cap V(\lambda,\nu)} \tau_-^4 r^4 |\underline{\alpha}_{34}(W)|^2 \\
&= \int_{\underline{C}(\nu)\cap V(\lambda,\nu)} \tau_-^4 r^4 |\nabla\!\!\!/ \underline{\beta}_3(W)|^2 + \int_{\underline{C}(\nu)\cap V(\lambda,\nu)} \tau_-^4 r^4 |\nabla\!\!\!/^2 \underline{\alpha}(W)|^2 \qquad (5.2.16)\\
&= \int_{\underline{C}(\nu)\cap V(\lambda,\nu)} \tau_-^4 |\underline{\alpha}(\hat{\mathcal{L}}_O^2 W)|^2 \leq \int_{\underline{C}(\nu)\cap V(\lambda,\nu)} Q(\hat{\mathcal{L}}_O^2 W)(\bar{K},\bar{K},T,e_3).
\end{aligned}
$$

Estimate for $\|\tau_-^4 \underline{\alpha}_{33}\|_{2,\underline{C}(\nu)\cap V(\lambda,\nu)}$

$$
\|\tau_-^4 \underline{\alpha}_{33}\|_{2,\underline{C}(\nu)\cap V(\lambda,\nu)} = \int_{\underline{C}(\nu)\cap V(\lambda,\nu)} \tau_-^8 |\underline{\alpha}_{33}(W)|^2 = \int_{\underline{C}(\nu)\cap V(\lambda,\nu)} \tau_-^8 |\underline{\alpha}_{44}(W)|^2
$$

$$
\begin{aligned}
&+ \int_{\underline{C}(\nu)\cap V(\lambda,\nu)} \tau_-^8 |\mathbf{D}_T \mathbf{D}_T \underline{\alpha}(W)|^2 + \int_{\underline{C}(\nu)\cap V(\lambda,\nu)} \tau_-^8 |(\mathbf{D}_T \underline{\alpha})_4(W)|^2 \\
&= \int_{\underline{C}(\nu)\cap V(\lambda,\nu)} \tau_-^8 |\underline{\alpha}_{44}(W)|^2 + \int_{\underline{C}(\nu)\cap V(\lambda,\nu)} \tau_-^8 |\underline{\alpha}(\hat{\mathcal{L}}_T^2 W)|^2 \\
&+ \int_{\underline{C}(\nu)\cap V(\lambda,\nu)} \tau_-^8 |\nabla\!\!\!/ \underline{\beta}(\hat{\mathcal{L}}_T W)|^2 = \int_{\underline{C}(\nu)\cap V(\lambda,\nu)} \tau_-^8 |\underline{\alpha}_{44}(W)|^2 \\
&+ \int_{\underline{C}(\nu)\cap V(\lambda,\nu)} \tau_-^8 |\underline{\alpha}(\hat{\mathcal{L}}_T^2 W)|^2 + \int_{\underline{C}(\nu)\cap V(\lambda,\nu)} \frac{\tau_-^8}{r^2} |\underline{\beta}(\hat{\mathcal{L}}_O \hat{\mathcal{L}}_T W)|^2 \\
&= \int_{\underline{C}(\nu)\cap V(\lambda,\nu)} \tau_-^8 |\underline{\alpha}_{44}(W)|^2 + \int_{\underline{C}(\nu)\cap V(\lambda,\nu)} \frac{\tau_-^8}{r^2} |\underline{\alpha}(\hat{\mathcal{L}}_S \hat{\mathcal{L}}_T W)|^2 \\
&+ \int_{\underline{C}(\nu)\cap V(\lambda,\nu)} \frac{\tau_-^8}{r^2} |\underline{\beta}(\hat{\mathcal{L}}_O \hat{\mathcal{L}}_T W)|^2 \le \int_{\underline{C}(\nu)\cap V(\lambda,\nu)} Q(\hat{\mathcal{L}}_O^2 W)(\bar{K}, \bar{K}, T, e_3) \qquad (5.2.17) \\
&+ \int_{\underline{C}(\nu)\cap V(\lambda,\nu)} Q(\hat{\mathcal{L}}_O \hat{\mathcal{L}}_T W)(\bar{K}, \bar{K}, \bar{K}, e_3) + \int_{\underline{C}(\nu)\cap V(\lambda,\nu)} Q(\hat{\mathcal{L}}_S \hat{\mathcal{L}}_T W)(\bar{K}, \bar{K}, \bar{K}, e_3).
\end{aligned}
$$

We end the appendix by discussing at length some of the more delicate estimates of Proposition 5.1.5. Detailed estimates for the other norms are easier and proceed along the same lines.

1.1 $\|r^4 \nabla\!\!\!/^2 \alpha\|_{2,C(\lambda)\cap V(\lambda,\nu)}$

Using Proposition 7.5.3 of [Ch-Kl] it follows that

$$
\begin{aligned}
\int_{C(\lambda)\cap V(\lambda,\nu)} r^8 |\nabla\!\!\!/^2 \alpha(W)|^2 &\le c \int_{C(\lambda)\cap V(\lambda,\nu)} r^6 |\mathcal{L}_O \nabla\!\!\!/ \alpha(W)|^2 \qquad (5.2.18) \\
&= c \int_{C(\lambda)\cap V(\lambda,\nu)} r^6 |\nabla\!\!\!/ \mathcal{L}_O \alpha(W)|^2 + c \int_{C(\lambda)\cap V(\lambda,\nu)} r^6 |[\mathcal{L}_O, \nabla\!\!\!/] \alpha(W)|^2.
\end{aligned}
$$

We use the relation (see (5.1.15))

$$
\mathcal{L}_O \alpha(W) \;=\; \alpha(\hat{\mathcal{L}}_O W) + S({}^{(O)}\pi, {}^{(O)}M) \cdot \alpha(W),
$$

where

$$
S({}^{(O)}\pi, {}^{(O)}M)_{abcd} = -\frac{1}{8} \mathrm{tr}\,{}^{(O)}\pi\, \delta_{ac}\delta_{bd} + \frac{1}{2} \delta_{ab}\, {}^{(O)}\mathbf{i}_{cd}
$$

to rewrite the first integral and estimate it as

$$
\begin{aligned}
&\int_{C(\lambda)\cap V(\lambda,\nu)} r^6 |\nabla\!\!\!/ \mathcal{L}_O \alpha(W)|^2 = \int_{C(\lambda)\cap V(\lambda,\nu)} r^6 |\nabla\!\!\!/ \alpha(\hat{\mathcal{L}}_O W)|^2 \\
&\quad + \int_{C(\lambda)\cap V(\lambda,\nu)} r^6 |\nabla\!\!\!/ [S({}^{(O)}\pi, {}^{(O)}M) \cdot \alpha(W)]|^2 \\
&\quad \le c \left(\int_{C(\lambda)\cap V(\lambda,\nu)} r^4 |\mathcal{L}_O \alpha(\hat{\mathcal{L}}_O W)|^2 + \int_{C(\lambda)\cap V(\lambda,\nu)} r^6 |\nabla\!\!\!/ S({}^{(O)}\pi, {}^{(O)}M)|^2 |\alpha(W)|^2 \right. \\
&\quad \left. + \int_{C(\lambda)\cap V(\lambda,\nu)} r^6 |S({}^{(O)}\pi, {}^{(O)}M)|^2 |\nabla\!\!\!/ \alpha(W)|^2 \right)
\end{aligned}
$$

$$
\begin{aligned}
&\le c\left(\int_{C(\lambda)\cap V(\lambda,\nu)} r^4|\alpha(\hat{\mathcal{L}}_O^2 W)|^2 + \int_{C(\lambda)\cap V(\lambda,\nu)} r^4|S({}^{(O)}\pi,{}^{(O)}M)|^2|\alpha(\hat{\mathcal{L}}_O W)|^2\right.\\
&+\int_{C(\lambda)\cap V(\lambda,\nu)} r^6|\nabla\!\!\!/ S({}^{(O)}\pi,{}^{(O)}M)|^2|\alpha(W)|^2\\
&\left.+\int_{C(\lambda)\cap V(\lambda,\nu)} r^6|S({}^{(O)}\pi,{}^{(O)}M)|^2|\nabla\!\!\!/\alpha(W)|^2\right)\\
&\le c\int_{C(\lambda)\cap V(\lambda,\nu)} r^4|\alpha(\hat{\mathcal{L}}_O^2 W)|^2 + c\frac{\epsilon_0^2}{r^2}\left(\int_{C(\lambda)\cap V(\lambda,\nu)} r^4|\alpha(\hat{\mathcal{L}}_O W)|^2\right.\\
&\left.+\int_{C(\lambda)\cap V(\lambda,\nu)} r^4|\alpha(W)|^2\right) \le c\int_{C(\lambda)\cap V(\lambda,\nu)} r^4|\alpha(\hat{\mathcal{L}}_O^2 W)|^2 + c\epsilon_0^2{\mathcal{R}_{[0]}}^2, \qquad (5.2.19)
\end{aligned}
$$

where we used the following estimates for $S({}^{(O)}\pi,{}^{(O)}M)$ which can be easily deduced from Chapter 3, Section 3.7, for $p\in[2,4]$:

$$
\begin{aligned}
&\sup_{\mathcal{K}}|rS({}^{(O)}\pi,{}^{(O)}M)| \le c\epsilon_0\\
&\sup_{\mathcal{K}}|r^{2-\frac{2}{p}}\nabla\!\!\!/ S({}^{(O)}\pi,{}^{(O)}M)|_{p,S} \le c\epsilon_0. \qquad (5.2.20)
\end{aligned}
$$

To estimate the second integral of (5.2.18), $\int_{C(\lambda)\cap V(\lambda,\nu)} r^6|[\mathcal{L}_O,\nabla\!\!\!/]\alpha(W)|^2$, we observe that (see (4.8.42) and Corollary 16.1.31 of [Ch-Kl])

$$
\begin{aligned}
([\nabla\!\!\!/, L_O]\alpha)_{abc} &= \left(\nabla\!\!\!/_b H_{da} + \nabla\!\!\!/_a H_{db} - \nabla\!\!\!/_d H_{ba}\right)\alpha_{dc}\\
&+ \left(\nabla\!\!\!/_b H_{da} + \nabla\!\!\!/_a H_{dc} - \nabla\!\!\!/_d H_{ca}\right)\alpha_{db}.
\end{aligned}
$$

Due to the estimates of Chapter 3, Section 3.7, the following inequalities hold, $p\in[2,4]$:

$$
|[\mathcal{L}_O,\nabla\!\!\!/]\alpha(W)|_{p,S}^2 \le c\frac{\epsilon_0^2}{r^2}|\alpha(W)|_{p,S}^2
$$

we obtain

$$
\int_{C(\lambda)\cap V(\lambda,\nu)} r^6|[\mathcal{L}_O,\nabla\!\!\!/]\alpha(W)|^2 \le c\epsilon_0^2\int_{C(\lambda)\cap V(\lambda,\nu)} r^4|\alpha(W)|^2 \le c\epsilon_0^2{\mathcal{R}_{[0]}}^2
$$

so that, finally,

$$
\|r^4\nabla\!\!\!/^2\alpha\|_{2,C(\lambda)\cap V(\lambda,\nu)}^2 \le c\int_{C(\lambda)\cap V(\lambda,\nu)} r^4|\alpha(\hat{\mathcal{L}}_O^2 W)|^2 + c\epsilon_0^2{\mathcal{R}_{[0]}}^2.
$$

1.2 $\|r^4\nabla\!\!\!/\alpha_4\|_{2,C(\lambda)\cap V(\lambda,\nu)}$ **.**

$$
\begin{aligned}
&\|r^4\nabla\!\!\!/\alpha_4\|_{2,C(\lambda)\cap V(\lambda,\nu)}^2 = \int_{C(\lambda)\cap V(\lambda,\nu)} r^8|\nabla\!\!\!/\alpha_4(W)|^2\\
&\le c\left(\int_{C(\lambda)\cap V(\lambda,\nu)} r^8|\nabla\!\!\!/\alpha_3(W)|^2 + \int_{C(\lambda)\cap V(\lambda,\nu)} r^8|\nabla\!\!\!/\mathbf{D}\!\!\!\!/_T\alpha(W)|^2\right.\\
&\left.+\int_{C(\lambda)\cap V(\lambda,\nu)} r^8|\nabla\!\!\!/(\frac{5}{2}\mathrm{tr}\chi + \frac{1}{2}\mathrm{tr}\underline{\chi})\alpha(W)|^2\right)
\end{aligned}
$$

$$\leq c\left(\int_{C(\lambda)\cap V(\lambda,\nu)} r^8|\nabla\alpha_3(W)|^2 + \int_{C(\lambda)\cap V(\lambda,\nu)} r^8|\nabla\mathbf{D}_T\alpha(W)|^2\right.$$
$$\left. + \int_{C(\lambda)\cap V(\lambda,\nu)} r^8|\nabla(\mathrm{tr}\chi, \mathrm{tr}\underline{\chi})|^2|\alpha(W)|^2 + \int_{C(\lambda)\cap V(\lambda,\nu)} r^8|(\mathrm{tr}\chi, \mathrm{tr}\underline{\chi})|^2|\nabla\alpha(W)|^2\right)$$
$$\leq c\left(\int_{C(\lambda)\cap V(\lambda,\nu)} r^8|\nabla\alpha_3(W)|^2 + \int_{C(\lambda)\cap V(\lambda,\nu)} r^8|\nabla\mathbf{D}_T\alpha(W)|^2\right.$$
$$\left. + \int_{C(\lambda)\cap V(\lambda,\nu)} r^6|\nabla\alpha(W)|^2 + \epsilon_0^2\int_{C(\lambda)\cap V(\lambda,\nu)} r^2|\alpha(W)|^2\right)$$
$$\leq c\left(\int_{C(\lambda)\cap V(\lambda,\nu)} r^8|\nabla\alpha_3(W)|^2 + \int_{C(\lambda)\cap V(\lambda,\nu)} r^8|\nabla\mathbf{D}_T\alpha(W)|^2 + \mathcal{R}_1{}^2 + \epsilon_0^2\mathcal{R}_{[0]}{}^2\right), \tag{5.2.21}$$

where we used the relation $\alpha_4 = -\alpha_3 + 2\mathbf{D}_T\alpha + (\frac{5}{2}\mathrm{tr}\chi + \frac{1}{2}\mathrm{tr}\underline{\chi})\alpha$ and the estimates of Chapter 3 for $\mathrm{tr}\chi$, $\mathrm{tr}\underline{\chi}$, $\nabla\mathrm{tr}\chi$, $\nabla\mathrm{tr}\underline{\chi}$. Using the Bianchi equation

$$\alpha_3 = \nabla\widehat{\otimes}\beta + \left[4\underline{\omega}\alpha - 3(\hat{\chi}\rho + {}^\star\hat{\chi}\sigma) + (\zeta + 4\eta)\widehat{\otimes}\beta\right],$$

and the estimates on the Ricci coefficients proved in Chapter 3 we obtain

$$\begin{aligned}\int_{C(\lambda)\cap V(\lambda,\nu)} r^8|\nabla\alpha_3(W)|^2 &\leq c\int_{C(\lambda)\cap V(\lambda,\nu)} r^8|\nabla^2\beta(W)|^2 + c\epsilon_0^2\mathcal{R}_{[1]}{}^2\\ &\leq c\mathcal{R}_2[\beta]^2 + c\epsilon_0^2\mathcal{R}_{[1]}{}^2.\end{aligned}$$

To estimate the last integral of (5.2.21) we recall that (see (5.1.24) and Proposition 5.1.1)

$$\mathbf{D}_T\alpha(W) = \mathcal{L}_T\alpha(W) + {}^{(T)}H\cdot\alpha(W)\,.$$

Moreover, using Lemma 5.1.1, we write

$$\begin{aligned}\nabla\mathcal{L}_T\alpha(W) &= \nabla\left(\alpha(\hat{\mathcal{L}}_T W) + G_1({}^{(T)}M, {}^{(T)}\pi)\alpha(W) + G_2({}^{(T)}P, {}^{(T)}Q)\beta(W)\right)\\ &= \nabla\alpha(\hat{\mathcal{L}}_T W + (\nabla G_1({}^{(T)}M, {}^{(T)}\pi))\alpha(W) + G_1({}^{(T)}M, {}^{(T)}\pi)(\nabla\alpha(W))\\ &+ \left(\nabla G_2({}^{(T)}P, {}^{(T)}Q)\right)\beta(W) + G_2({}^{(T)}P, {}^{(T)}Q)\nabla\beta(W)\end{aligned}$$

and, from this we obtain

$$\int_{C(\lambda)\cap V(\lambda,\nu)} r^8|\nabla\mathbf{D}_T\alpha(W)|^2 \leq$$
$$\leq \int_{C(\lambda)\cap V(\lambda,\nu)} r^6|\alpha(\hat{\mathcal{L}}_O\hat{\mathcal{L}}_T W)|^2 + \int_{C(\lambda)\cap V(\lambda,\nu)} r^8|{}^{(O)}G_1|^2|\alpha(\hat{\mathcal{L}}_T W)|^2$$
$$+ \int_{C(\lambda)\cap V(\lambda,\nu)} r^8|\nabla{}^{(T)}H|^2|\alpha(W)|^2 + \int_{C(\lambda)\cap V(\lambda,\nu)} r^8|{}^{(T)}H|^2|\nabla\alpha(W)|^2$$
$$+ \int_{C(\lambda)\cap V(\lambda,\nu)} r^8|\nabla{}^{(T)}G_1|^2|\alpha(W)|^2 + \int_{C(\lambda)\cap V(\lambda,\nu)} r^8|{}^{(T)}G_1|^2|\nabla\alpha(W)|^2$$
$$+ \int_{C(\lambda)\cap V(\lambda,\nu)} r^8|\nabla{}^{(T)}G_2|^2|\beta(W)|^2 + \int_{C(\lambda)\cap V(\lambda,\nu)} r^8|{}^{(T)}G_2|^2|\nabla\beta(W)|^2.$$

Using the estimates, previously proved,

$$\sup\left(|r^{(T)}H| + \sup|r^{(O)}G_1|\right) \le c\epsilon_0$$
$$\sup\left(|r^{(T)}P| + |r^{(T)}Q| + |r^{(T)}M| + |r^{(T)}\pi|\right) \le c\epsilon_0$$
$$|r^{2-\frac{2}{p}}\nabla\!\!\!/^{(T)}H|_{p,S} \le c\epsilon_0$$
$$|r^{2-\frac{2}{p}}\nabla\!\!\!/^{(T)}P|_{p,S} + |r^{2-\frac{2}{p}}\nabla\!\!\!/^{(T)}Q|_{p,S} + |r^{2-\frac{2}{p}}\nabla\!\!\!/^{(T)}M|_{p,S} + |r^{2-\frac{2}{p}}\nabla\!\!\!/^{(T)}\pi|_{p,S} \le c\epsilon_0,$$

for $p \in [2,4]$, we obtain

$$\int_{C(\lambda)\cap V(\lambda,\nu)} r^8|\nabla\!\!\!/\mathbf{D}\!\!\!/_T\alpha(W)|^2 \le \int_{C(\lambda)\cap V(\lambda,\nu)} r^6|\alpha(\hat{\mathcal{L}}_O\hat{\mathcal{L}}_T W)|^2 + c\epsilon_0^2{\mathcal{R}_{[1]}}^2,$$

and, finally,

$$\begin{aligned}
\|r^4\nabla\!\!\!/\alpha_4\|^2_{2,C(\lambda)\cap V(\lambda,\nu)} &\le \int_{C(\lambda)\cap V(\lambda,\nu)} r^6|\alpha(\hat{\mathcal{L}}_O\hat{\mathcal{L}}_T W)|^2 + c\mathcal{R}_2[\beta]^2 + c(1+\epsilon_0^2){\mathcal{R}_{[1]}}^2\\
&\le \int_{C(\lambda)\cap V(\lambda,\nu)} Q(\hat{\mathcal{L}}_O\hat{\mathcal{L}}_T W)(\bar{K},\bar{K},\bar{K},e_4) + c\mathcal{R}_2[\beta]^2 + c(1+\epsilon_0^2){\mathcal{R}_{[1]}}^2\\
&\le \int_{C(\lambda)\cap V(\lambda,\nu)} Q(\hat{\mathcal{L}}_O^2 W)(\bar{K},\bar{K},T,e_4) + \int_{C(\lambda)\cap V(\lambda,\nu)} Q(\hat{\mathcal{L}}_O\hat{\mathcal{L}}_T W)(\bar{K},\bar{K},\bar{K},e_4)\\
&\quad + c(1+\epsilon_0^2){\mathcal{R}_{[1]}}^2,
\end{aligned}$$

where we used the estimate of Proposition 5.1.5 for the $\mathcal{R}_2[\beta]^2$ term.

1.3 $\|\tau_- r^3\alpha_{44}\|_{2,C(\lambda)\cap V(\lambda,\nu)}$ **.** Starting from the relationship,[24]

$$\alpha_{44} = -\alpha_{34} - 2\mathbf{D}\!\!\!/_3\mathbf{D}\!\!\!/_T\alpha + 4\mathbf{D}\!\!\!/_T^2\alpha + \frac{5}{2}\mathrm{tr}\chi\alpha_3 + \frac{1}{2}\left[5\mathbf{D}\!\!\!/_4(\mathrm{tr}\chi\alpha) + \mathbf{D}\!\!\!/_4(\mathrm{tr}\underline{\chi}\alpha)\right],$$

it follows that

$$\begin{aligned}
\int_{C(\lambda)\cap V(\lambda,\nu)} \tau_-^2 r^6|\alpha_{44}(W)|^2 &\le c\Bigg(\int_{C(\lambda)\cap V(\lambda,\nu)} \tau_-^2 r^6|\alpha_{34}(W)|^2\\
&\quad + \int_{C(\lambda)\cap V(\lambda,\nu)} \tau_-^2 r^6|\mathbf{D}\!\!\!/_3\mathbf{D}\!\!\!/_T\alpha(W)|^2 + \int_{C(\lambda)\cap V(\lambda,\nu)} \tau_-^2 r^6|\mathbf{D}\!\!\!/_T^2\alpha(W)|^2\Bigg)\\
&\quad + \int_{C(\lambda)\cap V(\lambda,\nu)} \tau_-^2 r^4|\alpha_3(W)|^2 + \int_{C(\lambda)\cap V(\lambda,\nu)} \tau_-^2 r^4|\alpha_4(W)|^2 + c\epsilon_0^2{\mathcal{R}_{[0]}}^2\\
&\le c\Bigg(\int_{C(\lambda)\cap V(\lambda,\nu)} \tau_-^2 r^6|\alpha_{34}(W)|^2 + \int_{C(\lambda)\cap V(\lambda,\nu)} \tau_-^2 r^6|\mathbf{D}\!\!\!/_3\mathbf{D}\!\!\!/_T\alpha(W)|^2\\
&\quad + \int_{C(\lambda)\cap V(\lambda,\nu)} \tau_-^2 r^6|\mathbf{D}\!\!\!/_T^2\alpha(W)|^2\Bigg) + c(1+\epsilon_0^2){\mathcal{R}_{[1]}}^2.
\end{aligned} \tag{5.2.22}$$

24

$$\begin{aligned}
\alpha_{44} &= \mathbf{D}\!\!\!/_4\alpha_4 + \frac{7}{2}\mathrm{tr}\chi\alpha_4 = \mathbf{D}\!\!\!/_4[-\alpha_3 + 2\mathbf{D}\!\!\!/_T\alpha + \frac{1}{2}(5\mathrm{tr}\chi + \mathrm{tr}\underline{\chi})\alpha]\\
&= -\mathbf{D}\!\!\!/_4\alpha_3 - 2\mathbf{D}\!\!\!/_3\mathbf{D}\!\!\!/_T\alpha + 4\mathbf{D}\!\!\!/_T^2\alpha + \frac{1}{2}\left[5\mathbf{D}\!\!\!/_4(\mathrm{tr}\chi\alpha) + \mathbf{D}\!\!\!/_4(\mathrm{tr}\underline{\chi}\alpha)\right]\\
&= -\alpha_{34} - 2\mathbf{D}\!\!\!/_3\mathbf{D}\!\!\!/_T\alpha + 4\mathbf{D}\!\!\!/_T^2\alpha + \frac{5}{2}\mathrm{tr}\chi\alpha_3 + \frac{1}{2}\left[5\mathbf{D}\!\!\!/_4(\mathrm{tr}\chi\alpha) + \mathbf{D}\!\!\!/_4(\mathrm{tr}\underline{\chi}\alpha)\right]
\end{aligned}$$

Let us examine the second integral of (5.2.22), $\int_{C(\lambda)\cap V(\lambda,\nu)} \tau_-^2 r^6 |\mathbf{D}_3 \mathbf{D}_T \alpha(W)|^2$. From 5.1.24 and Proposition 5.1.1,

$$\mathbf{D}_T \alpha(W) = \mathcal{L}_T \alpha(W) + {}^{(T)}H \cdot \alpha(W)\,.$$

Moreover, using Lemma 5.1.1, we can write

$$\mathbf{D}_3 \mathcal{L}_T \alpha(W) = \mathbf{D}_3 \alpha(\hat{\mathcal{L}}_T W) + \mathbf{D}_3 \left(G_1({}^{(T)}M, {}^{(T)}\pi)\alpha(W), +G_2({}^{(T)}P, {}^{(T)}Q)\beta(W)\right)$$

and derive the inequality

$$\begin{aligned}
&\int_{C(\lambda)\cap V(\lambda,\nu)} \tau_-^2 r^6 |\mathbf{D}_3 \mathbf{D}_T \alpha(W)|^2 \le \int_{C(\lambda)\cap V(\lambda,\nu)} \tau_-^2 r^6 |\mathbf{D}_3 \alpha(\hat{\mathcal{L}}_T W)|^2 \\
&+c\epsilon_0^2 \left(\int_{C(\lambda)\cap V(\lambda,\nu)} \tau_-^2 r^4 |\mathbf{D}_3 \alpha(W)|^2 + \int_{C(\lambda)\cap V(\lambda,\nu)} r^4 |\alpha(W)|^2 \right) \\
&\le \int_{C(\lambda)\cap V(\lambda,\nu)} \tau_-^2 r^6 |\mathbf{D}_3 \alpha(\hat{\mathcal{L}}_T W)|^2 + c\epsilon_0^2 \mathcal{R}_{[1]}{}^2 \\
&\le c \int_{C(\lambda)\cap V(\lambda,\nu)} \tau_-^2 r^6 |\nabla\!\!\!/ \beta(\hat{\mathcal{L}}_T W)|^2 + c \int_{C(\lambda)\cap V(\lambda,\nu)} \tau_-^2 r^4 |\alpha(\hat{\mathcal{L}}_T W)|^2 \\
&+c\epsilon_0^2 \left(\mathcal{R}_0{}^2(\hat{\mathcal{L}}_T W) + \mathcal{R}_{[1]}{}^2 \right) \\
&\le c \int_{C(\lambda)\cap V(\lambda,\nu)} \tau_-^2 r^4 |\beta(\hat{\mathcal{L}}_O \hat{\mathcal{L}}_T W)|^2 + c \int_{C(\lambda)\cap V(\lambda,\nu)} \tau_-^2 r^4 |\alpha(\hat{\mathcal{L}}_T W)|^2 \\
&+c\epsilon_0^2 \left(\mathcal{R}_0{}^2(\hat{\mathcal{L}}_T W) + \mathcal{R}_{[1]}{}^2 \right)\,.
\end{aligned}$$

The third integral is estimated in a similar way, the main difference being that its main term, $\int_{C(\lambda)\cap V(\lambda,\nu)} \tau_-^2 r^6 |\mathbf{D}_4 \alpha(\hat{\mathcal{L}}_T W)|^2$, has to be estimated using the expression $\mathbf{D}_4 \alpha = \frac{1}{\tau_+} \mathbf{D}_S \alpha - 2\frac{u}{\tau_+} \mathbf{D}_3 \alpha$, in the following way:

$$\begin{aligned}
\int_{C(\lambda)\cap V(\lambda,\nu)} \tau_-^2 r^6 |\mathbf{D}_4 \alpha(\hat{\mathcal{L}}_T W)|^2 \;\le\; & \int_{C(\lambda)\cap V(\lambda,\nu)} \tau_-^2 r^4 |\mathbf{D}_S \alpha(\hat{\mathcal{L}}_T W)|^2 \\
+ & \int_{C(\lambda)\cap V(\lambda,\nu)} \tau_-^2 r^6 |\mathbf{D}_3 \alpha(\hat{\mathcal{L}}_T W)|^2.
\end{aligned}$$

The only term left to estimate is

$$\int_{C(\lambda)\cap V(\lambda,\nu)} \tau_-^2 r^4 |\mathbf{D}_S \alpha(\hat{\mathcal{L}}_T W)|^2 \le \int_{C(\lambda)\cap V(\lambda,\nu)} \tau_-^2 r^4 |\alpha(\hat{\mathcal{L}}_S \hat{\mathcal{L}}_T W)|^2 + \epsilon_0^2 \mathcal{R}_0{}^2(\hat{\mathcal{L}}_T W)\,.$$

Collecting all these estimates together we infer that

$$\begin{aligned}
&\int_{C(\lambda)\cap V(\lambda,\nu)} \tau_-^2 r^6 |\alpha_{44}(W)|^2 \\
&\le c \int_{C(\lambda)\cap V(\lambda,\nu)} \tau_-^2 r^6 |\alpha_{34}(W)|^2 + \int_{C(\lambda)\cap V(\lambda,\nu)} \tau_-^2 r^4 |\beta(\hat{\mathcal{L}}_O \hat{\mathcal{L}}_T W)|^2, \\
&+ \int_{C(\lambda)\cap V(\lambda,\nu)} \tau_-^2 r^4 |\alpha(\hat{\mathcal{L}}_S \hat{\mathcal{L}}_T W)|^2 + c(1+\epsilon_0^2) \left[\mathcal{R}_0{}^2(\hat{\mathcal{L}}_T W) + \mathcal{R}_{[1]}{}^2 \right]
\end{aligned}$$

which concludes the estimate.

6

The Error Estimates

In this chapter we assume the spacetime $\mathcal{K}$ is foliated by a double null canonical foliation that satisfies the assumptions

$$\mathcal{O} \leq \epsilon_0 \, , \; \mathcal{D} \leq \epsilon_0, \tag{6.0.1}$$

and we make use of the inequality proved in Theorem **M7**

$$\mathcal{R} \leq c \mathcal{Q}_{\mathcal{K}}^{\frac{1}{2}}. \tag{6.0.2}$$

The main result of the chapter is the proof of Theorem **M8**, which we restate below,

Theorem 3.7.10 (Theorem M8). *Under the assumptions (6.0.1) and (6.0.2) with ϵ_0 sufficiently small, the following estimate holds:*

$$\mathcal{Q}_{\mathcal{K}} \leq c_1 \mathcal{Q}_{\Sigma_0 \cap \mathcal{K}}, \tag{6.0.3}$$

with c_1 a constant independent from ϵ_0.

Remark: Observe that the assumptions of Theorem 3.7.10 stated in Chapter 3 imply the assumptions stated here.[1]

To prove this result we need to control the quantity

$$\mathcal{E}(u, \underline{u}) \equiv (\mathcal{Q} + \underline{\mathcal{Q}})(u, \underline{u}) - \mathcal{Q}_{\Sigma_0 \cap V(u,\underline{u})},$$

which we call the *error term* for all values of u and $\underline{u}$ on $\mathcal{K}$. Using the expression (see Proposition 3.2.3)

$$\begin{aligned} \mathbf{Div} P &= \mathbf{Div} Q_{\beta\gamma\delta} X^\beta Y^\gamma Z^\delta \\ &+ \frac{1}{2} Q^{\alpha\beta\gamma\delta} \left({}^{(X)}\pi_{\alpha\beta} Y_\gamma Z_\delta + {}^{(Y)}\pi_{\alpha\beta} Z_\gamma X_\delta + {}^{(Z)}\pi_{\alpha\beta} X_\gamma Y_\delta \right), \end{aligned} \tag{6.0.4}$$

[1] The assumption $\mathcal{R} \leq \epsilon_0$ stated in the version of this theorem, presented in Chapter 3, is needed to control the deformation tensors of the angular momentum vector fields; see Theorem 3.7.4.

and Stokes' theorem, it follows that

$$\int_{\underline{C}(\underline{u})\cap V(u,\underline{u})} Q(W)(X,Y,Z,e_3)+\int_{C(u)\cap V(u,\underline{u})} Q(W)(X,Y,Z,e_4)$$
$$-\int_{\Sigma_0\cap V(u,\underline{u})} Q(W)(X,Y,Z,T)$$
$$=\int_{V_{(u,\underline{u})}}\Big[\mathbf{Div}Q(W)_{\beta\gamma\delta}X^\beta Y^\gamma Z^\delta+\frac{1}{2}Q^{\alpha\beta\gamma\delta}(W)\left({}^{(X)}\pi_{\alpha\beta}Y_\gamma Z_\delta\right.$$
$$\left.+{}^{(Y)}\pi_{\alpha\beta}Z_\gamma X_\delta+{}^{(Z)}\pi_{\alpha\beta}X_\gamma Y_\delta\right)\Big]. \tag{6.0.5}$$

Therefore

$$\mathcal{E}(u,\underline{u})\equiv\mathcal{E}_1(u,\underline{u})+\mathcal{E}_2(u,\underline{u})$$

is a sum of terms like the right-hand side of (6.0.5) where W is replaced by $\hat{\mathcal{L}}_T W$, $\hat{\mathcal{L}}_O W$, $\hat{\mathcal{L}}_O\hat{\mathcal{L}}_T W$, $\hat{\mathcal{L}}_S\hat{\mathcal{L}}_T W$, $\hat{\mathcal{L}}_O^2 W$ and X, Y, Z take values in $\{T,\bar{K}\}$. $\mathcal{E}_1$ and $\mathcal{E}_2$ have the explicit expressions:[2]

$$\begin{aligned}\mathcal{E}_1(u,\underline{u}) &= \int_{V_{(u,\underline{u})}}\mathbf{Div}Q(\hat{\mathcal{L}}_T W)_{\beta\gamma\delta}(\bar{K}^\beta\bar{K}^\gamma\bar{K}^\delta)\\ &+\int_{V_{(u,\underline{u})}}\mathbf{Div}Q(\hat{\mathcal{L}}_O W)_{\beta\gamma\delta}(\bar{K}^\beta\bar{K}^\gamma T^\delta)\\ &+\frac{3}{2}\int_{V_{(u,\underline{u})}}Q(\hat{\mathcal{L}}_T W)_{\alpha\beta\gamma\delta}({}^{(\bar{K})}\pi^{\alpha\beta}\bar{K}^\gamma\bar{K}^\delta)\\ &+\int_{V_{(u,\underline{u})}}Q(\hat{\mathcal{L}}_O W)_{\alpha\beta\gamma\delta}({}^{(\bar{K})}\pi^{\alpha\beta}\bar{K}^\gamma T^\delta)\\ &+\frac{1}{2}\int_{V_{(u,\underline{u})}}Q(\hat{\mathcal{L}}_O W)_{\alpha\beta\gamma\delta}({}^{(T)}\pi^{\alpha\beta}\bar{K}^\gamma\bar{K}^\delta)\end{aligned} \tag{6.0.6}$$

$$\begin{aligned}\mathcal{E}_2(u,\underline{u}) &= \int_{V_{(u,\underline{u})}}\mathbf{Div}Q(\hat{\mathcal{L}}_O^2 W)_{\beta\gamma\delta}(\bar{K}^\beta\bar{K}^\gamma T^\delta)\\ &+\int_{V_{(u,\underline{u})}}\mathbf{Div}Q(\hat{\mathcal{L}}_O\hat{\mathcal{L}}_T W)_{\beta\gamma\delta}(\bar{K}^\beta\bar{K}^\gamma\bar{K}^\delta)\\ &+\int_{V_{(u,\underline{u})}}\mathbf{Div}Q(\hat{\mathcal{L}}_S\hat{\mathcal{L}}_T W)_{\beta\gamma\delta}(\bar{K}^\beta\bar{K}^\gamma\bar{K}^\delta)\\ &+\int_{V_{(u,\underline{u})}}Q(\hat{\mathcal{L}}_O^2 W)_{\alpha\beta\gamma\delta}({}^{(\bar{K})}\pi^{\alpha\beta}\bar{K}^\gamma T^\delta)\\ &+\frac{1}{2}\int_{V_{(u,\underline{u})}}Q(\hat{\mathcal{L}}_O^2 W)_{\alpha\beta\gamma\delta}({}^{(T)}\pi^{\alpha\beta}\bar{K}^\gamma\bar{K}^\delta)\end{aligned}$$

[2] Unlike Chapter 3, this chapter does not distinguish between the functions u, $\underline{u}$ and the values λ, ν they can assume, since here no ambiguity can arise. Moreover the $\mathcal{Q}$ integral norms are expressed in terms of a general Weyl tensor W instead of the curvature tensor $\mathbf{R}$ to remind the reader that these norms can be associated with a general Weyl field satisfying the Bianchi equations in a background spacetime; see Chapter 2. The reader can, anyway, identify W with $\mathbf{R}$.

$$+\frac{3}{2}\int_{V_{(u,\underline{u})}} Q(\hat{\mathcal{L}}_O\hat{\mathcal{L}}_T W)_{\alpha\beta\gamma\delta}({}^{(\bar{K})}\pi^{\alpha\beta}\bar{K}^\gamma\bar{K}^\delta) \tag{6.0.7}$$

$$+\frac{3}{2}\int_{V_{(u,\underline{u})}} Q(\hat{\mathcal{L}}_S\hat{\mathcal{L}}_T W)_{\alpha\beta\gamma\delta}({}^{(\bar{K})}\pi^{\alpha\beta}\bar{K}^\gamma\bar{K}^\delta).$$

The estimates of these terms are algebraically quite involved. The final result, however, is very simple. We shall show that

$$\mathcal{E}(u,\underline{u}) \leq c\epsilon_0 \mathcal{Q}_{\mathcal{K}} \tag{6.0.8}$$

with c an appropriate constant. This implies

$$\mathcal{Q}_{\mathcal{K}} \leq \frac{1}{1-c\epsilon_0}\mathcal{Q}_{\Sigma_0\cap\mathcal{K}}$$

which, for ϵ_0 sufficiently small, concludes the proof of the theorem. The next sections are devoted to the detailed estimates of the error terms required to prove (6.0.8).

Remark: The estimates of the spacetime integrals appearing in (6.0.6) and (6.0.7) are the most sensitive part of the proof of the Main Theorem. To understand how these estimates are made, we recall the discussion in Chapter 2 concerning global existence for nonlinear wave equations. To estimate the error terms appearing in the derivation of the energy estimates for the model problem (2.1.23), we had to introduce the commuting vector fields (refvectorfields), define the generalized energy norms (2.1.28) and use the global Sobolev inequalities (2.1.29) to derive decay estimates. These allowed us to prove (2.1.31), which implies the desired global existence result for $n > 3$. In dimension $n = 3$ we had, in addition, to rely on the special structure of nonlinear terms, called the "null condition." All these elements, except the last, were already incorporated in our discussion of the proof of the Main Theorem. To estimate the error terms (6.0.6), (6.0.7) we also need to use the special structure of these terms. Just as in the simple case of the null condition for the nonlinear wave equation, we have to make sure, by carefully decomposing all the terms appearing in the above integrals in terms of their null components, that the slowest decaying components are counterbalanced by terms that decay fast. For this reason we need to know the precise asymptotic behavior of all components of $W = \mathbf{R}$ and its derivatives as well as those of the various deformation tensors. The behavior of the null components of the deformation tensors depends crucially on that of the null connection coefficients.

6.1 Definitions and prerequisites

To estimate the first two integrals of $\mathcal{E}_1(u,\underline{u})$[3]

$$\int_{V_{(u,\underline{u})}} \mathbf{Div}Q(\hat{\mathcal{L}}_T W)_{\beta\gamma\delta}(\bar{K}^\beta\bar{K}^\gamma\bar{K}^\delta)\,,\ \int_{V_{(u,\underline{u})}} \mathbf{Div}Q(\hat{\mathcal{L}}_O W)_{\beta\gamma\delta}(\bar{K}^\beta\bar{K}^\gamma T^\delta)$$

[3]The following expressions are also used, with slight modifications, to estimate the first three integrals of $\mathcal{E}_2(u,\underline{u})$.

we have to compute explicitly $\mathbf{Div}Q(\hat{\mathcal{L}}_X W)$ with $X = T, O$. Denoting

$$D(X, W) \equiv \mathbf{Div}Q(\hat{\mathcal{L}}_X W),$$

it follows, by a straightforward calculation (see also [Ch-Kl] (8.1.3.c)) that

$$\begin{aligned} D(X, W)(\bar{K}, \bar{K}, T) &= \frac{1}{8}\tau_+^4(D(X, W)_{444} + D(X, W)_{344}) \\ &+\frac{1}{4}\tau_+^2\tau_-^2(D(X, W)_{344} + D(X, W)_{334}) \\ &+\frac{1}{8}\tau_-^4(D(X, W)_{334} + D(X, W)_{333}) \end{aligned} \tag{6.1.1}$$

$$\begin{aligned} D(X, W)(\bar{K}, \bar{K}, \bar{K}) &= \frac{1}{8}\tau_+^6 D(X, W)_{444} + \frac{3}{8}\tau_+^4\tau_-^2 D(X, W)_{344} \\ &+\frac{3}{8}\tau_+^2\tau_-^4 D(X, W)_{334} + \frac{1}{8}\tau_-^6 D(X, W)_{333} \end{aligned} \tag{6.1.2}$$

where

$$\begin{aligned} D(X, W)_{444} &= 4\alpha(\hat{\mathcal{L}}_X W) \cdot \Theta(X, W) - 8\beta(\hat{\mathcal{L}}_X W) \cdot \Xi(X, W) \\ D(X, W)_{443} &= 8\rho(\hat{\mathcal{L}}_X W)\Lambda(X, W) + 8\sigma(\hat{\mathcal{L}}_X W)K(X, W) \\ &\quad + 8\beta(\hat{\mathcal{L}}_X W) \cdot I(X, W) \\ D(X, W)_{334} &= 8\rho(\hat{\mathcal{L}}_X W)\underline{\Lambda}(X, W) - 8\sigma(\hat{\mathcal{L}}_X W)\underline{K}(X, W) \\ &\quad - 8\underline{\beta}(\hat{\mathcal{L}}_X W) \cdot \underline{I}(X, W) \\ D(X, W)_{333} &= 4\underline{\alpha}(\hat{\mathcal{L}}_X W) \cdot \underline{\Theta}(X, W) + 8\underline{\beta}(\hat{\mathcal{L}}_X W) \cdot \underline{\Xi}(X, W) \end{aligned} \tag{6.1.3}$$

$$\begin{aligned} &\Lambda(X, W)\,,\; K(X, W)\,,\; I(X, W)\,,\; \Theta(X, W)\,,\; \Xi(X, W) \\ &\underline{\Lambda}(X, W)\,,\; \underline{K}(X, W)\,,\; \underline{I}(X, W)\,,\; \underline{\Theta}(X, W)\,,\; \underline{\Xi}(X, W) \end{aligned}$$

are the null components of the Weyl current[4]

$$J(X, W)_{\beta\gamma\delta} \equiv D^\alpha(\hat{\mathcal{L}}_X W)_{\alpha\beta\gamma\delta}\,,$$

and[5]

$$\begin{aligned} &\Lambda(J) = \frac{1}{4}J_{434}\,,\; \underline{\Lambda}(J) = \frac{1}{4}J_{343}\,,\; \Xi(J)_a = \frac{1}{2}J_{44a}\,,\; \underline{\Xi}(J)_a = \frac{1}{2}J_{33a} \\ &I(J)_a = \frac{1}{2}J_{34a}\,,\; \underline{I}(J)_a = \frac{1}{2}J_{43a}\,,\; K(J) = \frac{1}{4}\in^{ab} J_{4ab}\,,\; \underline{K}(J) = \frac{1}{4}\in^{ab} J_{3ab} \\ &\Theta(J)_{ab} = J_{a4b} + J_{b4a} - (\delta^{cd} J_{c4d})\delta_{ab}\,,\; \underline{\Theta}(J)_{ab} = J_{a3b} + J_{b3a} - (\delta^{cd} J_{c3d})\delta_{ab} \end{aligned} \tag{6.1.4}$$

---

[4] If X is not a Killing or a conformal Killing vector field, $J(X, W)$ is different from zero even if W satisfies the homogeneous Bianchi equations.

[5] We remark also that $J_{a4b} = \Theta(J)_{ab} - \Lambda\delta_{ab} + K \in_{ab}$, $J_{a3b} = \underline{\Theta}(J)_{ab} - \underline{\Lambda}\delta_{ab} + \underline{K} \in_{ab}$, $J_{abc} = \in_{bc} \left({}^\star I(J)_a + {}^\star\underline{I}(J)_a\right)$.

where we used the relations

$$\begin{array}{llll} \Lambda(J^\star) &= K(J) &;\quad \underline{\Lambda}(J^\star) &= -\underline{K}(J) \\ K(J^\star) &= -\Lambda(J) &;\quad \underline{K}(J^\star) &= \underline{\Lambda}(J) \\ \Xi(J^\star) &= -{}^\star\Xi(J) &;\quad \underline{\Xi}(J^\star) &= {}^\star\underline{\Xi}(J) \\ I(J^\star) &= -{}^\star I(J) &;\quad \underline{I}(J^\star) &= {}^\star\underline{I}(J) \\ \Theta(J^\star) &= -{}^\star\Theta(J) &;\quad \underline{\Theta}(J^\star) &= {}^\star\underline{\Theta}(J), \end{array} \tag{6.1.5}$$

with J^* the Hodge dual of J, $J^*_{\beta\gamma\delta} = \frac{1}{2} J_{\beta\mu\nu}\epsilon^{\mu\nu}{}_{\gamma\delta}$.

Finally , $J(X, W)$ can be decomposed into three different parts.[6] [7]

$$J(X; W) = J^1(X; W) + J^2(X; W) + J^3(X; W)\ ,$$

where

$$\begin{aligned} J^1(X; W)_{\beta\gamma\delta} &= \frac{1}{2}{}^{(X)}\hat{\pi}^{\mu\nu}\mathbf{D}_\nu W_{\mu\beta\gamma\delta} \\ J^2(X; W)_{\beta\gamma\delta} &= \frac{1}{2}{}^{(X)}p_\lambda W^\lambda{}_{\beta\gamma\delta} \\ J^3(X; W)_{\beta\gamma\delta} &= \frac{1}{2}\left({}^{(X)}q_{\alpha\beta\lambda}W^{\alpha\lambda}{}_{\gamma\delta} + {}^{(X)}q_{\alpha\gamma\lambda}W^\alpha{}_\beta{}^\lambda{}_\delta + {}^{(X)}q_{\alpha\delta\lambda}W^\alpha{}_{\beta\gamma}{}^\lambda\right) \end{aligned} \tag{6.1.6}$$

and

$$\begin{aligned} {}^{(X)}p_\lambda &= \mathbf{D}^{\alpha}\,{}^{(X)}\hat{\pi}_{\alpha\gamma} \\ {}^{(X)}q_{\alpha\beta\gamma} &= \mathbf{D}^{\beta}\,{}^{(X)}\hat{\pi}_{\gamma\alpha} - \mathbf{D}^{\gamma}\,{}^{(X)}\hat{\pi}_{\beta\alpha} - \frac{1}{3}\left({}^{(X)}p p_\gamma g_{\alpha\beta} - {}^{(X)}p_\beta g_{\alpha\gamma}\right). \end{aligned} \tag{6.1.7}$$

It follows that the various factors $\Theta(X, W)$, $\Xi(X, W)$, $\Lambda(X, W), \dots, \underline{\Xi}(X, W)$ of (6.1.3) can also be decomposed into three parts, depending on which part of $J(X, W)$ they are connected to.

All these null components of the Weyl current can be explicitly written in terms of the null components of the Riemann tensor and its first derivatives, the null components of the traceless part of the deformation tensors and their derivatives ${}^{(X)}p$, ${}^{(X)}q$ which appear in the expressions of $J^1(X, W)$, $J^2(X, W)$, $J^3(X, W)$. Recalling the null decomposition of the deformation tensors (see (3.4.6))

$$\begin{array}{ll} {}^{(X)}\mathbf{i}_{ab} = {}^{(X)}\hat{\pi}_{ab}\ ; & {}^{(X)}\mathbf{j} = {}^{(X)}\hat{\pi}_{34} \\ {}^{(X)}\mathbf{m}_a = {}^{(X)}\hat{\pi}_{4a}\ ; & {}^{(X)}\underline{\mathbf{m}}_a = {}^{(X)}\hat{\pi}_{3a} \\ {}^{(X)}\mathbf{n} = {}^{(X)}\hat{\pi}_{44}\ ; & {}^{(X)}\underline{\mathbf{n}} = {}^{(X)}\hat{\pi}_{33}; \end{array}$$

the explicit expressions of the components of $J(X, W)$ are[8]

$$\underline{\Xi}(J^1) \ = \ \mathrm{Qr}\left[{}^{(X)}\mathbf{i};\ \not\nabla\underline{\alpha}\right] + \mathrm{Qr}\left[{}^{(X)}\mathbf{m};\ \underline{\alpha}_3\right] + \mathrm{Qr}\left[{}^{(X)}\underline{\mathbf{m}};\ \underline{\alpha}_4\right]$$

---

[6] See also Proposition 7.1.2 and (8.1.2b) of [Ch-Kl].

[7] To estimate $\mathcal{E}_2(u, \underline{u})$ it is necessary to consider also the divergence of the second Lie derivatives of the Weyl field $J(X, Y, W)_{\beta\gamma\delta} = D^\alpha(\hat{\mathcal{L}}_Y\hat{\mathcal{L}}_X W)_{\alpha\beta\gamma\delta}$. We will give their explicit expressions later on.

[8] See Proposition 8.1.4 of [Ch-Kl]. Qr[;] is a generic notation for any quadratic form with coefficients that depend only on the induced metric and area form of $S(u, \underline{u})$. We note also that the terms that are boxed below are in fact vanishing; we include them to emphasize the importance of the corresponding cancellations.

$$
\begin{aligned}
&+ \ \mathrm{Qr}\left[{}^{(X)}\underline{\mathbf{m}}\,;\ \not\nabla\underline{\beta}\right] + \mathrm{Qr}\left[{}^{(X)}\mathbf{j}\,;\ \underline{\beta}_3\right] + \mathrm{Qr}\left[{}^{(X)}\underline{\mathbf{n}}\,;\ \underline{\beta}_4\right] \\
&+ \ \mathrm{tr}\underline{\chi}\left(\mathrm{Qr}\left[{}^{(X)}\mathbf{m}\,;\ \underline{\alpha}\right] + \mathrm{Qr}\left[({}^{(X)}\mathbf{i},\, {}^{(X)}\mathbf{j})\,;\ \underline{\beta}\right] + \mathrm{Qr}\left[{}^{(X)}\underline{\mathbf{m}}\,;\ (\rho,\sigma)\right]\right) \\
&+ \ \mathrm{tr}\chi\left(\mathrm{Qr}\left[{}^{(X)}\underline{\mathbf{m}}\,;\ \underline{\alpha}\right] + \mathrm{Qr}\left[{}^{(X)}\underline{\mathbf{n}}\,;\ \underline{\beta}\right]\right) + \text{l.o.t.},
\end{aligned}
\tag{6.1.8}
$$

$$
\begin{aligned}
\underline{\Theta}(J^1) \ = \ &\mathrm{Qr}\left[{}^{(X)}\mathbf{m}\,;\ \not\nabla\underline{\alpha}\right] + \mathrm{Qr}\left[{}^{(X)}\mathbf{n}\,;\ \underline{\alpha}_3\right] + \mathrm{Qr}\left[{}^{(X)}\mathbf{j}\,;\ \underline{\alpha}_4\right] \\
+ \ &\mathrm{Qr}\left[{}^{(X)}\mathbf{i}\,;\ \not\nabla\underline{\beta}\right] + \mathrm{Qr}\left[{}^{(X)}\mathbf{m}\,;\ \underline{\beta}_3\right] + \mathrm{Qr}\left[{}^{(X)}\underline{\mathbf{m}}\,;\ \underline{\beta}_4\right] \\
+ \ &\mathrm{Qr}\left[{}^{(X)}\underline{\mathbf{m}}\,;\ \not\nabla(\rho,\sigma)\right] + \mathrm{Qr}\left[{}^{(X)}\mathbf{j}\,;\ (\rho_3,\sigma_3)\right] + \mathrm{Qr}\left[{}^{(X)}\underline{\mathbf{n}}\,;\ (\rho_4,\sigma_4)\right] \\
+ \ &\mathrm{tr}\underline{\chi}\left(\mathrm{Qr}\left[{}^{(X)}\mathbf{n}\,;\ \underline{\alpha}\right] + \mathrm{Qr}\left[{}^{(X)}\mathbf{m}\,;\ \underline{\beta}\right] + \mathrm{Qr}\left[({}^{(X)}\mathbf{i},\, {}^{(X)}\mathbf{j})\,;\ (\rho,\sigma)\right]\right. \\
+ \ &\left.\mathrm{Qr}\left[{}^{(X)}\underline{\mathbf{m}}\,;\ \beta\right]\right) \\
+ \ &\mathrm{tr}\chi\left(\boxed{\quad\mathrm{Qr}\left[{}^{(X)}\mathbf{i}\,;\ \underline{\alpha}\right]\quad} + \mathrm{Qr}\left[{}^{(X)}\mathbf{j}\,;\ \underline{\alpha}\right] + \mathrm{Qr}\left[{}^{(X)}\underline{\mathbf{m}}\,;\ \underline{\beta}\right]\right. \\
+ \ &\left.\mathrm{Qr}\left[{}^{(X)}\underline{\mathbf{n}}\,;\ (\rho,\sigma)\right]\right) + \text{l.o.t.},
\end{aligned}
\tag{6.1.9}
$$

$$
\begin{aligned}
\underline{\Lambda}(J^1) \ = \ &\mathrm{Qr}\left[{}^{(X)}\mathbf{i}\,;\ \not\nabla\underline{\beta}\right] + \mathrm{Qr}\left[{}^{(X)}\mathbf{m}\,;\ \underline{\beta}_3\right] + \mathrm{Qr}\left[{}^{(X)}\underline{\mathbf{m}}\,;\ \underline{\beta}_4\right] \\
+ \ &\mathrm{Qr}\left[{}^{(X)}\underline{\mathbf{m}}\,;\ \not\nabla(\rho,\sigma)\right] + \mathrm{Qr}\left[{}^{(X)}\mathbf{j}\,;\ (\rho_3,\sigma_3)\right] + \mathrm{Qr}\left[{}^{(X)}\underline{\mathbf{n}}\,;\ (\rho_4,\sigma_4)\right] \\
+ \ &\mathrm{tr}\underline{\chi}\left(\mathrm{Qr}\left[{}^{(X)}\mathbf{m}\,;\ \underline{\beta}\right] + \mathrm{Qr}\left[({}^{(X)}\mathbf{i},\, {}^{(X)}\mathbf{j})\,;\ (\rho,\sigma)\right] + \mathrm{Qr}\left[{}^{(X)}\underline{\mathbf{m}}\,;\ \beta\right]\right) \\
+ \ &\mathrm{tr}\chi\left(\mathrm{Qr}\left[({}^{(X)}\mathbf{i},\, {}^{(X)}\mathbf{j})\,;\ \underline{\alpha}\right] + \mathrm{Qr}\left[{}^{(X)}\underline{\mathbf{m}}\,;\ \underline{\beta}\right] + \mathrm{Qr}\left[{}^{(X)}\underline{\mathbf{n}}\,;\ (\rho,\sigma)\right]\right) \\
+ \ &\text{l.o.t.},
\end{aligned}
\tag{6.1.10}
$$

$$
\begin{aligned}
\underline{K}(J^1) \ = \ &\mathrm{Qr}\left[{}^{(X)}\mathbf{i}\,;\ \not\nabla\underline{\beta}\right] + \mathrm{Qr}\left[{}^{(X)}\mathbf{m}\,;\ \underline{\beta}_3\right] + \mathrm{Qr}\left[{}^{(X)}\underline{\mathbf{m}}\,;\ \underline{\beta}_4\right] \\
+ \ &\mathrm{Qr}\left[{}^{(X)}\underline{\mathbf{m}}\,;\ \not\nabla(\rho,\sigma)\right] + \mathrm{Qr}\left[{}^{(X)}\mathbf{j}\,;\ (\rho_3,\sigma_3)\right] + \mathrm{Qr}\left[{}^{(X)}\underline{\mathbf{n}}\,;\ (\rho_4,\sigma_4)\right] \\
+ \ &\mathrm{tr}\underline{\chi}\left(\mathrm{Qr}\left[{}^{(X)}\mathbf{m}\,;\ \underline{\beta}\right] + \mathrm{Qr}\left[({}^{(X)}\mathbf{i},\, {}^{(X)}\mathbf{j})\,;\ (\rho,\sigma)\right] + \mathrm{Qr}\left[{}^{(X)}\underline{\mathbf{m}}\,;\ \beta\right]\right) \\
+ \ &\mathrm{tr}\chi\left(\mathrm{Qr}\left[({}^{(X)}\mathbf{i},\, {}^{(X)}\mathbf{j})\,;\ \underline{\alpha}\right] + \mathrm{Qr}\left[{}^{(X)}\underline{\mathbf{m}}\,;\ \underline{\beta}\right] + \mathrm{Qr}\left[{}^{(X)}\underline{\mathbf{n}}\,;\ (\rho,\sigma)\right]\right) \\
+ \ &\text{l.o.t.},
\end{aligned}
\tag{6.1.11}
$$

$$
\begin{aligned}
\underline{I}(J^1) \ = \ &\mathrm{Qr}\left[{}^{(X)}\mathbf{m}\,;\ \not\nabla\underline{\beta}\right] + \mathrm{Qr}\left[{}^{(X)}\mathbf{n}\,;\ \underline{\beta}_3\right] + \mathrm{Qr}\left[{}^{(X)}\mathbf{j}\,;\ \underline{\beta}_4\right] \\
+ \ &\mathrm{Qr}\left[{}^{(X)}\mathbf{i}\,;\ \not\nabla(\rho,\sigma)\right] + \mathrm{Qr}\left[{}^{(X)}\mathbf{m}\,;\ (\rho_3,\sigma_3)\right] + \mathrm{Qr}\left[{}^{(X)}\underline{\mathbf{m}}\,;\ (\rho_4,\sigma_4)\right] \\
+ \ &\mathrm{tr}\underline{\chi}\left(\mathrm{Qr}\left[{}^{(X)}\mathbf{n}\,;\ \underline{\beta}\right] + \mathrm{Qr}\left[{}^{(X)}\mathbf{m}\,;\ (\rho,\sigma)\right] + \mathrm{Qr}\left[{}^{(X)}\mathbf{i}\,;\ \beta\right]\right) \\
+ \ &\mathrm{tr}\chi\left(\mathrm{Qr}\left[{}^{(X)}\mathbf{m}\,;\ \underline{\alpha}\right] + \mathrm{Qr}\left[({}^{(X)}\mathbf{i},\, {}^{(X)}\mathbf{j})\,;\ \underline{\beta}\right] + \mathrm{Qr}\left[{}^{(X)}\underline{\mathbf{m}}\,;\ (\rho,\sigma)\right]\right) \\
+ \ &\text{l.o.t.},
\end{aligned}
\tag{6.1.12}
$$

$$
\begin{aligned}
\Xi(J^1) &= \mathrm{Qr}\left[{}^{(X)}\mathbf{i}\,;\ \not\nabla\alpha\right] + \mathrm{Qr}\left[{}^{(X)}\underline{\mathbf{m}}\,;\ \alpha_4\right] + \mathrm{Qr}\left[{}^{(X)}\mathbf{m}\,;\ \alpha_3\right] \\
&+ \mathrm{Qr}\left[{}^{(X)}\mathbf{m}\,;\ \not\nabla\beta\right] + \mathrm{Qr}\left[{}^{(X)}\mathbf{j}\,;\ \beta_4\right] + \mathrm{Qr}\left[{}^{(X)}\mathbf{n}\,;\ \beta_3\right] \\
&+ \mathrm{tr}\chi\left(\mathrm{Qr}\left[{}^{(X)}\underline{\mathbf{m}}\,;\ \alpha\right] + \mathrm{Qr}\left[({}^{(X)}\mathbf{i},\ {}^{(X)}\mathbf{j})\,;\ \beta\right] + \mathrm{Qr}\left[{}^{(X)}\mathbf{m}\,;\ (\rho,\sigma)\right]\right) \\
&+ \mathrm{tr}\underline{\chi}\left(\mathrm{Qr}\left[{}^{(X)}\mathbf{m}\,;\ \alpha\right] + \mathrm{Qr}\left[{}^{(X)}\mathbf{n}\,;\ \beta\right]\right) + \mathrm{l.o.t.},
\end{aligned}
\tag{6.1.13}
$$

$$
\begin{aligned}
\Theta(J^1) &= \mathrm{Qr}\left[{}^{(X)}\underline{\mathbf{m}}\,;\ \not\nabla\alpha\right] + \mathrm{Qr}\left[{}^{(X)}\underline{\mathbf{n}}\,;\ \alpha_4\right] + \mathrm{Qr}\left[{}^{(X)}\mathbf{j}\,;\ \alpha_3\right] \\
&+ \mathrm{Qr}\left[{}^{(X)}\mathbf{i}\,;\ \not\nabla\beta\right] + \mathrm{Qr}\left[{}^{(X)}\underline{\mathbf{m}}\,;\ \beta_4\right] + \mathrm{Qr}\left[{}^{(X)}\mathbf{m}\,;\ \beta_3\right] \\
&+ \mathrm{Qr}\left[{}^{(X)}\mathbf{m}\,;\ \not\nabla(\rho,\sigma)\right] + \mathrm{Qr}\left[{}^{(X)}\mathbf{j}\,;\ (\rho_4,\sigma_4)\right] + \mathrm{Qr}\left[{}^{(X)}\mathbf{n}\,;\ (\rho_3,\sigma_3)\right] \\
&+ \mathrm{tr}\chi\left(\mathrm{Qr}\left[{}^{(X)}\underline{\mathbf{n}}\,;\ \alpha\right] + \mathrm{Qr}\left[{}^{(X)}\underline{\mathbf{m}}\,;\ \beta\right] + \mathrm{Qr}\left[({}^{(X)}\mathbf{i},\ {}^{(X)}\mathbf{j})\,;\ (\rho,\sigma)\right]\right. \\
&+ \left.\mathrm{Qr}\left[{}^{(X)}\mathbf{m}\,;\ \underline{\beta}\right]\right) + \mathrm{tr}\underline{\chi}\left(\boxed{\quad \mathrm{Qr}\left[{}^{(X)}\mathbf{i}\,;\ \alpha\right] \quad} + \mathrm{Qr}\left[{}^{(X)}\mathbf{j}\,;\ \alpha\right]\right. \\
&+ \left.\mathrm{Qr}\left[{}^{(X)}\mathbf{m}\,;\ \beta\right] + \mathrm{Qr}\left[{}^{(X)}\mathbf{n}\,;\ (\rho,\sigma)\right]\right) + \mathrm{l.o.t.},
\end{aligned}
\tag{6.1.14}
$$

$$
\begin{aligned}
\Lambda(J^1) &= \mathrm{Qr}\left[{}^{(X)}\mathbf{i}\,;\ \not\nabla\beta\right] + \mathrm{Qr}\left[{}^{(X)}\underline{\mathbf{m}}\,;\ \beta_4\right] + \mathrm{Qr}\left[{}^{(X)}\mathbf{m}\,;\ \beta_3\right] \\
&+ \mathrm{Qr}\left[{}^{(X)}\mathbf{m}\,;\ \not\nabla(\rho,\sigma)\right] + \mathrm{Qr}\left[{}^{(X)}\mathbf{j}\,;\ (\rho_4,\sigma_4)\right] + \mathrm{Qr}\left[{}^{(X)}\mathbf{n}\,;\ (\rho_3,\sigma_3)\right] \\
&+ \mathrm{tr}\chi\left(\mathrm{Qr}\left[{}^{(X)}\underline{\mathbf{m}}\,;\ \beta\right] + \mathrm{Qr}\left[({}^{(X)}\mathbf{i},\ {}^{(X)}\mathbf{j})\,;\ (\rho,\sigma)\right] + \mathrm{Qr}\left[{}^{(X)}\mathbf{m}\,;\ \underline{\beta}\right]\right) \\
&+ \mathrm{tr}\underline{\chi}\left(\mathrm{Qr}\left[({}^{(X)}\mathbf{i},\ {}^{(X)}\mathbf{j})\,;\ \alpha\right] + \mathrm{Qr}\left[{}^{(X)}\mathbf{m}\,;\ \beta\right] + \mathrm{Qr}\left[{}^{(X)}\mathbf{n}\,;\ (\rho,\sigma)\right]\right) \\
&+ \mathrm{l.o.t.},
\end{aligned}
\tag{6.1.15}
$$

$$
\begin{aligned}
K(J^1) &= \mathrm{Qr}\left[{}^{(X)}\mathbf{i}\,;\ \not\nabla\beta\right] + \mathrm{Qr}\left[{}^{(X)}\underline{\mathbf{m}}\,;\ \beta_4\right] + \mathrm{Qr}\left[{}^{(X)}\mathbf{m}\,;\ \beta_3\right] \\
&+ \mathrm{Qr}\left[{}^{(X)}\mathbf{m}\,;\ \not\nabla(\rho,\sigma)\right] + \mathrm{Qr}\left[{}^{(X)}\mathbf{j}\,;\ (\rho_4,\sigma_4)\right] + \mathrm{Qr}\left[{}^{(X)}\mathbf{n}\,;\ (\rho_3,\sigma_3)\right] \\
&+ \mathrm{tr}\chi\left(\mathrm{Qr}\left[{}^{(X)}\underline{\mathbf{m}}\,;\ \beta\right] + \mathrm{Qr}\left[({}^{(X)}\mathbf{i},\ {}^{(X)}\mathbf{j})\,;\ (\rho,\sigma)\right] + \mathrm{Qr}\left[{}^{(X)}\mathbf{m}\,;\ \underline{\beta}\right]\right) \\
&+ \mathrm{tr}\underline{\chi}\left(\mathrm{Qr}\left[({}^{(X)}\mathbf{i},\ {}^{(X)}\mathbf{j})\,;\ \alpha\right] + \mathrm{Qr}\left[{}^{(X)}\mathbf{m}\,;\ \beta\right] + \mathrm{Qr}\left[{}^{(X)}\mathbf{n}\,;\ (\rho,\sigma)\right]\right) \\
&+ \mathrm{l.o.t.},
\end{aligned}
\tag{6.1.16}
$$

$$
\begin{aligned}
I(J^1) &= \mathrm{Qr}\left[{}^{(X)}\underline{\mathbf{m}}\,;\ \not\nabla\beta\right] + \mathrm{Qr}\left[{}^{(X)}\underline{\mathbf{n}}\,;\ \beta_4\right] + \mathrm{Qr}\left[{}^{(X)}\mathbf{j}\,;\ \beta_3\right] \\
&+ \mathrm{Qr}\left[{}^{(X)}\mathbf{i}\,;\ \not\nabla(\rho,\sigma)\right] + \mathrm{Qr}\left[{}^{(X)}\underline{\mathbf{m}}\,;\ (\rho_4,\sigma_4)\right] + \mathrm{Qr}\left[{}^{(X)}\mathbf{m}\,;\ (\rho_3,\sigma_3)\right] \\
&+ \mathrm{tr}\chi\left(\mathrm{Qr}\left[{}^{(X)}\underline{\mathbf{n}}\,;\ \beta\right] + \mathrm{Qr}\left[{}^{(X)}\underline{\mathbf{m}}\,;\ (\rho,\sigma)\right] + \mathrm{Qr}\left[{}^{(X)}\mathbf{i}\,;\ \underline{\beta}\right]\right) \\
&+ \mathrm{tr}\underline{\chi}\left(\mathrm{Qr}\left[{}^{(X)}\underline{\mathbf{m}}\,;\ \alpha\right] + \mathrm{Qr}\left[({}^{(X)}\mathbf{i},\ {}^{(X)}\mathbf{j})\,;\ \beta\right] + \mathrm{Qr}\left[{}^{(X)}\mathbf{m}\,;\ (\rho,\sigma)\right]\right) \\
&+ \mathrm{l.o.t.}.
\end{aligned}
\tag{6.1.17}
$$

Remark: The terms which we denote by l.o.t. are cubic with respect to ${}^{(X)}\hat{\pi}$, W and the connection coefficients $\eta, \underline{\eta}, \omega, \underline{\omega}, \chi, \underline{\chi}$ and are linear with regard to each of them separately. They are manifestly of lower order by comparison to all other terms both in regard to their asymptotic behavior along the null outgoing hypersurfaces and to the order of differentiabilty relative to W. Hereafter we will disregard them.

The null decomposition of J^2 is given by

$$
\begin{aligned}
\underline{\Xi}(J^2) &= \mathrm{Qr}\left[{}^{(X)}\not{p};\ \underline{\alpha}\right] + \mathrm{Qr}\left[{}^{(X)}p_3;\ \underline{\beta}\right] \\
\underline{\Theta}(J^2) &= \mathrm{Qr}\left[{}^{(X)}p_4;\ \underline{\alpha}\right] + \mathrm{Qr}\left[{}^{(X)}\not{p};\ \underline{\beta}\right] + \mathrm{Qr}\left[{}^{(X)}p_3;\ (\rho,\sigma)\right] \\
\underline{\Lambda}(J^2) &= \mathrm{Qr}\left[{}^{(X)}\not{p};\ \underline{\beta}\right] + \mathrm{Qr}\left[{}^{(X)}p_3;\ (\rho,\sigma)\right] \\
\underline{K}(J^2) &= \mathrm{Qr}\left[{}^{(X)}\not{p};\ \underline{\beta}\right] + \mathrm{Qr}\left[{}^{(X)}p_3;\ (\rho,\sigma)\right] \\
\underline{I}(J^2) &= \mathrm{Qr}\left[{}^{(X)}p_4;\ \underline{\beta}\right] + \mathrm{Qr}\left[{}^{(X)}\not{p};\ (\rho,\sigma)\right] \qquad (6.1.18) \\
I(J^2) &= \mathrm{Qr}\left[{}^{(X)}p_3;\ \beta\right] + \mathrm{Qr}\left[{}^{(X)}\not{p};\ (\rho,\sigma)\right] \\
K(J^2) &= \mathrm{Qr}\left[{}^{(X)}\not{p};\ \beta\right] + \mathrm{Qr}\left[{}^{(X)}p_4;\ (\rho,\sigma)\right] \\
\Lambda(J^2) &= \mathrm{Qr}\left[{}^{(X)}\not{p};\ \beta\right] + \mathrm{Qr}\left[{}^{(X)}p_4;\ (\rho,\sigma)\right] \\
\Theta(J^2) &= \mathrm{Qr}\left[{}^{(X)}p_3;\ \alpha\right] + \mathrm{Qr}\left[{}^{(X)}\not{p};\ \beta\right] + \mathrm{Qr}\left[{}^{(X)}p_4;\ (\rho,\sigma)\right] \\
\Xi(J^2) &= \mathrm{Qr}\left[{}^{(X)}\not{p};\ \alpha\right] + \mathrm{Qr}\left[{}^{(X)}p_4;\ \beta\right],
\end{aligned}
$$

and the null decomposition of J^3 by

$$
\begin{aligned}
\underline{\Xi}(J^3) &= \mathrm{Qr}\left[\underline{\alpha};\ (I,\underline{I})({}^{(X)}q)\right] + \mathrm{Qr}\left[\underline{\beta};\ (\underline{K},\underline{\Lambda},\underline{\Theta})({}^{(X)}q)\right] + \mathrm{Qr}\left[(\rho,\sigma);\ \underline{\Xi}({}^{(X)}q)\right] \\
\underline{\Theta}(J^3) &= \mathrm{Qr}\left[\underline{\alpha};\ K({}^{(X)}q)\right] + \mathrm{Qr}\left[\underline{\alpha};\ \Lambda({}^{(X)}q)\right] + \boxed{\mathrm{Qr}\left[\underline{\alpha};\ \Theta({}^{(X)}q)\right]} \qquad (6.1.19) \\
&\quad + \mathrm{Qr}\left[\underline{\beta};\ (I,\underline{I})({}^{(X)}q)\right] + \mathrm{Qr}\left[(\rho,\sigma);\ \underline{\Theta}({}^{(X)}q)\right] + \boxed{\mathrm{Qr}\left[\beta;\ \underline{\Xi}({}^{(X)}q)\right]}
\end{aligned}
$$

$$
\begin{aligned}
\underline{\Lambda}(J^3) &= \mathrm{Qr}\left[\underline{\alpha};\ \Theta({}^{(X)}q)\right] + \boxed{\mathrm{Qr}\left[\underline{\beta};\ (I,\underline{I})({}^{(X)}q)\right]} \\
&\quad + \mathrm{Qr}\left[(\rho,\sigma);\ (\underline{K},\underline{\Lambda})({}^{(X)}q)\right] + \mathrm{Qr}\left[\beta;\ \underline{\Xi}({}^{(X)}q)\right] \\
\underline{K}(J^3) &= \mathrm{Qr}\left[\underline{\alpha};\ \Theta({}^{(X)}q)\right] + \boxed{\mathrm{Qr}\left[\underline{\beta};\ (I,\underline{I})({}^{(X)}q)\right]} \\
&\quad + \mathrm{Qr}\left[(\rho,\sigma);\ (\underline{K},\underline{\Lambda})({}^{(X)}q)\right] + \mathrm{Qr}\left[\beta;\ \underline{\Xi}({}^{(X)}q)\right] \qquad (6.1.20) \\
\underline{I}(J^3) &= \mathrm{Qr}\left[\underline{\alpha};\ \Xi({}^{(X)}q)\right] + \mathrm{Qr}\left[\underline{\beta};\ (K,\Lambda,\Theta)({}^{(X)}q)\right] \\
&\quad + \mathrm{Qr}\left[(\rho,\sigma);\ (I,\underline{I})({}^{(X)}q)\right] + \mathrm{Qr}\left[\beta;\ (\underline{K},\underline{\Lambda},\underline{\Theta})({}^{(X)}q)\right] \\
&\quad + \boxed{\mathrm{Qr}\left[\alpha;\ \underline{\Xi}({}^{(X)}q)\right]}
\end{aligned}
$$

$$
\begin{aligned}
I(J^3) &= \boxed{\mathrm{Qr}\left[\underline{\alpha};\ \Xi({}^{(X)}q)\right]} + \mathrm{Qr}\left[\underline{\beta};\ (K,\Lambda,\Theta)({}^{(X)}q)\right] \\
&\quad + \mathrm{Qr}\left[(\rho,\sigma);\ (I,\underline{I})({}^{(X)}q)\right] + \mathrm{Qr}\left[\beta;\ (\underline{K},\underline{\Lambda},\underline{\Theta})({}^{(X)}q)\right] \\
&\quad + \mathrm{Qr}\left[\alpha;\ \underline{\Xi}({}^{(X)}q)\right] \qquad (6.1.21) \\
K(J^3) &= \mathrm{Qr}\left[\alpha;\ \underline{\Theta}({}^{(X)}q)\right] + \boxed{\mathrm{Qr}\left[\beta;\ (I,\underline{I})({}^{(X)}q)\right]} \\
&\quad + \mathrm{Qr}\left[(\rho,\sigma);\ (K,\Lambda)({}^{(X)}q)\right] + \mathrm{Qr}\left[\underline{\beta};\ \Xi({}^{(X)}q)\right] \\
\Lambda(J^3) &= \mathrm{Qr}\left[\alpha;\ \underline{\Theta}({}^{(X)}q)\right] + \boxed{\mathrm{Qr}\left[\beta;\ (I,\underline{I})({}^{(X)}q)\right]}
\end{aligned}
$$

$$+\mathrm{Qr}\left[(\rho,\sigma)\,;\,(K,\Lambda)({}^{(X)}q)\right]+\mathrm{Qr}\left[\underline{\beta}\,;\,\Xi({}^{(X)}q)\right] \tag{6.1.22}$$

$$\begin{aligned}
\Theta(J^3) &= \mathrm{Qr}\left[\alpha\,;\,\underline{K}({}^{(X)}q)\right]+\mathrm{Qr}\left[\alpha\,;\,\underline{\Lambda}({}^{(X)}q)\right]+\boxed{\mathrm{Qr}\left[\alpha\,;\,\underline{\Theta}({}^{(X)}q)\right]}\\
&\quad+\mathrm{Qr}\left[\beta\,;\,(I,\underline{I})({}^{(X)}q)\right]+\mathrm{Qr}\left[(\rho,\sigma)\,;\,\Theta({}^{(X)}q)\right]+\boxed{\mathrm{Qr}\left[\underline{\beta}\,;\,\Xi({}^{(X)}q)\right]}\\
\Xi(J^3) &= \mathrm{Qr}\left[\alpha\,;\,(I,\underline{I})({}^{(X)}q)\right]+\mathrm{Qr}\left[\beta\,;\,(K,\Lambda,\Theta)({}^{(X)}q)\right]+\mathrm{Qr}\left[(\rho,\sigma)\,;\,\Xi({}^{(X)}q)\right].
\end{aligned} \tag{6.1.23}$$

The above expressions for the currents $J^2(X,W)$ and $J^3(X,W)$ depend on the null components of ${}^{(X)}p$ and ${}^{(X)}q$. They are:

$$\begin{aligned}
{}^{(X)}p_3 &= \not{d}\mathrm{iv}\,{}^{(X)}\underline{\mathbf{m}}-\frac{1}{2}(\not{\mathbf{D}}_4{}^{(X)}\underline{\mathbf{n}}+\not{\mathbf{D}}_3{}^{(X)}\mathbf{j})+(2\underline{\eta}+\eta-\zeta)\cdot{}^{(X)}\underline{\mathbf{m}}\\
&\quad-\hat{\chi}\cdot{}^{(X)}\mathbf{i}-\frac{1}{2}tr\chi\,(\mathrm{tr}{}^{(X)}\mathbf{i}+{}^{(X)}\mathbf{j})-\frac{1}{2}\mathrm{tr}\underline{\chi}\,{}^{(X)}\mathbf{n}-(\mathbf{D}_3\log\Omega),{}^{(X)}\mathbf{n}
\end{aligned} \tag{6.1.24}$$

$$\begin{aligned}
{}^{(X)}p_4 &= \not{d}\mathrm{iv}\,{}^{(X)}\mathbf{m}-\frac{1}{2}(\not{\mathbf{D}}_3{}^{(X)}\mathbf{n}+\not{\mathbf{D}}_4{}^{(X)}\mathbf{j})+(2\eta+\underline{\eta}+\zeta)\cdot{}^{(X)}\mathbf{m}\\
&\quad-\underline{\hat{\chi}}\cdot{}^{(X)}\mathbf{i}-\frac{1}{2}\mathrm{tr}\underline{\chi}\,(\mathrm{tr}{}^{(X)}\mathbf{i}+{}^{(X)}\mathbf{j})-\frac{1}{2}\mathrm{tr}\chi\,{}^{(X)}\underline{\mathbf{n}}-(\mathbf{D}_4\log\Omega){}^{(X)}\underline{\mathbf{n}},
\end{aligned} \tag{6.1.25}$$

$$\begin{aligned}
{}^{(X)}\not{p} &= \not{\nabla}_c{}^{(X)}\mathbf{i}-\frac{1}{2}(\not{\mathbf{D}}_4{}^{(X)}\underline{\mathbf{m}}+\not{\mathbf{D}}_3{}^{(X)}\mathbf{m})-\frac{1}{2}(\mathbf{D}_4\log\Omega){}^{(X)}\underline{\mathbf{m}}-\frac{1}{2}(\mathbf{D}_3\log\Omega){}^{(X)}\mathbf{m}\\
&\quad+\frac{1}{2}{}^{(X)}\mathbf{j}(\eta+\underline{\eta})+{}^{(X)}\mathbf{i}\cdot(\eta+\underline{\eta})-\frac{3}{4}\mathrm{tr}\chi\,{}^{(X)}\underline{\mathbf{m}}-\frac{3}{4}\mathrm{tr}\underline{\chi}\,{}^{(X)}\mathbf{m}-\frac{1}{2}\underline{\hat{\chi}}\cdot{}^{(X)}\mathbf{m}\\
&\quad-\frac{1}{2}\hat{\chi}\cdot{}^{(X)}\underline{\mathbf{m}}.
\end{aligned} \tag{6.1.26}$$

The null components of ${}^{(X)}q$ are expressed in the same notation as that used in (6.1.4) to denote the various null components:

$$\begin{aligned}
\Lambda({}^{(X)}q) &= \frac{1}{4}\left(\mathbf{D}_3{}^{(X)}\mathbf{n}-2(\mathbf{D}_3\log\Omega){}^{(X)}\mathbf{n}-4\eta\cdot{}^{(X)}\mathbf{m}\right)\\
&- \frac{1}{4}\left(\mathbf{D}_4{}^{(X)}\mathbf{j}-2\underline{\eta}\cdot{}^{(X)}\mathbf{m}\right)+\frac{2}{3}{}^{(X)}p_4\\
K({}^{(X)}q)_{ab} &= \frac{1}{2}\left(\not{\nabla}_a{}^{(X)}\mathbf{m}_b-\not{\nabla}_b{}^{(X)}\mathbf{m}_a\right)+\frac{1}{2}\left(\zeta_a{}^{(X)}\mathbf{m}_b-\zeta_b{}^{(X)}\mathbf{m}_a\right)\\
&- \frac{1}{2}\left(\hat{\chi}_{ac}{}^{(X)}\mathbf{i}_{cb}-\hat{\chi}_{bc}{}^{(X)}\mathbf{i}_{ca}\right)\\
\Xi({}^{(X)}q)_a &= \frac{1}{2}\not{\mathbf{D}}_4{}^{(X)}\mathbf{m}_a-\frac{1}{2}\not{\nabla}_a{}^{(X)}\mathbf{n}-\frac{1}{2}\underline{\eta}_a{}^{(X)}\mathbf{n}-\frac{1}{2}(\mathbf{D}_4\log\Omega){}^{(X)}\mathbf{m}_a\\
&+ \frac{1}{2}\mathrm{tr}\chi\,{}^{(X)}\mathbf{m}_a+\hat{\chi}_{ac}{}^{(X)}\mathbf{m}_c
\end{aligned} \tag{6.1.27}$$

$$
\begin{aligned}
I({}^{(X)}q)_a &= \frac{1}{2}\not{\mathbf{D}}_4{}^{(X)}\underline{\mathbf{m}}_a - \frac{1}{2}\not\nabla_a{}^{(X)}\mathbf{j} - \frac{1}{2}(\mathbf{D}_4\log\Omega){}^{(X)}\underline{\mathbf{m}}_a + \frac{1}{4}\mathrm{tr}\chi{}^{(X)}\underline{\mathbf{m}}_a \\
&+ \frac{1}{2}\hat{\chi}_{ac}{}^{(X)}\underline{\mathbf{m}}_c + \frac{1}{4}\mathrm{tr}\underline{\chi}{}^{(X)}\mathbf{m}_a + \frac{1}{2}\underline{\hat{\chi}}_{ac}{}^{(X)}\mathbf{m}_c - \frac{1}{2}\underline{\eta}_c{}^{(X)}\mathbf{i}_{ca} + \frac{3}{2}{}^{(X)}\not{p}_a \\
\Theta({}^{(X)}q)_{ab} &= 2\left(\not{\mathbf{D}}_4{}^{(X)}\mathbf{i}_{ab} - \frac{1}{2}\delta_{ab}\mathrm{tr}(\not{\mathbf{D}}_4{}^{(X)}\mathbf{i})\right) - \left(\not\nabla_a{}^{(X)}\mathbf{m}_b + \not\nabla_b{}^{(X)}\mathbf{m}_a - \delta_{ab}\not\nabla_c{}^{(X)}\mathbf{m}_c\right) \\
&- 2\left(\underline{\eta}_a{}^{(X)}\mathbf{m}_b + \underline{\eta}_b{}^{(X)}\mathbf{m}_a - \delta_{ab}\underline{\eta}_c{}^{(X)}\mathbf{m}_c\right) - \left(\zeta_a{}^{(X)}\mathbf{m}_b + \zeta_b{}^{(X)}\mathbf{m}_a - \delta_{ab}\zeta_c{}^{(X)}\mathbf{m}_c\right) \\
&+ \mathrm{tr}\chi{}^{(\hat{X})}\mathbf{i}_{ab} + \hat{\chi}_{ab}\mathrm{tr}{}^{(X)}\mathbf{i} + \underline{\hat{\chi}}_{ab}{}^{(X)}\mathbf{n} + \hat{\chi}_{ab}{}^{(X)}\mathbf{j},
\end{aligned}
$$

and the underlined quantities are obtained with the standard substitutions,

$$
\begin{aligned}
\underline{\Delta}({}^{(X)}q) &= \frac{1}{4}\left(\mathbf{D}_4{}^{(X)}\underline{\mathbf{n}} - 2(\mathbf{D}_4\log\Omega){}^{(X)}\underline{\mathbf{n}} - 4\underline{\eta}\cdot{}^{(X)}\underline{\mathbf{m}}\right) - \frac{1}{4}\left(\mathbf{D}_3{}^{(X)}\mathbf{j} - 2\eta\cdot{}^{(X)}\underline{\mathbf{m}}\right) + \frac{2}{3}{}^{(X)}p_3 \\
\underline{K}({}^{(X)}q)_{ab} &= \frac{1}{2}\left(\not\nabla_a{}^{(X)}\underline{\mathbf{m}}_b - \not\nabla_b{}^{(X)}\underline{\mathbf{m}}_a\right) - \frac{1}{2}\left(\zeta_a{}^{(X)}\underline{\mathbf{m}}_b - \zeta_b{}^{(X)}\underline{\mathbf{m}}_a\right) - \frac{1}{2}\left(\underline{\hat{\chi}}_{ac}{}^{(X)}\mathbf{i}_{cb} - \underline{\hat{\chi}}_{bc}{}^{(X)}\mathbf{i}_{ca}\right) \\
\underline{\Xi}({}^{(X)}q)_a &= \frac{1}{2}\not{\mathbf{D}}_3{}^{(X)}\underline{\mathbf{m}}_a - \frac{1}{2}\not\nabla_a{}^{(X)}\underline{\mathbf{n}} - \frac{1}{2}\eta_a{}^{(X)}\underline{\mathbf{n}} - \frac{1}{2}(\mathbf{D}_3\log\Omega){}^{(X)}\underline{\mathbf{m}}_a + \frac{1}{2}\mathrm{tr}\underline{\chi}{}^{(X)}\underline{\mathbf{m}}_a + \underline{\hat{\chi}}_{ac}{}^{(X)}\underline{\mathbf{m}}_c \\
\underline{I}({}^{(X)}q)_a &= \frac{1}{2}\not{\mathbf{D}}_3{}^{(X)}\mathbf{m}_a - \frac{1}{2}\not\nabla_a{}^{(X)}\mathbf{j} - \frac{1}{2}(\mathbf{D}_3\log\Omega){}^{(X)}\mathbf{m}_a + \frac{1}{4}\mathrm{tr}\underline{\chi}{}^{(X)}\mathbf{m}_a \\
&+ \frac{1}{2}\underline{\hat{\chi}}_{ac}{}^{(X)}\mathbf{m}_c + \frac{1}{4}\mathrm{tr}\chi{}^{(X)}\underline{\mathbf{m}}_a + \frac{1}{2}\hat{\chi}_{ac}{}^{(X)}\underline{\mathbf{m}}_c - \frac{1}{2}\eta_c{}^{(X)}\mathbf{i}_{ca} + \frac{3}{2}{}^{(X)}\not{p}_a \qquad (6.1.28) \\
\underline{\Theta}({}^{(X)}q)_{ab} &= 2\left(\not{\mathbf{D}}_3{}^{(X)}\mathbf{i}_{ab} - \frac{1}{2}\delta_{ab}\mathrm{tr}(\not{\mathbf{D}}_3{}^{(X)}\mathbf{i})\right) - \left(\not\nabla_a{}^{(X)}\underline{\mathbf{m}}_b + \not\nabla_b{}^{(X)}\underline{\mathbf{m}}_a - \delta_{ab}\not\nabla_c{}^{(X)}\underline{\mathbf{m}}_c\right) \\
&- 2\left(\eta_a{}^{(X)}\underline{\mathbf{m}}_b + \eta_b{}^{(X)}\underline{\mathbf{m}}_a - \delta_{ab}\eta_c{}^{(X)}\underline{\mathbf{m}}_c\right) + \left(\zeta_a{}^{(X)}\underline{\mathbf{m}}_b + \zeta_b{}^{(X)}\underline{\mathbf{m}}_a - \delta_{ab}\zeta_c{}^{(X)}\underline{\mathbf{m}}_c\right) \\
&+ \mathrm{tr}\underline{\chi}{}^{(X)}\mathbf{i}_{ab} + \underline{\hat{\chi}}_{ab}\mathrm{tr}{}^{(X)}\mathbf{i} + \hat{\chi}_{ab}{}^{(X)}\underline{\mathbf{n}} + \underline{\hat{\chi}}_{ab}{}^{(X)}\mathbf{j}.
\end{aligned}
$$

6.1.1 *Estimates for the T, S, $\bar{K}$ deformation tensors*

In Chapter 4 we proved the following results concerning the $\mathcal{O}$ norms (see Theorems 4.2.1, 4.2.2, 4.2.3) which, all together, prove Theorem **M1**:

$$
\begin{aligned}
&\mathcal{O}_0 + \underline{\mathcal{O}}_0 \le c(\mathcal{I}_0 + \mathcal{I}_* + \Delta_0) \\
&\mathcal{O}_{[1]} + \underline{\mathcal{O}}_{[1]} \le c(\mathcal{I}_0 + \mathcal{I}_* + \Delta_0 + \Delta_1) \\
&\mathcal{O}_{[2]} + \underline{\mathcal{O}}_{[2]} \le c(\mathcal{I}_0 + \mathcal{I}_* + \Delta_0 + \Delta_1 + \Delta_2) \\
&\mathcal{O}_3 + \underline{\mathcal{O}}_3 \le c(\mathcal{I}_0 + \mathcal{I}_* + \Delta_0 + \Delta_1 + \Delta_2).
\end{aligned}
$$

They justify, together with Theorem **M2**, assumptions (6.0.1) provided we choose $c(\mathcal{I}_0 + \mathcal{I}_* + \Delta_0 + \Delta_1 + \Delta_2) \le \epsilon_0$.

Based on these assumptions we now state a sequence of propositions concerning the components of deformation tensors associated with T, S, $\bar{K}$.

Proposition 6.1.1 *Under the assumptions (6.0.1), the following estimates hold for any $S \subset \mathcal{K}$ with $p \in [2, 4]$:*

$$
\begin{aligned}
&|r^{1-\frac{2}{p}}\tau_-{}^{(T)}\mathbf{i}|_{p,S} \leq c\epsilon_0\\
&|r^{1-\frac{2}{p}}\tau_-{}^{(T)}\mathbf{j}|_{p,S} \leq c\epsilon_0\\
&|r^{2-\frac{2}{p}}({}^{(T)}\mathbf{m},\ {}^{(T)}\underline{\mathbf{m}})|_{p,S} \leq c\epsilon_0\\
&|r^{2-\frac{2}{p}}{}^{(T)}\mathbf{n}|_{p,S} \leq c\epsilon_0\\
&|r^{1-\frac{2}{p}}\tau_-{}^{(T)}\underline{\mathbf{n}}|_{p,S} \leq c\epsilon_0
\end{aligned}
\tag{6.1.29}
$$

$$
\begin{aligned}
&|r^{2-\frac{2}{p}}\tau_-\nabla\!\!\!/{}^{(T)}\mathbf{i}|_{p,S} \leq c\epsilon_0\\
&|r^{2-\frac{2}{p}}\nabla\!\!\!/{}^{(T)}\mathbf{j}|_{p,S} \leq c\epsilon_0\\
&|r^{3-\frac{2}{p}}\nabla\!\!\!/({}^{(T)}\mathbf{m},\ {}^{(T)}\underline{\mathbf{m}})|_{p,S} \leq c\epsilon_0\\
&|r^{3-\frac{2}{p}}\nabla\!\!\!/{}^{(T)}\mathbf{n}|_{p,S} \leq c\epsilon_0\\
&|r^{2-\frac{2}{p}}\tau_-\nabla\!\!\!/{}^{(T)}\underline{\mathbf{n}}|_{p,S} \leq c\epsilon_0
\end{aligned}
\tag{6.1.30}
$$

$$
\begin{aligned}
&|r^{1-\frac{2}{p}}\tau_-^2 D\!\!\!\!/_3{}^{(T)}\mathbf{i}|_{p,S} \leq c\epsilon_0\\
&|r^{1-\frac{2}{p}}\tau_-^2 D\!\!\!\!/_3{}^{(T)}\mathbf{j}|_{p,S} \leq c\epsilon_0\\
&|r^{2-\frac{2}{p}}\tau_- D\!\!\!\!/_3({}^{(T)}\mathbf{m},\ {}^{(T)}\underline{\mathbf{m}})|_{p,S} \leq c\epsilon_0\\
&|r^{3-\frac{2}{p}} D\!\!\!\!/_3{}^{(T)}\mathbf{n}|_{p,S} \leq c\epsilon_0\\
&|r^{1-\frac{2}{p}}\tau_-^2 D\!\!\!\!/_3{}^{(T)}\underline{\mathbf{n}}|_{p,S} \leq c\epsilon_0.
\end{aligned}
\tag{6.1.31}
$$

$$
\begin{aligned}
&|r^{2-\frac{2}{p}}\tau_- D\!\!\!\!/_4{}^{(T)}\mathbf{i}|_{p,S} \leq c\epsilon_0\\
&|r^{2-\frac{2}{p}}\tau_- D\!\!\!\!/_4{}^{(T)}\mathbf{j}|_{p,S} \leq c\epsilon_0\\
&|r^{3-\frac{2}{p}} D\!\!\!\!/_4({}^{(T)}\mathbf{m},\ {}^{(T)}\underline{\mathbf{m}})|_{p,S} \leq c\epsilon_0\\
&|r^{3-\frac{2}{p}} D\!\!\!\!/_4{}^{(T)}\mathbf{n}|_{p,S} \leq c\epsilon_0\\
&\||r^{3-\frac{2}{p}} D\!\!\!\!/_4{}^{(T)}\underline{\mathbf{n}}|_{p,S} \leq c\epsilon_0.
\end{aligned}
\tag{6.1.32}
$$

Corollary 6.1.1 *Under the previous assumptions the following inequalities hold:*

$$
\begin{aligned}
&|r\tau_-{}^{(T)}\mathbf{i}|_{\infty,S} \leq c\epsilon_0\\
&|r\tau_-{}^{(T)}\mathbf{j}|_{\infty,S} \leq c\epsilon_0\\
&|r^2({}^{(T)}\mathbf{m},\ {}^{(T)}\underline{\mathbf{m}})|_{\infty,S} \leq c\epsilon_0\\
&|r^{2(T)}\mathbf{n}|_{\infty,S} \leq c\epsilon_0\\
&|r\tau_-{}^{(T)}\underline{\mathbf{n}}|_{\infty,S} \leq c\epsilon_0\\
&|r^{3-\frac{2}{p}} D\!\!\!\!/_3{}^{(T)}\mathbf{n}|_{\infty,S} \leq c\epsilon_0\\
&|r^{3-\frac{2}{p}} D\!\!\!\!/_4{}^{(T)}\underline{\mathbf{n}}|_{\infty,S} \leq c\epsilon_0.
\end{aligned}
\tag{6.1.33}
$$

Proposition 6.1.2 *Assuming the results of Theorems 4.2.1, 4.2.2 the following estimates hold for any $S \subset \mathcal{K}$ with $p \in [2, 4]$:*

$$
\begin{aligned}
&|r^{1-\frac{2}{p}}\frac{1}{\log r}{}^{(S)}\mathbf{i}|_{p,S} \leq c\epsilon_0\\
&|r^{1-\frac{2}{p}}\frac{1}{\log r}{}^{(S)}\mathbf{j}|_{p,S} \leq c\epsilon_0\\
&|r^{2-\frac{2}{p}}{}^{(S)}\mathbf{m}|_{p,S} \leq c\tau_-\epsilon_0\\
&|r^{1-\frac{2}{p}}{}^{(S)}\underline{\mathbf{m}}|_{p,S} \leq c\epsilon_0\\
&|r^{2-\frac{2}{p}}{}^{(S)}\mathbf{n}|_{p,S} \leq c\tau_-\epsilon_0\\
&|r^{-\frac{2}{p}}\tau_-{}^{(S)}\underline{\mathbf{n}}|_{p,S} \leq c\epsilon_0
\end{aligned}
\tag{6.1.34}
$$

$$
\begin{aligned}
&|r^{2-\frac{2}{p}}\nabla\!\!\!/\,{}^{(S)}\mathbf{i}|_{p,S} \leq c\epsilon_0\\
&|r^{2-\frac{2}{p}}\nabla\!\!\!/\,{}^{(S)}\mathbf{j}|_{p,S} \leq c\epsilon_0\\
&|r^{3-\frac{2}{p}}\nabla\!\!\!/\,{}^{(S)}\mathbf{m}|_{p,S} \leq c\tau_-\epsilon_0\\
&|r^{2-\frac{2}{p}}\nabla\!\!\!/\,{}^{(S)}\underline{\mathbf{m}}|_{p,S} \leq c\epsilon_0\\
&|r^{3-\frac{2}{p}}\nabla\!\!\!/\,{}^{(S)}\mathbf{n}|_{p,S} \leq c\tau_-\epsilon_0\\
&|r^{1-\frac{2}{p}}\tau_-\nabla\!\!\!/\,{}^{(S)}\underline{\mathbf{n}}|_{p,S} \leq c\epsilon_0
\end{aligned}
\tag{6.1.35}
$$

$$
\begin{aligned}
&|r^{1-\frac{2}{p}}\frac{\tau_-}{\log r}\mathcal{D}\!\!\!/_3{}^{(S)}\mathbf{i}|_{p,S} \leq c\epsilon_0\\
&|r^{1-\frac{2}{p}}\frac{\tau_-}{\log r}\mathcal{D}\!\!\!/_3{}^{(S)}\mathbf{j}|_{p,S} \leq c\epsilon_0\\
&|r^{2-\frac{2}{p}}\mathcal{D}\!\!\!/_3{}^{(S)}\mathbf{m}|_{p,S} \leq c\epsilon_0\\
&|r^{1-\frac{2}{p}}\tau_-\mathcal{D}\!\!\!/_3{}^{(S)}\underline{\mathbf{m}}|_{p,S} \leq c\epsilon_0\\
&|r^{-\frac{2}{p}}\tau_-^2\mathcal{D}\!\!\!/_3{}^{(S)}\underline{\mathbf{n}}|_{p,S} \leq c\epsilon_0
\end{aligned}
\tag{6.1.36}
$$

$$
\begin{aligned}
&|r^{2-\frac{2}{p}}\frac{\tau_-}{\log r}\mathcal{D}\!\!\!/_4{}^{(S)}\mathbf{i}|_{p,S} \leq c\epsilon_0\\
&|r^{2-\frac{2}{p}}\frac{\tau_-}{\log r}\mathcal{D}\!\!\!/_4{}^{(S)}\mathbf{j}|_{p,S} \leq c\epsilon_0\\
&|r^{3-\frac{2}{p}}\mathcal{D}\!\!\!/_4{}^{(S)}\mathbf{m}|_{p,S} \leq c\tau_-\epsilon_0\\
&|r^{2-\frac{2}{p}}\mathcal{D}\!\!\!/_4{}^{(S)}\underline{\mathbf{m}}|_{p,S} \leq c\epsilon_0\\
&|r^{3-\frac{2}{p}}\mathcal{D}\!\!\!/_4{}^{(S)}\mathbf{n}|_{p,S} \leq c\tau_-\epsilon_0,
\end{aligned}
\tag{6.1.37}
$$

and with $p \in [2, \infty]$,

$$
\begin{aligned}
&|r^{2-\frac{2}{p}}\mathcal{D}\!\!\!/_3{}^{(S)}\mathbf{n}|_{p,S} \leq c\epsilon_0\\
&|r^{1-\frac{2}{p}}\tau_-\mathcal{D}\!\!\!/_4{}^{(S)}\underline{\mathbf{n}}|_{p,S} \leq c\epsilon_0.
\end{aligned}
\tag{6.1.38}
$$

Proof: We examine only the estimate of $|r^{1-\frac{2}{p}}{}^{(S)}\mathbf{i}|_{p,S}$ to explain the logarithmic factor present in its estimate. The other estimates follow immediately from those relative to the connection coefficients. From the explicit expression (see (3.4.8)) of ${}^{(S)}\mathbf{i}$,

$$
{}^{(S)}\mathbf{i}_{ab} = \underline{u}\hat{\chi}_{ab} + u\hat{\underline{\chi}}_{ab} + \frac{1}{2}\delta_{ab}(\frac{1}{2}(\underline{u}\mathrm{tr}\chi + u\mathrm{tr}\underline{\chi}) + (\underline{u}\omega + u\underline{\omega}) - \frac{1}{\Omega}),
$$

it follows that the more delicate part to control is $\left[\frac{1}{2}(\underline{u}\mathrm{tr}\chi + u\mathrm{tr}\underline{\chi}) - \frac{1}{\Omega}\right]$, which we rewrite as

$$\begin{aligned}\frac{1}{2}(\underline{u}\mathrm{tr}\chi + u\mathrm{tr}\underline{\chi}) - \Omega^{-1} &= \Omega^{-1}\left[\frac{\Omega}{2}\left(\underline{u}\mathrm{tr}\chi + u\mathrm{tr}\underline{\chi}\right) - 1\right]\\ &= \Omega^{-1}\left[\frac{\Omega}{2}(\underline{u} - u)\mathrm{tr}\chi - 1\right] + \frac{u}{2}(\mathrm{tr}\chi + \mathrm{tr}\underline{\chi}).\end{aligned}$$

Using inequality (4.3.78) the second term can be bounded by $c\epsilon_0\frac{1}{r}$. The first term is estimated in the following way:

$$\begin{aligned}\left|\Omega^{-1}\left[\frac{\Omega}{2}(\underline{u} - u)\mathrm{tr}\chi - 1\right]\right| &\leq c\left(|\mathrm{tr}\chi||(\underline{u} - u) - 2r| + r\left|\mathrm{tr}\chi - \frac{2}{r}\right|\right)\\ &\leq c\epsilon_0\left(\frac{\log r}{r} + r\frac{1}{r^2}\right) \leq c\epsilon_0\frac{\log r}{r}\end{aligned}$$

using Lemma 4.1.8 and Proposition 4.3.6.

Corollary 6.1.2 *Under the previous assumptions the following inequalities hold for any* $S \subset \mathcal{K}$,

$$\begin{aligned}&|\tfrac{r}{\log r}{}^{(S)}\mathbf{i}|_{\infty,S} \leq c\epsilon_0\\ &|\tfrac{r}{\log r}{}^{(S)}\mathbf{j}|_{\infty,S} \leq c\epsilon_0\\ &|r^{2(S)}\mathbf{m}|_{\infty,S} \leq c\tau_-\epsilon_0\\ &|r^{(S)}\underline{\mathbf{m}}|_{\infty,S} \leq c\epsilon_0\\ &|r^{2(S)}\mathbf{n}|_{\infty,S} \leq c\tau_-\epsilon_0\\ &|\tau_-{}^{(S)}\underline{\mathbf{n}}|_{\infty,S} \leq c\epsilon_0.\end{aligned} \tag{6.1.39}$$

Remark: In the next propositions we provide estimates for the various components of the deformation tensor field ${}^{(K_0)}\pi$. The same estimates hold for the components of ${}^{(\bar{K})}\pi$.

Proposition 6.1.3 *Assume the results of Theorems 4.2.1, 4.2.2. Then the following estimates hold for any* $S \subset \mathcal{K}$ *and* $p \in [2, 4]$*:*

$$\begin{aligned}&|r^{-\frac{2}{p}}\tfrac{1}{\log r}{}^{(K_0)}\mathbf{i}|_{p,S} \leq c\epsilon_0\\ &|r^{-\frac{2}{p}}\tfrac{1}{\log r}{}^{(K_0)}\mathbf{j}|_{p,S} \leq c\epsilon_0\\ &|r^{2-\frac{2}{p}(K_0)}\mathbf{m}|_{p,S} \leq c\tau_-^2\epsilon_0\\ &|r^{-\frac{2}{p}(K_0)}\underline{\mathbf{m}}|_{p,S} \leq c\epsilon_0\\ &|r^{2-\frac{2}{p}(K_0)}\mathbf{n}|_{p,S} \leq c\tau_-^2\epsilon_0\\ &|r^{-\frac{2}{p}}\tau_-{}^{(K_0)}\underline{\mathbf{n}}|_{p,S} \leq cr\epsilon_0\end{aligned} \tag{6.1.40}$$

$$\begin{aligned}&|r^{1-\frac{2}{p}}\nabla\!\!\!\!/\,{}^{(K_0)}\mathbf{i}|_{p,S} \leq c\epsilon_0\\ &|r^{1-\frac{2}{p}}\nabla\!\!\!\!/\,{}^{(K_0)}\mathbf{j}|_{p,S} \leq c\epsilon_0\\ &|r^{3-\frac{2}{p}}\nabla\!\!\!\!/\,{}^{(K_0)}\mathbf{m}|_{p,S} \leq c\tau_-^2\epsilon_0\\ &|r^{1-\frac{2}{p}}\nabla\!\!\!\!/\,{}^{(K_0)}\underline{\mathbf{m}}|_{p,S} \leq c\epsilon_0\\ &|r^{3-\frac{2}{p}}\nabla\!\!\!\!/\,{}^{(K_0)}\mathbf{n}|_{p,S} \leq c\tau_-^2\epsilon_0\\ &|r^{-\frac{2}{p}}\tau_-\nabla\!\!\!\!/\,{}^{(K_0)}\underline{\mathbf{n}}|_{p,S} \leq c\epsilon_0\end{aligned} \tag{6.1.41}$$

$$
\begin{aligned}
&|r^{-\frac{2}{p}}\frac{\tau_-}{\log r}\mathbf{D}\!\!\!/_3{}^{(K_0)}\mathbf{i}|_{p,S} \le c\epsilon_0\\
&|r^{-\frac{2}{p}}\frac{\tau_-}{\log r}\mathbf{D}_3{}^{(K_0)}\mathbf{j}|_{p,S} \le c\epsilon_0\\
&|r^{2-\frac{2}{p}}\mathbf{D}\!\!\!/_3{}^{(K_0)}\mathbf{m}|_{p,S} \le c\tau_-\epsilon_0\\
&|r^{-\frac{2}{p}}\tau_-\mathbf{D}\!\!\!/_3{}^{(K_0)}\underline{\mathbf{m}}|_{p,S} \le c\epsilon_0\\
&|r^{-\frac{2}{p}}\tau_-^2\mathbf{D}\!\!\!/_3{}^{(K_0)}\underline{\mathbf{n}}|_{p,S} \le cr\epsilon_0
\end{aligned}
\tag{6.1.42}
$$

$$
\begin{aligned}
&|r^{1-\frac{2}{p}}\frac{\tau_-}{\log r}\mathbf{D}\!\!\!/_4{}^{(K_0)}\mathbf{i}|_{p,S} \le c\epsilon_0\\
&|r^{1-\frac{2}{p}}\frac{\tau_-}{\log r}\mathbf{D}\!\!\!/_4{}^{(K_0)}\mathbf{j}|_{p,S} \le c\epsilon_0\\
&|r^{3-\frac{2}{p}}\mathbf{D}\!\!\!/_4{}^{(K_0)}\mathbf{m}|_{p,S} \le c\tau_-\epsilon_0\\
&|r^{1-\frac{2}{p}}\mathbf{D}\!\!\!/_4{}^{(K_0)}\underline{\mathbf{m}}|_{p,S} \le c\epsilon_0\\
&|r^{3-\frac{2}{p}}\mathbf{D}\!\!\!/_4{}^{(K_0)}\mathbf{n}|_{p,S} \le c\tau_-^2\epsilon_0,
\end{aligned}
\tag{6.1.43}
$$

and with $p \in [2, \infty]$,

$$
\begin{aligned}
&|r^{2-\frac{2}{p}}\mathbf{D}\!\!\!/_3{}^{(K_0)}\mathbf{n}|_{p,S} \le c\tau_-\epsilon_0\\
&|r^{-\frac{2}{p}}\mathbf{D}\!\!\!/_4{}^{(K_0)}\underline{\mathbf{n}}|_{p,S} \le c\epsilon_0.
\end{aligned}
\tag{6.1.44}
$$

Proof: Again, the more delicate K_0-deformation tensor components to estimate are (see (3.4.9)):

$$
\begin{aligned}
{}^{(K_0)}\mathbf{i}_{ab} &= \underline{u}^2\hat{\chi}_{ab} + u^2\hat{\underline{\chi}}_{ab} + \frac{1}{2}\delta_{ab}\left(\frac{1}{2}(\underline{u}^2\mathrm{tr}\chi + u^2\mathrm{tr}\underline{\chi}) + (\underline{u}^2\omega + u^2\underline{\omega}) - \frac{\underline{u}+u}{\Omega}\right)\\
{}^{(K_0)}\mathbf{j} &= \frac{1}{2}(\underline{u}^2\mathrm{tr}\chi + u^2\mathrm{tr}\underline{\chi}) + (\underline{u}^2\omega + u^2\underline{\omega}) - \frac{\underline{u}+u}{\Omega}.
\end{aligned}
$$

From the expression of ${}^{(K_0)}\mathbf{i}$, the more delicate term is $\left[\frac{1}{2}(\underline{u}^2\mathrm{tr}\chi + u^2\mathrm{tr}\underline{\chi}) - \frac{1}{\Omega}(u+\underline{u})\right]$. Proceeding as in the case of ${}^{(S)}\mathbf{i}$, we estimate it, using inequality (4.3.78), Lemma 4.1.8 and Proposition 4.3.6,

$$
\begin{aligned}
\frac{1}{2}(\underline{u}^2\mathrm{tr}\chi + u^2\mathrm{tr}\underline{\chi}) - \frac{1}{\Omega}(u+\underline{u}) &= \frac{1}{2}(\underline{u}^2 - u^2)\mathrm{tr}\chi + \frac{1}{2}u^2(\mathrm{tr}\chi + \mathrm{tr}\underline{\chi}) - \frac{1}{\Omega}(u+\underline{u})\\
&= \frac{1}{2}u^2(\mathrm{tr}\chi + \mathrm{tr}\underline{\chi}) + (\underline{u}+u)\left[\frac{1}{2}(\underline{u}-u)\mathrm{tr}\chi - 2\right] + O(\epsilon_0)\\
&= O(r)\left[\frac{(\underline{u}-u)}{r} - 2\right] + \epsilon_0\left[1 + O(\tau_-^2 r^{-2})\right] = \epsilon_0\left[O(1) + O(\log r)\right].
\end{aligned}
$$

All the remaining estimates follow easily from the corresponding estimates of the connection coefficients.

Corollary 6.1.3 *Under the previous assumptions the following inequalities hold for any $S \subset \mathcal{K}$:*

$$
\begin{aligned}
&|\frac{1}{\log r}{}^{(K_0)}\mathbf{i}|_{\infty,S} \le c\epsilon_0\\
&|\frac{1}{\log r}{}^{(K_0)}\mathbf{j}|_{\infty,S} \le c\epsilon_0\\
&|r^{2\,(K_0)}\mathbf{m}|_{\infty,S} \le c\tau_-^2\epsilon_0\\
&|{}^{(K_0)}\underline{\mathbf{m}}|_{\infty,S} \le c\epsilon_0\\
&|r^{2\,(K_0)}\mathbf{n}|_{\infty,S} \le c\tau_-^2\epsilon_0\\
&|\tau_-{}^{(K_0)}\underline{\mathbf{n}}|_{\infty,S} \le cr\epsilon_0.
\end{aligned}
\tag{6.1.45}
$$

Proposition 6.1.4 *From Proposition 6.1.1 and from the explicit expressions of* ${}^{(T)}p_3$, ${}^{(T)}p_4$ *and* ${}^{(T)}\not{p}$, *we obtain the following estimates for any* $S \subset \mathcal{K}$ *with* $p \in [2,4]$*:*

$$
\begin{aligned}
&|r^{1-\frac{2}{p}}\tau_-^2\,{}^{(T)}p_3|_{p,S} \leq c\epsilon_0\\
&|r^{2-\frac{2}{p}}\tau_-\,{}^{(T)}p_4|_{p,S} \leq c\epsilon_0\\
&|r^{2-\frac{2}{p}}\tau_-\,{}^{(T)}\not{p}_a|_{p,S} \leq c\epsilon_0.
\end{aligned}
\tag{6.1.46}
$$

Proposition 6.1.5 *From Proposition 6.1.1 and from the explicit expressions of* $\not\nabla{}^{(T)}p_3$, $\not\nabla{}^{(T)}p_4$ *and* $\not\nabla{}^{(T)}\not{p}$, *we obtain the following estimates for any* $S \subset \mathcal{K}$ *with* $p \in [2,4]$*:*

$$
\begin{aligned}
&|r^{2-\frac{2}{p}}\tau_-^2\not\nabla{}^{(T)}p_3|_{p,S} \leq c\epsilon_0\\
&|r^{3-\frac{2}{p}}\tau_-\not\nabla{}^{(T)}p_4|_{p,S} \leq c\epsilon_0\\
&|r^{3-\frac{2}{p}}\tau_-\not\nabla{}^{(T)}\not{p}_a|_{p,S} \leq c\epsilon_0.
\end{aligned}
\tag{6.1.47}
$$

In addition to the estimates of Propositions 6.1.4, 6.1.5 we shall also need, for the T-deformation tensors, the following proposition.

Proposition 6.1.6 *From Proposition 6.1.1 and Propositions 4.3.10 and 4.4.2 the following estimates hold:*

$$
||\frac{1}{\sqrt{\tau_+}}\tau_-^2\,{}^{(T)}p_3||_{L_2(C\cap\mathcal{K})} \leq c\epsilon_0 \tag{6.1.48}
$$

$$
\begin{aligned}
&||\frac{1}{\sqrt{\tau_+}}\tau_-^2 r\not\nabla{}^{(T)}p_3||_{L_2(C\cap\mathcal{K})} \leq c\epsilon_0\\
&||\frac{1}{\sqrt{\tau_+}}\tau_-^2\hat{\mathcal{L}}_S\,{}^{(T)}p_3||_{L_2(C\cap\mathcal{K})} \leq c\epsilon_0.
\end{aligned}
$$

Proof: From the explicit expression of ${}^{(T)}p_3$ (see (6.1.24)) it follows immediate[illegible] that its more delicate term is $\mathbf{D}_3^2\log\Omega$. Then one refers to Propositions 4.3.10 and 4.4.2.

Proposition 6.1.7 *From Proposition 6.1.2 and from the explicit expressions of* ${}^{(S)}p_3$, ${}^{(S)}p_4$ *and* ${}^{(S)}\not{p}$, *we obtain the following estimates for any* $S \subset \mathcal{K}$ *with* $p \in [2,4]$*:*

$$
\begin{aligned}
&|r^{1-\frac{2}{p}}\frac{\tau_-}{\log r}\,{}^{(S)}p_3|_{p,S} \leq c\epsilon_0\\
&|r^{1-\frac{2}{p}}\frac{\tau_-}{\log r}\,{}^{(S)}p_4|_{p,S} \leq c\epsilon_0\\
&|r^{2-\frac{2}{p}}\,{}^{(S)}\not{p}_a|_{p,S} \leq c\epsilon_0.
\end{aligned}
\tag{6.1.49}
$$

The various components of ${}^{(T)}q$ and ${}^{(S)}q$ satisfy the following estimates.

Proposition 6.1.8 *From Proposition 6.1.1 and from the explicit expression of* ${}^{(T)}q$ *we obtain the following estimates for any* $S \subset \mathcal{K}$ *with* $p \in [2,4]$*:*

$$
\begin{aligned}
&|r^{2-\frac{2}{p}}\tau_-\Lambda({}^{(T)}q)|_{p,S} \le c\epsilon_0\\
&|r^{3-\frac{2}{p}}K({}^{(T)}q)|_{p,S} \le c\epsilon_0\\
&|r^{3-\frac{2}{p}}\Xi({}^{(T)}q)|_{p,S} \le c\epsilon_0\\
&|r^{2-\frac{2}{p}}\tau_-I({}^{(T)}q)|_{p,S} \le c\epsilon_0\\
&|r^{2-\frac{2}{p}}\tau_-\Theta({}^{(T)}q)|_{p,S} \le c\epsilon_0\\
&|r^{4-\frac{2}{p}}\nabla\!\!\!/ K({}^{(T)}q)|_{p,S} \le c\epsilon_0,
\end{aligned}
\tag{6.1.50}
$$

and

$$
\begin{aligned}
&|r^{1-\frac{2}{p}}\tau_-^2\underline{\Lambda}({}^{(T)}q)|_{p,S} \le c\epsilon_0\\
&|r^{2-\frac{2}{p}}\tau_-\underline{K}({}^{(T)}q)|_{p,S} \le c\epsilon_0\\
&|r^{2-\frac{2}{p}}\tau_-\underline{\Xi}({}^{(T)}q)|_{p,S} \le c\epsilon_0\\
&|r^{2-\frac{2}{p}}\tau_-\underline{I}({}^{(T)}q)|_{p,S} \le c\epsilon_0\\
&|r^{1-\frac{2}{p}}\tau_-^2\underline{\Theta}({}^{(T)}q)|_{p,S} \le c\epsilon_0\\
&|r^{3-\frac{2}{p}}\tau_-\nabla\!\!\!/ \underline{K}({}^{(T)}q)|_{p,S} \le c\epsilon_0.
\end{aligned}
\tag{6.1.51}
$$

Corollary 6.1.4 *Using the Sobolev Lemma 4.1.3, the sup norms for* $K({}^{(T)}q)$ *and* $\underline{K}({}^{(T)}q)$ *are bounded by*

$$
\begin{aligned}
&|r^3K({}^{(T)}q)|_{\infty,S} \le c\epsilon_0\\
&|r^2\tau_-\underline{K}({}^{(T)}q)|_{\infty,S} \le c\epsilon_0.
\end{aligned}
\tag{6.1.52}
$$

Remark: Recall that the norms associated with the vector field T with the slowest asymptotic behavior in r are

$$
\begin{aligned}
&|r^{1-\frac{2}{p}}\tau_-{}^{(T)}\mathbf{i}|_{p,S}\,,\ |r^{1-\frac{2}{p}}\tau_-{}^{(T)}\mathbf{j}|_{p,S}\,,\ |r^{1-\frac{2}{p}}\tau_-{}^{(T)}\mathbf{n}|_{p,S}\\
&|r^{1-\frac{2}{p}}\tau_-^2\mathbf{D}\!\!\!\!/_3{}^{(T)}\mathbf{i}|_{p,S}\,,\ |r^{1-\frac{2}{p}}\tau_-^2\mathbf{D}\!\!\!\!/_3{}^{(T)}\mathbf{j}|_{p,S}\,,\ |r^{1-\frac{2}{p}}\tau_-^2\mathbf{D}\!\!\!\!/_3{}^{(T)}\underline{\mathbf{n}}|_{p,S}\\
&|r^{1-\frac{2}{p}}\tau_-^{2\,(T)}p_3|_{p,S}\,,\ |r^{1-\frac{2}{p}}\tau_-^2\underline{\Lambda}({}^{(T)}q)|_{p,S}\,,\ |r^{1-\frac{2}{p}}\tau_-^2\underline{\Theta}({}^{(T)}q)|_{p,S}.
\end{aligned}
$$

Examining all these terms we observe that the slow decay of these quantities originates from the behavior of $\mathbf{D}_3\log\Omega$ or $\mathbf{D}_3^2\log\Omega$ on the last slice; see Subsection 3.5.5.

Proposition 6.1.9 *From Proposition 6.1.2 and from the explicit expressions of* ${}^{(S)}q$ *we obtain the following estimates for any* $S \subset \mathcal{K}$ *with* $p \in [2,4]$*:*

$$
\begin{aligned}
&|r^{2-\frac{2}{p}}\frac{\tau_-}{\log r}\Lambda({}^{(S)}q)|_{p,S} \le c\epsilon_0\\
&|r^{3-\frac{2}{p}}\frac{1}{\log r}K({}^{(S)}q)|_{p,S} \le c\epsilon_0\\
&|r^{3-\frac{2}{p}}\Xi({}^{(S)}q)|_{p,S} \le c\epsilon_0\\
&|r^{2-\frac{2}{p}}\frac{\tau_-}{\log r}I({}^{(S)}q)|_{p,S} \le c\epsilon_0\\
&|r^{2-\frac{2}{p}}\frac{\tau_-}{\log r}\Theta({}^{(S)}q)|_{p,S} \le c\epsilon_0\\
&|r^{4-\frac{2}{p}}\frac{1}{\log r}\nabla\!\!\!/ K({}^{(S)}q)|_{p,S} \le c\epsilon_0
\end{aligned}
\tag{6.1.53}
$$

and

$$
\begin{aligned}
&|r^{1-\frac{2}{p}}\frac{\tau_-^2}{\log r}\underline{\Lambda}({}^{(S)}q)|_{p,S}\leq c\epsilon_0\\
&|r^{2-\frac{2}{p}}\frac{\tau_-}{\log r}\underline{K}({}^{(S)}q)|_{p,S}\leq c\epsilon_0\\
&|r^{2-\frac{2}{p}}\tau_-\underline{\Xi}({}^{(S)}q)|_{p,S}\leq c\epsilon_0\\
&|r^{2-\frac{2}{p}}\frac{\tau_-}{\log r}\underline{I}({}^{(S)}q)|_{p,S}\leq c\epsilon_0\\
&|r^{1-\frac{2}{p}}\frac{\tau_-^2}{\log r}\underline{\Theta}({}^{(S)}q)|_{p,S}\leq c\epsilon_0\\
&|r^{3-\frac{2}{p}}\frac{\tau_-}{\log r}\nabla\!\!\!/\underline{K}({}^{(S)}q)|_{p,S}\leq c\epsilon_0.
\end{aligned}
\tag{6.1.54}
$$

6.1.2 *Estimates for the rotation deformation tensors*

We recall the result proved in Chapter 4, Section 4.6.

Corollary 4.7.1 *In $\mathcal{K}$, the following inequalities hold:*

$$
\begin{aligned}
&|r^{1-\frac{2}{p}}({}^{(O)}\mathbf{i},\,{}^{(O)}\mathbf{j},\,{}^{(O)}\underline{\mathbf{m}})|_{p,S}\leq c\epsilon_0\\
&|r^{2-\frac{2}{p}}\nabla\!\!\!/({}^{(O)}\mathbf{i},\,{}^{(O)}\mathbf{j},\,{}^{(O)}\underline{\mathbf{m}})|_{p,S}\leq c\epsilon_0\\
&|r^{2-\frac{2}{p}}\mathbf{D}\!\!\!/_4({}^{(O)}\mathbf{i},\,{}^{(O)}\mathbf{j},\,{}^{(O)}\underline{\mathbf{m}})|_{p,S}\leq c\epsilon_0\\
&|r^{1-\frac{2}{p}}\tau_-\mathbf{D}\!\!\!/_3({}^{(O)}\mathbf{i},\,{}^{(O)}\mathbf{j},\,{}^{(O)}\underline{\mathbf{m}})|_{p,S}\leq c\epsilon_0,
\end{aligned}
$$

where, for the first line, $p\in[2,\infty]$ and for the others, $p\in[2,4]$. Moreover,

$$
{}^{(O)}\mathbf{n},\,{}^{(O)}\underline{\mathbf{n}},\,{}^{(O)}\mathbf{m}=0\,.
$$

The following propositions provide estimates for the various components of ${}^{(O)}p$ and ${}^{(O)}q$. They are consequences of the estimates for the connection coefficients proved in Theorems 4.2.1, 4.2.2 and Corollary 4.7.1.

Proposition 6.1.10 *Based on Theorems 4.2.1, 4.2.2 and on Corollary 4.7.1, the following estimates hold for any $S\subset\mathcal{K}$ with $p\in[2,4]$:*

$$
\begin{aligned}
&|r^{1-\frac{2}{p}}\tau_-{}^{(O)}p_3|_{p,S}\leq c\epsilon_0\\
&|r^{2-\frac{2}{p}}({}^{(O)}p_4,\,{}^{(O)}\not{p})|_{p,S}\leq c\epsilon_0.
\end{aligned}
\tag{6.1.55}
$$

Proposition 6.1.11 *Based on Theorems 4.2.1, 4.2.2 and on Corollary 4.7.1, the following estimates hold:*

$$
\begin{aligned}
&||\hat{\mathcal{L}}_O{}^{(O)}p_3||_{L^2(\underline{C}(\underline{u})\cap V(u,\underline{u}))}\leq c\epsilon_0\\
&||\hat{\mathcal{L}}_O{}^{(O)}\not{p}||_{L^2(\underline{C}(\underline{u})\cap V(u,\underline{u}))}\leq c\epsilon_0\\
&||\frac{1}{\sqrt{\tau_+}}r\hat{\mathcal{L}}_O{}^{(O)}p_4||_{L^2(C(u)\cap V(u,\underline{u}))}\leq c\epsilon_0.
\end{aligned}
$$

Proof: It is enough to observe that the more delicate term for $\hat{\mathcal{L}}_O{}^{(O)}p_3$ is $\nabla\!\!\!/^{2(i)}Z_a$ and for $\hat{\mathcal{L}}_O{}^{(O)}\not{p}$ is $\nabla\!\!\!/^{2(i)}H_{ab}$ and then use Proposition 4.7.2. The last inequality is easier to obtain and we do not report it here.

Proposition 6.1.12 *Under the same assumptions as in Proposition 6.1.11 the following estimates hold:*

$$\begin{aligned}
&\sup_{\mathcal{K}} |r^{2-\frac{2}{p}}\Lambda({}^{(O)}q)|_{p,S} \leq c\epsilon_0\\
&\sup_{\mathcal{K}} |r^{3-\frac{2}{p}}K({}^{(O)}q)|_{p,S} \leq c\epsilon_0\\
&\sup_{\mathcal{K}} |r^{2-\frac{2}{p}}I({}^{(O)}q)|_{p,S} \leq c\epsilon_0\\
&\sup_{\mathcal{K}} |r^{2-\frac{2}{p}}\Theta({}^{(O)}q)|_{p,S} \leq c\epsilon_0\\
&\Xi({}^{(O)}q) = 0
\end{aligned} \tag{6.1.56}$$

and

$$\begin{aligned}
&\sup_{\mathcal{K}} |r^{1-\frac{2}{p}}\tau_-\underline{\Lambda}({}^{(O)}q)|_{p,S} \leq c\epsilon_0\\
&\sup_{\mathcal{K}} |r^{2-\frac{2}{p}}\underline{K}({}^{(O)}q)|_{p,S} \leq c\epsilon_0\\
&\sup_{\mathcal{K}} |r^{1-\frac{2}{p}}\underline{\Xi}({}^{(O)}q)|_{p,S} \leq c\epsilon_0\\
&\sup_{\mathcal{K}} |r^{2-\frac{2}{p}}\underline{I}({}^{(O)}q)|_{p,S} \leq c\epsilon_0\\
&\sup_{\mathcal{K}} |r^{1-\frac{2}{p}}\tau_-\underline{\Theta}({}^{(O)}q)|_{p,S} \leq c\epsilon_0
\end{aligned} \tag{6.1.57}$$

for $p \in [2, 4]$.

Proposition 6.1.13 *Under the same assumptions as in Proposition 6.1.11 the following estimates hold, the first with* $\delta > \epsilon > 0$*:*

$$\begin{aligned}
&\|\frac{1}{\sqrt{r^{1-2\epsilon}}}\hat{\mathcal{L}}_O\underline{\Xi}({}^{(O)}q)\|_{L^2(\underline{C}(\underline{u}')\cap V(u,\underline{u}'))} \leq c\frac{1}{u\sqrt{r^{1-2\delta}}}\epsilon_0\\
&\|\hat{\mathcal{L}}_O(I({}^{(O)}q), \underline{I}({}^{(O)}q))\|_{L^2(\underline{C}(\underline{u}')\cap V(u,\underline{u}'))} \leq c\epsilon_0\\
&\|\hat{\mathcal{L}}_O\underline{\Lambda}({}^{(O)}q)\|_{L^2(\underline{C}(\underline{u}')\cap V(u,\underline{u}'))} \leq c\epsilon_0\\
&\|\hat{\mathcal{L}}_O\underline{K}({}^{(O)}q)\|_{L^2(\underline{C}(\underline{u}')\cap V(u,\underline{u}'))} \leq c\epsilon_0\\
&\|\hat{\mathcal{L}}_O\underline{\Theta}({}^{(O)}q)\|_{L^2(\underline{C}(\underline{u}')\cap V(u,\underline{u}'))} \leq c\epsilon_0.
\end{aligned} \tag{6.1.58}$$

Proof: For the first inequality it is enough to observe that the more delicate term of $\hat{\mathcal{L}}_O\underline{\Xi}({}^{(O)}q)$ is $r\nabla\!\!\!\!/\,\mathbf{D}_3 Z$ and then use Proposition 4.7.2. An analogous argument holds for the remaining inequalities.

Remark: Some of the most delicate error terms appear in connection with the highest derivatives of the rotation deformation tensors ${}^{(O)}\hat{\pi}$. Indeed, as discussed in detail in Section 4.6, unlike all other deformation tensors, the second derivatives of ${}^{(O)}\hat{\pi}_{ab}$ and ${}^{(O)}\hat{\pi}_{a3}$ involve the third derivatives of the connection coefficients. This is why it is crucial to show that the $\mathcal{D}_2$ norms depend only on the second derivatives of the curvature tensor[9] and not on the third derivatives as it may appear from the structure equations. In the appendix to this chapter we recall precisely where the third derivatives of the connection coefficients appear.

---

[9] See Proposition 4.7.2.

6.2 The error terms $\mathcal{E}_1$

In Chapter 3, Subsection 3.3.4, we introduced above Σ_0 a narrow region $\mathcal{K}'_{\delta_0}$ called the initial layer region endowed with a different foliation, the initial layer foliation, which fits appropriately with the initial hypersurface Σ_0. The hypersurface Σ'_{δ_0} is the upper boundary of $\mathcal{K}'_{\delta_0}$. Moreover the Oscillation Lemma shows that we can define a hypersurface $\tilde{\Sigma}_{\tilde{\delta}_0}$ (see Corollary 4.1.2) associated with the double null canonical foliation at a distance $c\epsilon_0$ from Σ'_{δ_0}.

All the estimates made in this chapter are relative to the double null canonical foliation and the initial hypersurface is, in this case, $\tilde{\Sigma}_{\tilde{\delta}_0}$, which we call simply Σ_0 in the sequel.

To complete the proof of Theorem **M8** we have to estimate $\mathcal{Q}_{\tilde{\Sigma}_{\tilde{\delta}_0}\cap\mathcal{K}}$ in terms of $\mathcal{Q}_{\Sigma_0\cap\mathcal{K}}$. This is an immediate consequence of Theorem **M0** and the Oscillation Lemma.

6.2.1 *Estimate of* $\int_{V_{(u,\underline{u})}} \mathbf{Div}\, Q(\hat{\mathcal{L}}_T W)_{\beta\gamma\delta}(\bar{K}^\beta, \bar{K}^\gamma, \bar{K}^\delta)$

We must estimate the following four integrals:

$$B_1 \equiv \int_{V_{(u,\underline{u})}} \tau_+^6 D(T,W)_{444} \quad , \quad B_2 \equiv \int_{V_{(u,\underline{u})}} \tau_+^4\tau_-^2 D(T,W)_{344}$$
$$B_3 \equiv \int_{V_{(u,\underline{u})}} \tau_+^2\tau_-^4 D(T,W)_{334} \ , \ B_4 \equiv \int_{V_{(u,\underline{u})}} \tau_-^6 D(T,W)_{333}.$$

From equations 6.1.3, to estimate B_1 we have to control the integrals:

$$\begin{aligned}&\int_{V_{(u,\underline{u})}} \tau_+^6 \alpha(\hat{\mathcal{L}}_T W)\cdot\Theta(T,W)\\ &\int_{V_{(u,\underline{u})}} \tau_+^6 \beta(\hat{\mathcal{L}}_T W)\cdot\Xi(T,W);\end{aligned} \tag{6.2.1}$$

to estimate B_2, we have to control the integrals

$$\begin{aligned}&\int_{V_{(u,\underline{u})}} \tau_+^4\tau_-^2 \rho(\hat{\mathcal{L}}_T W)\Lambda(T,W)\\ &\int_{V_{(u,\underline{u})}} \tau_+^4\tau_-^2 \sigma(\hat{\mathcal{L}}_T W)K(T,W)\\ &\int_{V_{(u,\underline{u})}} \tau_+^4\tau_-^2 \beta(\hat{\mathcal{L}}_T W)\cdot I(T,W);\end{aligned} \tag{6.2.2}$$

to estimate B_3, we have to control the integrals:

$$\begin{aligned}&\int_{V_{(u,\underline{u})}} \tau_+^2\tau_-^4 \rho(\hat{\mathcal{L}}_T W)\underline{\Lambda}(T,W)\\ &\int_{V_{(u,\underline{u})}} \tau_+^2\tau_-^4 \sigma(\hat{\mathcal{L}}_T W)\underline{K}(T,W)\\ &\int_{V_{(u,\underline{u})}} \tau_+^2\tau_-^4 \underline{\beta}(\hat{\mathcal{L}}_T W)\cdot\underline{I}(T,W);\end{aligned} \tag{6.2.3}$$

and to estimate B_4, we have to control the integrals:

$$\begin{aligned}&\int_{V_{(u,\underline{u})}} \tau_-^6 \underline{\alpha}(\hat{\mathcal{L}}_T W)\cdot\underline{\Theta}(T,W)\\ &\int_{V_{(u,\underline{u})}} \tau_-^6 \underline{\beta}(\hat{\mathcal{L}}_T W)\cdot\underline{\Xi}(T,W).\end{aligned} \tag{6.2.4}$$

We estimate in detail the integrals appearing in B_1. Those relative to the other groups B_2, B_3, B_4 have lower weights in τ_+ and, therefore, are easier to treat.

Remark: The $\frac{1}{\log r}$ factors appearing in the various estimates of the deformation tensors relative to the S and K_0 vector fields appearing in Proposition 6.1.2, Corollary 6.1.2, Proposition 6.1.3, Corollary 6.1.3, Proposition 6.1.7 and Proposition 6.1.9 do not play any role in the subsequent estimates and, therefore, are hereafter disregarded.

Estimate of the B_1 integrals

From the decomposition $J(X;W) = J^1(X;W) + J^2(X;W) + J^3(X;W)$ (see (6.1.6)) it follows that

$$\Theta(T,W) = \Theta^{(1)}(T,W) + \Theta^{(2)}(T,W) + \Theta^{(3)}(T,W)$$
$$\Xi(T,W) = \Xi^{(1)}(T,W) + \Xi^{(2)}(T,W) + \Xi^{(3)}(T,W).$$

We write the two B_1 integrals as sums of three terms:

$$\int_{V_{(u,\underline{u})}} \tau_+^6 \alpha(\hat{\mathcal{L}}_T W)\cdot\Theta(T,W) = \sum_{i=1}^{3}\int_{V_{(u,\underline{u})}} \tau_+^6 \alpha(\hat{\mathcal{L}}_T W)\cdot\Theta^{(i)}(T,W)$$
$$\int_{V_{(u,\underline{u})}} \tau_+^6 \beta(\hat{\mathcal{L}}_T W)\cdot\Xi(T,W) = \sum_{i=1}^{3}\int_{V_{(u,\underline{u})}} \tau_+^6 \beta(\hat{\mathcal{L}}_T W)\cdot\Xi^{(i)}(T,W).$$

Proposition 6.2.1 *Under the assumptions (6.0.1) and (6.0.2) the following inequalities hold:*

$$\left|\int_{V_{(u,\underline{u})}} \tau_+^6 \alpha(\hat{\mathcal{L}}_T W)\cdot\Theta(T,W)\right| \le c\epsilon_0 \mathcal{Q}_{\mathcal{K}}$$
$$\left|\int_{V_{(u,\underline{u})}} \tau_+^6 \beta(\hat{\mathcal{L}}_T W)\cdot\Xi(T,W)\right| \le c\epsilon_0 \mathcal{Q}_{\mathcal{K}}.$$

Proof: We discuss in detail the first integral; the estimate of the second one is similar. Using the coarea formulas[10]

$$\int_{V(u,\underline{u})} F = \int_{u_0}^{u} du' \int_{C(u')\cap V(u,\underline{u})} F$$
$$\int_{V(u,\underline{u})} F = \int_{\underline{u}_0}^{\underline{u}} d\underline{u}' \int_{\underline{C}(\underline{u}')\cap V(u,\underline{u})} F \tag{6.2.5}$$

and the Schwartz inequality is written as

$$\left|\int_{V_{(u,\underline{u})}} \tau_+^6 \alpha(\hat{\mathcal{L}}_T W)\cdot\Theta(T,W)\right| \le c\int_{u_0}^{u} du' \left(\int_{C(u';[\underline{u}_0,\underline{u}])} \underline{u}'^6 |\alpha(\hat{\mathcal{L}}_T W)|^2\right)^{\frac{1}{2}},$$

[10]In the whole chapter $u_0 = u_0(\underline{u}) = u|_{\underline{C}(\underline{u})\cap\tilde{\Sigma}_{\tilde{\delta}_0}}$ and $\underline{u}_0 = \underline{u}_0(u) = \underline{u}|_{C(u)\cap\tilde{\Sigma}_{\tilde{\delta}_0}}$.

$$
\begin{aligned}
\left(\int_{C(u';[\underline{u}_0,\underline{u}])} \underline{u}'^6|\Theta(T,W)|^2\right)^{\frac{1}{2}} &\le c\mathcal{R}\left[\sum_{i=1}^{3}\int_{u_0}^{u} du'\left(\int_{C(u';[\underline{u}_0,\underline{u}])} \underline{u}'^6|\Theta^{(i)}(T,W)|^2\right)^{\frac{1}{2}}\right]\\
&\le c\mathcal{Q}_{\mathcal{K}}^{\frac{1}{2}}\left[\sum_{i=1}^{3}\int_{u_0}^{u} du'\left(\int_{C(u';[\underline{u}_0,\underline{u}])} \underline{u}'^6|\Theta^{(i)}(T,W)|^2\right)^{\frac{1}{2}}\right], \qquad (6.2.6)
\end{aligned}
$$

where we used the definition of $\mathcal{R}$ given in Chapter 3 and the inequality (6.0.2). The various terms on the right-hand side associated with the currents J^1, J^2, J^3 are estimated separately[11]. The result is obtained proving the following lemma,

Lemma 6.2.1 *Under the assumptions (6.0.1) and (6.0.2) the following inequalities hold:*

$$
\begin{aligned}
\left(\int_{C(u';[\underline{u}_0,\underline{u}])} \underline{u}'^6|\Theta^{(1)}(T,W)|^2\right)^{\frac{1}{2}} &\le c\epsilon_0\mathcal{Q}_{\mathcal{K}}^{\frac{1}{2}}\frac{1}{|u'|^{\frac{3}{2}}}\\
\left(\int_{C(u';[\underline{u}_0,\underline{u}])} \underline{u}'^6|\Theta^{(2)}(T,W)|^2\right)^{\frac{1}{2}} &\le c\epsilon_0\mathcal{Q}_{\mathcal{K}}^{\frac{1}{2}}\frac{1}{|u'|^{2}} \qquad (6.2.7)\\
\left(\int_{C(u';[\underline{u}_0,\underline{u}])} \underline{u}'^6|\Theta^{(3)}(T,W)|^2\right)^{\frac{1}{2}} &\le c\epsilon_0\mathcal{Q}_{\mathcal{K}}^{\frac{1}{2}}\frac{1}{|u'|^{2}}.
\end{aligned}
$$

Proof: All the various terms composing the first integral of (6.2.7) are estimated in the same way.[12] We discuss the first term

$$
\int_{C(u';[\underline{u}_0,\underline{u}])} \underline{u}'^6|{}^{(T)}\underline{\mathbf{m}}|^2|\nabla\!\!\!/\alpha(W)|^2.
$$

Using Corollary 6.1.1 for ${}^{(T)}\underline{\mathbf{m}}$,

$$
\begin{aligned}
\int_{C(u';[\underline{u}_0,\underline{u}])} \underline{u}'^6|{}^{(T)}\underline{\mathbf{m}}|^2|\nabla\!\!\!/\alpha(W)|^2 &\le c\frac{1}{u'^4}\left(\sup_{V(u,\underline{u})}|r^2\,{}^{(T)}\underline{\mathbf{m}}|\right)^2\int_{C(u';[\underline{u}_0,\underline{u}])}\tau_+^6|\nabla\!\!\!/\alpha(W)|^2\\
&\le c\frac{1}{u'^4}\left(\sup_{V(u,\underline{u})}|r^2\,{}^{(T)}\underline{\mathbf{m}}|\right)^2\mathcal{Q}_{\mathcal{K}} \le c\frac{1}{u'^4}\epsilon_0^2\mathcal{Q}_{\mathcal{K}}\,.
\end{aligned}
$$

To control the second integral of (6.2.7), recalling

$$
\Theta^{(2)}(T,W) = \mathrm{Qr}\left[{}^{(T)}p_3\,;\,\alpha\right] + \mathrm{Qr}\left[{}^{(T)}\not{p}\,;\,\beta\right] + \mathrm{Qr}\left[{}^{(T)}p_4\,;\,(\rho,\sigma)\right]\,,
$$

---

[11]The integrals depending on J^1 are estimated differently than those depending on J^2, J^3. The reason is that $\Theta^{(1)}(T,W)$ and $\Xi^{(1)}(T,W)$ (see 6.1.8,...,(6.1.17) are quadratic expressions depending linearly on the various components of the deformation tensor ${}^{(T)}\hat{\pi}$ and on the zero and first derivatives of the null Riemann components. Therefore, in this case, the components of the deformation tensor are estimated with their sup norms. On the other side, the terms $\Theta^{(2,3)}(T,W)$, $\Xi^{(2,3)}(T,W)$ associated with the J^2, J^3 currents (see (6.1.18),...,(6.1.23)) are quadratic expressions depending linearly on the deformation tensor ${}^{(T)}\hat{\pi}$, on its first derivatives and on the undifferentiated null Riemann components. Therefore, in this case, the first derivatives of the deformation tensor are estimated in the $|\cdot|_{p,S}$ norms with $p \in [2,4]$ and the Riemann components with their sup norms.

[12]In fact only the term depending on ρ produces the factor $u'^{-\frac{3}{2}}$; all the other ones behave better, as $O(u'^{-2})$.

we have to estimate the integrals

$$\begin{aligned}&\int_{C(u';[\underline{u}_0,\underline{u}])}\tau_+^6|{}^{(T)}p_3|^2|\alpha(W)|^2\\&\int_{C(u';[\underline{u}_0,\underline{u}])}\tau_+^6|{}^{(T)}\not{p}|^2|\beta(W)|^2\\&\int_{C(u';[\underline{u}_0,\underline{u}])}\tau_+^6|{}^{(T)}p_4|^2|(\rho,\sigma)(W)|^2.\end{aligned}\tag{6.2.8}$$

The first one is the more delicate, since ${}^{(T)}p_3$ has the slowest asymptotic decay (see (6.1.46))

$$\begin{aligned}\left(\int_{C(u';[\underline{u}_0,\underline{u}])}\tau_+^6|{}^{(T)}p_3|^2|\alpha(W)|^2\right)^{\frac12}&\leq\frac{1}{u'^2}\left(\int_{C(u';[\underline{u}_0,\underline{u}])}|\tau_-^2{}^{(T)}p_3|^2\tau_+^6|\alpha(W)|^2\right)^{\frac12}\\&\leq\frac{1}{u'^2}\left(\sup_{\mathcal{K}}\tau_+^{\frac72}|\alpha(W)|\right)\left(\int_{\underline{u}_0}^{\underline{u}}\frac{1}{\tau_+}|\tau_-^2{}^{(T)}p_3|^2_{p=2,S}\right)^{\frac12}\\&\leq c\frac{1}{u'^2}\mathcal{Q}_{\mathcal{K}}^{\frac12}\left(\int_{\underline{u}_0}^{\underline{u}}\frac{1}{\tau_+}|\tau_-^2{}^{(T)}p_3|^2_{p=2,S}\right)^{\frac12}\\&\leq c\epsilon_0\mathcal{Q}_{\mathcal{K}}^{\frac12}\frac{1}{u'^2},\end{aligned}\tag{6.2.9}$$

where we used Proposition 6.1.6 to estimate ${}^{(T)}p_3$ and the estimate for $\alpha(W)$ in (3.7.1), proved in Subsection 5.1.4. The estimates of the remaining integrals in (6.2.8) are easier and we do not report them here.

To control the third integral of (6.2.7), recalling that (see (6.1.23))

$$\begin{aligned}\Theta^{(3)}(T,W)&=\mathrm{Qr}\left[\alpha\,;\,\underline{K}({}^{(T)}q)\right]+\mathrm{Qr}\left[\alpha\,;\,\underline{\Lambda}({}^{(T)}q)\right]+\mathrm{Qr}\left[\beta\,;\,(I,\underline{I})({}^{(T)}q)\right]\\&+\ \mathrm{Qr}\left[(\rho,\sigma)\,;\,\Theta({}^{(T)}q)\right]\end{aligned}$$

we have to estimate the integrals

$$\begin{aligned}&\int_{C(u';[\underline{u}_0,\underline{u}])}\tau_+^6|(\underline{K}({}^{(T)}q),\underline{\Lambda}({}^{(T)}q))|^2|\alpha(W)|^2\\&\int_{C(u';[\underline{u}_0,\underline{u}])}\tau_+^6|(I({}^{(T)}q),\underline{I}({}^{(T)}q))|^2|\beta(W)|^2\\&\int_{C(u';[\underline{u}_0,\underline{u}])}\tau_+^6|\Theta({}^{(T)}q)|^2|(\rho,\sigma)(W)|^2.\end{aligned}\tag{6.2.10}$$

Again the worst asymptotic behavior is due to ${}^{(T)}p_3$ which is present in the explicit expression of $\underline{\Lambda}({}^{(T)}q)$ (see (6.1.28)). We write

$$\underline{\Lambda}({}^{(T)}q)=\tilde{\underline{\Lambda}}({}^{(T)}q)+\frac23{}^{(T)}p_3$$

and observe that, for $p \in [2, \infty]$, $\sup_{\mathcal{K}} |r^{2-\frac{2}{p}} \tau_- \tilde{\underline{\Lambda}}(^{(T)}q)|_{p,S} \leq c\epsilon_0$. Then, we have

$$\left(\int_{C(u';[\underline{u}_0,\underline{u}])} \tau_+^6 |(\underline{K}(^{(T)}q), \tilde{\underline{\Lambda}}(^{(T)}q))|^2 |\alpha(W)|^2 \right)^{\frac{1}{2}}$$
$$\leq c \frac{1}{u'^2} \left(\sup_{\mathcal{K}} |\tau_+^{\frac{7}{2}} \alpha(W)| \right) \left(\sup_{\mathcal{K}} |r^{2-\frac{2}{p}} \tau_- (K(^{(T)}q), \tilde{\underline{\Lambda}}(^{(T)}q))|_{p=2,S} \right)$$
$$\leq c\epsilon_0^2 \mathcal{Q}_{\mathcal{K}} \frac{1}{u'^2}. \tag{6.2.11}$$

The remaining part $\int_{C(u';[\underline{u}_0,\underline{u}])} \tau_+^6 |^{(T)}p_3|^2 |\alpha(W)|^2$ has already been estimated (see (6.2.9)). The estimates of the second and third integrals are the same as the first one and will be omitted.

The estimate of the second integral of Proposition 6.2.1 proceeds in a similar way. We have, for the part associated to J^1,

$$\left| \int_{V_{(u,\underline{u})}} \tau_+^6 \beta(\hat{\mathcal{L}}_T W) \cdot \Xi^{(1)}(T, W) \right|$$
$$\leq c \int_{u_0}^{u} du' \frac{1}{|u'|} \left(\int_{C(u';[\underline{u}_0,\underline{u}])} \underline{u}'^4 u'^2 |\beta(\hat{\mathcal{L}}_T W)|^2 \right)^{\frac{1}{2}} \left(\int_{C(u';[\underline{u}_0,\underline{u}])} \underline{u}'^8 |\Xi^{(1)}(T, W)|^2 \right)^{\frac{1}{2}}$$
$$\leq c \mathcal{Q}_{\mathcal{K}}^{\frac{1}{2}} \int_{u_0}^{u} du' \frac{1}{|u'|} \left(\int_{C(u';[\underline{u}_0,\underline{u}])} \underline{u}'^8 |\Xi^{(1)}(T, W)|^2 \right)^{\frac{1}{2}}, \tag{6.2.12}$$

where the first factor in the integrand is bounded by $\mathcal{Q}_{\mathcal{K}}^{\frac{1}{2}}$ according to (6.0.2) and the definition of $\mathcal{R}$. For the integrals associated with J^2, J^3 we proceed in a different way, using the second coarea formula in (6.2.5).

$$\left| \sum_{i=2}^{3} \int_{V_{(u,\underline{u})}} \tau_+^6 \beta(\hat{\mathcal{L}}_T W) \cdot \Xi^{(i)}(T, W) \right|$$
$$\leq c \int_{\underline{u}_0}^{\underline{u}} d\underline{u}' \left(\int_{\underline{C}(\underline{u}';[u_0,u])} \underline{u}'^6 |\beta(\hat{\mathcal{L}}_T W)|^2 \right)^{\frac{1}{2}} \left(\sum_{i=2}^{3} \int_{\underline{C}(\underline{u}';[u_0,u])} \underline{u}'^6 |\Xi^{(i)}(T, W)|^2 \right)^{\frac{1}{2}}$$
$$\leq c \mathcal{Q}_{\mathcal{K}}^{\frac{1}{2}} \sum_{i=2}^{3} \int_{\underline{u}_0}^{\underline{u}} d\underline{u}' \left(\int_{\underline{C}(\underline{u}';[u_0,u])} \underline{u}'^6 |\Xi^{(i)}(T, W)|^2 \right)^{\frac{1}{2}}, \tag{6.2.13}$$

where, again, the first factor in the integrand has been bounded by $\mathcal{Q}_{\mathcal{K}}^{\frac{1}{2}}$. The various terms on the right-hand side of (6.2.12) and (6.2.13), associated with the currents J^1, J^2, J^3, are estimated separately. The result is formulated in the next lemma.

Lemma 6.2.2 *Under the assumptions (6.0.1) and (6.0.2) the following inequalities hold:*

$$\left(\int_{C(u';[\underline{u}_0,\underline{u}])} \underline{u}'^8 |\Xi^{(1)}(T, W)|^2 \right)^{\frac{1}{2}} \leq c\epsilon_0 \mathcal{Q}_{\mathcal{K}}^{\frac{1}{2}} \frac{1}{|u'|}$$

$$\left(\int_{\underline{C}(\underline{u}';[u_0,u])} \underline{u}'^6|\Xi^{(2)}(T,W)|^2\right)^{\frac{1}{2}} \leq c\epsilon_0\mathcal{Q}_{\mathcal{K}}^{\frac{1}{2}}\frac{1}{|\underline{u}'|^{\frac{3}{2}}} \tag{6.2.14}$$

$$\left(\int_{\underline{C}(\underline{u}';[u_0,u])} \underline{u}'^6|\Xi^{(3)}(T,W)|^2\right)^{\frac{1}{2}} \leq c\epsilon_0\mathcal{Q}_{\mathcal{K}}^{\frac{1}{2}}\frac{1}{|\underline{u}'|^{\gamma}}$$

with $1 < \gamma < \frac{3}{2}$.

Proof: All the various terms appearing in the decomposition of $\Xi^{(1)}$ (see (6.1.13)) produce the same $|u'|^{-1}$ dependence. Let us consider, among them, the following one:

$$\int_{C(u';[\underline{u}_0,\underline{u}])} \underline{u}'^8|^{(T)}\mathbf{m}|^2|\beta_3(W)|^2.$$

We proceed as in Lemma 6.2.1, using Corollary 6.1.1 for ${}^{(T)}\mathbf{m}$ and inequality (6.0.2),

$$\int_{C(u';[\underline{u}_0,\underline{u}])} \underline{u}'^8|^{(T)}\mathbf{m}|^2|\beta_3(W)|^2 \leq c\left(\sup_{\mathcal{K}}|r^{2(T)}\mathbf{m}|^2\right)\int_{C(u';[\underline{u}_0,\underline{u}])} \underline{u}'^4|\beta_3|^2$$
$$\leq c\frac{1}{|u'|^2}\left(\sup_{\mathcal{K}}|r^{2(T)}\mathbf{m}|^2\right)\mathcal{Q}_1 \leq c\epsilon_0^2\mathcal{Q}_{\mathcal{K}}\frac{1}{|u'|^2}. \tag{6.2.15}$$

To estimate the second integral of (6.2.14), we recall (see (6.1.18))

$$\Xi^{(2)}(T,W) = \mathrm{Qr}\left[{}^{(X)}\not{p};\,\alpha\right] + \mathrm{Qr}\left[{}^{(X)}p_4;\,\beta\right],$$

we are led to examine the integrals

$$\int_{\underline{C}(\underline{u}';[u_0,u])} \underline{u}'^6|^{(T)}\not{p}|^2|\alpha(W)|^2$$
$$\int_{\underline{C}(\underline{u}';[u_0,u])} \underline{u}'^6|^{(T)}p_4|^2|\beta(W)|^2. \tag{6.2.16}$$

The first integral is estimated as follows, with the help of Proposition 6.1.4:

$$\int_{\underline{C}(\underline{u}';[u_0,u])} \underline{u}'^6|^{(T)}\not{p}|^2|\alpha(W)|^2 \leq \frac{1}{\underline{u}'^3}\left(\sup_{\mathcal{K}}|\tau_+^{\frac{7}{2}}\alpha(W)|^2\right)\int_{\underline{C}(\underline{u}';[u_0,u])} \frac{1}{u'^2}|r\tau_-{}^{(T)}\not{p}|^2$$
$$\leq c\mathcal{Q}_{\mathcal{K}}\frac{1}{\underline{u}'^3}\left(\sup_{\mathcal{K}}|r^{2-\frac{2}{p}}\tau_-{}^{(T)}\not{p}|^2_{p=2,S}\right)\int_{u_0}^{u} du'\frac{1}{u'^2} \leq c\epsilon_0^2\mathcal{Q}_{\mathcal{K}}\frac{1}{\underline{u}'^3}. \tag{6.2.17}$$

The second integral in (6.2.16) is estimated exactly as the previous one by substituting $|\tau_+^{\frac{7}{2}}\alpha(W)|$ with $|\tau_+^{\frac{7}{2}}\beta(W)|$ and $|r^{2-\frac{2}{p}}\tau_-{}^{(T)}\not{p}|_{p,S}$ with $|r^{2-\frac{2}{p}}\tau_-{}^{(T)}p_4|_{p,S}$.

To estimate the third integral of (6.2.14) since (see (6.1.23))

$$\begin{aligned}\Xi^{(3)}(T,W) &= \mathrm{Qr}\left[\alpha;\,(I,\underline{I})({}^{(X)}q)\right] + \mathrm{Qr}\left[\beta;\,(K,\Lambda,\Theta)({}^{(X)}q)\right]\\ &+ \mathrm{Qr}\left[(\rho,\sigma);\,\Xi({}^{(X)}q)\right],\end{aligned}$$

we have to control the following integrals:

$$\begin{aligned}
&\int_{\underline{C}(\underline{u}';[u_0,u])} \tau_+^6 \left| \left(I({}^{(T)}q), \underline{I}({}^{(T)}q)\right) \right|^2 |\alpha(W)|^2 \\
&\int_{\underline{C}(\underline{u}';[u_0,u])} \tau_+^6 \left| \left(K({}^{(T)}q), \Lambda({}^{(T)}q), \Theta({}^{(T)}q)\right) \right|^2 |\beta(W)|^2 \qquad (6.2.18)\\
&\int_{\underline{C}(\underline{u}';[u_0,u])} \tau_+^6 |\Xi({}^{(T)}q)|^2 |(\rho,\sigma)(W)|^2.
\end{aligned}$$

For the first integral we observe that the $|\cdot|_{p,S}$ norms with which we bound $I({}^{(T)}q)$ and $\underline{I}({}^{(T)}q)$ are the same as those used to control ${}^{(T)}p_4$ or ${}^{(T)}\not{p}$ (see (6.1.50), (6.1.51)). Therefore we proceed as in (6.2.17). The second integral can be bounded in the same way as the integral $\int_{V_{(u,\underline{u})}} \tau_+^6 |{}^{(T)}p_4|^2 |\beta(W)|^2$; see (6.2.16).

The third integral is controlled using the estimate of $\Xi({}^{(T)}q)$ in Proposition 6.1.8. We obtain, for any $\gamma < \frac{3}{2}$,

$$\begin{aligned}
&\int_{\underline{C}(\underline{u}';[u_0,u])} \tau_+^6 |\Xi({}^{(T)}q)|^2 |(\rho,\sigma)(W)|^2 \\
&\leq \left(\sup_{\mathcal{K}} |r^3(\rho,\sigma)(W)|^2\right) \int_{\underline{C}(\underline{u}';[u_0,u])} \tau_+^6 \frac{1}{r^{10}} |r^2 \Xi({}^{(T)}q)|^2 \\
&\leq c\frac{1}{\underline{u}'^{2\gamma}} \left(\sup_{\mathcal{K}} |r^{3-\frac{2}{p}} \Xi({}^{(T)}q)|^2_{p=2,S}\right) \mathcal{Q}_{\mathcal{K}} \int_{u_0}^{u} du' \frac{1}{u'^{4-2\gamma}} \leq c\epsilon_0^2 \mathcal{Q}_{\mathcal{K}} \frac{1}{\underline{u}'^{2\gamma}}.
\end{aligned}$$

Estimate of the B_2 integrals

The integrals in the B_2 group (see (6.2.2)), while similar to those we analyzed in the previous subsections, have a lower τ_+ weight and, therefore, are a little simpler. We will shortly show how they are estimated. However, we shall only analyze the integrals related to J^2, J^3, which, as we have shown in the previous discussion, are more delicate. Their estimates are collected in the following propositions.

Proposition 6.2.2 *Under the assumptions (6.0.1) and (6.0.2) the following inequalities hold:*

$$\begin{aligned}
&\int_{V_{(u,\underline{u})}} \tau_+^4 \tau_-^2 \rho(\hat{\mathcal{L}}_T W) \Lambda^{(2)}(T,W) \leq c\epsilon_0 \mathcal{Q}_{\mathcal{K}} \\
&\int_{V_{(u,\underline{u})}} \tau_+^4 \tau_-^2 \sigma(\hat{\mathcal{L}}_T W) K^{(2)}(T,W) \leq c\epsilon_0 \mathcal{Q}_{\mathcal{K}} \qquad (6.2.19)\\
&\int_{V_{(u,\underline{u})}} \tau_+^4 \tau_-^2 \beta(\hat{\mathcal{L}}_T W) \cdot I^{(2)}(T,W) \leq c\epsilon_0 \mathcal{Q}_{\mathcal{K}}.
\end{aligned}$$

Proposition 6.2.3 *Under the assumptions (6.0.1) and (6.0.2) the following inequalities hold:*

$$\int_{V_{(u,\underline{u})}} \tau_+^4 \tau_-^2 \rho(\hat{\mathcal{L}}_T W) \Lambda^{(3)}(T,W) \leq c\epsilon_0 \mathcal{Q}_{\mathcal{K}}$$

$$\int_{V_{(u,\underline{u})}} \tau_+^4 \tau_-^2 \sigma(\hat{\mathcal{L}}_T W) K^{(3)}(T,W) \le c\epsilon_0 \mathcal{Q}_{\mathcal{K}} \tag{6.2.20}$$

$$\int_{V_{(u,\underline{u})}} \tau_+^4 \tau_-^2 \beta(\hat{\mathcal{L}}_T W) \cdot I^{(3)}(T,W) \le c\epsilon_0 \mathcal{Q}_{\mathcal{K}}.$$

Proof of Proposition 6.2.2 The first and second integrals of (6.2.19) have the same structure (see (6.1.18)). Therefore we estimate only the first one. It decomposes into the following integrals (see (6.1.18)):

$$\int_{V_{(u,\underline{u})}} \tau_+^4 \tau_-^2 \rho(\hat{\mathcal{L}}_T W)^{(T)}\not{p}\beta(W)$$

$$\int_{V_{(u,\underline{u})}} \tau_+^4 \tau_-^2 \rho(\hat{\mathcal{L}}_T W)^{(T)}p_4(\rho,\sigma)(W). \tag{6.2.21}$$

We estimate the first one as follows:

$$\begin{aligned}
&\left|\int_{V_{(u,\underline{u})}} \tau_+^4 \tau_-^2 \rho(\hat{\mathcal{L}}_T W)^{(T)}\not{p}\beta(W)\right| \le c\int_{V_{(u,\underline{u})}} \frac{1}{\underline{u}'^{\frac{5}{2}}} |\tau_+^2 \tau_- \rho(\hat{\mathcal{L}}_T W)||r\tau_-{}^{(T)}\not{p}||\tau_+^{\frac{7}{2}}\beta(W)| \\
&\le c\left(\sup_{\mathcal{K}} |\tau_+^{\frac{7}{2}}\beta|\right)\left(\sup_{\mathcal{K}} \int_{\underline{C}(\underline{u}';[u_0,u])} \tau_+^4\tau_-^2|\rho(\hat{\mathcal{L}}_T W)|^2\right)^{\frac{1}{2}} \\
&\cdot \int_{\underline{u}_0}^{\underline{u}} \frac{d\underline{u}'}{\underline{u}'^{\frac{3}{2}}}\left(\int_{u_0}^{u} \frac{1}{u'^2}|r^{2-\frac{2}{p}}\tau_-{}^{(T)}\not{p}|^2_{p=2,S}\right)^{\frac{1}{2}} \qquad (6.2.22) \\
&\le c\left(\sup_{\mathcal{K}} |r^{2-\frac{2}{p}}\tau_-{}^{(T)}\not{p}|^2_{p=2,S}\right)^{\frac{1}{2}}\left(\sup_{\mathcal{K}} |\tau_+^{\frac{7}{2}}\beta|\right)\underline{\mathcal{Q}}_1^{\frac{1}{2}}\int_{\underline{u}_0}^{\underline{u}} \frac{d\underline{u}'}{\underline{u}'^{\frac{3}{2}}}\left(\int_{u_0}^{u}\frac{du'}{u'^2}\right)^{\frac{1}{2}} \le c\epsilon_0\mathcal{Q}_{\mathcal{K}}.
\end{aligned}$$

The estimate of the second integral in (6.2.21) is done exactly in the same way and we do not report it here.

The most sensitive term in the third integral of (6.2.19) is $\int_{V_{(u,\underline{u})}} \tau_+^4\tau_-^2\beta(\hat{\mathcal{L}}_T W)({}^{(T)}p_3)\beta(W)$. We obtain[13]

$$\begin{aligned}
&\left|\int_{V_{(u,\underline{u})}} \tau_+^4\tau_-^2\beta(\hat{\mathcal{L}}_T W)^{(T)}p_3\beta(W)\right| \\
&\le c\int d\underline{u}'\int_{\underline{C}(\underline{u}';[u_0,u])} |\tau_+^3\beta(\hat{\mathcal{L}}_T W)|\frac{1}{\tau_+^{\frac{5}{2}}}|\tau_-^{2}{}^{(T)}p_3||\tau_+^{\frac{7}{2}}\beta(W)| \qquad (6.2.23) \\
&\le c\left(\sup_{\mathcal{K}} |\tau_+^{\frac{7}{2}}\beta(W)|\right)\left(\sup_{\mathcal{K}}\int_{\underline{C}(\underline{u}';[u_0,u])} \tau_+^6|\beta(\hat{\mathcal{L}}_T W)|^2\right)^{\frac{1}{2}} \\
&\cdot \int \frac{d\underline{u}'}{\underline{u}'^{\frac{5}{2}}}\left(\int_{u_0}^{u} du'|r^{1-\frac{2}{p}}\tau_-^{2}{}^{(T)}p_3|^2_{p=2,S}\right)^{\frac{1}{2}} \\
&\le c\left(\sup_{\mathcal{K}} |r^{1-\frac{2}{p}}\tau_-^{2}{}^{(T)}p_3|^2_{p=2,S}\right)^{\frac{1}{2}}\left(\sup_{\mathcal{K}} |\tau_+^{\frac{7}{2}}\beta(W)|\right)\underline{\mathcal{Q}}_1^{\frac{1}{2}}\int\frac{d\underline{u}'}{\underline{u}'^{\frac{5}{2}}}\left(\int_{u_0}^{u} du'\right)^{\frac{1}{2}} \\
&\le c\epsilon_0\mathcal{Q}_{\mathcal{K}}.
\end{aligned}$$

[13]The other integral appearing in its decomposition (see (6.1.18)) is easier to treat.

Proof of Proposition 6.2.3: The first and second integrals of (6.2.20) are similar because ρ and σ behave in the same way and $\Lambda^{(3)}(T, W)$ and $K^{(3)}(T, W)$ have the same structure (see (6.1.22)). Therefore it is enough to estimate the first one. This amounts to controlling the three integrals

$$\begin{aligned}
&\int_{V_{(u,\underline{u})}} \tau_+^4 \tau_-^2 \rho(\hat{\mathcal{L}}_T W)\underline{\Theta}({}^{(T)}q)\alpha(W) \\
&\int_{V_{(u,\underline{u})}} \tau_+^4 \tau_-^2 \rho(\hat{\mathcal{L}}_T W)\big(K({}^{(T)}q), \Lambda({}^{(T)}q)\big)(\rho, \sigma)(W) \qquad (6.2.24)\\
&\int_{V_{(u,\underline{u})}} \tau_+^4 \tau_-^2 \rho(\hat{\mathcal{L}}_T W)\Xi({}^{(T)}q)\underline{\beta}(W).
\end{aligned}$$

The first integral is estimated in the following way:

$$\begin{aligned}
&\left|\int_{V_{(u,\underline{u})}} \tau_+^4 \tau_-^2 \rho(\hat{\mathcal{L}}_T W)(\underline{\Theta}({}^{(T)}q))\alpha(W)\right| \\
&\quad \le c\int_{\underline{u}_0}^{\underline{u}} d\underline{u}' \int_{\underline{C}(\underline{u}';[u_0,u])} |\tau_+^2 \tau_- \rho(\hat{\mathcal{L}}_T W)| \frac{1}{\tau_- \tau_+^{\frac{3}{2}}} |\tau_-^2 \underline{\Theta}({}^{(T)}q)||\tau_+^{\frac{7}{2}}\alpha(W)| \\
&\quad \le c\left(\sup_{\mathcal{K}} |r^{1-\frac{2}{p}}\tau_-^2 \underline{\Theta}({}^{(T)}q)|^2_{p=2,S}\right)^{\frac{1}{2}} \left(\sup_{\mathcal{K}} |\tau_+^{\frac{7}{2}}\alpha|\right) \underline{\mathcal{Q}}_1^{\frac{1}{2}} \int_{\underline{u}_0}^{\underline{u}} \frac{d\underline{u}'}{\underline{u}'^{\frac{3}{2}}} \left(\int_{u_0}^{u} \frac{du'}{u'^2}\right)^{\frac{1}{2}} \\
&\quad \le c\epsilon_0 \mathcal{Q}_{\mathcal{K}}. \qquad (6.2.25)
\end{aligned}$$

The norm estimates of $K({}^{(T)}q)$, $\Lambda({}^{(T)}q)$ are at least as good as that of ${}^{(T)}p_4$; see (6.1.50). Therefore the second integral in (6.2.24) is estimated as the second integral of (6.2.21) in Proposition 6.2.2. The estimate of the third integral proceeds as the previous one.

The estimate of the third integral of Proposition 6.2.3 does not differ from the previous ones and we do not report it here.

Estimate of the B_3, B_4 integrals

We recall the expressions for the B_3 integrals:

$$\begin{aligned}
&\textstyle\int_{V_{(u,\underline{u})}} \tau_+^2 \tau_-^4 \rho(\hat{\mathcal{L}}_T W)\underline{\Lambda}(T, W) \\
&\textstyle\int_{V_{(u,\underline{u})}} \tau_+^2 \tau_-^4 \sigma(\hat{\mathcal{L}}_T W)\underline{K}(T, W) \\
&\textstyle\int_{V_{(u,\underline{u})}} \tau_+^2 \tau_-^4 \underline{\beta}(\hat{\mathcal{L}}_T W)\cdot \underline{I}(T, W)
\end{aligned}$$

and those for the B_4 integrals:

$$\begin{aligned}
&\textstyle\int_{V_{(u,\underline{u})}} \tau_-^6 \underline{\alpha}(\hat{\mathcal{L}}_T W)\cdot \underline{\Theta}(T, W) \\
&\textstyle\int_{V_{(u,\underline{u})}} \tau_-^6 \underline{\beta}(\hat{\mathcal{L}}_T W)\cdot \underline{\Xi}(T, W).
\end{aligned}$$

The estimates of the various terms into which these integrals decompose are similar but easier than those for the integrals of groups B_2 and B_1, respectively. They are obtained with the obvious substitutions of the underlined quantities with those not underlined and vice versa.

The greater simplicity is due to the fact that, now, τ_-, although smaller than τ_+, plays an analogous role on the $\underline{C}$ null hypersurfaces. Moreover the factor $\mathbf{D}_3 \log \Omega$ with the slowest decay is now substituted by the better behaving factor $\mathbf{D}_4 \log \Omega$. Therefore we simply collect the final results in the next proposition.

Proposition 6.2.4 *Under the assumptions (6.0.1) and (6.0.2) the following inequalities hold:*

$$\int_{V_{(u,\underline{u})}} \tau_+^2 \tau_-^4 \rho(\hat{\mathcal{L}}_T W) \underline{\Lambda}(T, W) \leq c\epsilon_0 \mathcal{Q}_{\mathcal{K}}$$
$$\int_{V_{(u,\underline{u})}} \tau_+^2 \tau_-^4 \sigma(\hat{\mathcal{L}}_T W) \underline{K}(T, W) \leq c\epsilon_0 \mathcal{Q}_{\mathcal{K}}$$
$$\int_{V_{(u,\underline{u})}} \tau_+^2 \tau_-^4 \underline{\beta}(\hat{\mathcal{L}}_T W) \cdot \underline{I}(T, W) \leq c\epsilon_0 \mathcal{Q}_{\mathcal{K}}$$
$$\int_{V_{(u,\underline{u})}} \tau_-^6 \underline{\alpha}(\hat{\mathcal{L}}_T W) \cdot \underline{\Theta}(T, W) \leq c\epsilon_0 \mathcal{Q}_{\mathcal{K}}$$
$$\int_{V_{(u,\underline{u})}} \tau_-^6 \underline{\beta}(\hat{\mathcal{L}}_T W) \cdot \underline{\Xi}(T, W) \leq c\epsilon_0 \mathcal{Q}_{\mathcal{K}}.$$

6.2.2 *Estimate of* $\int_{V_{(u,\underline{u})}} Q(\hat{\mathcal{L}}_T W)_{\alpha\beta\gamma\delta}({}^{(\bar{K})}\pi^{\alpha\beta} \bar{K}^\gamma \bar{K}^\delta)$

Proposition 6.2.5 *Under the assumptions (6.0.1) and (6.0.2) the following inequality holds:*

$$\int_{V_{(u,\underline{u})}} |Q(\hat{\mathcal{L}}_T W)_{\alpha\beta\gamma\delta}({}^{(\bar{K})}\pi^{\alpha\beta} \bar{K}^\gamma \bar{K}^\delta)| \leq c\epsilon_0 \mathcal{Q}_{\mathcal{K}}. \tag{6.2.26}$$

Proof: We write the explicit expression of the integrand

$$\begin{aligned}
&{}^{(\bar{K})}\pi^{\alpha\beta} Q(\hat{\mathcal{L}}_T W)_{\alpha\beta\gamma\delta} \bar{K}^\gamma \bar{K}^\delta \\
&\quad = {}^{(\bar{K})}\pi^{\alpha\beta} \left\{ Q(\hat{\mathcal{L}}_T W)_{\alpha\beta 44} \tau_+^4 + 2Q(\hat{\mathcal{L}}_T W)_{\alpha\beta 43} \tau_+^2 \tau_-^2 + Q(\hat{\mathcal{L}}_T W)_{\alpha\beta 33} \tau_-^4 \right\}
\end{aligned}$$

where (see (3.4.9))

$$\begin{aligned}
&{}^{(\bar{K})}\pi^{\alpha\beta} Q(\hat{\mathcal{L}}_T W)_{\alpha\beta 44} \tau_+^4 \\
&= \frac{1}{16} \tau_+^4 \Big\{ 2|\alpha(\hat{\mathcal{L}}_T W)|^2 {}^{(\bar{K})}\underline{\mathbf{n}} + 4(|\rho(\hat{\mathcal{L}}_T W)|^2 + |\sigma(\hat{\mathcal{L}}_T W)|^2) {}^{(\bar{K})}\mathbf{n} + |\beta(\hat{\mathcal{L}}_T W)|^2 {}^{2(\bar{K})}\mathbf{j} \\
&- 8\alpha(\hat{\mathcal{L}}_T W) \cdot \beta(\hat{\mathcal{L}}_T W) \cdot {}^{(\bar{K})}\underline{\mathbf{m}} - 8\rho(\hat{\mathcal{L}}_T W) \beta(\hat{\mathcal{L}}_T W) \cdot {}^{(\bar{K})}\mathbf{m} \\
&+ 8\sigma(\hat{\mathcal{L}}_T W)^* \beta(\hat{\mathcal{L}}_T W) \cdot {}^{(\bar{K})}\mathbf{m} + 8(|\beta(\hat{\mathcal{L}}_T W)|^2) \mathrm{tr}^{(\bar{K})}\mathbf{i} \\
&+ \quad 8\rho(\hat{\mathcal{L}}_T W) \alpha(\hat{\mathcal{L}}_T W) \cdot {}^{(\bar{K})}\mathbf{i} - 8\sigma(\hat{\mathcal{L}}_T W)^* \alpha(\hat{\mathcal{L}}_T W) \cdot {}^{(\bar{K})}\mathbf{i} \Big\},
\end{aligned} \tag{6.2.27}$$

$$
\begin{aligned}
&{}^{(\bar{K})}\pi^{\alpha\beta}Q(\hat{\mathcal{L}}_T W)_{\alpha\beta 43}\tau_+^2\tau_-^2\\
&= \frac{1}{16}\tau_+^2\tau_-^2\Big\{4|\beta(\hat{\mathcal{L}}_T W)|^{2(\bar{K})}\underline{\mathbf{n}} + 4|\underline{\beta}(\hat{\mathcal{L}}_T W)|^{2(\bar{K})}\mathbf{n} + 4(|\rho(\hat{\mathcal{L}}_T W)|^2 + |\sigma(\hat{\mathcal{L}}_T W)|^2)^{(\bar{K})}\mathbf{j}\\
&- 4\rho(\hat{\mathcal{L}}_T W)\beta(\hat{\mathcal{L}}_T W)\cdot{}^{(\bar{K})}\underline{\mathbf{m}} + 4\sigma(\hat{\mathcal{L}}_T W)^*\beta(\hat{\mathcal{L}}_T W)\cdot{}^{(\bar{K})}\underline{\mathbf{m}}\\
&+ 4\rho(\hat{\mathcal{L}}_T W)\underline{\beta}(\hat{\mathcal{L}}_T W)\cdot{}^{(\bar{K})}\mathbf{m} + 4\sigma(\hat{\mathcal{L}}_T W)^*\underline{\beta}(\hat{\mathcal{L}}_T W)\cdot{}^{(\bar{K})}\mathbf{m}\\
&+ \quad 2(|\rho(\hat{\mathcal{L}}_T W)|^2 + \sigma(\hat{\mathcal{L}}_T W)|^2)\mathrm{tr}^{(\bar{K})}\mathbf{i} - 2(\beta(\hat{\mathcal{L}}_T W)\widehat{\otimes}\underline{\beta}(\hat{\mathcal{L}}_T W))\cdot{}^{(\bar{K})}\mathbf{i}\Big\},
\end{aligned}
\tag{6.2.28}
$$

$$
\begin{aligned}
&{}^{(\bar{K})}\pi^{\alpha\beta}Q(\hat{\mathcal{L}}_T W)_{\alpha\beta 33}\tau_-^4 =\\
&\frac{1}{16}\tau_-^4\Big\{4(\rho(\hat{\mathcal{L}}_T W)|^2 + \sigma(\hat{\mathcal{L}}_T W)|^2)^{(\bar{K})}\underline{\mathbf{n}} + 2|\underline{\alpha}(\hat{\mathcal{L}}_T W)|^{2(\bar{K})}\mathbf{n} + 4|\underline{\beta}(\hat{\mathcal{L}}_T W)|^{2(\bar{K})}\mathbf{j}\\
&+ 8(\underline{\alpha}(\hat{\mathcal{L}}_T W)\cdot\underline{\beta}(\hat{\mathcal{L}}_T W))\cdot{}^{(\bar{K})}\mathbf{m} - 8\rho(\hat{\mathcal{L}}_T W)\underline{\beta}(\hat{\mathcal{L}}_T W)\cdot{}^{(\bar{K})}\underline{\mathbf{m}}\\
&- 8\sigma(\hat{\mathcal{L}}_T W)^*\underline{\beta}(\hat{\mathcal{L}}_T W)\cdot{}^{(\bar{K})}\underline{\mathbf{m}} + 8(|\underline{\beta}(\hat{\mathcal{L}}_T W)|^2)\mathrm{tr}^{(\bar{K})}\mathbf{i}\\
&+ \quad 8\rho(\hat{\mathcal{L}}_T W)\underline{\alpha}(\hat{\mathcal{L}}_T W)\cdot{}^{(\bar{K})}\mathbf{i} + 8\sigma(\hat{\mathcal{L}}_T W)^*\underline{\alpha}(\hat{\mathcal{L}}_T W)\cdot{}^{(\bar{K})}\mathbf{i}\Big\}.
\end{aligned}
\tag{6.2.29}
$$

All factors have the same structure. They are cubic terms, quadratic in the null components of $\hat{\mathcal{L}}_T W$ and linear in the deformation tensor of $\bar{K}$. Therefore they are all estimated in the same way. Let us discuss explicitly the integral relative to the term $\tau_+^{4(\bar{K})}\underline{\mathbf{n}}|\alpha(\hat{\mathcal{L}}_T W)|^2$ and the one relative to the term $\tau_+^{4(\bar{K})}\mathbf{n}|\rho(\hat{\mathcal{L}}_T W)|^2$; see (6.2.27). For the first integral we obtain, using Corollary 6.1.3,

$$
\begin{aligned}
&\int_{V_{(u,\underline{u})}}\tau_+^4|\alpha(\hat{\mathcal{L}}_T W)|^2|{}^{(\bar{K})}\underline{\mathbf{n}}| \le c\int_{u_0}^{u}du'\int_{C(u';[\underline{u}_0,\underline{u}])}\tau_+^6|\alpha(\hat{\mathcal{L}}_T W)|^2\frac{1}{r^2}|{}^{(\bar{K})}\underline{\mathbf{n}}|\\
&\le c\left(\sup_{\mathcal{K}}|\frac{\tau_-}{r}{}^{(\bar{K})}\underline{\mathbf{n}}|\right)\left(\sup_{\mathcal{K}}\int_{C(u';[\underline{u}_0,\underline{u}])}\tau_+^6|\alpha(\hat{\mathcal{L}}_T W)|^2\right)\int_{u_0}^{u}du'\frac{1}{ru'} \le c\epsilon_0\mathcal{Q}_{\mathcal{K}}.
\end{aligned}
$$

The estimate of the second integral proceeds exactly as before:

$$
\begin{aligned}
&\int_{V_{(u,\underline{u})}}\tau_+^4|{}^{(\bar{K})}\mathbf{n}||\rho(\hat{\mathcal{L}}_T W)|^2 \le c\int_{u_0}^{u}du'\frac{1}{u'^2}\int_{C(u';[\underline{u}_0,\underline{u}])}|\frac{r^2}{\tau_-^2}{}^{(\bar{K})}\mathbf{n}|\tau_+^2\tau_-^4|\rho(\hat{\mathcal{L}}_T W)|^2\\
&\le c\left(\sup_{\mathcal{K}}|\frac{r^2}{\tau_-^2}{}^{(\bar{K})}\mathbf{n}|\right)\left(\sup_{\mathcal{K}}\int_{C(u';[\underline{u}_0,\underline{u}])}\tau_+^2\tau_-^4|\rho(\hat{\mathcal{L}}_T W)|^2\right)\int_{u_0}^{u}du'\frac{1}{u'^2} \le c\epsilon_0\mathcal{Q}_{\mathcal{K}}.
\end{aligned}
$$

6.2.3 *Estimate of* $\int_{V_{(u,\underline{u})}}\mathbf{Div}Q(\hat{\mathcal{L}}_O W)_{\beta\gamma\delta}(\bar{K}^\beta\bar{K}^\gamma T^\delta)$

We have to control the following integrals (see (6.1.1), (6.1.2), (6.1.3))

$$
\begin{aligned}
&\int_{V_{(u,\underline{u})}}\tau_+^4 D(O,W)_{444}\ ,\quad \int_{V_{(u,\underline{u})}}\tau_+^4 D(O,W)_{344}\\
&\int_{V_{(u,\underline{u})}}\tau_+^2\tau_-^2 D(O,W)_{344}\ ,\quad \int_{V_{(u,\underline{u})}}\tau_+^2\tau_-^2 D(O,W)_{334}\\
&\int_{V_{(u,\underline{u})}}\tau_-^4 D(O,W)_{334}\ ,\quad \int_{V_{(u,\underline{u})}}\tau_-^4 D(O,W)_{333}.
\end{aligned}
\tag{6.2.30}
$$

The integrals containing the weight τ_+^4 are the most sensitive terms. We estimate the first integral in the first line of (6.2.30), whose expression is

$$\int_{V_{(u,\underline{u})}} \tau_+^4 D(O,W)_{444} = \frac{1}{2}\int_{V_{(u,\underline{u})}} \tau_+^4 \alpha(\hat{\mathcal{L}}_O W)\cdot\Theta(O,W) - \int_{V_{(u,\underline{u})}} \tau_+^4 \beta(\hat{\mathcal{L}}_O W)\cdot\Xi(O,W). \tag{6.2.31}$$

Proposition 6.2.6 *Under the assumptions (6.0.1) and (6.0.2) the following inequalities hold:*

$$\left|\int_{V_{(u,\underline{u})}} \tau_+^4 \alpha(\hat{\mathcal{L}}_O W)\cdot\Theta(O,W)\right| \leq c\epsilon_0 \mathcal{Q}_{\mathcal{K}}$$

$$\left|\int_{V_{(u,\underline{u})}} \tau_+^4 \beta(\hat{\mathcal{L}}_O W)\cdot\Xi(O,W)\right| \leq c\epsilon_0 \mathcal{Q}_{\mathcal{K}},$$

where $\Theta(O,W) = \sum_{i=1}^3 \Theta(J^i(O,W))$, $\Xi(O,W) = \sum_{i=1}^3 \Xi(J^i(O,W))$.

Proof: Proceeding as in Subsection 6.2.1 we write

$$\begin{aligned}
&\left|\int_{V_{(u,\underline{u})}} \tau_+^4 \alpha(\hat{\mathcal{L}}_O W)\cdot\Theta(O,W)\right| \\
&\leq \left(\sup_{\mathcal{K}}\int_{C(u';[\underline{u}_0,\underline{u}])} \underline{u}'^4|\alpha(\hat{\mathcal{L}}_O W)|^2\right)^{\frac{1}{2}} \int_{u_0}^{u} du' \left(\int_{C(u';[\underline{u}_0,\underline{u}])} \underline{u}'^4|\Theta(O,W)|^2\right)^{\frac{1}{2}} \\
&\leq c\mathcal{Q}_{\mathcal{K}}^{\frac{1}{2}} \int_{u_0}^{u} du' \sum_{i=1}^{3}\left(\int_{C(u';[\underline{u}_0,\underline{u}])} \underline{u}'^4|\Theta^{(i)}(O,W)|^2\right)^{\frac{1}{2}}
\end{aligned} \tag{6.2.32}$$

$$\begin{aligned}
&\left|\int_{V_{(u,\underline{u})}} \underline{u}'^4 \beta(\hat{\mathcal{L}}_O W)\cdot\Xi(O,W)\right| \\
&\leq \left(\sup_{\mathcal{K}}\int_{C(u';[\underline{u}_0,\underline{u}])} \underline{u}'^4|\beta(\hat{\mathcal{L}}_O W)|^2\right)^{\frac{1}{2}} \int_{u_0}^{u} du' \left(\int_{C(u';[\underline{u}_0,\underline{u}])} \underline{u}'^4|\Xi(O,W)|^2\right)^{\frac{1}{2}} \\
&\leq c\mathcal{Q}_{\mathcal{K}}^{\frac{1}{2}} \int_{u_0}^{u} du' \sum_{i=1}^{3}\left(\int_{C(u';[\underline{u}_0,\underline{u}])} \underline{u}'^4|\Xi^{i}(O,W)|^2\right)^{\frac{1}{2}}.
\end{aligned} \tag{6.2.33}$$

The result is obtained by proving the next lemma.

Lemma 6.2.3 *Under the assumptions (6.0.1) and (6.0.2) the following inequalities hold:*

$$\left(\int_{C(u';[\underline{u}_0,\underline{u}])} \underline{u}'^4|\Theta^{(1)}(O,W)|^2\right)^{\frac{1}{2}} \leq c\epsilon_0 \mathcal{Q}_{\mathcal{K}}^{\frac{1}{2}}\frac{1}{|u'|^2}$$

$$\left(\int_{C(u';[\underline{u}_0,\underline{u}])} \underline{u}'^4|\Theta^{(2)}(O,W)|^2\right)^{\frac{1}{2}} \leq c\epsilon_0\mathcal{Q}_{\mathcal{K}}^{\frac{1}{2}}\frac{1}{|u'|^{\frac{3}{2}}} \tag{6.2.34}$$

$$\left(\int_{C(u';[\underline{u}_0,\underline{u}])} \underline{u}'^4|\Theta^{(3)}(O,W)|^2\right)^{\frac{1}{2}} \leq c\epsilon_0\mathcal{Q}_{\mathcal{K}}^{\frac{1}{2}}\frac{1}{|u'|^{\frac{3}{2}}}$$

$$\left(\int_{C(u';[\underline{u}_0,\underline{u}])} \underline{u}'^4|\Xi^{(1)}(O,W)|^2\right)^{\frac{1}{2}} \leq c\epsilon_0\mathcal{Q}_{\mathcal{K}}^{\frac{1}{2}}\frac{1}{|u'|^2}$$

$$\left(\int_{C(u';[\underline{u}_0,\underline{u}])} \underline{u}'^4|\Xi^{(2)}(O,W)|^2\right)^{\frac{1}{2}} \leq c\epsilon_0\mathcal{Q}_{\mathcal{K}}^{\frac{1}{2}}\frac{1}{|u'|^{\frac{3}{2}}} \tag{6.2.35}$$

$$\left(\int_{C(u';[\underline{u}_0,\underline{u}])} \underline{u}'^4|\Xi^{(3)}(O,W)|^2\right)^{\frac{1}{2}} \leq c\epsilon_0\mathcal{Q}_{\mathcal{K}}^{\frac{1}{2}}\frac{1}{|u'|^{\frac{3}{2}}}.$$

Proof: We start by estimating the integral in the first line, which is connected to the J^1 part of the current. From Corollary 4.7.1 and (6.1.14), we have

$$\begin{aligned}|\Theta^{(1)}(O,W)|^2 \;\leq\; & c\left(\sup_{\mathcal{K}}|r({}^{(O)}\mathbf{i},{}^{(O)}\mathbf{j},{}^{(O)}\underline{\mathbf{m}})|\right)^2\frac{1}{r^2}\Big[\big(|\nabla\!\!\!/\alpha|^2+|\alpha_3|^2+|\nabla\!\!\!/\beta|^2\\ &+|\beta_4|^2+|(\rho_4,\sigma_4)|^2+\frac{1}{r^2}\left(|\beta|^2+|(\rho,\sigma)|^2+|\alpha|^2\right)\Big]+\text{l.o.t.}\end{aligned}$$

Therefore

$$\begin{aligned}&\left(\int_{C(u';[\underline{u}_0,\underline{u}])} \underline{u}'^4|\Theta^{(1)}(O,W)|^2\right)^{\frac{1}{2}} \leq \cdot c\left(\sup_{\mathcal{K}}\left|r({}^{(O)}\mathbf{i},{}^{(O)}\mathbf{j},{}^{(O)}\underline{\mathbf{m}})\right|\right)\\ &\frac{1}{|u'|}\Bigg[\int_{C(u';[\underline{u}_0,\underline{u}])}\underline{u}'^2u'^2\left(|\nabla\!\!\!/\alpha|^2+|\alpha_3|^2+|\nabla\!\!\!/\beta|^2+|\beta_4|^2+|(\rho_4,\sigma_4)|^2\right)\\ &+u'^2\left(|\beta|^2+|(\rho,\sigma)|^2+|\alpha|^2\right)\Bigg]^{\frac{1}{2}} \leq c\left(\sup_{\mathcal{K}}\left|r({}^{(O)}\mathbf{i},{}^{(O)}\mathbf{j},{}^{(O)}\underline{\mathbf{m}})\right|\right)\\ &\cdot\frac{1}{|u'|^2}\Bigg[\left(\sup_{\mathcal{K}}\int_{C(u';[\underline{u}_0,\underline{u}])}Q(\hat{\mathcal{L}}_OW)(\bar{K},\bar{K},T,e_4)\right)^{\frac{1}{2}}\\ &+\left(\sup_{\mathcal{K}}\int_{\underline{C}(\underline{u}';[u_0,u])}Q(\hat{\mathcal{L}}_OW)(\bar{K},\bar{K},T,e_3)\right)^{\frac{1}{2}}+\sup_{\mathcal{K}\cap\Sigma_0}|r^3(\overline{\rho},\overline{\sigma})|\Bigg]\\ &\leq c\frac{1}{|u'|^2}\left(\sup_{\mathcal{K}}|r({}^{(O)}\mathbf{i},{}^{(O)}\mathbf{j},{}^{(O)}\underline{\mathbf{m}})|\right)\mathcal{Q}_{\mathcal{K}}^{\frac{1}{2}} \leq c\epsilon_0\mathcal{Q}_{\mathcal{K}}^{\frac{1}{2}}\frac{1}{|u'|^2}\end{aligned} \tag{6.2.36}$$

where we used the results of Proposition 5.1.3. The fourth integral of Lemma 6.2.3 is estimated recalling that[14] (see (6.1.13)),

$$|\Xi^1(O,W)|^2 \le c\frac{1}{r^2}\left(\sup_{\mathcal{K}}|r({}^{(O)}\mathbf{i},{}^{(O)}\mathbf{j},{}^{(O)}\underline{\mathbf{m}})|\right)^2\cdot \left[|\nabla\!\!\!/\alpha|^2+|\alpha_4|^2+|\beta_4|^2+\frac{1}{r^2}(|\alpha|^2+|\beta|^2)\right]+(\text{l.o.t}).$$

We obtain

$$\left(\int_{C(u';[\underline{u}_0,\underline{u}])}\underline{u}'^4|\Xi^1(O,W)|^2\right)^{\frac12}\le c\left(\sup_{\mathcal{K}}|r({}^{(O)}\mathbf{i},{}^{(O)}\mathbf{j},{}^{(O)}\underline{\mathbf{m}})|\right)\cdot \frac{1}{|u'|}\left[\int_{C(u';[\underline{u}_0,\underline{u}])}\underline{u}'^4\left(|\nabla\!\!\!/\alpha|^2+|\alpha_4|^2+|\beta_4|^2+\frac{1}{r^2}(|\alpha|^2+|\beta|^2)\right)\right]^{\frac12}.$$

The estimates of these terms proceed as in the previous case, using the results of Proposition 5.1.3.

We estimate the J^2 part only for the Θ integral in the second line of (6.2.34) since the estimate for the corresponding Ξ integral in the second line of (6.2.35), is done in the same way. Recalling the decomposition (see (6.1.18))

$$\Theta^{(2)}(O,W)=\mathrm{Qr}\left[{}^{(O)}p_3\,;\,\alpha\right]+\mathrm{Qr}\left[{}^{(O)}\not{p}\,;\,\beta\right]+\mathrm{Qr}\left[{}^{(O)}p_4\,;\,(\rho,\sigma)\right],$$

we write

$$\left(\int_{C(u';[\underline{u}_0,\underline{u}])}\underline{u}'^4|\Theta^{(2)}(O,W)|^2\right)^{\frac12} \tag{6.2.37}$$
$$\le c\left(\int_{C(u';[\underline{u}_0,\underline{u}])}\underline{u}'^4(|{}^{(O)}p_3|^2|\alpha(W)|^2+|{}^{(O)}\not{p}|^2|\beta(W)|^2+|{}^{(O)}p_4|^2|(\rho,\sigma)(W)|^2)\right)^{\frac12}$$

To estimate these integrals we have to control the $|\cdot|_{p,S}$ norms of ${}^{(O)}p_3$, ${}^{(O)}\not{p}$ and ${}^{(O)}p_4$. Using the estimates of Proposition 6.1.10, for any $p\in[2,4]$, we obtain

$$\left(\int_{C(u';[\underline{u}_0,\underline{u}])}\underline{u}'^4|{}^{(O)}p_3|^2|\alpha(W)|^2\right)^{\frac12}$$
$$\le c\left(\sup_{\mathcal{K}}|\tau_+^{\frac72}\alpha(W)|\right)\left(\int_{\underline{u}_0}^{\underline{u}}d\underline{u}'\frac{1}{\underline{u}'^3u'^2}|r^{1-\frac2p}u'{}^{(O)}p_3|^2_{p=2,S}\right)^{\frac12}$$
$$\le c\left(\sup_{\mathcal{K}}|r^{1-\frac2p}u'{}^{(O)}p_3|_{p=2,S}\right)\left(\sup_{\mathcal{K}}|\tau_+^{\frac72}\alpha(W)|\right)\frac{1}{|u'|}\left(\int_{\underline{u}_0}^{\underline{u}}d\underline{u}'\frac{1}{\underline{u}'^3}\right)^{\frac12}$$
$$\le c\epsilon_0\mathcal{Q}_{\mathcal{K}}^{\frac12}\frac{1}{|u'|^2}. \tag{6.2.38}$$

---

[14] In the following estimates we systematically neglect the (l.o.t.) terms.

The other integrals in (6.2.37) are estimated in the same way and we do not discuss them here.

The part associated with the J^3 current in the third line of (6.2.34) is estimated starting from the decomposition (see (6.1.23))

$$\begin{aligned}\Theta^{(3)}(O,W) &= \mathrm{Qr}\left[\alpha\,;\ \underline{K}(^{(O)}q)\right]+\mathrm{Qr}\left[\alpha\,;\ \underline{\Lambda}(^{(O)}q)\right]+\mathrm{Qr}\left[\beta\,;\ (I,\underline{I})(^{(O)}q)\right]\\ &\quad+\mathrm{Qr}\left[(\rho,\sigma)\,;\ \Theta(^{(O)}q)\right],\end{aligned}$$

from which we obtain the inequality

$$\begin{aligned}\left(\int_{C(u';[\underline{u}_0,\underline{u}])}\underline{u}'^4|\Theta^{(3)}(O,W)|^2\right)^{\frac{1}{2}} &\le c\left(\int_{C(u';[\underline{u}_0,\underline{u}])}\underline{u}'^4(|\underline{K}(^{(O)}q)|^2|\alpha(W)|^2\right.\\ &\left.+|\underline{\Lambda}(^{(O)}q)|^2|\alpha(W)|^2|+(I,\underline{I})(^{(O)}q)|^2|\beta(W)|^2+|\Theta(^{(O)}q)|^2|(\rho,\sigma)(W)|^2)\right)^{\frac{1}{2}}.\end{aligned}$$

To estimate these integrals we use the norm estimates of Proposition 6.1.12 for $\underline{K}(^{(O)}q)$, $\underline{\Lambda}(^{(O)}q)$, $(I,\underline{I})(^{(O)}q)$, $\Theta(^{(O)}q)$, with $p\in[2,4]$, which we recall here,

$$\begin{aligned}&\sup_{\mathcal{K}}|r^{2-\frac{2}{p}}(I(^{(O)}q),\underline{I}(^{(O)}q))|_{p,S}\le c\epsilon_0\ ,\quad \sup_{\mathcal{K}}|r^{2-\frac{2}{p}}\Theta(^{(O)}q)|_{p,S}\le c\epsilon_0\\ &\sup_{\mathcal{K}}|r^{1-\frac{2}{p}}\tau_-\underline{\Lambda}(^{(O)}q)|_{p,S}\le c\epsilon_0\ ,\quad \sup_{\mathcal{K}}|r^{2-\frac{2}{p}}\underline{K}(^{(O)}q)|_{p,S}\le c\epsilon_0\ .\end{aligned}$$

The estimates of all the integrals can be done in the same way. The factor of $|u'|^{-\frac{3}{2}}$ is due to the integral that depends on ρ, whereas the other estimates contain the better factor $|u'|^{-2}$. We just report the estimate of this term,

$$\begin{aligned}&\left(\int_{C(u';[\underline{u}_0,\underline{u}])}\underline{u}'^4|\Theta(^{(O)}q)|^2|\rho(W)|^2\right)^{\frac{1}{2}}\\ &\le c\left(\sup_{\mathcal{K}}|r^3\rho(W)|\right)\left(\int_{\underline{u}_0}^{\underline{u}}d\underline{u}'\frac{1}{\underline{u}'^4}|r^{2-\frac{2}{p}}\Theta(^{(O)}q)|^2_{p=2,S}\right)^{\frac{1}{2}}\\ &\le c\left(\sup_{\mathcal{K}}|r^{2-\frac{2}{p}}\Theta(^{(O)}q)|_{p=2,S}\right)\left(\sup_{\mathcal{K}}|r^3\rho(W)|\right)\left(\int_{\underline{u}_0}^{\underline{u}}d\underline{u}'\frac{1}{\underline{u}'^4}\right)^{\frac{1}{2}}\le c\epsilon_0\mathcal{Q}_{\mathcal{K}}^{\frac{1}{2}}\frac{1}{|u'|^{\frac{3}{2}}}.\end{aligned}$$

6.2.4 *Estimate of* $\int_{V_{(u,\underline{u})}}Q(\hat{\mathcal{L}}_O W)_{\alpha\beta\gamma\delta}(^{(\bar{K})}\pi^{\alpha\beta}\bar{K}^\gamma T^\delta)$

Proposition 6.2.7 *Under the assumptions (6.0.1) and (6.0.2) the following inequality holds:*

$$\int_{V_{(u,\underline{u})}}|Q(\hat{\mathcal{L}}_O W)_{\alpha\beta\gamma\delta}(^{(\bar{K})}\pi^{\alpha\beta}\bar{K}^\gamma T^\delta)|\le c\epsilon_0\mathcal{Q}_{\mathcal{K}}. \tag{6.2.39}$$

Proof: We write explicitly the various terms of the integrand

$$
\begin{aligned}
{}^{(\bar K)}\pi^{\alpha\beta} Q(\hat{\mathcal{L}}_O W)_{\alpha\beta\gamma\delta}\bar K^\gamma T^\delta &= {}^{(\bar K)}\pi^{\alpha\beta}\Big\{Q(\hat{\mathcal{L}}_O W)_{\alpha\beta 44}\tau_+^2 + Q(\hat{\mathcal{L}}_O W)_{\alpha\beta 43}\tau_+^2\\
&\quad + Q(\hat{\mathcal{L}}_O W)_{\alpha\beta 34}\tau_-^2 + Q(\hat{\mathcal{L}}_O W)_{\alpha\beta 33}\tau_-^2\Big\},
\end{aligned}
$$

where

$$
\begin{aligned}
&{}^{(\bar K)}\pi^{\alpha\beta} Q(\hat{\mathcal{L}}_O W)_{\alpha\beta 44}\tau_+^2 =\\
&\quad \frac{1}{16}\tau_+^2\Big\{2|\alpha(\hat{\mathcal{L}}_O W)|^{2\,(\bar K)}\underline{\mathbf{n}} + 4(|\rho(\hat{\mathcal{L}}_O W)|^2 + |\sigma(\hat{\mathcal{L}}_O W)|^2)^{(\bar K)}\mathbf{n} + |\beta(\hat{\mathcal{L}}_O W)|^{2\,(\bar K)}\mathbf{j}\\
&\quad - 8(\alpha(\hat{\mathcal{L}}_O W)\cdot\beta(\hat{\mathcal{L}}_O W))\cdot{}^{(\bar K)}\underline{\mathbf{m}} - 8\rho(\hat{\mathcal{L}}_O W)\beta(\hat{\mathcal{L}}_O W)\cdot{}^{(\bar K)}\mathbf{m}\\
&\quad + 8\sigma(\hat{\mathcal{L}}_O W){}^*\beta(\hat{\mathcal{L}}_O W)\cdot{}^{(\bar K)}\mathbf{m} + 8(|\beta(\hat{\mathcal{L}}_O W)|^2)\mathrm{tr}^{(\bar K)}\mathbf{i}\\
&\quad + 8\rho(\hat{\mathcal{L}}_O W)\alpha(\hat{\mathcal{L}}_O W)\cdot{}^{(\bar K)}\mathbf{i} - 8\sigma(\hat{\mathcal{L}}_O W){}^*\alpha(\hat{\mathcal{L}}_O W)\cdot{}^{(\bar K)}\mathbf{i}\Big\},
\end{aligned}
\tag{6.2.40}
$$

$$
\begin{aligned}
&{}^{(\bar K)}\pi^{\alpha\beta} Q(\hat{\mathcal{L}}_O W)_{\alpha\beta 43}\tau_+^2 =\\
&\quad \frac{1}{16}\tau_+^2\Big\{4|\beta(\hat{\mathcal{L}}_O W)|^{2\,(\bar K)}\underline{\mathbf{n}} + 4|\underline{\beta}(\hat{\mathcal{L}}_O W)|^{2\,(\bar K)}\mathbf{n} + 4(|\rho(\hat{\mathcal{L}}_O W)|^2 + |\sigma(\hat{\mathcal{L}}_O W)|^2)^{(\bar K)}\mathbf{j}\\
&\quad - 4\rho(\hat{\mathcal{L}}_O W)\beta(\hat{\mathcal{L}}_O W)\cdot{}^{(\bar K)}\underline{\mathbf{m}} + 4\sigma(\hat{\mathcal{L}}_O W){}^*\beta(\hat{\mathcal{L}}_O W)\cdot{}^{(\bar K)}\underline{\mathbf{m}}\\
&\quad + 4\rho(\hat{\mathcal{L}}_O W)\underline{\beta}(\hat{\mathcal{L}}_O W)\cdot{}^{(\bar K)}\mathbf{m} + 4\sigma(\hat{\mathcal{L}}_O W){}^*\underline{\beta}(\hat{\mathcal{L}}_O W)\cdot{}^{(\bar K)}\mathbf{m}\\
&\quad + 2(|\rho(\hat{\mathcal{L}}_O W)|^2 + \sigma(\hat{\mathcal{L}}_O W)|^2)\mathrm{tr}^{(\bar K)}\mathbf{i} - 2(\beta(\hat{\mathcal{L}}_O W)\widehat{\otimes}\underline{\beta}(\hat{\mathcal{L}}_O W))\cdot{}^{(\bar K)}\mathbf{i}\Big\},
\end{aligned}
\tag{6.2.41}
$$

$$
\begin{aligned}
&{}^{(\bar K)}\pi^{\alpha\beta} Q(\hat{\mathcal{L}}_O W)_{\alpha\beta 34}\tau_-^2 =\\
&\quad \frac{1}{16}\tau_-^2\Big\{4|\beta(\hat{\mathcal{L}}_O W)|^{2\,(\bar K)}\underline{\mathbf{n}} + 4|\underline{\beta}(\hat{\mathcal{L}}_O W)|^{2\,(\bar K)}\mathbf{n} + 4(|\rho(\hat{\mathcal{L}}_O W)|^2 + |\sigma(\hat{\mathcal{L}}_O W)|^2)^{(\bar K)}\mathbf{j}\\
&\quad - 4\rho(\hat{\mathcal{L}}_O W)\beta(\hat{\mathcal{L}}_O W)\cdot{}^{(\bar K)}\underline{\mathbf{m}} + 4\sigma(\hat{\mathcal{L}}_O W){}^*\beta(\hat{\mathcal{L}}_O W)\cdot{}^{(\bar K)}\underline{\mathbf{m}}\\
&\quad + 4\rho(\hat{\mathcal{L}}_O W)\underline{\beta}(\hat{\mathcal{L}}_O W)\cdot{}^{(\bar K)}\mathbf{m} + 4\sigma(\hat{\mathcal{L}}_O W){}^*\underline{\beta}(\hat{\mathcal{L}}_O W)\cdot{}^{(\bar K)}\mathbf{m}\\
&\quad + 2(|\rho(\hat{\mathcal{L}}_O W)|^2 + \sigma(\hat{\mathcal{L}}_O W)|^2)\mathrm{tr}^{(\bar K)}\mathbf{i} - 2(\beta(\hat{\mathcal{L}}_O W)\widehat{\otimes}\underline{\beta}(\hat{\mathcal{L}}_O W))\cdot{}^{(\bar K)}\mathbf{i}\Big\},
\end{aligned}
\tag{6.2.42}
$$

$$
\begin{aligned}
&{}^{(\bar K)}\pi^{\alpha\beta} Q(\hat{\mathcal{L}}_O W)_{\alpha\beta 33}\tau_-^2 =\\
&\quad \frac{1}{16}\tau_-^2\Big\{4(\rho(\hat{\mathcal{L}}_O W)|^2 + \sigma(\hat{\mathcal{L}}_O W)|^2)^{(\bar K)}\underline{\mathbf{n}} + 2|\underline{\alpha}(\hat{\mathcal{L}}_O W)|^{2\,(\bar K)}\mathbf{n} + 4|\underline{\beta}(\hat{\mathcal{L}}_O W)|^{2\,(\bar K)}\mathbf{j}\\
&\quad + 8(\underline{\alpha}(\hat{\mathcal{L}}_O W)\cdot\underline{\beta}(\hat{\mathcal{L}}_O W))\cdot{}^{(\bar K)}\mathbf{m} - 8\rho(\hat{\mathcal{L}}_O W)\underline{\beta}(\hat{\mathcal{L}}_O W)\cdot{}^{(\bar K)}\underline{\mathbf{m}}\\
&\quad - 8\sigma(\hat{\mathcal{L}}_O W){}^*\underline{\beta}(\hat{\mathcal{L}}_O W)\cdot{}^{(\bar K)}\underline{\mathbf{m}} + 8(|\underline{\beta}(\hat{\mathcal{L}}_O W)|^2)\mathrm{tr}^{(\bar K)}\mathbf{i}\\
&\quad + 8\rho(\hat{\mathcal{L}}_O W)\underline{\alpha}(\hat{\mathcal{L}}_O W)\cdot{}^{(\bar K)}\mathbf{i} + 8\sigma(\hat{\mathcal{L}}_O W){}^*\underline{\alpha}(\hat{\mathcal{L}}_O W)\cdot{}^{(\bar K)}\mathbf{i}\Big\},
\end{aligned}
\tag{6.2.43}
$$

All the integrals appearing in the decomposition of (6.2.39) can be treated in the same way. Explicitly, we discuss the integral associated with the term $\frac{1}{8}\tau_+^2|\alpha(\hat{\mathcal{L}}_O W)|^{2\,(\bar K)}\underline{\mathbf{n}}$ (see

(6.2.40)).

$$\int_{V_{(u,\underline{u})}} \tau_+^2|\alpha(\hat{\mathcal{L}}_O W)|^2|^{(\bar{K})}\underline{\mathbf{n}}| \leq c\int_{u_0}^{u} du' \int_{C(u';[\underline{u}_0,\underline{u}])} \tau_+^4 \frac{1}{ru'}|\alpha(\hat{\mathcal{L}}_O W)|^2|\frac{\tau_-}{r}{}^{(\bar{K})}\underline{\mathbf{n}}|$$

$$\leq c\left(\sup_{\mathcal{K}}|\tau_- r^{-1(\bar{K})}\underline{\mathbf{n}}|\right)\left(\sup_{\mathcal{K}}\int_{C(u';[\underline{u}_0,\underline{u}])}\tau_+^4|\alpha(\hat{\mathcal{L}}_O W)|^2\right)\left(\int_{u_0}^{u} du'\frac{1}{ru'}\right) \leq c\epsilon_0\mathcal{Q}_{\mathcal{K}}.$$

6.2.5 *Estimate of* $\int_{V_{(u,\underline{u})}} Q(\hat{\mathcal{L}}_O W)_{\alpha\beta\gamma\delta}({}^{(T)}\pi^{\alpha\beta}\bar{K}^\gamma\bar{K}^\delta)$

Proposition 6.2.8 *Under the assumptions (6.0.1) and (6.0.2) the following inequalities hold:*[15]

$$\int_{V_{(u,\underline{u})}} |Q(\hat{\mathcal{L}}_O W)_{\alpha\beta\gamma\delta}({}^{(T)}\pi^{\alpha\beta}\bar{K}^\gamma\bar{K}^\delta)| \leq c\epsilon_0\mathcal{Q}_{\mathcal{K}} \tag{6.2.44}$$

Proof: We write explicitly the various terms of

$$\begin{aligned}&{}^{(T)}\pi^{\alpha\beta}Q(\hat{\mathcal{L}}_O W)_{\alpha\beta\gamma\delta}\bar{K}^\gamma\bar{K}^\delta = \\ &\quad {}^{(T)}\pi^{\alpha\beta}\left\{Q(\hat{\mathcal{L}}_O W)_{\alpha\beta44}\tau_+^4 + 2Q(\hat{\mathcal{L}}_O W)_{\alpha\beta43}\tau_+^2\tau_-^2 + Q(\hat{\mathcal{L}}_O W)_{\alpha\beta33}\tau_-^4\right\}.\end{aligned}$$

$$\begin{aligned}&{}^{(T)}\pi^{\alpha\beta}Q(\hat{\mathcal{L}}_O W)_{\alpha\beta44}\tau_+^4 = \\ &\quad \frac{1}{16}\tau_+^4\Big\{2|\alpha(\hat{\mathcal{L}}_O W)|^{2(T)}\underline{\mathbf{n}} + 4(|\rho(\hat{\mathcal{L}}_O W)|^2 + |\sigma(\hat{\mathcal{L}}_O W)|^2)^{(T)}\mathbf{n} + |\beta(\hat{\mathcal{L}}_O W)|^{2(T)}\mathbf{j} \\ &\quad - 8(\alpha(\hat{\mathcal{L}}_O W)\cdot\beta(\hat{\mathcal{L}}_O W))\cdot{}^{(T)}\underline{\mathbf{m}} - 8\rho(\hat{\mathcal{L}}_O W)\beta(\hat{\mathcal{L}}_O W)\cdot{}^{(T)}\mathbf{m} \\ &\quad + 8\sigma(\hat{\mathcal{L}}_O W)^*\beta(\hat{\mathcal{L}}_O W)\cdot{}^{(T)}\mathbf{m} + 8(|\beta(\hat{\mathcal{L}}_O W)|^2)\mathrm{tr}^{(T)}\mathbf{i} \\ &\quad + 8\rho(\hat{\mathcal{L}}_O W)\alpha(\hat{\mathcal{L}}_O W)\cdot{}^{(T)}\mathbf{i} - 8\sigma(\hat{\mathcal{L}}_O W)^*\alpha(\hat{\mathcal{L}}_O W)\cdot{}^{(T)}\mathbf{i}\Big\}\end{aligned} \tag{6.2.45}$$

$$\begin{aligned}&{}^{(T)}\pi^{\alpha\beta}Q(\hat{\mathcal{L}}_O W)_{\alpha\beta43}\tau_+^2\tau_-^2 = \\ &\quad \frac{1}{16}\tau_+^2\tau_-^2\Big\{4|\beta(\hat{\mathcal{L}}_O W)|^{2(T)}\underline{\mathbf{n}} + 4|\underline{\beta}(\hat{\mathcal{L}}_O W)|^{2(T)}\mathbf{n} + 4(|\rho(\hat{\mathcal{L}}_O W)|^2 + |\sigma(\hat{\mathcal{L}}_O W)|^2)^{(T)}\mathbf{j} \\ &\quad - 4\rho(\hat{\mathcal{L}}_O W)\beta(\hat{\mathcal{L}}_O W)\cdot{}^{(T)}\underline{\mathbf{m}} + 4\sigma(\hat{\mathcal{L}}_O W)^*\beta(\hat{\mathcal{L}}_O W)\cdot{}^{(T)}\underline{\mathbf{m}} \\ &\quad + 4\rho(\hat{\mathcal{L}}_O W)\underline{\beta}(\hat{\mathcal{L}}_O W)\cdot{}^{(T)}\mathbf{m} + 4\sigma(\hat{\mathcal{L}}_O W)^*\underline{\beta}(\hat{\mathcal{L}}_O W)\cdot{}^{(T)}\mathbf{m} \\ &\quad + 2(|\rho(\hat{\mathcal{L}}_O W)|^2 + \sigma(\hat{\mathcal{L}}_O W)|^2)\mathrm{tr}^{(T)}\mathbf{i} - 2(\beta(\hat{\mathcal{L}}_O W)\widehat{\otimes}\underline{\beta}(\hat{\mathcal{L}}_O W))\cdot{}^{(T)}\mathbf{i}\Big\}\end{aligned} \tag{6.2.46}$$

$$\begin{aligned}&{}^{(T)}\pi^{\alpha\beta}Q(\hat{\mathcal{L}}_O W)_{\alpha\beta33}\tau_-^4 = \\ &\quad \frac{1}{16}\tau_-^4\Big\{4(\rho(\hat{\mathcal{L}}_O W)|^2 + \sigma(\hat{\mathcal{L}}_O W)|^2)^{(T)}\underline{\mathbf{n}} + 2|\underline{\alpha}(\hat{\mathcal{L}}_O W)|^{2(T)}\mathbf{n} + 4|\underline{\beta}(\hat{\mathcal{L}}_O W)|^{2(T)}\mathbf{j} \\ &\quad + 8(\underline{\alpha}(\hat{\mathcal{L}}_O W)\cdot\underline{\beta}(\hat{\mathcal{L}}_O W))\cdot{}^{(T)}\mathbf{m} - 8\rho(\hat{\mathcal{L}}_O W)\underline{\beta}(\hat{\mathcal{L}}_O W)\cdot{}^{(T)}\underline{\mathbf{m}} \\ &\quad - 8\sigma(\hat{\mathcal{L}}_O W)^*\underline{\beta}(\hat{\mathcal{L}}_O W)\cdot{}^{(T)}\underline{\mathbf{m}} + 8(|\underline{\beta}(\hat{\mathcal{L}}_O W)|^2)\mathrm{tr}^{(T)}\mathbf{i} \\ &\quad + 8\rho(\hat{\mathcal{L}}_O W)\underline{\alpha}(\hat{\mathcal{L}}_O W)\cdot{}^{(T)}\mathbf{i} + 8\sigma(\hat{\mathcal{L}}_O W)^*\underline{\alpha}(\hat{\mathcal{L}}_O W)\cdot{}^{(T)}\mathbf{i}\Big\}.\end{aligned} \tag{6.2.47}$$

[15] $\mathcal{Q}_{\mathcal{K}}$ can be replaced by $\mathcal{Q}_1 + \underline{\mathcal{Q}}_1$.

We use the estimates of Proposition 6.1.1 for the deformation tensor ${}^{(T)}\pi$. Note that all the integrals composing $\int_{V_{(u,\underline{u})}} |Q(\hat{\mathcal{L}}_O W)_{\alpha\beta\gamma\delta}({}^{(T)}\pi^{\alpha\beta}\bar{K}^\gamma\bar{K}^\delta)|$ are estimated in same way; and the estimates are of exactly the same type as those for $\int_{V_{(u,\underline{u})}} |Q(\hat{\mathcal{L}}_O W)_{\alpha\beta\gamma\delta}({}^{(\bar{K})}\pi^{\alpha\beta}\bar{K}^\gamma T^\delta)|$. Therefore we do not report them here.

6.3 The error terms $\mathcal{E}_2$

The remarks at the beginning of Section 6.2 also apply to the estimates of $\mathcal{E}_2(u,\underline{u})$ and we do not repeat them here.

$\mathcal{E}_2(u,\underline{u})$ collects the error terms associated with the integrals of $\mathcal{Q}_2$ and $\underline{\mathcal{Q}}_2$,

$$\begin{aligned}\mathcal{E}_2(u,\underline{u}) = &\int_{V_{(u,\underline{u})}} |\mathbf{Div}\, Q(\hat{\mathcal{L}}_O^2 W)_{\beta\gamma\delta}(\bar{K}^\beta\bar{K}^\gamma T^\delta)| + \int_{V_{(u,\underline{u})}} |\mathbf{Div}\, Q(\hat{\mathcal{L}}_O\hat{\mathcal{L}}_T W)_{\beta\gamma\delta}(\bar{K}^\beta\bar{K}^\gamma\bar{K}^\delta)|\\
&+\int_{V_{(u,\underline{u})}} |\mathbf{Div}\, Q(\hat{\mathcal{L}}_S\hat{\mathcal{L}}_T W)_{\beta\gamma\delta}(\bar{K}^\beta\bar{K}^\gamma\bar{K}^\delta)| + \int_{V_{(u,\underline{u})}} |Q(\hat{\mathcal{L}}_O^2 W)_{\alpha\beta\gamma\delta}({}^{(\bar{K})}\pi^{\alpha\beta}\bar{K}^\gamma T^\delta)|\\
&+\frac{1}{2}\int_{V_{(u,\underline{u})}} |Q(\hat{\mathcal{L}}_O^2 W)_{\alpha\beta\gamma\delta}({}^{(T)}\pi^{\alpha\beta}\bar{K}^\gamma\bar{K}^\delta)| + \frac{3}{2}\int_{V_{(u,\underline{u})}} |Q(\hat{\mathcal{L}}_O\hat{\mathcal{L}}_T W)_{\alpha\beta\gamma\delta}({}^{(\bar{K})}\pi^{\alpha\beta}\bar{K}^\gamma\bar{K}^\delta)|\\
&+\frac{3}{2}\int_{V_{(u,\underline{u})}} |Q(\hat{\mathcal{L}}_S\hat{\mathcal{L}}_T W)_{\alpha\beta\gamma\delta}({}^{(\bar{K})}\pi^{\alpha\beta}\bar{K}^\gamma\bar{K}^\delta)|.\end{aligned}$$

Decomposing these integrals as we have before leads to a very large number of terms. Nevertheless it is easy to see, with the experience accumulated in the previous sections, how to estimate all of them.

Remark: As far as asymptotic behavior is concerned most terms can be treated like the corresponding ones in the previous section. The main new complication arising here is that of the presence of higher order derivatives. In particular in order to control the second-order derivatives of ${}^{(O)}\pi$, we had to work hard to ensure that they could be estimated in terms of only two derivatives of the curvature rather than the three derivatives that seem necessary upon first analysis. To stress this fact we shall concentrate here mainly on the terms involving the highest derivatives.

We start by controlling the integrals

$$\begin{aligned}&\int_{V_{(u,\underline{u})}} |\mathbf{Div}\, Q(\hat{\mathcal{L}}_O^2 W)_{\beta\gamma\delta}(\bar{K}^\beta\bar{K}^\gamma T^\delta)|\\
&\int_{V_{(u,\underline{u})}} |Q(\hat{\mathcal{L}}_O^2 W)_{\alpha\beta\gamma\delta}({}^{(\bar{K})}\pi^{\alpha\beta}\bar{K}^\gamma T^\delta)|\\
&\int_{V_{(u,\underline{u})}} |Q(\hat{\mathcal{L}}_O^2 W)_{\alpha\beta\gamma\delta}({}^{(T)}\pi^{\alpha\beta}\bar{K}^\gamma\bar{K}^\delta)|,\end{aligned}$$

which are needed to prove the boundedness of

$$\int_C Q(\hat{\mathcal{L}}_O^2 W)(\bar{K},\bar{K},T,e_4), \quad \int_{\underline{C}} Q(\hat{\mathcal{L}}_O^2 W)(\bar{K},\bar{K},T,e_3).$$

6.3.1 Estimate of $\int_{V_{(u,\underline{u})}} \mathbf{Div} Q(\hat{\mathcal{L}}_O^2 W)_{\beta\gamma\delta}(\bar{K}^\beta \bar{K}^\gamma T^\delta)$

By a straightforward calculation[16]

$$\begin{aligned}\mathbf{Div} Q(\hat{\mathcal{L}}_O^2 W)_{\beta\gamma\delta} &= (\hat{\mathcal{L}}_O^2 W)_{\beta\ \delta}^{\ \mu\ \nu} J(O,O;W)_{\mu\gamma\nu} + (\hat{\mathcal{L}}_O^2 W)_{\beta\ \gamma}^{\ \mu\ \nu} J(O,O;W)_{\mu\delta\nu}\\ &+ {}^*(\hat{\mathcal{L}}_O^2 W)_{\beta\ \delta}^{\ \mu\ \nu} J(O,O;W)^*{}_{\mu\gamma\nu} + {}^*(\hat{\mathcal{L}}_O^2 W)_{\beta\ \gamma}^{\ \mu\ \nu} J(O,O;W)^*{}_{\mu\delta\nu}\end{aligned}$$

where[17]

$$J(O,O;W) = J^0(O,O;W) + \frac{1}{2}\left(J^1(O,O;W) + J^2(O,O;W) + J^3(O,O;W)\right) \quad (6.3.1)$$

and

$$\begin{aligned}J^0(O,O;W) &= \hat{\mathcal{L}}_O J(O;W)\\ J^1(O,O;W) &= J^1(O;\hat{\mathcal{L}}_O W)\\ J^2(O,O;W) &= J^2(O;\hat{\mathcal{L}}_O W)\\ J^3(O,O;W) &= J^3(O;\hat{\mathcal{L}}_O W).\end{aligned} \quad (6.3.2)$$

Recallthe equation (see (6.1.1))

$$\begin{aligned}\mathbf{Div} Q(\hat{\mathcal{L}}_O^2 W)(\bar{K},\bar{K},T) &= \frac{1}{8}\tau_+^4 (D(O,O;W)_{444} + D(O,O;W)_{344})\\ &+ \frac{1}{4}\tau_+^2\tau_-^2 (D(O,O;W)_{344} + D(O,O;W)_{334})\\ &+ \frac{1}{8}\tau_-^4 (D(O,O;W)_{334} + D(O,O;W)_{333}),\end{aligned}$$

where (see (6.1.3))

$$\begin{aligned}D(O,O;W)_{444} &= 4\alpha(\hat{\mathcal{L}}_O^2 W)\cdot\Theta(O,O;W) - 8\beta(\hat{\mathcal{L}}_O^2 W)\cdot\Xi(O,O;W)\\ D(O,O;W)_{443} &= 8\rho(\hat{\mathcal{L}}_O^2 W)\Lambda(O,O;W) + 8\sigma(\hat{\mathcal{L}}_O^2 W)K(O,O;W)\\ &+ 8\beta(\hat{\mathcal{L}}_O^2 W)\cdot I(O,O;W)\\ D(O,O;W)_{334} &= 8\rho(\hat{\mathcal{L}}_O^2 W)\underline{\Lambda}(O,O;W) - 8\sigma(\hat{\mathcal{L}}_O^2 W)\underline{K}(O,O;W)\\ &- 8\underline{\beta}(\hat{\mathcal{L}}_O^2 W)\cdot\underline{I}(O,O;W)\\ D(O,O;W)_{333} &= 4\underline{\alpha}(\hat{\mathcal{L}}_O^2 W)\cdot\underline{\Theta}(O,O;W) + 8\underline{\beta}(\hat{\mathcal{L}}_O^2 W)\cdot\underline{\Xi}(O,O;W).\end{aligned} \quad (6.3.3)$$

The terms $\Theta(O,O;W),\dots,\underline{\Xi}(O,O;W)$[18] have the following structure:

$$X(O,O;W) = X^0(O,O;W) + \frac{1}{2}\left(X^1(O,O;W) + X^2(O,O;W) + X^3(O,O;W)\right),$$

[16] See [Ch-Kl], Propositions 7.1.1, 7.1.2.

[17] See also [Ch-Kl], (8.1.2d).

[18] They are the null components of $J(O,O;W)$.

and the upper indices 0, 1, 2, 3 refer to the various parts of the current $J(O, O; W)$ (see (6.3.2)). Therefore

$$X^i(O, O; W) = X^i(O, \hat{\mathcal{L}}_O W), \tag{6.3.4}$$

and

$$X^0(O, O; W) = \frac{1}{2}\left[\hat{\mathcal{L}}_O X^1(O; W) + \hat{\mathcal{L}}_O X^2(O; W) + \hat{\mathcal{L}}_O X^3(O; W)\right]. \tag{6.3.5}$$

We have to control the integrals

$$\begin{aligned}
&\int_{V_{(u,\underline{u})}} \tau_+^4 D(O, O; W)_{444}\,, \int_{V_{(u,\underline{u})}} \tau_+^4 D(O, O; W)_{443}\\
&\int_{V_{(u,\underline{u})}} \tau_+^2\tau_-^2 D(O, O; W)_{334}\,, \int_{V_{(u,\underline{u})}} \tau_-^4 D(O, O; W)_{333}.
\end{aligned} \tag{6.3.6}$$

The more sensitive terms are the first two containing the highest weight in τ_+. We look at the estimate of the first one[19],

$$\begin{aligned}
\int_{V_{(u,\underline{u})}} \tau_+^4 D(O, O; W)_{444} &= \frac{1}{2}\int_{V_{(u,\underline{u})}} \tau_+^4 \alpha(\hat{\mathcal{L}}_O^2 W)\cdot\Theta(O, O; W)\\
&\quad - \int_{V_{(u,\underline{u})}} \tau_+^4 \beta(\hat{\mathcal{L}}_O^2 W)\cdot\Xi(O, O; W).
\end{aligned}$$

The other terms are simpler and can be treated similarly. The two integrals on the right-hand side are estimated in the next proposition, in a way similar to Proposition 6.2.6,

Proposition 6.3.1 *Under the assumptions (6.0.1) and (6.0.2) the following inequalities hold:*

$$\begin{aligned}
&\left|\int_{V_{(u,\underline{u})}} \tau_+^4 \alpha(\hat{\mathcal{L}}_O^2 W)\cdot\Theta(O, O; W)\right| \le c\epsilon_0\mathcal{Q}_{\mathcal{K}}\\
&\left|\int_{V_{(u,\underline{u})}} \tau_+^4 \beta(\hat{\mathcal{L}}_O^2 W)\cdot\Xi(O, O; W)\right| \le c\epsilon_0\mathcal{Q}_{\mathcal{K}}.
\end{aligned}$$

Proof: We start by looking at the first integral; equation (6.3.4) implies that the terms in the integrand associated with $J^1(O, O; W)$, $J^2(O, O; W)$, $J^3(O, O; W)$ are estimated exactly as the corresponding terms of Proposition 6.2.6 by substituting W with $\hat{\mathcal{L}}_O W$ and by observing that, in the estimates analogous to those of Lemma 6.2.3, $\mathcal{Q}_1^{\frac{1}{2}}$ and $\underline{\mathcal{Q}}_1^{\frac{1}{2}}$ are replaced by $\mathcal{Q}_2^{\frac{1}{2}}$ and $\underline{\mathcal{Q}}_2^{\frac{1}{2}}$.

We still have to control

$$\int_{V_{(u,\underline{u})}} \tau_+^4 \alpha(\hat{\mathcal{L}}_O^2 W)\cdot\Theta^0(O, O; W)\,.$$

[19] Both terms have the same weight, but $D(O, O; W)_{443}$ has lower signature (see (3.1.24)). Therefore the second one is potentially more dangerous, yet it is easy to check that it does not, in fact, introduce any additional difficulties.

For this integral we have to estimate

$$\Theta^0(O,O;W)=\frac{1}{2}\Big[\hat{\mathcal{L}}_O\Theta^1(O;W)+\hat{\mathcal{L}}_O\Theta^2(O;W)+\hat{\mathcal{L}}_O\Theta^3(O;W)\Big].$$

The more delicate parts are those associated with $\hat{\mathcal{L}}_O\Theta^2(O;W)$ and $\hat{\mathcal{L}}_O\Theta^3(O;W)$; therefore we start by considering

$$\int_{V_{(u,\underline{u})}}\tau_+^4\alpha(\hat{\mathcal{L}}_O^2W)\cdot\hat{\mathcal{L}}_O\Theta^2(O;W). \tag{6.3.7}$$

Recalling the expression of $\Theta^2(O;W)$ (see (6.1.18))

$$\Theta^2(O;W)=\mathrm{Qr}\big[{}^{(O)}p_3\,;\,\alpha\big]+\mathrm{Qr}\big[{}^{(O)}\not{p};\,\beta\big]+\mathrm{Qr}\big[{}^{(O)}p_4\,;\,(\rho,\sigma)\big],$$

we find that $\hat{\mathcal{L}}_O\Theta^2(O;W)$ has the following expression:

$$\begin{aligned}\hat{\mathcal{L}}_O\Theta^2(O;W)&=\mathrm{Qr}\Big[\hat{\mathcal{L}}_O{}^{(O)}p_3\,;\,\alpha(W)\Big]+\mathrm{Qr}\Big[\hat{\mathcal{L}}_O{}^{(O)}\not{p};\,\beta(W)\Big]+\mathrm{Qr}\Big[\hat{\mathcal{L}}_O{}^{(O)}p_4\,;\,(\rho,\sigma)\Big]\\&+\ \mathrm{Qr}\Big[{}^{(O)}p_3\,;\,\hat{\mathcal{L}}_O\alpha(W)\Big]+\mathrm{Qr}\Big[{}^{(O)}\not{p};\,\hat{\mathcal{L}}_O\beta(W)\Big]+\mathrm{Qr}\Big[{}^{(O)}p_4\,;\,\hat{\mathcal{L}}_O(\rho,\sigma)\Big].\end{aligned} \tag{6.3.8}$$

As noted in the remark after Proposition 6.1.13 and in the discussion in the appendix to this chapter, the dependence on the third derivatives of the connection coefficients appears in $\hat{\mathcal{L}}_O{}^{(O)}p_3$ and $\hat{\mathcal{L}}_O{}^{(O)}\not{p}$.[20] Therefore we concentrate on

$$\mathrm{Qr}\Big[\hat{\mathcal{L}}_O{}^{(O)}p_3\,;\,\alpha(W)\Big]+\mathrm{Qr}\Big[\hat{\mathcal{L}}_O{}^{(O)}\not{p};\,\beta(W)\Big],$$

and check that the corresponding integrals in (6.3.7)

$$\begin{aligned}&\int_{V_{(u,\underline{u})}}\tau_+^4\alpha(\hat{\mathcal{L}}_O^2W)\mathrm{Qr}\Big[\hat{\mathcal{L}}_O{}^{(O)}\not{p};\,\beta(W)\Big]\\&\int_{V_{(u,\underline{u})}}\tau_+^4\alpha(\hat{\mathcal{L}}_O^2W)\mathrm{Qr}\Big[\hat{\mathcal{L}}_O{}^{(O)}p_3\,;\,\alpha(W)\Big],\end{aligned} \tag{6.3.9}$$

satisfy the following lemma.

Lemma 6.3.1 *Under the assumptions (6.0.1) and (6.0.2) the following inequalities hold:*

$$\int_{V_{(u,\underline{u})}}\tau_+^4\alpha(\hat{\mathcal{L}}_O^2W)Qr\Big[\hat{\mathcal{L}}_O{}^{(O)}\not{p};\,\beta(W)\Big]\le c\mathcal{Q}_2\left(\sup_{\mathcal{K}}\int_{\underline{C}(\underline{u};[u_0(\underline{u}),u])}|\hat{\mathcal{L}}_O{}^{(O)}\not{p}|^2\right)^{\frac{1}{2}}$$

$$\int_{V_{(u,\underline{u})}}\tau_+^4\alpha(\hat{\mathcal{L}}_O^2W)Qr\Big[\hat{\mathcal{L}}_O{}^{(O)}p_3\,;\,\alpha(W)\Big]\le c\mathcal{Q}_2\left(\sup_{\mathcal{K}}\int_{\underline{C}(\underline{u};[u_0(\underline{u}),u])}|\hat{\mathcal{L}}_O{}^{(O)}p_3|^2\right)^{\frac{1}{2}}.$$

[20]From the discussion in the appendix to this chapter (see Subsection 6.4.1) it follows that $\hat{\mathcal{L}}_O{}^{(O)}p_4$ does not depend on the third derivatives of the connection coefficients.

We postpone its proof until Subsection 6.3.2.

With the help of this lemma and Proposition 6.1.11, we obtain

$$\int_{V_{(u,\underline{u})}} \tau_+^4 \alpha(\hat{\mathcal{L}}_O^2 W) \mathrm{Qr}\left[\hat{\mathcal{L}}_O {}^{(O)}\not{p};\ \beta(W)\right] \le c\epsilon_0 \mathcal{Q}_{\mathcal{K}}$$
$$\int_{V_{(u,\underline{u})}} \tau_+^4 \alpha(\hat{\mathcal{L}}_O^2 W) \mathrm{Qr}\left[\hat{\mathcal{L}}_O {}^{(O)}p_3\,;\ \alpha(W)\right] \le c\epsilon_0 \mathcal{Q}_{\mathcal{K}}, \tag{6.3.10}$$

which proves Proposition 6.3.1 for these particular terms. The contribution to the integral (6.3.7) due to $\mathrm{Qr}\left[\hat{\mathcal{L}}_O {}^{(O)}p_4\,;\ (\rho,\sigma)\right]$ is easier to treat and we do not discuss it here.

The contributions to the integral (6.3.7) due to the terms present in the second line of (6.3.8) are

$$\int_{V_{(u,\underline{u})}} \tau_+^4 \alpha(\hat{\mathcal{L}}_O^2 W) \mathrm{Qr}\left[{}^{(O)}\not{p};\ \hat{\mathcal{L}}_O \beta(W)\right]$$
$$\int_{V_{(u,\underline{u})}} \tau_+^4 \alpha(\hat{\mathcal{L}}_O^2 W) \mathrm{Qr}\left[{}^{(O)}p_3\,;\ \hat{\mathcal{L}}_O \alpha(W)\right] \tag{6.3.11}$$
$$\int_{V_{(u,\underline{u})}} \tau_+^4 \alpha(\hat{\mathcal{L}}_O^2 W) \mathrm{Qr}\left[{}^{(O)}p_4\,;\ \hat{\mathcal{L}}_O (\rho,\sigma)(W)\right],$$

and turn out to be easier to control. A short discussion on their estimates is given at the end of this section.

The estimate of the integral

$$\int_{V_{(u,\underline{u})}} \tau_+^4 \alpha(\hat{\mathcal{L}}_O^2 W) \cdot \hat{\mathcal{L}}_O \Theta^3(O;\, W) \tag{6.3.12}$$

is made precisely as that for the integral (6.3.7). Again it follows (see (6.1.23)) that the contributions to $\Theta^3(O;\, W) \equiv \Theta(J^3(O;\, W))$ which depend on the second derivatives of the connection coefficients are

$$\mathrm{Qr}\left[\alpha\,;\ \underline{K}({}^{(O)}q)\right]\ ,\ \mathrm{Qr}\left[\alpha\,;\ \underline{\Lambda}({}^{(O)}q)\right]\ ,\ \mathrm{Qr}\left[\beta\,;\ (I,\underline{I})({}^{(O)}q)\right],$$

and therefore, the more delicate integrals to control are

$$\int_{V_{(u,\underline{u})}} \tau_+^4 \alpha(\hat{\mathcal{L}}_O^2 W) \mathrm{Qr}\left[\hat{\mathcal{L}}_O \underline{K}({}^{(O)}q)\,;\ \alpha(W)\right]$$
$$\int_{V_{(u,\underline{u})}} \tau_+^4 \alpha(\hat{\mathcal{L}}_O^2 W) \mathrm{Qr}\left[\hat{\mathcal{L}}_O \underline{\Lambda}({}^{(O)}q)\,;\ \alpha(W)\right]$$
$$\int_{V_{(u,\underline{u})}} \tau_+^4 \alpha(\hat{\mathcal{L}}_O^2 W) \mathrm{Qr}\left[\hat{\mathcal{L}}_O \underline{I}({}^{(O)}q)\,;\ \alpha(W)\right].$$

These are estimated exactly as those in (6.3.9)[21] and we do not report them here.

[21] See the discussion in the appendix to this chapter, Subsection 6.4.1, about the dependence on the connection coefficients of $\underline{K}({}^{(O)}q)$, $\underline{\Lambda}({}^{(O)}q)$, $\underline{I}({}^{(O)}q)$.

The previous estimates have shown that the more delicate terms to estimate are those depending on the third-order derivatives of the connection coefficients. So far we have discussed the estimates for the terms that depend on $\nabla\!\!\!/^3\eta$, $\nabla\!\!\!/^3\underline{\eta}$ and $\nabla\!\!\!/^3\chi$. Next we examine the terms that depend on $\nabla\!\!\!/^3\underline{\omega}$. Therefore we consider explicitly only those parts of the integrals (6.3.6) that depend on $\nabla\!\!\!/^3\underline{\omega}$.[22] The factor $\nabla\!\!\!/^2\underline{\omega}$ appears in the expression

$$\underline{\Xi}({}^{(O)}q)_a = \frac{1}{2}\mathbf{D}\!\!\!/_3{}^{(O)}\underline{\mathbf{m}}_a - \frac{1}{2}(\mathbf{D}_3\log\Omega)^{(O)}\underline{\mathbf{m}}_a + \frac{1}{2}\mathrm{tr}\underline{\chi}^{(O)}\underline{\mathbf{m}}_a + \hat{\underline{\chi}}_{ac}{}^{(O)}\underline{\mathbf{m}}_c$$

through $\mathbf{D}\!\!\!/_3{}^{(O)}\underline{\mathbf{m}}$. The terms of $J^3(O;W)$ that depend on $\underline{\Xi}({}^{(O)}q)$ are (see (6.1.19), (6.1.21))

$$\begin{aligned}
\underline{\Xi}^3(O;W) &= \mathrm{Qr}\left[(\rho,\sigma)\,;\,\underline{\Xi}({}^{(O)}q)\right]+\cdots\\
\underline{\Lambda}^3(O;W) &= \mathrm{Qr}\left[\beta\,;\,\underline{\Xi}({}^{(O)}q)\right]+\cdots\\
\underline{K}^3(O;W) &= \mathrm{Qr}\left[\beta\,;\,\underline{\Xi}({}^{(O)}q)\right]+\cdots\\
I^3(O;W) &= \mathrm{Qr}\left[\alpha\,;\,\underline{\Xi}({}^{(O)}q)\right]+\cdots\,.
\end{aligned}\tag{6.3.13}$$

These $J^3(O;W)$ components are present in (see (6.1.3))

$$\begin{aligned}
D(O,W)_{443} &= 8\beta(\hat{\mathcal{L}}_O W)\cdot I^3(O,W)+\cdots\\
D(O,W)_{334} &= 8\rho(\hat{\mathcal{L}}_O W)\underline{\Lambda}^3(O,W) - 8\sigma(\hat{\mathcal{L}}_O W)\underline{K}^3(O,W)+\cdots\\
D(O,W)_{333} &= 8\underline{\beta}(\hat{\mathcal{L}}_O W)\cdot\underline{\Xi}^3(O,W)+\cdots\,.
\end{aligned}\tag{6.3.14}$$

In the error term $\int_{V_{(u,\underline{u})}}\mathbf{Div}Q(\hat{\mathcal{L}}_O^2 W)_{\beta\gamma\delta}\bar{K}^\beta\bar{K}^\gamma T^\delta$ we have to consider the terms $D(O,O;W)_{443}$, $D(O,O;W)_{334}$, $D(O,O;W)_{333}$. The contributions to these terms coming from $\hat{\mathcal{L}}_O J^3(O;W)$ are:[23]

$$\begin{aligned}
D(O,O;W)_{443} &= 8\beta(\hat{\mathcal{L}}_O^2 W)\cdot I(O,O;W)+\cdots\\
D(O,O;W)_{334} &= 8\rho(\hat{\mathcal{L}}_O^2 W)\underline{\Lambda}(O,O;W) - 8\sigma(\hat{\mathcal{L}}_O^2 W)\underline{K}(O,O;W)+\cdots\\
D(O,O;W)_{333} &= 8\underline{\beta}(\hat{\mathcal{L}}_O^2 W)\cdot\underline{\Xi}(O,O;W)+\cdots\,,
\end{aligned}$$

and from $I(O,O;W)$, $\underline{\Lambda}(O,O;W)$, $\underline{K}(O,O;W)$, $\underline{\Xi}(O,O;W)$ we select the parts containing $\hat{\mathcal{L}}_O\underline{\Xi}({}^{(O)}q)$,

$$\begin{aligned}
\underline{\Xi}(O,O;W) &= \mathrm{Qr}\left[(\rho,\sigma)(W)\,;\,\hat{\mathcal{L}}_O\underline{\Xi}({}^{(O)}q)\right]+\cdots\\
\underline{\Lambda}(O,O;W) &= \mathrm{Qr}\left[\beta(W)\,;\,\hat{\mathcal{L}}_O\underline{\Xi}({}^{(O)}q)\right]+\cdots\\
\underline{K}(O,O;W) &= \mathrm{Qr}\left[\beta(W)\,;\,\hat{\mathcal{L}}_O\underline{\Xi}({}^{(O)}q)\right]+\cdots\\
I(O,O;W) &= \mathrm{Qr}\left[\alpha(W)\,;\,\hat{\mathcal{L}}_O\underline{\Xi}({}^{(O)}q)\right]+\cdots\,.
\end{aligned}$$

---

[22] The term $\nabla\!\!\!/^3\underline{\omega}$ does not appear in the integrals examined up to now.

[23] $\hat{\mathcal{L}}_O J^3(O;W)$ appears in the current $J^0(O,O;W) = \hat{\mathcal{L}}_O J^3(O;W)+\cdots$.

In conclusion the terms that depend on $\nabla\!\!\!\!/^3\underline{\omega}$ are

$$D(O,O;W)_{443}=8\beta(\hat{\mathcal{L}}_O^2W)\cdot\mathrm{Qr}\left[\alpha(W)\,;\ \hat{\mathcal{L}}_O\underline{\Xi}({}^{(O)}q)\right]+\cdots$$
$$D(O,O;W)_{334}=8\rho(\hat{\mathcal{L}}_O^2W)\cdot\mathrm{Qr}\left[\beta(W)\,;\ \hat{\mathcal{L}}_O\underline{\Xi}({}^{(O)}q)\right]+\cdots$$
$$D(O,O;W)_{333}=8\underline{\beta}(\hat{\mathcal{L}}_O^2W)\cdot\mathrm{Qr}\left[(\rho,\sigma)(W)\,;\ \hat{\mathcal{L}}_O\underline{\Xi}({}^{(O)}q)\right]+\cdots.$$

They appear in the following integrals

$$\begin{aligned}\int_{V_{(u,\underline{u})}}\tau_+^4(D(O,O;W)_{443}&=\int_{V_{(u,\underline{u})}}\tau_+^4\beta(\hat{\mathcal{L}}_O^2W)\cdot\mathrm{Qr}\left[\alpha(W)\,;\ \hat{\mathcal{L}}_O\underline{\Xi}({}^{(O)}q)\right]+\cdots\\ \int_{V_{(u,\underline{u})}}\tau_+^2\tau_-^2(D(O,O;W)_{334}&=\int_{V_{(u,\underline{u})}}\tau_+^2\tau_-^2\rho(\hat{\mathcal{L}}_O^2W)\cdot\mathrm{Qr}\left[\beta(W)\,;\ \hat{\mathcal{L}}_O\underline{\Xi}({}^{(O)}q)\right]+\cdots\\ \int_{V_{(u,\underline{u})}}\tau_-^4(D(O,O;W)_{333}&=\int_{V_{(u,\underline{u})}}\tau_-^4\underline{\beta}(\hat{\mathcal{L}}_O^2W)\cdot\mathrm{Qr}\left[(\rho,\sigma)(W)\,;\ \hat{\mathcal{L}}_O\underline{\Xi}({}^{(O)}q)\right]+\cdots.\end{aligned}\tag{6.3.15}$$

The neglected terms either have been estimated already in Proposition 6.3.1 or are easier to control. The integrals on the right-hand side are estimated in the following proposition.

Proposition 6.3.2 *Under the assumptions (6.0.1) and (6.0.2) the following inequalities hold:*

$$\begin{aligned}&\int_{V_{(u,\underline{u})}}\tau_+^4\beta(\hat{\mathcal{L}}_O^2W)\cdot Qr\left[\alpha(W)\,;\ \hat{\mathcal{L}}_O\underline{\Xi}({}^{(O)}q)\right]\le c\epsilon_0\mathcal{Q}_{\mathcal{K}}\\ &\int_{V_{(u,\underline{u})}}\tau_+^2\tau_-^2\rho(\hat{\mathcal{L}}_O^2W)Qr\left[\beta(W)\,;\ \hat{\mathcal{L}}_O\underline{\Xi}({}^{(O)}q)\right]\le c\epsilon_0\mathcal{Q}_{\mathcal{K}}\\ &\int_{V_{(u,\underline{u})}}\tau_-^4\underline{\beta}(\hat{\mathcal{L}}_O^2W)\cdot Qr\left[(\rho,\sigma)(W)\,;\ \hat{\mathcal{L}}_O\underline{\Xi}({}^{(O)}q)\right]\le c\epsilon_0\mathcal{Q}_{\mathcal{K}}.\end{aligned}\tag{6.3.16}$$

Proof: The proof is based on the following lemma, whose proof is postponed until the next subsection.

Lemma 6.3.2 *Under the assumptions (6.0.1) and (6.0.2) the following inequalities hold with $\epsilon>0$:*

$$\int_{V_{(u,\underline{u})}}\tau_+^4\beta(\hat{\mathcal{L}}_O^2W)\cdot Qr\left[\alpha(W)\,;\ \hat{\mathcal{L}}_O\underline{\Xi}({}^{(O)}q)\right]\le$$
$$c\mathcal{Q}_{\mathcal{K}}\left(\sup_{\mathcal{K}}\int_{\underline{C}(\underline{u};[u_0(\underline{u}),u])}\frac{1}{r^{1-2\epsilon}}|\hat{\mathcal{L}}_O\underline{\Xi}({}^{(O)}q)|^2\right)^{\frac{1}{2}}$$
$$\int_{V_{(u,\underline{u})}}\tau_+^2\tau_-^2\rho(\hat{\mathcal{L}}_O^2W)Qr\left[\beta(W)\,;\ \hat{\mathcal{L}}_O\underline{\Xi}({}^{(O)}q)\right]\le$$
$$c\mathcal{Q}_{\mathcal{K}}\left(\sup_{\mathcal{K}}\int_{\underline{C}(\underline{u};[u_0(\underline{u}),u])}\frac{1}{r^{1-2\epsilon}}|\hat{\mathcal{L}}_O\underline{\Xi}({}^{(O)}q)|^2\right)^{\frac{1}{2}}$$

$$\int_{V_{(u,\underline{u})}} \tau_-^4 \underline{\beta}(\hat{\mathcal{L}}_O^2 W)\cdot Qr\left[(\rho,\sigma)(W)\,;\,\hat{\mathcal{L}}_O \underline{\Xi}({}^{(O)}q)\right]$$
$$\leq c\mathcal{Q}_{\mathcal{K}}\left(\sup_{\mathcal{K}}\int_{\underline{C}(\underline{u};[u_0(\underline{u}),u])}\frac{1}{r^{1-2\epsilon}}|\hat{\mathcal{L}}_O\underline{\Xi}({}^{(O)}q)|^2\right)^{\frac{1}{2}}.$$

Proposition 6.3.2 is an immediate consequence of Proposition 6.1.13 and Lemma 6.3.2.

6.3.2 *Proof of Lemma 6.3.1 and Lemma 6.3.2*

To prove Lemma 6.3.1 we estimate the first integral of (6.3.9),

$$\int_{V_{(u,\underline{u})}} \tau_+^4\alpha(\hat{\mathcal{L}}_O^2 W)\mathrm{Qr}\left[\hat{\mathcal{L}}_O{}^{(O)}\not{p};\,\beta(W)\right] \leq \int_{V_{(u,\underline{u})}} \tau_+^4|\alpha(\hat{\mathcal{L}}_O^2 W)||\hat{\mathcal{L}}_O{}^{(O)}\not{p}||\beta(W)|$$
$$\leq \left(\int_{V_{(u,\underline{u})}} \tau_+^{2\gamma}|\alpha(\hat{\mathcal{L}}_O^2 W)|^2\right)^{\frac{1}{2}}\left(\int_{V_{(u,\underline{u})}} \tau_+^{2\sigma}|\hat{\mathcal{L}}_O{}^{(O)}\not{p}|^2|\beta(W)|^2\right)^{\frac{1}{2}}, \tag{6.3.17}$$

with $\gamma+\sigma=4$. The first factor satisfies the inequality

$$\int_{V_{(u,\underline{u})}} \tau_+^{2\gamma}|\alpha(\hat{\mathcal{L}}_O^2 W)|^2 \leq c\mathcal{Q}_2, \tag{6.3.18}$$

with $2\gamma<3$, which implies $2\sigma>5$. In fact, from the results of Chapter 5, see Proposition 5.1.5,

$$\begin{aligned}
\int_{V_{(u,\underline{u})}} \tau_+^{2\gamma}|\alpha(\hat{\mathcal{L}}_O^2 W)|^2 &= \int_{u_0(\underline{u})}^{u} du'\int_{C(u';[\underline{u}_0(u'),\underline{u}])} \tau_+^{2\gamma}|\alpha(\hat{\mathcal{L}}_O^2 W)|^2\\
&\leq \int_{u_0(\underline{u})}^{u} du'\frac{1}{u'^{4-2\gamma}}\left(\sup_{V(u,\underline{u})}\int_{C(u';[\underline{u}_0(u'),\underline{u}])} \tau_+^4|\alpha(\hat{\mathcal{L}}_O^2 W)|^2\right)\\
&\leq c\left(\sup_{\mathcal{K}}\int_{C(u')} Q(\hat{\mathcal{L}}_O^2 W)(\bar{K},\bar{K},T,e_4)\right) \leq c\mathcal{Q}_2.
\end{aligned}$$

The second factor satisfies

$$\begin{aligned}
\int_{V_{(u,\underline{u})}} \tau_+^{2\sigma}|\hat{\mathcal{L}}_O{}^{(O)}\not{p}|^2|\beta(W)|^2 &\leq \left(\sup|r^{\frac{7}{2}}\beta(W)|\right)^2\int_{V_{(u,\underline{u})}} \tau_+^{-(7-2\sigma)}|\hat{\mathcal{L}}_O{}^{(O)}\not{p}|^2\\
&\leq c\mathcal{Q}_{\mathcal{K}}\int_{V_{(u,\underline{u})}} \tau_+^{-(7-2\sigma)}|\hat{\mathcal{L}}_O{}^{(O)}\not{p}|^2.
\end{aligned} \tag{6.3.19}$$

We decompose the integral in the following way, writing $7-2\sigma=1+\tau$ with $\tau\in(0,1)$,

$$\begin{aligned}
\int_{V_{(u,\underline{u})}} \tau_+^{-(7-2\sigma)}|\hat{\mathcal{L}}_O{}^{(O)}\not{p}|^2 &\leq c\int_{\underline{u}_0(u)}^{\underline{u}} d\underline{u}'\int_{\underline{C}(\underline{u}';[u_0(\underline{u}'),u])} \underline{u}'^{-(1+\tau)}|\hat{\mathcal{L}}_O{}^{(O)}\not{p}|^2\\
&\leq c\int_{\underline{u}_0(u)}^{\underline{u}} d\underline{u}'\frac{1}{\underline{u}'^{(1+\tau)}}\left(\int_{\underline{C}(\underline{u}';[u_0(\underline{u}'),u])} |\hat{\mathcal{L}}_O{}^{(O)}\not{p}|^2\right)\\
&\leq c\left(\sup_{\mathcal{K}}\int_{\underline{C}(\underline{u}';[u_0(\underline{u}'),u])} |\hat{\mathcal{L}}_O{}^{(O)}\not{p}|^2\right) \leq c\epsilon_0,
\end{aligned} \tag{6.3.20}$$

where we have used, in the last inequality, the second estimate of Proposition 6.1.11. Inequalities (6.3.20) and (6.3.18) together prove the first estimate of the lemma. The estimate of the second line of (6.3.9) proceeds in the same way, but it relies on the first inequality of Proposition 6.1.11.

To prove Lemma 6.3.2 we prove only the first inequality of the lemma; the others are proved in the same way. The proof is similar to but easier than that of Lemma 6.3.1,

$$
\begin{aligned}
&\int_{V_{(u,\underline{u})}} \tau_+^4 \beta(\hat{\mathcal{L}}_O^2 W)\cdot \mathrm{Qr}\left[\alpha(W)\,;\ \hat{\mathcal{L}}_O \underline{\Xi}({}^{(O)}q)\right]\\
&\le \left(\sup_{V(u,\underline{u})}\int_{\underline{C}(\underline{u};[u_0,u])} \underline{u}'^4|\beta(\hat{\mathcal{L}}_O^2 W)|^2\right)^{\frac12}\int_{\underline{u}_0}^{\underline{u}} d\underline{u}'\left(\int_{\underline{C}(\underline{u}';[u_0,u])}\underline{u}'^4|\alpha(W)|^2||\hat{\mathcal{L}}_O\underline{\Xi}({}^{(O)}q)|^2\right)^{\frac12}\\
&\le c\mathcal{Q}_{\mathcal{K}}^{\frac12}\left(\sup_{\mathcal{K}}|r^{\frac72}\alpha(W)|\right)\int_{\underline{u}_0}^{\underline{u}} d\underline{u}'\left(\int_{\underline{C}(\underline{u}';[u_0,u])}\frac{1}{\underline{u}'^3}|\hat{\mathcal{L}}_O\underline{\Xi}({}^{(O)}q)|^2\right)^{\frac12}\\
&\le c\mathcal{Q}_{\mathcal{K}}\int_{\underline{u}_0}^{\underline{u}} d\underline{u}'\frac{1}{\underline{u}'^{1+\epsilon}}\left(\int_{\underline{C}(\underline{u};[u_0,u])}\frac{1}{r^{1-2\epsilon}}|\hat{\mathcal{L}}_O\underline{\Xi}({}^{(O)}q)|^2\right)^{\frac12}, \qquad (6.3.21)
\end{aligned}
$$

with $\epsilon > 0$, where in the last line we used Proposition 6.1.13.

We shall now provide a sketch of the estimate of the second integral of (6.3.11).[24]. We use the estimates (6.1.55), valid for any $p \in [2,4]$,

$$
\sup_{\mathcal{K}}|r^{1-\frac2p}\tau_-{}^{(O)}\mathrm{p}_3|_{p,S}\le c\epsilon_0\ ,\quad \sup_{\mathcal{K}}|r^{2-\frac2p}({}^{(O)}\mathrm{p}_4,{}^{(O)}\not{\mathrm{p}})|_{p,S}\le c\epsilon_0\ ,
$$

and obtain

$$
\begin{aligned}
&\int_{C(u';[\underline{u}_0,\underline{u}])}\underline{u}'^4|{}^{(O)}\mathrm{p}_3|^2|\hat{\mathcal{L}}_O\alpha(W)|^2\le c\int_{\underline{u}_0}^{\underline{u}} d\underline{u}'\underline{u}'^4\int_{S(u',\underline{u}')}|{}^{(O)}\mathrm{p}_3|^2|\hat{\mathcal{L}}_O\alpha(W)|^2\\
&\le c\int_{\underline{u}_0}^{\underline{u}} d\underline{u}'\underline{u}'^4\frac{1}{r^2\tau_-^2}|r^{1-\frac2p}\tau_-{}^{(O)}\mathrm{p}_3|_{p=4,S}^2|r^{1-\frac2p}\hat{\mathcal{L}}_O\alpha(W)|_{p=4,S}^2\\
&\le c\left(\sup_{\mathcal{K}}|r^{1-\frac2p}\tau_-{}^{(O)}\mathrm{p}_3|_{p=4,S}\right)^2\int_{\underline{u}_0}^{\underline{u}} d\underline{u}'\underline{u}'^4\frac{1}{r^2\tau_-^2}|r^{1-\frac2p}\hat{\mathcal{L}}_O\alpha(W)|_{p=4,S}^2\\
&\le c\epsilon_0^2\int_{\underline{u}_0}^{\underline{u}} d\underline{u}'\underline{u}'^4\frac{1}{r^7\tau_-^2}|r^{\frac72-\frac2p}\hat{\mathcal{L}}_O\alpha(W)|_{p=4,S}^2\le c\epsilon_0^2\mathcal{Q}_{\mathcal{K}}\left(\int_{\underline{u}_0}^{\underline{u}} d\underline{u}'\underline{u}'^4\frac{1}{r^7\tau_-^2}\right)\\
&\le c\epsilon_0^2\mathcal{Q}_{\mathcal{K}}\frac{1}{u'^4}\le c\epsilon_0^2\mathcal{Q}_{\mathcal{K}}. \qquad (6.3.22)
\end{aligned}
$$

Exactly the same argument can be used for the other integrals in (6.3.11) and for those associated with the current $J^3(O;W)$.

[24]Compare the way this integral is estimated with the estimate of the second integral of (6.2.34) in Lemma 6.2.3; see inequality (6.2.38)

6.3.3 *Estimate of* $\int_{V_{(u,\underline{u})}} Q(\hat{\mathcal{L}}_O^2 W)_{\alpha\beta\gamma\delta}({}^{(\bar{K})}\pi^{\alpha\beta}\bar{K}^\gamma T^\delta)$

Proposition 6.3.3 *Under the assumptions (6.0.1) and (6.0.2) the following inequality holds:*

$$\int_{V_{(u,\underline{u})}} |Q(\hat{\mathcal{L}}_O^2 W)_{\alpha\beta\gamma\delta}({}^{(\bar{K})}\pi^{\alpha\beta}\bar{K}^\gamma T^\delta)| \leq c\epsilon_0 \mathcal{Q}_{\mathcal{K}}. \tag{6.3.23}$$

Proof: These estimates are exactly of the same type as those obtained for

$$\int_{V_{(u,\underline{u})}} |Q(\hat{\mathcal{L}}_O W)_{\alpha\beta\gamma\delta}({}^{(\bar{K})}\pi^{\alpha\beta}\bar{K}^\gamma T^\delta)|$$

with the obvious substitution in the first and second factors of $\mathcal{Q}_1^{\frac{1}{2}}$ and $\underline{\mathcal{Q}}_1^{\frac{1}{2}}$ with $\mathcal{Q}_2^{\frac{1}{2}}$ or $\underline{\mathcal{Q}}_2^{\frac{1}{2}}$.

6.3.4 *Estimate of* $\int_{V_{(u,\underline{u})}} Q(\hat{\mathcal{L}}_O^2 W)_{\alpha\beta\gamma\delta}({}^{(T)}\pi^{\alpha\beta}\bar{K}^\gamma \bar{K}^\delta)$

Proposition 6.3.4 *Under the assumptions (6.0.1) and (6.0.2) the following inequality holds:*

$$\int_{V_{(u,\underline{u})}} |Q(\hat{\mathcal{L}}_O^2 W)_{\alpha\beta\gamma\delta}({}^{(T)}\pi^{\alpha\beta}\bar{K}^\gamma \bar{K}^\delta)| \leq c\epsilon_0 \mathcal{Q}_{\mathcal{K}}. \tag{6.3.24}$$

Proof: These estimates are exactly of the same type as those obtained for

$$\int_{V_{(u,\underline{u})}} |Q(\hat{\mathcal{L}}_O W)_{\alpha\beta\gamma\delta}({}^{(\bar{K})}\pi^{\alpha\beta}\bar{K}^\gamma T^\delta)|$$

with the obvious substitution in the first and second factors of $\mathcal{Q}_1^{\frac{1}{2}}$ and $\underline{\mathcal{Q}}_1^{\frac{1}{2}}$ with $\mathcal{Q}_2^{\frac{1}{2}}$ and $\underline{\mathcal{Q}}_2^{\frac{1}{2}}$.

This completes the control of the error terms associated to the integrals

$$\int_C Q(\hat{\mathcal{L}}_O^2 W)(\bar{K}, \bar{K}, T, e_4)\ , \ \int_{\underline{C}} Q(\hat{\mathcal{L}}_O^2 W)(\bar{K}, \bar{K}, T, e_3)\ .$$

6.3.5 *Estimate of* $\int_{V_{(u,\underline{u})}} \mathbf{Div} Q(\hat{\mathcal{L}}_O \hat{\mathcal{L}}_T W)_{\beta\gamma\delta}\bar{K}^\beta \bar{K}^\gamma \bar{K}^\delta$

We recall the equation (see [Ch-Kl], Propositions 7.1.1, 7.1.2),

$$\begin{aligned}\mathbf{Div} Q(\hat{\mathcal{L}}_O \hat{\mathcal{L}}_T W)_{\beta\gamma\delta} &= (\hat{\mathcal{L}}_O \hat{\mathcal{L}}_T W)_{\beta\ \delta}^{\ \mu\ \nu} J(T, O; W)_{\mu\gamma\nu} + (\hat{\mathcal{L}}_O \hat{\mathcal{L}}_T W)_{\beta\ \gamma}^{\ \mu\ \nu} J(T, O; W)_{\mu\delta\nu} \\ &\quad + {}^*(\hat{\mathcal{L}}_O \hat{\mathcal{L}}_T W)_{\beta\ \delta}^{\ \mu\ \nu} J(T, O; W)^*{}_{\mu\gamma\nu} + {}^*(\hat{\mathcal{L}}_O \hat{\mathcal{L}}_T W)_{\beta\ \gamma}^{\ \mu\ \nu} J(T, O; W)^*{}_{\mu\delta\nu},\end{aligned}$$

where

$$J(T, O; W) = J^0(T, O; W) + \frac{1}{2}\left(J^1(T, O; W) + J^2(T, O; W) + J^3(T, O; W)\right)$$

and

$$J^0(T,O;W)=\hat{\mathcal{L}}_O J(T;W)$$
$$J^i(T,O;W)=J^i(O;\hat{\mathcal{L}}_T W)\quad ,i\in\{1,2,3\}.$$

We see that the terms associated with the current $J^1(T,O;W)=J^1(O;\hat{\mathcal{L}}_T W)$ are the same as those of

$$\int_{V_{(u,\underline{u})}}\mathbf{Div}Q(\hat{\mathcal{L}}_O W)_{\beta\gamma\delta}(\bar{K}^\beta\bar{K}^\gamma T^\delta)$$

with $\hat{\mathcal{L}}_O W$ replaced by $\hat{\mathcal{L}}_O\hat{\mathcal{L}}_T W$ and $J^1(O;W)$ replaced by $J^1(O;\hat{\mathcal{L}}_T W)$. Considering the equation (see (6.1.3))

$$D^1(T,O;W)_{444}=4\alpha(\hat{\mathcal{L}}_O\hat{\mathcal{L}}_T W)\cdot\Theta^1(O,\hat{\mathcal{L}}_T W)-8\beta(\hat{\mathcal{L}}_O\hat{\mathcal{L}}_T W),\cdot\Xi^1(O,\hat{\mathcal{L}}_T W)\,,$$

it follows that we have to control the integrals

$$\int_{V_{(u,\underline{u})}}\tau_+^6\alpha(\hat{\mathcal{L}}_O\hat{\mathcal{L}}_T W)\cdot\Theta^1(O,\hat{\mathcal{L}}_T W)$$
$$\int_{V_{(u,\underline{u})}}\tau_+^6\beta(\hat{\mathcal{L}}_O\hat{\mathcal{L}}_T W)\cdot\Xi^1(O,\hat{\mathcal{L}}_T W).\qquad(6.3.25)$$

The main difference with respect to Proposition 6.2.6 is that the Riemann components present in the terms $\Theta^1(O,\hat{\mathcal{L}}_T W)$ and $\Xi^1(O,\hat{\mathcal{L}}_T W)$ (see (6.1.13), (6.1.14)) appear as first derivatives along the tangential and null directions of the null components of $\hat{\mathcal{L}}_T W$. We conclude that terms like $\mathbf{D}_3\mathbf{D}_3 W$, $\mathbf{D}_3\mathbf{D}_4 W$, $\mathbf{D}_4\mathbf{D}_3 W$, $\mathbf{D}_4\mathbf{D}_4 W$ are present.[25] In particular we have to control the integrals

$$\int\frac{du'}{u'}\left(\int_{C(u')}r^6|\mathbf{D}_T\alpha(\hat{\mathcal{L}}_T W)|^2\right)^{\frac{1}{2}}$$
$$\int\frac{d\underline{u}'}{\underline{u}'}\left(\int_{\underline{C}(\underline{u}')}r^6|\mathbf{D}_T\underline{\alpha}(\hat{\mathcal{L}}_T W)|^2\right)^{\frac{1}{2}}.$$

The terms associated with $J^2(T,O;W)$ and $J^3(T,O;W)$ are treated like the corresponding terms of $\int_{V_{(u,\underline{u})}}\mathbf{Div}Q(\hat{\mathcal{L}}_O W)_{\beta\gamma\delta}(\bar{K}^\beta\bar{K}^\gamma T^\delta)$ with the obvious modifications.

Finally, to control the terms associated with the current $J^0(T,O;W)=\hat{\mathcal{L}}_O J(T;W)$ we have to look, carefully, at the following integrals:

$$\int_{V_{(u,\underline{u})}}\tau_+^6\alpha(\hat{\mathcal{L}}_O\hat{\mathcal{L}}_T W)(\hat{\mathcal{L}}_O{}^{(T)}p_3)\alpha(W)$$

[25] Observe that the terms in $\Theta^1(O,\hat{\mathcal{L}}_T W)$, $\Xi^1(O,\hat{\mathcal{L}}_T W)$ involve, in particular $\mathbf{D}_3\mathbf{D}_3\underline{\alpha}$ and $\mathbf{D}_4\mathbf{D}_4\alpha$ which do not appear in the Bianchi equations. These terms have been treated in Chapter 5 by expressing them in terms of $\mathbf{D}_T\alpha(\hat{\mathcal{L}}_T W)$ or $\mathbf{D}_T\underline{\alpha}(\hat{\mathcal{L}}_T W)+$ [easier terms].

$$\int_{V_{(u,\underline{u})}} \tau_+^6 \alpha(\hat{\mathcal{L}}_O \hat{\mathcal{L}}_T W)(\hat{\mathcal{L}}_O {}^{(T)}\not{p})\beta(W)$$

$$\int_{V_{(u,\underline{u})}} \tau_+^6 \alpha(\hat{\mathcal{L}}_O \hat{\mathcal{L}}_T W)(\hat{\mathcal{L}}_O {}^{(T)}p_4)(\rho,\sigma)(W).$$

Their estimates are summarized in the following proposition.

Proposition 6.3.5 *Under the assumptions (6.0.1) and (6.0.2) the following inequalities hold:*

$$\begin{aligned}
&\int_{V_{(u,\underline{u})}} \tau_+^6 \alpha(\hat{\mathcal{L}}_O \hat{\mathcal{L}}_T W)(\hat{\mathcal{L}}_O {}^{(T)}p_3)\alpha(W) \leq c\epsilon_0 \mathcal{Q}_{\mathcal{K}}^{\frac{1}{2}}\\
&\int_{V_{(u,\underline{u})}} \tau_+^6 \alpha(\hat{\mathcal{L}}_O \hat{\mathcal{L}}_T W)(\hat{\mathcal{L}}_O {}^{(T)}\not{p})\beta(W) \leq c\epsilon_0 \mathcal{Q}_{\mathcal{K}}^{\frac{1}{2}} \qquad (6.3.26)\\
&\int_{V_{(u,\underline{u})}} \tau_+^6 \alpha(\hat{\mathcal{L}}_O \hat{\mathcal{L}}_T W)(\hat{\mathcal{L}}_O {}^{(T)}p_4)(\rho,\sigma)(W) \leq c\epsilon_0 \mathcal{Q}_{\mathcal{K}}^{\frac{1}{2}}.
\end{aligned}$$

Proof: The first term is the more delicate one. It has to be treated like the term $\int_{V_{(u,\underline{u})}} \tau_+^6 \alpha(\hat{\mathcal{L}}_T W){}^{(T)}p_3\alpha(W)$ (see (6.2.9)) using, in this case, the following bound of Proposition 6.1.6,

$$\left(\sup_{u\in\mathcal{K}} \int_{\underline{u}_0}^{\underline{u}} \frac{1}{\tau_+(u,\underline{u}')} |r\tau_-^2 \not{\nabla}{}^{(T)}p_3|^2_{p=2,S}(u,\underline{u}')d\underline{u}'\right)^{\frac{1}{2}} \leq c\epsilon_0\,.$$

The second and third integrals in (6.3.26) are treated in the same way, but in this case, it is enough to use only the estimates of Proposition 6.1.5.[26]

6.3.6 *Estimate of* $\int_{V_{(u,\underline{u})}} Q(\hat{\mathcal{L}}_O \hat{\mathcal{L}}_T W)_{\alpha\beta\gamma\delta}({}^{(\bar{K})}\pi^{\alpha\beta}\bar{K}^\gamma \bar{K}^\delta)$

Proposition 6.3.6 *Under the assumptions (6.0.1) and (6.0.2) the following inequality holds:*

$$\int_{V_{(u,\underline{u})}} |Q(\hat{\mathcal{L}}_O \hat{\mathcal{L}}_T W)_{\alpha\beta\gamma\delta}({}^{(\bar{K})}\pi^{\alpha\beta}\bar{K}^\gamma \bar{K}^\delta)| \leq c\epsilon_0 \mathcal{Q}_{\mathcal{K}}. \qquad (6.3.27)$$

Proof: The estimate of this term proceeds exactly like the estimate of $\int_{V_{(u,\underline{u})}} |Q(\hat{\mathcal{L}}_T W)_{\alpha\beta\gamma\delta}({}^{(\bar{K})}\pi^{\alpha\beta}\bar{K}^\gamma \bar{K}^\delta)|$. The final result is the same with the obvious substitutions of the factors $\mathcal{Q}_1^{\frac{1}{2}}$ and $\underline{\mathcal{Q}}_1^{\frac{1}{2}}$ with $\mathcal{Q}_2^{\frac{1}{2}}$ and $\underline{\mathcal{Q}}_2^{\frac{1}{2}}$.

6.3.7 *Estimate of* $\int_{V_{(u,\underline{u})}} \mathbf{Div} Q(\hat{\mathcal{L}}_S \hat{\mathcal{L}}_T W)_{\beta\gamma\delta}(\bar{K}^\beta \bar{K}^\gamma \bar{K}^\delta)$

We recall the equation[27] (see [Ch-Kl], Propositions 7.1.1, 7.1.2)

$$\mathbf{Div} Q(\hat{\mathcal{L}}_S \hat{\mathcal{L}}_T W)_{\beta\gamma\delta} = (\hat{\mathcal{L}}_S \hat{\mathcal{L}}_T W)_{\beta\ \delta}^{\ \mu\ \nu} J(T,S;W)_{\mu\gamma\nu} + (\hat{\mathcal{L}}_S \hat{\mathcal{L}}_T W)_{\beta\ \gamma}^{\ \mu\ \nu} J(T,S;W)_{\mu\delta\nu}$$

---

[26]This is due to the better asymptotic behavior of ${}^{(T)}p_4$, ${}^{(T)}\not{p}$ with respect to ${}^{(T)}p_3$; see the remark after Proposition 6.1.5.

[27]The estimate of this term is similar to that of $\int_{V_{(u,\underline{u})}} |\mathbf{Div} Q(\hat{\mathcal{L}}_O \hat{\mathcal{L}}_T W)_{\beta\gamma\delta}(\bar{K}^\beta \bar{K}^\gamma \bar{K}^\delta)|$.

$$+{}^*(\hat{\mathcal{L}}_S\hat{\mathcal{L}}_T W)_{\beta\ \delta}^{\ \mu\ \nu} J(T,S;W)^*{}_{\mu\gamma\nu} + {}^*(\hat{\mathcal{L}}_S\hat{\mathcal{L}}_T W)_{\beta\ \gamma}^{\ \mu\ \nu} J(T,S;W)^*{}_{\mu\delta\nu},$$

where

$$J(T,S;W) = J^0(T,S;W) + \frac{1}{2}\left(J^1(T,S;W) + J^2(T,S;W) + J^3(T,S;W)\right)$$

and

$$J^0(T,S;W) = \hat{\mathcal{L}}_S J(T;W)\,,\ \ J^1(T,S;W) = J^1(S;\hat{\mathcal{L}}_T W)$$
$$J^2(T,S;W) = J^2(S;\hat{\mathcal{L}}_T W)\,,\ \ J^3(T,S;W) = J^3(S;\hat{\mathcal{L}}_T W).$$

Proceeding as in Subsection 6.3.5 we analyze the integrals

$$\int_{V_{(u,\underline{u})}} \tau_+^6 D(T,S;W)_{444}\,,\ \int_{V_{(u,\underline{u})}} \tau_+^4\tau_-^2 D(T,S;W)_{344}$$
$$\int_{V_{(u,\underline{u})}} \tau_+^2\tau_-^4 D(T,S;W)_{334}\,,\ \int_{V_{(u,\underline{u})}} \tau_-^6 D(T,S;W)_{333}.$$

We examine the first of these integrals since the others are controlled in the same way and are, in fact, even easier. We recall (see (6.1.3)) the equation

$$D(T,S;W)_{444} = 4\alpha(\hat{\mathcal{L}}_S\hat{\mathcal{L}}_T W)\cdot\Theta(T,S;W) - 8\beta(\hat{\mathcal{L}}_S\hat{\mathcal{L}}_T W)\cdot\Xi(T,S;W), \quad (6.3.28)$$

where

$$\Theta(T,S;W) \ = \ \Theta^0(T,S;W) + \frac{1}{2}\left(\Theta^1(T,S;W) + \Theta^2(T,S;W) + \Theta^3(T,S;W)\right)$$

and the indices 0, 1, 2, 3 refer to the various parts of the current $J(T,S;W)$.[28] Therefore, for $i\in\{1,2,3\}$, $\Theta^i(T,S;W) = \Theta^i(S,\hat{\mathcal{L}}_T W)$. We consider first the terms with $i = 1,2,3$ and in particular the first term of (6.3.28)

$$\int_{V_{(u,\underline{u})}} \tau_+^6\alpha(\hat{\mathcal{L}}_S\hat{\mathcal{L}}_T W)\cdot\Theta^i(T,S;W)\,.$$

Proceeding as in the case of $\int_{V_{(u,\underline{u})}} \tau_+^6\alpha(\hat{\mathcal{L}}_T W)\cdot\Theta^i(T,W)$ (see (6.2.6)) we have

$$\int_{V_{(u,\underline{u})}} \tau_+^6\alpha(\hat{\mathcal{L}}_S\hat{\mathcal{L}}_T W)\cdot\Theta^i(S,\hat{\mathcal{L}}_T W)$$
$$\leq c\int_{u_0}^{u} du'\left(\int_{C(u';[\underline{u}_0,\underline{u}])} \underline{u}'^6|\alpha(\hat{\mathcal{L}}_S\hat{\mathcal{L}}_T W)|^2\right)^{\frac{1}{2}}\left(\int_{C(u';[\underline{u}_0,\underline{u}])} \underline{u}'^6|\Theta^i(S,\hat{\mathcal{L}}_T W)|^2\right)^{\frac{1}{2}}.$$

The first integral factor is estimated by (see (3.5.1))

$$\sup_{\mathcal{K}}\left(\int_{C(u';[\underline{u}_0,\underline{u}])} \underline{u}'^6|\alpha(\hat{\mathcal{L}}_S\hat{\mathcal{L}}_T W)|^2\right)^{\frac{1}{2}} \leq c\mathcal{Q}_2^{\frac{1}{2}}. \quad (6.3.29)$$

[28] The corresponding expressions hold for $\Xi(T,S;W)$.

To estimate

$$\int_{u_0}^{u} du' \left(\int_{C(u';[\underline{u}_0,\underline{u}])} \underline{u}'^6 |\Theta^i(S, \hat{\mathcal{L}}_T W)|^2 \right)^{\frac{1}{2}},$$

we compare it with the integral

$$\int_{u_0}^{u} du' \left(\int_{C(u';[\underline{u}_0,\underline{u}])} \underline{u}'^6 |\Theta^i(T, W)|^2 \right)^{\frac{1}{2}},$$

which has been estimated in Subsection 6.2.1.[29] Proceeding in the same way we obtain

$$\int_{V_{(u,\underline{u})}} \tau_+^6 \alpha(\hat{\mathcal{L}}_S \hat{\mathcal{L}}_T W) \cdot \Theta^1(T, S; W) \le c\epsilon_0 \mathcal{Q}_2^{\frac{1}{2}} (\mathcal{Q}_1 + \mathcal{Q}_2)^{\frac{1}{2}} \le c\epsilon_0 \mathcal{Q}_{\mathcal{K}}.$$

Operating in an analogous way for the second term we derive

$$\int_{V_{(u,\underline{u})}} \tau_+^6 \beta(\hat{\mathcal{L}}_S \hat{\mathcal{L}}_T W) \cdot \Xi^1(S, T; W) \le c\epsilon_0 \mathcal{Q}_2^{\frac{1}{2}} (\mathcal{Q}_1 + \mathcal{Q}_2)^{\frac{1}{2}} \le c\epsilon_0 \mathcal{Q}_{\mathcal{K}}.$$

In the case of the currents $J^2(T, S; W)$ and $J^3(T, S; W)$, we have to proceed somewhat differently than in the corresponding proof of Subsection 6.2.1. This is due to the fact that we have to deal with the null components of $\hat{\mathcal{L}}_T W$ which cannot be estimated in the sup norm. We shall sketch here the estimates for $\Theta^{(2)}(S, T; W)$; those for $\Theta^{(3)}(S, T; W)$ are obtained in the same way. From

$$\begin{aligned} \Theta^{(2)}(S, T; W) &= \mathrm{Qr}\left[{}^{(S)}p_3\, ;\, \alpha(\hat{\mathcal{L}}_T W)\right] + \mathrm{Qr}\left[{}^{(S)}\not{p};\, \beta(\hat{\mathcal{L}}_T W)\right] \\ &+ \mathrm{Qr}\left[{}^{(S)}p_4\, ;\, (\rho, \sigma)(\hat{\mathcal{L}}_T W)\right], \end{aligned}$$

the following integrals have to be estimated:

$$\begin{aligned} &\int_{V_{(u,\underline{u})}} \tau_+^6 \alpha(\hat{\mathcal{L}}_S \hat{\mathcal{L}}_T W)({}^{(S)}p_3)\alpha(\hat{\mathcal{L}}_T W) \\ &\int_{V_{(u,\underline{u})}} \tau_+^6 \alpha(\hat{\mathcal{L}}_S \hat{\mathcal{L}}_T W)({}^{(S)}\not{p})\beta(\hat{\mathcal{L}}_T W) \qquad (6.3.30) \\ &\int_{V_{(u,\underline{u})}} \tau_+^6 \alpha(\hat{\mathcal{L}}_S \hat{\mathcal{L}}_T W)({}^{(S)}p_4)(\rho, \sigma)(\hat{\mathcal{L}}_T W). \end{aligned}$$

---

[29]The main difference is that now the deformation tensors and their derivatives refer to the vector field S instead of T and the Riemann null components and their derivatives are relative to $\hat{\mathcal{L}}_T W$ instead of W. Because the estimates for the deformation tensor relative to S : ${}^{(S)}\hat{\pi}$ are worse than those for the deformation tensor ${}^{(T)}\hat{\pi}$, by a factor r or τ_- and, at the same time, the estimates for the null Riemann components of $\hat{\mathcal{L}}_T W$ are better than those relative to the null components of W by a factor r, we easily conclude that the the term $\int_{V_{(u,\underline{u})}} \tau_+^6 \alpha(\hat{\mathcal{L}}_S \hat{\mathcal{L}}_T W) \cdot \Theta^1(T, S; W)$ is under control.

To estimate the first integral of (6.3.30) we write

$$\int_{V_{(u,\underline{u})}} \tau_+^6 \alpha(\hat{\mathcal{L}}_S\hat{\mathcal{L}}_T W)^{(S)}p_3\,\alpha(\hat{\mathcal{L}}_T W)$$

$$\leq \int \frac{du'}{u'}\left(\int_{C(u';[\underline{u}_0,\underline{u}])} \tau_+^6|\alpha(\hat{\mathcal{L}}_S\hat{\mathcal{L}}_T W)|^2\right)^{\frac{1}{2}}\left(\int_{C(u';[\underline{u}_0,\underline{u}])} |\tau_-{}^{(S)}p_3|^2\tau_+^6|\alpha(\hat{\mathcal{L}}_T W)|^2\right)^{\frac{1}{2}}$$

$$\leq \left(\sup_{\mathcal{K}}\int_{C(u';[\underline{u}_0,\underline{u}])} Q(\hat{\mathcal{L}}_S\hat{\mathcal{L}}_T W)(\bar{K},\bar{K},\bar{K},e_4)\right)^{\frac{1}{2}} \cdot \tag{6.3.31}$$

$$\cdot\int\frac{du'}{u'}\left[\int_{\underline{u}_0}^{\underline{u}} d\underline{u}'\left(\int_{S(u',\underline{u}')}|\tau_+^{\frac{1}{2}}\tau_-{}^{(S)}p_3|^4\right)^{\frac{1}{2}}\left(\int_{S(u',\underline{u}')}\tau_+^{10}|\alpha(\hat{\mathcal{L}}_T W)|^4\right)^{\frac{1}{2}}\right]^{\frac{1}{2}}$$

$$\leq c\mathcal{Q}_2^{\frac{1}{2}}\int\frac{du'}{u'}\left[\int_{\underline{u}_0}^{\underline{u}} d\underline{u}'|r^{1-\frac{2}{p}}\tau_-{}^{(S)}p_3|^2_{p=4,S}\left(\int_{S(u',\underline{u}')}\tau_+^{10}|\alpha(\hat{\mathcal{L}}_T W)|^4\right)^{\frac{1}{2}}\right]^{\frac{1}{2}}.$$

Using Proposition 6.1.7, (see (6.1.49)) and the estimate proved in Chapter 5,

$$\left(\int_{S(u',\underline{u}')}\tau_+^{16}|\alpha(\hat{\mathcal{L}}_T W)|^4\right)^{\frac{1}{4}} \tag{6.3.32}$$

$$\leq c\left(\int_{C(u')} r^6|\alpha(\hat{\mathcal{L}}_T W)|^2 + r^8|\nabla\!\!\!/\alpha(\hat{\mathcal{L}}_T W)|^2 r^8|\mathbf{D}\!\!\!\!/_4\alpha(\hat{\mathcal{L}}_T W)|^2\right)^{\frac{1}{2}} \leq c\mathcal{Q}_{\mathcal{K}}^{\frac{1}{2}},$$

we have

$$\int_{V_{(u,\underline{u})}} \tau_+^6\alpha(\hat{\mathcal{L}}_S\hat{\mathcal{L}}_T W)({}^{(S)}p_3)\alpha(\hat{\mathcal{L}}_T W)$$

$$\leq c\left(\sup|r^{1-\frac{2}{p}}\tau_-{}^{(S)}p_3|_{p=4,S}\right)\mathcal{Q}_2^{\frac{1}{2}}\mathcal{Q}_{\mathcal{K}}^{\frac{1}{2}}\int\frac{du'}{u'}\left(\int_{\underline{u}_0}^{\underline{u}} d\underline{u}'\frac{1}{\underline{u}'^3}\right)^{\frac{1}{2}}$$

$$\leq c\left(\sup|r^{1-\frac{2}{p}}\tau_-{}^{(S)}p_3|_{p=4,S}\right)\mathcal{Q}_{\mathcal{K}} \leq c\epsilon_0\mathcal{Q}_{\mathcal{K}}, \tag{6.3.33}$$

which completes the estimate. The remaining integrals in (6.3.30) are estimated in the same way. We are left with the estimate of the part associated with $J^0(T,S;W) = \hat{\mathcal{L}}_S J(T;W)$. From

$$\Theta^0(T,S;W) = \frac{1}{2}\left(\hat{\mathcal{L}}_S\Theta^1(T;W) + \hat{\mathcal{L}}_S\Theta^2(T;W) + \hat{\mathcal{L}}_S\Theta^3(T;W)\right),$$

we have to control the integrals

$$\int_{V_{(u,\underline{u})}} \tau_+^6\alpha(\hat{\mathcal{L}}_S\hat{\mathcal{L}}_T W)\hat{\mathcal{L}}_S\Theta^1(T;W)$$

$$\int_{V_{(u,\underline{u})}} \tau_+^6\alpha(\hat{\mathcal{L}}_S\hat{\mathcal{L}}_T W)\hat{\mathcal{L}}_S\Theta^2(T;W) \tag{6.3.34}$$

$$\int_{V_{(u,\underline{u})}} \tau_+^6\alpha(\hat{\mathcal{L}}_S\hat{\mathcal{L}}_T W)\hat{\mathcal{L}}_S\Theta^3(T;W).$$

We write, using the estimates proved in Chapter 6,

$$\begin{aligned}
&\int_{V_{(u,\underline{u})}} \tau_+^6 \alpha(\hat{\mathcal{L}}_S \hat{\mathcal{L}}_T W) \hat{\mathcal{L}}_S \Theta^i(T;W) \\
&\leq \int du' \left(\int_{C(u';[\underline{u}_0,\underline{u}])} \tau_+^6 |\alpha(\hat{\mathcal{L}}_S \hat{\mathcal{L}}_T W)|^2 \right)^{\frac{1}{2}} \left(\int_{C(u';[\underline{u}_0,\underline{u}])} \tau_+^6 |\hat{\mathcal{L}}_S \Theta^i(T;W)|^2 \right)^{\frac{1}{2}} \\
&\leq c \mathcal{Q}_2^{\frac{1}{2}} \int du' \left(\int_{C(u';[\underline{u}_0,\underline{u}])} \tau_+^6 |\hat{\mathcal{L}}_S \Theta^i(T;W)|^2 \right)^{\frac{1}{2}} . \qquad (6.3.35)
\end{aligned}$$

Recall that $\Theta^i(T;W)$ consists of a sum of terms that are products of a deformation tensor, or its first derivative, and a null Riemann component, or its first derivative; see (6.1.8),...,(6.1.23). Let us first consider

$$\begin{aligned}
\Theta(J^1) &= \text{Qr}\left[{}^{(T)}\underline{\mathbf{m}} ; \nabla\!\!\!/ \alpha\right] + \text{Qr}\left[{}^{(T)}\underline{\mathbf{n}} ; \alpha_4\right] + \text{Qr}\left[{}^{(T)}\mathbf{j} ; \alpha_3\right] \\
&+ \text{Qr}\left[{}^{(T)}\mathbf{i} ; \nabla\!\!\!/ \beta\right] + \text{Qr}\left[{}^{(T)}\underline{\mathbf{m}} ; \beta_4\right] + \text{Qr}\left[{}^{(T)}\mathbf{m} ; \beta_3\right] \\
&+ \cdots .
\end{aligned}$$

The Lie derivative $\hat{\mathcal{L}}_S$ can operate on both factors. When it operates on the null Riemann components it adds a derivative.[30] Therefore in $\hat{\mathcal{L}}_S \Theta(J^1)$ the null Riemann components can be twice differentiated, but multiplied by undifferentiated deformation tensors. In this case the estimates are made by taking the sup norms for the deformation tensors and the $L^2(C)$ or $L^2(\underline{C})$ norms for the twice differentiated Riemann components.

In the case of

$$\Theta(J^2) = \text{Qr}\left[{}^{(X)}p_3 ; \alpha\right] + \text{Qr}\left[{}^{(X)}\not{p} ; \beta\right] + \text{Qr}\left[{}^{(X)}p_4 ; (\rho,\sigma)\right],$$

when $\hat{\mathcal{L}}_S$ operates on the Riemann components, we estimate the integrals by taking the $L^p(S)$ norms both for the first derivatives of the Riemann components and for the ${}^{(T)}p_3$, ${}^{(T)}p_3$, ${}^{(T)}\not{p}$ terms which depend on the first derivatives of the deformation tensors. A similar procedure applies in the case of $\Theta(J^3)$. Therefore the estimate of this error term, when $\hat{\mathcal{L}}_S$ operates on the Riemann components, does not produce complications since the asymptotic behavior is not changed and the derivatives involved are only the zero and first derivatives of the deformation tensors and at most the first derivatives for the Riemann components.

We now consider the case in which $\hat{\mathcal{L}}_S$ operates on the deformation tensors. This case is simple for the $\Theta(J^1)$ part since the only change is that the deformation tensor is replaced by its first derivatives which depend on the first derivatives of the connection coefficients, we already know how to control.

When $\hat{\mathcal{L}}_S$ operates on the deformation tensor parts of $\Theta(J^2)$ and $\Theta(J^3)$ the situation is more delicate. In this case the Riemann tensor components are not differentiated and can be estimated with the sup norms, but when $\hat{\mathcal{L}}_S$ operates on ${}^{(T)}p_3$, ${}^{(T)}p_3$, ${}^{(T)}\not{p}$ terms

[30] Without changing the asymptotic behavior of the component.

that depend on the second derivatives of the connection coefficients, this requires careful control of their norms. The same happens when $\hat{\mathcal{L}}_S$ operates on the ${}^{(T)}q$ factors of $\Theta(J^3)$.

Let us consider, as an example, the integral of (6.3.35) for $i = 2$. This amounts to controlling the terms

$$\begin{aligned}
&\int du'\left(\int_{C(u';[\underline{u}_0,\underline{u}])}\tau_+^6|\hat{\mathcal{L}}_S{}^{(T)}p_3|^2|\alpha(W)|^2\right)^{\frac{1}{2}}\\
&\int du'\left(\int_{C(u';[\underline{u}_0,\underline{u}])}\tau_+^6|\hat{\mathcal{L}}_S{}^{(T)}\not{p}|^2|\beta(W)|^2\right)^{\frac{1}{2}}\\
&\int du'\left(\int_{C(u';[\underline{u}_0,\underline{u}])}\tau_+^6|\hat{\mathcal{L}}_S{}^{(T)}p_4|^2|(\rho,\sigma)(W)|^2\right)^{\frac{1}{2}}.
\end{aligned}\tag{6.3.36}$$

To estimate the first term we write

$$\begin{aligned}
&\int du'\left(\int_{C(u';[\underline{u}_0,\underline{u}])}\tau_+^6|\hat{\mathcal{L}}_S{}^{(T)}p_3|^2|\alpha(W)|^2\right)^{\frac{1}{2}}\\
&\leq c\left(\sup|\tau_+^{\frac{7}{2}}\alpha(W)|\right)\int du'\frac{1}{\tau_-^2}\left(\int_{C(u';[\underline{u}_0,\underline{u}])}\frac{1}{\tau_+}|\tau_-^2\hat{\mathcal{L}}_S{}^{(T)}p_3|^2\right)^{\frac{1}{2}}\\
&\leq c\epsilon_0\mathcal{Q}_{\mathcal{K}}^{\frac{1}{2}},
\end{aligned}\tag{6.3.37}$$

where the last inequality was derived using Proposition 6.1.6. The estimates of the remaining two integrals are easier due to the better asymptotic behavior of ${}^{(T)}\not{p}$ and ${}^{(T)}p_4$ compared to ${}^{(T)}p_3$. In this case it is enough to use Proposition 6.1.5. In fact we have, for the last integral,

$$\begin{aligned}
&\int du'\left(\int_{C(u';[\underline{u}_0,\underline{u}])}\tau_+^6|\hat{\mathcal{L}}_S{}^{(T)}p_4|^2|(\rho(W),\sigma(W))|^2\right)^{\frac{1}{2}}\\
&\leq c\left(\sup|\tau_+^3(\rho(W),\sigma(W))|\right)\int du'\frac{1}{\tau_-}\left(\int_{C(u';[\underline{u}_0,\underline{u}])}\frac{1}{\tau_+^2}|\tau_+\tau_-\hat{\mathcal{L}}_S{}^{(T)}p_4|^2\right)^{\frac{1}{2}}\\
&\leq c\epsilon_0\mathcal{Q}_{\mathcal{K}}^{\frac{1}{2}}\int du'\frac{1}{\tau_-}\left(\int d\underline{u}'\frac{1}{\tau_+^2}\right)^{\frac{1}{2}}\leq c\epsilon_0\mathcal{Q}_{\mathcal{K}}^{\frac{1}{2}}.
\end{aligned}\tag{6.3.38}$$

6.3.8 Estimate of $\int_{V_{(u,\underline{u})}}Q(\hat{\mathcal{L}}_S\hat{\mathcal{L}}_T W)_{\alpha\beta\gamma\delta}({}^{(\bar{K})}\pi^{\alpha\beta}\bar{K}^\gamma\bar{K}^\delta)$

The estimate of this integral is made exactly in the same way as the estimate of $\int_{V_{(u,\underline{u})}}|Q(\hat{\mathcal{L}}_T W)_{\alpha\beta\gamma\delta}({}^{(\bar{K})}\pi^{\alpha\beta}\bar{K}^\gamma\bar{K}^\delta)|$ with the obvious substitution of $\mathcal{Q}_1$ with $\mathcal{Q}_2$.

6.4 Appendix

6.4.1 The third-order derivatives of the connection coefficients

The third-order derivatives of the connection coefficients From the evolution equations, in Section 4.6, of ${}^{(i)}F$, ${}^{(i)}H_{ab}$, ${}^{(i)}Z_a$ and from the explicit expressions of the rotation deformation tensors (see (4.6.10) and (4.6.11)) it follows that ${}^{(O)}\underline{\mathbf{m}}_a$ depends on $\nabla\eta$ and $\nabla\underline{\eta}$, ${}^{(O)}\mathbf{i}_{ab}$ depends on $\nabla\chi$ and ${}^{(O)}\mathbf{j}$ depends on $\nabla\log\Omega$. From the expression (see (6.1.24))

$$\begin{aligned} {}^{(O)}p_3 &= \mathrm{div}\,{}^{(O)}\underline{\mathbf{m}} - \frac{1}{2}\mathbf{D}_3{}^{(O)}\mathbf{j} + (2\underline{\eta} + \eta - \zeta)\cdot{}^{(O)}\underline{\mathbf{m}} - \hat{\chi}\cdot{}^{(O)}\mathbf{i} \\ &\quad -\frac{1}{2}\mathrm{tr}\chi\,(\mathrm{tr}{}^{(O)}\mathbf{i} + {}^{(O)}\mathbf{j}), \end{aligned}$$

it follows that ${}^{(O)}p_3$ depends on the second derivatives of the connection coefficients $\nabla^2\eta$ and $\nabla^2\underline{\eta}$ through $\mathrm{div}\,{}^{(O)}\underline{\mathbf{m}}$,[31] which in turn depend on the first derivatives of the Riemann tensor.[32] Therefore it follows immediately that $\hat{\mathcal{L}}_O{}^{(O)}p_3$ depends on $\nabla^3\eta$, $\nabla^3\underline{\eta}$ and through them, on the second derivatives of the Riemann tensor. From

$$\begin{aligned} {}^{(O)}\not{p} &= \nabla_c{}^{(O)}\mathbf{i} - \frac{1}{2}\mathbf{D}_4{}^{(O)}\underline{\mathbf{m}} - \frac{1}{2}(\mathbf{D}_4\log\Omega){}^{(O)}\underline{\mathbf{m}} + \frac{1}{2}{}^{(O)}\mathbf{j}(\eta + \underline{\eta}) \\ &\quad + {}^{(O)}\mathbf{i}\cdot(\eta + \underline{\eta}) - \frac{3}{4}\mathrm{tr}\chi\,{}^{(O)}\underline{\mathbf{m}} - \frac{1}{2}\hat{\chi}\cdot{}^{(O)}\underline{\mathbf{m}}, \end{aligned}$$

we see that ${}^{(O)}\not{p}$ depends, through the term $\nabla_c{}^{(O)}\mathbf{i}$, on $\nabla^2\chi$.[33] This implies that $\hat{\mathcal{L}}_O{}^{(O)}\not{p}$ depends on the third derivative $\nabla^3\chi$, and, through it, on the second derivatives of the Riemann tensor. From

$${}^{(O)}p_4 = -\frac{1}{2}\mathbf{D}_4{}^{(O)}\mathbf{j} - \hat{\underline{\chi}}\cdot{}^{(O)}\mathbf{i} - \frac{1}{2}\mathrm{tr}\underline{\chi}\,(\mathrm{tr}{}^{(O)}\mathbf{i} + {}^{(O)}\mathbf{j}) - \frac{1}{2}\mathrm{tr}\chi\,{}^{(O)}\underline{\mathbf{n}} - (\mathbf{D}_4\log\Omega){}^{(O)}\underline{\mathbf{n}},$$

we see that ${}^{(O)}p_4$ does not depend on second-order derivatives of connection coefficients. An analogous argument holds for the derivatives of the various components of ${}^{(O)}q$. From the explicit expressions (6.1.27) and (6.1.28), in the case $X = O$, we obtain

1. $\Lambda({}^{(O)}q) = -\frac{1}{4}\mathbf{D}_4{}^{(O)}\mathbf{j} + \frac{2}{3}{}^{(O)}p_4$ does not depend on second derivatives of connection coefficients;

2. $K({}^{(O)}q)_{ab} = -\frac{1}{2}\left(\hat{\chi}_{ac}{}^{(O)}\mathbf{i}_{cb} - \hat{\chi}_{bc}{}^{(O)}\mathbf{i}_{ca}\right)$ does not depend on second derivatives of connection coefficients;

---

[31] Because ∇Z also depends on ∇H, it follows that through the dependence on $\mathrm{div}\,{}^{(O)}\underline{\mathbf{m}}$ we have also a dependence on $\nabla^2\chi$ and finally $\hat{\mathcal{L}}_O{}^{(O)}p_3$ depends on $\nabla^3(\eta, \underline{\eta}, \chi)$.

[32] In fact through $\mathbf{D}_3{}^{(O)}\mathbf{j}$ there is also a dependence on $\nabla\underline{\omega}$. This is, nevertheless, not harmful since the tangential derivatives of ${}^{(O)}p_3$, $\nabla^3\eta$ and $\nabla^3\underline{\eta}$ depend on the second derivatives of the Riemann tensor, whereas $\nabla^2\underline{\omega}$ depends only on the first derivative, as follows from Proposition 4.4.1.

[33] It seems that, through the term $\mathbf{D}_4{}^{(O)}\underline{\mathbf{m}}$, ${}^{(O)}\not{p}$ also depends on $\nabla^2\omega$. This is not true because $\mathbf{D}_4{}^{(O)}\underline{\mathbf{m}}$ is proportional to $\frac{d}{d\underline{u}}(\Omega{}^{(i)}Z_a)$ which (see Proposition 4.6.2) is proportional to $\nabla_a{}^{(i)}F$, ${}^{(i)}H_{ab}$ and $L_{{}^{(i)}O}(\zeta + \eta)$, and therefore depends on $\nabla\eta$, $\nabla\underline{\eta}$ and $\nabla\chi$, all first derivatives of the connection coefficients.

3. $I({}^{(O)}q)_a = \frac{1}{2}\mathbf{D}\!\!\!/_4{}^{(O)}\underline{\mathbf{m}}_a - \frac{1}{2}\nabla\!\!\!\!/_a{}^{(O)}\mathbf{j} - \frac{1}{2}(\mathbf{D}_4 \log\Omega)^{(O)}\underline{\mathbf{m}}_a + \frac{1}{4}\mathrm{tr}\chi{}^{(O)}\underline{\mathbf{m}}_a$

$$+\frac{1}{2}\hat{\chi}_{ac}{}^{(O)}\underline{\mathbf{m}}_c - \frac{1}{2}\underline{\eta}_c{}^{(O)}\mathbf{i}_{ca} + \frac{3}{2}{}^{(O)}\not{p}_a$$

depends, through ${}^{(O)}\not{p}$, on the second derivatives of Ricci coefficient $\nabla\!\!\!\!/^2\chi$;

4. $\Theta({}^{(O)}q)_{ab} = 2\left(\mathbf{D}\!\!\!/_4{}^{(O)}\mathbf{i}_{ab} - \frac{1}{2}\delta_{ab}\mathrm{tr}(\mathbf{D}\!\!\!/_4{}^{(O)}\mathbf{i})\right) + \mathrm{tr}\chi{}^{(\hat{O})}\mathbf{i}_{ab} + \hat{\chi}_{ab}\mathrm{tr}^{(O)}\mathbf{i} + \hat{\chi}_{ab}{}^{(O)}\mathbf{j}$

does not depend on the second derivatives of the connection coefficients;[34].

5. $\underline{\Lambda}({}^{(O)}q) = -\underline{\eta}\cdot{}^{(O)}\underline{\mathbf{m}} - \frac{1}{4}\left(\mathbf{D}_3{}^{(O)}\mathbf{j} - 2\eta\cdot{}^{(O)}\underline{\mathbf{m}}\right) + \frac{2}{3}{}^{(O)}p_3$ depends, through ${}^{(O)}p_3$, on $\nabla\!\!\!\!/^2\eta$ and $\nabla\!\!\!\!/^2\underline{\eta}$;[35]

6. $\underline{K}({}^{(O)}q)_{ab} = \frac{1}{2}\left(\nabla\!\!\!\!/_a{}^{(O)}\underline{\mathbf{m}}_b - \nabla\!\!\!\!/_b{}^{(O)}\underline{\mathbf{m}}_a\right) - \frac{1}{2}\left(\zeta_a{}^{(O)}\underline{\mathbf{m}}_b - \zeta_b{}^{(O)}\underline{\mathbf{m}}_a\right)$
$-\frac{1}{2}\left(\underline{\hat{\chi}}_{ac}{}^{(O)}\mathbf{i}_{cb} - \underline{\hat{\chi}}_{bc}{}^{(O)}\mathbf{i}_{ca}\right)$
depends, through $\nabla\!\!\!\!/_a{}^{(O)}\underline{\mathbf{m}}$, on $\nabla\!\!\!\!/^2\eta$ and $\nabla\!\!\!\!/^2\underline{\eta}$;

7. $\underline{\Xi}({}^{(O)}q)_a = \frac{1}{2}\mathbf{D}\!\!\!/_3{}^{(O)}\underline{\mathbf{m}}_a - \frac{1}{2}(\mathbf{D}_3\log\Omega)^{(O)}\underline{\mathbf{m}}_a + \frac{1}{2}\mathrm{tr}\underline{\chi}{}^{(O)}\underline{\mathbf{m}}_a + \underline{\hat{\chi}}_{ac}{}^{(O)}\underline{\mathbf{m}}_c$ depends, through $\mathbf{D}\!\!\!/_3{}^{(O)}\underline{\mathbf{m}}$, on $\nabla\!\!\!\!/^2\underline{\omega}$;

8. $\underline{I}({}^{(O)}q)_a = -\frac{1}{2}\nabla\!\!\!\!/_a{}^{(X)}\mathbf{j} + \frac{1}{4}\mathrm{tr}\chi{}^{(O)}\underline{\mathbf{m}}_a + \frac{1}{2}\hat{\chi}_{ac}{}^{(O)}\underline{\mathbf{m}}_c - \frac{1}{2}\eta_c{}^{(O)}\mathbf{i}_{ca} + \frac{3}{2}{}^{(O)}\not{p}_a$ depends, through ${}^{(O)}\not{p}$, on $\nabla\!\!\!\!/^2\chi$;

9. $$\underline{\Theta}({}^{(O)}q)_{ab} = 2\left(\mathbf{D}\!\!\!/_3{}^{(O)}\mathbf{i}_{ab} - \frac{1}{2}\delta_{ab}\mathrm{tr}(\mathbf{D}\!\!\!/_3{}^{(O)}\mathbf{i})\right) - \left(\nabla\!\!\!\!/_a{}^{(O)}\underline{\mathbf{m}}_b + \nabla\!\!\!\!/_b{}^{(O)}\underline{\mathbf{m}}_a - \delta_{ab}\nabla\!\!\!\!/_c{}^{(O)}\underline{\mathbf{m}}_c\right)$$
$$-2\left(\eta_a{}^{(O)}\underline{\mathbf{m}}_b + \eta_b{}^{(O)}\underline{\mathbf{m}}_a - \delta_{ab}\eta_c{}^{(O)}\underline{\mathbf{m}}_c\right) + \left(\zeta_a{}^{(O)}\underline{\mathbf{m}}_b + \zeta_b{}^{(O)}\underline{\mathbf{m}}_a - \delta_{ab}\zeta_c{}^{(O)}\underline{\mathbf{m}}_c\right)$$
$$+\mathrm{tr}\underline{\chi}{}^{(O)}\mathbf{i}_{ab} + \underline{\hat{\chi}}_{ab}\mathrm{tr}^{(O)}\mathbf{i} + \underline{\hat{\chi}}_{ab}{}^{(O)}\mathbf{j}$$

depends, through $\nabla\!\!\!\!/^{(O)}\underline{\mathbf{m}}$, on $\nabla\!\!\!\!/^2\eta$ and $\nabla\!\!\!\!/^2\underline{\eta}$.

[34] The term $\mathbf{D}\!\!\!/_4{}^{(O)}\mathbf{i}$ can be expressed, using the structure equations, in terms of first derivatives of connection coefficients.
[35] The term $\mathbf{D}_3{}^{(O)}\mathbf{j}$ gives a dependence on $\nabla\!\!\!\!/\underline{\omega}$ which, as discussed before, is harmless.

7

The Initial Hypersurface and the Last Slice

7.1 Initial hypersurface foliations

7.1.1 Some general properties of a foliation of Σ_0

Let the function $w(p)$ define a foliation on Σ_0. Its leaves are

$$S_0(\nu) = \{p \in \Sigma_0 | w(p) = \nu\}.$$

We define on $\Sigma_0 \backslash K$ a moving orthonormal frame $\{\tilde{N}, e_A\}$[1] adapted to this foliation, where $A \in \{1, 2\}$ and $\tilde{N}^i = \frac{1}{|\partial w|} g^{ij} \partial_j w$ is the unit vector field defined on Σ_0, normal to each $S_0(\nu)$. The metric on Σ_0 can be written, in adapted coordinates $\{w, \phi^a\}$, as

$$g(\cdot, \cdot) = a^2 dw^2 + \gamma_{ab} d\phi^a d\phi^b, \tag{7.1.1}$$

and in these coordinates, $\tilde{N} = \frac{1}{a}\frac{\partial}{\partial w}$, $a^{-2} = |\partial w|^2$.

The second fundamental form associated with the leaves of this foliation is

$$\theta_{ij} = \Pi_i^l \Pi_j^s \nabla_l \tilde{N}_s = \nabla_i \tilde{N}_j - \tilde{N}_i \nabla_{\tilde{N}} \tilde{N}_j,$$

where $\Pi_s^l = (\delta_s^l - \tilde{N}^l \tilde{N}_s)$ is the projection over TS_0 and ∇ the covariant derivative relative to Σ_0. A simple computation shows that, in adapted coordinates,

$$\theta_{ab} = \frac{1}{2a} \partial_w \gamma_{ab} \, .$$

Moreover the adapted moving frame satisfies the following equations, where $\nabla\!\!\!/_A \equiv \nabla\!\!\!/_{e_A}$,

$$\nabla_{\tilde{N}} \tilde{N} = (a^{-1} \nabla\!\!\!/_A a) e_A$$

---

[1] In this chapter unlike in the rest of the book, we use the capital letters $A, B, C, \ldots$ for an orthonormal basis tangent to S, $e_A = e_A^b \frac{\partial}{\partial \phi_b}$. We use the small ones $a, b, c, \ldots$ as coordinate indices.

$$\begin{aligned}
\nabla_A \tilde{N} &= \theta_{AB} e_B \\
\nabla_{\tilde{N}} e_A &= \nabla\!\!\!/_{\tilde{N}} e_A + (a^{-1} \nabla\!\!\!/_A a) \tilde{N} \\
\nabla_B e_A &= \nabla\!\!\!/_B e_A - \theta_{AB} \tilde{N}.
\end{aligned} \tag{7.1.2}$$

7.1.2 The structure equations on Σ_0

We start by writing the Gauss and the Codazzi–Mainardi equations[2]

$$\begin{aligned}
{}^{(3)}R_{abcd} &= {}^{(2)}R_{abcd} - \theta_{ac}\theta_{bd} + \theta_{ad}\theta_{bc} \\
{}^{(3)}R_{wabc} &= -a \left(\nabla\!\!\!/_b \theta_{ca} - \nabla\!\!\!/_c \theta_{ab} \right).
\end{aligned} \tag{7.1.3}$$

Contracting the Gauss equation with respect to the indices b, d, we obtain[3] the *Contracted Gauss equation*

$${}^{(3)}R_{ac} = {}^{(2)}R_{ac} + {}^{(3)}R^{w}{}_{awc} - \mathrm{tr}\theta\theta_{ac} + \theta_{ad}\theta^{d}{}_{c}, \tag{7.1.4}$$

where ${}^{(3)}R_{ac}$ is the Ricci tensor of Σ_0 and ${}^{(2)}R_{ac}$ is that associated with $S_0(\nu)$. The explicit computation of ${}^{(3)}R^{w}{}_{awc}$ gives

$${}^{(3)}R^{w}{}_{awc} = -a^{-1} \nabla\!\!\!/_a \nabla\!\!\!/_c a - \nabla\!\!\!/_{\tilde{N}} \theta_{ac} - \theta_{ad}\theta^{d}{}_{c}, \tag{7.1.5}$$

which, substituted in (7.1.4), gives

$${}^{(3)}R_{ac} = {}^{(2)}R_{ac} - a^{-1} \nabla\!\!\!/_a \nabla\!\!\!/_c a - \nabla\!\!\!/_{\tilde{N}} \theta_{ac} - \mathrm{tr}\theta\theta_{ac}. \tag{7.1.6}$$

Contracting (7.1.5) with respect to the indices a, c, we obtain

$${}^{(3)}R_{\tilde{N}\tilde{N}} = -a^{-1} \Delta\!\!\!/ a - \nabla\!\!\!/_{\tilde{N}} \mathrm{tr}\theta - |\theta|^2. \tag{7.1.7}$$

Moreover contracting (7.1.4) with respect to the indices a, c, we obtain

$${}^{(3)}R - 2\,{}^{(3)}R_{\tilde{N}\tilde{N}} = {}^{(2)}R - (\mathrm{tr}\theta)^2 + |\theta|^2, \tag{7.1.8}$$

where ${}^{(3)}R$ is the scalar curvature of Σ_0 and ${}^{(2)}R$ is that of S_0. Finally contracting the Codazzi–Mainardi equation with respect to the a, c indices, we obtain the *Contracted Codazzi-Mainardi equations*

$${}^{(3)}R_{\tilde{N}b} = -\nabla\!\!\!/_b \mathrm{tr}\theta + \nabla\!\!\!/^c \theta_{cb}. \tag{7.1.9}$$

[2] Here the index w denotes the corresponding coordinate w and a, b are associated with ϕ^a, ϕ^b. In arbitrary coordinates equations (7.1.3) become

$$\begin{aligned}
{}^{(3)}R_{rspq} \Pi^r_i \Pi^s_j \Pi^p_k \Pi^q_l &= {}^{(2)}R_{ijkl} - \left(\theta_{ik}\theta_{jl} - \theta_{il}\theta_{jk} \right) \\
{}^{(3)}R_{srpq} \tilde{N}^s \Pi^r_k \Pi^p_i \Pi^q_j &= -\left(\nabla\!\!\!/_i \theta_{jk} - \nabla\!\!\!/_j \theta_{ki} \right)
\end{aligned}$$

[3] In arbitrary coordinates it has the form

$${}^{(3)}R_{rp} \Pi^r_i \Pi^p_k = {}^{(2)}R_{ik} + {}^{(3)}R_{iskq} \tilde{N}^s \tilde{N}^q - \mathrm{tr}\theta\theta_{ik} + \theta_{is}\theta^{s}{}_{k}.$$

Decomposing θ into its traceless and trace parts, $\theta = \hat{\theta} + \frac{1}{2}\gamma \mathrm{tr}\theta$, and using equations (7.1.6), (7.1.7) and (7.1.8) we obtain the evolution equations for $\mathrm{tr}\theta$ and $\hat{\theta}$.

Evolution equations:

$$\begin{aligned} \nabla\!\!\!\!/_{\tilde{N}}\hat{\theta}_{ac} + \mathrm{tr}\theta\,\hat{\theta}_{ac} &= -a^{-1}\widehat{\nabla\!\!\!\!/_a\nabla\!\!\!\!/_c}a - \left[{}^{(3)}R_{ac} + 2^{-1}\gamma_{ac}({}^{(3)}R_{\tilde{N}\tilde{N}} - {}^{(3)}R)\right] \\ \nabla_{\tilde{N}}\mathrm{tr}\theta + \frac{1}{2}\mathrm{tr}\theta^2 &= -a^{-1}\triangle\!\!\!\!/ a - |\hat{\theta}|^2 - {}^{(3)}R_{\tilde{N}\tilde{N}}, \end{aligned} \tag{7.1.10}$$

where $\widehat{\nabla\!\!\!\!/_a\nabla\!\!\!\!/_c} = \nabla\!\!\!\!/_a\nabla\!\!\!\!/_c - \frac{1}{2}\gamma_{ac}\triangle\!\!\!\!/$.

Definition 7.1.1 *We denote by $\widehat{{}^{(3)}R}$ the traceless part of the Ricci tensor ${}^{(3)}R$ with respect to the metric of Σ_0, g_{ij},*

$$\widehat{{}^{(3)}R} = {}^{(3)}R - \frac{1}{3}g_{ij}\,{}^{(3)}R, \tag{7.1.11}$$

where ${}^{(3)}R$ is the scalar curvature of Σ_0. The various components of the traceless part of the Ricci tensor in the frame adapted to the foliation are defined as[4]

$$S_{AB} \equiv \widehat{{}^{(3)}R}_{AB}\ ,\ \ P_A \equiv \widehat{{}^{(3)}R}_{A\tilde{N}}\ ,\ \ Q \equiv \widehat{{}^{(3)}R}_{\tilde{N}\tilde{N}}. \tag{7.1.12}$$

It follows easily that the traceless part of S, $\hat{S}_{AB} = S_{AB} - \frac{1}{2}\delta_{AB}\mathrm{tr}S$ satisfies

$$\hat{S}_{AB} = {}^{(3)}R_{AB} + \frac{1}{2}\delta_{AB}\left({}^{(3)}R_{\tilde{N}\tilde{N}} - {}^{(3)}R\right), \tag{7.1.13}$$

which allows us to rewrite the evolution equation for $\hat{\theta}$ (7.1.10) as

$$\nabla\!\!\!\!/_{\tilde{N}}\hat{\theta}_{ac} + \mathrm{tr}\theta\,\hat{\theta}_{ac} = -a^{-1}\widehat{\nabla\!\!\!\!/_a\nabla\!\!\!\!/_c}a - \hat{S}_{ac}. \tag{7.1.14}$$

The second Bianchi identities and the definition[5]

$$B_{ij} = (\mathrm{curl}\ \widehat{{}^{(3)}R})_{ij} \tag{7.1.15}$$

imply that the components of $\widehat{{}^{(3)}R}$, S, P, Q, satisfy the following equations (see [Ch-Kl], Chapter 5), assuming the adapted frame with $a = 1$ and Fermi transported,[6]

$$\begin{aligned} \mathrm{d\!\!\!/iv}\,P &= \frac{1}{6}\nabla_N R - \nabla_N Q - \frac{3}{2}\mathrm{tr}\theta\,Q + \hat{S}\cdot\hat{\theta} \\ \mathrm{cu\!\!\!/rl}\,P &= B_{NN} + \hat{\theta}\wedge\hat{S} \\ \nabla\!\!\!\!/_N P + \mathrm{tr}\theta\,P &= \frac{1}{12}\nabla\!\!\!\!/_a R + {}^*B\!\!\!\!/_N + \nabla\!\!\!\!/ Q - 2\hat{\theta}\cdot P \\ \mathrm{d\!\!\!/iv}\,\hat{S} &= \left(\frac{1}{12}\nabla\!\!\!\!/ R - {}^*B\!\!\!\!/_N\right) - \frac{1}{12}\nabla\!\!\!\!/ Q + (\hat{\theta}\cdot P) - \frac{1}{2}\mathrm{tr}\theta\,P \\ \nabla\!\!\!\!/_N\hat{S} + \frac{1}{2}\mathrm{tr}\theta\,\hat{S} &= {}^*B\!\!\!\!/ + \frac{1}{2}\nabla\!\!\!\!/\hat{\otimes}P + \frac{3}{2}\hat{\theta}\,Q, \end{aligned} \tag{7.1.16}$$

where $\nabla\!\!\!\!/_N X$ is the projection of $\nabla_N X$ on TS_0, $B\!\!\!\!/$ is the S-tangent symmetric 2-tensor $B\!\!\!\!/_{AB} = B_{AB}$ and ${}^*B\!\!\!\!/_{ab} = \epsilon_a{}^c\epsilon_b{}^d B\!\!\!\!/_{cd}$.

---

[4] In arbitrary coordinates we have $S_{ij} = \Pi_i^l\Pi_j^t\,\widehat{{}^{(3)}R}_{lt}\ ,\ \ P_i = \Pi_i^l\,\widehat{{}^{(3)}R}_{l\tilde{N}}\ ,\ \ Q = \widehat{{}^{(3)}R}_{\tilde{N}\tilde{N}}$.

[5] $(\mathrm{curl}\ \widehat{{}^{(3)}R})_{ij} \equiv \epsilon_j{}^{ls}\nabla_l\left({}^{(3)}R_{is} - \frac{1}{3}g_{is}\,{}^{(3)}R\right)$.

[6] $\nabla\!\!\!\!/_N e_A = 0$.

7.1.3 *The construction of the background foliation of Σ_0*

We start with the following theorem.

Theorem 7.1.1 *Assume that, given $\varepsilon > 0$, the initial data are such that $J_K(\Sigma_0, g, k) \leq \varepsilon^2$ is bounded. There exists a global geodesic foliation on $\Sigma_0 \backslash K$ with lapse function $a = 1$, such that the following inequalities hold:*[7]

$$\inf_{\Sigma_0\backslash K} r\mathrm{tr}\theta \leq c\varepsilon, \ \sup_{\Sigma_0\backslash K} r\mathrm{tr}\theta \leq c\varepsilon, \ \inf_{\Sigma_0\backslash K} r^2\tilde{K} \leq c\varepsilon, \ \sup_{\Sigma_0\backslash K} r^2\tilde{K} \leq c\varepsilon$$

$$\sup_{\Sigma_0\backslash K} \frac{r^2}{1+|\log r|}|\hat{\theta}| \leq c\varepsilon, \ \sup_{\Sigma_0\backslash K} r^2(\mathrm{tr}\theta - \overline{\mathrm{tr}\theta}) \leq c\varepsilon. \tag{7.1.17}$$

Also

$$\begin{aligned}
&|r^{\frac{7}{2}-\frac{2}{p}}\hat{S}|_{p,S_0} + |r^{\frac{9}{2}-\frac{2}{p}}\nabla\!\!\!/\hat{S}|_{p,S_0} \leq c\varepsilon \\
&|r^{\frac{7}{2}-\frac{2}{p}}P|_{p,S_0} + |r^{\frac{9}{2}-\frac{2}{p}}\nabla\!\!\!/ P|_{p,S_0} \leq c\varepsilon \\
&|r^{\frac{7}{2}-\frac{2}{p}}(Q-\overline{Q})|_{p,S_0} + |r^{\frac{9}{2}-\frac{2}{p}}\nabla\!\!\!/ Q|_{p,S_0} \leq c\varepsilon.
\end{aligned} \tag{7.1.18}$$

Proof: The proof of Theorem 7.1.1 follows by a simple adaptation of the proof of Proposition 5.0.1 in [Ch-Kl], Chapter 5. Observe that one can also prove additional estimates for the derivatives of $\mathrm{tr}\theta$ and $\hat{\theta}$ as well as for $\hat{S}$, P and $(Q-\overline{Q})$ up to second order (see also (7.1.18)).

Remark: We can choose K such that ∂K coincides with a leaf of the background foliation.

The results of Theorem 7.1.1 and assumption $J_K(\Sigma_0, g, k) \leq \varepsilon^2$ allow us to control on Σ_0 both the connection coefficients and the various components of the four-dimensional Riemann tensor. To achieve this result we use the following relationships for the four-dimensional Riemann tensor:

$$\begin{aligned}
&{}^{(3)}R_{\tilde{N}\tilde{N}} = \rho' + k^2_{\tilde{N}\tilde{N}} + \sum_A |k_{e_A\tilde{N}}|^2 \\
&{}^{(3)}R_{e_A\tilde{N}} = -\frac{1}{2}(\beta'_A + \underline{\beta}'_A) - \frac{1}{2}(\chi' + \underline{\chi}')_{AC}\zeta'_C + (\omega' + \underline{\omega}')\zeta'_A \\
&{}^{(3)}R_{e_Ae_B} = \frac{1}{4}(\alpha'_{AB} + \underline{\alpha}'_{AB}) - \frac{1}{2}\delta_{AB}\rho'_B + \frac{1}{4}(\chi' + \underline{\chi}')_{AC}(\chi' + \underline{\chi}')_{CB} + \zeta'_A\zeta'_B \\
&0 = \frac{1}{4}(\alpha'_{AB} - \underline{\alpha}'_{AB}) + \frac{1}{2}\epsilon_{AB}\sigma' + \nabla_{\tilde{N}}k_{BA} - \nabla_B k_{\tilde{N}A} \\
&0 = \frac{1}{4}(\beta'_A - \underline{\beta}'_A) + \nabla_{\tilde{N}}k_{A\tilde{N}} - \nabla_A k_{\tilde{N}\tilde{N}}.
\end{aligned} \tag{7.1.19}$$

The first three relations come from the Gauss equation,

$${}^{(3)}R^{\mu}_{\ \nu\rho\sigma} = {}^{(4)}R^{\tau}_{\ \gamma\lambda\zeta}\Pi^{\mu}_{\tau}\Pi^{\gamma}_{\nu}\Pi^{\lambda}_{\rho}\Pi^{\zeta}_{\sigma} - k^{\mu}_{\ \rho}k_{\nu\sigma} + k^{\mu}_{\ \sigma}k_{\nu\rho}, \tag{7.1.20}$$

---

[7] We denote here by $\tilde{K}$ the Gauss curvature of $S_0(\nu)$ to distinguish it from the compact set K.

where $\Pi^\mu_\nu = (\delta^\mu_\nu - T^\mu_0 T_{0\nu})$ is the projection on $T\Sigma_0$, and the remaining two come from the Codazzi–Mainardi equation,

$$^{(4)}R_{T_0\gamma\lambda\zeta} = -\nabla_\lambda k_{\zeta\gamma} + \nabla_\zeta k_{\lambda\gamma}. \tag{7.1.21}$$

For the connection coefficients the relations involved are

$$\begin{aligned}
\zeta'_A &= \frac{1}{2} g(\mathbf{D}_A e_4, e_3) = k_{A\tilde{N}} \\
\chi'_{AB} &= g(\mathbf{D}_A e_4, e_B) = -k_{AB} + \theta_{AB} \\
\underline{\chi}'_{AB} &= g(\mathbf{D}_A e_3, e_B) = -k_{AB} - \theta_{AB} \\
\omega' + \underline{\omega}' &= k_{\tilde{N}\tilde{N}} \\
2\omega' &= -\mathbf{D}_4 \log a \ , \ 2\underline{\omega}' = -\mathbf{D}_3 \log a. \\
2\Omega' &= a
\end{aligned} \tag{7.1.22}$$

7.1.4 *The construction of the canonical foliation of* Σ_0

We start by giving the motivation for the introduction of the canonical foliation on $\Sigma_0 \backslash K$. Let us consider the evolution equation for $\mathrm{tr}\theta$ on Σ_0 (see (7.1.10))

$$\nabla_{\tilde{N}} \mathrm{tr}\theta + \frac{1}{2}\mathrm{tr}\theta^2 = -a^{-1} \Delta a - |\hat{\theta}|^2 - {}^{(3)}R_{\tilde{N}\tilde{N}}. \tag{7.1.23}$$

Expressing the spacetime curvature tensor $\mathbf{R}_{\alpha\beta\gamma\delta}$ relative to the null pair $\{e_4 = T_0 + \tilde{N}, e_3 = T_0 - \tilde{N}\}$[8] and using the Gauss equation (see (7.1.20)), we obtain

$$^{(3)}R_{\tilde{N}\tilde{N}} = \rho + k^2_{\tilde{N}\tilde{N}} + \sum_A |k_{e_A\tilde{N}}|^2. \tag{7.1.24}$$

Using this equation we rewrite the evolution equation (7.1.23) as

$$\nabla_{\tilde{N}} \mathrm{tr}\theta + \frac{1}{2}\mathrm{tr}\theta^2 = -(\Delta \log a + \rho) + \left[-|\nabla \log a|^2 - |\hat{\theta}|^2 + g(k) \right], \tag{7.1.25}$$

where $g(k) \equiv k^2_{\tilde{N}\tilde{N}} + \sum_a |k_{e_A\tilde{N}}|^2$. Observe that the right-hand side of (7.1.25) depends through ρ on the second derivatives of the metric g, which implies that we can estimate, at most, two angular derivatives of $\mathrm{tr}\theta$. To do better we have to modify the $(\Delta \log a + \rho)$ term. This leads to the following definition.

Definition 3.3.1: *We say that a foliation is canonical on* $\Sigma_0 \backslash K$ *if it is defined by a function* $\underline{u}_{(0)}(p)$ *that is a solution of the initial slice problem,*

$$\begin{aligned}
|\nabla \underline{u}_{(0)}| &= a^{-1} \ , \ \underline{u}_{(0)}|_{\partial K} = \nu_0 \\
\Delta \log a &= -(\rho - \overline{\rho}) \ , \ \overline{\log a} = 0.
\end{aligned} \tag{7.1.26}$$

The leaves of the canonical foliation are denoted by

$$S_{(0)}(\nu) = \{p \in \Sigma_0 | \underline{u}_{(0)}(p) = \nu\}.$$

Moreover the initial leaf $S_0(\nu_0) = \partial K$ *of the background foliation is also the initial leaf,* $S_{(0)}(\nu_0)$*, of the canonical foliation.*

---

[8] T_0 is the vector field defined on Σ_0 orthonormal to Σ_0.

7.1.5 *Proof of Theorem* M3

Let us now recall the statement of Theorem **M3**.

Theorem 3.3.1 (Theorem M3) *Consider an initial data set that satisfies the exterior global smallness condition $J_K(\Sigma_0, g, k) \le \varepsilon^2$ with ε sufficiently small. There exists a canonical foliation on $\Sigma_0 \backslash K$ such that the following estimates hold:*

$$\mathcal{O}_{[3]}(\Sigma_0 \backslash K) \le c\varepsilon \ , \ \underline{\mathcal{O}}_{[3]}(\Sigma_0 \backslash K) \le c\varepsilon. \tag{7.1.27}$$

Proof of Theorem M3, part I: The proof consists of a local existence argument followed by a continuation argument. The local existence part is of the same type, but easier than the local part of the proof of the last slice problem presented in [Ch-Kl], Chapter 6. The global extension argument is far simpler than the corresponding global argument given in Chapter 14 of [Ch-Kl]. Both the local existence and the extension argument are based on Theorem 7.1.1.

7.2 The initial hypersurface connection estimates

In this section we continue the proof Theorem **M3**. We assume that the initial hypersurface $\Sigma_0 \backslash K$ is endowed with a canonical foliation and prove that we can estimate the $\Sigma_0 \backslash K$ norms of the connection coefficients and their derivatives up to third order in terms of the initial data norm $J_K(\Sigma_0, g, k)$. We shall in fact prove slightly stronger estimates expressed in the following definitions of the connection coefficient norms $\mathcal{O}'(\Sigma_0 \backslash K)$ and $\underline{\mathcal{O}}'(\Sigma_0 \backslash K)$.[9]

$$\begin{aligned}
\mathcal{O}'^{\infty}_{[0]}(\Sigma_0 \backslash K) &\equiv \mathcal{O}'^{\infty}_{0}(\Sigma_0 \backslash K) + \sup_{\Sigma_0 \backslash K} |r^{\frac{5}{2}}(\overline{\mathrm{tr}\chi'} - \frac{2}{r})| + \sup_{\Sigma_0 \backslash K} |r(\Omega' - \frac{1}{2})| \\
\underline{\mathcal{O}}'^{\infty}_{[0]}(\Sigma_0 \backslash K) &\equiv \underline{\mathcal{O}}'^{\infty}_{0}(\Sigma_0 \backslash K) + \sup_{\Sigma_0 \backslash K} |r^{\frac{5}{2}}(\overline{\mathrm{tr}\underline{\chi}'} + \frac{2}{r})| \\
\mathcal{O}'_{[1]}(\Sigma_0 \backslash K) &\equiv \left[\mathcal{O}'_1(\Sigma_0 \backslash K) + \sup_{p\in[2,4]} \mathcal{O}'^{p,S}_0(\Sigma_0 \backslash K)(\mathbf{D}'_3 \underline{\omega}') \right] + \mathcal{O}'^{\infty}_{[0]}(\Sigma_0 \backslash K) \\
\underline{\mathcal{O}}'_{[1]}(\Sigma_0 \backslash K) &\equiv \left[\underline{\mathcal{O}}'_1(\Sigma_0 \backslash K) + \sup_{p\in[2,4]} \mathcal{O}'^{p,S}_0(\Sigma_0 \backslash K)(\mathbf{D}'_4 \omega') \right] + \underline{\mathcal{O}}'^{\infty}_{[0]}(\Sigma_0 \backslash K) \qquad (7.2.1) \\
\mathcal{O}'_{[2]}(\Sigma_0 \backslash K) &\equiv \left[\mathcal{O}'_2(\Sigma_0 \backslash K) + \sup_{p\in[2,4]} \left(\mathcal{O}'^{p,S}_1(\Sigma_0 \backslash K)(\mathbf{D}'_3 \underline{\omega}') + \mathcal{O}'^{p,S}_0(\Sigma_0 \backslash K)(\mathbf{D}'^2_3 \underline{\omega}') \right) \right] \\
&\quad + \mathcal{O}'_{[1]}(\Sigma_0 \backslash K) \\
\underline{\mathcal{O}}'_{[2]}(\Sigma_0 \backslash K) &\equiv \left[\underline{\mathcal{O}}'_2(\Sigma_0 \backslash K) + \sup_{p\in[2,4]} \left(\mathcal{O}'^{p,S}_1(\Sigma_0 \backslash K)(\mathbf{D}'_4 \omega') + \mathcal{O}'^{p,S}_0(\Sigma_0 \backslash K)(\mathbf{D}'^2_4 \omega') \right) \right] \\
&\quad + \underline{\mathcal{O}}'_{[1]}(\Sigma_0 \backslash K), \\
\mathcal{O}'_{[3]}(\Sigma_0 \backslash K) &\equiv \mathcal{O}'_3(\Sigma_0 \backslash K) + \mathcal{O}'_{[2]}(\Sigma_0 \backslash K)
\end{aligned}$$

[9] Observe that here all the connection coefficients $\chi', \underline{\chi}'$... are primed because they are the restrictions on Σ_0 of the connection coefficients relative to the initial layer foliation; see Subsection 3.3.4.

where for $q \leq 2$,

$$\mathcal{O}'^{p,S}_{q}(\Sigma_0\backslash K)(X) \equiv \sup_{\Sigma_0\backslash K} \mathcal{O}'^{p,S}_{q}(\Sigma_0\backslash K)(X)(\nu). \tag{7.2.2}$$

The $\mathcal{O}'^{p,S}_{q}(\Sigma_0\backslash K)(X)(\nu)$ norms presented in (7.2.1) are listed below.[10]

$$\begin{aligned}
&\mathcal{O}'^{p,S}_{q}(\Sigma_0\backslash K)(\hat{\chi}')(\nu) = |r^{(\frac{5}{2}+q-\frac{2}{p})}\nabla'^{q}\hat{\chi}'|_{p,S_{(0)}(\nu)}\\
&\mathcal{O}'^{p,S}_{q}(\Sigma_0\backslash K)(\underline{\hat{\chi}}')(\nu) = |r^{(\frac{5}{2}+q-\frac{2}{p})}\nabla'^{q}\underline{\hat{\chi}}'|_{p,S_{(0)}(\nu)}\\
&\mathcal{O}'^{p,S}_{q}(\Sigma_0\backslash K)(\mathrm{tr}\chi')(\nu) = |r^{(\frac{5}{2}+q-\frac{2}{p})}\nabla'^{q}(\mathrm{tr}\chi' - \overline{\mathrm{tr}\chi'})|_{p,S_{(0)}(\nu)}\\
&\mathcal{O}'^{p,S}_{q}(\Sigma_0\backslash K)(\mathrm{tr}\underline{\chi}')(\nu) = |r^{(\frac{5}{2}+q-\frac{2}{p})}\nabla'^{q}(\mathrm{tr}\underline{\chi}' - \overline{\mathrm{tr}\underline{\chi}'})|_{p,S_{(0)}(\nu)}\\
&\mathcal{O}'^{p,S}_{q}(\Sigma_0\backslash K)(\eta')(\nu) = |r^{(\frac{5}{2}+q-\frac{2}{p})}\nabla'^{q}\eta'|_{p,S_{(0)}(\nu)}\\
&\mathcal{O}'^{p,S}_{q}(\Sigma_0\backslash K)(\underline{\eta}')(\nu) = |r^{(\frac{5}{2}+q-\frac{2}{p})}\nabla'^{q}\underline{\eta}'|_{p,S_{(0)}(\nu)}\\
&\mathcal{O}'^{p,S}_{q}(\Sigma_0\backslash K)(\underline{\omega}')(\nu) = |r^{(\frac{5}{2}+q-\frac{2}{p})}\nabla'^{q}\underline{\omega}'|_{p,S_{(0)}(\nu)}\\
&\mathcal{O}'^{p,S}_{q}(\Sigma_0\backslash K)(\mathbf{D}'_3\underline{\omega}')(\nu) = |r^{(\frac{7}{2}+q-\frac{2}{p})}\nabla'^{q}\mathbf{D}'_3\underline{\omega}'|_{p,S_{(0)}(\nu)}\\
&\mathcal{O}'^{p,S}_{q}(\Sigma_0\backslash K)(\mathbf{D}'^2_3\underline{\omega}')(\nu) = |r^{(\frac{9}{2}+q-\frac{2}{p})}\nabla'^{q}\mathbf{D}'^2_3\underline{\omega}'|_{p,S_{(0)}(\nu)}\\
&\mathcal{O}'^{p,S}_{q}(\Sigma_0\backslash K)(\omega')(\nu) = |r^{(\frac{5}{2}+q-\frac{2}{p})}\nabla'^{q}\omega'|_{p,S_{(0)}(\nu)}\\
&\mathcal{O}'^{p,S}_{q}(\Sigma_0\backslash K)(\mathbf{D}'_4\omega')(\nu) = |r^{(\frac{7}{2}+q-\frac{2}{p})}\nabla^{q}\mathbf{D}'_3\omega'|_{p,S_{(0)}(\nu)}\\
&\mathcal{O}'^{p,S}_{q}(\Sigma_0\backslash K)(\mathbf{D}'^2_4\omega')(\nu) = |r^{(\frac{9}{2}+q-\frac{2}{p})}\nabla'^{q}\mathbf{D}'^2_4\omega'|_{p,S_{(0)}(\nu)}.
\end{aligned} \tag{7.2.3}$$

Finally we introduce the norms $\mathcal{O}'_3(\Sigma_0\backslash K)$ and $\underline{\mathcal{O}}'_3(\Sigma_0\backslash K)$ on the initial slice. They are defined in the following way:

$$\begin{aligned}
\mathcal{O}'_3(\Sigma_0\backslash K) &= \mathcal{O}'_3(\Sigma_0\backslash K)(\mathrm{tr}\chi') + \mathcal{O}'_3(\Sigma_0\backslash K)(\underline{\omega}')\\
\underline{\mathcal{O}}'_3(\Sigma_0\backslash K) &= \mathcal{O}'_3(\Sigma_0\backslash K)(\mathrm{tr}\underline{\chi}') + \mathcal{O}'_3(\Sigma_0\backslash K)(\omega'),
\end{aligned} \tag{7.2.4}$$

where

$$\begin{aligned}
\mathcal{O}'_3(\mathrm{tr}\underline{\chi}') &= \sup_{\nu\in[\nu_0,\nu_*]}\left(r^{\frac{1}{2}}(\nu)||r^3\nabla'^3\mathrm{tr}\underline{\chi}'||_{L^2(\Sigma_0\backslash K)}\right)\\
\mathcal{O}'_3(\omega') &= \sup_{\nu\in[\nu_0,\nu_*]}\left(r^{\frac{1}{2}}(\nu)||r^3\nabla'^3\omega'||_{L^2(\Sigma_0\backslash K)}\right)\\
\mathcal{O}'_3(\mathrm{tr}\chi') &= \sup_{\nu\in[\nu_0,\nu_*]}\left(r^{\frac{1}{2}}(\nu)||r^3\nabla'^3\mathrm{tr}\chi'||_{L^2(\Sigma_0\backslash K)}\right)\\
\mathcal{O}'_3(\omega') &= \sup_{\nu\in[\nu_0,\nu_*]}\left(r^{\frac{1}{2}}(\nu)||r^3\nabla'^3\underline{\omega}'||_{L^2(\Sigma_0\backslash K)}\right).
\end{aligned} \tag{7.2.5}$$

---

[10]Observe that these norms differ from those defined on the whole $\mathcal{K}$ (see (3.5.22,. . ., (3.5.28)) by a factor $r^{\frac{1}{2}}$.

Proof of Theorem M3, part II:

We sketch the proof of the theorem below.

1. We recall the evolution equation satisfied by $\mathrm{tr}\theta$, the Codazzi–Mainardi equation for $\hat{\theta}$ (7.1.9), and the elliptic equation satisfied by a, relative to the canonical foliation:

$$\begin{aligned} &\nabla_{\tilde{N}}\mathrm{tr}\theta + \frac{1}{2}\mathrm{tr}\theta^2 = -\overline{\rho} + \left[-|\nabla\!\!\!/ \log a|^2 - |\hat{\theta}|^2 + g(k)\right] \\ &\nabla\!\!\!/^c\hat{\theta}_{cb} = \frac{1}{2}\nabla\!\!\!/_b\mathrm{tr}\theta + P_b \\ &\Delta\!\!\!/ \log a = -(\rho - \overline{\rho}) \ , \ \overline{\log a} = 0. \end{aligned} \tag{7.2.6}$$

Observe that the quantities $\hat{S}$, P, $(Q - \overline{Q})$ (see (7.1.12)), which decompose the Ricci tensor relative to $\Sigma_0 \backslash K$, are expressed relative to the canonical foliation.

2. Using equations (7.2.6) we proceed precisely as in the proof of Theorem 7.1.1 and deduce the following results:

$$\begin{aligned} &|r^{\frac{7}{2}-\frac{2}{p}}\hat{S}|_{p,S_{(0)}} + |r^{\frac{9}{2}-\frac{2}{p}}\nabla\!\!\!/\hat{S}|_{p,S_{(0)}} + |r^{\frac{9}{2}-\frac{2}{p}}\nabla_{\tilde{N}}\hat{S}|_{p,S_{(0)}} \leq c\varepsilon \\ &|r^{\frac{7}{2}-\frac{2}{p}}P|_{p,S_{(0)}} + |r^{\frac{9}{2}-\frac{2}{p}}\nabla\!\!\!/ P|_{p,S_{(0)}} + |r^{\frac{9}{2}-\frac{2}{p}}\nabla_{\tilde{N}}P|_{p,S_{(0)}} \leq c\varepsilon \\ &|r^{\frac{7}{2}-\frac{2}{p}}(Q - \overline{Q})|_{p,S_{(0)}} + |r^{\frac{9}{2}-\frac{2}{p}}\nabla\!\!\!/ Q|_{p,S_{(0)}} + |r^{\frac{9}{2}-\frac{2}{p}}Q_{\tilde{N}}|_{p,S_{(0)}} \leq c\varepsilon \\ &||r^4\nabla\!\!\!/\nabla_{\tilde{N}}\hat{S}||_{L^2(\Sigma_0\backslash K)} + ||r^4\nabla^2_{\tilde{N}}\hat{S}||_{L^2(\Sigma_0\backslash K)} \leq c\varepsilon \\ &||r^4\nabla\!\!\!/\nabla_{\tilde{N}}P||_{L^2(\Sigma_0\backslash K)} + ||r^4\nabla^2_{\tilde{N}}P||_{L^2(\Sigma_0\backslash K)} \leq c\varepsilon \\ &||r^4\nabla_{\tilde{N}}Q_{\tilde{N}}||_{L^2(\Sigma_0\backslash K)} + ||r^4\nabla\!\!\!/ Q_{\tilde{N}}||_{L^2(\Sigma_0\backslash K)} + ||r^4\nabla\!\!\!/^2 Q||_{L^2(\Sigma_0\backslash K)} \leq c\varepsilon. \end{aligned} \tag{7.2.7}$$

Remark: In the process of proving this result we also prove the same results for $\hat{\theta}$, $\mathrm{tr}\theta$ and their derivatives up to second order as in Theorem 7.1.1 (see (7.1.17)) and [Ch-Kl], Chapter 5. In fact, we will do better in what follows.

3. Using the curvature estimates established above we use the equations (7.2.6) to obtain the estimates (7.1.27). We sketch the main ideas of the proof below.

(a) The first important observation is that the curvature term, $\overline{\rho}$, in the third equation of (7.2.6) is constant on the $S_{(0)}$ surfaces. Therefore we can eliminate it by taking tangential derivatives. This leads to a coupled system between the evolution equation for $\nabla\!\!\!/\mathrm{tr}\theta$ and the Codazzi equation for $\hat{\theta}$. This system is similar to but simpler than the system for $\nabla\!\!\!/\mathrm{tr}\chi$ and $\nabla\!\!\!/\hat{\chi}$ studied in Chapter 4, Subsection 4.3.1 and can be treated as that case was. Therefore we derive the following estimates:

$$\begin{aligned} &|r^{\frac{5}{2}-\frac{2}{p}}\hat{\theta}|_{p,S_{(0)}} \leq c\varepsilon \ , \ |r^{\frac{7}{2}-\frac{2}{p}}\nabla\!\!\!/\hat{\theta}|_{p,S_{(0)}} \leq c\varepsilon \\ &|r^{\frac{5}{2}-\frac{2}{p}}(\mathrm{tr}\theta - \overline{\mathrm{tr}\theta})|_{p,S_{(0)}} \leq c\varepsilon \ , \ |r^{2-\frac{2}{p}}(\overline{\mathrm{tr}\theta} - \frac{2}{r})|_{p,S_{(0)}} \leq c\varepsilon \ , \\ &|r^{\frac{7}{2}-\frac{2}{p}}\nabla\!\!\!/\mathrm{tr}\theta|_{p,S_{(0)}} \leq c.\varepsilon \end{aligned} \tag{7.2.8}$$

(b) We can also obtain the estimates for up to two more derivatives for these quantities. For $p \in [2, 4]$, we have

$$|r^{\frac{9}{2}-\frac{2}{p}}\nabla\!\!\!/^2\mathrm{tr}\theta|_{p,S_{(0)}} \leq c\varepsilon \ , \ |r^{\frac{9}{2}-\frac{2}{p}}\nabla\!\!\!/^2\hat{\theta}|_{p,S_{(0)}} \leq c\varepsilon$$

$$r^{\frac{1}{2}}||r^3 \nabla\!\!\!\!/^3 \mathrm{tr}\theta||_{L^2(\Sigma_0\setminus K)} \leq c\varepsilon. \tag{7.2.9}$$

We can also obtain the estimate for the scalar function a for $q = 0, 1, 2,$

$$|r^{(\frac{3}{2}+q)-\frac{2}{p}} \nabla\!\!\!\!/^q \log a|_{p,S_{(0)}} \leq c\varepsilon, \tag{7.2.10}$$

due to the elliptic character of the equation satisfied by $\log a$.

4. The final step consists in using these results to estimate the various connection coefficients χ', $\underline{\chi}'$, ζ', ω' and $\underline{\omega}'$. We consider the future-directed unit vector normal to Σ_0, T_0, and the null frame adapted to the canonical foliation, which we denote by $\{e'_4, e'_3, e'_A\}$ where $e'_4 = T_0 + \tilde{N}$, $e'_3 = T_0 - \tilde{N}$. The connection coefficients we consider here are those computed with respect to this null frame.[11] The following expressions hold (see also (7.1.22)):

$$\begin{aligned}
2\Omega' &= a \\
\zeta'_A &= \frac{1}{2} g(\mathbf{D}_{A'} e'_4, e'_3) = -g(\mathbf{D}_{e'_A} T_0, \tilde{N}) = k(e'_A, \tilde{N}) \\
\chi'_{AB} &= g(\mathbf{D}'_A e'_4, e'_B) = -k_{AB} + \theta_{AB} \\
\underline{\chi}'_{AB} &= g(\mathbf{D}'_A e'_3, e'_B) = -k_{AB} - \theta_{AB} \\
2\omega' &= -\mathbf{D}_{e'_4} \log a \ , \ 2\underline{\omega}' = -\mathbf{D}_{e'_3} \log a.
\end{aligned} \tag{7.2.11}$$

In view of the estimates for k implicit in the assumption $J_K(\Sigma_0, \mathbf{g}, k) \leq \varepsilon^2$, we obtain the following result:

$$\mathcal{O}'_{[3]}(\Sigma_0\setminus K) \leq c\varepsilon \ , \ \underline{\mathcal{O}}'_{[3]}(\Sigma_0\setminus K) \leq c\varepsilon. \tag{7.2.12}$$

7.2.1 *Proof of Lemma 3.7.1*

We recall the statement of the lemma.

Lemma 3.7.1: *If $J_K(\Sigma_0, g, k)$ is sufficiently small, the following inequality holds:*

$$\mathcal{Q}_{\Sigma_0 \cap \mathcal{K}} \leq c J_K(\Sigma_0, g, k) \ .$$

Proof: The proof of this Lemma is straightforward. We sketch here only the main steps. From definitions 3.5.9, 3.5.10, 3.5.12, it follows that we have to estimate on Σ_0 the L^2 norms of the various components of the Riemann tensors $\mathbf{R}$, $\hat{\mathcal{L}}_0\mathbf{R}$, $\hat{\mathcal{L}}_T\mathbf{R}$, $\hat{\mathcal{L}}_0^2\mathbf{R}$, $\hat{\mathcal{L}}_S\hat{\mathcal{L}}_T\mathbf{R}$. In (7.1.19) these Riemann components are written in terms of the Ricci tensor of Σ_0 and of quadratic expressions in k. Using the previous Theorem **M3**, (see (7.2.7)) and the conditions on k imposed by the smallness of J_K, we can easily estimate these quantities. The main simplification is that, on Σ_0, all these Riemann components have the same asymptotic behavior as $r \to \infty$. More precisely we have $O(r^{-\frac{7}{2}})$ for the terms associated with $\mathbf{R}$, $\hat{\mathcal{L}}_0\mathbf{R}$, $\hat{\mathcal{L}}_0^2\mathbf{R}$ and stronger for $\hat{\mathcal{L}}_T\mathbf{R}$ and $\hat{\mathcal{L}}_S\hat{\mathcal{L}}_T\mathbf{R}$.[12]

---

[11] Observe that, as discussed in Chapters 3 and 4, this frame is the restriction to Σ_0 of the null frame relative to the initial layer foliation.

[12] With only the exception of Q (see (7.1.12)) where we have to subtract its average $\overline{Q}$.

7.3 The last slice foliation

7.3.1 Construction of the canonical foliation of $\underline{C}_*$

The construction of the canonical foliation on $\underline{C}_*$ is based on the proof of Theorem **M6**.

Theorem M6: *Assume given on $\underline{C}_*$ a foliation we call "background foliation" not necessarily canonical, whose connection coefficients and null curvature components satisfy the inequalities:*

$$\mathcal{R}'(\underline{C}_*) \equiv \mathcal{R}_{[2]}{}'(\underline{C}_*) + \underline{\mathcal{R}}_{[2]}{}'(\underline{C}_*) \leq \epsilon_0'$$
$$\mathcal{O}'(\underline{C}_*) \equiv \underline{\mathcal{O}}_{[2]}{}'(\underline{C}_*) + \mathcal{O}_{[2]}{}'(\underline{C}_*) \leq \epsilon_0', \tag{7.3.1}$$

where $\mathcal{R}_{[2]}{}'(\underline{C}_*)$, $\underline{\mathcal{R}}_{[2]}{}'(\underline{C}_*)$, $\underline{\mathcal{O}}_{[2]}{}'(\underline{C}_*)$, $\mathcal{O}_{[2]}{}'(\underline{C}_*)$ are the norms introduced in Chapter 3, Section 3.5, restricted to $\underline{C}_*$, relative to the background foliation. Then there exists a canonical foliation on $\underline{C}_*$ relative to which we have

$$\mathcal{R}(\underline{C}_*) \equiv \mathcal{R}_{[2]}(\underline{C}_*) + \underline{\mathcal{R}}_{[2]}(\underline{C}_*) \leq c\epsilon_0'$$
$$\mathcal{O}(\underline{C}_*) \equiv \underline{\mathcal{O}}_{[2]}(\underline{C}_*) + \mathcal{O}_{[3]}(\underline{C}_*) \leq c\epsilon_0'. \tag{7.3.2}$$

In addition it can be shown that these two foliations remain close to each other in a sense that can be made precise.

Proof: We divide the proof into four parts.

1. We prove that the canonical foliation exists locally on $\underline{C}_*$ close to $S_*(\lambda_1)$.

2. Given a canonical foliation on $\underline{C}_*$, such that $\mathcal{R}(\underline{C}_*)$ is sufficiently small, we show that the following inequality holds:

$$\mathcal{O}(\underline{C}_*) \leq c\left(\mathcal{I}_0 + \mathcal{R}(\underline{C}_*)\right). \tag{7.3.3}$$

This is the content of Theorem **M4** whose proof is given in Section 7.4.

3. We compare the norms $\mathcal{R}(\underline{C}_*)$ and the norms $\mathcal{R}'(\underline{C}_*)$ and establish the following inequality which we write, schematically, as

$$\mathcal{R}(\underline{C}_*) \leq \mathcal{R}'(\underline{C}_*) + c\big(\mathcal{O}(\underline{C}_*) + \mathcal{O}'(\underline{C}_*)\big)\mathcal{R}'(\underline{C}_*). \tag{7.3.4}$$

See the proof in the appendix.

4. Combining (2) and (3) and using assumptions (7.3.1) we deduce

$$\begin{aligned}\mathcal{O}(\underline{C}_*) &\leq c\Big(\mathcal{I}_0 + \mathcal{R}'(\underline{C}_*) + c\big(\mathcal{O}(\underline{C}_*) + \mathcal{O}'(\underline{C}_*)\big)\mathcal{R}'(\underline{C}_*)\Big)\\ &\leq c\left(\mathcal{I}_0 + \mathcal{R}'(\underline{C}_*) + \mathcal{O}(\underline{C}_*)\epsilon_0' + O(\epsilon_0'^2)\right).\end{aligned} \tag{7.3.5}$$

Therefore, if ϵ_0' is sufficiently small, we obtain the result of the theorem for $\mathcal{O}(\underline{C}_*)$ and using again inequality (7.3.5) in (7.3.4) we also derive the estimate for $\mathcal{R}(\underline{C}_*)$.

The proof of the local existence result, (1), is sketched in the appendix to this chapter (see Subsection 7.7.2) and proved in detail elsewhere (see [Ni]). The continuation argument, (2), to extend the foliation to the whole $\underline{C}_*$ is based on the a priori estimates (7.3.3), proved in the next section.

The next corollary specifies in which sense the two foliations on $\underline{C}_*$ are near.

Corollary 7.3.1 *Under the assumptions of Theorem* **M6**, *let S' be a leaf on $\underline{C}_*$ of the background foliation,*[13] *and let $u(p) = u_*(p)$ be the function defined on $\underline{C}_*$ whose level surfaces are the leaves of the canonical foliation of $\underline{C}_*$. Then*

$$\sup_{(p,p')\in S'} |u(p) - u(p')| \leq c\epsilon'_0. \tag{7.3.6}$$

Proof: It is an immediate consequence of the proof of Theorem **M6** and we do not report it here.

7.4 The last slice connection estimates

In this section we give the proof of Theorem **M4**. We shall in fact prove slightly stronger estimates which are encompassed in the following definitions of the connection coefficient norms $\mathcal{O}^*$ and $\underline{\mathcal{O}}^*$ on $\underline{C}_*$.

7.4.1 $\mathcal{O}$ norms on the last slice

The norms we introduce here are slightly different from the corresponding ones defined in $\mathcal{K}$; in fact they are a little stronger. It is possible to prove their boundedness due to the fact that $\underline{C}_*$ is endowed with a canonical foliation. Their expressions are:

$$\begin{aligned}
\mathcal{O}^{*\infty}_{[0]} &\equiv \mathcal{O}^{*\infty}_0 + \sup_{\underline{C}_*}|r^2(\overline{\mathrm{tr}\chi} - \frac{2}{r})| + \sup_{\underline{C}_*}|r(\Omega - \frac{1}{2})|\\
\underline{\mathcal{O}}^{*\infty}_{[0]} &\equiv \underline{\mathcal{O}}^{*\infty}_0 + \sup_{\underline{C}_*}|r\tau_-(\overline{\mathrm{tr}\underline{\chi}} + \frac{2}{r})|\\
\mathcal{O}^*_{[1]} &\equiv \left[\mathcal{O}^*_1 + \sup_{p\in[2,4]} \mathcal{O}^{*p,S}_0(\mathbf{D}_3\underline{\omega})\right] + \mathcal{O}^{*\infty}_{[0]}\\
\underline{\mathcal{O}}^*_{[1]} &\equiv \underline{\mathcal{O}}^*_1 + \underline{\mathcal{O}}^{*\infty}_{[0]}\\
\mathcal{O}^*_{[2]} &= \left[\mathcal{O}^*_2 + \sup_{p\in[2,4]} \left(\mathcal{O}^{*p,S}_1(\mathbf{D}_3\underline{\omega}) + \mathcal{O}^{*p,S}_0(\mathbf{D}^2_3\underline{\omega})\right)\right] + \mathcal{O}^*_{[1]}\\
\underline{\mathcal{O}}^*_{[2]} &= \underline{\mathcal{O}}^*_2 + \underline{\mathcal{O}}^*_{[1]}\ , \quad \mathcal{O}^*_{[3]} = \mathcal{O}^*_3 + \mathcal{O}^*_{[2]},
\end{aligned} \tag{7.4.1}$$

where for $q \leq 2$,

$$\begin{aligned}
\mathcal{O}^*_q &= \sup_{p\in[2,4]} \mathcal{O}^{*p}_q\ , \underline{\mathcal{O}}^*_q = \sup_{p\in[2,4]} \underline{\mathcal{O}}^{*p}_q\\
\mathcal{O}^{*p}_q &= \mathcal{O}^{*p,S}_q(\mathrm{tr}\chi) + \mathcal{O}^{*p,S}_q(\hat{\chi}) + \mathcal{O}^{*p,S}_q(\eta) + \mathcal{O}^{*p,S}_q(\underline{\omega})\\
\underline{\mathcal{O}}^{*p}_q &= \mathcal{O}^{*p,S}_q(\mathrm{tr}\underline{\chi}) + \mathcal{O}^{*p,S}_q(\underline{\hat{\chi}}) + \mathcal{O}^{*p,S}_q(\underline{\eta}) + \mathcal{O}^{*p,S}_q(\omega)
\end{aligned} \tag{7.4.2}$$

and

$$\mathcal{O}^{*p,S}_q(X) = \sup_{\underline{C}_*} \mathcal{O}^{*p,S}_q(X).$$

[13]In the Oscillation Lemma this result is needed only for a small distance from $\underline{C}_* \cap \Sigma_0$.

The $\mathcal{O}^{*p,S}_{q}$ norms, present in (7.4.1), are listed below:[14]

$$
\begin{aligned}
\mathcal{O}^{*p,S}_{q}(\hat{\chi}) &= |r^{(2+q-\frac{2}{p})}\tau_-^{\frac{1}{2}}\nabla^q\hat{\chi}|_{p,S(\lambda,\nu_*)}\\
\mathcal{O}^{*p,S}_{q}(\underline{\hat{\chi}}) &= |r^{(1+q-\frac{2}{p})}\tau_-^{\frac{3}{2}}\nabla^q\underline{\hat{\chi}}|_{p,S(\lambda,\nu_*)}\\
\mathcal{O}^{*p,S}_{q}(\mathrm{tr}\chi) &= |r^{(2+q-\frac{2}{p})}\tau_-^{\frac{1}{2}}\nabla^q(\mathrm{tr}\chi-\overline{\mathrm{tr}\chi})|_{p,S(\lambda,\nu_*)}\\
\mathcal{O}^{*p,S}_{q}(\mathrm{tr}\underline{\chi}) &= |r^{(2+q-\frac{2}{p})}\tau_-^{\frac{1}{2}}\nabla^q(\mathrm{tr}\underline{\chi}-\overline{\mathrm{tr}\underline{\chi}})|_{p,S(\lambda,\nu_*)}\\
\mathcal{O}^{*p,S}_{q}(\eta) &= |r^{(2+q-\frac{2}{p})}\tau_-^{\frac{1}{2}}\nabla^q\eta|_{p,S(\lambda,\nu_*)}\\
\mathcal{O}^{*p,S}_{q}(\underline{\eta}) &= |r^{(2+q-\frac{2}{p})}\tau_-^{\frac{1}{2}}\nabla^q\underline{\eta}|_{p,S(\lambda,\nu_*)}\\
\mathcal{O}^{*p,S}_{q}(\underline{\omega}) &= |r^{(1+q-\frac{2}{p})}\tau_-^{\frac{3}{2}}\nabla^q\underline{\omega}|_{p,S(\lambda,\nu_*)}\\
\mathcal{O}^{*p,S}_{q}(\mathbf{D}_3\underline{\omega}) &= |r^{(1+q-\frac{2}{p})}\tau_-^{\frac{5}{2}}\nabla^q\mathbf{D}_3\underline{\omega}|_{p,S(\lambda,\nu_*)}\\
\mathcal{O}^{*p,S}_{q}(\mathbf{D}_3^2\underline{\omega}) &= |r^{(1+q-\frac{2}{p})}\tau_-^{\frac{7}{2}}\nabla^q\mathbf{D}_3^2\underline{\omega}|_{p,S(\lambda,\nu_*)}.
\end{aligned}
\tag{7.4.3}
$$

Finally we introduce the $\mathcal{O}^*_3$ norm on the final slice,

$$
\mathcal{O}^*_3 = \mathcal{O}^*_3(\mathrm{tr}\chi) + \mathcal{O}^*_3(\underline{\omega}), \tag{7.4.4}
$$

where

$$
\begin{aligned}
\mathcal{O}^*_3(\mathrm{tr}\chi) &= \sup_{\lambda\in[\lambda_0,\lambda_1]}\left(r^{\frac{1}{2}}(\lambda,\nu)||r^3\nabla^3\mathrm{tr}\chi||_{L^2(\underline{C}_*(\nu_*,[\lambda_1,\lambda]))}\right)\\
\mathcal{O}^*_3(\underline{\omega}) &= \sup_{\lambda\in[\lambda_0,\lambda_1]}\left(r^{\frac{1}{2}}(\lambda,\nu)||r^3\nabla^3\underline{\omega}||_{L^2(\underline{C}_*(\nu_*,[\lambda_1,\lambda]))}\right).
\end{aligned}
\tag{7.4.5}
$$

Using these norms the improved version of Theorem **M4** is the following.

Theorem M4: *Let a canonical foliation be given on $\underline{C}_*$, relative to which*

$$
\mathcal{R}_{[2]} + \underline{\mathcal{R}}_{[2]} \le \Delta\,.
$$

Moreover assume

$$
\mathcal{O}_{[2]}(\underline{C}_*\cap\Sigma_0) + \mathcal{O}_3(\Sigma_0) + \underline{\mathcal{O}}_{[3]}(\Sigma_0) \le \mathcal{I}_0\,.
$$

If Δ, $\mathcal{I}_0$ are sufficiently small, then the following estimate holds:

$$
\underline{\mathcal{O}}^*_{[2]} + \mathcal{O}^*_{[3]} \le c(\mathcal{I}_0+\Delta)\,.
$$

The proof of this theorem is divided between Proposition 7.4.1 and Proposition 7.4.2.

[14] Observe that all these norms are different from those defined on the whole $\mathcal{K}$; (see (3.5.22),..., (3.5.28)) by a factor $\tau_-^{\frac{1}{2}}$. Moreover, we do not need, on $\underline{C}_*$, norms relative to ω.

Proposition 7.4.1 *Assume that, relative to a canonical foliation on the last slice, $\underline{C}_*$, we have*

$$\begin{aligned}\mathcal{R}_0^\infty + \underline{\mathcal{R}}_0^\infty &\le \Delta_0\\ \mathcal{R}_1^S + \underline{\mathcal{R}}_1^S &\le \Delta_1.\end{aligned} \tag{7.4.6}$$

Moreover assume that

$$\underline{\mathcal{O}}^*_{[1]}(\underline{C}_* \cap \Sigma_0) + \mathcal{O}^*_{[1]}(\underline{C}_* \cap \Sigma_0) \le \mathcal{I}_0. \tag{7.4.7}$$

If Δ_0, Δ_1 and $\mathcal{I}_0$ are sufficiently small, the following inequality holds:

$$\underline{\mathcal{O}}^*_{[1]} + \mathcal{O}^*_{[1]} \le c\,(\mathcal{I}_0 + \Delta_0 + \Delta_1)\,. \tag{7.4.8}$$

Proposition 7.4.2 *Assume that, relative to a canonical foliation on the last slice, $\underline{C}_*$, we have*

$$\begin{aligned}\mathcal{R}_0^\infty + \underline{\mathcal{R}}_0^\infty &\le \Delta_0\\ \mathcal{R}_1^S + \underline{\mathcal{R}}_1^S &\le \Delta_1\\ \mathcal{R}_2 + \underline{\mathcal{R}}_2 &\le \Delta_2.\end{aligned} \tag{7.4.9}$$

Moreover assume[15]

$$\mathcal{O}^*_{[2]}(\underline{C}_* \cap \Sigma_0) + \underline{\mathcal{O}}^*_{[2]}(\underline{C}_* \cap \Sigma_0) \le \mathcal{I}_0. \tag{7.4.10}$$

Then if Δ_0, Δ_1, Δ_2 and $\mathcal{I}_0$ are sufficiently small, the following inequalities hold:

$$\begin{aligned}\underline{\mathcal{O}}^*_{[2]} + \mathcal{O}^*_{[2]} &\le c\,(\mathcal{I}_0 + \Delta_0 + \Delta_1)\\ \mathcal{O}^*_3 &\le c\,(\mathcal{I}_0 + \Delta_0 + \Delta_1 + \Delta_2)\,.\end{aligned} \tag{7.4.11}$$

Proof of Proposition 7.4.1: The proposition is proved by a bootstrap argument similar to the one of Theorem 4.2.1.

1. We prove that, assuming

$$\mathcal{O}^{*\infty}_{[0]} \le \Gamma_0\,,\ \underline{\mathcal{O}}^{*\infty}_{[0]} \le \Gamma_0, \tag{7.4.12}$$

with Γ_0 sufficiently small,[16], the following inequalities hold:

$$\begin{aligned}\underline{\mathcal{O}}^*_0 + \mathcal{O}^*_0 &\le c\,(\mathcal{I}_0 + \Delta_0)\\ \underline{\mathcal{O}}^*_1 + \mathcal{O}^*_1 &\le c\,(\mathcal{I}_0 + \Delta_0)\,.\end{aligned} \tag{7.4.13}$$

Also,

$$\sup_{p\in[2,4]}\left(\sup_{\underline{C}_*}|r^{1-\frac{2}{p}}\tau_-(\overline{\mathrm{tr}\underline{\chi}} + \frac{2}{r})|_{p,S}\right) \le c\,(\mathcal{I}_0 + \Delta_0)$$

[15] Observe that the assumptions (7.4.10) follow from the global initial data conditions (3.6.3); see Theorem 3.3.1.

[16] Γ_0 must be such that $\Gamma_0^2 < (\Delta_0 + \mathcal{I}_0) < \Gamma_0$.

$$\sup_{p\in[2,4]}\left(\sup_{\underline{C}_*}|r^{2-\frac{2}{p}}(\overline{\mathrm{tr}\chi}-\frac{2}{r})|_{p,S}\right)\leq c\,(\mathcal{I}_0+\Delta_0) \tag{7.4.14}$$

$$\sup_{p\in[2,4]}\left(\sup_{\underline{C}_*}|r^{1-\frac{2}{p}}(\Omega-\frac{1}{2})|_{p,S}+\mathcal{O}^{*p,S}_0(\underline{\omega})\right)\leq c\,(\mathcal{I}_0+\Delta_0)$$

$$\sup_{p\in[2,4]}\left(\mathcal{O}^{*p,S}_1(\underline{\omega})+\mathcal{O}^{*p,S}_0(\mathbf{D}_3\underline{\omega})\right)\leq c\,(\mathcal{I}_0+\Delta_0+\Delta_1)\,.$$

2. From the estimates (7.4.13), (7.4.14), using the Sobolev inequality of Lemma 4.1.3, it is possible to control $\underline{\mathcal{O}}^{*\infty}_{[0]}$ and $\mathcal{O}^{*\infty}_{[0]}$. We obtain

$$\underline{\mathcal{O}}^{*\infty}_{[0]}(\underline{C}_*)+\mathcal{O}^{*\infty}_{[0]}(\underline{C}_*)\leq c\,(\mathcal{I}_0+\Delta_0)\,. \tag{7.4.15}$$

3. To complete the bootstrap argument we consider the portion of $\underline{C}_*$,

$$\underline{C}_*([\lambda_1,\lambda_2))=\{p\in\underline{C}_*|u(p)\in[\lambda_1,\lambda_2)\}\,,$$

where the following inequality holds:[17]

$$\underline{\mathcal{O}}^{*\infty}_{[0]}(\underline{C}_*([\lambda_1,\lambda_2)))+\mathcal{O}^{*\infty}_{[0]}(\underline{C}_*([\lambda_1,\lambda_2)))\leq\Gamma_0.$$

Using the result in (2), assuming $\mathcal{I}_0+\Delta_0$ sufficiently small, it follows that in this portion of $\underline{C}_*$ a better inequality holds, namely

$$\underline{\mathcal{O}}^{*\infty}_{[0]}(\underline{C}_*([\lambda_1,\lambda_2)))+\mathcal{O}^{*\infty}_{[0]}(\underline{C}_*([\lambda_1,\lambda_2)))\leq\frac{\Gamma_0}{2},$$

so that the region can be made larger. To avoid a contradiction one has to conclude that it coincides with the whole $\underline{C}_*$.

Repeating step (1) again, we also prove that inequalities (7.4.13) and (7.4.14) hold on the whole $\underline{C}_*$, and therefore the inequality

$$\underline{\mathcal{O}}^*_{[1]}+\mathcal{O}^*_{[1]}\leq c\,(\mathcal{I}_0+\Delta_0+\Delta_1)$$

which completes the proposition.

Corollary 7.4.1 *The previous result implies the following inequality; see also Corollary 4.3.1,*

$$\sup_{\underline{C}_*}|r\tau_-(\mathrm{tr}\chi+\mathrm{tr}\underline{\chi})|\leq c(\mathcal{I}_0+\mathcal{I}_*+\Delta_0). \tag{7.4.16}$$

7.4.2 Implementation of Proposition 7.4.1

Part (1): Under the bootstrap assumption (7.4.12) we prove the first estimate of (7.4.13). We start by controlling the norms $\mathcal{O}^{p,S}_q(\underline{C}_*)(\eta)$ for $q=0,1$, $p\in[2,4]$, using the Hodge system

$$\begin{aligned}\not{\mathrm{div}}\,\eta&=\frac{1}{2}(\hat{\chi}\cdot\hat{\underline{\chi}}-\overline{\hat{\chi}\cdot\hat{\underline{\chi}}})-(\rho-\overline{\rho})\\ \not{\mathrm{curl}}\,\eta&=\frac{1}{2}\hat{\underline{\chi}}\wedge\hat{\chi}+\sigma.\end{aligned} \tag{7.4.17}$$

[17] That the interval $[\lambda_1,\lambda_2)$ is not empty is guaranteed from the assumptions (7.4.7).

Observe that this Hodge system is derived from the Hodge system (4.3.34) with the help of the relation $\mu = \overline{\mu}$ defining the canonical foliation (see (3.3.9)).

We use the estimate of $\hat{\chi}$ and $\hat{\underline{\chi}}$ contained in the bootstrap assumption (7.4.12) and the assumptions on $\rho - \overline{\rho}$ and σ contained in the first line of (7.4.6). All this implies the inequality

$$\mathcal{O}_0^{*p,S}(\eta) + \mathcal{O}_1^{*p,S}(\eta) \leq c(\Delta_0 + \Gamma_0^2) \leq c(\mathcal{I}_0 + \Delta_0). \tag{7.4.18}$$

We estimate next the norms $\mathcal{O}_0^{*p,S}(\mathrm{tr}\chi)$ and $\mathcal{O}_0^{*p,S}(\hat{\chi})$. For them we use the evolution equations for $\mathrm{tr}\chi$ and $\hat{\chi}$ along $\underline{C}_*$ (see (3.1.46)):

$$\begin{aligned}
&\frac{d}{du}(\Omega\mathrm{tr}\chi) + \frac{1}{2}\Omega\mathrm{tr}\underline{\chi}(\Omega\mathrm{tr}\chi) = 2\Omega^2(|\eta|^2 - \frac{1}{2}\overline{\hat{\chi}\hat{\underline{\chi}}} + \overline{\rho}) \\
&\frac{d}{du}(\Omega\hat{\chi}_{ab}) + \frac{1}{2}\Omega\mathrm{tr}\underline{\chi}(\Omega\hat{\chi}_{ab}) = -\frac{1}{2}\Omega\hat{\underline{\chi}}_{ab}(\Omega\mathrm{tr}\chi) + \Omega^2(\nabla\!\!\!/\widehat{\otimes}\eta - \eta\widehat{\otimes}\eta)_{ab}.
\end{aligned} \tag{7.4.19}$$

Using the bootstrap assumption (7.4.12) for χ and for $\underline{\chi}$, the estimate for $\overline{\rho}$ and the previous result for η and $\nabla\!\!\!/\eta$, we estimate the right-hand side of the equations (7.4.19). Applying the evolution Lemma 4.1.5 we obtain immediately that

$$\begin{aligned}
&|r^{1-\frac{2}{p}}\mathrm{tr}\chi|_{p,S_*} \leq |r^{1-\frac{2}{p}}\mathrm{tr}\chi|_{p,\underline{C}_*\cap\Sigma_0} + \frac{1}{r}c\left(\Gamma_0\left(\mathcal{I}_0 + \Delta_0\right) + \Gamma_0^2\right) \\
&|r^{2-\frac{2}{p}}\tau_-^{\frac{1}{2}}\hat{\chi}|_{p,S_*} \leq |r^{2-\frac{2}{p}}\tau_-^{\frac{1}{2}}\hat{\chi}|_{p,\underline{C}_*\cap\Sigma_0} + c\left(\Gamma_0\left(\mathcal{I}_0 + \Delta_0\right) + \Gamma_0^2\right),
\end{aligned}$$

and using the assumption (7.4.10) for the connection coefficients on Σ_0, we obtain

$$\begin{aligned}
&|r^{1-\frac{2}{p}}\mathrm{tr}\chi|_{p,S_*} \leq c\left((\mathcal{I}_0 + \Delta_0) + \Gamma_0\left(\mathcal{I}_0 + \Delta_0\right) + \Gamma_0^2\right) \leq c\left(\mathcal{I}_0 + \Delta_0\right) \\
&|r^{2-\frac{2}{p}}\tau_-^{\frac{1}{2}}\hat{\chi}|_{p,S_*} \leq c\left((\mathcal{I}_0 + \Delta_0) + \Gamma_0\left(\mathcal{I}_0 + \Delta_0\right) + \Gamma_0^2\right) \leq c\left(\mathcal{I}_0 + \Delta_0\right).
\end{aligned} \tag{7.4.20}$$

To control the norms relative to $\underline{\eta}$, $\mathrm{tr}\chi - \overline{\mathrm{tr}\chi}$ and $\mathrm{tr}\underline{\chi} - \overline{\mathrm{tr}\underline{\chi}}$, we first need to estimate $(\mathbf{D}_3 \log\Omega - \overline{\mathbf{D}_3 \log\Omega})$ and $\nabla\!\!\!/\mathbf{D}_3 \log\Omega$. This is achieved in the following way.

The function $\log\Omega$ satisfies, on the last slice (see (3.3.12)), the elliptic equation

$$\Delta\!\!\!/\log\Omega = \frac{1}{2}\left[\mathrm{d}\!\!\!/\mathrm{iv}\,\underline{\eta} + \frac{1}{2}(\hat{\chi}\cdot\hat{\underline{\chi}} - \overline{\hat{\chi}\cdot\hat{\underline{\chi}}}) - (\rho - \overline{\rho})\right].$$

Differentiating this equation with respect to $\mathbf{D}_3$ we infer that $\mathbf{D}_3 \log\Omega$ satisfies the following elliptic equation (see Lemma 7.7.4):

$$\Delta\!\!\!/(\Omega\mathbf{D}_3 \log\Omega) = \mathrm{d}\!\!\!/\mathrm{iv}\,F_1 + G_1 - \overline{G_1}, \tag{7.4.21}$$

where

$$\begin{aligned}
&F_1 = \Omega\underline{\beta} + \tilde{F}_1\ ,\quad \tilde{F}_1 = \left(\frac{3}{2}\Omega\eta\cdot\hat{\underline{\chi}} + \frac{1}{4}\Omega\eta\mathrm{tr}\underline{\chi}\right), \\
&G_1 = H + \frac{1}{4}\Omega\mathbf{D}_3(\hat{\chi}\cdot\hat{\underline{\chi}}) - \frac{1}{2}(\Omega\mathrm{tr}\underline{\chi})(\rho - \overline{\rho}) + \frac{1}{4}(\Omega\mathrm{tr}\underline{\chi})(\hat{\chi}\cdot\hat{\underline{\chi}} - \overline{\hat{\chi}\cdot\hat{\underline{\chi}}}).
\end{aligned}$$

Remark: The term in $\tilde{F}_1$

$$\frac{1}{4}\Omega\eta\mathrm{tr}\underline{\chi} = \frac{1}{2r}\Omega\eta + \frac{1}{4}\Omega\eta\left(\mathrm{tr}\underline{\chi} - \frac{2}{r}\right)$$

contains a linear term; nevertheless, $\mathrm{d\!/\!iv}(\Omega\eta\mathrm{tr}\underline{\chi}) = \Omega\mathrm{tr}\chi\,\mathrm{d\!/\!iv}\,\eta$ plus quadratic terms that can be estimated by Γ_0^2. Thus using again the relation $\mu = \overline{\mu}$,

$$\mathrm{d\!/\!iv}(\Omega\eta\mathrm{tr}\underline{\chi}) = (\Omega\mathrm{tr}\chi)\left(\frac{1}{2}(\hat{\chi}\cdot\underline{\hat{\chi}} - \overline{\hat{\chi}\cdot\underline{\hat{\chi}}}) - (\rho - \overline{\rho})\right) + O(\Gamma_0^2) \leq c(\Delta_0 + \Gamma_0^2)\,.$$

Therefore applying Proposition 4.1.3 to (7.4.21) we obtain for $p \in [2, 4]$,

$$\begin{aligned}&|r^{1-\frac{2}{p}}\tau_-^{\frac{3}{2}}(\mathbf{D\!\!\!/}_3\log\Omega - \overline{\mathbf{D\!\!\!/}_3\log\Omega})|_{p,S} \leq c(\Delta_0 + \Gamma_0^2)\\ &|r^{2-\frac{2}{p}}\tau_-^{\frac{3}{2}}\nabla\!\!\!\!/\,\mathbf{D\!\!\!/}_3\log\Omega|_{p,S} \leq c(\Delta_0 + \Gamma_0^2).\end{aligned} \tag{7.4.22}$$

To estimate $\mathcal{O}_0^{*p,S}(\mathrm{tr}\chi)$ (see (7.4.3)), we have to control $\mathrm{tr}\chi - \mathrm{tr}\underline{\chi}$. It is simple to show that if $\tilde{W} = \mathrm{tr}\chi - \overline{\mathrm{tr}\chi}$, $\underline{\tilde{W}} = \mathrm{tr}\underline{\chi} - \overline{\mathrm{tr}\underline{\chi}}$, its evolutions equation is (see also Proposition 4.3.14)

$$\frac{d}{du}\tilde{W} + \frac{1}{2}\Omega\mathrm{tr}\underline{\chi}\tilde{W} = -\frac{1}{2}\underline{\tilde{W}}(\overline{\Omega\mathrm{tr}\chi}) - \frac{1}{2}\overline{\underline{\tilde{W}}(\Omega\mathrm{tr}\chi)} + F - \overline{F}, \tag{7.4.23}$$

where $F = 2\Omega^2(|\eta|^2 - \frac{1}{2}\overline{\hat{\chi}\underline{\hat{\chi}}})$. To use this evolution equation we need an estimate on $\underline{C}_*$ of $\underline{\tilde{W}}$. We observe that $\underline{V} = \Omega\mathrm{tr}\underline{\chi} - \overline{\Omega\mathrm{tr}\underline{\chi}}$ satisfies the following evolution equation (see (4.3.114) and Proposition 4.3.15:[18]

$$\begin{aligned}\frac{d}{du}\underline{V} + \Omega\mathrm{tr}\underline{\chi}\,\underline{V} &= \frac{1}{2}\underline{V}^2 - \overline{\underline{V}^2} + 2(\Omega\mathbf{D}_3\log\Omega)\underline{V} - \left[\Omega^2|\underline{\hat{\chi}}|^2 - \overline{\Omega^2|\underline{\hat{\chi}}|^2}\right]\\ &\quad + 2(\overline{\Omega\mathrm{tr}\underline{\chi}})\left(\Omega\mathbf{D}_3\log\Omega - \overline{\Omega\mathbf{D}_3\log\Omega}\right)\\ &\quad - \overline{2(\Omega\mathrm{tr}\underline{\chi})\left(\Omega\mathbf{D}_3\log\Omega - \overline{\Omega\mathbf{D}_3\log\Omega}\right)}.\end{aligned}$$

Applying the evolution Lemma 4.1.5 and Gronwall's lemma, and using the estimate for $\underline{\hat{\chi}}$ in the bootstrap assumption (7.4.12) and the estimate (see (7.4.22))

$$|r^{1-\frac{2}{p}}\tau_-^{\frac{3}{2}}(\Omega\mathbf{D}_3\log\Omega - \overline{\Omega\mathbf{D}_3\log\Omega})|_{p,S_*} \leq c\,(\mathcal{I}_0 + \Delta_0)\,,$$

and, finally, the assumptions for the connection coefficients on Σ_0,[19] (see (7.4.7)) we obtain the following estimate for $\underline{\tilde{W}}$:

$$|r^{1-\frac{2}{p}}\tau_-^{\frac{3}{2}}\underline{\tilde{W}}|_{p,S_*} \leq c\left((\mathcal{I}_0 + \Delta_0) + \Gamma_0\,(\mathcal{I}_0 + \Delta_0) + \Gamma_0^2\right) \leq c\,(\mathcal{I}_0 + \Delta_0)\,. \tag{7.4.24}$$

---

[18]In Proposition 4.3.15 the estimate of $\underline{\tilde{W}}$ depends also on $\mathcal{I}_*$ due to the factor $\underline{\omega} - \overline{\underline{\omega}}$. Here the dependence on $\mathcal{I}_*$ is absent since, on $\underline{C}_*$, $\underline{\omega} - \overline{\underline{\omega}}$ is estimated differently; see the proof of Proposition 7.4.4.

[19]$\underline{\mathcal{O}}^*_{[1]}(\underline{C}_* \cap \Sigma_0) + \underline{\mathcal{O}}^*_{[1]}(\underline{C}_* \cap \Sigma_0) \leq \mathcal{I}_0$.

Therefore the right-hand side of (7.4.23) can be bounded in the $|\cdot|_{p,S}$ norm as

$$\left|\frac{1}{2}\underline{\tilde{W}(\overline{\Omega \mathrm{tr}\chi})} + \frac{1}{2}\overline{\underline{\tilde{W}(\Omega \mathrm{tr}\chi)}} - (F - \overline{F})\right|_{p,S_*} \leq \frac{1}{r^2 \tau_-^{\frac{3}{2}}} c(\Gamma_0)\,(\mathcal{I}_0 + \Delta_0)\ .$$

Applying the evolution Lemma 4.1.5 and Gronwall's lemma to the evolution equation (7.4.23) for $\tilde{W} = \mathrm{tr}\chi - \overline{\mathrm{tr}\chi}$, and using assumptions (7.4.7) relative to Σ_0, we conclude that

$$\begin{aligned} |r^{2-\frac{2}{p}} \tau_-^{\frac{1}{2}} (\mathrm{tr}\chi - \overline{\mathrm{tr}\chi})|_{p,S_*} &\leq |r^{2-\frac{2}{p}} \tau_-^{\frac{1}{2}} (\mathrm{tr}\chi - \overline{\mathrm{tr}\chi})|_{p,\underline{C}_* \cap \Sigma_0} + c(\Gamma_0)\,(\mathcal{I}_0 + \Delta_0) \\ &\leq c\,(\mathcal{I}_0 + \Delta_0)\,, \end{aligned} \tag{7.4.25}$$

which, together with the inequalities (7.4.20) completes the proof of the first inequality of (7.4.13) for the connection coefficients that are not underlined. Inequality (7.4.24) allows us also to conclude that

$$\begin{aligned} |r^{1-\frac{2}{p}} \tau_-^{\frac{3}{2}} (\mathrm{tr}\underline{\chi} - \overline{\mathrm{tr}\underline{\chi}})|_{p,S_*} &\leq |r^{1-\frac{2}{p}} \tau_-^{\frac{3}{2}} (\mathrm{tr}\underline{\chi} - \overline{\mathrm{tr}\underline{\chi}})|_{p,\underline{C}_* \cap \Sigma_0} + c(\Gamma_0)\,(\mathcal{I}_0 + \Delta_0) \\ &\leq c\,(\mathcal{I}_0 + \Delta_0)\,. \end{aligned} \tag{7.4.26}$$

We shall now estimate the angular derivatives $\nabla\!\!\!/ \mathrm{tr}\chi$, $\nabla\!\!\!/ \hat{\chi}$ and $\nabla\!\!\!/ \eta$. The last term has already been estimated (see (7.4.18)). To estimate $\nabla\!\!\!/ \mathrm{tr}\chi$ we take the tangential derivative of the first equation of (7.4.19). The estimate then proceeds in the same way as that for $\mathrm{tr}\chi$. Observe that because the foliation on the last slice is canonical, the dependence on the Riemann components disappears, since $\nabla\!\!\!/ \overline{\rho} = 0$. We obtain the result for $p \in [2, 4]$,

$$|r^{3-2/p} \tau_-^{\frac{1}{2}} \nabla\!\!\!/ \mathrm{tr}\chi|_{p,S_*} \leq c(\mathcal{I}_0 + \Delta_0). \tag{7.4.27}$$

To estimate $\nabla\!\!\!/ \hat{\chi}$ we can proceed in the same way, differentiating tangentially the evolution equation of $\hat{\chi}$ along $\underline{C}_*$ (see (7.4.19)). Observe that the right-hand side of this evolution equation for $\nabla\!\!\!/ \hat{\chi}$ does not contain curvature terms. We obtain the estimate

$$|r^{3-2/p} \tau_-^{\frac{1}{2}} \nabla\!\!\!/ \hat{\chi}|_{p,S_*} \leq c(\mathcal{I}_0 + \Delta_0). \tag{7.4.28}$$

Remark: It is interesting to note that the estimate of $\nabla\!\!\!/ \hat{\chi}$ could also be obtained from the Hodge system,

$$\mathrm{d\!\!/iv}\,\hat{\chi} = \frac{1}{2} \nabla\!\!\!/ \mathrm{tr}\chi - \zeta \cdot \chi - \beta\,,$$

(see (4.3.13)) by applying Proposition 4.1.3 to it. To achieve this result, however, we need a better estimate for ζ. This is done below.

To estimate ζ we consider its evolution equation along the null-incoming hypersurfaces (see (3.1.46))

$$\mathbf{D\!\!\!/}_3 \zeta + 2\underline{\chi} \cdot \zeta - \mathbf{D}_3 \nabla\!\!\!/ \log\Omega = -\underline{\beta}.$$

Due to the "better" estimate for $\nabla\!\!\!/ \mathbf{D}_3 \log\Omega$ (see (7.4.22)) it follows immediately that, on $\underline{C}_*$, ζ satisfies the inequality

$$|r^{2-2/p} \tau_-^{\frac{1}{2}} \zeta|_{p,S_*} \leq c(\mathcal{I}_0 + \Delta_0). \tag{7.4.29}$$

This estimate for ζ and the analogous one for its tangential derivative, obtained exactly in the same way, together with the estimates for η and $\nabla\eta$ (see (7.4.18)) allows us also to conclude that $\underline{\eta}$ and $\nabla\underline{\eta}$ satisfy on $\underline{C}_*$ the same estimate as η and $\nabla\eta$,

$$\mathcal{O}_0^{p,S}(\underline{C}_*)(\underline{\eta})+\mathcal{O}_1^{p,S}(\underline{C}_*)(\underline{\eta})\leq c(\Delta_0+\Gamma_0^2)\leq c(\mathcal{I}_0+\Delta_0). \tag{7.4.30}$$

To control $\nabla\mathrm{tr}\underline{\chi}$ we derive the evolution equation for $\mathrm{tr}\underline{\chi}$ along the incoming "cones," and using again the bootstrap assumption (7.4.12) for $\hat{\underline{\chi}}$ and $\nabla\hat{\underline{\chi}}$ and the estimate (7.4.47) for $\nabla\underline{\omega}$, we obtain

$$\begin{aligned} |r^{3-\frac{2}{p}}\tau_-^{\frac{1}{2}}\nabla\mathrm{tr}\underline{\chi}|_{p,S_*} &\leq |r^{3-\frac{2}{p}}\tau_-^{\frac{1}{2}}\nabla\mathrm{tr}\underline{\chi}|_{p,\underline{C}_*\cap\Sigma_0}+c\left(\Gamma_0(\mathcal{I}_0+\Delta_0)+\Gamma_0^2\right) \\ &\leq c\left((\mathcal{I}_0+\Delta_0)+\Gamma_0(\mathcal{I}_0+\Delta_0)+\Gamma_0^2\right)\leq c\,(\mathcal{I}_0+\Delta_0)\,. \end{aligned} \tag{7.4.31}$$

To complete the proof of the inequalities in (7.4.13) we are left with estimating, on $\underline{C}_*$, $\hat{\underline{\chi}}$ and $\nabla\hat{\underline{\chi}}$. These estimates are obtained by applying Proposition 4.1.3 to the Hodge system

$$\mathrm{d}\!\!/\mathrm{iv}\,\hat{\underline{\chi}}=\frac{1}{2}\nabla\mathrm{tr}\underline{\chi}+\zeta\cdot\underline{\chi}-\underline{\beta}\,,$$

which gives

$$\begin{aligned} |r^{1-\frac{2}{p}}\tau_-^{\frac{3}{2}}\hat{\underline{\chi}}|_{p,S_*} &\leq c\,((\mathcal{I}_0+\Delta_0)+\Gamma_0(\mathcal{I}_0+\Delta_0))\leq c\,(\mathcal{I}_0+\Delta_0) \\ |r^{2-\frac{2}{p}}\tau_-^{\frac{3}{2}}\nabla\hat{\underline{\chi}}|_{p,S_*} &\leq c\,((\mathcal{I}_0+\Delta_0)+\Gamma_0(\mathcal{I}_0+\Delta_0))\leq c\,(\mathcal{I}_0+\Delta_0)\,. \end{aligned} \tag{7.4.32}$$

To complete the proof of part (1) we still have to prove the inequalities (7.4.14). We obtain the estimates for $\overline{\mathrm{tr}\chi}$ and $\overline{\mathrm{tr}\underline{\chi}}$, $p\in[2,4]$,

$$\begin{aligned} |r^{2-\frac{2}{p}}(\overline{\mathrm{tr}\chi}-\frac{2}{r})|_{p,S_*} &\leq c(\Gamma_0)\left((\mathcal{I}_0+\Delta_0)+\Gamma_0^2\right) \\ |r^{1-\frac{2}{p}}\tau_-(\overline{\mathrm{tr}\underline{\chi}}+\frac{2}{r})|_{p,S_*} &\leq c(\Gamma_0)\left((\mathcal{I}_0+\Delta_0)+\Gamma_0^2\right), \end{aligned} \tag{7.4.33}$$

by starting again from the evolution equations for $\mathrm{tr}\chi$ and $\mathrm{tr}\underline{\chi}$ along $\underline{C}_*$ and using the estimate (7.2.8) for $\left(\overline{\mathrm{tr}\theta}-\frac{2}{r}\right)$ at $\Sigma_0\cap\underline{C}_*$, proved in Theorem **M3**.

The remaining estimates of (7.4.14) are obtained in the following way. We control $\Delta\!\!\!/\log\Omega$ from the estimates we already have of $\mathrm{d}\!\!/\mathrm{iv}\,\eta$ and $\mathrm{d}\!\!/\mathrm{iv}\,\underline{\eta}$. From it we control $\log\Omega$, recalling that on $\underline{C}_*$, $\overline{\log\Omega}=0$, and its first and second tangential derivatives. This result is collected in the following proposition.

Proposition 7.4.3 *Under the bootstrap assumptions (7.4.12) and the assumptions (7.4.6) for the Riemann components, we have for any $p\geq 2$,*

$$\begin{aligned} |r^{3-2/p}\tau_-^{\frac{1}{2}}\nabla^2\log\Omega|_{p,S_*} &\leq c|r^{3-2/p}\tau_-^{\frac{1}{2}}\Delta\!\!\!/\log\Omega|_{p,S_*}\leq c(\Gamma_0)(\mathcal{I}_0+\Delta_0) \\ |r^{2-2/p}\tau_-^{\frac{1}{2}}\nabla\log\Omega|_{p,S_*} &\leq c|r^{3-2/p}\tau_-^{\frac{1}{2}}\Delta\!\!\!/\log\Omega|_{p,S_*}\leq c(\Gamma_0)(\mathcal{I}_0+\Delta_0) \\ |r^{1-2/p}\tau_-^{\frac{1}{2}}\log 2\Omega|_{p,S_*} &\leq c|r^{3-2/p}\tau_-^{\frac{1}{2}}\Delta\!\!\!/\log\Omega|_{p,S_*}\leq c(\Gamma_0)(\mathcal{I}_0+\Delta_0). \end{aligned} \tag{7.4.34}$$

To complete part (1) we still have to control $\mathcal{O}_0^{p,S}(\mathbf{D}_3\underline{\omega})(\underline{C}_*)$. This is discussed in the next subsection together with the proof of Proposition 7.4.2.

Proof of Corollary 7.4.1: The estimate of $(\mathrm{tr}\chi + \mathrm{tr}\underline{\chi})$ is an easy consequence of the previous estimates (7.4.25), (7.4.26) and (7.4.33). The final result is

$$\sup_{\underline{C}_*} |r\tau_-(\mathrm{tr}\chi + \mathrm{tr}\underline{\chi})| \leq c\,(\mathcal{I}_0 + \Delta)\,.$$

7.4.3 Implementation of Proposition 7.4.2

The proof of Proposition 7.4.2 is similar to that of Proposition 7.4.1. We use, in fact, for the norms associated with $\chi, \underline{\chi}$ the evolution equations obtained by differentiating the previous ones.

The estimates for the second tangential derivatives $\nabla\!\!\!/^2\eta$ are obtained by differentiating the Hodge system (7.4.17) and those for $\nabla\!\!\!/^2\underline{\eta}$ by differentiating twice the evolution equation for ζ and taking into account the estimate for $\nabla\!\!\!/^2\mathbf{D}\!\!\!\!/_3\log\Omega$.

The control of the norms $\mathcal{O}_0^*(\mathbf{D}\!\!\!\!/_3^2\underline{\omega})$, $\mathcal{O}_1^*(\mathbf{D}\!\!\!\!/_3\underline{\omega})$ and $\mathcal{O}_2^*(\underline{\omega})$ is obtained, via elliptic estimates, in Propositions 7.4.4, 7.4.5. These propositions allow us to control the norm $\mathcal{O}_0^{*p,S}(\mathbf{D}_3\underline{\omega})$ also, completing, therefore, the proof of Proposition 7.4.1.

We omit the details concerning the estimates for the second angular derivatives of $\mathrm{tr}\chi, \hat{\chi}, \mathrm{tr}\underline{\chi}, \underline{\hat{\chi}}, \eta, \underline{\eta}$ and we sketch below the estimate for the third derivatives of $\mathrm{tr}\chi$ and $\underline{\omega}$.

Estimate for $\mathcal{O}_3^*(\mathrm{tr}\chi)$:
The estimate of $||r^3\nabla\!\!\!/^3\mathrm{tr}\chi||_{L^2(\underline{C}_*\cap V(u,\underline{u}_*))}$ proceeds in a way similar to the estimate of $\nabla\!\!\!/\mathcal{U}\!\!\!/$ in Proposition 4.5.1 where $\mathcal{U}\!\!\!/ \equiv \mathrm{d}\!\!\!/\mathrm{iv}\,\mathcal{U}\!\!\!/ + \Omega^{-1}\mathrm{tr}\chi\rho$, but it is simpler due to the canonical foliation of $\underline{C}_*$. We start with the evolution equation for $\nabla\!\!\!/^3\mathrm{tr}\chi$ obtained by deriving three times tangentially the evolution equation along $\underline{C}_*$ for $\mathrm{tr}\chi$,

$$\frac{d}{du}(\mathrm{tr}\chi) + \frac{1}{2}\Omega\mathrm{tr}\underline{\chi}(\mathrm{tr}\chi) = \mathcal{X}_0, \tag{7.4.35}$$

where

$$\mathcal{X}_0 = -\Omega(\mathbf{D}_3\log\Omega)\mathrm{tr}\chi + 2\Omega|\eta|^2 + \Omega\left(2\overline{\rho} - \overline{\hat{\chi}\underline{\hat{\chi}}}\right), \tag{7.4.36}$$

satisfies, from the results of Proposition 7.4.1

$$|\mathcal{X}_0| = O(r^{-3}) + O(r^{-2}\tau_-^{-\frac{3}{2}}). \tag{7.4.37}$$

The evolution equation for $\nabla\!\!\!/\mathrm{tr}\chi$ can be written as

$$\frac{d}{du}(\nabla\!\!\!/\mathrm{tr}\chi) + \Omega\mathrm{tr}\underline{\chi}(\nabla\!\!\!/\mathrm{tr}\chi) + (\mathbf{D}_3\log\Omega)(\nabla\!\!\!/\mathrm{tr}\chi) = -\underline{\hat{\chi}}(\nabla\!\!\!/\mathrm{tr}\chi) + \mathcal{X}_1, \tag{7.4.38}$$

where

$$\mathcal{X}_1 = \left[-\frac{1}{2}(\nabla\!\!\!/\mathrm{tr}\underline{\chi})\mathrm{tr}\chi - (\nabla\!\!\!/\mathbf{D}_3\log\Omega)\mathrm{tr}\chi + 4\eta\cdot\nabla\!\!\!/\eta + \nabla\!\!\!/\log\Omega(\mathbf{D}_3\log\Omega)\right] \tag{7.4.39}$$

and

$$|\mathcal{X}_1| = O(r^{-4}\tau_-^{-\frac{1}{2}}) + O(r^{-3}\tau_-^{-\frac{3}{2}}). \tag{7.4.40}$$

Iterating the procedure we obtain for $\nabla\!\!\!/^3\mathrm{tr}\chi$ the evolution equation

$$\frac{d}{du}(\nabla\!\!\!/^3\mathrm{tr}\chi) + 2\Omega\mathrm{tr}\underline{\chi}(\nabla\!\!\!/^3\mathrm{tr}\chi) + (\mathbf{D}_3\log\Omega)(\nabla\!\!\!/^3\mathrm{tr}\chi) = -\hat{\underline{\chi}}(\nabla\!\!\!/^3\mathrm{tr}\chi), +\mathcal{X}_3 \tag{7.4.41}$$

where $\mathcal{X}_3$ is a term, up to fourth order in the connection coefficients up to second tangential derivatives and linear in the Riemann components up to first derivatives. Its asymptotic behavior is

$$|r^{-\frac{2}{p}}\mathcal{X}_3|_{p,S} = O(r^{-6}\tau_-^{-\frac{1}{2}}) + O(r^{-5}\tau_-^{-\frac{3}{2}}). \tag{7.4.42}$$

Therefore, applying the evolution Lemma 4.1.5 and the Gronwall Lemma we obtain

$$\begin{aligned} |r^{4-\frac{2}{p}}\nabla\!\!\!/^3\mathrm{tr}\chi|_{p=2,S}(u,\underline{u}_*) &\le c|r^{4-\frac{2}{p}}\nabla\!\!\!/^3\mathrm{tr}\chi|_{p=2,\underline{C}_*\cap\Sigma_0} + \int_{u_0(\underline{u}_*)}^{u} du'|r^{4-\frac{2}{p}}\mathcal{X}_3|_{p=2,S} \\ &\le c|r^{4-\frac{2}{p}}\nabla\!\!\!/^3\mathrm{tr}\chi|_{p=2,\underline{C}_*\cap\Sigma_0} + c\frac{(\mathcal{I}_0+\Delta)}{r(u,\underline{u}_*)u^{\frac{1}{2}}}. \end{aligned} \tag{7.4.43}$$

From this expression we obtain, with $\epsilon > 0$,

$$\begin{aligned} &||r^3\nabla\!\!\!/^3\mathrm{tr}\chi||_{L^2(\underline{C}_*\cap V(u,\underline{u}_*))} \\ &\le c\left(\int_{u_0(\underline{u}_*)}^{u}|r^3\nabla\!\!\!/^3\mathrm{tr}\chi|^2_{p=2,\underline{C}_*\cap\Sigma_0}\right)^{\frac{1}{2}} + c(\mathcal{I}_0+\Delta)\left(\int_{u_0(\underline{u}_*)}^{u}\frac{1}{r(u,\underline{u}_*)^2u}\right)^{\frac{1}{2}} \\ &\le c|u_0(\underline{u}_*)-u|^{\frac{1}{2}}\left(|r^3\nabla\!\!\!/^3\mathrm{tr}\chi|_{p=2,\underline{C}_*\cap\Sigma_0}\right) + c(\mathcal{I}_0+\Delta)\frac{1}{r^{1-\epsilon}(u,\underline{u}_*)u^{\epsilon}}. \end{aligned} \tag{7.4.44}$$

The estimate of the first term is connected to the estimate of $\nabla\!\!\!/^3\mathrm{tr}\theta$ on Σ_0. In fact, on Σ_0, $\mathrm{tr}\chi = \mathrm{tr}\theta - k_{\tilde{N},\tilde{N}}$ and the second fundamental form is controlled from the assumption on J_K. Therefore we are left with estimating $|r^3\nabla\!\!\!/^3\mathrm{tr}\theta|_{p=2,\underline{C}_*\cap\Sigma_0}$. Observing that $\underline{C}_*\cap\Sigma_0$ is, by construction, a leaf of the canonical foliation on Σ_0, we can integrate the evolution equation for $\mathrm{tr}\theta$, (7.1.25), after differentiating it tangentially three times,

$$\nabla_{\tilde{N}}(\nabla\!\!\!/^3\mathrm{tr}\theta) + \frac{5}{2}\mathrm{tr}\theta(\nabla\!\!\!/^3\mathrm{tr}\theta) = -2\hat{\theta}\cdot\nabla\!\!\!/^3\hat{\theta} + \mathcal{P}_3, \tag{7.4.45}$$

where $\mathcal{P}_3$ depends on the second fundamental form up to third derivatives and on terms with lower order tangential derivatives. Taking into account that, due to strong asymptotic flatness, $\lim_{r\to\infty}|r^{\frac{11}{2}}\nabla\!\!\!/^3\mathrm{tr}\theta| = 0$, and recalling the estimate (7.2.9), we conclude that

$$|r^{\frac{11}{2}-\frac{2}{p}}\nabla\!\!\!/^3\mathrm{tr}\theta|_{p=2,C_*\cap\Sigma_0} \le c(\mathcal{I}_0+\Delta).$$

This implies that

$$|r^3\nabla\!\!\!/^3\mathrm{tr}\theta|_{p=2,C_*\cap\Sigma_0} \le c\frac{(\mathcal{I}_0+\Delta)}{r^{\frac{3}{2}}(u_0,(\underline{u}_*),\underline{u}_*)}, \tag{7.4.46}$$

which, when substituted into (7.4.44), completes the estimate of $r^{\frac{1}{2}}||r^3\nabla\!\!\!/^3\mathrm{tr}\chi||_{L^2(\underline{C}_*\cap V(u,\underline{u}_*))}$.

Estimate for $\mathcal{O}_3^*(\omega)$: This estimate is a corollary of the elliptic estimates proved in the next subsection.

The elliptic estimates on the last slice

Proposition 7.4.4 *Under the assumptions of Proposition 7.4.1 there exists a generic constant c such that for $p \in [2,4]$ on $\underline{C}_*$,*[20][21]

$$
\begin{aligned}
&|r^{1-\frac{2}{p}}\tau_-^{\frac{3}{2}}\Omega\mathbf{D}_3\log\Omega|_{p,S}(u,\underline{u}_*) \leq c(\mathcal{I}_0+\Delta_0)\\
&|r^{2-\frac{2}{p}}\tau_-^{\frac{3}{2}}\nabla\!\!\!/(\Omega\mathbf{D}_3\log\Omega)|_{p,S}(u,\underline{u}_*) \leq c_0(\mathcal{I}_0+\Delta_0)\\
&|r^{3-\frac{2}{p}}\tau_-^{\frac{3}{2}}\nabla\!\!\!/^2(\Omega\mathbf{D}_3\log\Omega)|_{p,S}(u,\underline{u}_*) \leq c_0(\mathcal{I}_0+\Delta_0+\Delta_1).
\end{aligned}
\tag{7.4.47}
$$

Proposition 7.4.5 *Under the assumptions of Proposition 7.4.2 there exists a generic constant c such that*

$$
\begin{aligned}
&|r^{1-\frac{2}{p}}\tau_-^{\frac{5}{2}}(\Omega\mathbf{D}_3)^2\log\Omega|_{p,S}(u,\underline{u}_*) \leq c(\mathcal{I}_0+\Delta_0)\\
&|r^{2-\frac{2}{p}}\tau_-^{\frac{5}{2}}\nabla\!\!\!/(\Omega\mathbf{D}_3)^2\log\Omega)|_{p,S}(u,\underline{u}_*) \leq c(\mathcal{I}_0+\Delta_0+\Delta_1)\\
&|r^{1-\frac{2}{p}}\tau_-^{\frac{7}{2}}(\Omega\mathbf{D}_3)^3\log\Omega|_{p,S_*}(u,\underline{u}_*) \leq c(\mathcal{I}_0+\Delta_0+\Delta_1).
\end{aligned}
\tag{7.4.48}
$$

The proofs of these propositions are in the appendix to this chapter.

7.5 The last slice rotation deformation estimates

In this section we prove Theorem **M5** which we recall here.

Theorem M5: *Assume that, relative to a canonical foliation on $\underline{C}_*$, we have*

$$\mathcal{R}_{[2]}+\underline{\mathcal{R}}_{[2]} \leq \Delta$$

$$\mathcal{O}_{[2]}(\underline{C}_*\cap\Sigma_0)+\mathcal{O}_3(\Sigma_0\backslash K)+\underline{\mathcal{O}}_{[3]}(\Sigma_0\backslash K) \leq c\mathcal{I}_0\,.$$

If Δ, $\mathcal{I}_0$ are sufficiently small, the following estimate holds:

$$\mathcal{D}(\underline{C}_*) \leq c(\mathcal{I}_0+\Delta)\,.$$

The proof of this theorem is divided into two propositions.

Proposition 7.5.1 *Under the assumptions of Proposition 7.4.2, as well as*

$$\mathcal{D}_{[1]}(\underline{C}_*\cap\Sigma_0) \leq c(\mathcal{I}_0+\Delta) \tag{7.5.1}$$

$$\left(\int_{u_0(\underline{u}')}^{u} du'|r^{2-\frac{2}{p}}\nabla\!\!\!/^2 H|^2_{p=2,\underline{C}_*\cap\Sigma_0}\right)^{\frac{1}{2}} \leq c(\mathcal{I}_0+\Delta),$$

the following estimates hold for $p \in [2,4]$:

$$
\begin{aligned}
&|r^{-1\,(i)}O|_{p,S_*} \leq c\,(\mathcal{I}_0+\Delta) \quad , \quad |\nabla\!\!\!/^{(i)}O|_{p,S_*} \leq c\,(\mathcal{I}_0+\Delta)\\
&|r^{(i)}H_{ab}|_{p,S_*} \leq c\,(\mathcal{I}_0+\Delta) \quad , \quad |r^{2-\frac{2}{p}}\nabla\!\!\!/^{(i)}H_{ab}|_{p,S_*} \leq c\,(\mathcal{I}_0+\Delta)\,,
\end{aligned}
\tag{7.5.2}
$$

---

[20] We stress here that the proof of this proposition does not require the completion of Proposition 7.4.1.

[21] Observe that this result for $\Omega\mathbf{D}_3\log\Omega$ on $\underline{C}_*$ is stronger than the one that holds on $\mathcal{K}$ (see Propositions 4.3.4).

and

$$||r\nabla\!\!\!\!/^2 H||_{L^2(\underline{C}_*\cap V(u,\underline{u}_*))} =\leq c\,(\mathcal{I}_0+\Delta)\,. \tag{7.5.3}$$

Proposition 7.5.2 *Assume that, relative to a canonical foliation on $\Sigma_0\backslash K$,*

$$\mathcal{R}_{[2]}+\underline{\mathcal{R}}_{[2]}\leq\Delta$$
$$\mathcal{O}_{[2]}(\underline{C}_*\cap\Sigma_0)+\mathcal{O}_{[3]}(\Sigma_0\backslash K)+\underline{\mathcal{O}}_{[3]}(\Sigma_0\backslash K)\leq c\mathcal{I}_0.$$

Then, if Δ and $\mathcal{I}_0$ are sufficiently small,

$$\mathcal{D}_{[1]}(\underline{C}_*\cap\Sigma_0)\leq c(\mathcal{I}_0+\Delta) \tag{7.5.4}$$
$$\left(\int_{u_0(\underline{u}')}^{u} du'|r^{2-\frac{2}{p}}\nabla\!\!\!\!/^2 H|^2_{p=2,\underline{C}_*\cap\Sigma_0}\right)^{\frac12}\leq c(\mathcal{I}_0+\Delta).$$

Proof of Proposition 7.5.1: We first recall the construction of the rotation vector fields on $\underline{C}_*$. We start with the vector fields defined in $S_*(\lambda_1)=\underline{C}_*\cap\Sigma_0$ where we have already defined the rotation group (see [Ch-Kl], Chapter 3 and the proof of Proposition 7.5.2). This is achieved in a similar way as for the extension discussed in Chapter 4, Subsection 4.6.1. Let $q\in S_*(\lambda)$ be an arbitrary point of $\underline{C}_*$. Since $S_*(\lambda)$ is diffeomorphic, via $\underline{\phi}_{\Delta}$, to $S_*(\lambda_1)$, with $\Delta=\lambda-\lambda_1$, there exists a point $p\in S_*(\lambda_1)$ such that $q=\underline{\phi}_{\Delta}(p)$. We define the element O_* of the rotation group operating on $q\in\underline{C}_*$ as[22]

$$(O_*;q)\equiv\underline{\phi}_{\Delta}(O;p),$$

where $(O_*;q)$ is a point of $S_*(\lambda)$ and $(O;p)$ is the point of $S_*(\lambda_1)$ obtained by applying O to the point p. This extension of the action of the rotation group to the whole $\underline{C}_*$ satisfies

$$O_*=\underline{\phi}_t^{-1}O_*\underline{\phi}_t,$$

which implies that the generators, ${}^{(i)}O_*$, satisfy

$$[\underline{N},{}^{(i)}O_*]=0.$$

From the previous definitions we can easily check that

$$[{}^{(i)}O_*,{}^{(j)}O_*]=\epsilon_{ijk}{}^{(k)}O_*\ ,\ i,j,k\in\{1,2,3\}.$$

In conclusion the generators ${}^{(i)}O_*$, defined on the whole $\underline{C}_*$, tangent to $S_*(\lambda)$ at each point, satisfy

$$\begin{aligned}&[{}^{(i)}O_*,{}^{(j)}O_*]=\epsilon_{ijk}{}^{(k)}O_*\\&[\underline{N},{}^{(i)}O_*]=0\\&g({}^{(i)}O_*,e_3)=g({}^{(i)}O_*,e_4)=0.\end{aligned} \tag{7.5.5}$$

Moreover, as $\underline{N}=\Omega e_3$, it follows that

$$[{}^{(i)}O_*,e_3]={}^{(i)}Fe_3, \tag{7.5.6}$$

where ${}^{(i)}F\equiv-{}^{(i)}O_{*c}(\nabla\!\!\!\!/_c\log\Omega)\ ,\ {}^{(i)}O_{*c}\equiv g({}^{(i)}O_*,e_c)$.

---

[22] At the differential level the extension is defined as: ${}^{(i)}O_*\equiv\underline{\phi}_{\Delta *}{}^{(i)}O$.

Proposition 7.5.3 *The quantities* ${}^{(i)}O_{*a}$ *and* $(\nabla\!\!\!/\, {}^{(i)}O_*)_{ab}$ *satisfy the evolution equations*

$$\frac{d}{du}{}^{(i)}O_{*b} = \Omega\underline{\chi}_{bc}{}^{(i)}O_{*c} \tag{7.5.7}$$

$$\frac{d}{du}(\nabla\!\!\!/\, {}^{(i)}O_*)_{ab} = \Omega\Big[\hat{\underline{\chi}}_{bc}(\nabla\!\!\!/\, {}^{(i)}O_*)_{ac} - \hat{\underline{\chi}}_{ac}(\nabla\!\!\!/\, {}^{(i)}O_*)_{cb} + {}^{(i)}O_{*c}(\underline{\chi}_{cb}\eta_a - \underline{\chi}_{ca}\eta_b)$$
$$+ {}^{(i)}O_{*c}R_{3abc} - {}^{(i)}O_{*c}\underline{\chi}_{cb}\zeta_a + \underline{\chi}_{ab}({}^{(i)}O_{*c}\eta_c) + {}^{(i)}O_{*c}(\nabla\!\!\!/_a\underline{\chi})_{cb}\Big]. \tag{7.5.8}$$

Proof: From $[\underline{N}, {}^{(i)}O_*] = 0$ we infer that

$$\Omega\mathbf{D}_3{}^{(i)}O_* = {}^{(i)}O_{*c}(\nabla\!\!\!/_c \log\Omega)\underline{N} + \Omega\mathbf{D}_{{}^{(i)}O_*}e_3, \tag{7.5.9}$$

and choosing a moving frame satisfying $\mathbf{D}\!\!\!\!/_3 e_b = 0$, we obtain (7.5.7).

To obtain an evolution equation for $\nabla\!\!\!/_a{}^{(i)}O_{*b} \equiv (\nabla\!\!\!/\, {}^{(i)}O_*)_{ab}$ we start from equation (7.5.9), which we rewrite as

$$\mathbf{D}_3{}^{(i)}O_* = {}^{(i)}O_{*c}\underline{\chi}_{cb}e_b + {}^{(i)}O_{*c}\eta_c e_3.$$

Using the commutation relations proved in the appendix to Chapter 4 (see Proposition 4.8.1), we obtain

$$\frac{d}{du}(\nabla\!\!\!/\, {}^{(i)}O_*)_{ab} = \Omega\Big[\hat{\underline{\chi}}_{bc}(\nabla\!\!\!/\, {}^{(i)}O_*)_{ac} - \hat{\underline{\chi}}_{ac}(\nabla\!\!\!/\, {}^{(i)}O_*)_{cb} + {}^{(i)}O_{*c}(\underline{\chi}_{cb}\eta_a - \underline{\chi}_{ca}\eta_b)$$
$$+ {}^{(i)}O_{*c}R_{3abc} - {}^{(i)}O_{*c}\underline{\chi}_{cb}\zeta_a + \underline{\chi}_{ab}({}^{(i)}O_{*c}\eta_c) + {}^{(i)}O_{*c}(\nabla\!\!\!/_a\underline{\chi})_{cb}\Big].$$

Using the evolution equations (7.5.7) and (7.5.8), we obtain immediately the estimates (7.5.2) for ${}^{(i)}O_*$. To estimate ${}^{(i)}Z_a$ and ${}^{(i)}H_{ab}$ we recall their last slice expressions (see (4.6.11)):

$${}^{(i)}Z_a \equiv \frac{1}{4}\big(g(\mathbf{D}_a{}^{(i)}O_*, e_3) + g(\mathbf{D}_3{}^{(i)}O_*, e_a)\big) = \frac{1}{4}{}^{(i)}\pi_{3a}$$
$${}^{(i)}H_{ab} \equiv \frac{1}{2}\big(g(\nabla\!\!\!/_a{}^{(i)}O_*, e_b) + g(\nabla\!\!\!/_b{}^{(i)}O_*, e_a)\big) = \frac{1}{2}{}^{(i)}\pi_{ab}.$$

An elementary calculation shows that on $\underline{C}_*$, ${}^{(i)}Z_a = 0$. To prove the remaining estimates of (7.5.2) and (7.5.3) we need the evolution equation of ${}^{(i)}H_{ab}$ and of its tangential derivatives along $\underline{C}_*$, up to second order. These evolution equations can be obtained as in Chapter 4, Subsection 4.8.4, with the obvious modifications. The final result is

$$\frac{d}{du}{}^{(i)}H_{ab} = -\Omega\left(\hat{\underline{\chi}}_{ac}{}^{(i)}H_{cb} + \hat{\underline{\chi}}_{bc}{}^{(i)}H_{ca}\right) + \Omega{}^{(i)}O_c(\nabla\!\!\!/_c\underline{\chi})_{ab} + \Omega\underline{\chi}_{ab}(\nabla\!\!\!/_c\log\Omega){}^{(i)}O_c$$
$$+\Omega\left(\hat{\underline{\chi}}_{bc}\nabla\!\!\!/_a{}^{(i)}O_c + \hat{\underline{\chi}}_{ac}\nabla\!\!\!/_b{}^{(i)}O_c\right)$$
$$\frac{d}{du}(\nabla\!\!\!/_c H_{ab}) + \frac{1}{2}\Omega\mathrm{tr}\underline{\chi}(\nabla\!\!\!/_c H_{ab}) = -\Omega\hat{\underline{\chi}}_{ad}(\nabla\!\!\!/ H)_{cdb} + \hat{\underline{\chi}}_{ad}[(\nabla\!\!\!/ H)_{bdc} - (\nabla\!\!\!/ H)_{dbc}]$$
$$+\hat{\underline{\chi}}_{bd}[(\nabla\!\!\!/ H)_{adc} - (\nabla\!\!\!/ H)_{dac}] + \underline{\mathcal{H}}_1. \tag{7.5.10}$$

From these evolution equations the estimates in the last line of (7.5.2) follow immediately.

To prove inequality (7.5.3), which involves up to third derivatives for the connection coefficients, we need the evolution equation for the second tangential derivative of ${}^{(i)}H$ along the null-incoming hypersurface. Proceeding as in Chapter 4, with the obvious modifications, we obtain an evolution equation of the form

$$\frac{d}{du}(\nabla\!\!\!/^2 H) + \Omega \mathrm{tr}\underline{\chi}(\nabla\!\!\!/^2 H) = \underline{\hat{\chi}}(\nabla\!\!\!/^2 H) + (\mathcal{L}_O \nabla\!\!\!/^2 \underline{\chi}) + \underline{\mathcal{H}}_2, \tag{7.5.11}$$

where $\underline{\mathcal{H}}_2$ collects all the quadratic error terms that do not depend on third order derivatives of the connection coefficients. From this equation it is immediate to prove that

$$|r^{2-\frac{2}{p}} \nabla\!\!\!/^2 H|_{p=2,S}(u', \underline{u}_*) \leq c\left(|r^{2-\frac{2}{p}} \nabla\!\!\!/^2 H|_{p=2,\underline{C}_*\cap\Sigma_0} + \frac{1}{r(u', \underline{u}_*)u'}\left(\mathcal{I}_0 + \Delta\right)\right),$$

where the second factor comes from the integration along $\underline{C}_*$ and Proposition 7.4.2. Substituting this estimate into the integral in (7.5.4) we obtain the expected result, once we prove the inequality

$$\int_{u_0(\underline{u}*)}^{u} du' |r^{2-\frac{2}{p}} \nabla\!\!\!/^2 H|^2_{p=2,\underline{C}_*\cap\Sigma_0} \leq c\left(\mathcal{I}_0 + \Delta\right)^2. \tag{7.5.12}$$

Proof of Proposition 7.5.2: We prove only the second inequality of (7.5.4). The first can be proved in the same way, but is easier. To control the left-hand side of (7.5.12) let us observe that

$$\begin{aligned} \int_{u_0(\underline{u}')}^{u} du' |r^{2-\frac{2}{p}} \nabla\!\!\!/^2 H|^2_{p=2,\underline{C}_*\cap\Sigma_0} &= |u_0(\underline{u}') - u| |r^{2-\frac{2}{p}} \nabla\!\!\!/^2 H|^2_{p=2,\underline{C}_*\cap\Sigma_0} \\ &\leq c r_0(\lambda_1) |r^{2-\frac{2}{p}} \nabla\!\!\!/^2 H|^2_{p=2,\underline{C}_*\cap\Sigma_0}, \end{aligned} \tag{7.5.13}$$

where $r_0(\lambda_1) \equiv \left[\frac{1}{4\pi}|S_*(\lambda_1)|\right]^{\frac{1}{2}}$ is the radius of of $S_*(\lambda_1) = S_{(0)}(\nu_*) = \underline{C}_* \cap \Sigma_0$ (see Definition 3.3.1 and Definition 3.3.8). To control $|r^{2-\frac{2}{p}} \nabla\!\!\!/^2 H|_{p=2,\underline{C}_*\cap\Sigma_0}$ we have, therefore, to control $|r^{2-\frac{2}{p}} \nabla\!\!\!/^2 H|_{p=2,S_{(0)}(\nu)}$ on Σ_0. ${}^{(i)}H_{ab}$ on Σ_0 has the expression

$${}^{(i)}H_{ab} \equiv \frac{1}{2}\left(g(\nabla\!\!\!/_a {}^{(i)}O, e_b) + g(\nabla\!\!\!/_b {}^{(i)}O, e_a)\right), \tag{7.5.14}$$

where ${}^{(i)}O = {}^{(i)}O_{\Sigma_0}$ are the generators of the rotation group defined on Σ_0 and $g = \mathbf{g}|_{\Sigma_0}$ is the metric restricted to Σ_0. To define the rotation group on Σ_0, we use exactly the same procedure used in [Ch-Kl], Chapter 3, where the function $u(p)$, defining the foliation, is $u(p) = u_{(0)}(p)$; see Definition 3.3.1. The strategy consists in using the global initial data conditions to define the rotation group at the spacelike infinity of Σ_0 and to extend it to the whole Σ_0 using the diffeomorphism generated by the vector field $N \equiv \frac{\partial}{\partial u}$. The final result is that the ${}^{(i)}O$'s satisfy the relations:[23]:

$$\begin{aligned} [{}^{(i)}O, {}^{(j)}O] &= \epsilon_{ijk} {}^{(k)}O \\ [N, {}^{(i)}O] &= 0 \\ g({}^{(i)}O, N) &= 0. \end{aligned} \tag{7.5.15}$$

[23]The relation between $N = \frac{\partial}{\partial u}$ and $\tilde{N} = \frac{1}{a}\frac{\partial}{\partial u}$ on Σ_0 is of the same type as the relation between N and e_3 on $\underline{C}_*$.

On Σ_0, in the coordinates $\{u, \theta, \phi\}$, the metric is written as

$$g(\cdot,\cdot) = a^2 du^2 + \gamma_{ab} d\phi^a d\phi^b$$

and $\{\phi^a\} = \{\theta, \phi\}$. The adapted moving frame, Fermi propagated along Σ_0, is denoted $\{\tilde{N}, e_A\}$, with $A \in \{1, 2\}$ and $\tilde{N} = \frac{1}{a}\frac{\partial}{\partial u}$.

From the commutation relation $[N, {}^{(i)}O] = 0$ and the properties of the adapted moving frame we obtain $(\nabla_N {}^{(i)}O)_A = \theta_{AB} {}^{(i)}O_B$, which can be rewritten as

$$\frac{d}{du} {}^{(i)}O_A = a\theta_{AB} {}^{(i)}O_B. \tag{7.5.16}$$

To obtain the evolution equation for $\nabla_A {}^{(i)}O_B \equiv (\nabla {}^{(i)}O)_{AB}$, we derive tangentially the equation $(\nabla_N {}^{(i)}O)_A = \theta_{AB} {}^{(i)}O_B$ and use the commutation relations $[\nabla\!\!\!\!/_{\tilde{N}}, \nabla\!\!\!\!/]$, which have the following expression (see the appendix to Chapter 4),

$$\begin{aligned} ([\nabla\!\!\!\!/_{\tilde{N}}, \nabla\!\!\!\!/]V)_{AB} &= -\theta_{AC}(\nabla\!\!\!\!/ V)_{CB} - (a^{-1}\nabla\!\!\!\!/_A a)\theta_{AC}V_C + \theta_{AB}(a^{-1}\nabla\!\!\!\!/_C a)V_C \\ &\quad +(a^{-1}\nabla\!\!\!\!/_A a)(\nabla\!\!\!\!/_{\tilde{N}} V)_B + \tilde{N}^j e^s_A e^t_B [\nabla_j, \nabla_s]V_t, \end{aligned}$$

where V is a vector field tangent to $S_{(0)}$. Choosing $V = {}^{(i)}O$ we obtain

$$\begin{aligned} \frac{d}{du}(\nabla_A {}^{(i)}O)_B &= a\Big[\hat{\theta}_{BC}(\nabla_A {}^{(i)}O)_C - \hat{\theta}_{AC}(\nabla_C {}^{(i)}O)_B + {}^{(i)}O_C\left(\theta_{CB}(a^{-1}\nabla\!\!\!\!/_A a)\right. \\ &\quad \left.-\theta_{CA}(a^{-1}\nabla\!\!\!\!/_B a)\right) + {}^{(i)}O_C(\nabla\!\!\!\!/_A\theta)_{BC} + \theta_{AB}(a^{-1}\nabla\!\!\!\!/_C a){}^{(i)}O_C \\ &\quad +\tilde{N}^j e^s_A [\nabla_j, \nabla_s]{}^{(i)}O_B\Big], \end{aligned} \tag{7.5.17}$$

and from it, immediately, the evolution equation for ${}^{(i)}H$ on Σ_0,

$$\frac{d}{du} {}^{(i)}H_{AB} = a\left(\hat{\theta}_{BC} {}^{(i)}H_{CA} - \hat{\theta}_{AC} {}^{(i)}H_{CB}\right) + \mathcal{H}_0 \tag{7.5.18}$$

where

$$\begin{aligned} \mathcal{H}_0 &= a\Big[-\left(\hat{\theta}_{BC}(\nabla_C {}^{(i)}O)_A + \hat{\theta}_{AC}(\nabla_C {}^{(i)}O_B)\right) + \frac{1}{2}{}^{(i)}O_C\left((\nabla\!\!\!\!/_A\theta)_{BC}\right. \\ &\quad \left.+(\nabla\!\!\!\!/_B\theta)_{AC}\right) + \theta_{AB}(a^{-1}\nabla\!\!\!\!/_C a){}^{(i)}O)_C + \frac{1}{2}\left({}^{(3)}\mathbf{R}_{BC\tilde{N}A} + {}^{(3)}\mathbf{R}_{AC\tilde{N}B}\right){}^{(i)}O_C\Big]. \end{aligned} \tag{7.5.19}$$

From the results of Section 7.2, the estimates for ${}^{(i)}O$ and $\nabla\!\!\!\!/ {}^{(i)}O$ obtained using the evolution equations (7.5.16), (7.5.17) and the properties of the rotation group at the spacelike infinity of Σ_0, it is immediate to infer that $\mathcal{H}_0 = O(\frac{1}{r^2})$. Differentiating twice tangentially this evolution equation and using the commutation relations of $[\nabla\!\!\!\!/_{\tilde{N}}, \nabla\!\!\!\!/]$ we obtain the evolution equation for $(\nabla\!\!\!\!/^2_A {}^{(i)}H)$,

$$\frac{d}{du}(\nabla\!\!\!\!/^2 {}^{(i)}H)_{AB} + a\mathrm{tr}\theta(\nabla\!\!\!\!/^2 {}^{(i)}H)_{AB} = a\left(\hat{\theta}_{BC}(\nabla\!\!\!\!/^2 {}^{(i)}H)_{CA} - \hat{\theta}_{AC}(\nabla\!\!\!\!/^2 {}^{(i)}H)_{CB}\right) + \mathcal{H}_{2AB},$$

where $\mathcal{H}_2$ depends on the second derivatives of the Riemann components. Applying the evolution Lemma 4.1.5 and the Gronwall Lemma to $|r^{2-\frac{2}{p}}\nabla\!\!\!\!/^2 {}^{(i)}H|_{p=2,S_{(0)}}$, we obtain

$$|r^{2-\frac{2}{p}}\nabla\!\!\!\!/^2 {}^{(i)}H|_{p=2,S_{(0)}}(\underline{u}_*) \leq c\int_{\underline{u}_*}^{\infty} |r^{2-\frac{2}{p}}\mathcal{H}_2|_{p=2,S_{(0)}}, \tag{7.5.20}$$

recalling that $\lim_{\underline{u}\to\infty} |r^{2-\frac{2}{p}} \nabla\!\!\!\!/^{(i)} H|_{p=2,S}(\underline{u}) = 0$. The integral on the right-hand side can be estimated using the explicit expression of $\mathcal{H}_2$ and the estimates of Section 7.2 by $r_0(\lambda_1)^{-\frac{1}{2}}(\mathcal{I}_0 + \Delta)$, which, when substituted into (7.5.13) gives

$$\int_{u_0(\underline{u}')}^{u} du' |r^{2-\frac{2}{p}} \nabla\!\!\!\!/^2 H|^2_{p=2,\underline{C}_*\cap\Sigma_0} \leq c(\mathcal{I}_0 + \Delta)^2, \tag{7.5.21}$$

which completes the proof of Proposition 7.5.2.

7.6 The extension argument

In this section we present the proof of Theorem **M9** which we recall below.

Theorem M9: *Consider the spacetime $\mathcal{K}(\lambda_0, \nu_*)$ together with its double null (canonical) foliation given by the functions u and $\underline{u}$ such that*

1. *The norms $\mathcal{Q}, \mathcal{O}, \mathcal{R}$ are sufficiently small*

$$\mathcal{Q} \leq \epsilon_0' \,, \; \mathcal{O} \leq \epsilon_0' \,, \; \mathcal{R} \leq \epsilon_0' \,.$$

2. *The initial conditions on Σ_0 are such that*

$$\mathcal{O}(\Sigma_0[\nu_*, \nu_* + \delta]) \leq \epsilon_0' \,,$$

where $\Sigma_0[\nu_*, \nu_* + \delta] \equiv \{p \in \Sigma_0 | u_{(0)}(p) \in [\nu_*, \nu_* + \delta]\}$.

Then we can extend the spacetime $\mathcal{K}(\lambda_0, \nu_*)$ and the double null foliation $\{u, \underline{u}\}$ to a larger spacetime $\mathcal{K}(\lambda_0, \nu_* + \delta)$, with δ sufficiently small, such that the extended norms, denoted by $\mathcal{O}', \mathcal{R}'$ satisfy

$$\mathcal{O}' \leq c\epsilon_0' \,, \; \mathcal{R}' \leq c\epsilon_0' \,.$$

Proof: By an adapted version of the classical local existence theorem, starting with initial data in the annulus $\mathcal{A}_0 = \{p \in \Sigma_0 | u_{(0)}(p) \in [\nu_*, \nu_* + \delta]\}$ provided δ is sufficiently small, we can construct a solution of the Einstein equations in its future domain of dependence, which we denote as $\mathcal{K}_0^{(\delta)}$,

$$\mathcal{K}_0^{(\delta)} \equiv \mathcal{K}(\lambda_1, \nu_* + \delta) \,.$$

The boundary of $\mathcal{K}_0^{(\delta)}$ consists of the annulus $\mathcal{A}_0 \subset \Sigma_0$ and two null hypersurfaces, one incoming, given by the portion,[24]

$$\underline{C}_{**}(\lambda_1 - \sigma) \equiv \underline{C}(\nu_* + \delta; [\lambda_1 - \sigma, \lambda_1]),$$

of the null hypersurface $\underline{C}_{**} \equiv \underline{C}(\nu_* + \delta)$, where

$$\lambda_1 - \sigma \equiv u|_{\underline{C}(\nu_*+\delta)\cap\Sigma_0} \,, \; \lambda_1 = u|_{\underline{C}(\nu_*)\cap\Sigma_0} \,,$$

[24]Recall that $u < 0$; therefore λ varies inside $\mathcal{K} \equiv \mathcal{K}(\lambda_0, \nu_*)$ in the interval $[\lambda_1, \lambda_0]$.

and one outgoing, given by the portion of the null-outgoing hypersurface initiating at $S_*(\lambda_1) = \underline{C}(\nu_*) \cap \Sigma_0$ and contained in $\mathcal{K}(\lambda_0, \nu_* + \delta)$, which we denote as $C^*(\delta)$. We can endow the region $\mathcal{K}_0^{(\delta)}$ with a double null foliation $\{u, \underline{u}\}$, where u and $\underline{u}$ are incoming and outgoing solutions of the eikonal equation with, as initial data, the function $u_{(0)}$ restricted to $\mathcal{A}_0$. It is trivial to see that, provided δ is sufficiently small, relative to this double null foliation we have

$$\mathcal{O}' \leq \frac{3}{2}\epsilon_0' \,,\; \mathcal{R}' \leq \frac{3}{2}\epsilon_0'. \tag{7.6.1}$$

Using a nonstandard version of the local existence theorem (see the discussion in Remark 3 below) we extend the spacetime $\mathcal{K}_0^{(\delta)} \cup \mathcal{K}(\lambda_1 + \sigma, \nu_*)$ to the future domain of dependence of $C^*(\delta) \cup \underline{C}_*(\lambda_1 + \sigma)$, where

$$\underline{C}_*(\lambda_1 + \sigma) \equiv \underline{C}(\nu_*; [\lambda_1, \lambda_1 + \sigma]), \tag{7.6.2}$$

provided σ is sufficiently small. Moreover starting with the foliation induced on $C^*(\delta)$ and on $\underline{C}_*(\lambda_1 + \sigma)$ we can extend the double null foliation in this region in such a way that $\mathcal{O}'$ and $\mathcal{R}'$ satisfy

$$\mathcal{O}' \leq 2\epsilon_0' \,,\; \mathcal{R}' \leq 2\epsilon_0'. \tag{7.6.3}$$

Denote by $\overline{\sigma}$ the supremum of all the values of σ for which this extension can be done in such a way that $\mathcal{O}'$ and $\mathcal{R}'$ satisfy

$$\mathcal{O}' \leq c_0\epsilon_0' \,,\; \mathcal{R}' \leq c_0\epsilon_0', \tag{7.6.4}$$

where the constant c_0 will be specified later on. If $\overline{\sigma} = \lambda_0 - \lambda_1$ the proof is completed. Otherwise let us consider the spacetime $\mathcal{K}(\lambda_1 + \overline{\sigma}, \nu_* + \delta)$ where $\mathcal{O}'$ and $\mathcal{R}'$ satisfy (7.6.4). By using a somewhat simplified[25] version of the apriori estimates developed in Chapters 4,5,6, we show that, in fact, $\mathcal{O}'$ and $\mathcal{R}'$ are strictly less than $c_0\epsilon_0'$ and thus reach a contradiction if we choose c_0 sufficiently large. This is accomplished in the following steps:

Using a variant of the methods of Chapter 4 we show that inside the region[26]

$$\tilde{\Delta}(\lambda_1 + \sigma, \nu_* + \delta) \equiv \mathcal{K}(\lambda_1 + \sigma, \nu_* + \delta) \setminus \left(\mathcal{K}_0^{(\delta)} \cup \mathcal{K}(\lambda_1 + \sigma, \nu_*)\right), \tag{7.6.5}$$

$\mathcal{O}'$ can be bounded as

$$\mathcal{O}'|_{\tilde{\Delta}} \leq c\left(\mathcal{O}'|_{C^*(\delta)} + \mathcal{O}'|_{\underline{C}_*(\lambda_1+\sigma)} + \mathcal{R}'|_{\tilde{\Delta}}\right) \leq c(3\epsilon_0' + \mathcal{R}'). \tag{7.6.6}$$

By the comparison argument of Chapter 5 we know that $\mathcal{R}'$ can be bounded by $c\mathcal{Q}^{\frac{1}{2}}$ where $\mathcal{Q}$ is the quantity defined in Chapter 3, Subsection 3.5.1, relative to the vector fields S, T, K_0 and the rotation vector fields ${}^{(i)}O$, defined in $\tilde{\Delta}(\lambda_1 + \sigma, \nu_* + \delta)$.

Remark 1: The vector fields S, T, K_0 are defined, as before, with the help of the extended functions u and $\underline{u}$ defined in $\tilde{\Delta}(\lambda_1 + \sigma, \nu_* + \delta)$. The rotation vector fields ${}^{(i)}O$ are defined

[25] Simplified with respect to the length of the interval in which u varies.

[26] Recalling definition 3.7.23, we have $\Delta(\lambda_1 + \sigma, \nu_*) = \tilde{\Delta}(\lambda_1 + \sigma, \nu_* + \delta) \cup \mathcal{K}_0^{(\delta)}$.

in the same way as in Chapter 4 by an extension argument starting from $\underline{C}_*$. With the help of the diffeomorphism ϕ_t along $C(u)$, we are extending in the future direction.

To complete the argument we must apply again, but in a slightly different situation, Theorem **M8**. Therefore we prove that, in the region $\tilde{\Delta}(\lambda_1+\sigma, \nu_*+\delta)$, $\mathcal{Q}$ is bounded by a constant multiple of its restriction to $C^*(\delta)\cup\underline{C}_*(\lambda_1+\sigma)$. Because, on the other hand, $\mathcal{Q}^{\frac{1}{2}}|_{C^*(\delta)\cup\underline{C}_*(\lambda_1+\sigma)}$ is bounded by

$$\mathcal{Q}^{\frac{1}{2}}|_{C^*(\delta)\cup\underline{C}_*(\lambda_1+\sigma)} = \mathcal{Q}^{\frac{1}{2}}|_{C^*(\delta)} + \mathcal{Q}^{\frac{1}{2}}|_{\underline{C}_*(\lambda_1+\sigma)} \le 4\epsilon_0', \tag{7.6.7}$$

it follows that $\mathcal{Q}|^{\frac{1}{2}}_{\tilde{\Delta}} \le 4c\epsilon_0'$ and $\mathcal{R}|'_{\Delta} \le 4c^2\epsilon_0'$ which implies, from (7.6.6), that

$$\mathcal{O}'|_{\tilde{\Delta}} \le c\left(\mathcal{O}'|_{C^*(\delta)} + \mathcal{O}'|_{\underline{C}_*(\lambda_1+\sigma)} + \mathcal{R}'|_{\tilde{\Delta}}\right) \le (3c\epsilon_0' + 4c^3\epsilon_0'). \tag{7.6.8}$$

Choosing $c_0 > 8c^3$ completes the proof of the theorem.

Remark 2: The only argument that is somewhat different from those developed in Chapters 4, 5, 6 is the one relative to the $\mathcal{O}$ norms. The estimates needed above to control the $\mathcal{O}$ and $\mathcal{D}$ norms in $\tilde{\Delta}$ are somewhat different from those of Chapter 4. Indeed in Chapter 4 the quantities that are not underlined, $\mathrm{tr}\chi$, $\hat{\chi}$ and η, were estimated using the evolution equations along $C(u)$ moving backward in time starting from $\underline{C}_*$. Now, on the other hand, we start from $\underline{C}_*$ and move forward in time for a very short interval of size δ. The smallness of the interval makes this procedure straightforward.

Remark 3: The nonstandard local existence theorem that we introduced above does not seem to exist in the literature. Although we do not prove it here, we sketch a possible approach to the proof.

Using the Einstein equations written relative to a double null foliation (see Subsection 3.1.7), we can first prove an adapted version of the Cauchy–Kowaleski theorem assuming that the spacetimes $\mathcal{K}_0^{(\delta)}$ and $\mathcal{K}(\lambda_1+\sigma, \nu_*)$ are real analytic. Once we have that, in order to get rid of the analyticity assumption, we propose the approach outlined in [Kl-Ni] which is based on a priori estimates similar, but far simpler, than the ones described in Chapters 4, 5, 6.

An alternative approach would be to make use of Rendall's solution, [Ren], of the characteristic Cauchy problem. More precisely. we will need an adaptation of his approach to the H^k category; see also [Mu] and [Do]. Starting with $\hat{\chi}$ on $C^*(\delta)$ and $\underline{\hat{\chi}}$ on $\underline{C}_*(\lambda_1+\sigma)$ and assuming that they are sufficiently differentiable, say $\hat{\chi}\in H^k(C^*(\delta))$ and $\underline{\hat{\chi}}\in H^k(\underline{C}_*(\lambda_1+\sigma))$ for k sufficiently large, this H^k variant of Rendall's result should allow us to construct a spacetime $\tilde{\Delta}(\lambda_1+\tilde{\sigma}, \nu_*+\tilde{\delta})$, with $\tilde{\sigma}$ and $\tilde{\delta}$ depending on the H^k norms of $\hat{\chi}$ and $\underline{\hat{\chi}}$. Due to the loss of derivatives inherent in the characteristic Cauchy problem, to apply this result we need a degree of smoothness for $\hat{\chi}$ and $\underline{\hat{\chi}}$ incompatible with our setup. This loss of derivatives can be attributed to the fact that in the characteristic Cauchy problem one treats the data $\hat{\chi}, \underline{\hat{\chi}}$ (or $\hat{\gamma}$ and Ω (see discussion in Subsection 3.1.7) as arbitrary). In reality, however, because our $\hat{\chi}, \underline{\hat{\chi}}$ are induced by the spacetimes $\mathcal{K}_0^{(\delta)}$ and $\mathcal{K}(\lambda_1+\sigma, \nu_*)$ they satisfy additional equations. In particular, this means that we do not just know $\hat{\chi}, \underline{\hat{\chi}}$ on $C^*(\delta)$, $\underline{C}_*(\lambda_1+\sigma)$ but also their derivatives $\mathbf{D}\!\!\!/_3\hat{\chi}, \mathbf{D}\!\!\!/_4\underline{\hat{\chi}}$. Of course we cannot in general prescribe both $\hat{\chi}$, $\hat{\underline{\chi}}$ and $\not{D}_3\hat{\chi}$, $\not{D}_4\hat{\underline{\chi}}$; in our case, however, these quantities satisfy on $C^*(\delta)$, $\underline{C}_*(\lambda_1+\sigma)$ compatibility relations induced by the structure equations.

To have the expected regularity in the solution of the characteristic Cauchy problem, one needs to appropriately approximate our $\hat{\chi}$, $\hat{\underline{\chi}}$, taking into account the compatibility relations mentioned above, induced by the spacetimes $\mathcal{K}_0^{(\delta)}$ and $\mathcal{K}(\lambda_1+\sigma, \nu_*)$ by a smooth sequence $\hat{\chi}_n$, $\hat{\underline{\chi}}_n$ and one needs to associate with them the spacetimes $E_n = \tilde{\Delta}(\lambda_1 + \tilde{\sigma}_n, \nu_* + \tilde{\delta}_n)$ constructed by the variant of Rendall's result mentioned above. Once this is done we can apply a vastly simplified version of the a priori estimates described in Chapters 3–7 to show that these spacetimes can be extended to values of $\tilde{\sigma}$, $\tilde{\delta}$ independent of n and can pass to the limit. Since the details of this argument are not very relevant to this book we plan to present them in a separate publication.

Once theorems **M1–M9** have been proved it follows immediately that the proof of Theorem **M0**, which describes the properties of the initial layer foliation can be derived using similar steps in a much simpler local situation. In this case, in fact, we do not have to look for a canonical foliation on the last slice since both null-incoming and null-outgoing hypersurfaces are based on the initial slice canonical foliation of Σ_0. Moreover the weights of the various integral norms are not important since the initial layer region has a finite and small height. The previous theorems can, therefore, be easily applied to this case to prove Theorem **M0**.

7.7 Appendix

7.7.1 Comparison between different foliations

We discuss here how to compare different foliations associated with different solutions of the eikonal equation and how to prove that, under appropriate conditions, they stay near each other.

Consider two double null foliations $\{u', \underline{u}\}$ and $\{u, \underline{u}\}$ with common null-incoming hypersurfaces $\underline{C}(\nu)$. We denote by $C'(\lambda)$, $C(\lambda)$ the null-outgoing hypersurfaces $u' = \lambda$, $u=\lambda$.

In the application of this result to the proof of the Main Theorem, in particular Step 6, we need to assume that the $\{u', \underline{u}\}$ foliation is globally defined and small, that is, $\mathcal{O}' \leq \epsilon_0'$ and that the foliation $\{u, \underline{u}\}$ is defined in a neighborhood Δ of $\underline{C}_{**}$; see (3.7.23). We can assume also that $\mathcal{O}$ is sufficiently small in Δ.

We want to establish a quantitative relationship between the two foliations in Δ. Associated with the null hypersurfaces of these foliations we introduce the null geodesic vector fields

$$L' = -g^{\mu\nu}\partial_\nu u' \frac{\partial}{\partial x^\mu}\ ,\ L = -g^{\mu\nu}\partial_\nu u \frac{\partial}{\partial x^\mu}$$
$$\underline{L} = -g^{\mu\nu}\partial_\nu \underline{u} \frac{\partial}{\partial x^\mu}, \tag{7.7.1}$$

and the corresponding spacetime lapse functions Ω and Ω' (see Definition 3.1.12),

$$g^{\mu\nu}\partial_\mu u' \partial_\nu \underline{u} = -(2\Omega'^2)^{-1}\ ,\ g^{\mu\nu}\partial_\mu u \partial_\nu \underline{u} = -(2\Omega^2)^{-1}. \tag{7.7.2}$$

Associated with the two double null foliations are two different double null integrable S-foliations whose leaves are

$$S'(\lambda,\nu)=C'(\lambda)\cap\underline{C}(\nu)\,,\quad S(\lambda,\nu)=C(\lambda)\cap\underline{C}(\nu)\,.$$

Starting from the geodesic vector fields we associate with these foliations two adapted null frames $\{\hat{e}'_4,\hat{e}'_3,e'_a\}$ and $\{\hat{e}_4,\hat{e}_3,e_a\}$ where[27]

$$\begin{aligned}&\hat{e}'_4=4\Omega'^2L'\,,\ \hat{e}'_3=\underline{L}\,,\ \ e'_a\text{ tangent to }S'(\lambda,\nu)\\&\hat{e}_4=4\Omega^2L'\,,\ \hat{e}_3=\underline{L}\,,\ \ e_a\text{ tangent to }S(\lambda,\nu).\end{aligned}\tag{7.7.3}$$

The two null frames are related in the following way:

$$\begin{aligned}\hat{e}_4&=\hat{e}'_4+\left[4\Omega^2\Omega'^2(-2\mathbf{g}(L,L'))\right]\hat{e}'_3+2\left[4\Omega^2\Omega'^2(-2\mathbf{g}(L,L'))\right]^{\frac12}\hat{\sigma}_ae'_a\\\hat{e}_3&=\hat{e}'_3\\e_a&=e'_a+\left[4\Omega^2\Omega'^2(-2\mathbf{g}(L,L'))\right]^{\frac12}\hat{\sigma}_a\hat{e}'_3,\end{aligned}\tag{7.7.4}$$

where $|\hat{\sigma}|^2=1$. Moreover

$$\begin{aligned}\hat{e}'_4&=\hat{e}_4+\left[4\Omega^2\Omega'^2(-2\mathbf{g}(L,L'))\right]\hat{e}_3-2\left[4\Omega^2\Omega'^2(-2\mathbf{g}(L,L'))\right]^{\frac12}\hat{\sigma}_ae_a\\\hat{e}'_3&=\hat{e}_3\\e'_a&=e_a-\left[4\Omega^2\Omega'^2(-2\mathbf{g}(L,L'))\right]^{\frac12}\hat{\sigma}_a\hat{e}_3.\end{aligned}\tag{7.7.5}$$

These formulas follow immediately from the fact that both frames are null frames and from the relation $\mathbf{g}(\hat{e}_4,\hat{e}'_4)=16\Omega^2\Omega'^2\mathbf{g}(L,L')$.

How much the foliations are near to one another is controlled by the term

$$\Theta\equiv\left[4\Omega^2\Omega'^2(-2\mathbf{g}(L,L'))\right].\tag{7.7.6}$$

To estimate $\mathbf{g}(L,L')$ we start from its expression

$$\mathbf{g}(L,L')=g^{\mu\nu}\partial_\mu u\partial_\nu u'=\frac{1}{2}g^{\mu\nu}\partial_\mu(u-u')\partial_\nu(u'-u),\tag{7.7.7}$$

and express the right-hand side of (7.7.7) using a specific choice of coordinates. We choose $\{v,\underline{u},\omega^a\}$ as coordinates, where v is the affine parameter of the null-incoming geodesic curves along the hypersurfaces $\underline{C}(\underline{u})$. It is an easy computation to write the explicit expression of the metric,

$$\mathbf{g}(\cdot,\cdot)=X^2d\underline{u}^2-\left(dvd\underline{u}+d\underline{u}dv\right)-X_a\left(d\underline{u}d\omega^a+d\omega^ad\underline{u}\right)+\gamma_{ab}d\omega^ad\omega^b,$$

where, as in Subsection 3.1.6 (see equation (3.1.62)), $N=\frac{\partial}{\partial\underline{u}}+X$ and

$$\frac{\partial}{\partial v}X^a=Z^a=2\gamma^{ab}\zeta_b\,.$$

[27]The null vector fields chosen here are not normalized null pairs in the sense of Definition 3.1.13. In fact we have $\hat{e}_4=2\Omega\hat{N}=2N$ and $\hat{e}_3=\underline{L}=(2\Omega)^{-1}\underline{\hat{N}}$ and the same for the primed ones.

The components of the inverse metric are

$$g^{vv}=0\,,\ g^{\underline{u}\,\underline{u}}=0\,,\ g^{v\underline{u}}=-1\,,\ g^{vd}=X^d\,,\ g^{\underline{u}d}=0\,,\ g^{ab}=\gamma^{ab}\,.$$

Using these coordinates we observe that, along the $\underline{C}(\nu)$ null hypersurfaces common to both foliations, u and u' satisfy

$$\begin{aligned}u(p)=u(v,\underline{u},\omega)&=\int^v\frac{du}{dv}dv'=\int^v(2\Omega^2)^{-1}(v',\underline{u},\omega)dv'\\ u'(p)=u'(v,\underline{u},\omega)&=\int^v\frac{du'}{dv}dv'=\int^v(2\Omega'^2)^{-1}(v',\underline{u},\omega)dv'.\end{aligned}\tag{7.7.8}$$

Therefore

$$\begin{aligned}2\mathbf{g}(L,L')&=g^{v\underline{u}}\left[\partial_v(u-u')\partial_{\underline{u}}(u'-u)+\partial_{\underline{u}}(u-u')\partial_v(u'-u)\right]\\&\quad+g^{vd}\left[\partial_v(u-u')\partial_d(u'-u)+\partial_d(u-u')\partial_v(u'-u)\right]\\&\quad+\gamma^{ab}\partial_a(u-u')\partial_b(u'-u)\\&=-\left[\partial_v(u-u')\partial_{\underline{u}}(u'-u)+\partial_{\underline{u}}(u-u')\partial_v(u'-u)\right]\\&\quad-X^d\left[\partial_v(u-u')\partial_d(u'-u)+\partial_d(u-u')\partial_v(u'-u)\right]\\&\quad+\gamma^{ab}\partial_a(u-u')\partial_b(u'-u).\end{aligned}\tag{7.7.9}$$

In the chosen coordinates the right-hand side of (7.7.9) becomes

$$\begin{aligned}2\mathbf{g}(L,L')=&-\Bigg[\partial_v\left(\int^v\left[(2\Omega^2)^{-1}-(2\Omega'^2)^{-1}\right](v',\underline{u},\omega)dv'\right)\\&\qquad\partial_{\underline{u}}\left(\int^v\left[(2\Omega'^2)^{-1}-(2\Omega^2)^{-1}\right](v',\underline{u},\omega)dv'\right)\\&\quad+\partial_{\underline{u}}\left(\int^v\left[(2\Omega^2)^{-1}-(2\Omega'^2)^{-1}\right](v',\underline{u},\omega)dv'\right)\\&\qquad\partial_v\left(\int^v\left[(2\Omega'^2)^{-1}-(2\Omega^2)^{-1}\right](v',\underline{u},\omega)dv'\right)\Bigg]\\&-X^d\Bigg[\partial_v\left(\int^v\left[(2\Omega^2)^{-1}-(2\Omega'^2)^{-1}\right](v',\underline{u},\omega)dv'\right)\\&\qquad\partial_d\left(\int^v\left[(2\Omega'^2)^{-1}-(2\Omega^2)^{-1}\right](v',\underline{u},\omega)dv'\right)\\&\quad+\partial_d\left(\int^v\left[(2\Omega^2)^{-1}-(2\Omega'^2)^{-1}\right](v',\underline{u},\omega)dv'\right)\\&\qquad\partial_v\left(\int^v\left[(2\Omega'^2)^{-1}-(2\Omega^2)^{-1}\right](v',\underline{u},\omega)dv'\right)\Bigg]\\&\quad\gamma^{ab}\partial_a\left(\int^v\left[(2\Omega^2)^{-1}-(2\Omega'^2)^{-1}\right](v',\underline{u},\omega)dv'\right)\\&\qquad\partial_b\left(\int^v\left[(2\Omega'^2)^{-1}-(2\Omega^2)^{-1}\right](v',\underline{u},\omega)dv'\right).\end{aligned}$$

Computing these terms explicitly we obtain

$$\mathbf{g}(L,L')=-\frac{(\Omega'^2-\Omega^2)}{2\Omega^2\Omega'^2}\left\{\int_0^v\left(\frac{1}{\Omega^2}\partial_{\underline{u}}\log\Omega-\frac{1}{\Omega'^2}\partial_{\underline{u}}\log\Omega'\right)\right.$$

$$
\begin{aligned}
&\quad +X^d\int_0^v\left(\frac{1}{\Omega^2}\partial_d\log\Omega-\frac{1}{\Omega'^2}\partial_d\log\Omega'\right)\\
&- \gamma^{ab}\int_0^v\left(\frac{1}{\Omega^2}\partial_a\log\Omega-\frac{1}{\Omega'^2}\partial_a\log\Omega'\right)\left(\frac{1}{\Omega^2}\partial_b\log\Omega-\frac{1}{\Omega'^2}\partial_b\log\Omega'\right)\Bigg\}\\
&\equiv [I]+[II]+[III]. \qquad (7.7.10)
\end{aligned}
$$

Writing this expression in terms of the null frames (7.7.4), (7.7.5), we obtain the following expressions:

$$
\begin{aligned}
[I]+[II] &= -\frac{(\Omega'^2-\Omega^2)}{2\Omega^2\Omega'^2}\Bigg\{\int_0^v\left(\frac{1}{2\Omega^2}\partial_{\hat{e}_4}\log\Omega-\frac{1}{2\Omega'^2}\partial_{\hat{e}'_4}\log\Omega'\right)\\
&+\int_0^v\left(\frac{\Theta}{2\Omega'^2}\partial_{\hat{e}'_3}\log\Omega'+\frac{\Theta^{\frac12}}{\Omega'^2}\hat{\sigma}_a\partial_{e'_a}\log\Omega'\right)\Bigg\}\\
&+\frac{(\Omega'^2-\Omega^2)}{2\Omega^2\Omega'^2}\Bigg[\int_0^v\left(\frac{1}{\Omega^2}X^d\partial_d\log\Omega-\frac{1}{\Omega'^2}X^d\partial_d\log\Omega'\right)\\
&-X^d\int_0^v\left(\frac{1}{\Omega^2}\partial_d\log\Omega-\frac{1}{\Omega'^2}\partial_d\log\Omega'\right)\Bigg]\\
&= -\frac{(\Omega'^2-\Omega^2)}{2\Omega^2\Omega'^2}\Bigg\{\int_0^v\left(\frac{1}{2\Omega^2}\partial_{\hat{e}_4}\log\Omega-\frac{1}{2\Omega'^2}\partial_{\hat{e}'_4}\log\Omega'\right)\\
&+\int_0^v\left(\frac{\Theta}{2\Omega'^2}\partial_{\hat{e}'_3}\log\Omega'+\frac{\Theta^{\frac12}}{\Omega'^2}\hat{\sigma}_a\partial_{e'_a}\log\Omega'\right)\Bigg\}+\frac{(\Omega'^2-\Omega^2)}{2\Omega^2\Omega'^2}\\
&\cdot\Bigg[\int_0^v\left(\frac{1}{\Omega^2}\theta^d(X)\partial_{e_d}\log\Omega-\frac{1}{\Omega'^2}\theta^d(X)\partial_{e'_d}\log\Omega'-\frac{1}{\Omega'^2}\Theta^{\frac12}(\theta^d(X)\hat{\sigma}_d)\partial_{\hat{e}'_3}\log\Omega'\right)\\
&-X^c\int_0^v\left(\frac{1}{\Omega^2}\theta^d_c\partial_{e_d}\log\Omega-\frac{1}{\Omega'^2}\theta^d_c\partial_{e'_d}\log\Omega'-\frac{1}{\Omega'^2}\Theta^{\frac12}(\theta^d_c\hat{\sigma}_d)\partial_{\hat{e}'_3}\log\Omega'\right)\Bigg], \qquad (7.7.11)
\end{aligned}
$$

and

$$
[III]=-\frac{(\Omega'^2-\Omega^2)}{2\Omega^2\Omega'^2}\gamma^{ab}\int_0^v\frac{1}{\Omega^2}\theta^d_a\partial_{e_d}\log\Omega\int_0^v\frac{1}{\Omega'^2}\theta^d_b\left(\partial_{e'_d}\log\Omega'+\Theta^{\frac12}(\theta^d_b\hat{\sigma}_d)\partial_{\hat{e}'_3}\log\Omega'\right)
\qquad (7.7.12)
$$

where θ^d is the 1-form associated with the vector fields e_d, $d\in\{1,2\}$. Putting together (7.7.10) and (7.7.12) we obtain

$$
\begin{aligned}
&\Theta(v,\underline{u},\omega)\\
&=4(\Omega'^2-\Omega^2)\Bigg\{\int_0^v\left(\frac{1}{2\Omega^2}\partial_{\hat{e}_4}\log\Omega-\frac{1}{2\Omega'^2}\partial_{\hat{e}'_4}\log\Omega'\right)\\
&\quad+\int_0^v\left(\frac{\Theta}{2\Omega'^2}\partial_{\hat{e}'_3}\log\Omega'+\frac{\Theta^{\frac12}}{\Omega'^2}\hat{\sigma}_a\partial_{e'_a}\log\Omega'\right)\\
&\quad-\Bigg[\int_0^v\left(\frac{1}{\Omega^2}\theta^d(X)\partial_{e_d}\log\Omega-\frac{1}{\Omega'^2}\theta^d(X)\partial_{e'_d}\log\Omega'-\frac{1}{\Omega'^2}\Theta^{\frac12}(\theta^d(X)\hat{\sigma}_d)\partial_{\hat{e}'_3}\log\Omega'\right)
\end{aligned}
$$

$$-X^c\int_0^v\left(\frac{1}{\Omega^2}\theta^d_c\partial_{e_d}\log\Omega-\frac{1}{\Omega'^2}\theta^d_c\partial_{e'_d}\log\Omega'-\frac{1}{\Omega'^2}\Theta^{\frac{1}{2}}(\theta^d_c\hat{\sigma}_d)\partial_{\hat{e}'_3}\log\Omega'\right)\Bigg]$$
$$+\gamma^{ab}\int_0^v\frac{1}{\Omega^2}\theta^d_a\partial_{e_d}\log\Omega\int_0^v\frac{1}{\Omega'^2}\theta^d_b\left(\partial_{e'_d}\log\Omega'+\Theta^{\frac{1}{2}}(\theta^d_b\hat{\sigma}_d)\partial_{\hat{e}'_3}\log\Omega'\right)\Bigg\}, \tag{7.7.13}$$

and from this expression we have immediately the following lemma.

Lemma 7.7.1 *Assume the spacetime lapse function Ω is bounded; then, if $\mathcal{O}'_{[0]}$ is sufficiently small, we have*[28]

$$|r^2\Theta|\le c(\mathcal{O}_{[0]}+\mathcal{O}'_{[0]})(1+\mathcal{O}'_{[0]}). \tag{7.7.14}$$

Proof: Taking the sup of Θ along $\underline{C}(\nu)$ and noting that $\mathcal{O}'_{[0]}$ is small and recalling that the connection coefficients ω' and $\eta',\underline{\eta}'$ have the appropriate decay, the result follows immediately. The next corollary is an immediate consequence of this lemma.

Corollary 7.7.1 *Assume the spacetime lapse function Ω is bounded; then, if $\mathcal{O}'_{[0]}$ is sufficiently small we have*

$$\mathcal{R}\le\mathcal{R}'+c\mathcal{O}_{[2]}(1+\mathcal{O}'_{[2]})\mathcal{R}'. \tag{7.7.15}$$

7.7.2 *Proof of the local existence part of Theorem* **M6**

We recall the equations that define the canonical foliation on the last slice (see (3.3.12)):

$$\begin{aligned}&\frac{du_*}{dv}=(2\Omega^2)^{-1};\quad u_*|_{\underline{C}_*\cap\Sigma_0}=\lambda_1\\&\Delta\!\!\!/\log\Omega=\frac{1}{2}\mathrm{d\!\!/iv}\,\underline{\eta}+\frac{1}{2}\left(\mathbf{K}-\overline{\mathbf{K}}+\frac{1}{4}(\mathrm{tr}\chi\,\mathrm{tr}\underline{\chi}-\overline{\mathrm{tr}\chi\,\mathrm{tr}\underline{\chi}})\right)\\&\overline{\log 2\Omega}=0.\end{aligned} \tag{7.7.16}$$

We rewrite these equations with respect to the following null frame

$$\underline{L}=\frac{1}{2\Omega}\underline{\hat{N}}\,,\ L^*=2\Omega\hat{N}\,,\ e''_A=e_A\,,$$

where $\{\hat{N},\underline{\hat{N}}\}$ is the normalized null pair associated with the canonical foliation. The quantities that refer to the null frame $\{\underline{L},L^*,e''_A\}$ will be denoted with a double prime, for instance $\chi'',\zeta'',\rho''\dots$, and the following relations hold between primed and unprimed quantities

$$\begin{aligned}&\zeta''_A=\zeta_A-\nabla\!\!\!/_A\log\Omega\,,\ \underline{\eta}''_A=\underline{\eta}_A=-\zeta''\,,\ \eta''_A=\eta_A\\&\underline{\chi}''_{AB}=\frac{1}{2\Omega}\underline{\chi}_{AB}\,,\ \chi''_{AB}=2\Omega\chi_{AB}\,,\ \rho''=\rho.\end{aligned} \tag{7.7.17}$$

[28] The decay is stronger on $\underline{C}_*$.

The equations (7.7.16) become

$$\frac{du_*}{dv}=(2\Omega^2)^{-1},\quad u_*|_{S'_*(0)}=\lambda_1,$$
$$\Delta\log 2\Omega=\frac{1}{2}\mathrm{div}\,\underline{\eta}''+\frac{1}{2}\left[(\frac{1}{2}\hat{\chi}''\hat{\underline{\chi}}''-\frac{1}{2}\overline{\hat{\chi}''\hat{\underline{\chi}}''})-(\rho''-\overline{\rho''})\right],$$
$$\overline{\log 2\Omega}=0, \tag{7.7.18}$$

where we write $S_0=S_*(\lambda_1)=\underline{C}_*\cap\Sigma_0$.

To solve this system of equations we make the following preliminary steps.

1. We observe that we can replace the given background foliation on $\underline{C}_*$ by the geodesic foliation that we define below. This can be easily done locally near S_0. The geodesic foliation is defined by the level surfaces of the affine parameter v,

$$S'(\tau)=\left\{p\in\underline{C}_*|v(p)=\tau\in[0,v_1]\right\},$$

where $S'(0)=S_0$ and v_1 is defined later on.

2. Associated with this new background foliation we define a null frame adapted to it,

$$\{\underline{L},N',e'_A\},$$

with the e'_A vector fields, Fermi transported along $\underline{C}_*$.

3. We choose (v,ω) as coordinates of a point $p\in\underline{C}_*$ where $\omega=(\theta,\phi)$ are the angular coordinates of S_0.[29] The vector fields e'_A can be expressed in the form

$$e'_A|_p=e'^{\,a}_A\frac{\partial}{\partial\omega^a}\bigg|_{\omega(p)}.$$

We denote by $\gamma(v)$ the restriction of the metric g on the two-dimensional surfaces $S'(\tau)\subset\underline{C}_*$,

$$\gamma(v,\omega)(\cdot,\cdot)=\mathbf{g}(p)|_{S'(\tau)}(\cdot,\cdot).$$

The null geodesics on $\underline{C}_*$ define a family of maps $\{\psi_v\}$ between $S'(0)$ and $S'(v)$. In our adapted coordinates they are given by

$$S'(0):p_0\equiv(0,\omega)\to p=\psi_v(p_0)=(v,\omega)\in S'(v).$$

Therefore the metrics $\{\gamma(v,\cdot)\}$ can be thought as a family of metrics on $S'(0)$,

$$\gamma(v,\omega)=\gamma_{ab}(v;\omega)d\omega^a d\omega^b.$$

4. We consider the class of foliations defined through the functions $W(\lambda,\omega)$,[30]

$$^{(W)}F:[\lambda_1,\lambda_2]\times S_0\to\underline{C}_*$$
$$^{(W)}F(\lambda,\omega)=(W(\lambda,\omega),\omega),\quad W(\lambda_1,\omega)=0. \tag{7.7.19}$$

[29]Let p be a point $\in\underline{C}_*$. There exists a null geodesic λ starting at $p_0\in S_0$ such that $p=\lambda(\overline{v},p_0)$. Then $(v(p),\omega(p))=(\overline{v},\theta(p_0),\phi(p_0))$.

[30]The function $W(\lambda,\omega)$ must have some appropriate properties to define a foliation. In particular W must have no critical points and, for any fixed λ, the level surfaces of $W(\lambda,\omega)$ must be diffeomorphic to S^2.

The leaves of the ${}^{(W)}F$ foliations are the two-dimensional surfaces

$$S_{(W)}(\lambda) \equiv \{p \in \underline{C}_* | (v(p), \omega(p)) = (W(\lambda, \omega), \omega)\} \tag{7.7.20}$$

and $S_{(W)}(\lambda_1) = S'(0)$. Observe that the background geodesic foliation corresponds to $W_0(\lambda, \omega) = \lambda - \lambda_1$.

Once we have introduced this space of foliations, we define an appropriate norm on it and construct a transformation such that its fixed point will be the solution of the system (7.7.18). This is achieved through the following steps.

1. Observe that the vector fields

$$\frac{\partial}{\partial \omega^a} + \frac{\partial W}{\partial \omega^a} \frac{\partial}{\partial v} \tag{7.7.21}$$

are tangent to $S_{(W)}(\lambda)$ for every λ. Using them we define the orthonormal frame $\{{}^{(W)}e_A\}$, adapted to the $W(\lambda, \cdot)$ foliation, as

$${}^{(W)}e_A = e'^{\,a}_A \left(\frac{\partial}{\partial \omega^a} + \frac{\partial W}{\partial \omega^a} \frac{\partial}{\partial v} \right) \equiv e'_A + (\partial'_A W)\underline{L}. \tag{7.7.22}$$

2. We construct a null frame adapted to $S_{(W)}(\lambda)$

$$\{\underline{L}, {}^{(W)}N, {}^{(W)}e_A\} .$$

The relation of this null frame with the background one, $\{\underline{L}, N', e'_A\}$, is given by[31]

$$\begin{aligned} {}^{(W)}N &= N' + (\partial' W)^2 \underline{L} + 2(\partial'_B W) e'_B \\ {}^{(W)}e_A &= e'_A + (\partial'_A W)\underline{L} \end{aligned} \tag{7.7.23}$$

where

$$(\partial' W)^2 = \sum_A (\partial'_A W)^2 \equiv \sum_A (\partial_{e'_A} W)^2.$$

The connection coefficients and the relevant null curvature component relative to this $S_{(W)}$ foliation can be expressed in terms of those relative to the background foliation:[32]

$$\begin{aligned}
&{}^{(W)}\underline{\chi}_{AB} = \underline{\chi}'_{AB} \\
&{}^{(W)}\zeta_A = \zeta'_A - (\partial'_C W)\underline{\chi}'_{C,A} \\
&{}^{(W)}\underline{\eta}_A = \underline{\eta}'_A + (\partial'_C W)\underline{\chi}'_{C,A} = -{}^{(W)}\zeta_A \\
&{}^{(W)}\chi_{AB} = \chi'_{AB} + (\partial' W)^2 \underline{\chi}'_{AB} + 2\left[(\partial_{e'_B} W)\zeta'_A + (\partial_{e'_A} W)\zeta'_B\right] \\
&\qquad -2(\partial'_C W)\left[(\partial'_B W)\underline{\chi}'_{CA} + (\partial'_A W)\underline{\chi}'_{CB}\right] + 2\left(\nabla\!\!\!\!/'(\nabla\!\!\!\!/' W)|_{S_{W(\lambda)}}\right)_{AB} \\
&{}^{(W)}\rho = \rho' - \frac{1}{2}(\partial'_C W)\underline{\beta}'_C + \frac{1}{4}(\partial'_B W)(\partial'_C W)\underline{\alpha}'_{BC}.
\end{aligned} \tag{7.7.24}$$

---

[31] These relations are the same as those in Subsection 7.7.1. Here, however, we have a more refined control over W.

[32] Repeated capital indices are to be summed over. We use the notation $\chi'_{AB} \equiv \chi'(e'_A, e'_B)$, ${}^{(W)}\chi_{AB} \equiv {}^{(W)}\chi({}^{(W)}e_A, {}^{(W)}e_B)$ for all connection coefficients.

3. We introduce the nonlinear map $\mathcal{A}$ whose fixed point will be the solution of the system (7.7.18). We denote with $||G||_{L^p(S_0)}$ the norm:

$$||G||_{L^p(S_0)} = \left(\int_{S_0} |G|^p d\mu_0\right)^{\frac{1}{p}}, \tag{7.7.25}$$

where $d\mu_0$ is the measure on S_0. On the space function $C^0(I; L^p(S_0))$ with $I = [0, \lambda_2]$,[33] we introduce the nonlinear map

$$\mathcal{A} : W(\lambda, \omega) \rightarrow \widetilde{W}(\lambda, \omega) \equiv \mathcal{A}(W)(\lambda, \omega),$$

defined through the following steps.

(a) We consider the portion of the $\underline{C}_*$ null hypersurface

$$\underline{C}_*(I; W) = \{p \in \underline{C}_* | p \in S_{(W)}(\lambda); \lambda \in I\}. \tag{7.7.26}$$

(b) Given W we consider on $\underline{C}_*(I; W)$ the null frame

$$\{\underline{L}, {}^{(W)}N, {}^{(W)}e_A\},$$

the associated connection coefficients ${}^{(W)}\zeta$, ${}^{(W)}\underline{\eta}$, ${}^{(W)}\chi$, ${}^{(W)}\underline{\chi}$ and the curvature component ${}^{(W)}\rho - \overline{{}^{(W)}\rho}$ where the average is made with respect to S_W.

(c) On the two-dimensional surfaces $S_{(W)}(\lambda)$ of $\underline{C}_*(I; W)$ we solve the elliptic equation

$$\begin{aligned} {}^{(W)}\not\Delta \log 2{}^{(W)}\Omega &= \frac{1}{2}{}^{(W)}\not{\mathrm{div}}\,{}^{(W)}\underline{\eta} + \frac{1}{2}\left[\left(\frac{1}{2}{}^{(W)}\hat{\chi}\,{}^{(W)}\underline{\hat{\chi}} - \frac{1}{2}\overline{{}^{(W)}\hat{\chi}\,{}^{(W)}\underline{\hat{\chi}}}\right) - \left({}^{(W)}\rho - \overline{{}^{(W)}\rho}\right)\right] \\ \overline{\log 2{}^{(W)}\Omega} &= 0, \end{aligned} \tag{7.7.27}$$

where for any given λ, ${}^{(W)}\not\Delta = {}^{(W)}\not\nabla_A {}^{(W)}\not\nabla_A$ is the intrinsic Laplacian relative to the surface $S_{(W)}(\lambda)$.

(d) We define the nonlinear map

$$\mathcal{A}(W)(\lambda, \omega) = \int_0^\lambda 2\left({}^{(W)}\Omega|_{S_W(\lambda')}(\omega)\right)^2 d\lambda'. \tag{7.7.28}$$

To prove the local existence of a canonical foliation we have to show that the transformation $\mathcal{A}(W) = \widetilde{W}$ has a fixed point W_*,

$$\mathcal{A}(W_*) = W_*.$$

Indeed given W_* we can define implicitly $u_* = u_*(v, \omega)$ according to[34]

$$u_*(W_*(\lambda, \omega), \omega) = \lambda. \tag{7.7.29}$$

---

[33] Here we use $\lambda \equiv \lambda - \lambda_1$.

[34] Note that since the background foliation is equivariant with respect to the vector field $\underline{L}$, the (local) foliation ${}^{\mathcal{A}(W)}S(\lambda) = \{p \in \underline{C}_* | \mathcal{A}(W)(p) = \lambda\}$ is equivariant with respect to the vector field ${}^{(\mathcal{A}(W))}\underline{N} \equiv 2({}^{(W)}\Omega)^2 \underline{L}$.

Then,

$$1=\frac{du_*}{d\lambda}=\frac{du_*}{dv}\frac{dW_*}{d\lambda}=\frac{du_*}{dv}2({}^{(W_*)}\Omega)^2,$$

which implies

$$\frac{du_*}{dv}=\frac{1}{2({}^{(W_*)}\Omega)^2},$$

as desired. The portion of $\underline{C}_*$ endowed with this canonical foliation is

$$\underline{C}_*(I;W_*)=\{p\in\underline{C}_*|p\in S_{(W_*)}(\lambda);\lambda\in I\}\,.$$

In the following we consider a simplified version of the map $\mathcal{A}(W)$ close to the previous one. To obtain it we rewrite (7.7.28) as

$$\mathcal{A}(W)(\lambda,\omega)=\frac{\lambda}{2}+\int_0^\lambda\left(2\left({}^{(W)}\Omega|_{S_{(W)}(\lambda')}(\omega)\right)^2-\frac{1}{2}\right)d\lambda'.$$

Assuming $\left|2\left({}^{(W)}\Omega|_{S_{(W)}(\lambda')}(\omega)\right)^2-\frac{1}{2}\right|$ is small for any function W and any $\lambda\in I$, we expand $2\left({}^{(W)}\Omega|_{S_{(W)}(\lambda')}(\omega)\right)^2-\frac{1}{2}$ in terms of $\log{}^{(W)}\Omega$ and consider only the lowest order term of the expansion in $\log 2{}^{(W)}\Omega$,[35]

$$\begin{aligned}\mathcal{A}(W)(\lambda,\omega)&=\frac{\lambda}{2}+\int_0^\lambda\left(\log 2{}^{(W)}\Omega|_{S_{(W)}(\lambda')}\right)(\omega)d\lambda'\\&=\frac{\lambda}{2}+\int_0^\lambda\left(\log 2{}^{(W)}\Omega\right)(W(\lambda',\omega),\omega)d\lambda',\end{aligned}\tag{7.7.30}$$

where the last line follows, since by definition

$${}^{(W)}\Omega|_{S_W(\lambda)}(\omega)={}^{(W)}\Omega(W(\lambda,\omega),\omega)\,.$$

(e) We look for a fixed point of the map (7.7.30) in the space

$$\mathcal{E}\equiv\cap_{p=2}^4C^1(I;W_2^p(S_0)),\tag{7.7.31}$$

with the Sobolev norms $||\cdot||_{W_k^p(S_0)}$ defined by

$$||G||_{W_k^p(S_0)}\equiv\left(\sum_{l=0}^k\int_{S_0}|{}^{(0)}\nabla^l G|^pd\mu_0\right)^{\frac{1}{p}}<\infty,\tag{7.7.32}$$

where ${}^{(0)}\nabla$ denotes the covariant derivative with respect to S_0.

---

[35]We consider only the first term of the expansion

$$2\Omega^2-\frac{1}{2}=\log 2\Omega+\frac{1}{2}\sum_{k=2}^\infty\frac{2^k}{k!}(\log 2\Omega)^k\,.$$

It will be clear from the proof that the result obtained using this "approximate" map can be immediately extended to the exact map if the portion of the null hypersurface, $\underline{C}_*(I)$, does not differ too much from a portion of a null cone in the Minkowski spacetime.

Definition 7.7.1 *In $\mathcal{E}$ the closed set $\mathcal{K}_{\delta_0,\sigma_1}$ is defined such that $W \in \mathcal{K}_{\delta_0,\sigma_1}$ if, for any $p \in [2,4]$,*[36]

$$\begin{cases} W(\lambda = 0, \omega) = 0 \\ \sup_{\lambda \in I} ||{}^{(0)}\nabla^{0,1,2}\left(W(\lambda,\cdot) - \frac{\lambda}{2}\right)||_{L^p(S_0)} \leq \delta_0 \\ \sup_{\lambda \in I} ||\partial_\lambda {}^{(0)}\nabla^{0,1,2}\left(W(\lambda,\cdot) - \frac{\lambda}{2}\right)||_{L^p(S_0)} \leq \sigma_1 . \end{cases} \tag{7.7.33}$$

We can finally state a precise form of the local existence part of Theorem **M6**.

Theorem 7.7.2 *Assume that, with ϵ_0' sufficiently small,*

$$\mathcal{R}_{[2]}{}'(\underline{C}_*) + \underline{\mathcal{R}}_{[2]}{}'(\underline{C}_*) \leq \epsilon_0'$$
$$\underline{\mathcal{O}}_{[2]}{}'(\underline{C}_*) + \mathcal{O}_{[2]}{}'(\underline{C}_*) \leq \epsilon_0',$$

relative to the background foliation. There exist positive constants $|I|, \delta_0, \sigma_1$ such that $\mathcal{A}$ has a fixed point in the set $\mathcal{K}_{\delta_0,\sigma_1}$.

Proof: We need to show that:

1. $\mathcal{A}$ maps $\mathcal{K}_{\delta_0,\sigma_1}$, subset of $\cap_{p=2}^4 C^1(I; W_2^p(S_0))$, into itself.
2. $\mathcal{A}$ is a contraction on $\mathcal{K}_{\delta_0,\sigma_1}$.

We shall omit the proof here; see [Ni].

We now restrict the comparison between different foliations discussed in Subsection 7.7.1 to the specific case where the two foliations are the background and the canonical foliation on $\underline{C}_*$. We recall that the existence of the canonical foliation has been proved in Theorem **M6**. As discussed in Subsection 7.7.1 and as required by the proof of the Oscillation Lemma (see also Lemma 4.8.2), we have to control the quantity $(-2\mathbf{g}(L',L))$ in the initial portion of $\underline{C}_*$. The proof we present here is different from the one discussed in Subsection 7.7.1. [37] The result is expressed in the following lemma.

Lemma 7.7.2 *Assume on $\underline{C}_*(I)$*[38]

$$\mathcal{O}'(\underline{C}_*(I)) \leq \epsilon_0' \ , \ \mathcal{O}^*(\underline{C}_*(I)) \leq \epsilon_0'. \tag{7.7.34}$$

Assume also that

$$|r'^{\frac{5}{2}}\eta'|_{\Sigma_0} \leq \epsilon_0' \ , \ |\mathbf{g}(L',L)|_{\underline{C}_* \cap \Sigma_0} = 0. \tag{7.7.35}$$

Then, on $\underline{C}_*(I)$, the following inequality holds:

$$|r^2 \tau_- \mathbf{g}(L',L)| \leq c\epsilon_0'. \tag{7.7.36}$$

[36] Conditions (7.7.33) imply also that the following norms are bounded

$$\sup_{\lambda \in I} |{}^{(0)}\nabla^{0,1} W|_{L^\infty(S_0)} \ , \ \sup_{\lambda \in I} |\partial_\lambda {}^{(0)}\nabla^{0,1} W|_{L^\infty(S_0)} .$$

[37] The proof used here can be easily adapted to prove Lemma 7.7.1.

[38] $\mathcal{O}^*(\underline{C}_*)$ has the same expression as $\mathcal{O}(\underline{C}_*)$, with all the norms of the connection coefficients norms replaced by those defined in (7.4.1), (7.4.3), (7.4.4).

Proof: The proof is similar to, but easier than, the one for the estimate of $\mathbf{g}(L', L)$ on Σ'_{δ_0}. The evolution equation along $\underline{C}_*$ for $\mathbf{g}(L', L)$ is

$$\frac{d}{du'}\mathbf{g}(L', L) = 4\frac{\Omega'}{\Omega}(\Omega\underline{\omega}'+\Omega'\underline{\omega})\mathbf{g}(L', L)+\frac{\Omega'}{\Omega}(-2\mathbf{g}(L', L))^{\frac{1}{2}}\hat{\sigma}\cdot(\eta'-\eta), \quad (7.7.37)$$

which is obtained by a direct computation. Applying now Gronwall's Lemma we obtain

$$|\mathbf{g}(L', L)|_{\underline{C}_*}(u', \underline{u}_*) \leq c\int_{u'_0}^{u'} du''\left(|\mathbf{g}(L', L)|^{\frac{1}{2}}|\eta' - \eta|\right)(u'', \underline{u}_*). \quad (7.7.38)$$

Since, due to the assumptions of the lemma, $|\eta' - \eta| = O(r'^{-2}\tau_-^{-\frac{1}{2}})$, the lemma is proved by mimicking the argument used to complete the proof of Lemma 4.8.2.

7.7.3 *Proof of Propositions 7.4.4, 7.4.5*

Proof: The proof of these propositions is a consequence of the next two lemmas.

Lemma 7.7.3 *Let us consider on S a solution in the sense of distributions of the equation*[39]

$$\Delta\!\!\!/ u = G,$$

where G satisfies the condition: $\overline{G} = 0$. *Moreover let us assume that*

$$G \in W^{-2,p}(S),$$

where an element of $W^{-2,p}(S)$ *is a bounded linear functional on the space of the test functions* $C^\infty(S)$ *such that the following inequality holds:*

$$| < G, \phi > | \leq C\left(|\nabla\!\!\!/^2\phi|_{L^q(S)} + r^{-1}|\nabla\!\!\!/\phi|_{L^q(S)} + r^{-2}|\phi|_{L^q(S)}\right),$$

with q the number conjugate to p and with C a constant independent of ϕ. *Introducing the norm* $|G|_{W^{-2,p}(S)}$ *as the infimum of the possible constants C, and noting that the solution u is an element of* $L^p(S)$, *we find that the following inequality holds:*

$$|u - \overline{u}| \leq c|G|_{W^{-2,p}(S)},$$

with a constant c depending on ${k_m}^{-1}$ *and* k_M.

Lemma 7.7.4 *Assume the last slice is endowed with a canonical foliation. Then* $\Omega\mathbf{D}_3\log\Omega$, $(\Omega\mathbf{D}_3)^2\log\Omega$, $(\Omega\mathbf{D}_3)^3\log\Omega$ *satisfy the following elliptic equations:*

$$\Delta\!\!\!/(\Omega\mathbf{D}_3\log\Omega) = \mathrm{d}\!\!\!/\mathrm{iv}\, F_1 + G_1 - \overline{G_1} \quad (7.7.39)$$

$$\Delta\!\!\!/((\Omega\mathbf{D}_3)^2\log\Omega) = \mathrm{d}\!\!\!/\mathrm{iv}\, F_2 + G_2 - \overline{G_2} \quad (7.7.40)$$

$$\Delta\!\!\!/((\Omega\mathbf{D}_3)^3\log\Omega) = \mathrm{d}\!\!\!/\mathrm{iv}\, F_3 + G_3 - \overline{G_3}, \quad (7.7.41)$$

[39]Here S is a two-dimensional compact surface such that $k_m > 0$, where $k_m = \min_S r^2K$, $k_M = \max_S r^2K$ and K is its Gauss curvature.

where

$$
\begin{aligned}
F_1 &= \Omega\underline{\beta} + \tilde{F}_1 \ , \quad \tilde{F}_1 = \left(\frac{3}{2}\Omega\eta\cdot\underline{\hat{\chi}} + \frac{1}{4}\Omega\eta\mathrm{tr}\underline{\chi}\right)\\
G_1 &= H + \frac{1}{4}\Omega\mathbf{D}_3(\hat{\chi}\cdot\underline{\hat{\chi}}) - \frac{1}{2}(\Omega\mathrm{tr}\underline{\chi})(\rho-\overline{\rho}) + \frac{1}{4}(\Omega\mathrm{tr}\underline{\chi})(\hat{\chi}\cdot\underline{\hat{\chi}} - \overline{\hat{\chi}\cdot\underline{\hat{\chi}}})\\
H &= \frac{\Omega}{2}\left(\frac{3}{2}\mathrm{tr}\underline{\chi}\rho + \frac{1}{2}\hat{\chi}\cdot\underline{\alpha} + \eta\cdot\underline{\beta}\right)
\end{aligned} \tag{7.7.42}
$$

$$
\begin{aligned}
F_2 &= \Omega\,\mathrm{d\!\!/iv}\,\underline{\alpha} + \widetilde{F}_2\\
\widetilde{F}_2 &= (\mathbf{D}_3\Omega\underline{\beta} - \Omega\,\mathrm{d\!\!/iv}\,\underline{\alpha}) + \Omega\left(\mathbf{D}_3\widetilde{F}_1 + 2\nabla\!\!\!/(\Omega\mathbf{D}_3\log\Omega)\cdot\underline{\hat{\chi}} - \underline{\hat{\chi}}\cdot F_1 + \frac{1}{2}\mathrm{tr}\underline{\chi}F_1\right)\\
G_2 &= \Omega\mathbf{D}_3G_1 + (\Omega\mathrm{tr}\underline{\chi})(G_1 - \overline{G_1})
\end{aligned} \tag{7.7.43}
$$

$$
\begin{aligned}
F_3 &= \Omega\,\mathrm{d\!\!/iv}\,\mathbf{D}_3\underline{\alpha} + \widetilde{F}_3\\
\widetilde{F}_3 &= (\mathbf{D}_3\Omega\,\mathrm{d\!\!/iv}\,\underline{\alpha} - \Omega\,\mathrm{d\!\!/iv}\,\mathbf{D}_3\underline{\alpha})\\
&\quad + \Omega\left(\mathbf{D}_3\widetilde{F}_2 + 2\nabla\!\!\!/[(\Omega D_3)^2\log\Omega]\cdot\underline{\hat{\chi}} - \underline{\hat{\chi}}\cdot F_2 + \frac{1}{2}\mathrm{tr}\underline{\chi}F_2\right)\\
G_3 &= \Omega\mathbf{D}_3G_2 + (\Omega\mathrm{tr}\underline{\chi})(G_2 - \overline{G_2}).
\end{aligned} \tag{7.7.44}
$$

Proof of Lemma 7.7.3: Let ϕ be a function $\in \mathcal{C}^\infty(S)$ and let ψ be a solution of $\Delta\!\!\!/\psi = \phi$. Then it is easy to prove that ψ satisfies the following bound for any $q > 0$

$$
|\nabla\!\!\!/^2\psi|_{L^q(S)} + r^{-1}|\nabla\!\!\!/\psi|_{L^q(S)} + r^{-2}|\psi - \overline{\psi}|_{L^q(S)} \le c|\phi|_{L^q(S)}.
$$

Therefore

$$
\begin{aligned}
|<u-\overline{u},\phi>| &= |<u-\overline{u},\Delta\!\!\!/\psi>| = |<\Delta\!\!\!/u,\psi>|\\
&= |<G,\psi>| = |<G,\psi-\overline{\psi}>|\\
&\le |G|_{W^{-2,p}(S)}\left(|\nabla\!\!\!/^2\psi|_{L^q(S)} + r^{-1}|\nabla\!\!\!/\psi|_{L^q(S)} + r^{-2}|\psi-\overline{\psi}|_{L^q(S)}\right)\\
&\le c|G|_{W^{-2,p}(S)}|\phi|_{L^q(S)},
\end{aligned} \tag{7.7.45}
$$

so that

$$
|<u-\overline{u},\phi>| \le c|G|_{W^{-2,p}(S)}|\phi|_{L^q(S)},
$$

for any $\phi \in \mathcal{C}^\infty(S)$, and consequently,

$$
|u-\overline{u}|_{L^p(S)} \le c|G|_{W^{-2,p}(S)}\ .
$$

Proof of Lemma 7.7.4: The proof of this lemma is postponed to the end of the appendix.

Proof of Proposition 7.4.4:

We apply Lemma 7.7.3 to equation (7.7.39) with

$$\begin{aligned} u &= \Omega \mathbf{D}_3 \log \Omega \\ G &= \not{d}\mathrm{iv}\, F_1 + (G_1 - \overline{G_1}) = \not{d}\mathrm{iv}\, \Omega \underline{\beta} + \not{d}\mathrm{iv}\, \tilde{F}_1 + (G_1 - \overline{G_1}) \\ &= \not{d}\mathrm{iv}\, \Omega \underline{\beta} + \not{d}\mathrm{iv} \left(\frac{3}{2} \Omega \eta \cdot \hat{\underline{\chi}} + \frac{1}{4} \Omega \eta \mathrm{tr} \underline{\chi} \right) + (G_1 - \overline{G_1}). \end{aligned}$$

The only derivative of the Riemann component is $\not{d}\mathrm{iv}\, \Omega \underline{\beta}$. All the other terms of G are controlled from the sup norm estimates of the undifferentiated Riemann components and the estimates of the connection coefficients and their first derivatives on the last slice. In particular, assuming $\mathcal{I}_0$, Δ_0 sufficiently small,

$$\begin{aligned} &|r^{3-\frac{2}{p}} \tau_-^{\frac{3}{2}} (G_1 - \overline{G_1})|_{p,S} \leq c\,(\mathcal{I}_0 + \Delta_0) \ , \quad \forall p \geq 2, \\ &|r^{3-\frac{2}{p}} \tau_-^{\frac{3}{2}} \not{d}\mathrm{iv}\, \Omega \underline{\beta}|_{p,S} \leq c \Delta_1 \ , \quad \forall p \in [2, 4], \end{aligned} \tag{7.7.46}$$

and

$$|r^{4-\frac{2}{p}} \tau_-^{\frac{1}{2}} \not{d}\mathrm{iv}\, \tilde{F}_1|_{p,S} \leq c\,(\mathcal{I}_0 + \Delta_0) \ , \quad \forall p \geq 2. \tag{7.7.47}$$

The last estimate is true on the last slice due to its canonical foliation. In fact from (7.7.42) it follows that

$$\tilde{F}_1 = \left(\frac{3}{2} \Omega \eta \cdot \hat{\underline{\chi}} + \frac{1}{4} \Omega \eta \mathrm{tr} \underline{\chi} \right),$$

and to control $|r^{4-\frac{2}{p}} \tau_-^{\frac{1}{2}} \not{d}\mathrm{iv}\, \tilde{F}_1|_{p,S}$, recalling the estimate for $\hat{\underline{\chi}}$, we have to control $|r^{4-\frac{2}{p}} \tau_-^{\frac{1}{2}} \Omega \not{\nabla} \eta \mathrm{tr} \underline{\chi}|_{p,S}$. In fact from (3.3.12) we obtain (see also (4.3.37))

$$\begin{aligned} \frac{1}{2} \not{d}\mathrm{iv}\, \eta = \frac{1}{2} \not{d}\mathrm{iv}\, \zeta + \frac{1}{2} \not{\Delta} \log \Omega &= \frac{1}{2} \left(K - \overline{K} + \frac{1}{4} (\mathrm{tr}\chi \mathrm{tr}\underline{\chi} - \overline{\mathrm{tr}\chi \mathrm{tr}\underline{\chi}}) \right) \\ &= \frac{1}{2} \left(\frac{1}{2} \hat{\chi} \hat{\underline{\chi}} - \frac{1}{2} \overline{\hat{\chi} \hat{\underline{\chi}}} - (\rho - \overline{\rho}) \right), \end{aligned} \tag{7.7.48}$$

and from this expression the required estimate for $\not{d}\mathrm{iv}\, \eta$ follows. Therefore

$$\begin{aligned} | < G, \psi > | &\leq \left| \int_S (\not{d}\mathrm{iv}\, \Omega \underline{\beta}) \psi \right| + \left| \int_S (\not{d}\mathrm{iv}\, \tilde{F}) \psi \right| + \left| \int_S (G_1 - \overline{G_1}) \psi \right| \\ &\leq \int_S |\Omega \underline{\beta}| |\not{\nabla} \psi| + \int_S |\not{d}\mathrm{iv}\, \tilde{F}_1| |\psi| + \int_S |G_1 - \overline{G_1}| |\psi| \\ &\leq \frac{1}{|r^{2-\frac{2}{p}} \tau_-^{\frac{3}{2}}|} |r^{2-\frac{2}{p}} \tau_-^{\frac{3}{2}} \Omega \underline{\beta}|_{p,S} |\not{\nabla} \psi|_{q,S} + \frac{1}{|r^{4-\frac{2}{p}} \tau_-^{\frac{1}{2}}|} |r^{4-\frac{2}{p}} \tau_-^{\frac{1}{2}} \not{d}\mathrm{iv}\, \tilde{F}_1|_{p,S} |\psi|_{q,S} \\ &+ \frac{1}{|r^{3-\frac{2}{p}} \tau_-^{\frac{3}{2}}|} |r^{3-\frac{2}{p}} \tau_-^{\frac{3}{2}} (G_1 - \overline{G_1})|_{p,S} |\psi|_{q,S} \end{aligned}$$

$$\leq c\frac{1}{|r^{1-\frac{2}{p}}\tau_-^{\frac{3}{2}}|}\sup_{\underline{C}_*}\Bigg[\Big(|r^{2-\frac{2}{p}}\tau_-^{\frac{3}{2}}\Omega\underline{\beta}|_{p,S}+|r^{4-\frac{2}{p}}\tau_-^{\frac{1}{2}}\nabla\tilde{F}_1|_{p,S}\Big) + |r^{3-\frac{2}{p}}\tau_-^{\frac{3}{2}}(G_1-\overline{G_1})|_{p,S}\Bigg]\left(\frac{1}{r}|\nabla\psi|_{q,S}+\frac{1}{r^2}|\psi|_{q,S}\right) \tag{7.7.49}$$

which implies

$$|G|_{W^{-2,p}(S)} \leq c_0\frac{1}{|r^{1-\frac{2}{p}}\tau_-^{\frac{3}{2}}|}\sup_{\underline{C}_*}\Bigg[\Big(|r^{2-\frac{2}{p}}\tau_-^{\frac{3}{2}}\Omega\underline{\beta}|_{p,S}+|r^{4-\frac{2}{p}}\tau_-^{\frac{1}{2}}\tilde{F}_1|_{p,S}\Big) + |r^{3-\frac{2}{p}}\tau_-^{\frac{3}{2}}(G_1-\overline{G_1})|_{p,S}\Bigg] \leq c_0\frac{1}{|r^{1-\frac{2}{p}}\tau_-^{\frac{3}{2}}|}(\mathcal{I}_0+\Delta_0)\,.$$

From it we obtain for any $p\geq 2$,

$$|\Omega\mathbf{D}_3\log\Omega-\overline{\Omega\mathbf{D}_3\log\Omega}|_{p,S}\leq c_0\frac{1}{|r^{1-\frac{2}{p}}\tau_-^{\frac{3}{2}}|}(\mathcal{I}_0+\Delta_0). \tag{7.7.50}$$

Considering $\nabla(\Omega\mathbf{D}_3\log\Omega)$ and $\nabla^2(\Omega\mathbf{D}_3\log\Omega)$ and proceeding in the same way we find that for $p\geq 2$,[40]

$$|\nabla(\Omega\mathbf{D}_3\log\Omega)|_{p,S}\leq c\frac{1}{|r^{2-\frac{2}{p}}\tau_-^{\frac{3}{2}}|}(\mathcal{I}_0+\Delta_0), \tag{7.7.51}$$

and for $p\in[2,4]$,

$$|\nabla^2(\Omega\mathbf{D}_3\log\Omega)|_{p,S}\leq c_0\frac{1}{|r^{3-\frac{2}{p}}\tau_-^{\frac{3}{2}}|}(\mathcal{I}_0+\Delta_0+\Delta_1). \tag{7.7.52}$$

Finally, since on the last slice $\overline{\log 2\Omega}=0$, we have

$$\begin{aligned}
0=\frac{\Omega}{2}\mathbf{D}_3\overline{\log 2\Omega} &= \frac{1}{2}\frac{d}{du}\frac{1}{|S|}\int_S\log 2\Omega\\
&= -\frac{1}{2}\frac{1}{|S|^2}\left(\frac{d|S|}{du}\right)\int_S\log 2\Omega+\frac{1}{2}\frac{1}{|S|}\frac{d}{du}\int_S\log 2\Omega\\
&= \frac{1}{2}\left[-\frac{1}{|S|}\left(\frac{d|S|}{du}\right)\overline{\log 2\Omega}+\overline{(\Omega\mathrm{tr}\underline{\chi})\log 2\Omega}+\overline{\Omega\mathbf{D}_3\log 2\Omega}\right]\\
&= \overline{(\Omega\mathrm{tr}\underline{\chi})\log 2\Omega+\Omega\mathbf{D}_3\log 2\Omega}\ .
\end{aligned}$$

Therefore, recalling the inequality[41]

$$|\overline{f}|_{p,S}\leq|f|_{p,S}, \tag{7.7.53}$$

---

[40] From it $\sup_{\underline{C}_*}|r\tau_-^{\frac{3}{2}}\Omega\mathbf{D}_3\log\Omega|\leq c(\mathcal{I}_0+\Delta_0)$ follows.

[41] It follows immediately from the Hölder inequality. In fact

$$|\overline{f}|_{p,S}=\left(\int_S\frac{1}{|S|^p}\left(\int_S f\right)^p\right)^{\frac{1}{p}}=|S|^{\frac{1}{p}-1}\int_S f\leq|S|^{\frac{1}{p}-1}|1|_{q,S}|f|_{p,S}=|S|^{\frac{1}{p}+\frac{1}{q}-1}|f|_{p,S}=|f|_{p,S}\,.$$

and Proposition 7.4.3, we have

$$|\overline{\Omega\mathbf{D}_3\log 2\Omega}|_{p,S_*} \leq |\overline{(\Omega\mathrm{tr}\underline{\chi})\log 2\Omega}|_{p,S_*} \leq c\frac{1}{r}|(\log 2\Omega - \overline{\log 2\Omega})|_{p,S_*}, \tag{7.7.54}$$

and

$$|r^{1-\frac{2}{p}}\tau_-\overline{\Omega\mathbf{D}_3\log 2\Omega}|_{p,S_*} \leq c\frac{\tau_-^{\frac{1}{2}}}{r}|r^{1-\frac{2}{p}}\tau_-^{\frac{1}{2}}(\log 2\Omega - \overline{\log 2\Omega})|_{p,S_*} \leq c\frac{\tau_-^{\frac{1}{2}}}{r}(\mathcal{I}_0 + \Delta_0),$$

so that, finally,

$$|r^{1-\frac{2}{p}}\tau_-^{\frac{3}{2}}\overline{\Omega\mathbf{D}_3\log 2\Omega}|_{p,S_*} \leq c(\mathcal{I}_0 + \Delta_0), \tag{7.7.55}$$

which completes the proof of Proposition 7.4.4.

Proof of Proposition 7.4.5

To apply Lemma 7.7.3 to this case we use the following definitions,

$$\begin{aligned} u &= \Omega\mathbf{D}_3(\Omega\mathbf{D}_3\log\Omega) \\ G &= \mathrm{d\!\!/iv}\left(\Omega\mathbf{D}_3 F_1 + 2\Omega(\nabla\!\!\!\!/(\Omega\mathbf{D}_3\log\Omega))\cdot\hat{\underline{\chi}} - \Omega\hat{\underline{\chi}}\cdot F_1 + \frac{1}{2}\Omega\mathrm{tr}\underline{\chi}F_1\right) + (G_2 - \overline{G_2}) \\ &\equiv \mathrm{d\!\!/iv}\,\mathrm{d\!\!/iv}\,\underline{\alpha} + \mathrm{d\!\!/iv}\,\tilde{F}_2 + (G_2 - \overline{G_2}). \end{aligned}$$

Therefore we have to control the integrals

$$\int_S \mathrm{d\!\!/iv}\,\mathrm{d\!\!/iv}\,\underline{\alpha}\psi\ ,\ \int_S \mathrm{d\!\!/iv}\,\tilde{F}_2\psi\ ,\ \int_S (G_2 - \overline{G_2})\psi\ .$$

The following estimates hold for any $p \geq 2$:

$$\begin{aligned} \left|\int_S \mathrm{d\!\!/iv}\,\mathrm{d\!\!/iv}\,\underline{\alpha}\psi\right| &\leq c\int_S |\underline{\alpha}||\nabla\!\!\!\!/^2\psi| \leq c\frac{1}{|r^{1-\frac{2}{p}}\tau_-^{\frac{5}{2}}|}\left(|r^{1-\frac{2}{p}}\tau_-^{\frac{5}{2}}\underline{\alpha}|_{p,S}\right)|\nabla\!\!\!\!/^2\psi|_{q,S} \\ \left|\int_S \mathrm{d\!\!/iv}\,\tilde{F}_2\psi\right| &\leq c\int_S |\tilde{F}_2||\nabla\!\!\!\!/\psi| \leq c\frac{1}{|r^{3-\frac{2}{p}}\tau_-^{\frac{3}{2}}|}|r^{3-\frac{2}{p}}\tau_-^{\frac{3}{2}}\tilde{F}_2|_{p,S}|\nabla\!\!\!\!/\psi|_{q,S} \\ &\leq c\frac{1}{|r^{2-\frac{2}{p}}\tau_-^{\frac{3}{2}}|}|r^{3-\frac{2}{p}}\tau_-^{\frac{3}{2}}\tilde{F}_2|_{p,S}\left(\frac{1}{r}|\nabla\!\!\!\!/\psi|_{q,S}\right). \end{aligned} \tag{7.7.56}$$

Clearly the last estimate is the appropriate one since a long but simple analysis of all the terms composing $\tilde{F}_2$ allows us to conclude that for any $p \geq 2$,

$$|r^{3-\frac{2}{p}}\tau_-^{\frac{3}{2}}\tilde{F}_2|_{p,S} \leq c(\mathcal{I}_* + \Delta_0). \tag{7.7.57}$$

To estimate the third integral $\int_S (G_2 - \overline{G_2})\psi$ we have to examine the structure of the term $G_2 = \Omega\mathbf{D}_3 G_1 + (\Omega\mathrm{tr}\underline{\chi})(G_1 - \overline{G_1})$. We already have the estimate of $(G_1 - \overline{G_1})$; therefore, we have only to investigate the term $\Omega\mathbf{D}_3 G_1$. Extracting the terms where $\mathbf{D}_3$ operates on the Riemann components we can write

$$\mathbf{D}_3 G_1 = O\left(\frac{1}{r}\right)\mathrm{d\!\!/iv}\,\underline{\beta} + O\left(\frac{1}{r^2}\right)\mathrm{d\!\!/iv}\,\underline{\alpha} + O\left(\frac{1}{r^2}\right)\rho + O\left(\frac{1}{r^3}\right)(\underline{\beta} + \underline{\alpha}) + \widetilde{\mathbf{D}_3 G_1},$$

where $\widetilde{\mathbf{D}_3 G_1}$ is the part of $\mathbf{D}_3 G_1$ that does not depend explicitly on the Riemann components and for which it is easy to prove that the following bounds hold for any $p \geq 2$:

$$|r^{3-\frac{2}{p}}\tau_-^{\frac{5}{2}}\widetilde{\mathbf{D}_3 G_1}|_{p,S} \leq c(\mathcal{I}_* + \Delta_0)\ .$$

Therefore we write

$$\begin{aligned}
G_2 &= \Omega\mathbf{D}_3 G_1 + (\Omega\mathrm{tr}\underline{\chi})(G_1 - \overline{G_1}) = \left(O\left(\frac{1}{r}\right)\mathrm{d\!\!/iv}\,\underline{\beta} + O\left(\frac{1}{r^2}\right)\mathrm{d\!\!/iv}\,\underline{\alpha}\right)\\
&+ \left[\left(O\left(\frac{1}{r^2}\right)\rho + O\left(\frac{1}{r^3}\right)(\underline{\beta} + \underline{\alpha})\right) + \widetilde{\mathbf{D}_3 G_1} + (\Omega\mathrm{tr}\underline{\chi})(G_1 - \overline{G_1})\right]\\
&\equiv \left(O\left(\frac{1}{r}\right)\mathrm{d\!\!/iv}\,\underline{\beta} + O\left(\frac{1}{r^2}\right)\mathrm{d\!\!/iv}\,\underline{\alpha}\right) + \tilde{G}_2,
\end{aligned}$$

and it is easy to prove by collecting all the previous estimates that

$$|r^{3-\frac{2}{p}}\tau_-^{\frac{5}{2}}\tilde{G}_2|_{p,S} \leq c(\mathcal{I}_* + \Delta_0),$$

for any $p \geq 2$. Combining all these results we obtain

$$\begin{aligned}
&\left|\int_S (G_2 - \overline{G_2})\psi\right|\\
&\leq c\int_S \left|\left(O\left(\frac{1}{r}\right)\underline{\beta}\right)\right||\nabla\!\!\!/\psi| + c\int_S \left|\left(O\left(\frac{1}{r^2}\right)\underline{\alpha}\right)\right||\nabla\!\!\!/\psi| + c\int_S |\tilde{G}_2 - \overline{\tilde{G}_2}||\psi|\\
&\leq c\frac{1}{|r^{3-\frac{2}{p}}\tau_-^{\frac{5}{2}}|}|r^{3-\frac{2}{p}}\tau_-^{\frac{5}{2}}(\tilde{G}_2 - \overline{\tilde{G}_2})|_{p,S}|\psi|_{q,S} + c\frac{1}{|r^{3-\frac{2}{p}}\tau_-^{\frac{3}{2}}|}|r^{2-\frac{2}{p}}\tau_-^{\frac{3}{2}}\underline{\beta}|_{p,S}|\nabla\!\!\!/\psi|_{q,S}\\
&\quad + c\frac{1}{|r^{3-\frac{2}{p}}\tau_-^{\frac{5}{2}}|}|r^{1-\frac{2}{p}}\tau_-^{\frac{5}{2}}\underline{\alpha}|_{p,S}|\nabla\!\!\!/\psi|_{q,S}\\
&\leq c\frac{1}{|r^{1-\frac{2}{p}}\tau_-^{\frac{5}{2}}|}\left[|r^{3-\frac{2}{p}}\tau_-^{\frac{5}{2}}(\tilde{G}_2 - \overline{\tilde{G}_2})|_{p,S} + \frac{1}{r}\left(|r^{2-\frac{2}{p}}\tau_-^{\frac{3}{2}}\underline{\beta}|_{p,S} + |r^{1-\frac{2}{p}}\tau_-^{\frac{5}{2}}\underline{\alpha}|_{p,S}\right)\right]\\
&\quad \cdot\left(\frac{1}{r}|\nabla\!\!\!/\psi|_{q,S} + \frac{1}{r^2}|\psi|_{q,S}\right).
\end{aligned} \tag{7.7.58}$$

Using Lemma 7.7.3 and the analogue of (7.7.54) we conclude

$$\begin{aligned}
&|(\Omega D_3(\Omega D_3 \log\Omega))|_{p,S}\\
&\quad\leq c\frac{1}{|r^{1-\frac{2}{p}}\tau_-^{\frac{5}{2}}|}\left\{\left[|r^{3-\frac{2}{p}}\tau_-^{\frac{5}{2}}(\tilde{G}_2 - \overline{\tilde{G}_2})|_{p,S} + \frac{1}{r}\left(|r^{2-\frac{2}{p}}\tau_-^{\frac{3}{2}}\underline{\beta}|_{p,S} + |r^{1-\frac{2}{p}}\tau_-^{\frac{5}{2}}\underline{\alpha}|_{p,S}\right)\right]\right.\\
&\quad\left. + \frac{1}{r}\left(|r^{3-\frac{2}{p}}\tau_-^{\frac{3}{2}}\tilde{F}_2|_{p,S}\right) + \frac{1}{\tau_-^{\frac{1}{2}}}\left(|r^{1-\frac{2}{p}}\tau_-^{\frac{5}{2}}\underline{\alpha}|_{p,S}\right)\right\}\\
&\quad\cdot\left(\frac{1}{r^2}|\nabla\!\!\!/^2\psi|_{q,S} + \frac{1}{r}|\nabla\!\!\!/\psi|_{q,S} + \frac{1}{r^2}|\psi|_{q,S}\right).
\end{aligned} \tag{7.7.59}$$

Therefore

$$|r^{1-\frac{2}{p}}\tau_-^{\frac{5}{2}}(\Omega D_3(\Omega D_3 \log\Omega))|_{p,S} \leq c|G|_{W^{-p,S}} \tag{7.7.60}$$

and, for any $p \geq 2$,

$$|G|_{W^{-p,S}} \leq \Bigg\{\Bigg[|r^{3-\frac{2}{p}}\tau_-^{\frac{5}{2}}(\tilde{G}_2 - \overline{\tilde{G}_2})|_{p,S} + \frac{1}{r}\left(|r^{2-\frac{2}{p}}\tau_-^{\frac{3}{2}}\underline{\beta}|_{p,S} + |r^{1-\frac{2}{p}}\tau_-^{\frac{5}{2}}\underline{\alpha}|_{p,S}\right)\Bigg]$$
$$+\frac{1}{r}\left(|r^{3-\frac{2}{p}}\tau_-^{\frac{3}{2}}\tilde{F}_2|_{p,S}\right) + \frac{1}{\tau_-^{\frac{1}{2}}}\left(|r^{1-\frac{2}{p}}\tau_-^{\frac{5}{2}}\underline{\alpha}|_{p,S}\right)\Bigg\} \leq c(\mathcal{I}_* + \Delta_0) \tag{7.7.61}$$

Proceeding in a similar way we obtain, for any $p \in [2, 4]$,

$$|r^{2-\frac{2}{p}}\tau_-^{\frac{5}{2}}\nabla\!\!\!/(\Omega\mathbf{D}_3(\Omega\mathbf{D}_3 \log\Omega))|_{p,S} \leq c(\mathcal{I}_* + \Delta_0 + \Delta_1)\,.$$

Estimate for $(\Omega\mathbf{D}_3)^3 \log\Omega$

The proof goes basically as in the case of $(\Omega\mathbf{D}_3)^2 \log\Omega$; we stress only the main differences. From the explicit expressions of F_3 and G_3 we have to examine the dependence on the various Riemann components. In a symbolic way we can write

$$\begin{aligned}
F_3 &= \text{d}\!\!\!/\text{iv}\,\mathbf{D}_3\underline{\alpha} + O(\frac{1}{r})\text{d}\!\!\!/\text{iv}\,\underline{\alpha} + O(\frac{1}{r^2})\mathbf{D}_3\underline{\alpha}\\
&= \text{d}\!\!\!/\text{iv}\,\underline{\alpha}(\hat{\mathcal{L}}_T W) + \text{d}\!\!\!/\text{iv}\,\text{d}\!\!\!/\text{iv}\,\underline{\beta} + O(\frac{1}{r})\text{d}\!\!\!/\text{iv}\,\underline{\alpha} + O(\frac{1}{r^2})(\underline{\alpha}(\hat{\mathcal{L}}_T W) + \text{d}\!\!\!/\text{iv}\,\underline{\beta})\\
G_3 &= O(\frac{1}{r})\text{d}\!\!\!/\text{iv}\,\text{d}\!\!\!/\text{iv}\,\underline{\alpha} + O(\frac{1}{r^2})\text{d}\!\!\!/\text{iv}\,\underline{\alpha}(\hat{\mathcal{L}}_T W) + O(\frac{1}{r^2})\text{d}\!\!\!/\text{iv}\,\text{d}\!\!\!/\text{iv}\,,\underline{\beta}
\end{aligned} \tag{7.7.62}$$

where we reported just the dependence on the various Riemann components associated with the Weyl tensor W except when explicitly indicated. The factors $O(\frac{1}{r})$, $O(\frac{1}{r^2})$ in front of some of them are to remind us of the asymptotic behavior of the factors multiplying the various components.

It is now easy to apply to this case Lemma 7.7.3 and obtain the thesis of the lemma.[42]

Proof of equation (7.7.39) of Lemma 7.7.4

We write

$$\Delta\!\!\!/\mathbf{D}_3 \log\Omega = \mathbf{D}_3\Delta\!\!\!/ \log\Omega + [\Delta\!\!\!/, \mathbf{D}_3]\log\Omega.$$

Using Proposition 4.8.1 for $[\Delta\!\!\!/, \mathbf{D}_3]\log\Omega$ we obtain

$$\Delta\!\!\!/\mathbf{D}_3 \log\Omega \;=\; \mathbf{D}_3\Delta\!\!\!/ \log\Omega$$

---

[42]Note that the control of $|r^{1-\frac{2}{p}}\tau_-^{\frac{7}{2}}\hat{\mathcal{L}}_T\underline{\alpha}|_{p,S}$ required by the assumptions of the lemma is provided by the control of

$$\int_{\underline{C}(\underline{u})} Q(\hat{\mathcal{L}}_O\hat{\mathcal{L}}_T W)(\bar{K}, \bar{K}, \bar{K}, e_3) \text{ and } \int_{\underline{C}(\underline{u})} Q(\hat{\mathcal{L}}_T W)(\bar{K}, \bar{K}, \bar{K}, e_3),$$

whereas we cannot obtain the analogous bound for $|\nabla\!\!\!/\hat{\mathcal{L}}_T\underline{\alpha}|_{p,S}$ by proceeding in the same way.

$$-\Big\{-\eta_a \underline{\chi}_{ab} \nabla\!\!\!\!/_b \log\Omega + \mathrm{tr}\underline{\chi}\,\eta_b \nabla\!\!\!\!/_b \log\Omega - 2\underline{\chi}_{ab}(\nabla\!\!\!\!/\nabla\!\!\!\!/ \log\Omega)_{ab}$$
$$- \zeta_a \underline{\chi}_{ab} \nabla\!\!\!\!/_b \log\Omega - (\mathrm{d}\!\!\!/\mathrm{iv}\,\underline{\chi})_b \nabla\!\!\!\!/_b \log\Omega - \underline{\beta}_b \nabla\!\!\!\!/_b \log\Omega\Big\}$$
$$-\Big[(\nabla\!\!\!\!/_a \log\Omega)(\mathbf{D}_3 \nabla\!\!\!\!/ \log\Omega)_a - \zeta_a(\mathbf{D}_3\nabla\!\!\!\!/ \log\Omega)_a + \eta_a \nabla\!\!\!\!/_a \mathbf{D}_3 \log\Omega$$
$$+(\Delta\!\!\!\!/ \log\Omega)\mathbf{D}_3\log\Omega + \zeta_a(\nabla\!\!\!\!/_a \log\Omega)\mathbf{D}_3\log\Omega\Big]. \qquad (7.7.63)$$

A long but easy computation allows us to rewrite this equation as

$$\Delta\!\!\!\!/(\Omega\mathbf{D}_3\log\Omega) = \Omega\mathbf{D}_3(\Delta\!\!\!\!/ \log\Omega) + \Omega\Big[2\underline{\chi}_{ab}(\nabla\!\!\!\!/\nabla\!\!\!\!/ \log\Omega)_{ab} + (\nabla\!\!\!\!/_b \log\Omega)(\nabla\!\!\!\!/_a \log\Omega)\underline{\chi}_{ab}$$
$$+ (\nabla\!\!\!\!/_b \log\Omega)(\mathrm{d}\!\!\!/\mathrm{iv}\,\underline{\chi})_b + (\nabla\!\!\!\!/_b \log\Omega)\left(\eta_a\underline{\chi}_{ab} - \eta_b \mathrm{tr}\underline{\chi} + \underline{\beta}_b\right)\Big]. \qquad (7.7.64)$$

Recalling the structure equation (see (3.1.47))

$$\nabla\!\!\!\!/ \mathrm{tr}\underline{\chi} - \mathrm{d}\!\!\!/\mathrm{iv}\,\underline{\chi} + \zeta\cdot\underline{\chi} - \zeta\,\mathrm{tr}\underline{\chi} + \underline{\beta} = 0,$$

we obtain

$$(\eta_a\underline{\chi}_{ab} - \eta_b\mathrm{tr}\underline{\chi} + \underline{\beta}_b) = (\nabla\!\!\!\!/_a \log\Omega)\underline{\chi}_{ab} - (\nabla\!\!\!\!/_b \log\Omega)\mathrm{tr}\underline{\chi} + (\mathrm{d}\!\!\!/\mathrm{iv}\,\underline{\chi})_b - \nabla\!\!\!\!/_b \mathrm{tr}\underline{\chi},$$

so that the [ ] part of the previous equation, (7.7.64), becomes:

$$\Omega[\ \ ] = 2\mathrm{d}\!\!\!/\mathrm{iv}\,(\Omega\nabla\!\!\!\!/ \log\Omega\cdot\underline{\chi}) - (\nabla\!\!\!\!/_b \log\Omega)(\nabla\!\!\!\!/_b \Omega\mathrm{tr}\underline{\chi}),$$

and therefore,

$$\Delta\!\!\!\!/(\Omega\mathbf{D}_3\log\Omega) = \Omega\mathbf{D}_3(\Delta\!\!\!\!/ \log\Omega) + \Big[2\mathrm{d}\!\!\!/\mathrm{iv}\,(\Omega\nabla\!\!\!\!/ \log\Omega\cdot\underline{\chi}) - (\nabla\!\!\!\!/_b \log\Omega)(\nabla\!\!\!\!/_b \Omega\mathrm{tr}\underline{\chi})\Big].$$

Because the foliation we chose on the last slice $\underline{C}_*$ is canonical, $\log\Omega$ satisfies the elliptic equation (see (3.3.12))

$$\Delta\!\!\!\!/ \log\Omega = \frac{1}{2}\left[\mathrm{d}\!\!\!/\mathrm{iv}\,\underline{\eta} + \frac{1}{2}(\hat{\chi}\cdot\underline{\hat{\chi}} - \overline{\hat{\chi}\cdot\underline{\hat{\chi}}}) - (\rho - \overline{\rho})\right],$$

so that

$$\Omega\mathbf{D}_3(\Delta\!\!\!\!/ \log\Omega) = \frac{\Omega}{2}\left[\mathbf{D}_3(\mathrm{d}\!\!\!/\mathrm{iv}\,\underline{\eta}) - (\mathbf{D}_3\rho - \mathbf{D}_3\overline{\rho}) + \frac{1}{2}\left(\mathbf{D}_3(\hat{\chi}\cdot\underline{\hat{\chi}}) - \mathbf{D}_3\overline{(\hat{\chi}\cdot\underline{\hat{\chi}})}\right)\right].$$

Recalling the evolution equation of $\underline{\eta}$ along the $\underline{C}$ null hypersurfaces

$$\mathbf{D}\!\!\!\!/_3\underline{\eta} = (\eta - \underline{\eta})\cdot\underline{\chi} + \underline{\beta}$$

a long but simple computation gives[43]

$$\begin{aligned}\frac{\Omega}{2}\mathbf{D}_3(\not{\mathrm{d}}\mathrm{iv}\,\underline{\eta}) &= \frac{1}{2}\Big[\not{\mathrm{d}}\mathrm{iv}\,(\Omega\underline{\beta}+\Omega(\eta-\underline{\eta})\cdot\underline{\chi})-(\not{\mathrm{d}}\mathrm{iv}\,\Omega\underline{\eta}\cdot\underline{\chi})+(\not\nabla\Omega\mathrm{tr}\underline{\chi})\cdot\underline{\eta}\Big]\\ &= \frac{1}{2}\Big[\not{\mathrm{d}}\mathrm{iv}\,(\Omega\underline{\beta}+\Omega\eta\cdot\underline{\chi}-2\Omega\underline{\eta}\cdot\underline{\chi})\Big]+\frac{1}{2}(\not\nabla\Omega\mathrm{tr}\underline{\chi})\cdot\underline{\eta},\end{aligned}\tag{7.7.65}$$

where we used again the structure equation

$$\not\nabla\mathrm{tr}\underline{\chi}-\not{\mathrm{d}}\mathrm{iv}\,\underline{\chi}+\zeta\cdot\underline{\chi}-\zeta\,\mathrm{tr}\underline{\chi}+\underline{\beta}=0\,.$$

From it

$$\begin{aligned}\Omega\mathbf{D}_3(\not\Delta\log\Omega) &= \frac{1}{2}\Big[\not{\mathrm{d}}\mathrm{iv}\,(\Omega\underline{\beta}+\Omega\eta\cdot\underline{\chi}-2\Omega\underline{\eta}\cdot\underline{\chi})\Big]+\frac{1}{2}(\not\nabla\Omega\mathrm{tr}\underline{\chi})\cdot\underline{\eta}\\ &+\left[-\frac{\Omega}{2}\mathbf{D}_3(\rho-\overline{\rho})+\frac{\Omega}{4}\left(\mathbf{D}_3(\hat{\chi}\cdot\hat{\underline{\chi}})-\mathbf{D}_3\overline{(\hat{\chi}\cdot\hat{\underline{\chi}})}\right)\right],\end{aligned}\tag{7.7.66}$$

so that finally

$$\begin{aligned}\not\Delta(\Omega\mathbf{D}_3\log\Omega) &= \not{\mathrm{d}}\mathrm{iv}\left(\frac{1}{2}\Omega\underline{\beta}+\frac{3}{2}\Omega\eta\cdot\underline{\chi}\right)-\frac{1}{2}\eta\cdot(\not\nabla\Omega\mathrm{tr}\underline{\chi})\\ &+\left[-\frac{\Omega}{2}\mathbf{D}_3(\rho-\overline{\rho})+\frac{\Omega}{4}\left(\mathbf{D}_3(\hat{\chi}\cdot\hat{\underline{\chi}})-\mathbf{D}_3\overline{(\hat{\chi}\cdot\hat{\underline{\chi}})}\right)\right].\end{aligned}\tag{7.7.67}$$

Recalling that on the last slice (see (3.3.10)),

$$\not{\mathrm{d}}\mathrm{iv}\,\eta=-\not{\mathrm{d}}\mathrm{iv}\,\underline{\eta}+2\not\Delta\log\Omega=\left[\tfrac{1}{2}(\hat{\chi}\cdot\hat{\underline{\chi}}-\overline{\hat{\chi}\cdot\hat{\underline{\chi}}})-(\rho-\overline{\rho})\right],$$

we have

$$\begin{aligned}-\frac{1}{2}\eta_b(\not\nabla_b\Omega\mathrm{tr}\underline{\chi}) &= -\frac{1}{2}\not{\mathrm{d}}\mathrm{iv}\,(\Omega\mathrm{tr}\underline{\chi}\eta)+\frac{1}{2}(\not{\mathrm{d}}\mathrm{iv}\,\eta)\Omega\mathrm{tr}\underline{\chi}\\ &= -\frac{1}{2}\not{\mathrm{d}}\mathrm{iv}\,(\Omega\mathrm{tr}\underline{\chi}\eta)+\frac{1}{2}\Omega\mathrm{tr}\underline{\chi}\left[\frac{1}{2}(\hat{\chi}\cdot\hat{\underline{\chi}}-\overline{\hat{\chi}\cdot\hat{\underline{\chi}}})-(\rho-\overline{\rho})\right].\end{aligned}\tag{7.7.68}$$

Moreover from

$$\begin{aligned}\frac{\Omega}{2}\mathbf{D}_3\overline{\rho} &= \frac{1}{2}\frac{d}{du}\left(\frac{1}{|S|}\int_S\rho\right)=-\frac{1}{2}\frac{1}{|S|^2}\left(\frac{d|S|}{du}\right)\int_S\rho+\frac{1}{2}\frac{1}{|S|}\frac{d}{du}\int_S\rho\\ &= \frac{1}{2}\left\{-\frac{1}{|S|}\left(\frac{d|S|}{du}\right)\overline{\rho}+\overline{(\Omega\mathrm{tr}\underline{\chi})\rho}+\overline{\Omega\mathbf{D}_3\rho}\right\},\end{aligned}$$

---

[43]More explicitly

$$\begin{aligned}\mathbf{D}_3(\not{\mathrm{d}}\mathrm{iv}\,\underline{\eta}) &= \not{\mathrm{d}}\mathrm{iv}\,(\mathbf{D}_3\underline{\eta})+(\not\nabla_a\log\Omega)(\mathbf{D}_3\underline{\eta})_a-(\eta_a\underline{\chi}_{ab}-\eta_b\mathrm{tr}\underline{\chi}+\underline{\beta}_b)\underline{\eta}_b-\underline{\chi}_{ab}(\not\nabla_b\underline{\eta})_a\\ &= \not{\mathrm{d}}\mathrm{iv}\,(\underline{\beta}+(\eta-\underline{\eta})\cdot\underline{\chi})+(\not\nabla_a\log\Omega)(\underline{\beta}+(\eta-\underline{\eta})\cdot\underline{\chi})_a-(\eta_a\underline{\chi}_{ab}-\eta_b\mathrm{tr}\underline{\chi}+\underline{\beta}_b)\underline{\eta}_b-\underline{\chi}_{ab}(\not\nabla_b\underline{\eta})_a\\ &= \frac{1}{\Omega}\not{\mathrm{d}}\mathrm{iv}\,(\Omega\underline{\beta}+\Omega(\eta-\underline{\eta})\cdot\underline{\chi})-\Big((\not\nabla_a\log\Omega)\underline{\chi}_{ab}+(\not{\mathrm{d}}\mathrm{iv}\,\underline{\chi})_b-(\not\nabla_b\log\Omega)\mathrm{tr}\underline{\chi}-\not\nabla_b\mathrm{tr}\underline{\chi}\Big)\underline{\eta}_b-\underline{\chi}_{ab}(\not\nabla_b\underline{\eta})_a\\ &= \frac{1}{\Omega}\not{\mathrm{d}}\mathrm{iv}\,(\Omega\underline{\beta}+\Omega(\eta-\underline{\eta})\cdot\underline{\chi})-\frac{1}{\Omega}(\not{\mathrm{d}}\mathrm{iv}\,\Omega\underline{\chi})_b\underline{\eta}_b-\underline{\chi}_{ab}(\not\nabla_a\underline{\eta})_b+\frac{1}{\Omega}(\not\nabla(\Omega\mathrm{tr}\underline{\chi}))\underline{\eta}_b\\ &= \frac{1}{\Omega}\Big[\not{\mathrm{d}}\mathrm{iv}\,(\Omega\underline{\beta}+\Omega(\eta-\underline{\eta})\cdot\underline{\chi})-(\not{\mathrm{d}}\mathrm{iv}\,\Omega\underline{\eta}\cdot\underline{\chi})+(\not\nabla_b\Omega\mathrm{tr}\underline{\chi})\underline{\eta}_b\Big].\end{aligned}$$

we have

$$\frac{\Omega}{2}\mathbf{D}_3(\rho-\overline{\rho})=\frac{1}{2}\left(\Omega\mathbf{D}_3\rho-\overline{\Omega\mathbf{D}_3\rho}\right)-\frac{1}{2}\left(\overline{(\Omega\mathrm{tr}\underline{\chi})\rho}-\overline{(\Omega\mathrm{tr}\underline{\chi})}\overline{\rho}\right). \tag{7.7.69}$$

Using the Bianchi equation for ρ (see (3.2.8)),

$$\mathbf{D}_3\rho=-\left(\frac{3}{2}\mathrm{tr}\underline{\chi}\rho+\frac{1}{2}\hat{\chi}\cdot\underline{\alpha}+\eta\cdot\underline{\beta}+\nabla\!\!\!/\log\Omega\cdot\underline{\beta}\right)-\mathrm{d}\!\!\!/\mathrm{iv}\,\underline{\beta}$$

we obtain

$$-\frac{\Omega}{2}\mathbf{D}_3(\rho-\overline{\rho})=\frac{1}{2}\mathrm{d}\!\!\!/\mathrm{iv}\,(\Omega\underline{\beta})+(H-\overline{H}), \tag{7.7.70}$$

where

$$H\equiv\frac{\Omega}{2}\left(\frac{3}{2}\mathrm{tr}\underline{\chi}\rho+\frac{1}{2}\hat{\chi}\cdot\underline{\alpha}+\eta\cdot\underline{\beta}\right). \tag{7.7.71}$$

In the same way

$$\begin{aligned}\frac{\Omega}{4}\left(\mathbf{D}_3(\hat{\chi}\cdot\underline{\hat{\chi}})-\mathbf{D}_3\overline{(\hat{\chi}\cdot\underline{\hat{\chi}})}\right) &= \frac{1}{4}\left(\Omega\mathbf{D}_3(\hat{\chi}\cdot\underline{\hat{\chi}})-\overline{\Omega\mathbf{D}_3(\hat{\chi}\cdot\underline{\hat{\chi}})}\right)\\ &- \frac{1}{4}\left(\overline{(\Omega\mathrm{tr}\underline{\chi})(\hat{\chi}\cdot\underline{\hat{\chi}})}-\overline{(\Omega\mathrm{tr}\underline{\chi})}\,\overline{(\hat{\chi}\cdot\underline{\hat{\chi}})}\right),\end{aligned} \tag{7.7.72}$$

and finally, collecting all these equations together,

$$\Delta\!\!\!/(\Omega\mathbf{D}_3\log\Omega)=\mathrm{d}\!\!\!/\mathrm{iv}\left(\Omega\underline{\beta}+\frac{3}{2}\Omega\eta\cdot\underline{\hat{\chi}}+\frac{1}{4}\Omega\eta\mathrm{tr}\underline{\chi}\right)+G_1-\overline{G_1}, \tag{7.7.73}$$

where[44]

$$\begin{aligned}G_1 &= H+\frac{1}{4}\Omega\mathbf{D}_3(\hat{\chi}\cdot\underline{\hat{\chi}})-\frac{1}{2}(\Omega\mathrm{tr}\underline{\chi})(\rho-\overline{\rho})\\ &+ \frac{1}{4}(\Omega\mathrm{tr}\underline{\chi})(\hat{\chi}\cdot\underline{\hat{\chi}}-\overline{\hat{\chi}\cdot\underline{\hat{\chi}}}).\end{aligned} \tag{7.7.74}$$

Denoting

$$\mathrm{d}\!\!\!/\mathrm{iv}\,F_1\equiv\mathrm{d}\!\!\!/\mathrm{iv}\left(\Omega\underline{\beta}+\frac{3}{2}\Omega\eta\cdot\underline{\hat{\chi}}+\frac{1}{4}\Omega\eta\mathrm{tr}\underline{\chi}\right) \tag{7.7.75}$$

then the final equation is

$$\Delta\!\!\!/(\Omega\mathbf{D}_3\log\Omega)=\mathrm{d}\!\!\!/\mathrm{iv}\,F_1+G_1-\overline{G_1}. \tag{7.7.76}$$

---

[44]From the expression of H it seems that G_1 depends on $\underline{\alpha}$, but in fact the $\underline{\alpha}$ present in H is cancelled by the $\underline{\alpha}$ with the opposite sign appearing in $\frac{1}{4}\Omega\mathbf{D}_3(\hat{\chi}\cdot\underline{\hat{\chi}})$.

Proof of equation (7.7.40) of Lemma 7.7.4

From

$$\begin{aligned}\not\Delta(\Omega\mathbf{D}_3(\Omega\mathbf{D}_3\log\Omega)) &= \Omega(\not\Delta\mathbf{D}_3(\Omega\mathbf{D}_3\log\Omega)) + (\not\Delta\Omega)\mathbf{D}_3(\Omega\mathbf{D}_3\log\Omega)\\ &+ 2\Omega(\not\nabla\log\Omega)\cdot\not\nabla(\mathbf{D}_3\Omega\mathbf{D}_3\log\Omega),\end{aligned} \tag{7.7.77}$$

it follows that

$$\begin{aligned}\frac{1}{\Omega}\not\Delta(\Omega\mathbf{D}_3(\Omega\mathbf{D}_3\log\Omega)) &= \not\Delta\mathbf{D}_3(\Omega\mathbf{D}_3\log\Omega) + \left[\frac{\not\Delta\Omega}{\Omega}\mathbf{D}_3(\Omega\mathbf{D}_3\log\Omega)\right.\\ &\left.+ 2(\not\nabla\log\Omega)\not\nabla\mathbf{D}_3(\Omega\mathbf{D}_3\log\Omega)\right].\end{aligned} \tag{7.7.78}$$

From the previous lemma (see (7.7.63)) we have

$$\not\Delta\mathbf{D}_3(\Omega\mathbf{D}_3\log\Omega) = \mathbf{D}_3\not\Delta(\Omega\mathbf{D}_3\log\Omega) - [\cdot]_{(\Omega D_3\log\Omega)} - \{\cdot\}_{(\Omega D_3\log\Omega)}, \tag{7.7.79}$$

where $\{\cdot\}_{(\Omega D_3\log\Omega)}$ and $[\cdot]_{(\Omega D_3\log\Omega)}$ have the same expressions as in (7.7.63) with $\log\Omega$ substituted with $\Omega\mathbf{D}_3\log\Omega$,

$$\begin{aligned}\{\cdot\}_{(\Omega D_3\log\Omega)} &= \Big\{-\eta_a\underline{\chi}_{ab}\not\nabla_b(\Omega\mathbf{D}_3\log\Omega) + \mathrm{tr}\underline{\chi}\eta_b\not\nabla_b(\Omega\mathbf{D}_3\log\Omega)\\ &-2\underline{\chi}_{ab}(\not\nabla\not\nabla(\Omega\mathbf{D}_3\log\Omega))_{ab} - \zeta_a\underline{\chi}_{ab}\not\nabla_b(\Omega\mathbf{D}_3\log\Omega)\\ &-(\not{\mathrm{div}}\,\underline{\chi})_b\not\nabla_b(\Omega\mathbf{D}_3\log\Omega) - \underline{\beta}_b\not\nabla_b(\Omega\mathbf{D}_3\log\Omega)\Big\},\end{aligned} \tag{7.7.80}$$

and

$$\begin{aligned}[\cdot]_{(\Omega D_3\log\Omega)} &= \big[(\not\nabla_a\log\Omega)(\mathbf{D}_3\not\nabla_a(\Omega\mathbf{D}_3\log\Omega)) - \zeta_a(\mathbf{D}_3\not\nabla_a(\Omega\mathbf{D}_3\log\Omega))\\ &+ \eta_a\not\nabla_a\mathbf{D}_3(\Omega\mathbf{D}_3\log\Omega) + (\not\Delta\log\Omega)\mathbf{D}_3(\Omega\mathbf{D}_3\log\Omega)\\ &+ \zeta_a(\not\nabla_a\log\Omega)\mathbf{D}_3(\Omega\mathbf{D}_3\log\Omega)\big].\end{aligned} \tag{7.7.81}$$

Observe that a simple use of commutation relations allows us to rewrite $[\cdot]_{(\Omega D_3\log\Omega)}$ in the following way, which will be used subsequently,

$$\begin{aligned}[\cdot]_{(\Omega D_3\log\Omega)} &= \left[2(\not\nabla_a\log\Omega)(\not\nabla_a\mathbf{D}_3(\Omega\mathbf{D}_3\log\Omega)) + \frac{\not\Delta\Omega}{\Omega}\mathbf{D}_3(\Omega\mathbf{D}_3\log\Omega)\right.\\ &- \left.\underline{\chi}_{ab}(\not\nabla_a\log\Omega)\not\nabla_b(\Omega\mathbf{D}_3\log\Omega) + \zeta_a\underline{\chi}_{ab}\not\nabla_b(\Omega\mathbf{D}_3\log\Omega)\right].\end{aligned} \tag{7.7.82}$$

Substituting the expression for $\not\Delta\mathbf{D}_3(\Omega\mathbf{D}_3\log\Omega)$ (see (7.7.79)) into (7.7.78) we obtain

$$\begin{aligned}\frac{1}{\Omega}\not\Delta(\Omega\mathbf{D}_3(\Omega\mathbf{D}_3\log\Omega)) &= \mathbf{D}_3\not\Delta(\Omega\mathbf{D}_3\log\Omega) + \left[\frac{\not\Delta\Omega}{\Omega}\mathbf{D}_3(\Omega\mathbf{D}_3\log\Omega)\right.\\ &+ \left.2(\not\nabla_a\log\Omega)\not\nabla_a\mathbf{D}_3(\Omega\mathbf{D}_3\log\Omega) - [\cdot]_{(\Omega D_3\log\Omega)}\right] - \{\cdot\}_{(\Omega D_3\log\Omega)}\end{aligned} \tag{7.7.83}$$

A long but easy computation on exactly the same lines as in the previous case allows us to rewrite the last equation as

$$\begin{aligned}
\not\Delta(\Omega\mathbf{D}_3(\Omega\mathbf{D}_3\log\Omega)) &= \Omega\mathbf{D}_3\not\Delta(\Omega\mathbf{D}_3\log\Omega) + \Omega\Big[2\underline{\chi}_{ab}(\not\nabla\not\nabla(\Omega\mathbf{D}_3\log\Omega))_{ab} \\
&\quad + (\not\nabla_b\log\Omega)\not\nabla_a(\Omega\mathbf{D}_3\log\Omega)\underline{\chi}_{ab} + (\not\nabla_b(\Omega\mathbf{D}_3\log\Omega))(\not{\mathrm{div}}\,\underline{\chi})_b \\
&\quad + (\not\nabla_b(\Omega\mathbf{D}_3\log\Omega))\left(\eta_a\underline{\chi}_{ab} - \eta_b\mathrm{tr}\underline{\chi} + \underline{\beta}_b\right)\Big] \\
&\equiv \Omega\mathbf{D}_3\not\Delta(\Omega\mathbf{D}_3\log\Omega) + \Omega[IV]_{(\Omega D_3\log\Omega)}, \qquad (7.7.84)
\end{aligned}$$

where[45]

$$\begin{aligned}
\Omega[IV]_{(\Omega D_3\log\Omega)} &= \Omega\Big[2\underline{\chi}_{ab}(\not\nabla\not\nabla(\Omega\mathbf{D}_3\log\Omega))_{ab} + (\not\nabla_b\log\Omega)\not\nabla_a(\Omega\mathbf{D}_3\log\Omega)\underline{\chi}_{ab} \\
&\quad + (\not\nabla_b(\Omega\mathbf{D}_3\log\Omega))(\not{\mathrm{div}}\,\underline{\chi})_b + (\not\nabla_b(\Omega\mathbf{D}_3\log\Omega))\left(\eta_a\underline{\chi}_{ab} - \eta_b\mathrm{tr}\underline{\chi} + \underline{\beta}_b\right)\Big] \\
&= \Omega\Big[2\underline{\chi}_{ab}(\not\nabla\not\nabla(\Omega\mathbf{D}_3\log\Omega))_{ab} + 2(\not\nabla_b\log\Omega)\not\nabla_a(\Omega\mathbf{D}_3\log\Omega)\underline{\chi}_{ab} \\
&\quad + 2(\not\nabla_b(\Omega\mathbf{D}_3\log\Omega))(\not{\mathrm{div}}\,\underline{\chi})_b - (\not\nabla_b(\Omega\mathbf{D}_3\log\Omega))\left((\not\nabla_b\log\Omega)\mathrm{tr}\underline{\chi} + \not\nabla_b\mathrm{tr}\underline{\chi}\right)\Big] \\
&= 2\not{\mathrm{div}}\,(\Omega\,(\not\nabla(\Omega\mathbf{D}_3\log\Omega))\cdot\underline{\chi}) - (\not\nabla_b(\Omega\mathbf{D}_3\log\Omega))(\not\nabla_b\Omega\mathrm{tr}\underline{\chi}). \qquad (7.7.85)
\end{aligned}$$

Finally we obtain

$$\begin{aligned}
\not\Delta(\Omega\mathbf{D}_3(\Omega\mathbf{D}_3\log\Omega)) &= \Omega\mathbf{D}_3\left[\not{\mathrm{div}}\,\Omega\underline{\beta} + \not{\mathrm{div}}\,\tilde{F}_1 + (G_1 - \overline{G_1})\right] \\
&+ \not{\mathrm{div}}\,(2\Omega\,(\not\nabla(\Omega\mathbf{D}_3\log\Omega))\cdot\underline{\chi}) - (\not\nabla_b(\Omega\mathbf{D}_3\log\Omega))(\not\nabla_b\Omega\mathrm{tr}\underline{\chi}) \\
&= \Omega\mathbf{D}_3\left[\not{\mathrm{div}}\,F_1 + (G_1 - \overline{G_1})\right] \\
&+ \not{\mathrm{div}}\,(2\Omega\,(\not\nabla(\Omega\mathbf{D}_3\log\Omega))\cdot\underline{\chi}) - (\not\nabla_b(\Omega\mathbf{D}_3\log\Omega))(\not\nabla_b\Omega\mathrm{tr}\underline{\chi}),
\end{aligned}$$
(7.7.86)

where $\not{\mathrm{div}}\,F_1 \equiv \not{\mathrm{div}}\,\Omega\underline{\beta} + \not{\mathrm{div}}\,\tilde{F}_1$ and $\tilde{F}_1 = \left(\frac{3}{2}\Omega\eta\cdot\hat{\underline{\chi}} + \frac{1}{4}\Omega\eta\mathrm{tr}\underline{\chi}\right)$. Repeating the previous computation, we have

$$\begin{aligned}
\mathbf{D}_3\not{\mathrm{div}}\,F_1 &= \not{\mathrm{div}}\,(\mathbf{D}_3F_1) + (\not\nabla_a\log\Omega)(\mathbf{D}_3F_1)_a + (-\eta_a\underline{\chi}_{ab} + \eta_b\mathrm{tr}\underline{\chi} - \underline{\beta}_b)F_{1b} - \underline{\chi}_{ab}(\not\nabla_bF_1)_a \\
&= \frac{1}{\Omega}\not{\mathrm{div}}\,(\Omega\mathbf{D}_3F_1) + \left[(-\eta_a\hat{\underline{\chi}}_{ab} + \frac{1}{2}\eta_b\mathrm{tr}\underline{\chi} - \underline{\beta}_b)F_{1b} - \underline{\chi}_{ab}(\not\nabla_bF_1)_a\right],
\end{aligned}$$

and using the previous structure equation again we obtain

$$\Omega\mathbf{D}_3\not{\mathrm{div}}\,F_1 = \not{\mathrm{div}}\,(\Omega\mathbf{D}_3F_1 - \Omega\underline{\chi}\cdot F_1) + (\not\nabla_b\Omega\mathrm{tr}\underline{\chi})F_{1b}. \qquad (7.7.87)$$

Inserting this expression into the equation for $\not\Delta(\Omega\mathbf{D}_3(\Omega\mathbf{D}_3\log\Omega))$ (see (7.7.86)) it is easy to conclude that

$$\not\Delta(\Omega\mathbf{D}_3(\Omega\mathbf{D}_3\log\Omega)) = \not{\mathrm{div}}\,(2\Omega\not\nabla(\Omega\mathbf{D}_3\log\Omega)\cdot\underline{\chi}) + \not{\mathrm{div}}\,(\Omega\mathbf{D}_3F_1 - \Omega\underline{\chi}\cdot F_1)$$

---

[45] We used the structure equation

$$\eta_a\underline{\chi}_{ab} - \eta_b\mathrm{tr}\underline{\chi} + \underline{\beta}_b = (\not\nabla_a\log\Omega)\underline{\chi}_{ab} - (\not\nabla_b\log\Omega)\mathrm{tr}\underline{\chi} + (\not{\mathrm{div}}\,\underline{\chi})_b - \not\nabla_b\mathrm{tr}\underline{\chi}\,.$$

$$+ \quad \left(F_{1b} - \nabla\!\!\!/_b(\Omega \mathbf{D}_3 \log\Omega)\right)(\nabla\!\!\!/_b \Omega \mathrm{tr}\underline{\chi}) + \Omega \mathbf{D}_3(G_1 - \overline{G_1}). \tag{7.7.88}$$

Observing that

$$\begin{aligned}
&\left(F_{1b} - \nabla\!\!\!/_b(\Omega \mathbf{D}_3 \log\Omega)\right)(\nabla\!\!\!/_b \Omega \mathrm{tr}\underline{\chi}) \\
&= \mathrm{d}\!\!\!/\mathrm{iv}\left((F_{1b} - \nabla\!\!\!/_b(\Omega \mathbf{D}_3 \log\Omega))(\Omega \mathrm{tr}\underline{\chi})\right) - \Omega \mathrm{tr}\underline{\chi}\,(\mathrm{d}\!\!\!/\mathrm{iv} F_1 - \Delta\!\!\!\!/(\Omega \mathbf{D}_3 \log\Omega)) \\
&= \mathrm{d}\!\!\!/\mathrm{iv}\left((F_{1b} - \nabla\!\!\!/_b(\Omega \mathbf{D}_3 \log\Omega))(\Omega \mathrm{tr}\underline{\chi})\right) + (\Omega \mathrm{tr}\underline{\chi})(G_1 - \overline{G_1}),
\end{aligned}$$

the final result is

$$\begin{aligned}
\Delta\!\!\!\!/(\Omega \mathbf{D}_3(\Omega \mathbf{D}_3 \log\Omega)) &= \mathrm{d}\!\!\!/\mathrm{iv}\Big[(\Omega \mathbf{D}_3 F_1 - \Omega\underline{\chi}\cdot F_1) + 2\Omega(\nabla\!\!\!/(\Omega \mathbf{D}_3 \log\Omega))\cdot\underline{\chi} \\
&\quad + F_1 \Omega \mathrm{tr}\underline{\chi} - (\nabla\!\!\!/(\Omega \mathbf{D}_3 \log\Omega))\Omega \mathrm{tr}\underline{\chi}\Big] \\
&\quad + \left\{\Omega \mathbf{D}_3(G_1 - \overline{G_1}) + (\Omega \mathrm{tr}\underline{\chi})(G_1 - \overline{G_1})\right\} \\
&\equiv \mathrm{d}\!\!\!/\mathrm{iv} F_2 + G_2 - \overline{G_2},
\end{aligned} \tag{7.7.89}$$

where

$$\begin{aligned}
F_2 &= \Omega \mathbf{D}_3 F_1 + 2\Omega(\nabla\!\!\!/(\Omega \mathbf{D}_3 \log\Omega))\cdot\hat{\underline{\chi}} - \Omega\hat{\underline{\chi}}\cdot F_1 + \frac{1}{2}\Omega \mathrm{tr}\underline{\chi} F_1 \\
G_2 &= \Omega \mathbf{D}_3 G_1 + (\Omega \mathrm{tr}\underline{\chi})(G_1 - \overline{G_1}).
\end{aligned}$$

Therefore the final equation has the same structure as the one for $\Omega \mathbf{D}_3 \log\Omega$:

$$\Delta\!\!\!\!/(\Omega \mathbf{D}_3(\Omega \mathbf{D}_3 \log\Omega)) = \mathrm{d}\!\!\!/\mathrm{iv} F_2 + G_2 - \overline{G_2}, \tag{7.7.90}$$

and obviously $\overline{\mathrm{d}\!\!\!/\mathrm{iv} F_2 + G_2 - \overline{G_2}} = 0$.

Proof of equation (7.7.41) of Lemma 7.7.4

Proceeding as in the previous case we have the following expression:

$$\begin{aligned}
\Delta\!\!\!\!/((\Omega \mathbf{D}_3)^3 \log\Omega)) &= \Omega \Delta\!\!\!\!/\mathbf{D}_3[(\Omega \mathbf{D}_3)^2 \log\Omega] + \Delta\!\!\!\!/\Omega)\mathbf{D}_3[(\Omega \mathbf{D}_3)^2 \log\Omega] \\
&+ 2\Omega(\nabla\!\!\!/ \log\Omega)\cdot\nabla\!\!\!/\mathbf{D}_3[(\Omega \mathbf{D}_3)^2 \log\Omega].
\end{aligned} \tag{7.7.91}$$

As before

$$\begin{aligned}
\Delta\!\!\!\!/\mathbf{D}_3[(\Omega \mathbf{D}_3)^2 \log\Omega] &= \mathbf{D}_3\Delta\!\!\!\!/[(\Omega \mathbf{D}_3)^2 \log\Omega] - [\cdot]_{[(\Omega D_3)^2 \log\Omega]} \\
&\quad -\{\cdot\}_{[(\Omega D_3)^2 \log\Omega]},
\end{aligned} \tag{7.7.92}$$

where (see (7.7.82))

$$\begin{aligned}
[\cdot]_{[(\Omega D_3)^2 \log\Omega]} &= \Big[2(\nabla\!\!\!/ \log\Omega)\nabla\!\!\!/\mathbf{D}_3[(\Omega \mathbf{D}_3)^2 \log\Omega] + \left(\frac{\Delta\!\!\!\!/\Omega}{\Omega}\right)\mathbf{D}_3[(\Omega \mathbf{D}_3)^2 \log\Omega] \\
&\quad -\underline{\chi}_{ab}(\nabla\!\!\!/_a \log\Omega)\nabla\!\!\!/_b[(\Omega \mathbf{D}_3)^2 \log\Omega] + \zeta_a \underline{\chi}_{ab}\nabla\!\!\!/_b[(\Omega \mathbf{D}_3)^2 \log\Omega]\Big],
\end{aligned}$$

and

$$
\begin{aligned}
\{\cdot\}_{[(\Omega D_3)^2 \log \Omega]} &= \Big\{-\eta_a \underline{\chi}_{ab} \nabla\!\!\!/_b[(\Omega\mathbf{D}_3)^2 \log \Omega] + \mathrm{tr}\underline{\chi}\,\eta_b \nabla\!\!\!/_b[(\Omega\mathbf{D}_3)^2 \log \Omega] \\
&\quad - 2\underline{\chi}_{ab}(\nabla\!\!\!/\nabla\!\!\!/[(\Omega\mathbf{D}_3)^2 \log \Omega])_{ab} - \zeta_a \underline{\chi}_{ab} \nabla\!\!\!/_b[(\Omega\mathbf{D}_3)^2 \log \Omega] \\
&\quad - (\mathrm{d}\!\!\!/\mathrm{iv}\,\underline{\chi})_b \nabla\!\!\!/_b[(\Omega\mathbf{D}_3)^2 \log \Omega] - \underline{\beta}_b \nabla\!\!\!/_b[(\Omega\mathbf{D}_3)^2 \log \Omega]\Big\}.
\end{aligned}
$$

Substituting the expression for $\Delta\!\!\!/\mathbf{D}_3[(\Omega\mathbf{D}_3)^2 \log \Omega]$ into (7.7.91) we obtain

$$
\begin{aligned}
\frac{1}{\Omega}\Delta\!\!\!/((\Omega\mathbf{D}_3)^3 \log \Omega) &= \mathbf{D}_3\Delta\!\!\!/[(\Omega\mathbf{D}_3)^2 \log \Omega] + \Big[\underline{\chi}_{ab}(\nabla\!\!\!/_a \log \Omega)\nabla\!\!\!/_b[(\Omega\mathbf{D}_3)^2 \log \Omega] \\
&\quad - \zeta_a \underline{\chi}_{ab} \nabla\!\!\!/_b[(\Omega\mathbf{D}_3)^2 \log \Omega]\Big] - \{\cdot\}_{[(\Omega D_3)^2 \log \Omega]}. \qquad (7.7.93)
\end{aligned}
$$

A long but easy computation on the same lines as for the previous case allows us to rewrite the equation as

$$
\begin{aligned}
\Delta\!\!\!/((\Omega\mathbf{D}_3)^3 \log \Omega)) &= \Omega\mathbf{D}_3\Delta\!\!\!/[(\Omega\mathbf{D}_3)^2 \log \Omega] + \Omega\Big[\underline{\chi}_{ab}(\nabla\!\!\!/_a \log \Omega)\nabla\!\!\!/_b[(\Omega\mathbf{D}_3)^2 \log \Omega] \\
&\quad + 2\underline{\chi}_{ab}(\nabla\!\!\!/\nabla\!\!\!/[(\Omega\mathbf{D}_3)^2 \log \Omega])_{ab} + (\mathrm{d}\!\!\!/\mathrm{iv}\,\underline{\chi})_b(\nabla\!\!\!/_b[(\Omega\mathbf{D}_3)^2 \log \Omega]) \\
&\quad + (\nabla\!\!\!/_b[(\Omega\mathbf{D}_3)^2 \log \Omega])\left(\eta_a \underline{\chi}_{ab} - \eta_b \mathrm{tr}\underline{\chi} + \underline{\beta}_b\right)\Big] \\
&\equiv \Omega\mathbf{D}_3\Delta\!\!\!/[(\Omega\mathbf{D}_3)^2 \log \Omega] + \Omega[IV]_{[(\Omega\mathbf{D}_3)^2 \log \Omega]}. \qquad (7.7.94)
\end{aligned}
$$

After some easy but long computations in which we used the structure equation $\eta_a \underline{\chi}_{ab} - \eta_b \mathrm{tr}\underline{\chi} + \underline{\beta}_b = (\nabla\!\!\!/_a \log \Omega)\underline{\chi}_{ab} - (\nabla\!\!\!/_b \log \Omega)\mathrm{tr}\underline{\chi} + (\mathrm{d}\!\!\!/\mathrm{iv}\,\underline{\chi})_b - \nabla\!\!\!/_b \mathrm{tr}\underline{\chi}$, we obtain (see 7.7.85))

$$
\Omega[IV]_{[(\Omega D_3)^2 \log \Omega]} = \mathrm{d}\!\!\!/\mathrm{iv}\,(2\Omega[(\Omega\mathbf{D}_3)^2 \log \Omega]) \cdot \underline{\chi}) - (\nabla\!\!\!/_b[(\Omega\mathbf{D}_3)^2 \log \Omega])\nabla\!\!\!/_b(\Omega \mathrm{tr}\underline{\chi}).
$$

Finally, repeating the previous steps, we conclude that

$$
\Delta\!\!\!/((\Omega\mathbf{D}_3)^3 \log \Omega)) = \mathrm{d}\!\!\!/\mathrm{iv}\, F_3 + G_3 - \overline{G_3}, \qquad (7.7.95)
$$

where

$$
\begin{aligned}
F_3 &= \Omega\mathbf{D}_3 F_2 + 2\Omega(\nabla\!\!\!/[(\Omega\mathbf{D}_3)^2 \log \Omega]) \cdot \hat{\underline{\chi}} - \Omega\hat{\underline{\chi}} \cdot F_2 + \frac{1}{2}\Omega \mathrm{tr}\underline{\chi} F_2 \\
G_3 &= \Omega\mathbf{D}_3 G_2 + (\Omega \mathrm{tr}\underline{\chi})(G_2 - \overline{G_2}), \qquad (7.7.96)
\end{aligned}
$$

and obviously $\overline{\mathrm{d}\!\!\!/\mathrm{iv}\, F_3 + G_3 - \overline{G_3}} = 0$.

8

Conclusions

In this chapter we derive the most important consequences of our Main Theorem. In particular, we give a rigorous derivation of the Bondi mass law and the precise asymptotic formula for the optical function u expressed in terms of t and r. Due to the construction of our spacetime based on the double null foliation which allows us, in particular, to give a straightforward definition of the null-outgoing infinity, the derivation of these results is simpler and more intuitive than the one in [Ch-Kl]. In addition our approach allows us to give a simple derivation of the connection between the Bondi mass and the ADM mass. Before embarking on the main topic of this chapter it helps to summarize some of the main relevant features of our proof of the Main Theorem.

Our spacetime has been constructed together with a double null foliation generated by the level hypersurfaces, $C(\lambda)$, $\underline{C}(\nu)$ of the optical functions $u, \underline{u}$. Associated with these null hypersurfaces we have the two dimensional surfaces $S(\lambda, \nu) = C(\lambda) \cap \underline{C}(\nu)$, the adapted null frames (see Definition 3.1.13), and the connection coefficients χ, $\underline{\chi}$, η, $\underline{\eta}$,ω, $\underline{\omega}$, which satisfy the structure equations (3.1.46), (3.1.47) and (3.1.48). We have also decomposed the curvature tensor relative to the adapted null frames into its null components α, β, ρ, σ, $\underline{\beta}$, $\underline{\alpha}$ (see (3.1.20)). The boundedness of the $\mathcal{R}$ norms implies in particular the uniform decay of these components (see (3.7.1)):

$$
\begin{aligned}
&\sup_{\mathcal{K}} r^{7/2}|\alpha| \leq C_0\,, \ \sup_{\mathcal{K}} r|\tau_-|^{\frac{5}{2}}|\underline{\alpha}| \leq C_0\\
&\sup_{\mathcal{K}} r^{7/2}|\beta| \leq C_0\,, \ \sup_{\mathcal{K}} r^2|\tau_-|^{\frac{3}{2}}|\underline{\beta}| \leq C_0\\
&\sup_{\mathcal{K}} r^{3}|\rho| \leq C_0\,, \ \sup_{\mathcal{K}} r^3|\tau_-|^{\frac{1}{2}}|(\rho-\overline{\rho}, \sigma)| \leq C_0.
\end{aligned}
\tag{8.0.1}
$$

In Theorem 8.5.2 of Section 8.5, we show that the $\overline{\rho}$ component, whose decay is $O(r^{-3})$, is intimately tied to M, the ADM mass of Σ_0. In fact the following relation is proved:

$$\overline{\rho}(\lambda,\nu) = -\frac{2M}{r^3} + O\left(\frac{1}{r^3\lambda}\right) .$$

The boundedness of the $\mathcal{O}$ norms implies pointwise estimates for the connection coefficients and their first derivatives. Based on the properties of the canonical foliation we were in fact able to prove somewhat stronger estimates for the connection coefficients in Chapters 4. These stronger estimates were not relevant[1] in the proof of the Main Theorem but play a fundamental role in this chapter. We repeat here those estimates that will be used in the sequel and we refer the reader to Chapter 4 (see in particular Subsection 4.3.16) for their detailed proofs:

$$\begin{aligned}
&|r^{2-2/p}\tau_-^{\frac{1}{2}}\hat{\chi}|_{p,S} \le C_0 \ , \quad |r^{3-2/p}\tau_-^{\frac{1}{2}}\nabla\!\!\!\!/\,\hat{\chi}|_{p,S} \le C_0\\
&|r^{1-2/p}\tau_-^{\frac{3}{2}}\underline{\hat{\chi}}|_{p,S} \le C_0 \ , \quad |r^{2-2/p}\tau_-^{\frac{3}{2}}\nabla\!\!\!\!/\,\underline{\hat{\chi}}|_{p,S} \le C_0\\
&|r^{2-\frac{2}{p}}\tau_-^{\frac{1}{2}}\left(\Omega\mathrm{tr}\chi - \overline{\Omega\mathrm{tr}\chi}\right)|_{p,S} \le C_0 \ , \ |r^{2-\frac{2}{p}}\tau_-^{\frac{1}{2}}\left(\Omega\mathrm{tr}\underline{\chi} - \overline{\Omega\mathrm{tr}\underline{\chi}}\right)|_{p,S} \le C_0\\
&|r^{3-2/p}\tau_-^{\frac{1}{2}}\nabla\!\!\!\!/\,\mathrm{tr}\chi|_{p,S} \le C_0 \ , \quad |r^{3-2/p}\tau_-^{\frac{1}{2}}\nabla\!\!\!\!/\,\mathrm{tr}\underline{\chi}|_{p,S} \le C_0\\
&|r^{2-2/p}\tau_-^{\frac{1}{2}}\eta|_{p,S} \le C_0 \ , \quad |r^{2-2/p}\tau_-^{\frac{1}{2}}\underline{\eta}|_{p,S} \le C_0\\
&|r^{3-2/p}\tau_-^{\frac{1}{2}}\nabla\!\!\!\!/\,\eta|_{p,S} \le C_0 \ , \quad |r^{3-2/p}\tau_-^{\frac{1}{2}}\nabla\!\!\!\!/\,\underline{\eta}|_{p,S} \le C_0\\
&|r^{2-\frac{2}{p}}\tau_-^{\frac{3}{2}}(\Omega\nabla\!\!\!\!/\,\mathbf{D}_3\log\Omega)|_{p,S} \le C_0 \ , \quad |r^{3-\frac{2}{p}}\tau_-^{\frac{1}{2}}(\Omega\nabla\!\!\!\!/\,\mathbf{D}_4\log\Omega)|_{p,S} \le C_0\\
&|r^{1-\frac{2}{p}}\tau_-^{\frac{3}{2}}(\Omega\mathbf{D}_3\log\Omega - \overline{\Omega\mathbf{D}_3\log\Omega})|_{p,S} \le C_0\\
&|r^{2-\frac{2}{p}}\tau_-^{\frac{1}{2}}(\Omega\mathbf{D}_4\log\Omega - \overline{\Omega\mathbf{D}_4\log\Omega})|_{p,S} \le C_0,
\end{aligned} \tag{8.0.2}$$

where C_0 is a constant depending on the initial data.

Observe that the quantities ω, $\underline{\omega}$, $\mathbf{D}_4\omega$, $\mathbf{D}_3\underline{\omega}$, $(\overline{\Omega\mathrm{tr}\chi} - \frac{1}{r})$ and $(\overline{\Omega\mathrm{tr}\underline{\chi}} + \frac{1}{r})$ do not have the $\sqrt{\tau_-}$ improvement factor manifest in all the other terms. This is due to their relation (in the structure equations) to the ρ component of the curvature tensor, or more precisely, to its $\overline{\rho}$ part, which is tied to the ADM mass as explained above. Note also that these connection coefficients are the only nontrivial ones in Schwarzschild spacetime. From this perspective our main result can be interpreted as a stability of the external region of the Schwarzschild spacetime. One of the results proved in this chapter further justifies this conclusion. In fact we show, in Proposition 8.6.1, that on any null-outgoing hypersurface $C(\lambda)$ the following relation holds:

$$\frac{dr}{dt} = 1 - \frac{2M}{r} + O\left(\frac{1}{r^2}\right), \tag{8.0.3}$$

[1] But, of course, the fact that the double null foliation is canonical was.

where t is the time function $t(p) = \frac{1}{2}(\underline{u}+u)(p)$ introduced in Chapter 3; see Proposition 3.3.1.[2] This proves that the null-outgoing hypersurfaces converge asymptotically to the null-outgoing Schwarzschild cones.

As mentioned above our double null foliation approach allows us to define the null-outgoing infinity by simply taking the limit of the null-incoming hypersurfaces $\underline{C}(\nu)$ as $\nu \to \infty$. This approach not only simplifies significantly the derivation of the main conclusions in [Ch-Kl] but also allows us to connect the null-outgoing infinity $\mathcal{J}^+$ to the spacelike infinity i_0. In particular we are able to connect the Bondi mass to the ADM mass M.[3]. In fact introducing the Bondi mass (see Definition 8.5.2) as the limit of the Hawking mass $m(\lambda, \nu)$ for $\nu \to \infty$, we prove in Theorem 8.5.2 that

$$\lim_{\lambda \to -\infty} M_B(\lambda) = M.$$

8.1 The spacetime null infinity

8.1.1 The existence of a global optical function

The Main Theorem provides us with a family of optical functions ${}^{(\nu_*)}u(p)$ that are outgoing solutions of the eikonal equation with initial data on $\underline{C}(\nu_*)$. The following corollary allows us to conclude the existence of a global optical function u with initial data at null infinity.

Corollary 8.1.1 *Under the assumptions of the Main Theorem, the following limit holds in $(\mathcal{M}, \mathbf{g})$:*

$$u(p) = \lim_{\nu_* \to \infty} {}^{(\nu_*)}u(p). \tag{8.1.1}$$

Proof: We show that $\{{}^{(\nu_*)}u(p)\}$ forms a Cauchy sequence. We first prove that for $\tilde{\epsilon}$ arbitrary small, it is possible to choose $\overline{\nu}_*$ such that, given $\nu_{*,2} > \nu_{*,1} > \overline{\nu}_*$, we have $\left|{}^{(\nu_{*,2})}u - {}^{(\nu_{*,1})}u\right|_{\underline{C}_*(\nu_{*,1})} \leq \tilde{\epsilon}$ or, in view of our definition ${}^{(\nu_{*,1})}u(p)|_{\underline{C}_*(\nu_{*,1})} = {}^{(\nu_{*,1})}u_*(p)$,

$$\sup_{p \in \underline{C}_*(\nu_{*,1})} \left|{}^{(\nu_{*,2})}u(p) - {}^{(\nu_{*,1})}u_*(p)\right| \leq \tilde{\epsilon}, \tag{8.1.2}$$

where ${}^{(\nu_{*,1})}u_*(p)$ is the solution of the last slice problem on $\underline{C}_*(\nu_{*,1})$ (see Definition 3.3.2). Once (8.1.2) is proved we can conclude that in any spacetime region $\mathcal{K}(\lambda_0, \nu_1) \subset \mathcal{M}$ the difference between ${}^{(\nu_{*,2})}u(p)$ and ${}^{(\nu_{*,1})}u(p)$ tends to zero as $\nu_{*,1} \to \infty$. This implies the convergence of ${}^{(\nu_*)}u(p)$ to $u(p)$, which proves the result.

---

[2] Note that although we do not use the maximal spacelike foliation, we do have a spacelike foliation at our disposal and, therefore, a global time function. Both are in fact provided by Proposition 3.3.1 and the spacelike hypersurfaces are such that each surface $S(\lambda, \nu)$ is immersed in the hypersurface $\tilde{\Sigma}_t = \{p \in \mathcal{M} | t(p) = t\}$, where $t = \frac{1}{2}(\nu + \lambda)$.

[3] Analogously, using these null-incoming hypersurfaces, the derivation of the asymptotic rotational symmetry of the spacetime is more straightforward than in [Ch-Kl]. In fact, as described in Subsection 3.4.1, the angular momentum vector fields on $\mathcal{M}$ are defined starting from the last slice null hypersurface $\underline{C}(\nu_*)$. On the other hand, the angular momentum vector fields on the last slice are defined starting from $\underline{C}(\nu_*) \cap \Sigma_0$ so that the connection between the limits is easily established as $\nu_* \to \infty$.

To prove (8.1.2) we choose $\overline{\nu}_*$ sufficiently large such that the initial data norm in $\Sigma_{0ext}(\nu_{*,1})$, the Σ_0 external region outside $\underline{C}_*(\nu_{*,1}) \cap \Sigma_0$, is of order $O(\tilde{\epsilon})$,

$$\left[\sup_{\Sigma_{0ext}(\nu_{*,1})}\left((d_0^2+1)^3|\mathrm{Ric}|^2\right)+\int_{\Sigma_{0ext}(\nu_{*,1})}\sum_{l=0}^{3}(d_0^2+1)^{l+1}|\nabla^l k|^2\right.$$
$$\left.+\int_{\Sigma_0}\sum_{l=0}^{1}(d_0^2+1)^{l+3}|\nabla^l B|^2\right]\leq\tilde{\epsilon}^2 .$$

The restriction of ${}^{(\nu_{*,2})}u$ induces a foliation on $\underline{C}_*(\nu_{*,1})$. Alhough strictly speaking this is not a background foliation[4] we can nevertheless deform it in the neighborhood of $\underline{C}_*(\nu_{*,1}) \cap \Sigma_0$, generating only errors of order $O(\tilde{\epsilon})$, such that it becomes one. More precisely, we can assume that $v(p) = {}^{(\nu_{*,2})}u|_{\underline{C}_*(\nu_{*,1})}(p)$ induces a background foliation on $\underline{C}_*(\nu_{*,1})$ which satisfies the assumptions of Theorem 3.3.2. Since ${}^{(\nu_{*,1})}u_*$ is the canonical foliation of $\underline{C}_*(\nu_{*,1})$, we infer from 3.3.12 that

$$\frac{d\,{}^{(\nu_{*,1})}u_*}{dv}(p) = (2\,{}^{(\nu_{*,1})}\Omega)^{-2}, \tag{8.1.3}$$

where ${}^{(\nu_{*,1})}\Omega$ satisfies the elliptic equation

$$\not{\Delta}\log 2\,{}^{(\nu_{*,1})}\Omega = \frac{1}{2}\not{\mathrm{div}}\,{}^{(\nu_{*,1})}\underline{\eta} + \left[\frac{1}{2}{}^{(\nu_{*,1})}\hat{\chi}\,{}^{(\nu_{*,1})}\hat{\underline{\chi}} - \overline{{}^{(\nu_{*,1})}\hat{\chi}\,{}^{(\nu_{*,1})}\hat{\underline{\chi}}}) - ({}^{(\nu_{*,1})}\rho - \overline{{}^{(\nu_{*,1})}\rho})\right]$$
$$\overline{\log 2\,{}^{(\nu_{*,1})}\Omega} = 0. \tag{8.1.4}$$

Integrating equation (8.1.3) along $\underline{C}_*(\nu_{*,1})$ we obtain, apart from higher order corrections,

$${}^{(\nu_{*,1})}u_*(p) - {}^{(\nu_{*,2})}u(p)|_{\underline{C}_*(\nu_{*,1})} = \int_0^{\lambda(p)}\left(\frac{1}{(2\,{}^{(\nu_{*,1})}\Omega)^2} - 1\right), \tag{8.1.5}$$

from which we infer

$$\left|{}^{(\nu_{*,1})}u_*(p) - {}^{(\nu_{*,2})}u(p)|_{\underline{C}_*(\nu_{*,1})}\right| \leq c\int_0^{\lambda(p)}\left|\log({}^{(\nu_{*,1})}\Omega)\right|. \tag{8.1.6}$$

In view of the elliptic equation (8.1.4) satisfied by $\log({}^{(\nu_{*,1})}\Omega)$, and using the estimates for the connection coefficients on the last slice proved in Chapter 7, Theorem **M4**, we obtain

$$\begin{aligned}|\log({}^{(\nu_{*,1})}\Omega)| &\leq r^2\left(|\not{\mathrm{div}}\,{}^{(\nu_{*,1})}\underline{\eta}| + |{}^{(\nu_{*,1})}\hat{\chi}||{}^{(\nu_{*,1})}\hat{\underline{\chi}}| + |({}^{(\nu_{*,1})}\rho - \overline{{}^{(\nu_{*,1})}\rho})|\right)\\ &= O\left(\frac{1}{r\lambda^{\frac{1}{2}}}\right).\end{aligned} \tag{8.1.7}$$

Therefore, integrating the right-hand side of (8.1.6) we obtain,

$$\left|{}^{(\nu_{*,1})}u_*(p) - {}^{(\nu_{*,2})}u(p)|_{\underline{C}_*(\nu_{*,1})}\right| \leq c\frac{1}{r^{\frac{1}{2}}} \leq \tilde{\epsilon}, \tag{8.1.8}$$

since $r = r(\lambda, \overline{\nu}_*)$ can be made sufficiently large, provided we choose $\overline{\nu}_*$ appropriately large.

---

[4]This requires some extra work. Between the leaves defined on $\underline{C}_*(\nu_{*,1})$ by the restriction of ${}^{(\nu_{*,2})}u(p)$ there exists one that is "near" of order $\tilde{\epsilon}$ to $\underline{C}_*(\nu_{*,1}) \cap \Sigma_0$. The proof of this requires the use of the Oscillation Lemma (see Lemma 4.1.6) and has to take into account that $\overline{\nu}_*$ is chosen sufficiently large.

8.1.2 *The null-outgoing infinity $\mathcal{J}^+$*

To define the null-outgoing infinity $\mathcal{J}^+$, we start by defining in the global spacetime $(\mathcal{M}, \mathbf{g})$ constructed in the Main Theorem a family of diffeomorphisms $\psi(\lambda, \nu)$ such that

$$\psi(\lambda, \nu) : S^2 \to S(\lambda, \nu). \tag{8.1.9}$$

These diffeomorphisms are associated with the diffeomorphisms ϕ_ν and $\underline{\phi}_\lambda$ generated by the null equivariant vector fields $N, \underline{N}$ introduced in Subsection 3.1.4, Lemma 3.1.1, as follows:

According to the results of the previous subsection we already have the two optical functions $u(p) = \lim_{\nu_* \to \infty} {}^{(\nu_*)}u(p)$ and $\underline{u}(p)$ defining a double null canonical foliation in $\mathcal{M}$. Associated with it we define the function Ω (see Subsection 3.1.4) through the relation

$$2\Omega^2 = -(g^{\rho\sigma}\partial_\rho u \partial_\sigma \underline{u})^{-1} \tag{8.1.10}$$

and the null geodesic vector fields,

$$L^\rho \equiv -g^{\rho\mu}\partial_\mu u \text{ and } \underline{L}^\rho \equiv -g^{\rho\mu}\partial_\mu \underline{u}, \tag{8.1.11}$$

as well as (see Lemma 3.1.2) the null-outgoing vector field $N = 2\Omega^2 L$ and the null-incoming vector field $\underline{N} = 2\Omega^2 \underline{L}$, equivariant relative to the double null integral S-foliation $\{S(\lambda, \nu) = C(\lambda) \cap \underline{C}(\nu)\}$. Using N and $\underline{N}$ we consider the diffeomorphisms ϕ_t and $\underline{\phi}_s$ generated by them (see Definition 3.1.14) and recall that they map a leaf of the double null integral S-foliation $\{S(\lambda, \nu)\}$ into another leaf of the same foliation.

The diffeomorphism $\psi(\lambda, \nu) : S^2 \to S(\lambda, \nu)$ is defined by

$$\psi(\lambda, \nu) = \underline{\phi}_{(\lambda - \lambda_0)} \circ \phi_{(\nu - \nu_0)}, \tag{8.1.12}$$

where $\nu_0 = \underline{u}|_{C(\lambda_0) \cap \Sigma_0}$, $\lambda_0 = u|_{C(\lambda_0) \cap \Sigma_0}$. Here we have identified $S_{(0)}(\nu_0)$ with the topological sphere S^2 through a diffeomorphism we have not written explicitly.
Given the diffeomorphism $\psi(\lambda, \nu)$ we can map the $S(\lambda, \nu)$-tangent tensor fields defined on $\mathcal{M}$ to tensor fields defined on S^2 with the help of the pull-back map $\psi^*(\lambda, \nu)$. Therefore, given an $S(\lambda, \nu)$-tangent p-covariant tensor field ω, we define the p-covariant tensor field $\tilde{\omega}$ on S^2 by the relation

$$\tilde{\omega}(\lambda, \nu) \equiv \psi^*(\lambda, \nu)(r^{-p}\omega). \tag{8.1.13}$$

This allows us to introduce a precise definition of the null-outgoing infinite limit of ω.

Definition 8.1.1 *We say that the $S(\lambda, \nu)$-tangent p-covariant tensor field ω has W as its null-outgoing infinite limit along $C(\lambda)$,*

$$\lim_{C(\lambda), \nu \to \infty} \omega = W,$$

if the following limit exists,

$$W(\lambda) = \lim_{\nu \to \infty} \tilde{\omega}(\lambda, \nu) = \lim_{\nu \to \infty} \psi^*(\lambda, \nu)(r^{-p}\omega). \tag{8.1.14}$$

In this case W is a p-covariant tensor field on S^2.

Because we are interested in studying the null-outgoing infinite limit of some of the structure equations we need also an explicit expression for $\frac{\partial}{\partial\lambda}W$. To obtain it we need the following lemma.

Lemma 8.1.1 *The following relations hold:*

$$\begin{aligned}\frac{\partial}{\partial\lambda}\psi^*(\lambda,\nu)\omega &= \psi^*(\lambda,\nu)(\mathcal{L}_{\underline{N}}\omega)\\ \frac{\partial}{\partial\nu}\psi^*(\lambda,\nu)\omega &= \psi^*(\lambda,\nu)(\mathcal{L}_V\omega),\end{aligned}\tag{8.1.15}$$

where $V = \underline{\phi}^{-1}_{*(\lambda-\lambda_0)}N$ *is the vector field generating the 1-parameter diffeomorphisms* $\underline{\phi}^{-1}_{(\lambda-\lambda_0)}\circ\phi_h\circ\underline{\phi}_{(\lambda-\lambda_0)}$.

Proof: Let $p\in S(\lambda,\nu)$ and $p_0=\psi^{-1}(\lambda,\nu)(p)\in S^2$. Then

$$(\frac{\partial}{\partial\lambda}\psi^*(\lambda,\nu)\omega)\big|_{p_0} = \lim_{h\to 0}\frac{1}{h}\Big[\big(\psi^*(\lambda+h,\nu)\omega\big)_{p_0} - \big(\psi^*(\lambda,\nu)\omega\big)_{p_0}\Big].\tag{8.1.16}$$

Since

$$(\psi^*(\lambda+h,\nu)\omega)_{p_0} = \phi^*_{(\nu-\nu_0)}\underline{\phi}^*_{(\lambda+h-\lambda_0)}\omega = \psi^*(\lambda,\nu)(\underline{\phi}^*_h\omega)_p,\tag{8.1.17}$$

we have

$$\begin{aligned}(\frac{\partial}{\partial\lambda}\psi^*(\lambda,\nu)\omega)\big|_{p_0} &= \lim_{h\to 0}\frac{1}{h}\Big[\big(\psi^*(\lambda,\nu)\underline{\phi}^*_h\omega\big)_{p_0} - \big(\psi^*(\lambda,\nu)\omega\big)_{p_0}\Big]\\ &= \psi^*(\lambda,\nu)\left(\lim_{h\to 0}\frac{1}{h}\Big[(\underline{\phi}^*_h\omega)_p-\omega_p\Big]\right) = \psi^*(\lambda,\nu)\left(\mathcal{L}_{\underline{N}}\omega_p\right)\end{aligned}\tag{8.1.18}$$

To prove the second relation we write

$$(\frac{\partial}{\partial\nu}\psi^*(\lambda,\nu)\omega)\big|_{p_0} = \lim_{h\to 0}\frac{1}{h}\Big[\big(\psi^*(\lambda,\nu+h)\omega\big)_{p_0} - \big(\psi^*(\lambda,\nu)\omega\big)_{p_0}\Big].\tag{8.1.19}$$

Because

$$(\psi^*(\lambda,\nu+h)\omega)_{p_0} = \psi^*(\lambda,\nu)\left((\underline{\phi}^*_{(\lambda-\lambda_0)})^{-1}\phi^*_h\underline{\phi}^*_{(\lambda-\lambda_0)}\omega\right)_p,\tag{8.1.20}$$

we obtain

$$\begin{aligned}(\frac{\partial}{\partial\nu}\psi^*(\lambda,\nu)\omega)\big|_{p_0} &= \psi^*(\lambda,\nu)\lim_{h\to 0}\frac{1}{h}\left[\left((\underline{\phi}^*_{(\lambda-\lambda_0)})^{-1}\phi^*_h\underline{\phi}^*_{(\lambda-\lambda_0)}\omega\right)_p-\omega_p\right]\\ &= \psi^*(\lambda,\nu)(\mathcal{L}_V\omega|_p),\end{aligned}\tag{8.1.21}$$

where V is the vector field generating the 1-parameter diffeomorphisms $\underline{\phi}^{-1}_{(\lambda-\lambda_0)}\circ\phi_h\circ\underline{\phi}_{(\lambda-\lambda_0)}$.

From Lemma 8.1.1 it follows that

$$\frac{\partial}{\partial\lambda}W(\lambda) = \lim_{C(\lambda),\nu\to\infty}\mathcal{L}_{\underline{N}}\omega = \lim_{\nu\to\infty}\psi^*(\lambda,\nu)(r^{-p}\mathcal{L}_{\underline{N}}\omega).\tag{8.1.22}$$

8.1.3 The null-outgoing limit of the metric

Let $\gamma = \mathbf{g}|_{S(\lambda,\nu)}$ be the induced metric on $S(\lambda, \nu)$. Define $\tilde{\gamma}(\lambda, \nu)$ the Riemannian metric on S^2,

$$\tilde{\gamma}(\lambda, \nu) = \psi^*(\lambda, \nu)(r^{-2}\gamma). \tag{8.1.23}$$

Lemma 8.1.2 *$\tilde{\gamma}$ satisfies the following equation on S^2,*

$$\frac{\partial \tilde{\gamma}}{\partial \lambda} = \left(\Omega \mathrm{tr}\underline{\chi} - \overline{\Omega \mathrm{tr}\underline{\chi}}\right)\tilde{\gamma} + 2\Omega \widetilde{\underline{\hat{\chi}}}. \tag{8.1.24}$$

Proof:

$$\begin{aligned}
\frac{\partial \tilde{\gamma}}{\partial \lambda} &= \psi^*(\lambda, \nu)(\mathcal{L}_{\underline{N}} r^{-2}\gamma) = \psi^*(\lambda, \nu)\left\{-2r^{-1}\underline{N}(r)r^{-2}\gamma + r^{-2}\mathcal{L}_{\underline{N}}\gamma\right\} \\
&= -(\overline{\Omega \mathrm{tr}\underline{\chi}})\tilde{\gamma} + r^{-2}\psi^*(\lambda, \nu)(2\Omega\underline{\chi}) = -(\overline{\Omega \mathrm{tr}\underline{\chi}})\tilde{\gamma} + r^{-2}\psi^*(\lambda, \nu)(\gamma\Omega\mathrm{tr}\underline{\chi} + 2\Omega\underline{\hat{\chi}}) \\
&= \left(\Omega \mathrm{tr}\underline{\chi} - \overline{\Omega \mathrm{tr}\underline{\chi}}\right)\tilde{\gamma} + 2\Omega r^{-2}\psi^*(\lambda, \nu)(\underline{\hat{\chi}}) = \left(\Omega \mathrm{tr}\underline{\chi} - \overline{\Omega \mathrm{tr}\underline{\chi}}\right)\tilde{\gamma} + 2\Omega\widetilde{\underline{\hat{\chi}}}.
\end{aligned}$$

Using this lemma we prove the following proposition.

Proposition 8.1.1 *The metric $\tilde{\gamma}_{(\lambda,\nu)}$ converges as $\nu \to \infty$ to a metric $\tilde{\gamma}_\infty$ on S^2,*

$$\lim_{\nu\to\infty} \tilde{\gamma}_{(\lambda,\nu)} = \tilde{\gamma}_\infty. \tag{8.1.25}$$

Moreover $\tilde{\gamma}_\infty$ has Gauss curvature 1, is independent from λ, and can be considered as the standard metric on S^2.

Proof: We start by comparing $\tilde{\gamma}_{(\lambda,\nu)}$ to $\tilde{\gamma}_{(\lambda_0(\nu),\nu)}$. To do this we choose an orthonormal basis $\{E_A\}$ on $(S^2, \tilde{\gamma}_{(\lambda_0(\nu),\nu)})$ such that the matrix $\tilde{\gamma}_{(\lambda,\nu)}(E_A, E_B)$ is diagonal, with smallest eigenvalue $\Lambda_-(\lambda, \nu)$ and highest eigenvalue $\Lambda_+(\lambda, \nu)$.

Proceeding as in Section 3.3 of [Ch-Kl] we denote

$$\mu_{\tilde{\gamma}}(\lambda, \nu) \equiv \sqrt{\Lambda_-\Lambda_+}\ , \quad \nu_{\tilde{\gamma}}(\lambda, \nu) \equiv \sqrt{\frac{\Lambda_+}{\Lambda_-}},$$

and prove, under appropriate conditions on the connection coefficients listed below and satisfied in view of the Main Theorem, that $\mu_{\tilde{\gamma}}(\lambda, \nu)$ and $\nu_{\tilde{\gamma}}(\lambda, \nu)$ can be bounded by their values at $(\lambda_0(\nu), \nu)$ plus lower order corrections that go to zero as $\nu \to \infty$. Therefore

$$\begin{aligned}
\lim_{\nu\to\infty} \mu_{\tilde{\gamma}}(\lambda, \nu) &= \lim_{\nu\to\infty} \mu_{\tilde{\gamma}}(\lambda_0(\nu), \nu) \\
\lim_{\nu\to\infty} \nu_{\tilde{\gamma}}(\lambda, \nu) &= \lim_{\nu\to\infty} \nu_{\tilde{\gamma}}(\lambda_0(\nu), \nu).
\end{aligned} \tag{8.1.26}$$

This means that, for $\nu \to \infty$, the metric $\tilde{\gamma}_{(\lambda,\nu)}$ converges to the metric $\tilde{\gamma}_{(\lambda_0(\nu),\nu)}$ in the sense that its eigenvalues converge to $\lim_{\nu\to\infty} \Lambda_-(\lambda_0(\nu), \nu)$ and $\lim_{\nu\to\infty} \Lambda_+(\lambda_0(\nu), \nu)$, respectively. Moreover the connection $\Gamma(\lambda, \nu)$ of the metric $\tilde{\gamma}_{(\lambda,\nu)}$ is bounded by the connection of the metric $\tilde{\gamma}_{(\lambda_0(\nu),\nu)}$ plus correction terms (see the remark below). Thus in the $\nu \to \infty$ limit we have

$$\lim_{\nu\to\infty} |\Gamma(\lambda, \nu) - \Gamma(\lambda_0(\nu), \nu)| = 0, \tag{8.1.27}$$

with the pointwise norm taken relative to the metric $\tilde{\gamma}_{(\lambda_0(\nu),\nu)}$.

Similarly, using the diffeomorphism on Σ_0 generated by the gradient flow of the canonical function $u_{(0)}$ (see Definition 3.3.1) we can connect $\Lambda_-(\lambda_0(\nu),\nu)$ and $\Lambda_+(\lambda_0(\nu),\nu)$ to the corresponding eigenvalues of the rescaled metric $r^{-2}\gamma_{(\lambda_0(\nu),\nu)}$. In view of the initial conditions (see Definition 3.6.1) it is easily seen that this rescaled metric tends to the standard metric at spacelike infinity as $\nu \to \infty$, and therefore

$$\lim_{\nu\to\infty} \Lambda_-(\lambda_0(\nu),\nu) = \lim_{\nu\to\infty} \Lambda_+(\lambda_0(\nu),\nu) = 1. \tag{8.1.28}$$

We have, therefore proved that the limit $\lim_{\nu\to\infty} \tilde{\gamma}_{(\lambda,\nu)} = \tilde{\gamma}_\infty$ exists. Moreover, in view of the boundedness of the pointwise norms (see 8.0.2)

$$|r^2\tau_-^{\frac{1}{2}}\left(\Omega\mathrm{tr}\underline{\chi} - \overline{\Omega\mathrm{tr}\underline{\chi}}\right)|\ ,\ \ |r\tau_-^{\frac{3}{2}}\hat{\underline{\chi}}|, \tag{8.1.29}$$

we deduce from (8.1.24) that

$$\lim_{\nu\to\infty}\left|\frac{\partial\tilde{\gamma}}{\partial\lambda}\right|_{\tilde{\gamma}_\infty} = 0. \tag{8.1.30}$$

Thus $\tilde{\gamma}_\infty$ does not depend on λ.

Remark: To prove the limit of the connection in (8.1.27) we need the boundedness of $|r^3\tau_-^{\frac{1}{2}}\nabla\!\!\!/\mathrm{tr}\underline{\chi}|$, $|r\tau_-^{\frac{3}{2}}\hat{\underline{\chi}}|$, $|r^2\tau_-^{\frac{1}{2}}\eta|$, $|r^2\tau_-^{\frac{1}{2}}\underline{\eta}|$ and the fact that $r^2K(\lambda,\nu)$ tends as $\nu\to\infty$ to the Gauss curvature of the spacelike infinity surface: $\lim_{\nu\to\infty} S(\lambda_0(\nu),\nu)$. This boundedness is provided by the estimates (8.0.1) and (8.0.2).

8.1.4 The null-outgoing infinite limit of the $S(\lambda,\nu)$-orthonormal frame

Let e_a be an orthonormal basis for the tangent space to $S(\lambda,\nu)$. Using the diffeomorphism $\psi(\lambda,\nu)$ introduced above, we define a basis on S^2 as

$$\tilde{E}_a|_{p_0} = \psi_*{}^{-1}(\lambda,\nu)(re_a|_p), \tag{8.1.31}$$

where $p = \psi(\lambda,\nu)(p_0)$ and p_0 is a point on S^2.

Lemma 8.1.3 *The frame $\{\tilde{E}_a(\lambda,\nu)\}$ converges as $\nu\to\infty$ to a frame orthonormal with respect to $\tilde{\gamma}_\infty$.*

Proof: The frame $\{\tilde{E}_a(\lambda,\nu)\}$ is orthonormal with respect to the metric $\tilde{\gamma}(\lambda,\nu) = \psi^*(\lambda,\nu)(r^{-2}\gamma)$. Deriving $\tilde{E}_a(\lambda,\nu)$ with repect to ν, we obtain

$$\begin{aligned}
\frac{\partial\tilde{E}_a|_{p_0}}{\partial\nu} &= \lim_{h\to 0}\frac{1}{h}\left[\psi_*{}^{-1}(\lambda,\nu+h)(re_a) - \psi_*{}^{-1}(\lambda,\nu)(re_a)\right]\Big|_{p_0} \\
&= \psi_*^{-1}(\lambda,\nu)\lim_{h\to 0}\frac{1}{h}\left[\left((\underline{\phi}_{\lambda-\lambda_0}\circ\phi_h\circ\underline{\phi}^{-1}_{\lambda-\lambda_0})(re_a)\right)|_p - (re_a)|_p\right] \\
&= \psi_*^{-1}(\lambda,\nu)(\mathcal{L}_V(re_a)|_p),
\end{aligned} \tag{8.1.32}$$

where $V = \underset{-*(\lambda-\lambda_0)}{\phi^{-1}} N$. From the definition of the $|\cdot|_{\tilde{\gamma}_\infty}$ and the definition of the Lie derivative it is evident that

$$\left|\frac{\partial \tilde{E}_a}{\partial \nu}\right|_{\tilde{\gamma}_\infty} \leq \frac{1}{r}|\mathcal{L}_N(re_a)|_\gamma + O(\frac{1}{r^2}) \tag{8.1.33}$$

We further assume that the frame $\{e_a\}$ is Fermi transported along $C(\lambda)$; that is, $\not{\mathcal{L}}_N e_a = -\Omega\chi_{ab}e_b$. An explicit calculation gives

$$\not{\mathcal{L}}_N(re_a) = -\frac{1}{2}(\Omega\text{tr}\chi - \overline{\Omega\text{tr}\chi})re_a - \Omega\hat{\chi}_{AB}re_B, \tag{8.1.34}$$

from which $|\mathcal{L}_N(re_a)|_\gamma = |\not{\mathcal{L}}_N(re_a)|_\gamma \leq cr^{-1}$ and finally

$$\left|\frac{\partial \tilde{E}_a}{\partial \nu}\right|_{\tilde{\gamma}_\infty} = O(r^{-2}). \tag{8.1.35}$$

This implies that $\{\tilde{E}_a(\lambda, \nu)\}$ converges as $\nu \to \infty$ to a frame orthonormal with respect to $\tilde{\gamma}_\infty$.

With this definition of the $\{\tilde{E}_a(\lambda, \nu)\}$ frame, given ω an $S(\lambda, \nu)$-tangent p-covariant tensor field on $\mathcal{M}$ and $\tilde{\omega}$ its rescaled pull back on S^2, the following relation holds:

$$\tilde{\omega}(\tilde{E}_{a_1}, ...\tilde{E}_{a_p}) = \omega(e_{a_1}....e_{a_p}). \tag{8.1.36}$$

Therefore we have proved the following lemma.

Lemma 8.1.4 *An $S(\lambda, \nu)$-tangent p-covariant tensor field ω has a null-outgoing limit, in the sense of Definition 8.1.1, if and only if $\lim_{\nu\to\infty} \omega|_{S(\lambda,\nu)}(e_{a_1}....e_{a_p})$ exists.*

8.2 The behavior of the curvature tensor at the null-outgoing infinity

To examine the behavior of the various components of the Riemann tensor moving toward the future null infinity along an null-outgoing hypersurface $C(\lambda)$ we recall that, as discussed in Section 8.1, a covariant p-tensor w defined on $S(\lambda, \nu)$ has a null infinity limit along $C(\lambda)$ if the following limit exists for any λ,[5]

$$\lim_{C(\lambda);\nu\to\infty} \psi^*(\lambda, \nu)(r^{-p}w) = \lim_{C(\lambda);\nu\to\infty} \tilde{w}(\lambda, \nu) \equiv W(\lambda). \tag{8.2.1}$$

Moreover, Lemma 8.1.4 shows that it is equivalent to prove that the following limit exists,

$$\lim_{C(\lambda);\nu\to\infty} w(p)(e_{a_1}, ..., e_{a_p}). \tag{8.2.2}$$

Using these results we prove the following proposition.

[5] Any λ means in fact any $\lambda \leq \lambda_0$.

Proposition 8.2.1 *The null components of the Riemann tensor have the following future null-outgoing infinity limits:*[6]

$$\lim_{C(\lambda);\nu\to\infty} r\underline{\alpha} = \underline{A}(\lambda,\omega) \quad , \quad \lim_{C(\lambda);\nu\to\infty} r^2\underline{\beta} = \underline{B}(\lambda,\omega) \tag{8.2.3}$$
$$\lim_{C(\lambda);\nu\to\infty} r^3\rho = P(\lambda,\omega) \quad , \quad \lim_{C(\lambda);\nu\to\infty} r^3\sigma = Q(\lambda,\omega),$$

where $\underline{A}(\lambda,\omega)$, $\underline{B}(\lambda,\omega)$, $P(\lambda,\omega)$, $Q(\lambda,\omega)$ *satisfy the estimates*

$$|\underline{A}(\lambda,\omega)| \le c(1+|\lambda|)^{-\frac{5}{2}} \quad ; \quad |\underline{B}(\lambda,\omega)| \le c(1+|\lambda|)^{-\frac{3}{2}} \tag{8.2.4}$$
$$|(P-\overline{P})(\lambda,\omega)| \le c(1+|\lambda|)^{-\frac{1}{2}} \;;\; |(Q-\overline{Q})(\lambda,\omega)| \le c(1+|\lambda|)^{-\frac{1}{2}}.$$

Remark: The limits $\lim_{\lambda\to\lambda_0}\overline{P}$ and $\lim_{\lambda\to\lambda_0}\overline{Q}$ will be discussed in Section 8.5.

Proof: The results of Proposition 8.2.1 are, basically, the same as those of conclusion (17.0.1) in [Ch-Kl]. We sketch its proof for completeness.

Using the Bianchi equation for $\underline{\alpha}$ (see (3.2.8)), and assuming the frame $\{e_a\}$ is Fermi transported along the null-outgoing hypersurfaces we have the evolution equation for $\underline{\alpha}_{ab} = \underline{\alpha}(e_a,e_b)$,

$$\frac{\partial\underline{\alpha}_{ab}}{\partial\nu} + \frac{1}{2}\Omega\mathrm{tr}\chi\,\underline{\alpha}_{ab} = \Omega\Big\{-(\nabla\!\!\!/\widehat{\otimes}\underline{\beta})_{ab} + \Big[4\omega\underline{\alpha}_{ab} - 3(\hat{\underline{\chi}}_{ab}\rho - {}^{\star}\hat{\underline{\chi}}_{ab}\sigma) + \big((\zeta-4\underline{\eta})\widehat{\otimes}\underline{\beta}\big)_{ab}\Big]\Big\}.$$

Recalling that $\frac{\partial}{\partial\nu}r = \frac{1}{2}r\overline{\Omega\mathrm{tr}\chi}$ (see (4.1.30)), from the previous equation we derive

$$\frac{\partial(r\underline{\alpha}_{ab})}{\partial\nu} = f(\nu,\lambda,\cdot)(r\underline{\alpha}_{ab}) + F(\nu,\lambda,\cdot), \tag{8.2.5}$$

where

$$f(\nu,\lambda,\omega^a) = \Big[-\frac{1}{2}(\Omega\mathrm{tr}\chi - \overline{\Omega\mathrm{tr}\chi}) + 4\Omega\omega\Big] \tag{8.2.6}$$
$$F(\nu,\lambda,\omega^a) = r\Omega\Big\{-(\nabla\!\!\!/\widehat{\otimes}\underline{\beta})_{ab} + \Big[-3(\hat{\underline{\chi}}_{ab}\rho - {}^{\star}\hat{\underline{\chi}}_{ab}\sigma) + \big((\zeta-4\underline{\eta})\widehat{\otimes}\underline{\beta}\big)_{ab}\Big]\Big\}.$$

From (8.2.5) we easily obtain, omitting the dependence on the angular variables,

$$|r\underline{\alpha}_{ab}(\nu,\lambda) - r\underline{\alpha}_{ab}(\nu',\lambda)| \le \int_{\nu'}^{\nu}\big(|f||r\underline{\alpha}| + |F|\big)(\nu'',\lambda)d\nu''. \tag{8.2.7}$$

From the Main Theorem, (see (8.0.1) and (8.0.2)), $f(\nu,\lambda) = O(r^{-2}(\lambda,\nu))$ and $\sup_{\mathcal{M}} r|\lambda|^{\frac{5}{2}}|\underline{\alpha}| \le C_0$. Therefore

$$\int_{\nu'}^{\nu}|fr\underline{\alpha}|(\nu'',\lambda)d\nu'' \le C_0\frac{1}{r(\lambda,\nu')}\frac{1}{\lambda^{\frac{5}{2}}}. \tag{8.2.8}$$

[6] ω denotes the angular variables of S^2.

In view of the results of the Main Theorem, the $\mathcal{R}$ norms are uniformly bounded by a constant C_0. This implies immediately the following estimate for $\int_{\nu'}^{\nu} F(\nu', \lambda)d\nu'$, taking into account only the principal terms of $F(\nu, \lambda, \omega^a)$, $r\Omega(\nabla\!\!\!/\widehat{\otimes}\underline{\beta})_{ab}$,

$$\left|\int_{\nu'}^{\nu} F(\nu'', \lambda)d\nu''\right| \leq c\int_{\nu'}^{\nu} |r(\nabla\!\!\!/\widehat{\otimes}\underline{\beta})_{ab}|d\nu'' \tag{8.2.9}$$

$$\leq c\left(\int_{\nu'}^{\nu} |r\nabla\!\!\!/\underline{\beta}|^2 r^2 d\nu''\right)^{\frac{1}{2}}\left(\int_{\nu'}^{\nu} r^{-2}d\nu''\right)^{\frac{1}{2}} \leq \frac{1}{r(\lambda, \nu')^{\frac{1}{2}}}\frac{1}{\lambda^2}.$$

Choosing ν' sufficiently large we can make the right-hand side of (8.2.8) and (8.2.9) arbitrarily small, which proves the existence of the limit for $r\underline{\alpha}(\lambda, \nu)$. Then, using (8.2.5), the Gronwall inequality and the estimate for $\underline{\alpha}$ on Σ_0 (see (7.1.19)), we obtain

$$\lim_{\nu\to\infty} |r\underline{\alpha}_{ab}|(\nu, \lambda) \leq c\left(|r\underline{\alpha}_{ab}|(\nu_0(\lambda)) + \int_{\nu_0(\lambda)}^{\infty} F(\nu', \lambda)d\nu'\right) \leq c\frac{1}{\lambda^{\frac{5}{2}}},$$

which proves the proposition. The limits for $\underline{\beta}$, ρ, σ are obtained in the same way using the corresponding Bianchi equations (see 3.2.8), and we do not present them here.[7]

According to the Penrose hypothesis of smooth conformal compactification [Pe1], [Pe2], [Ne-Pe1], [Ne-Pe2], α and β should also have null-outgoing infinity limits. It is in fact known that smooth compactification implies

$$\lim_{C(\lambda);\nu\to\infty} r^5\alpha = A(\lambda, \omega) \quad , \quad \lim_{C(\lambda);\nu\to\infty} r^4\beta = B(\lambda, \omega). \tag{8.2.10}$$

These results are, however, out of reach with our methods. In this work, as well as in [Ch-Kl], we have only been able to prove the boundedness of $r^{\frac{7}{2}}\alpha$ and $r^{\frac{7}{2}}\beta$. This, by itself, does not exclude the possibility that α, β have better bounds under, possibly, more stringent conditions on the initial data.[8]

The issue of smooth conformal compactification has drawn a lot of attention in the last twenty years. In particular H. Friedrich is one of the main promoters of the idea that there must exist an important class of data that leads to a smooth compactification [Fr1]–[Fr4]. On the other hand, a lot of evidence has been accumulated suggesting that one cannot expect smoothness of the compactification for arbitrary initial data or for physically relevant ones. In particular, D. Christodoulou has shown, under some reasonable "physical" assumptions concerning the past null infinity $\mathcal{J}^-$ that the following limits hold:[9]

$$\begin{aligned}\alpha &= r^{-4}A_1(\theta, \phi) + r^{-5}\log r A_2(\lambda; \theta, \phi) + r^{-5}A_3(\lambda; \theta, \phi)\\ \beta &= r^{-4}\log r B_1(\theta, \phi) + r^{-4}B_2(\lambda; \theta, \phi).\end{aligned} \tag{8.2.11}$$

---

[7] More precisely, due to the presence of $\nabla\!\!\!/\hat{\otimes}\underline{\beta}$ in the evolution equation for $\underline{\alpha}$, one has to prove the existence of the limits $\lim_{\nu\to\infty} |r\underline{\alpha}_{ab}|_{p,S}(\lambda, \nu)$, $\lim_{\nu\to\infty} |r^2|\nabla\!\!\!/\underline{\alpha}_{ab}|_{p,S}(\lambda, \nu)$ and deduce the result from them.

[8] Note that we do not show that the quantities $r^{\frac{7}{2}}\alpha$, $r^{\frac{7}{2}}\beta$ have a null-outgoing limit. Indeed when we try to implement the same strategy for α and β as for the other quantities we encounter the following difficulties: in the case of α, the Bianchi equations (see (3.2.8)) do not contain an evolution equation for α along the e_4 direction. On the other hand, for β we have an evolution equation along $C(\lambda)$, which suggests that $r^4\beta$ has a null-outgoing limit. To prove this, however, we would need to control the quantity $\int_{\nu'}^{\nu} |r^4\nabla\!\!\!/\alpha|$ whose boundedness is not at our disposal.

[9] It is assumed that there is no incoming radiation from $\mathcal{J}^-$ and that the outgoing radiation has the structure suggested by the quadrupole approximation of the gravitational radiation produced by N accelerated point masses. D. Christodoulou, private communication; see also [Ch 8].

This is in agreement with the polyhomogeneous expansions suggested by the work of many authors; see [Kr] and the references therein.

8.3 The behavior of the connection coefficients at the null-outgoing infinity

Proposition 8.3.1 *The following null-outgoing infinity limits hold:*

$$\lim_{C(\lambda);\nu\to\infty}\Omega=\frac{1}{2}\,,\quad \lim_{C(\lambda);\nu\to\infty}r^2\underline{\omega}=-\frac{\nu P}{4}$$
$$\lim_{C(\lambda);\nu\to\infty}r\mathrm{tr}\chi=2\,,\quad \lim_{C(\lambda);\nu\to\infty}r\mathrm{tr}\underline{\chi}=-2. \tag{8.3.1}$$

Proof: From the estimates (8.0.2) proved in the Main Theorem it follows that if $\nu_1\geq\nu_2\geq M>0$, then $|\log 2\Omega(\nu_1,\lambda)-\log 2\Omega(\nu_2,\lambda)|$ can be made arbitrarily small. In fact

$$\begin{aligned}|\log 2\Omega(\nu_1,\lambda)-\log 2\Omega(\nu_2,\lambda)| &\leq c\int_{\nu_2}^{\nu_1}|(\Omega\mathbf{D}_4\log\Omega)|(\nu',\lambda)d\nu'\\ &\leq c\left|\frac{1}{r(\lambda,\nu_1)}-\frac{1}{r(\lambda,\nu_2)}\right|.\end{aligned} \tag{8.3.2}$$

Therefore $\lim_{C(\lambda);\nu\to\infty}\log 2\Omega(\nu,\lambda)$ does exist. To prove that it is equal to zero we observe first that the relation $\triangle\log\Omega=\frac{1}{2}\nabla\!\!\!/(\eta+\underline{\eta})$ implies that $|\log 2\Omega-\overline{\log 2\Omega}|$ goes to zero as ν goes to infinity (see (4.3.64)). Subsequently we look at the evolution equation of $\overline{\log 2\Omega}$ along the $\underline{C}(\nu)$ null hypersurfaces, with the same procedure as in Lemma 4.3.4, and we obtain the final result estimating, first, $\overline{\log 2\Omega(\lambda,\nu)}$ in terms of $\overline{\log 2\Omega(\lambda,\nu)}|_{\underline{C}(\nu)\cap\Sigma_0}=\overline{\log 2\Omega(\lambda_0(\nu),\nu)}$ plus correction terms which go to zero as $\nu\to\infty$, and second, recalling that $\overline{\log 2\Omega(\lambda,\nu)}|_{\underline{C}(\nu)\cap\Sigma_0}$ goes to zero as $\nu\to\infty$.

To prove the limit for $\underline{\omega}$, we observe that it satisfies the evolution equation (4.3.59),

$$\frac{\partial}{\partial\nu}\Omega\underline{\omega}=-\frac{1}{2}(\Omega\hat{\underline{F}}-\Omega^2\rho). \tag{8.3.3}$$

As $\lim_{C(\lambda);\nu\to\infty}\underline{\omega}=0$, integrating (8.3.3) we obtain

$$(\Omega\underline{\omega})(\lambda,\nu)=\frac{1}{2}\int_{\nu}^{\infty}\Omega\hat{\underline{F}}-\frac{1}{2}\int_{\nu}^{\infty}\Omega^2\rho, \tag{8.3.4}$$

and multiplying both sides by r^2,

$$(r^2\Omega\underline{\omega})(\lambda,\nu)=\frac{r^2(\lambda,\nu)}{2}\int_{\nu}^{\infty}\Omega\hat{\underline{F}}-\frac{r^2(\lambda,\nu)}{2}\int_{\nu}^{\infty}\Omega^2\rho. \tag{8.3.5}$$

From the estimates (8.0.2), validated in the Main Theorem, it is immediate to see that

$$\int_{\nu}^{\infty}\Omega\hat{\underline{F}}(\lambda,\nu')d\nu'=O(r^{-3}). \tag{8.3.6}$$

Therefore, performing the limit $\nu \to \infty$, it follows that

$$\lim_{\nu\to\infty} (r^2\Omega\underline{\omega})(\lambda,\nu) = -\lim_{\nu\to\infty} \frac{r^2(\lambda,\nu)}{2}\int_\nu^\infty \Omega^2\rho \tag{8.3.7}$$
$$= -\lim_{\nu\to\infty}\left(\frac{r^2(\lambda,\nu)}{2}\int_\nu^\infty \frac{\Omega^2}{r^3}(r^3\rho - P)\right) - P\lim_{\nu\to\infty}\left(\frac{r^2(\lambda,\nu)}{2}\int_\nu^\infty \frac{\Omega^2}{r^3}\right).$$

From the results of Proposition 8.2.1 and from $\lim_{C(\lambda);\nu\to\infty}\Omega = \frac{1}{2}$ it follows immediately that the first integral in the right-hand side goes to zero, while $\lim_{\nu\to\infty}\frac{r^2(\lambda,\nu)}{2}\int_\nu^\infty \frac{\Omega^2}{r^3} = \frac{1}{8}$ so that

$$\lim_{C(\lambda);\nu\to\infty} (r^2\underline{\omega})(\lambda,\nu) = -\frac{P}{4}, \tag{8.3.8}$$

which proves the result.

The proof of the limits for $r\mathrm{tr}\chi$ and $r\mathrm{tr}\underline{\chi}$ proceeds exactly in the same way. First, by looking at their evolution equations along $C(\lambda)$ and using the estimates from the Main Theorem one proves that these limits do exist. Then using the corresponding evolution equations along the $\underline{C}(\nu)$ null hypersurfaces it is possible to connect these limits to the limits of the corresponding quantities on Σ_0, thus obtaining the result.

Remark: Observe that the limit for $r^2\underline{\omega}$ gives a stronger decay than the estimate provided by the Main Theorem; see Proposition 4.3.4 and Proposition 7.4.4. Once this limit is proved one can obtain a better estimate also for $\nabla\!\!\!\!/\,\underline{\omega}$.

Proposition 8.3.2 *The connection coefficients $\hat{\chi}$, $\hat{\underline{\chi}}$ and ζ have the null-outgoing infinity limits*

$$\lim_{C(\lambda);\nu\to\infty} r^2\hat{\chi} = X(\lambda,\cdot)\,, \quad \lim_{C(\lambda);\nu\to\infty} r\hat{\underline{\chi}} = \underline{X}(\lambda,\cdot)$$
$$\lim_{C(\lambda);\nu\to\infty} r^2\zeta = Z(\lambda,\cdot) \tag{8.3.9}$$

and

$$|X(\lambda,\cdot)| \le c(1+|\lambda|)^{-\frac{1}{2}}\,, \quad |\underline{X}(\lambda,\cdot)| \le c(1+|\lambda|)^{-\frac{3}{2}}$$
$$|Z(\lambda,\cdot)| \le c(1+|\lambda|)^{-\frac{1}{2}}. \tag{8.3.10}$$

Proof: We start by looking at the null infinity limit of $\hat{\chi}$. As for the underlined null components of the curvature tensor, we look at the evolution equation along the null-outgoing hypersurfaces $C(\lambda)$ for $\hat{\chi}$ starting with initial data on Σ_0 and use it for estimating the limit $\nu \to \infty$.[10] The evolution equation for $\hat{\chi}$ (see (3.1.46)) can be written as

$$\frac{\partial}{\partial\nu}\hat{\chi}_{ab} + \Omega\mathrm{tr}\chi\,\hat{\chi}_{ab} + 2\Omega\omega\hat{\chi}_{ab} = -\Omega\alpha_{ab}, \tag{8.3.11}$$

[10] Observe that our procedure here differs from that of Chapter 4, where the integration took place with the initial conditions given on the last slice.

and from it,

$$\frac{\partial(r^2\hat{\chi}_{ab})}{\partial\nu}=h(\nu,\lambda,\cdot)(r^2\hat{\chi}_{ab})-\Omega r^2\alpha_{ab}, \tag{8.3.12}$$

where

$$h(\nu,\lambda,\omega^a)=-(\Omega\text{tr}\chi-\overline{\Omega\text{tr}\chi})+2\Omega\omega. \tag{8.3.13}$$

In view of the Main Theorem results (see (8.0.1), (8.0.2)) we easily check the pointwise bounds,

$$|r^{\frac{7}{2}}\alpha|\leq C_0\,,\ |r^2h|\leq C_0\,,\ |r^2\tau_-^{\frac{1}{2}}\hat{\chi}|\leq C_0. \tag{8.3.14}$$

Thus by integration of (8.3.12), with $\nu_1\geq\nu_2\geq M>0$, we have

$$\left|r^2\hat{\chi}_{ab}(\nu_1,\lambda)-r^2\hat{\chi}_{ab}(\nu_2,\lambda)\right|\leq c\left(\int_{\nu_2}^{\nu_1}|r^2\alpha_{ab}|(\lambda,\nu')d\nu'+\int_{\nu_2}^{\nu_1}O\left(\frac{1}{r^2\lambda^{\frac{1}{2}}}\right)\right).$$

Since the right-hand side can be made arbitrarily small as $M\to\infty$, we conclude that the following limit exists:

$$\lim_{C(\lambda);\nu\to\infty}r^2\hat{\chi}=X(\lambda,\cdot). \tag{8.3.15}$$

To estimate the behavior of $X(\lambda,\cdot)$ with respect to λ, we use the evolution equation for $\hat{\chi}$ along the null-incoming hypersurfaces $\underline{C}(\nu)$ written as

$$\frac{\partial}{\partial\lambda}\hat{\chi}_{ab}=-\left[\frac{1}{2}\Omega\text{tr}\underline{\chi}+2\Omega\underline{\omega}\right]\hat{\chi}_{ab}-\frac{1}{2}\Omega\text{tr}\chi\hat{\underline{\chi}}_{ab}+\Omega(\nabla\!\!\!/\widehat{\otimes}\eta+\eta\widehat{\otimes}\eta), \tag{8.3.16}$$

from which we derive

$$\frac{\partial}{\partial\lambda}(r\hat{\chi}_{ab})=t(\nu,\lambda,\omega^a)(r\hat{\chi}_{ab})+T(\nu,\lambda,\omega^a), \tag{8.3.17}$$

where

$$\begin{aligned}\underline{t}(\nu,\lambda,\omega^a)&=\left[-\frac{1}{2}(\Omega\text{tr}\underline{\chi}-\overline{\Omega\text{tr}\underline{\chi}})+2\Omega\underline{\omega}\right]\\ \underline{T}(\nu,\lambda,\omega^a)&=-\frac{1}{2}\Omega\text{tr}\chi(r\hat{\underline{\chi}}_{ab})+\Omega r(\nabla\!\!\!/\widehat{\otimes}\eta+\eta\widehat{\otimes}\eta).\end{aligned} \tag{8.3.18}$$

Integrating, multiplying by r, applying the Gronwall inequality and denoting $\lambda_0(\nu)=u|_{\Sigma_0\cap\underline{C}(\nu)}$, we obtain

$$\begin{aligned}|r^2\hat{\chi}_{ab}|(\nu,\lambda)&\leq c\left(|r^2\hat{\chi}_{ab}|(\lambda_0(\nu),\nu)+\int_{\lambda_0(\nu)}^{\lambda}|r^2(\nabla\!\!\!/\widehat{\otimes}\eta+\eta\widehat{\otimes}\eta)|(\lambda',\nu)d\lambda'\right)\\ &\leq c|r^2\hat{\chi}_{ab}|(\lambda_0(\nu),\nu)+c\int_{\lambda_0(\nu)}^{\lambda}(|r^2\nabla\!\!\!/\eta|+r^2|\eta|^2)(\lambda',\nu)d\lambda'\\ &\leq c|r^2\hat{\chi}_{ab}|(\lambda_0(\nu),\nu)+c(1+|\lambda|)^{-\frac{1}{2}}\end{aligned} \tag{8.3.19}$$

where the last inequality follows from the estimates (8.0.2). Taking the limit as ν goes to infinity and recalling the asymptotic behavior of $|r^2\hat{\chi}_{ab}|$ on Σ_0 (see (7.2.3)) we infer that $|X(\lambda,\cdot)|\leq c(1+|\lambda|)^{-\frac{1}{2}}$.

We proceed in the same way for $\underline{\hat{\chi}}$. Consider the evolution equation for $\underline{\hat{\chi}}$ along $C(\lambda)$, which we write as

$$\frac{\partial}{\partial\nu}\underline{\hat{\chi}}_{ab}=-\left[\frac{1}{2}\Omega\mathrm{tr}\chi+2\Omega\omega\right]\underline{\hat{\chi}}_{ab}-\frac{1}{2}\Omega\mathrm{tr}\underline{\chi}\,\hat{\chi}_{ab}+\Omega(\nabla\widehat{\otimes}\underline{\eta}+\underline{\eta}\widehat{\otimes}\underline{\eta}). \tag{8.3.20}$$

This also implies

$$\frac{\partial}{\partial\nu}(r\underline{\hat{\chi}}_{ab})=q(\nu,\lambda,\omega^a)(r\underline{\hat{\chi}}_{ab})+Q(\nu,\lambda,\omega^a), \tag{8.3.21}$$

where

$$\begin{aligned}q(\nu,\lambda,\omega^a)&=\left[-\frac{1}{2}(\Omega\mathrm{tr}\chi-\overline{\Omega\mathrm{tr}\chi})+2\Omega\omega\right]\\ Q(\nu,\lambda,\omega^a)&=-\frac{1}{2}\Omega\mathrm{tr}\underline{\chi}(r\hat{\chi}_{ab})+\Omega r(\nabla\widehat{\otimes}\underline{\eta}+\underline{\eta}\widehat{\otimes}\underline{\eta}).\end{aligned} \tag{8.3.22}$$

In view of the Main Theorem results (see (8.0.1), (8.0.2)) we easily check the pointwise bounds

$$|r^2q|\leq C_0\ ,\ |r^3\tau_-^{\frac{1}{2}}(\nabla\widehat{\otimes}\underline{\eta}+\underline{\eta}\widehat{\otimes}\underline{\eta})|\leq C_0\ ,\ |r\tau_-^{\frac{3}{2}}\underline{\hat{\chi}}|\leq C_0. \tag{8.3.23}$$

Proceeding as before, we obtain

$$\begin{aligned}|r\underline{\hat{\chi}}_{ab}(\nu_1,\lambda)-r\underline{\hat{\chi}}_{ab}(\nu_2,\lambda)|&\leq c\int_{\nu_2}^{\nu_1}|r(\nabla\widehat{\otimes}\underline{\eta}+\underline{\eta}\widehat{\otimes}\underline{\eta})|(\lambda,\nu')d\nu' \qquad (8.3.24)\\ &+\int_{\nu_2}^{\nu_1}O\left(\frac{1}{r^2\lambda^{\frac{3}{2}}}\right). \qquad (8.3.25)\end{aligned}$$

Letting ν_1, ν_2 going to ∞, as before, we infer that $r\underline{\hat{\chi}}$ has a limit

$$\lim_{C(\lambda);\nu\to\infty}r\underline{\hat{\chi}}=\underline{X}(\lambda,\cdot)\,.$$

Moreover, recalling the asymptotic behavior of $|r\underline{\hat{\chi}}_{ab}|$ on Σ_0, derived from the boundedness of the norms (7.2.3), it is immediate to prove, proceeding as in the case of $X(\lambda,\cdot)$, that $\underline{X}$ behaves as $O(\lambda^{-\frac{3}{2}})$. Therefore

$$|\underline{X}(\lambda,\cdot)|\leq c(1+|\lambda|)^{-\frac{3}{2}}\,.$$

To prove the limit for $r^2\zeta$ we look at its evolution equation (see (3.1.46)) along the null-incoming hypersurfaces $\underline{C}(\nu)$, which we rewrite as

$$\begin{aligned}\frac{\partial}{\partial\lambda}\zeta+\Omega\mathrm{tr}\underline{\chi}\zeta&=-\Omega\underline{\hat{\chi}}\cdot\zeta+\Omega\nabla\mathbf{D}_3\log\Omega-\Omega[\nabla,\mathbf{D}_3]\log\Omega-\Omega\underline{\beta}\\ &=-\Omega\underline{\hat{\chi}}\cdot\zeta-2\Omega\nabla\underline{\omega}-\Omega\Big(-2\underline{\omega}\nabla\log\Omega-\underline{\chi}\cdot(\eta+\underline{\eta})\Big)-\Omega\underline{\beta}.\end{aligned} \tag{8.3.26}$$

From it we derive

$$\frac{\partial}{\partial\lambda}(r^2\zeta_a) = p(\nu,\lambda,\omega^b)(r^2\zeta_a) + P_a(\nu,\lambda,\omega^b) - 2\Omega r^2 \nabla\!\!\!/_a \underline{\omega} - \Omega r^2 \underline{\beta}_a, \tag{8.3.27}$$

where

$$p(\nu,\lambda,\omega^b) = -(\Omega \text{tr}\underline{\chi} - \overline{\Omega \text{tr}\underline{\chi}})$$
$$P_a(\nu,\lambda,\omega^b) = -\Omega(\hat{\underline{\chi}}\cdot r^2\zeta)_a - 2\left[2\Omega r^2 \underline{\omega} \nabla\!\!\!/_a \log\Omega + \Omega r^2 \underline{\chi}_{ac}(\eta + \underline{\eta})_c\right].$$

Integrating along $\underline{C}(\nu)$ we obtain

$$\begin{aligned}(r^2\zeta)_a(\lambda,\nu) &= (r^2\zeta)_a(\lambda_0(\nu),\nu) + \int_{\lambda_0(\nu)}^{\lambda} p(\nu,\lambda,\omega^b)(r^2\zeta_a) + \int_{\lambda_0(\nu)}^{\lambda} P_a(\nu,\lambda,\omega^b) \\ &\quad -2\int_{\lambda_0(\nu)}^{\lambda} \Omega r^2 \nabla\!\!\!/_a \underline{\omega} - \int_{\lambda_0(\nu)}^{\lambda} \Omega r^2 \underline{\beta}_a. \end{aligned}\tag{8.3.28}$$

In view of the Main Theorem results (see (8.0.1), (8.0.2)) we easily check the pointwise bounds

$$|p(\nu,\lambda,\omega^b)(r^2\zeta_a)| \le c r^{-2}\tau_-^{-1}, \quad |P(\nu,\lambda,\omega^b)| \le c r^{-1}\tau_-^{-\frac{1}{2}}.$$

Taking the $\nu \to \infty$ limit of (8.3.28) the first two integrals of the right-hand side go to zero and we are left with

$$\begin{aligned}\lim_{\nu\to\infty}(r^2\zeta)_a(\lambda,\nu) &= \lim_{\nu\to\infty}(r^2\zeta)_a(\lambda_0(\nu),\nu) - 2\int_{-\infty}^{\lambda} \lim_{\nu\to\infty} \Omega r^2 \nabla\!\!\!/_a \underline{\omega} \\ &\quad - \int_{-\infty}^{\lambda} \lim_{\nu\to\infty} \Omega r^2 \underline{\beta}_a. \end{aligned}\tag{8.3.29}$$

We assume now that $r^2\nabla\!\!\!/\underline{\omega} = O(r^{-1}\tau_-^{-\frac{1}{2}})$. This result is proved in Proposition 8.3.3. Moreover from the results of Chapter 7 (see Section 7.2) it follows immediately that $\lim_{\nu\to\infty}(r^2\zeta)_a(\lambda_0(\nu),\nu) = 0$. Therefore the limit in (8.3.29) becomes, recalling (8.2.3),

$$\lim_{\nu\to\infty}(r^2\zeta)(\lambda,\nu) = -\frac{1}{2}\int_{-\infty}^{\lambda} \lim_{\nu\to\infty} r^2\underline{\beta} = -\frac{1}{2}\int_{-\infty}^{\lambda} \underline{B}(\lambda,\omega) \equiv Z(\lambda,\cdot) \tag{8.3.30}$$

and $|Z(\lambda,\cdot)| \le c(1+|\lambda|)^{-\frac{1}{2}}$, which completes the proof of the proposition.

Proposition 8.3.3 *The first derivatives of the connection coefficients $\nabla\!\!\!/\underline{\omega}$, $\nabla\!\!\!/\text{tr}\chi$ have the following null-outgoing infinity limits:*

$$\lim_{C(\lambda);\nu\to\infty} r^2\nabla\!\!\!/\underline{\omega} = 0, \quad \lim_{C(\lambda);\nu\to\infty} r^3\nabla\!\!\!/\text{tr}\chi = \frac{1}{2}\mathcal{V} - 2Z = \tilde{\nabla\!\!\!/}H, \tag{8.3.31}$$

where $\mathcal{V}$ and H are defined through

$$\mathcal{V} = \lim_{C(\lambda);\nu\to\infty} r^3\mathcal{U}, \quad H = \lim_{C(\lambda);\nu\to\infty} r^2(\text{tr}\chi - \frac{2}{r}), \tag{8.3.32}$$

where $\mathcal{U}$ has been defined in equation (4.3.5). Moreover

$$r^2\nabla\!\!\!/\underline{\omega} = O(r^{-1}\tau_-^{-\frac{1}{2}}) \tag{8.3.33}$$

Proof: The proof is similar to the one for $\underline{\omega}$ in the previous proposition. First we observe that $\lim_{C(\lambda);\nu\to\infty}(r\nabla\!\!\!/\underline{\omega}) = 0$ as follows from the $L^p(S)$ norm estimates for $\nabla\!\!\!/\underline{\omega}$ and $\nabla\!\!\!/^2\underline{\omega}$. Then we look at the evolution equation (4.3.80) satisfied by $\nabla\!\!\!/\underline{\omega}$, which can be rewritten as

$$\frac{\partial}{\partial\nu}(\nabla\!\!\!/\underline{\omega}) + \frac{\Omega\mathrm{tr}\chi}{2}(\nabla\!\!\!/\underline{\omega}) = -\Omega\hat{\chi}(\nabla\!\!\!/\underline{\omega}) + 2\Omega\left(\omega\nabla\!\!\!/\underline{\omega} - \underline{\omega}\nabla\!\!\!/\omega\right) + \frac{1}{2}\underline{\hat{H}} - \frac{1}{2}\nabla\!\!\!/\rho \qquad (8.3.34)$$

where $\underline{\hat{H}}$ satisfies $|r^{-\frac{2}{p}}\underline{\hat{H}}|_{p,S} \leq C_0 r^{-5}$; see Proposition 4.3.8. This allows us to write

$$\frac{\partial}{\partial\nu}(r\nabla\!\!\!/\underline{\omega}) = -\frac{1}{2}\left(\Omega\mathrm{tr}\chi - \overline{\Omega\mathrm{tr}\chi}\right)(r\nabla\!\!\!/\underline{\omega}) - \frac{1}{2}r\nabla\!\!\!/\rho + O\left(\frac{1}{r^3\tau_-}\right) = -\frac{1}{2}r\nabla\!\!\!/\rho + O\left(\frac{1}{r^3\tau_-}\right),$$

and from it, using that $\lim_{C(\lambda);\nu\to\infty}(r\nabla\!\!\!/\underline{\omega}) = 0$,

$$(r^2\nabla\!\!\!/\underline{\omega})(\lambda,\nu) = -\frac{1}{2}r(\lambda,\nu)\int_\nu^\infty r\nabla\!\!\!/\rho(\lambda,\nu') - \frac{1}{2}r(\lambda,\nu)\int_\nu^\infty O\left(\frac{1}{r^3\tau_-}\right), \qquad (8.3.35)$$

which implies that

$$(r^2\nabla\!\!\!/\underline{\omega})(\lambda,\nu) = O\big(r^{-1}\tau_-^{-\frac{1}{2}}\big). \qquad (8.3.36)$$

As before we can estimate the limit of the right-hand side as $\nu\to\infty$. Due to the fact that the $\nabla\!\!\!/\rho$, in (8.3.35), behaves asymptotically as $O(r^{-4}\tau_-^{-\frac{1}{2}})$, we conclude that it goes to zero, proving the first part of the proposition.[11]

To prove the existence of the limit for $r^3\nabla\!\!\!/\mathrm{tr}\chi$ we have to follow an argument similar to the one used in Chapter 4, Proposition 4.3.1. Starting from the evolution equation of $\psi\!\!\!/ = \Omega^{-1}\big(\nabla\!\!\!/\mathrm{tr}\chi + \mathrm{tr}\chi\zeta\big)$, equation (4.3.6),

$$\frac{\partial}{\partial\nu}\psi\!\!\!/ + \frac{3}{2}\Omega\mathrm{tr}\chi\psi\!\!\!/ = F\!\!\!/,$$

we derive

$$\frac{\partial}{\partial\nu}r^3\psi\!\!\!/ = s(\nu,\lambda,\omega^b)r^3\psi\!\!\!/ + S(\nu,\lambda,\omega^b) - \mathrm{tr}\chi r^3\beta, \qquad (8.3.37)$$

where

$$s(\nu,\lambda,\omega^b) = -\frac{3}{2}(\Omega\mathrm{tr}\chi - \overline{\Omega\mathrm{tr}\chi})$$
$$S(\nu,\lambda,\omega^b) = -\Omega(\underline{\hat{\chi}}\cdot r^3\psi\!\!\!/) + r^3\left[-\nabla\!\!\!/|\hat{\chi}|^2 - \eta|\hat{\chi}|^2 + \mathrm{tr}\chi\ \hat{\chi}\cdot\underline{\eta}\right].$$

Proceeding as in Proposition 8.3.2 for $r^2\chi$ we find first that the limit of $r^3\psi\!\!\!/$ for $\nu\to\infty$ does exist

$$\lim_{C(\lambda);\nu\to\infty} r^3\psi\!\!\!/ = -V\!\!\!/. \qquad (8.3.38)$$

[11]Observe that in this case we do not have a pointwise limit as $\nu\to\infty$ for $r^3\nabla\!\!\!/\underline{\omega}$ due to the fact that its evolution equation involves the first tangential derivatives of ρ which does not have a pointwise limit.

In fact from

$$(r^3\underline{\not{\Psi}})(\lambda,\nu_1)-(r^3\underline{\not{\Psi}})(\lambda,\nu_2) = \int_{\nu_2}^{\nu_1}\big(s(\nu,\lambda,\omega^b)r^3\underline{\not{\Psi}}(\lambda,\nu)+S(\nu,\lambda,\omega^b)\big) - \int_{\nu_2}^{\nu_1}\mathrm{tr}\chi r^3\beta(\lambda,\nu), \tag{8.3.39}$$

using the Main Theorem results, the right-hand side of (8.3.39) can be made smaller than an arbitrary $\epsilon > 0$, provided $\nu_1 \geq \nu_2 \geq N$ for N sufficiently large.

To obtain the λ behavior of $\not{V}(\lambda,\cdot)$ we proceed as before writing the evolution equation along $\underline{C}(\nu)$ for $\underline{\not{\Psi}}$. Proceeding as for $r^2\hat{\chi}$ (see (8.3.19)) we infer that $|\not{V}(\lambda,\cdot)| \leq c(1+|\lambda|)^{-\frac{1}{2}}$. From knowledge of the null-outgoing infinity limits of $r^3\underline{\not{\Psi}}$ and $r^3\mathrm{tr}\chi\zeta$ we deduce the limit of $r^3\not{\nabla}\mathrm{tr}\chi$,

$$\lim_{C(\lambda);\nu\to\infty} r^3\not{\nabla}\mathrm{tr}\chi = \frac{1}{2}\not{V}-2Z. \tag{8.3.40}$$

Finally to connect $\frac{1}{2}\not{V}-2Z$ with $\tilde{\not{\nabla}}H$ we have to prove the relation

$$H=\lim_{C(\lambda);\nu\to\infty} r^2(\mathrm{tr}\chi-\frac{2}{r}).$$

To achieve this result we look at the evolution equation of $r^2(\overline{\Omega^{-1}\mathrm{tr}\chi}-\frac{4}{r})$ along $C(\lambda)$. Proceeding analogously to Proposition 4.3.6 we obtain,

$$\frac{\partial}{\partial\nu}(\overline{\Omega^{-1}\mathrm{tr}\chi})+\frac{1}{2}\Omega\mathrm{tr}\chi(\overline{\Omega^{-1}\mathrm{tr}\chi}) = -\overline{|\hat{\chi}|^2}+\frac{1}{2}V(\overline{\Omega^{-1}\mathrm{tr}\chi})+\frac{1}{2}\overline{V(\Omega^{-1}\mathrm{tr}\chi)}, \tag{8.3.41}$$

where $V \equiv \big(\Omega\mathrm{tr}\chi-\overline{\Omega\mathrm{tr}\chi}\big)$. Moreover (see (4.1.30))

$$\frac{\partial}{\partial\nu}\frac{1}{r} = -\frac{1}{r^2}\frac{\partial r}{\partial\nu} = -\frac{1}{2r}\overline{\Omega\mathrm{tr}\chi} = -\frac{1}{2}\Omega\mathrm{tr}\chi\frac{1}{r}+\frac{1}{2r}V. \tag{8.3.42}$$

Putting together (8.3.41) and (8.3.42) and denoting $\Psi=\big(\overline{\Omega^{-1}\mathrm{tr}\chi}-\frac{4}{r}\big)$, we have

$$\frac{\partial}{\partial\nu}\Psi+\frac{1}{2}\Omega\mathrm{tr}\chi\Psi = -\overline{|\hat{\chi}|^2}+\frac{1}{2}V\Psi+\frac{1}{2}\overline{V\Psi} \tag{8.3.43}$$

and

$$\frac{\partial}{\partial\nu}r\Psi = -r\overline{|\hat{\chi}|^2}+\frac{1}{2}\overline{Vr\Psi}. \tag{8.3.44}$$

Proceeding as before this implies that $\lim_{C(\lambda);\nu\to\infty} r\Psi = 0$ and we can write

$$(r^2\Psi)(\lambda,\nu) = r(\lambda,\nu)\int_\nu^\infty\left(\overline{r|\hat{\chi}|^2}-\frac{1}{2}\overline{Vr\Psi}\right). \tag{8.3.45}$$

Observing that the integrand in (8.3.46) is $O(r^{-3})$ we conclude, taking the limit $\nu\to\infty$, that $\lim_{C(\lambda);\nu\to\infty} r^2\big(\overline{\Omega^{-1}\mathrm{tr}\chi}-\frac{4}{r}\big)=0$ and, as the limit of Ω is $\frac{1}{2}$,

$$\lim_{C(\lambda);\nu\to\infty} r^2\big(\overline{\mathrm{tr}\chi}-\frac{2}{r}\big)=0. \tag{8.3.46}$$

To prove the existence of $\lim_{C(\lambda);\nu\to\infty} r^2\big(\mathrm{tr}\chi - \frac{2}{r}\big)$ we are left to prove the existence of $\lim_{C(\lambda);\nu\to\infty} r\nabla\!\!\!/ r^2\big(\mathrm{tr}\chi - \frac{2}{r}\big)$. This is has been obtained in (8.3.40) and, therefore, denoting $H = \lim_{C(\lambda);\nu\to\infty} r^2\big(\mathrm{tr}\chi - \frac{2}{r}\big)$ we can write

$$\lim_{C(\lambda);\nu\to\infty} r^3\nabla\!\!\!/\mathrm{tr}\chi = \tilde{\nabla\!\!\!/}H. \tag{8.3.47}$$

8.4 The null-outgoing infinity limit of the structure equations

We show in this section that some of the structure equations have some limit equations when $\nu \to \infty$, involving $X(\lambda,\cdot)$, $\underline{X}(\lambda,\cdot)$ and the null infinity limit of the null Riemann tensor components.

Proposition 8.4.1 *The following equations are satisfied by the null-outgoing infinity limit of the connection coefficients and of the null Riemann components,*

$$\widetilde{\mathrm{d}\!\!\!/\mathrm{iv}}\,\underline{X} = \underline{B}\ ,\quad \frac{\partial}{\partial\lambda}\underline{X} = -\frac{1}{2}\underline{A}$$
$$\frac{\partial}{\partial\lambda}X = -\frac{1}{2}\underline{X}\ ,\quad \widetilde{\mathrm{d}\!\!\!/\mathrm{iv}}\,X = \frac{1}{2}\tilde{\nabla\!\!\!/}H + Z. \tag{8.4.1}$$

Proof: Let us consider the structure equation (see (3.1.47))

$$\nabla\!\!\!/\mathrm{tr}\underline{\chi} - \mathrm{d}\!\!\!/\mathrm{iv}\,\underline{\chi} + \zeta\cdot\underline{\chi} - \zeta\,\mathrm{tr}\underline{\chi} = -\underline{\beta}\,.$$

Multiplying it by r^2 we obtain

$$r\mathrm{d}\!\!\!/\mathrm{iv}\,(r\hat{\underline{\chi}})_a = \frac{r^2}{2}(\nabla\!\!\!/_a\mathrm{tr}\underline{\chi}) + r^2(\zeta\cdot\underline{\chi})_a - r^2(\zeta\,\mathrm{tr}\underline{\chi})_a + r^2\underline{\beta}_a, \tag{8.4.2}$$

and taking the limit $\nu\to\infty$, recalling the estimates for the connection coefficients and for the Riemann null components provided by the Main Theorem, we obtain, denoting with $\widetilde{\mathrm{d}\!\!\!/\mathrm{iv}}$ the divergence on S^2 relative to the $\tilde{\gamma}_\infty$ metric,

$$\widetilde{\mathrm{d}\!\!\!/\mathrm{iv}}\,\underline{X} = \underline{B}. \tag{8.4.3}$$

Let us now consider the structure equation

$$\mathbf{D}\!\!\!\!/_3\hat{\underline{\chi}} + \mathrm{tr}\underline{\chi}\,\hat{\underline{\chi}} - (\mathbf{D}_3\log\Omega)\hat{\chi} = -\underline{\alpha},$$

which we rewrite as

$$\frac{\partial}{\partial\lambda}\hat{\underline{\chi}}_{ab} + \Omega\mathrm{tr}\underline{\chi}\,\hat{\underline{\chi}}_{ab} + 2\Omega\underline{\omega}\hat{\underline{\chi}}_{ab} = -\Omega\underline{\alpha}_{ab}. \tag{8.4.4}$$

Multiplying it by r and recalling the definition and asymptotic properties of h (see (8.3.13) and (8.0.2))

$$\frac{\partial(r\hat{\underline{\chi}}_{ab})}{\partial\lambda} = h(\nu,\lambda,\cdot)(r^2\hat{\underline{\chi}}_{ab}) - \Omega(r\underline{\alpha}_{ab}). \tag{8.4.5}$$

Therefore, taking the $\nu \to \infty$ limit, we obtain

$$\frac{\partial}{\partial\lambda}\underline{X} = -\frac{1}{2}\underline{A}. \tag{8.4.6}$$

Finally, from the structure equation

$$\frac{\partial}{\partial\lambda}\hat{\chi}_{ab} = -\left[\frac{1}{2}\Omega\mathrm{tr}\underline{\chi} + 2\Omega\underline{\omega}\right]\hat{\chi}_{ab} - \frac{1}{2}\Omega\mathrm{tr}\chi\hat{\underline{\chi}}_{ab} - \Omega(\nabla\!\!\!/\widehat{\otimes}\eta - \eta\widehat{\otimes}\eta), \tag{8.4.7}$$

proceeding exactly in the same way, multiplying by r^2 and taking the limit $\nu \to \infty$ we obtain

$$\lim_{C(\lambda);\nu\to\infty}\frac{\partial}{\partial\lambda}(r^2\hat{\chi}_{ab}) = -\frac{1}{2}\lim_{C(\lambda);\nu\to\infty}(r\hat{\underline{\chi}}_{ab}), \tag{8.4.8}$$

which implies

$$\frac{\partial}{\partial\lambda}X = -\frac{1}{2}\underline{X}. \tag{8.4.9}$$

To prove the last relation of (8.4.1) we recall the structure equation (see (3.1.47))

$$\mathrm{d}\!\!\!/\mathrm{iv}\,\hat{\chi} = \frac{1}{2}\nabla\!\!\!/\mathrm{tr}\chi - \zeta\cdot\hat{\chi} + \frac{1}{2}\zeta\,\mathrm{tr}\chi - \beta = 0.$$

Multiplying this equation by r^3 and performing the limit $\nu \to \infty$ we obtain

$$\lim_{C(\lambda);\nu\to\infty} r\,\mathrm{d}\!\!\!/\mathrm{iv}\,r^2\hat{\chi} = \widetilde{\mathrm{d}\!\!\!/\mathrm{iv}}\,X = \frac{1}{2}\tilde{\nabla\!\!\!/}H + Z. \tag{8.4.10}$$

8.5 The Bondi mass

Definition 8.5.1 *The Hawking mass enclosed by a 2-surface $S(\lambda,\nu)$ is given (see[Ch-Kl]) by the expression*

$$m(\lambda,\nu) = \frac{r(\lambda,\nu)}{2}\left(1 + \frac{1}{16\pi}\int_{S(\lambda,\nu)}\mathrm{tr}\chi\,\mathrm{tr}\underline{\chi}\right). \tag{8.5.1}$$

Recalling the definition of the mass aspect function (see (3.3.6))

$$\underline{\mu}(\lambda,\nu) = \mathbf{K} + \frac{1}{4}\mathrm{tr}\chi\,\mathrm{tr}\underline{\chi} - \mathrm{d}\!\!\!/\mathrm{iv}\,\underline{\eta} = -\mathrm{d}\!\!\!/\mathrm{iv}\,\underline{\eta} + \frac{1}{2}\hat{\chi}\cdot\hat{\underline{\chi}} - \rho, \tag{8.5.2}$$

and using the Gauss–Bonnet theorem, we can express the Hawking mass in the form[12]

$$\begin{aligned} m(\lambda,\nu) &= \frac{r(\lambda,\nu)}{2}\left(1 + \frac{1}{4\pi}\int_{S(\lambda,\nu)}(\underline{\mu} - \mathbf{K} + \mathrm{d}\!\!\!/\mathrm{iv}\,\underline{\eta})\right) \\ &= \frac{r(\lambda,\nu)}{8\pi}\int_{S(\lambda,\nu)}\underline{\mu} = \frac{r(\lambda,\nu)}{8\pi}\int_{S(\lambda,\nu)}\left(\frac{1}{2}\hat{\chi}\cdot\hat{\underline{\chi}} - \rho\right). \end{aligned} \tag{8.5.3}$$

[12]In most of the equations we omit the dependence on the angular variables, except where strictly needed.

Proposition 8.5.1 *The Hawking mass $m(\lambda, \nu)$ satisfies*

$$\frac{\partial}{\partial \nu} m(\lambda, \nu) = O(r^{-2}). \tag{8.5.4}$$

Proof: From Definition 8.5.1 we have

$$\begin{aligned}
\frac{\partial}{\partial \nu} m(\lambda, \nu) &= \frac{1}{2}\frac{\partial r}{\partial \nu}\left(1 + \frac{1}{16\pi}\int_{S(\lambda,\nu)} \mathrm{tr}\chi\,\mathrm{tr}\underline{\chi}\right) + \frac{r}{32\pi}\frac{\partial}{\partial \nu}\int_{S(\lambda,\nu)} \mathrm{tr}\chi\,\mathrm{tr}\underline{\chi} \\
&= \frac{\partial r}{\partial \nu}\frac{1}{8\pi}\int_{S(\lambda,\nu)} \underline{\mu} + \frac{r}{32\pi}\frac{\partial}{\partial \nu}\int_{S(\lambda,\nu)} \mathrm{tr}\chi\,\mathrm{tr}\underline{\chi} \\
&= \frac{\overline{\Omega \mathrm{tr}\chi}}{2}\frac{r}{8\pi}\int_{S(\lambda,\nu)} \underline{\mu} + \frac{r}{32\pi}\int_{S(\lambda,\nu)}\left(\frac{\partial}{\partial \nu}(\mathrm{tr}\chi\,\mathrm{tr}\underline{\chi}) + \Omega \mathrm{tr}\chi^2 \mathrm{tr}\underline{\chi}\right) \\
&= \frac{r}{16\pi}\int_{S(\lambda,\nu)}\left(\overline{\Omega \mathrm{tr}\chi}\underline{\mu} + \frac{1}{2}\Omega \mathrm{tr}\chi^2 \mathrm{tr}\underline{\chi} + \frac{1}{2}\frac{\partial}{\partial \nu}(\mathrm{tr}\chi\,\mathrm{tr}\underline{\chi})\right).
\end{aligned} \tag{8.5.5}$$

From the structure equations (3.1.46) it is easy to derive

$$\begin{aligned}
\frac{\partial}{\partial \nu}\mathrm{tr}\chi\,\mathrm{tr}\underline{\chi} &= -\Omega(\mathrm{tr}\chi)^2\mathrm{tr}\underline{\chi} - \Omega|\hat{\chi}|^2\mathrm{tr}\underline{\chi} + 2\Omega \mathrm{tr}\chi|\eta|^2 + 2\Omega \mathrm{tr}\chi\left(\mathrm{d}\!\!/\mathrm{iv}\,\underline{\eta} - \frac{1}{2}\hat{\chi}\cdot\underline{\hat{\chi}} + \rho\right) \\
&= -2\Omega \mathrm{tr}\chi\underline{\mu} - \Omega(\mathrm{tr}\chi)^2\mathrm{tr}\underline{\chi} - \Omega|\hat{\chi}|^2\mathrm{tr}\underline{\chi} + 2\Omega \mathrm{tr}\chi|\eta|^2.
\end{aligned} \tag{8.5.6}$$

Plugging this relation into (8.5.5) we obtain

$$\frac{\partial}{\partial \nu} m(\lambda, \nu) = \frac{r}{16\pi}\int_{S(\lambda,\nu)}\left[(\overline{\Omega \mathrm{tr}\chi} - \Omega \mathrm{tr}\chi)\underline{\mu} + \Omega \mathrm{tr}\chi|\eta|^2 - \frac{1}{2}\Omega \mathrm{tr}\underline{\chi}|\hat{\chi}|^2\right], \tag{8.5.7}$$

and due to the estimates (8.0.1) and (8.0.2), the right-hand side of (8.5.7) is $O(r^{-2})$, which proves the proposition.

From the expression of the Hawking mass in equation (8.5.3) and the existence of the null-outgoing infinity limit of $r^2\hat{\chi}$, $r\underline{\hat{\chi}}$ and $r^3\rho$ proved in Propositions 8.2.1 and 8.3.2, it follows immediately that $m(\lambda, \nu)$ has a limit as $\nu \to \infty$, uniform in λ. We can therefore introduce the Bondi mass (see [Bo-Bu-Me]) in the following way.[13]

Definition 8.5.2 *The Bondi mass relative to the null-outgoing hypersurface $C(\lambda)$ is*

$$M_B(\lambda) = \lim_{\nu\to\infty} m(\lambda, \nu). \tag{8.5.8}$$

A corollary to Proposition 8.5.1 is as follows.

Corollary 8.5.1 *On any $C(\lambda)$ the following relation holds:*

$$m(\lambda, \nu) = M_B(\lambda) + O(r^{-1}) \tag{8.5.9}$$

Proof: It follows immediately by integrating the right-hand side of (8.5.4).

Observe that, on any $\underline{C}(\nu)$, λ varies in the interval $[\lambda_0(\nu), \lambda_0]$, where $\lambda_0(\nu) = u|_{\underline{C}(\nu)\cap\Sigma_0}$. As $\lim_{\nu\to\infty}\lambda_0(\nu) = -\infty$, $M_B(\lambda)$ is defined in the interval $(-\infty, \lambda_0]$ and we have the following result.

[13] For a detailed discussion about the Bondi mass, see [Wa2], Chapter 11.

Theorem 8.5.2 *The Bondi mass has the following limit*

$$\lim_{\lambda\to-\infty} M_B(\lambda) = M, \tag{8.5.10}$$

where M, defined in the global initial data conditions (see Definition 3.6.1) is the ADM energy on Σ_0.

Proof: From equation (8.5.3) and the definition of the Bondi mass it follows immediately that

$$M_B(\lambda) = \frac{1}{8\pi}\int_{S^2} \big(X\cdot \underline{X} - P\big)(\lambda,\cdot) \tag{8.5.11}$$

where the integration is relative to the standard volume element of S^2. The asymptotic behavior in λ of $X(\lambda,\cdot)$ and $\underline{X}(\lambda,\cdot)$, proved in Proposition 8.3.2, implies that

$$M_B(\lambda) = -\frac{1}{8\pi}\int_{S^2} P(\lambda,\cdot) + O\left(\frac{1}{\lambda^2}\right) \tag{8.5.12}$$

and

$$M_B(-\infty) = -\frac{1}{2}\lim_{\lambda\to-\infty}\overline{P}(\lambda) = -\frac{1}{2}\lim_{\lambda\to-\infty}\left(\lim_{\nu\to\infty}(r^3\overline{\rho})(\lambda,\nu)\right). \tag{8.5.13}$$

We express $(r^3\overline{\rho})(\lambda,\nu)$ using its evolution equation along $\underline{C}(\nu)$ (see Subsection 5.1.4))

$$\begin{aligned}(r^3\overline{\rho})(\lambda,\nu) &= (r^3\overline{\rho})(\lambda_0(\nu),\nu) + \frac{1}{8\pi}\int_{\lambda_0(\nu)}^{\lambda}\int_{S(\lambda,\nu)} r(\overline{\Omega \mathrm{tr}\underline{\chi}} - \Omega\mathrm{tr}\underline{\chi})(\rho-\overline{\rho})\\ &-\frac{1}{4\pi}\int_{\lambda_0(\nu)}^{\lambda}\int_{S(\lambda,\nu)} \Omega r\left((\frac{3}{2}\eta - \frac{1}{2}\underline{\eta})\cdot\underline{\beta} + \frac{1}{2}\hat{\chi}\cdot\underline{\alpha} - \zeta\cdot\underline{\beta}\right)\end{aligned} \tag{8.5.14}$$

and recalling the results of Propositions 8.2.1 and 8.3.2, we can write

$$\begin{aligned}\overline{P}(\lambda) &= \lim_{\nu\to\infty}(r^3\overline{\rho})(\lambda_0(\nu),\nu) - \frac{1}{8\pi}\int_{-\infty}^{\lambda}\int_{S^2} X\underline{A}(\lambda,\cdot)\\ &= \lim_{\nu\to\infty}(r^3\overline{\rho})(\lambda_0(\nu),\nu) + O\left(\frac{1}{\lambda^2}\right),\end{aligned} \tag{8.5.15}$$

where the last integral in (8.5.15) has been estimated observing that $X = O\left(\lambda^{-\frac{1}{2}}\right)$ and $\underline{A} = O\left(\lambda^{-\frac{5}{2}}\right)$. Computing explicitly the asymptotic expression of ρ on Σ_0 we obtain (see Subsection 7.1.3)

$$M_B(-\infty) = -\frac{1}{2}\lim_{\nu\to\infty}(r^3\overline{\rho})(\lambda_0(\nu),\nu) = E_{ADM} = M, \tag{8.5.16}$$

which proves the proposition.

We are now ready to give a rigorous derivation of the *Bondi mass formula*.

Theorem 8.5.3 *The following equation is satisfied in the null-outgoing infinity limit,*

$$\frac{\partial M_B(\lambda)}{\partial\lambda} = -\frac{1}{32\pi}\int_{S^2} |\underline{X}(\lambda,\cdot)|^2. \tag{8.5.17}$$

Proof: To prove equation (8.5.17) we first differentiate with respect to λ the Hawking mass $m(\lambda,\nu)$ and subsequently take the limit $\nu\to\infty$. Proceeding as in the derivation of (8.5.5)

$$\begin{aligned}\frac{\partial}{\partial\lambda}m(\lambda,\nu) &= \frac{\partial r}{\partial\lambda}\frac{1}{8\pi}\int_{S(\lambda,\nu)}\underline{\mu} + \frac{r}{32\pi}\frac{\partial}{\partial\lambda}\int_{S(\lambda,\nu)}\mathrm{tr}\chi\,\mathrm{tr}\underline{\chi} \\ &= \frac{r}{16\pi}\int_{S(\lambda,\nu)}\left(\overline{\Omega\mathrm{tr}\underline{\chi}}\,\underline{\mu} + \frac{1}{2}\Omega\mathrm{tr}\underline{\chi}^2\mathrm{tr}\chi + \frac{1}{2}\frac{\partial}{\partial\lambda}(\mathrm{tr}\chi\,\mathrm{tr}\underline{\chi})\right).\end{aligned} \tag{8.5.18}$$

From the structure equations (3.1.46), it is easy to obtain

$$\begin{aligned}\frac{\partial}{\partial\lambda}\mathrm{tr}\chi\,\mathrm{tr}\underline{\chi} &= -\Omega(\mathrm{tr}\underline{\chi})^2\mathrm{tr}\chi - \Omega|\hat{\underline{\chi}}|^2\mathrm{tr}\chi + 2\Omega\mathrm{tr}\underline{\chi}|\eta|^2 + 2\Omega\mathrm{tr}\underline{\chi}\left(\mathrm{d}\!\!/\mathrm{iv}\,\eta - \frac{1}{2}\hat{\underline{\chi}}\cdot\hat{\chi} + \rho\right) \\ &= -2\Omega\mathrm{tr}\underline{\chi}\mu - \Omega(\mathrm{tr}\underline{\chi})^2\mathrm{tr}\chi - \Omega|\hat{\underline{\chi}}|^2\mathrm{tr}\chi + 2\Omega\mathrm{tr}\chi|\eta|^2,\end{aligned} \tag{8.5.19}$$

and equation (8.5.18) can be rewritten as

$$\frac{\partial}{\partial\lambda}m(\lambda,\nu) = \frac{r}{16\pi}\int_{S(\lambda,\nu)}\left(\overline{\Omega\mathrm{tr}\underline{\chi}}\,\underline{\mu} - \Omega\mathrm{tr}\underline{\chi}\mu - \frac{1}{2}\Omega\mathrm{tr}\chi|\hat{\underline{\chi}}|^2 + \Omega\mathrm{tr}\chi|\eta|^2\right). \tag{8.5.20}$$

Using the previous results we see that the integrand of (8.5.20) admits a limit for $\nu\to\infty$, uniform in λ. Moreover the only term in the right-hand side not converging to zero is $-(32\pi)^{-1}r\int_{S(\lambda,\nu)}\Omega\mathrm{tr}\chi|\hat{\underline{\chi}}|^2$. Thus we conclude, recalling the estimates (8.0.1), (8.0.2), and the limits proved in Propositions 8.3.1 and 8.3.2 that

$$\frac{\partial}{\partial\lambda}M_B(\lambda) = -\frac{1}{32\pi}\int_{S^2}|\underline{X}(\lambda,\cdot)|^2,$$

which proves our result. The right-hand side of this expression, $\frac{1}{32\pi}\int_{S^2}|\underline{X}(\lambda,\cdot)|^2$ is therefore the energy carried away to infinity by gravitational radiation in a given direction, per unit solid angle.

Using equation (8.5.15) we can complete Proposition 8.2.1 proving the following lemma:

Lemma 8.5.1 *The following limits hold:*

$$\begin{aligned}\lim_{\lambda\to\lambda_0}\overline{P} &= -2M + \frac{1}{4\pi}\int_{-\infty}^{\lambda_0}\int_{S^2}\left(X\frac{\partial}{\partial\lambda}\underline{X}\right)(\lambda_0,\cdot) \\ \lim_{\lambda\to\lambda_0}\overline{Q} &= \frac{1}{4\pi}\int_{S^2}\left(\mathrm{c\!\!/url}\,Z - \frac{1}{2}\underline{X}\wedge X\right)(\lambda_0,\cdot).\end{aligned} \tag{8.5.21}$$

Proof: From equations (8.5.15) and (8.4.6) it follows immediately that

$$\begin{aligned}\overline{P}(\lambda_0) &= -2M - \frac{1}{8\pi}\int_{-\infty}^{\lambda_0}\int_{S^2}(X\underline{A})(\lambda_0,\cdot) \\ &= -2M + \frac{1}{4\pi}\int_{-\infty}^{\lambda_0}\int_{S^2}\left(X\frac{\partial}{\partial\lambda}\underline{X}\right)(\lambda_0,\cdot).\end{aligned} \tag{8.5.22}$$

The second limit follows immediately from the structure equation

$$\sigma = \mathrm{c}\not{\mathrm{u}}\mathrm{rl}\,\zeta - \frac{1}{2}\hat{\underline{\chi}} \wedge \hat{\chi}\ ,$$

(see (3.1.47) and also Chapter 5, Subsection 5.1.4). Multiplying it by r^3, taking the limit $\nu \to \infty$ and using the results of Proposition 8.3.2 we obtain

$$\overline{Q}(\lambda_0) \;=\; \frac{1}{4\pi}\int_{S^2}\left(\widetilde{\mathrm{c}\not{\mathrm{u}}\mathrm{rl}}\, Z - \frac{1}{2}\underline{X} \wedge X\right)(\lambda_0, \cdot), \tag{8.5.23}$$

denoting with $\widetilde{\mathrm{c}\not{\mathrm{u}}\mathrm{rl}}$ the curl on S^2 relative to the $\tilde{\gamma}_\infty$ metric.

8.6 Asymptotic behavior of null-outgoing hypersurfaces

In this section we show[14] that as $\nu \to \infty$, the null-outgoing hypersurfaces $C(\lambda)$ approach the null-outgoing cones of the Schwarzschild spacetime with ADM mass, $M = M_B(-\infty)$. In particular we show that they diverge logarithmically from the standard position of the null-outgoing cones in Minkowski spacetime.

Proposition 8.6.1 *On any null-outgoing hypersurface $C(\lambda)$ the following relation holds:*

$$\frac{dr}{dt} = 1 - \frac{2M}{r} + O\left(\frac{1}{r^2}\right). \tag{8.6.1}$$

Proof: We first recall the definition of the global time function in the spacetime $\mathcal{M}$, see Proposition 3.3.1,

$$t(\lambda, \nu) = \frac{1}{2}(\lambda + \nu)\ .$$

Recall also that $r = r(\lambda, \nu)$ is defined by the formula

$$r(\lambda, \nu) = (4\pi)^{-\frac{1}{2}}|S(\lambda, \nu)|^{\frac{1}{2}}\ .$$

Computing $\frac{d}{dt}r$ on a null hypersurface $C(\lambda)$ we obtain (see (4.1.30))

$$\frac{d}{dt}r|_{C(\lambda)} = 2\frac{\partial}{\partial\nu}r = r\overline{\Omega\mathrm{tr}\chi} = 1 + r(\overline{\Omega\mathrm{tr}\chi} - \frac{1}{r}). \tag{8.6.2}$$

To obtain an explicit relation between $r(\overline{\Omega\mathrm{tr}\chi} - \frac{1}{r})$ and the Bondi mass, we express this quantity as an integral along the null-incoming hypersurface $\underline{C}(\nu)$,

$$\begin{aligned}\frac{1}{4\pi r(\lambda,\nu)}\int_{S(\lambda,\nu)}(\Omega\mathrm{tr}\chi - \frac{1}{r}) &= \frac{1}{4\pi r(\lambda_0(\nu),\nu)}\int_{S(\lambda_0(\nu),\nu)}(\Omega\mathrm{tr}\chi - \frac{1}{r})\\ &+\frac{1}{4\pi}\int_{\lambda_0(\nu)}^{\lambda}\frac{\partial}{\partial\lambda}\left(\frac{1}{r}\int_S(\Omega\mathrm{tr}\chi - \frac{1}{r})\right)(\lambda', \nu).\end{aligned} \tag{8.6.3}$$

[14] See (17.0.6) of [Ch-Kl]. The result is in fact slightly stronger due to the fact that, here, M is the ADM mass.

Using Lemma 3.1.3 we have

$$\begin{aligned}
\frac{1}{4\pi}\int_{\lambda_0(\nu)}^{\lambda}\frac{\partial}{\partial\lambda}\left(\frac{1}{r}\int_S(\Omega\mathrm{tr}\chi-\frac{1}{r})\right)&=\frac{1}{4\pi}\int_{\lambda_0(\nu)}^{\lambda}\left\{-\frac{1}{r^2}\left(\frac{\partial}{\partial\lambda}r\right)\int_{S(\lambda,\nu)}(\Omega\mathrm{tr}\chi-\frac{1}{r})\right.\\
&\quad\left.+\frac{1}{r}\int_{S(\lambda,\nu)}\left(\frac{\partial}{\partial\lambda}(\Omega\mathrm{tr}\chi-\frac{1}{r})+\Omega\mathrm{tr}\underline{\chi}(\Omega\mathrm{tr}\chi-\frac{1}{r})\right)\right\}\\
&=\frac{1}{4\pi}\int_{\lambda_0(\nu)}^{\lambda}\left\{\frac{1}{r}\int_{S(\lambda,\nu)}\left[\frac{\partial}{\partial\lambda}(\Omega\mathrm{tr}\chi-\frac{1}{r})+\frac{\Omega\mathrm{tr}\underline{\chi}}{2}(\Omega\mathrm{tr}\chi-\frac{1}{r})\right]\right.\\
&\quad\left.+\frac{1}{2r}\int_{S(\lambda,\nu)}(\Omega\mathrm{tr}\underline{\chi}-\overline{\Omega\mathrm{tr}\underline{\chi}})(\Omega\mathrm{tr}\chi-\frac{1}{r})\right\} \qquad (8.6.4)\\
&=\frac{1}{4\pi}\int_{\lambda_0(\nu)}^{\lambda}\frac{1}{r}\int_{S(\lambda,\nu)}\left[\frac{\partial}{\partial\lambda}(\Omega\mathrm{tr}\chi-\frac{1}{r})+\frac{\Omega\mathrm{tr}\underline{\chi}}{2}(\Omega\mathrm{tr}\chi-\frac{1}{r})\right]+O\left(\frac{1}{r^3(\lambda,\nu)}\right)
\end{aligned}$$

where the estimate of the last term uses the boundedness of $r^2|\Omega\mathrm{tr}\underline{\chi}-\overline{\Omega\mathrm{tr}\underline{\chi}}|$ and of $r^2|(\Omega\mathrm{tr}\chi-\frac{1}{r})|$ implicit in the bounds for the $\mathcal{O}$ norms proved in Theorem **M1**. With the help of the structure equations (3.1.46)

$$\begin{aligned}
\left[\frac{\partial}{\partial\lambda}(\Omega\mathrm{tr}\chi-\frac{1}{r})+\frac{\Omega\mathrm{tr}\underline{\chi}}{2}(\Omega\mathrm{tr}\chi-\frac{1}{r})\right]&=\Omega\left[\left(-\hat{\underline{\chi}}\cdot\hat{\chi}+2\rho\right)+2\,\mathrm{d\!\!/iv}\,\zeta+2\Delta\!\!\!/\log\Omega\right.\\
&\quad\left.+2|\zeta|^2+4\zeta\cdot\nabla\!\!\!\!/\log\Omega+2|\nabla\!\!\!\!/\log\Omega|^2\right]. \qquad (8.6.5)
\end{aligned}$$

Using once more the estimates for the connection coefficients implicit in the bounds for the $\mathcal{O}$ norms provided by the Main Theorem, we write

$$\frac{\partial}{\partial\lambda}\left(\frac{1}{r}\int_{S(\lambda,\nu)}(\Omega\mathrm{tr}\chi-\frac{1}{r})\right)=\frac{2}{r}\int_{S(\lambda,\nu)}\left(-\frac{1}{2}\hat{\underline{\chi}}\cdot\hat{\chi}+\rho\right)+O\left(\frac{1}{r^3}\right). \qquad (8.6.6)$$

Therefore, from (8.6.4) and using (8.5.3) we have,

$$\begin{aligned}
\frac{1}{4\pi}\int_{\lambda_0(\nu)}^{\lambda}\frac{\partial}{\partial\lambda}\left(\frac{1}{r}\int_{S(\lambda,\nu)}(\Omega\mathrm{tr}\chi-\frac{1}{r})\right)&=2\int_{\lambda_0(\nu)}^{\lambda}\frac{1}{r^2}\left(-\frac{r}{4\pi}\int_{S(\lambda',\nu)}\left(\frac{1}{2}\hat{\underline{\chi}}\cdot\hat{\chi}-\rho\right)\right)\\
&=-2\int_{\lambda_0(\nu)}^{\lambda}\frac{1}{r^2(\lambda',\nu)}m(\lambda',\nu)+O\left(\frac{1}{r^2}\right) \qquad (8.6.7)
\end{aligned}$$

Recalling that from Corollary 8.5.1, $m(\lambda,\nu)=M_B(\lambda)+O(r^{-1})$, we write

$$\begin{aligned}
\frac{d}{dt}r&=\frac{1}{4\pi r(\lambda_0(\nu),\nu)}\int_{S(\lambda_0(\nu),\nu)}(\Omega\mathrm{tr}\chi-\frac{1}{r})+\frac{1}{4\pi}\int_{\lambda_0(\nu)}^{\lambda}\frac{\partial}{\partial\lambda}\left(\frac{1}{r}\int_S(\Omega\mathrm{tr}\chi-\frac{1}{r})\right)\\
&=\frac{1}{4\pi r(\lambda_0(\nu),\nu)}\int_{S(\lambda_0(\nu),\nu)}(\Omega\mathrm{tr}\chi-\frac{1}{r})-2\int_{\lambda_0(\nu)}^{\lambda}\frac{1}{r^2(\lambda',\nu)}M_B(\lambda')+O\left(\frac{1}{r^2}\right).
\end{aligned} \qquad (8.6.8)$$

Using the Bondi masss formula, equation 8.5.17, we write

$$-2\int_{\lambda_0(\nu)}^{\lambda}\frac{1}{r^2(\lambda',\nu)}M_B(\lambda')=-2\int_{\lambda_0(\nu)}^{\lambda}\frac{1}{r^2(\lambda',\nu)}\left(M_B(-\infty)+\int_{-\infty}^{\lambda'}\frac{\partial M_B}{\partial\lambda}(\lambda'')\right)$$

$$
\begin{aligned}
&= -2\int_{\lambda_0(\nu)}^{\lambda} \frac{1}{r^2(\lambda',\nu)}\left(M_B(-\infty) - \frac{1}{32\pi}\int_{-\infty}^{\lambda'}\int_{S^2} |\underline{X}(\lambda'',\cdot)|^2\right)\\
&= -2\int_{\lambda_0(\nu)}^{\lambda} \frac{1}{r^2(\lambda',\nu)} M_B(-\infty) + O\left(\frac{1}{r^2\lambda}\right) \qquad (8.6.9)
\end{aligned}
$$

Therefore

$$
\begin{aligned}
\frac{d}{dt}r &= \frac{1}{4\pi r(\lambda_0(\nu),\nu)}\int_{S(\lambda_0(\nu),\nu)} (\Omega\mathrm{tr}\chi - \frac{1}{r}) - 2M_B(-\infty)\int_{\lambda_0(\nu)}^{\lambda}\frac{1}{r^2(\lambda',\nu)}d\lambda' + O\left(\frac{1}{r^2}\right)\\
&= \frac{1}{4\pi r(\lambda_0(\nu),\nu)}\int_{S(\lambda_0(\nu),\nu)} (\Omega\mathrm{tr}\chi - \frac{1}{r}) - 2M_B(-\infty)\left(\frac{1}{r(\lambda,\nu)} - \frac{1}{r(\lambda_0(\nu),\nu)}\right) + O\left(\frac{1}{r^2}\right),
\end{aligned}
$$

where we have used the relation $d\lambda = -(1+\frac{c\varepsilon}{r})dr$, which follows from Lemma 4.1.8. To complete the proof we write the first term in the last line of (8.6.8) as an integral on Σ_0, recalling the relations $\lambda = u|_{\Sigma_0\cap C(\lambda)}$, $\nu_0(\lambda) = \underline{u}|_{\Sigma_0\cap C(\lambda)}$ and $\lambda_0(\nu) = u|_{\Sigma_0\cap \underline{C}(\nu)}$,

$$
\begin{aligned}
&\frac{1}{4\pi r(\lambda_0(\nu),\nu)}\int_{S(\lambda_0(\nu),\nu)} (\Omega\mathrm{tr}\chi - \frac{1}{r}) \qquad (8.6.10)\\
&\quad = \int_{-\infty}^{\lambda_0(\nu)} \frac{\partial}{\partial\lambda}\left(\frac{1}{4\pi r(\lambda,\nu_0(\lambda))}\int_{S(\lambda,\nu_0(\lambda))}(a\mathrm{tr}\theta - \frac{1}{r})\right)d\lambda + O\left(\frac{1}{r^2}\right),
\end{aligned}
$$

where the $O\left(\frac{1}{r^2}\right)$ term originates from the integration of the terms due to $(\Omega\mathrm{tr}\chi - a\mathrm{tr}\theta)$; see Subsections 3.3.1, 7.1.3 and 7.1.4.

Repeating the computation in (8.6.4) with $\tilde{N}$ the unit vector field along Σ_0 normal to the canonical foliation $\{S_0(\nu)\}$ and taking into account equation (7.2.6),

$$
\nabla_{\tilde{N}}\mathrm{tr}\theta + \frac{1}{2}(\mathrm{tr}\theta)^2 = -\overline{\rho} + \left[-|\nabla\!\!\!/ \log a|^2 - |\hat{\theta}|^2 + g(k)\right]
$$

and $\nabla_{\tilde{N}}r = \frac{\overline{a\mathrm{tr}\theta}}{2}$ as well as the estimates of the norms $\mathcal{O}(\Sigma_0\backslash K)$ we write

$$
\begin{aligned}
&\frac{\partial}{\partial\lambda}\left(\frac{1}{4\pi r(\lambda,\nu_0(\lambda))}\int_{S(\lambda,\nu_0(\lambda))}(a\mathrm{tr}\theta - \frac{1}{r})\right) = -a\tilde{N}\left(\frac{1}{4\pi r(\lambda,\nu_0(\lambda))}\int_{S(\lambda,\nu_0(\lambda))}(a\mathrm{tr}\theta - \frac{1}{r})\right)\\
&= \frac{1}{4\pi r^2}\left(\nabla_{\tilde{N}}r\right)\int_{S(\lambda,\nu_0(\lambda))}(a\mathrm{tr}\theta - \frac{1}{r})\\
&\quad - \frac{1}{4\pi r}\int_{S(\lambda,\nu_0(\lambda))}\left(\nabla_{\tilde{N}}(a\mathrm{tr}\theta - \frac{1}{r}) + a\mathrm{tr}\theta(a\mathrm{tr}\theta - \frac{1}{r})\right) + O\left(\frac{1}{r^3(\lambda,\nu_0(\lambda))}\right)\\
&= -\frac{1}{4\pi r}\int_{S(\lambda,\nu_0(\lambda))}\left[\nabla_{\tilde{N}}(a\mathrm{tr}\theta - \frac{1}{r}) + \frac{a\mathrm{tr}\theta}{2}(a\mathrm{tr}\theta - \frac{1}{r})\right]\\
&\quad + \frac{1}{8\pi r}\int_{S(\lambda,\nu_0(\lambda))}(a\mathrm{tr}\theta - \overline{(a\mathrm{tr}\theta)})(a\mathrm{tr}\theta - \frac{1}{r}) + O\left(\frac{1}{r^3(\lambda,\nu_0(\lambda))}\right) \qquad (8.6.11)\\
&= -\frac{1}{4\pi r}\int_{S(\lambda,\nu_0(\lambda))}\left[\nabla_{\tilde{N}}(a\mathrm{tr}\theta - \frac{1}{r}) + \frac{a\mathrm{tr}\theta}{2}(a\mathrm{tr}\theta - \frac{1}{r})\right] + O\left(\frac{1}{r^3(\lambda,\nu_0(\lambda))}\right)\\
&= \frac{1}{4\pi r}\int_{S(\lambda,\nu_0(\lambda))}\overline{\rho} + O\left(\frac{1}{r^3(\lambda,\nu_0(\lambda))}\right) = r(\lambda,\nu_0(\lambda))\overline{\rho} + O\left(\frac{1}{r^3(\lambda,\nu_0(\lambda))}\right).
\end{aligned}
$$

The factor a in the first line of (8.6.11) has been written as one plus a correction term which gives a contribution to the integral of order $O(r^{-3})$.

Plugging this result into (8.6.10) we obtain

$$\frac{1}{4\pi r(\lambda_0(\nu),\nu)}\int_{S(\lambda_0(\nu),\nu)}(\Omega\mathrm{tr}\chi-\frac{1}{r})$$
$$=\int_{-\infty}^{\lambda_0(\nu)}\left(\frac{1}{r^2}r^3\overline{\rho}\right)(\lambda,\nu_0(\lambda))d\lambda+O\left(\frac{1}{r^2(\lambda,\nu_0(\lambda))}\right). \tag{8.6.12}$$

Recalling the Global initial data conditions (see Definition 3.6.1), an explicit computation similar to the one in Subsection 5.1.4, Lemma 5.1.2, but done on Σ_0 using the estimates (8.0.2) gives[15]

$$r^3\overline{\rho}=-2M+O(r^{-1}). \tag{8.6.13}$$

Therefore

$$\frac{1}{4\pi r(\lambda_0(\nu),\nu)}\int_{S(\lambda_0(\nu),\nu)}(\Omega\mathrm{tr}\chi-\frac{1}{r})=(\lim_{\lambda\to-\infty}r^3\overline{\rho})\int_{-\infty}^{\lambda_0(\nu)}\frac{1}{r^2}d\lambda$$
$$+\int_{-\infty}^{\lambda_0(\nu)}\frac{1}{r^2}\left(r^3\overline{\rho}-(\lim_{\lambda\to-\infty}r^3\overline{\rho})\right)d\lambda+O\left(\frac{1}{r^2}\right)$$
$$=-\frac{2M}{r(\lambda_0(\nu),\nu)}+O\left(\frac{1}{r^2}\right). \tag{8.6.14}$$

The constant M above is the ADM mass associated to the initial data and concide, as proved in Proposition 8.5.2, with the Bondi mass for $\lambda\to-\infty$, $M=M_B(-\infty)$. Using this relation, equation (8.6.8) can be written as

$$\frac{d}{dt}r(\lambda,\nu)=1-\frac{2M}{r(\lambda,\nu)}+O\left(\frac{1}{r^2(\lambda,\nu)}\right), \tag{8.6.15}$$

completing the proof of Proposition 8.6.1.

Remark: From equation (8.6.15) we obtain immediately for r the implicit expression

$$r=t-2M\log(r-2M)+c, \tag{8.6.16}$$

and defining $r^*=r+2M\log(r-2M)$ we conclude that the null hypersurfaces $C(\lambda)$ tend asymptotically to the Schwarzschild null-outgoing cones $C_S(\lambda)\equiv\{p\in\mathcal{M}|t(p)-r^*(p)=\lambda\}$. Therefore on any null hypersurfaces $C(\lambda)$ we have, as $\nu\to\infty$,

$$r=t-2M\log t+O(1)\ .$$

[15] Observe that in (8.6.13) to prove that the correction term is $O(r^{-1})$ we have to use the "improved" estimates (8.0.2); see the discussion at the beginning of this chapter.

References

[Ad] R. E. Adams, *Sobolev Spaces*, Academic Press, N.Y., 1975.

[An-Mon] L. Andersson, V. Moncrief, On the global evolution problem in 3+1 gravity, *J. Geom. Phys.*, under preparation.

[Ar-De-Mi] R. Arnowitt, S. Deser, C. Misner, Coordinate invariance and energy expressions in general relativity, *Phys. Rev.* **122** (1961), 997–1006.

[Ba] R. Bartnik, Existence of maximal surfaces in asymptotically flat spacetimes, *Comm. Math. Phys.* **94** (1984), 155–175.

[Ba-Ch1] H. Bahouri and J. Y. Chemin, Equations d'ondes quasilinéaires et estimation de Strichartz, *Amer. J. Math.* **121** (1999), 1337–1777.

[Bel] L. Bel, Introduction d'un tenseur du quatrième ordre, *C. R. Acad. Sci. Paris* **247** (1959), 1094–1096.

[Bl-D] L. Blanchet, T. Damour, Hereditary effects in gravitational radiation, *Phys. Rev.* **D46** (1992), 4304–4319.

[Bo-Bu-Me] H. Bondi, M. G. J. van der Burg, A. W. K. Metzner, Gravitational waves in General Relativity VII. Waves from Axi-symmetric Isolated Systems, *Proc. Roy. Soc. Lond.* **A269** (1962), 21–52.

[Br-D-Is-M] P. R. Brady, S. Droz, W. Israel, S. M. Morsink, Covariant double-null dynamics:(2+2)-splitting of the Einstein equations, *Classical Quantum Gravity*, **13** (1996), 2211–2230.

[Br1] Y. Choquet-Bruhat, Théorème d'existence pour certain systèmes d'équations aux dériveés partielles non linéaires, *Acta Matematica* **88** (1952), 141–225.

[Br2] Y. Choquet-Bruhat, Solutions C^∞ d'équations hyperboliques non linéaires, *C. R. Acad. Sci. Paris* **272** (1968), 386–388.

[Br3] Y. Choquet-Bruhat, Un théorème d'instabilité pour certains équations hyperboliques non linéaires, *C. R. Acad. Sci. Paris* **276A** (1973), 281.

[Br-Ch2] Y. Choquet-Bruhat, D. Christodoulou, Elliptic systems in $H_{s,\delta}$ spaces on manifolds which are euclidean at infinity, *Acta Math.* **145** (1981), 129–150.

[Br-Ge] Y. Choquet-Bruhat, R. P. Geroch, Global aspects of the Cauchy problem in General Relativity, *Comm. Math. Phys.* **14** (1969), 329–335.

[Br-Y] Y. Choquet-Bruhat, J. York, The Cauchy Problem, in A. Held, ed., *General Relativity and Gravitation*, Vol. 1, Plenum, N. Y., 1980, pp. 99–172.

[Ch1] D. Christodoulou, Solutions globales des équations de champ de Yang–Mills, *C. R. Acad. Sci. Paris* **293 Series A** (1981), 39–42.

[Ch2] D. Christodoulou, Global solutions for nonlinear hyperbolic equations for small data, *Comm. Pure Appl. Math.* **39** (1986), 267–282.

[Ch3] D. Christodoulou, The formation of black holes and singularities in spherically symmetric gravitational collapse, *Comm. Pure and Appl. Math.* **XLIV** (1991), 339–373.

[Ch4] D. Christodoulou, Examples of naked singularity formation in the gravitational collapse of a scalar field, *Ann. Math.* **140** (1994), 607–653.

[Ch5] D. Christodoulou, The instability of naked singularity formation in the gravitational collapse of a scalar field, *Ann. Math.* **149** (1999), 149–183.

[Ch6] D. Christodoulou, On the global inital value problem and the issue of singularities, *Classical Quantum Gravity* **13** (1999), A23–A35.

[Ch7] D. Christodoulou, The Stability of Minkowski Spacetime, *Proceedings of the International Congress of Mathematicians Kyoto 1990*, (1990), 1114–1121.

[Ch8] D. Christodoulou, The global initial value problem in general relativity, Lecture given at the ninth Marcel Grossmann meeting (Rome July 2-8,2000), submitted to World Scientific, on August 1, 2001.

[Ch-Kl1] D. Christodoulou, S. Klainerman, Asymptotic properties of linear field equations in Minkowski space, *Comm. Pure Appl. Math.* **XLIII** (1990), 137–199.

[Ch-Kl] D. Christodoulou, S. Klainerman, *The global nonlinear stability of the Minkowski space*, Princeton Mathematical Series, 41, 1993.

[Do] M. Dossa, Espaces de Sobolev non isotropes, à poids et problèmes de Cauchy quasi-linéaires sur un conoide caractéristique, Ann. Inst. H. Poincaré **66**: n.1, (1997), 37–107.

[Ch-Mu] D. Christodoulou, N.'O'Murchadha, The boost problem in General Relativity, *Comm. Math. Phys.* **80** (1981), 271–300.

[Cha] I. Chavel, *Riemannian Geometry: A Modern Introduction*, Cambridge University Press, 108, 1993.

[Chr] P. T. Chrúsciel, On the uniqueness in the large of solutions to the Einstein equations (strong cosmic censorship), *Mathematical Aspects of Classical Field Theory*, (Seattle, WA, 1991), Amer. Math. Soc., Providence, RI, 1992, pp. 235–273.

[Ea-Mon] D. Eardley, V. Moncrief, The global existence problem and cosmic censorship in general relativity, *Gen. Rel. Gravit.* **13** (1981), 887–892.

[F-Ms1] A. Fisher, J. E. Marsden, The Einstein evolution equations as a first-order quasi-linear symmetric hyperbolic system. I, *Comm. Math. Phys.* **28** (1972), 1–38.

[F-Ms2] A. Fisher, J. E. Marsden, General relativity, partial differential equations and dynamical systems, *AMS Proc. Symp. Pure Math.* **23** (1973), 309–327.

[Fr1] H. Friedrich, Cauchy problems for the conformal vacuum field equations in General Relativity, *Comm. Math. Phys.* **91** (1983), 445–472.

[Fr2] H.!Friedrich, *Existence and structure of past asymptotically simple solutions of Einstein's field equations with positive cosmological constant, J. Geom. Phys.* **3** (1986), 101–117.

[Fr3] H. Friedrich, On the global existence and the asymptotic behavior of solutions to the Einstein-Maxwell–Yang–Mills equations, *J. Differential Geom.* **34**: no.2 (1991), 275–345.

[Fr4] H. Friedrich, Hyperbolic reductions for Einstein equations, *Classical Quantum Gravity* **13** (1996), 1451–1469.

[Fr-Re] H. Friedrich, A. Rendall, The Cauchy problem for the Einstein equations, arXiv:gr-qc/0002074 (22 Feb 2000).

[Ge] R. P. Geroch, The domain of dependence, *J. Math. Phys.* **11** (1970), 437–439.

[Haw-El] S. W. Hawking, G. F. R. Hellis, *The Large Scale Structure of Spacetime*, Cambridge Monographs on Mathematical Physics, 1973.

[Ho] L. Hörmander, *Lectures on Nonlinear Hyperbolic Equations*, Mathematics and Applications 26, Springer-Verlag, 1987.

[Hu-Ka-Ms] T. J. R.Hughes, T. Kato, J. E. Marsden, *Well-posed quasi-linear second-order hyperbolic systems with applications to nonlinear elastodynamics and general relativity*, Arch. Rational Mech. Anal. **63**: no. 3 (1976), 273–294.

[John1] F. John, *Formation of Singularities in Elastic Waves*, Lecture Notes in Physics, Springer-Verlag, Berlin, 1984, pp. 190–214.

[John2] F. John, *Nonlinear wave equations, formation of singularities*, Amer. Math. Soc., Providence, 1990.

[Kl1] S. Klainerman, *Long time behavior of solutions to nonlinear wave equations*, *Proceedings of the Intern. Congress of Mathematicians*, Warsaw, 1982.

[Kl2] S. Klainerman, *The null condition and global existence to nonlinear wave equations*, Lect. Appl. Math. **23** (1986), 293–326.

[Kl3] S. Klainerman, Remarks on the global Sobolev inequalities in Minkowski Space, *Comm. Pure Appl. Math.* **40** (1987), 111–117.

[Kl4] S. Klainerman, Uniform decay estimates and the Lorentz invariance of the classical wave equation. *Comm. Pure Appl. Math.* **38** (1985), 321–332.

[Kl5] S. Klainerman, A commuting vector field approach to Strichartz type inequalities and applications to quasilinear wave equations, *IMRN* to appear.

[Kl-Ni] S. Klainerman, F.Nicolò, On local and global aspects of the Cauchy problem in General Relativity, *Classical Quantum Gravity* **16** (1999), R73–R157.

[Kl-Se] S. Klainerman, S. Selberg, Bilinear Estimates and Applications to Nonlinear Wave Equations, *Communications in Contemporary Mathematics*, 2002.

[Kl-Rodn1] S. Klainerman, I. Rodnianski, Improved local well posedness for quasilinear wave equations in dimension three, to appear in *Duke Math. Journ.*

[Kl-Rodn2] S. Klainerman, I. Rodnianski Rough solutions of the Einstein vacuum equations, arXiv:math.AP/0109173, submitted to *Annals Math.*

[Kl-Rodn3] S. Klainerman, I. Rodnianski *The causal structure of microlocalized, rough, Einstein metrics*, arXiv:math.AP/0109174, submitted to Annals of Math.

[Kl-Rodn] S. Klainerman, I.Rodnianski The causal structure of microlocalized, rough, Einstein metrics, *C. R. Acad. Sci. Paris* **Ser.I334** (2002), 125–130.

[Kr] J. A. V. Kroon, Polyhomogeneity and zero-rest-mass fields with applications to Newman-Penrose constants, *Classical Quantum Gravity* **17**: no.3 (2000), 605–621.

[Le] J. Leray, *Lectures on Hyperbolic Equations*, Institute for Advanced Study, Notes, 1953.

[Mu] H. Muller Zum Hagen, Characteristic initial value problem for hyperbolic systems of second order differential equations, *Ann. Inst. H. Poincaré* 53, n.2, (1990), 159–216.

[Ne-Pe1] E. T. Newman, R.Penrose, An approach to gravitational radiation by a method of spin coefficients, *J. Math. Phys.* **3** (1962), 566–578.

[Ne-Pe2] E. T. Newman, R. Penrose, New conservation laws for zero rest-mass fields in asymptotically flat space-time, *Proc. Roy. Soc. Lond.* **A305** (1968), 175–204.

[Ni] F. Nicolò, Canonical foliation on a null hypersurface, To appear.

[Pe1] R. Penrose, Conformal Treatment of Infinity, *Relativity, Groups, and Topology*, B. deWitt and C. deWitt, eds., Gordon and Breach, 1963.

[Pe2] R. Penrose, Zero rest mass fields including gravitation: asymptotic behavior, *Proc. Roy. Soc. Lond.* **A284** (1962), 159–203.

[Pe3] R. Penrose, Gravitational collapse and spacetime singularities, *Phys. Rev. Lett.* **14** (1965), 57–59.

[Ren] A. D. Rendall, Reduction of the characteristic initial value problem to the Cauchy problem and its applications to the Einstein equations, *Proc. Roy. Soc. Lond.* **A427** (1990), 221–239.

[Se] H. Seifert, *Kausal Lorentzraume*, Doctoral thesis, 1968.

[Sc-Yau1] R. Schoen, S. T. Yau, Proof of the positive mass theorem I, *Comm. Math. Phys.* **65** (1979), 45–76.

[Sc-Yau2] R. Schoen, S. T. Yau, Proof of the positive mass theorem II, *Comm. Math. Phys.* **79** (1981), 231–260.

[Sp] M. Spivak, *A Comprehensive Introduction to Differential Geometry*, Publish or Perish, Inc., Wilmington, 1970.

[Ta] D. Tataru Strichartz estimates for operators with nonsmooth coefficients, to appear in *Journ. of A.M.S.*

[Wa1] R. Wald, *Gravitational collapse and cosmic censorship.* (1997), gr-qc/9710068.

[Wa2] R. Wald, *General Relativity*, University of Chicago Press, 1984.

[W] T. Wolff, *Recent work connected with the Kakeya problem*, in Prospects in Mathematics, H. Rossi, ed. AMS, 1998.

Index